AIR POLLUTION RESEARCH REPORTS

PHYSICO-CHEMICAL BEHAVIOUR OF ATMOSPHERIC POLLUTANTS

Proceedings of the Fifth European Symposium held in Varese (Italy) from
25 to 28 September 1989 and organised within the framework of the
Concerted Action COST 611, with the incorporation of a joint
EUROTRAC-COST 611 Workshop on LACTOZ and HALIPP

This is report no 23 in the Air Pollution Report Series of the
Environmental Research Programme of the Commission of the European
Communities, Directorate-General for Science, Research and Development.

Commission of the European Communities

PHYSICO-CHEMICAL BEHAVIOUR OF ATMOSPHERIC POLLUTANTS

Edited by

G. RESTELLI

Joint Research Centre, Environment Institute, Ispra

and

G. ANGELETTI

Commission of the European Communities,
Directorate-General Science, Research and Development

KLUWER ACADEMIC PUBLISHERS

DORDRECHT / BOSTON / LONDON

Library of Congress Cataloging in Publication Data

Physico-chemical behaviour of atmospheric pollutants : proceedings of
  the fifth European symposium held in Varese, Italy from 25 to 28
  September 1989 and organised within the framework of the Concerted
  Action COST 611 / edited by G. Restelli and G. Angeletti.
       p.   cm. -- (Air pollution research reports : no. 30) (EUR ; 12542)
  English and French.
  At head of title: Commission of the European Communities.
  "Fifth European Symposium on Physico-Chemcial Behaviour of
Atmospheric Pollutants"--Pref.
  Includes bibliographical references.
  ISBN-13: 978-94-010-6743-0   (alk. paper)
  1. Air--Pollution--Congresses. 2. Pollutants--Measurement-
-Congresses. 3. Environmental chemistry--Congresses. 4. Acid
deposition--Congresses. 5. Photochemical oxidants--Congresses.
I. Restelli, G. II. Angeletti, g., 1943-   . III. Commission of
the European Communities. IV. European Symposium on Physico
-Chemical Behaviour of Atmospheric Pollutants (5th : 1989 : Varese,
Italy) V. Series: Air pollution research report ; 23. VI. Series:
EUR (Series) ; 12542.
TD881.P485   1990
628.5'3--dc20   90-4181

ISBN -13: 978-94-010-6743-0   e-ISBN-13: 978-94-009-0567-2
DOI 10.1007/978-94-009-0567-2

Publication arrangements by
Commission of the European Communities
Directorate-General Telecommunications, Information Industries and Innovation,
Scientific and Technical Communications Unit, Luxembourg

EUR 12542 EN-FR
© 1990 ECSC, EEC, EAEC, Brussels and Luxembourg

Softcover reprint of the hardcover 1st edition 1990

LEGAL NOTICE
Neither the Commission of the European Communities nor any person acting on behalf of the
Commission is responsible for the use which might be made of the following information.

Published by Kluwer Academic Publishers,
P.O. Box 17, 3300 AA Dordrecht, The Netherlands.

Kluwer Academic Publishers incorporates the publishing programmes of
D. Reidel, Martinus Nijhoff, Dr W. Junk and MTP Press.

Sold and distributed in the U.S.A. and Canada
by Kluwer Academic Publishers,
101 Philip Drive, Norwell, MA 02061, U.S.A.

In all other countries, sold and distributed
by Kluwer Academic Publishers Group,
P.O. Box 322, 3300 AH Dordrecht, The Netherlands.

Printed on acid-free paper

PREFACE

In October 1979 the First European Symposium on Physico-Chemical Behaviour of Atmospheric Pollutants was held in Ispra (Italy); 83 scientists attended the conference contributing 44 papers.
Ten years later, the Fifth European Symposium on Physico-Chemical Behaviour of Atmospheric Pollutants, organized as for the previous Symposia in the framework of the Concerted Action *COST 611, was held in Varese (Italy) from 25 to 28 September 1989.
This Volume contains the oral papers and the posters presented at this Symposium.
Participation at this Conference is more than doubled of that in 1979 in terms of scientists (185) and contributed papers (110). This simple comparison demonstrates once more the growing attention of the scientific community to the problems related to the pollution of the atmosphere.
During these years, important new issues have arisen (global pollution/climatic changes) while old ones have been reviewed due to new experimental evidence (depletion of stratospheric ozone).
The Symposium offered the best opportunity for a review of the current studies and technical progress achieved in the various sectors of the Concerted Action since the Fourth Symposium held in Stresa (Italy) in September 1986.
In 1987 the scientific programme and the operational structures of the COST 611 Project were revised. The Project is now structured into three Working Parties:

1. Development of Analytical Methods to measure Trace Components of the Atmosphere.
2. Atmospheric Chemical and Photochemical Processes.
3. Field measurements and their interpretation.

The Contents of this Volume have been organized accordingly. Close working relationships have been established between this Project and the environmental Eureka Project Eurotrac; in particular the subprojects Halipp and Lactoz are jointly coordinated by COST 611 and by Eurotrac. Papers related to these subprojects and presented at the Symposium in the frame of the respective Discussion Meetings appear as special Sessions in these Proceedings.

A special Session has been also dedicated to the papers pertinent to the "1988 Polarstern Cruise ANT VII/1" to underline the character of a major International/European collaborative project in the field.
We believe that these Proceedings give a largely complete overview of current activities and trends in this field, in Europe.

Ispra/Brussels, November 1989

G. RESTELLI   **G. ANGELETTI**

* COST 611 : Scientific and Technical Cooperation among European Community Member Countries and the Non-Member Countries Finland, Norway, Sweden and Switzerland in the field of "Physico-Chemical Behaviour of Atmospheric Pollutants".

CONTENTS

SESSION II/A - ATMOSPHERIC CHEMICAL AND PHOTOCHEMICAL PROCESSES

SESSION I

DEVELOPMENT OF ANALYTICAL

METHODS TO MEASURE

TRACE COMPONENTS

OF THE ATMOSPHERE.

SUMMARY OF SESSION I
DEVELOPMENT OF ANALYTICAL METHODS TO MEASURE TRACE COMPONENTS OF THE
ATMOSPHERE.

Ivo ALLEGRINI
C.N.R. - Ist. Inquinamento Atmosferico,
Via Salaria Km. 29,300 - CP10
00016 Monterotondo stazione, Roma, Italy

The activities related to Working Party 1 has been addressed to a large number of interesting investigations which cover the most relevant researches in the field of tropospheric chemistry. Fourteen oral presentations and ten posters emphasize the results obtained by the application of analytical techniques for the measurement of trace components, which swing from classic wet-chemistry procedures to highly sophisticated instrumental approaches.

The need for a better understanding of the processes governing the dry deposition of gaseous and particulate compounds has resulted into the development and application of many procedures based upon the use of diffusion denuders. The unique ability of denuders to modulate the sampling efficiency of different compounds has been applied to an interesting series of automatic instruments characterized by a rather low response time. Diffusional properties of trace components have also been employed for passive devices, which might find interesting applications for large scale screening and measurement of species relevant to acid deposition. Work on denuders and diffusional samplers should continue in order to identify and correct for possible interferents which might impair the measurement of the components at very low concentration levels such as those encountered in remote locations.

Diffusion properties have also been applied to the characterization of airborne particulate matter, resulting in an aerosol size spectrometer which is able to cover the most important size region for atmospheric pollution studies. As it is well known, most atmospheric species evolve, through complex chemical reactions, to aerosol phase which is removed from the atmosphere by either wet and dry deposition. Thus, observations on aerosol might be very useful for the development and the application of receptor oriented models. Measurements of the optical properties and heavy metals content are equally important for reaching a such important goal.

A peculiar system of suspended particles in atmosphere is that relevant to fogs and clouds. Although interesting studies have been carried out in laboratory and in field studies, much work remains to be done for the characterization of interstitial air and trace compounds in liquid phase, especially for the analysis of peroxides and organic compounds. Even for the measurement of the liquid water content, it seems that there is not much agreement between devices currently used for such evaluation. As the sampling step is probably the determining factor for the analysis of fogs and clouds, it is strongly recommended to develop and field test some dedicated impactors with well characterized penetration efficiencies.

Many papers have been specifically addressed to the improvement of the quality of observations. As many compounds of interest for acid deposition and related processes are present in the atmosphere at very low concentration levels, highly precise and accurate measurements have to be performed. In addition, if organic compounds are excluded, the major scientific and technical attention is given to a relatively small number of different species, which are therefore evaluated in many institutions and laboratories. In order to achieve intercomparability of the data, proper field intercomparison exercises and Quality Assurance/Quality Control protocols have to be performed and elaborated. In this context results from the intercomparison on nitric acid in gas phase and nitrates in particulate matter carried out in Rome on Sept. 1988, have been reported and discussed, with a major emphasis on the definition and characterization of the penetration functions shown by the species of interest and by the potential interferents.

Since these exercises provide significant improvement in the measurement of a given compounds, they are strongly recommended if better data sets need to be obtained. In addition, as it has been shown for the intercomparison carried out in Rome, species and potential interferents are measured simultaneously in a very intense way so that the exercise might turn easily into a field experiment which, at least in principle, give rise to significant advancements in the understanding of chemical processes in atmosphere. Thus, many exercises have been deviced for the very near future as a significant part of the COST 611 action. Ammonia and volatile organics (VOC) are the species for which the top priority has been established.

Quality of data sets may also be improved through the development of calibration procedures for instrumentation and methods. These also include the preparation of standard atmospheres of pure compounds relevant to tropospheric chemistry which can be used to test analytical methods prior to an extensive field use. Of course, standard atmospheres might also be conveniently used for the calibration of instruments. Within this respect, the calibration of PAN analyzers is strongly recommended.

A substantial number of papers reported measurements on trace components of environmental significance such as those related to halogenated organic compounds in marine environments. Since oceans and seas appear to be important sources of several species of interest for tropospheric chemistry, activities related to the analysis of naturally emitted compounds should be continued and encouraged.

Research and developments on instrumentation has resulted in a series of very important papers. Either the scientific work and the discussion highlighted the need for a better scientific understanding about complex instruments prior to their use in the field. In most instances, instruments developed to a prototype level are put in the international market, especially if they are intended to be used for the measurement of species taken into account by the national legislation, i.e., the primary pollutants listed in air quality criteria. Since users are usually not very familiar with the scientific background of the instrument, the results are very often poor in quality and significance.

Instruments which are based on long range measurements over extended pathlength such as Differential Optical Absorption Spectrometers should be thoroughly investigated as to their responses against point samplers measuring the same compounds. Such a need is very impulsive especially for those species which are recognized to play an important role in tropospheric chemistry as formaldehyde, nitrous acid, nitrogen dioxide and hydrogen peroxides. Thus, properly designed experiments aimed to such intercomparison should be planned for the near future.

Spectral properties of pollutant molecules have been largely used for the development of specific instruments such as those which make use of Tunable Diode Lasers (TDLAS) for the measurement of trace gases in the infrared region.   As this instrument requires that the air is sampled through an adsorption cell, investigations are devised in order to verify that filtration and transfer of the air into the cell do not induce change in the chemical composition and that they do preserve the physico-chemical representativeness of the sample.   Also for such applications the development of suitable standards and testing procedure might be very useful.

Papers related to WP1 and WP3 which have been presented during the Symposium have shown that a major need for airborne instrumentation is felt.   Such instruments are intended to give important answers to compulsory questions related to the regional budget of atmospheric pollutants including their deposition and chemical evolution.   Therefore, they should be automatic, fast and technological rugged in order to be easily carried out on aircrafts, in addition to superior sensitivity. Although the technical development of such instruments is partly covered by the current activities carried out by COST 611, it is strongly recommended that participating research groups spend efforts in order to actively cooperate with institutions which are in charge for the development of advanced instrumentation for pollution monitoring.

As a conclusion, the critical examination of the work carried out reveals that many important goals have been reached and that several scientific objectives have been met.   The high standard of the work suggest that activities related to Working Party 1 will certainly contribute to a substantial advancement in understanding of the physico-chemical processes which take place in the troposphere.

DETERMINATION OF NH$_3$, HNO$_3$ AND NH$_4$NO$_3$ IN AMBIENT AIR BY AUTOMATED THERMODENUDER SYSTEMS AND A WET ANNULAR DENUDER SYSTEM.

M.P. Keuken, A. Wayers-IJpelaan, R.P. Otjes and J. Slanina.

Netherlands Energy Research Foundation (ECN), P.O. Box 1,
1755 ZG Petten, The Netherlands.

SUMMARY

Automated denuder systems have been developed for the measurement of air concentrations of NH$_3$, HNO$_3$ and NH$_4$NO$_3$. The response time is in the order of one hour based on a sampling time of about 20 minutes. The limit of detection for NH$_3$, HNO$_3$ and NH$_4$NO$_3$ is 0.1 μg.m^{-3}. During a field intercomparison at Rome on 19-23 September 1988 for the measurement of HNO$_3$ and nitrate-containing aerosol in ambient air, it was demonstrated that these novel techniques show good agreement with conventional techniques. Advantages, such as automation and the relatively short sampling time are demonstrated by the observed strong diurnal fluctuation of the HNO$_3$ concentration.
As the reproducibility is better than 5%, these denuder systems can be used for dry deposition flux measurements by the gradient method. This will be tested during a field experiment from July 1989 to May 1990 in The Netherlands for the NH$_3$ flux over a forest by two thermodenuders at 30 and 20 m height.

1. INTRODUCTION

As a consequence of heavy traffic, thermal powerplants and intensive agricultural activities in The Netherlands there are high emissions of NO$_x$ and NH$_3$. Especially, effects related to the deposition of nitrogen-containing gases and aerosols, such as NH$_3$, HNO$_3$, NH$_4$NO$_3$ and (NH$_4$)$_2$SO$_4$ on vegetation and drinking-water resources have received increased attention. About 70% of the yearly potential acid deposition in The Netherlands (1987) is accounted for by dry deposition of: NH$_3$ 30%, HNO$_3$ 3%, HNO$_2$ 1%, NOx 5%, SO$_2$ 25%, HCl 4% and NH$_4^+$-containing aerosols 5% (1).

Only for the measurement of air concentrations of NO$_x$ and SO$_2$ are relatively low cost, direct detection systems available. In this paper sampling techniques for the measurement of air concentrations of NH$_3$, HNO$_3$, HNO$_2$, HCl and NH$_4^+$-containing aerosols are evaluated.

2. INSTRUMENTATION

By using denuder/filter pack combinations problems encountered with filter techniques, such as interaction between gases and aerosols, are minimized and information is obtained on the chemical speciation of air pollutants. At ECN a configuration is in use, consisting of two denuders and a filter pack in series (2). In the first, NaF-coated, denuder HNO$_3$ and HCl are collected, while ammonia is retained in the second,

H_3PO_4-coated, denuder. Aerosols, such as NH_4NO_3 and $(NH_4)_2SO_4$, are sampled by a filter pack which contains a PTFE filter, a NaF-coated paper filter and a H_3PO_4-coated paper filter in series. After sampling the denuder and filters are extracted with water and the solutions are analyzed for NO_3^-, Cl^-, SO_4^{2-} and NH_4^+. At a sample flow of 5 l.min^{-1} and a sampling time of 4 hours the limit of detection of gases and aerosols is 0.1 $\mu g.m^{-3}$.

Air concentrations of HNO_3, NH_3 and NH_4NO_3 at Petten The Netherlands from December 1988 to March 1989 (n=34) were measured by the denuder/filter pack combination and a filter pack. The results for HNO_3 are shown in Figure 1.

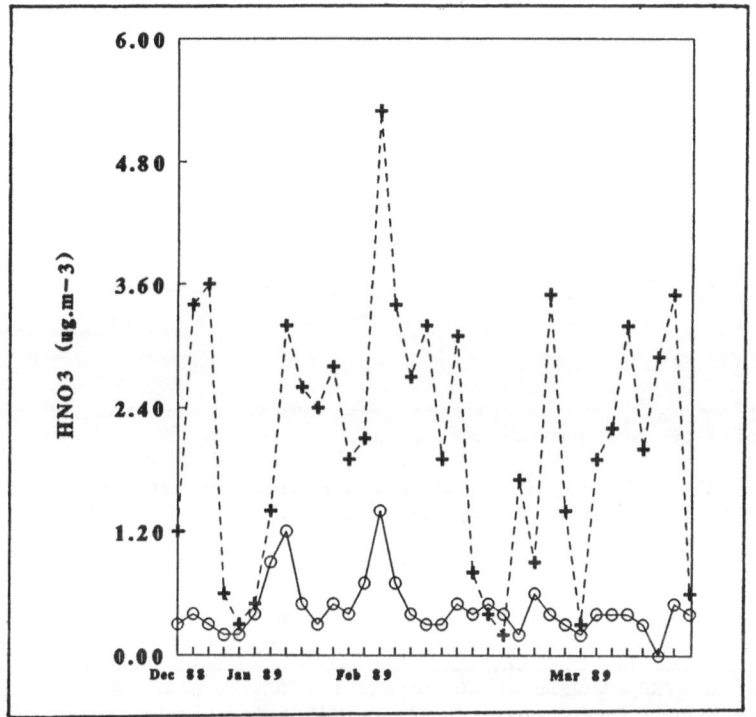

Figure 1: Air concentrations of HNO_3 at Petten, The Netherlands measured by a denuder (0) and a filter pack (+).

Average air concentrations, obtained by denuder/filter pack and filter pack, are: 0.5 and 2.1 $\mu g.m^{-3}$ HNO_3, 1.2 and 1.9 $\mu g.m^{-3}$ NH_3 and 3.3 and 2.3 $\mu g.m^{-3}$ NH_4NO_3, respectively. These results demonstrate the interferences on the sampling of HNO_3 and NH_3 by filter packs due to evaporation of NH_4NO_3 from the PTFE filter, which results in an overestimation of HNO_3 and NH_3 air concentrations .

Advantages of the denuder/filter pack method are the relative simple and low cost equipment. However, disadvantages are the labour-intensive coating and extraction procedures involved, and the time delay between sampling and analysis. The use of thermodenuders, based on thermal desorption of the collected species and on-line detection by a sensitive monitor of the liberated gases, eliminates these problems.

At ECN an annular thermodenuder coated with V_2O_5 is used to collect NH_3 (3). After sampling the denuder is heated at 700 °C, NH_3 is converted to NO_x, which is detected by a NO_x-monitor. The apparatus is shown in Figure 2.

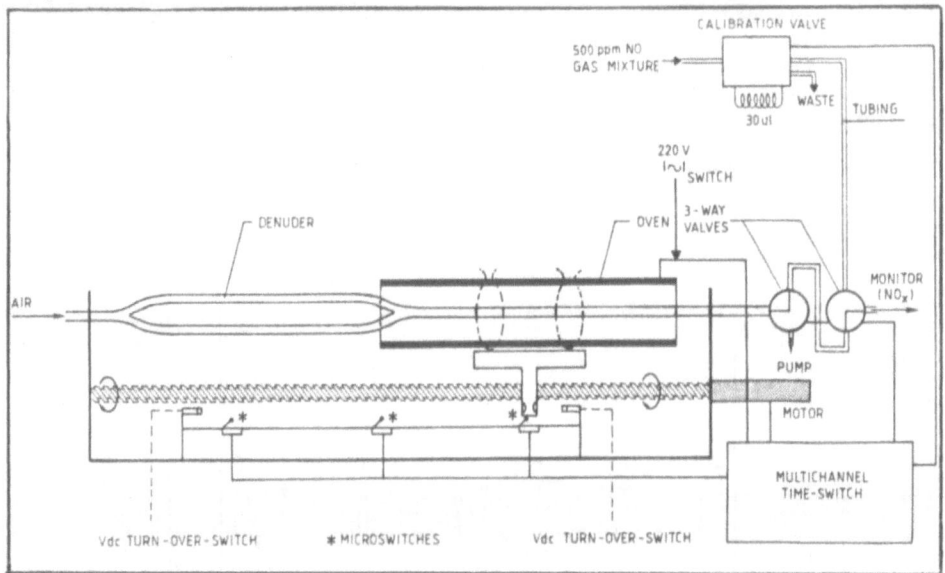

Figure 2: Automated NH_3 thermodenuder setup.

The sampling flow is 10 $1.min^{-1}$ and the limit of detection is 0.1 $\mu g.m^{-3}$ NH_3, obtained by using a sampling time of 10 minutes.

At ECN an annular thermodenuder coated with $MgSO_4$ have been developed along the lines of research performed in the past in cooperation with the University of Dortmund (F.R.G.) (4). Nitric acid is collected at ambient temperature in the first part of the denuder while ammonium nitrate is sampled at 140 °C in the second part. This device is constructed as the NH_3-thermodenuder but with a longer length of the annulus: the HNO_3/NH_4NO_3-denuder 40 cm and the NH_3-denuder 20 cm. After sampling, the second (oven- position 1) and first (oven-position 2 and 3) part of the HNO_3/NH_4NO_3-denuder are subsequently heated at 700 °C and the liberated NO_x is detected by a NO_x monitor. This is illustrated in Figure 3.

The detection limit of HNO_3 and NH_4NO_3 is 0.1 $\mu g.m^{-3}$ obtained by using a sampling flow of 5 $1.min^{-1}$ and a sampling time of 30 minutes.

An important disadvantage of thermodenuders is the rather expensive equipment, which includes the NO_x monitor, especially, if several components are measured simultaneously. ECN has developed a "wet annular denuder" for the simultaneous sampling of gases such as SO_2, HCl, NH_3, HNO_3 and HNO_2 (5). The limits of detection for HNO_3 and NH_3 are 0.1 $\mu g.m^{-3}$. Details of this novel technique are described in an accompaning paper in this volume.

(*) The thermodenuder systems are commercially available at van Essen Instruments (Delft, The Netherlands).

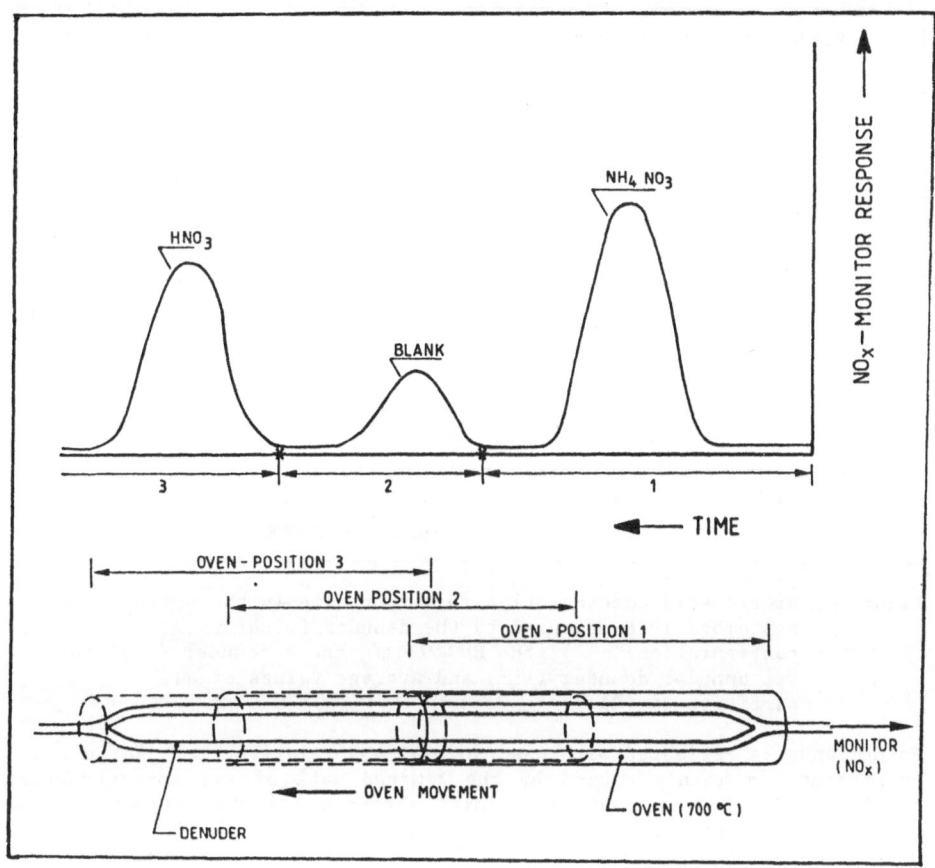

Figure 3: Response of the NO$_x$ monitor as a function of the position of
the oven during thermal desorption of the HNO$_3$/NH$_4$NO$_3$
thermodenuder.

During a field intercomparison experiment from 19 to 23 September
1988 at Rome, Italy, the denuder/filter pack combination, the
thermodenuder and the wet annular denuder were used for the determination
of HNO$_3$ in ambient air. The results given in Figure 4 show good agreement
between these techniques.

Remarkable are the diurnal fluctuations of the HNO$_3$ concentrations
and the relatively high values up to 8 μg.m^{-3} HNO$_3$ at Rome compared to
values in The Netherlands. This shows the advantage of sampling
techniques based on a relative short sampling period of about 30 minutes,
such as the thermodenuder and the wet annular denuder.

3. DRY DEPOSITION FLUX MEASUREMENTS
The dry deposition of a component may be calculated from the
product of its dry deposition velocity and air concentration. The yearly
average dry deposition flux of nitrogen-containing components in The

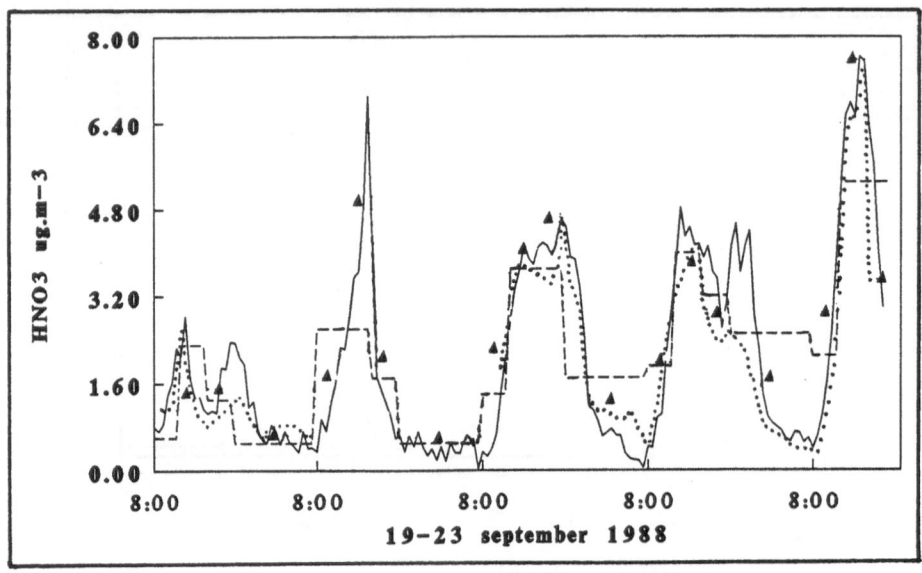

Figure 4: Nitric acid concentration in air at Rome in the period 19-23
September 1988 measured by the denuder/filter pack
configuration (- -), the HNO_3/NH_4NO_3-thermodenuder (..), the
wet annular denuder (---) and average values of all
participants (▲).

Netherlands is estimated within a range of 40% relative. This large
uncertainty is mainly caused by the limited data of air concentrations
for NH_3, HNO_2, HNO_3 and NH_4^+-containing aerosols <u>and</u> the estimated value
of their dry deposition velocities (e.g. SO_2 0.9 cm.s^{-1}, NH_3 1.5 cm.s^{-1},
HNO_3 2.0 cm.s^{-1}, HNO_2 0.9 cm.s^{-1}, NH_4NO_3 0.1 cm.s^{-1} and $(NH_4)_2SO_4$ 0.1
cm.s^{-1}).

As no fast detection systems - with a response time in the order of
seconds - are available for nitrogen-containing gases and aerosols
(except for NO_x), eddy-correlation techniques cannot be used to determine
the dry deposition velocity of these gases. Tower-based measurements of
the air concentration and meteorological parameters at two (or more)
heights has been used to calculate the dry deposition flux by means of
micrometeorological procedures, such as the Bowen ratio method. As a
consequence of the present available inadequate analytical equipment,
relatively few and time restricted (weeks to months) field experiments
have been performed in relation with dry deposition flux of
nitrogen-containing components. Dry deposition velocities depend on many
factors, such as meteorological conditions, type of vegetation, wetness
of vegetation and reactivity of the component. Therefore, results
obtained at one location during relatively short field experiments may
not be representative for another location.

Automated thermodenuders and the wet annular denuder with their
sampling times in the order of minutes and reproducibility better than 5
%, seem promising instruments for dry deposition flux measurements of

NH_3, HNO_3, HNO_2, HCl and NH_4NO_3 based on gradient measurements. Especially, the dry deposition flux of NH_3 in The Netherlands is of importance for effects of air pollution and its measurement by the thermodenuder is discussed below.

4. NH_3 GRADIENT MEASUREMENTS

Ambient air measurements (n=493) at Petten, The Netherlands for NH_3 by two parallel V_2O_5-coated thermodenuders resulted in a coefficient of correlation of 0.9962 in the range of 0.1 - 29 $\mu g.m^{-3}$ NH_3 with a linear regression of 0.97 and an intercept of 0.06 $\mu g.m^{-3}$ NH_3. This demonstrates the potential application of automated annular denuder systems for dry deposition flux measurements based on the gradient method. Two sampling methods were investigated to sample air at two heights (e.g. the gradient method over a forest involves tower-based measurements at 20 and 30 m height):
1. sampling at turbulent flow conditions by a flow of 2 $m^3.min^{-1}$ with a tube of 30 m length and 7 cm ID, which results in a residence time of about 0.3 s
2. sampling at laminar flow conditions by a flow of 10 $1.min^{-1}$ with a NaOH-coated tube of 30 m length and 1 cm ID.

It was observed that both sampling methods showed interferences either by sorption/desorption of NH_4NO_3 resulting in variation of the NH_3 air concentration as a function of temperature and relative humidity (1) or sorption of NH_3 on the wall of the sample tube resulting in strong tailing (2).

Starting July 1989 until May 1990 at two locations in The Netherlands gradient measurements of NH_3 are performed with V_2O_5-coated thermodenuders placed at 0.5 and 2 m and 20 and 30 m over heather and a forest, respectively. The NH_3 gradient is measured automatically during this period every hour with a sampling time of 10 minutes.

Acknowledgements-Research on the development of the thermodenuder and the wet annular denuder was funded by both the Dutch Ministry of Economic Affairs and the European Community (Contract N°: EV4V-00883-C).

REFERENCES
(1) Slanina J., Keuken M.P., Woittiez J., Mallant R., Janssen A.J. and Das H.A. (1989) Research on atmospheric acidification. ECN-1989.
(2) Keuken M.P., Wayers-IJpelaan A., Möls J.J., Otjes R.P. and Slanina J. (1989) The determination of ammonia in ambient air by an automated thermodenuder system. Atmos. Environ. (in press).
(3) Keuken M.P. (1989) Determination of acid-deposition-related compounds in the lower atmosphere. ECN-218.
(4) Klockow D., Niessner R., Malejczyk M., Kiendl H., vom Berg B., Keuken M.P., Wayers-IJpelaan A. and Slanina J. (1989) Determination of nitric acid and ammonium nitrate by means of a computer-controlled thermodenuder systems. Atmos. Environ. (in press).
(5) Keuken M.P., Schoonebeek C.A.M., van Wensveen-Louter A. and Slanina J. (1988) Simultaneous sampling of NH_3, HNO_3, HCl, SO_2 and H_2O_2 in ambient air by a wet annular denuder system. Atmos. Environ. 22, 2541-2548.

<u>NOUVEAU DISPOSITIF INERTIEL ET DIFFUSIONNEL POUR LA MESURE
DE LA GRANULOMETRIE DES AEROSOLS</u>

D. BOULAUD, M. DIOURI*, C. COMPER, G. MADELAINE
Laboratoire de Physique et Métrologie des Aérosols
IPSN/DPT/SPIN/LEPA
Commissariat à l'Energie Atomique
BP 6, 92265 FONTENAY AUX ROSES CEDEX, FRANCE

<u>Résumé</u>

On décrit les performances d'un nouveau dispositif inertiel et
diffusionnel (SDI 2000) permettant de couvrir une dynamique de 2000
sur un domaine de dimension compris entre 0,0075 et 15 µm. Le
principe de réalisation choisi est l'association en série d'un
impacteur en cascade avec une batterie de lits de billes de verre.
Après une description succincte de l'appareil, les performances sont
données et le SDI 2000 est comparé aux méthodes fondées sur les
propriétés électriques des aérosols (Analyseur Electrique d'Aérosol
et Analyseur Différentiel de Mobilité Electrique). Enfin, on donne
un exemple d'application concernant la mesure des aérosols émis par
un moteur diesel.

1. INTRODUCTION

L'objectif fixé pour ce nouveau Spectromètre Diffusionnel et
Inertiel (SDI 2000) est de couvrir un domaine de dimension correspondant
à une dynamique de 2000 dans le domaine 0,0075 à 15 µm et de fournir une
distribution en dimension massique correspondant au comportement
aérodynamique réel de l'aérosol mesuré. Pour ce faire, on a choisi de
caractériser d'une part, sont comportement inertiel pour les dimensions
supérieures à 0,3 µm et d'autre part, son comportement diffusionnel pour
les dimensions inférieures à cette valeur.

Pour atteindre cet objectif, le principe de réalisation adopté est
d'associer en série un impacteur en cascade et une batterie de diffusion
composée de canaux disposés en parrallèle. Pour accéder à la distribution
en dimension massique d'un aérosol on effectue donc, premièrement, une
sélection inertielle en recueillant les masses déposées sur les
différents étages de l'impacteur et, deuxièmement, une sélection
diffusionnelle en recueillant les masses déposées sur les filtres
placées en aval des différents canaux où l'aérosol aura diffusé.

Le fonctionnement des impacteurs en cascade est maintenant assez
bien maîtrisé et notre choix s'est porté sur l'Andersen Mark II. Le
débit nominal de cet appareil est fixé à 28,3 1/min ce qui nous fixe le
débit de passage dans chaque canal une fois adopté le nombre de canaux

---

* Laboratoire de Physique des Aérosols et Transferts des Contaminations,
  Université PARIS XII et Université d'OUJDA, Maroc

composant la batterie de diffusion.

Les principales contraintes qui pèsent sur le choix d'un type de batterie de diffusion pouvant être utilisé dans le SDI 2000 sont essentiellement de trois ordres :

1. connaissance des lois de diffusion,
2. risque de colmatage du milieu diffusif lors du prélèvement d'aérosol hautement concentré,
3. coût du matériau constituant le milieu diffusif et de la réalisation de la batterie de diffusion.

Les deux modèles de batterie de diffusion les plus répandus sont les canaux cylindriques ou parallélépipédiques et les grilles. Pour notre part, nous avons choisi le lit granulaire constitué de billes de verre sphériques qui est un milieu couramment utilisé pour traiter les problèmes de filtration des effluents gazeux mais qui a notre connaissance, n'a jamais été utilisé comme milieu diffusif dans une batterie de diffusion. Ce média qui permet de répondre aux contraintes 2. et 3. a nécessité une étude expérimentale des lois de diffusion d'un aérosol dans un lit granulaire afin de lever l'incertitude sur la connaissance de ces lois (1).

2. DESCRIPTION ET FONCTIONNEMENT DU SDI 2000

On trouvera une description complète de cet appareil dans plusieurs références (2 ; 3). Néanmoins on décrit succinctement les principaux éléments du SDI 2000 et on rappelle brièvement son fonctionnement.

La figure 1 représente le schéma de principe du SDI 2000. La partie inertielle est constituée par l'impacteur en cascade Andersen Mark II comprenant huit étages et dont les diamètres de coupure s'échelonnent de 0,35 à 7,5 µm pour le débit nominal de 28,3 l/min. La partie diffusionnelle est constituée par six tubes en parallèle de 20 cm de long et 4 cm de diamètre qui contiennent les lits granulaires de différentes profondeurs. Le diamètre des billes varie de 1 à 5 mm selon l'efficacité de collection souhaitée. Le sixième canal reste vide et sert de canal de référence. En aval de ces canaux six filtres recueillent les particules ayant traversées l'impacteur et les différents lits granulaires. Le débit total est celui correspondant au débit nominal de l'impacteur et le débit dans chaque canal est maintenu constant par l'intermédiaire d'orifices critiques.

La durée d'échantillonnage et le volume d'air prélevé dépendent de la concentration de l'aérosol et de la méthode d'analyse utilisée pour déterminer les masses recueillies sur les cibles de collection de l'impacteur et les filtres placés en aval de la batterie de diffusion.

A partir de ces données on restitue la granulométrie de l'aérosol, à partir d'un algorithme fondé sur l'inversion numérique de l'intégrale de Fredholm, en utilisant la méthode itérative non linéaire introduite par S. TWOMEY (4).

3. PERFORMANCES DU SDI 2000

L'étude des performances du SDI 2000 a été centrée essentiellement sur l'influence de la nature de la surface de collection de l'impacteur vis-à-vis du piégeage des particules les plus fines. L'aérosol d'essai choisi est celui produit par un générateur d'uranine utilisé pour le contrôle des filtres de très haute efficacité selon la norme AFNOR NFX-44011.

Notre mesure de référence est constituée par l'utilisation de plaques de verre enduites de graisse comme surfaces de collection dans l'impacteur (2). La figure 2 représente un exemple d'un résultat de mesure, on a transcrit d'une part, l'histogramme restitué à partir du traitement des données selon la méthode de TWOMEY et d'autre part, l'ajustement de cet histogramme par une loi log-normale à partir du diamètre moyen massique et de l'écart type géométrique obtenus. On notera que l'histogramme tronqué pour des particules de diamètre supérieur à 0,3 µm est dû au fait que dans le générateur d'aérosols d'uranine on a une cascade de deux buses qui assurent le piégeage des plus grosses gouttelettes. La reproductibilité de ces mesures est très bonne car sur plus de dix essais identiques les variations sur le diamètre moyen massique est l'écart type géométrique n'excèdent pas 5 %.

L'influence de la surface de collection dans l'impacteur a été étudiée à partir de cinq types de surfaces : plaques de verre et plaques métalliques avec ou sans graisse et filtres.

La présence de graisse sur les surfaces de collection est recommandée pour le piégeage des grosses particules (> 1 µm) qui ne peuvent pas ainsi rebondir ou être réentrainées (2). Dans le cas des particules plus fines (< 1 µm) cet effet est encore présent car on constate bien que le diamètre moyen massique déterminé avec des plaques sans graisse est toujours plus petit. Les variations induites sur le diamètre moyen massique restent faibles de l'ordre de - 10 % mais significatives.

Comme on peut l'observer sur la figure 3 les effets les plus importants sont obtenus avec les filtres. Dans ce cas les variations induites sur le diamètre moyen massique et l'écart type sont respectivement de l'ordre de + 30 % et + 40 %. L'explication probable de ce phénomène est que les filtres du fait de leur non planéité et de leur perméabilité à l'air piègent une fraction importante des particules les plus fines qui se retrouvent alors comptabilisées dans l'impacteur plutôt que dans la batterie de diffusion. Il est donc fortement recommandé d'utiliser soit des plaques de verre ou des plaques métalliques enduites de graisses, le petit écart observé entre ces deux supports provenant sans doute d'une légère différence dans la distance séparant l'ajutage d'accélération et la surface de collection.

4. COMPARAISON DU SDI 2000 AVEC DES METHODES FONDEES SUR LES PROPRIETES ELECTRIQUES DES AEROSOLS

Le générateur d'aérosols évoqué au paragraphe précédent est encore utilisé comme source de particules pour cette comparaison, et on fait varier la granulométrie des aérosols en modifiant son fonctionnement.

Les appareils comparés avec le SDI 2000 sont l'EAA (Analyseur Electrique d'Aérosols) et le DMPS (Analyseur Différentiel de Mobilité Electrique).

Les principaux résultats de cette comparaison sont transcrits sur les figures 4 et 5 représentant respectivement les comparaisons portant sur les diamètres moyens massiques et les masses produites par le générateur par unité de temps. On constate donc sur la figure 4 que les diamètres déterminés par les appareils électriques sont toujours plus faibles, l'écart devenant important lorsque le diamètre mesuré au SDI dépasse 0,5 µm.

L'écart observé pour les valeurs de diamètres aux environs de 0,2 µm est faible mais il s'accentue au fur et à mesure que le diamètre mesuré au SDI croît. La principale raison de ce décalage croissant réside sans doute dans la mauvaise prise en compte des particules les plus grosses par les appareils électriques, en particulier pour le DMPS

du fait de la présence d'un impacteur à l'entrée de la prise
d'échantillon. Cette hypothèse semble confirmée lorsque l'on compare les
mesures des masses générées par unité de temps (figure 5). En effet, si
l'accord est très bon pour les masses générées les plus faibles (en
absence de grosses particules), l'écart se creuse en particulier pour le
DMPS lorsque les masses générées augmentent (présence de plus en plus de
grosses particules).

5. MESURE DES AEROSOLS EMIS PAR UN MOTEUR DIESEL

Une des applications particulièrement intéressante du SDI 2000
concerne la caractérisation de la distribution en dimension massique des
aérosols émis par les moteurs diesel, car leur diamètre moyen est
souvent compris entre 0,1 et 0,5 µm. Ce domaine de valeur correspond
d'une part, à la limite inférieure d'utilisation des impacteurs et
d'autre part, à la limite supérieure d'utilisation des batteries de
diffusion, ce qui implique que le choix de l'un ou l'autre de ces
dispositifs induira un manque de résolution dans la partie haute ou
basse de la granulométrie. L'emploi des méthodes électriques pose des
problèmes délicats de dilution car à l'échappement ces aérosols sont
très concentrés. De plus, dans ce cas, comme on a pu le constater
ci-dessus l'accès à la fraction supermicronique de l'aérosol est limité.

A partir de ces considérations, une série de mesures a été
effectuée avec le SDI 2000, sur les aérosols émis par les moteurs
diesel. La description du détail des conditions opératoires sortirait du
cadre de cet exposé et on se reportera aux articles détaillés (5).
Néanmoins nous avons représenté sur la figure 6 un exemple représentatif
d'une distribution en dimension massique obtenue. On constate sur cette
figure qu'il apparaît bien un mode significatif pour un diamètre compris
entre 0,2 et 0,3 µm mais que la fraction supermicronique n'est pas
négligeable ce qui limite l'utilisation des méthodes électriques dans ce
cas.

REFERENCES

(1) DIOURI M., BOULAUD D., et MADELAINE G., (1986). Collection of fine
    particles by granular bed. In Aerosols Formation and Reactivity,
    p. 842, Pergamon Press.
(2) DIOURI M., (1987). Contribution à l'étude du comportement
    aérodynamique des aérosols. Mise au point du SDI 2000.
    Rapport CEA-R-5412, Thèse d'état de l'Université PARIS XII.
(3) BOULAUD D., DIOURI M., (1988). A new inertial and diffusional device
    (SDI 2000). J. of Aerosol Science, 19, (7), 927-930.
(4) TWOMEY S., (1977). In introduction to the mathematics of inversion
    in remote sensing and indirect measurements. Elsevier, Amsterdam.
(5) MARDUEL J.L., BOULAUD D., GAUDICHET A., (1987). Emissions diesel :
    analyse des polluants non réglementés. In Pollution de l'air par
    les transports, Pollution atmosphérique, n° spécial, 113-123.

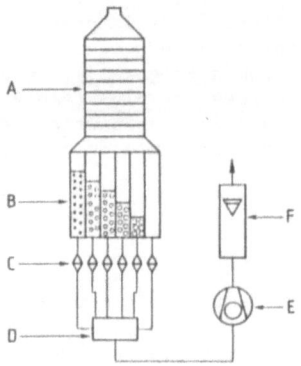

A : impacteur en cascade
B : lits granulaires
C : porte-filtres
D : volume avec 6 orifices critiques
E : pompe
F : débit mètre

Figure 1 - Schéma de principe
du   SDI

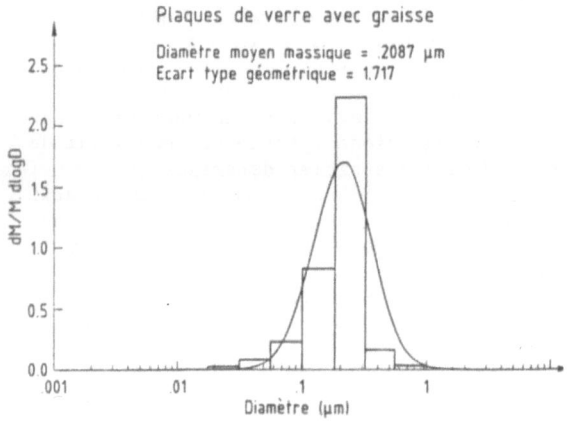

Figure 2
Exemple de résultats
obtenus avec le SDI

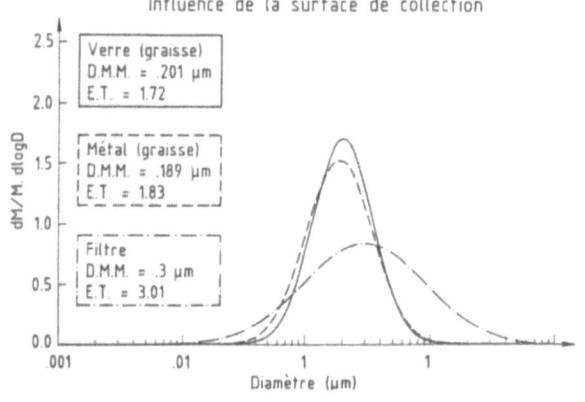

Figure 3
Influence de la surface
de collection dans
l'impacteur

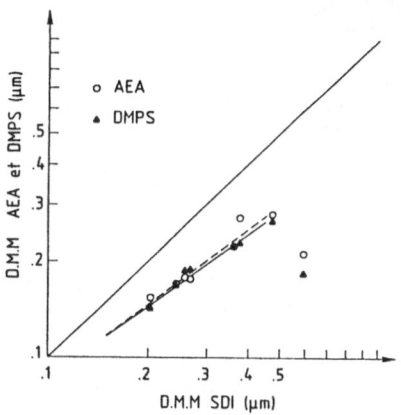

Comparaison des diamètres

○ AEA
▲ DMPS

Figure 4
Comparaison entre les diamètres
moyens massiques obtenus avec le
SDI et les méthodes électriques

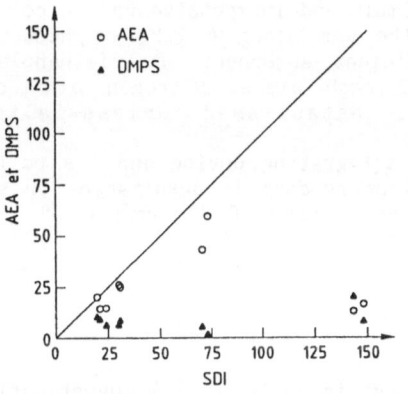

Comparaison des masses générées
(mg/h)

○ AEA
▲ DMPS

Figure 5
Comparaison des masses générées
mesurées avec le SDI et les
méthodes électriques

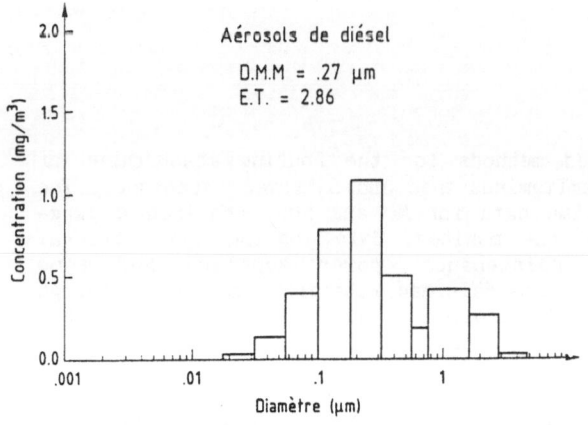

Aérosols de diésel

D.M.M = .27 μm
E.T. = 2.86

Figure 6

Exemple de résultats
obtenus avec le SDI
concernant les aérosols
émis par un moteur diesel

A PASSIVE DIFFUSION TUBE SAMPLER FOR THE MEASUREMENT OF ATMOSPHERIC NO2. A NEW APPROACH.

D.H.F. Atkins
Environment Institute - Chemistry Division
Joint Research Centre - Ispra (VA) ITALY

ABSTRACT

Methods currently used for the determination of NO_2 in the atmosphere require power supplies and some protection from the environment and, in the case of the generally accepted chemiluminescence analyser, a large initial capital outlay for the purchase of the instrument, data logger and retrieval systems as well as regular skilled maintenance in use. These disadvantages have hindered the accumulation of a data base of NO_2 concentrations similar to that available for SO_2 in some countries.

Recently a passive diffusion tube sampler has been validated for ambient NO_2 sampling. This simple and inexpensive device collects by molecular diffusion of the gas along a tube of accurately known dimensions to an efficient absorbent of triethanolamine coated onto stainless steel mesh discs. Nitrogen dioxide is determined by the well established Griess-Saltzman spectrophotometric method.

The sampler described is an integrating device and its reliance on diffusion for NO_2 collection renders it unsuitable for short term measurements. Its collection rate of "72 cm^3 hr^{-1}" gives, however a limit of detection of about 200 ppb hours allowing concentrations down to 1-2 ppb to be measured with sampling periods of one week.

In this paper the preparation, analysis and performance of the sampler are discussed and data are presented for its validation in the field against the chemiluminiscence method. Finally examples are included of its use in environmental investigations, some of which could hardly have been contemplated with hitherto available methods.

1. INTRODUCTION

The two most widely used methods for the routine measurement of atmospheric NO_2 are the chemiluminescence and Saltzman techniques. The former, which gives real time data for NO and NO_2, requires a large initial capital outlay for the monitor, data logging and retrieval systems, regular skilled maintenance, power supplies and some protection from the weather. The Saltzman (chemical) method, although simpler, also requires power supplies protection from weather and regular maintenance. For these reasons neither method is attractive for survey work.

Recently a new sampler for NO_2 has been introduced which, with its low cost and independence from power and maintenance requirements is particularly convenient for surveys allowing them, if necessary, to be carried out on a hitherto impractically large scale. The sampler is a passive device which samples by molecular diffusion and is based on that developed by Palmes (1) in 1976 for NO2 measurements at the ppm level in occupational hygiene studies. The demonstration of this sampler's suitability for determinations at the ppb level in the outdoor environment is the subject of this paper.

2. THE NITROGEN DIOXIDE PASSIVE SAMPLER

The sampler (Fig. 1) consists of a plastic, usually acrylic, tube of accurately known dimensions approximately 7cm long and 1 cm internal diameter with ends machined to accommodate closely fitting polythene caps. Two stainless steel mesh discs coated with triethanolamine, an efficient absorbent for NO2, are held in position at one end of the tube by a polythene cap which is usually coloured for reference. The other end of the sampler is sealed with a white cap. In use the sampler is mounted vertically in positions of unrestricted air movement with the coloured cap uppermost. The lower, white cap is removed at the onset of sampling, allowing NO2 to be transported by molecular diffusion up the tube to the absorbent, and replaced at the end. The nitrogen dioxide collected is measured as nitrite by the well established Griess-Saltzman spectrophotometric analytical method.

The sampling rate is determined by Fick's Law (Fig. 2) the amount Q collected in time t being given by

$$Q = \frac{-D(C_1 - C_2)At}{L}$$

where D is the diffusion coefficient of NO2 in air, C_1 and C_2 are the concentrations of NO2 in the atmosphere and at the surface of the absorbent respectively and A and L are the cross-sectional area and length of the sampler body.

With an efficient absorbent C_2 may be taken as zero when the equation becomes

$$Q = \frac{-DC_1At}{L}$$

As the diffusion coefficient and tube dimensions are all constants, which may be calculated or measured, the sampler may be regarded as an absolute device, independent of standards for use in the field and only requiring reference materials for the determination in the laboratory of the NO2 collected. For the sampler shown in Fig. 1 the collection rate is 72 cm^3 hr^{-1}.

Sampler preparation is simple. An aqueous solution of triethanolamine, just sufficient to wet the stainless steel discs, is added by micropipette, the tubes are then sealed and stored in a refrigerator until required. The sampler blank, derived from reagents and the pick-up of NO2 during tube preparation, corresponds to an amount of nitrite equivalent to approximately 100 ppb hours exposure. This gives an overall limit of detection of about 200 ppb and allows NO2 concentrations down to almost 1 ppb to be determined with the one week sampling periods often employed.

3. VALIDATION OF THE SAMPLER

The following aspects of sampler performance have been considered (2).

1. Precision
2. The effect of long sampling periods on sampler performance.
3. Accuracy when compared with other sampling methods.
4. The effect of wind speed.
5. The effect of temperature and humidity.
6. Interferences.
7. Breakdown of the photochemical state.

3.1 Precision

Batches of 10 samplers were exposed over an area of about 1 m^2 in a region of homogeneous NO$_2$ distribution for periods of one week. The precision (defined as a coefficent of variation and equal to the standard deviation as a percentage of the mean) is shown in Table 1 for sampling periods between one and eight weeks. For sampling periods of up to four weeks the coefficent of variation is of the order of 6% and apparently independent of NO$_2$ concentration above 5 ppb. Below 5 ppb the precision is poorer although the figure of 15% shown is heavily weighted by one unexplained very poor result. The true figure is probably of the order of 10%. For long (eight week) sampling periods the coefficent is about 12%.

3.2 The effect of long sampling periods on sampler performance

Batches of samplers were exposed for up to eight weeks side by side with samplers exposed for consecutive periods of one week over the same interval. The results, expressed as a ratio between the extended period and consecutive weeks is shown in Table 2. A fall in sampler efficiency of about 1% per week is apparent.

3.3 Accuracy

Accuracy was investigated in comparisons with chemiluminescence analysers and the Saltzman method and urban at rural sites in the UK and Italy. All results (Fig. 2) are for average weekly concentrations. Agreement between the techniques was excellent.

3.4 The effect of wind speed

Air moving across the open end of the sampler during use will generate turbulance in the open end, effectively shortening the diffusion length and causing the sampler to give a spuriously high reading. An attempt to quantify the effect was made by calculating the ratio

$$\frac{NO_2 \text{ measured by diffusion tube sampler}}{NO_2 \text{ measured by chemiluminescence analyser}}$$

over 19 weeks for which average wind speed data were available (Fig. 3). No significant change in the ratio was found for average wind speeds between 1 and 4.5 m sec^{-1} and it was assumed that the effect was too small to be revealed by this approach and must be presumed to lie within the limits of precision quoted earlier.

3.5 The efect of temperature and humidity

The kinetic theory of gases indicates that the diffusion coefficient, and hence the sampler collection rate, increase by about 0.2% °C^{-1}. Allowance may be made for this during sampling if necessary. The triethanolamine absorbent melts at 21 °C and it has been suggested its collection efficency falls by 15% between 27° and 15°C as a result of this change of phase.

No effect from the transition on sampler performance has ever been observed and it should be noted that triethanolamine, which is very hygroscopic, is likely to be present on the discs as an aqueous solution over this range of temperature.

No effect on sampler performance when compared with chemiluminescence analysers has been found with relative humidities between 20% and 90%.

3.6 Interferences

The analytical method used for NO_2 determination is specific and species such as SO_2 and HNO_3 collected by the sampler do not interfere. Compounds such as HNO_2 and PAN, containing nitrite or giving it on hydrolysis, clearly interfere. Neither compound is present in the atmosphere in sufficient concentration to be of importance in the applications discussed in this paper. Both compounds are more reactive than NO_2 and should they be present it is likely that they would deposit on the walls of the sampler without reaching the absorbent. Should this occur, removal of the coated discs from the sampler before analysis would overcome the problem.

3.7 Breakdown of the photochemical state

Although air in the sampler body is isolated from the atmosphere its chemistry is not frozen. The light transmission characteristics of the acrylic tube used for sampling showed some attenuation in the photochemically important region possibly leading to changes in the photochemical equilibrium. This information, together with ozone deposition data, was used to model the system with reactants at typical urban and rural levels. The model showed an increase in NO_2 of about 11% in rural areas and 6% in urban. The photochemical reaction is only of importance during daylight and even in the more sensitive rural areas the effect would be difficult to detect. A practical check on the effect was made by exposing the acrylic samplers, where some light attenuation took place, side by side with samplers coated to exclude all light. No difference in the performance of the two systems was detected.

4. ADVANTAGES AND DISADVANTAGES OF THE NO2 SAMPLER

Advantages
1. Precise and accurate.
2. No maintenance, power or protection from weather required.
3. Easy to prepare, handle and analyse allowing them to be used with a high grid density.
4. No calibration required in the field.
5. Inexpensive (£ 0-40 or 700Lit) and sampler components may be used repeatedly.

Disadvantages
1. Unsuitable for short sampling periods unless levels high.
2. Only average concentrations over the sampling periods.
3. No real time data.
4. Errors may occur with rapidly fluctuating concentrations.

5. APPLICATIONS

As we have seen the simplicity and low cost of the samplers allows them to be used with a high grid density if required. Some applications where this unique advantage has been used will be considered in the remainder of this paper.

The EEC Directive on air quality standards for NO_2 requires monitoring to be carried out in areas where the risk to the public is greatest. Their identification presents many problems. Surveys of NO_2 to assist in their identification were carried out in several UK towns chosen for their large population and high traffic density. In one city, Glasgow (3), about 180 samplers were installed across the city in locations designated as background, that is at least 50 m from large NO_x sources such as busy roads. In these situations changes of NO_2 concentration with distance are relatively small and data from them may be regarded as locally representative and used for mapping NO_2 distribution. Weekly sampling was carried out for the six month period from winter to mid-summer. The high density of sampling points allowed accurate maps of NO_2 distribution to be prepared Average NO_2 concentrations were found to be generally higher to the north of the River Clyde with the maximum (> 35 ppb) the north of the City Centre. A chemiluminescence monitor, as required by the Directive, has been installed in this area (Fig. 5).

A second application, demonstrating the spatial resolution which may be obtained, was in survey around a long established NO_2 monitoring station (Minster House) in the Victoria area of central London (4). About 100 samplers were installed in background situations in a circle of approximatly 0.5 Km radius around the site.

Sampling was carried out for two periods of two weeks. The data from one of them is mapped in Fig. 5. In both periods Minster House was found to be situated in the area of highest average concentration.

Finally a survey has been made across the whole of Great Britain with sampling at 50 sites in the UK Acid Rain Review Groups secondary network (5). The results for the year June 1987 to May 1988 are shown in Fig. 6. Average concentrations are lowest in the north and west (< 8ppb) with the highest concentrations in central and south east England. Data from this survey have been used to validate computer models of NO_2 distribution and in the assessment of acid deposition.

REFERENCES
(1)   PALMES, E.D., GUNNISON, A.I., DIMATTIO, J. and TOMCZYK, C. Personal sampler for nitrogen dioxide. Am. Ind. Hyg. Assoc. J. 37, 570-577 (1976).
(2)   ATKINS, D.H.F., SANDALLS, J., LAW, D.V., HOUGH, A.M. and STEVENSON, K. The measurement of nitrogen dioxide in the outdoor environment using simple diffusion tube samplers. United Kingdom Atomic Energy authority Report AERE R 12133 (1986).
(3)   ATKINS, D.H.F., LAW D.V. and NEATE, M. A survey of nitrogen dioxide in the City of Glasgow and surrounding districts. United Kingdom Atomic Energy Authority Report AERE R 12447 (1989)
(4)   ATKINS, D.H.F., and LAW D.V. Nitrogen dioxide survey in the Victoria area of London using passive diffusion tube samplers. United Kingdom Atomic Energy Authority Report AERE 12107 (1989).
(5)   ATKINS, D.H.F., LAW, D.V., and SANDALLS, R. The distribution of nitrogen dioxide in the United Kingdom. A first report. United Kingdom Atomic Energy Authority Report AERE R 13304 (1989).

TABLE 1

PRECISION OF NO$_2$ SAMPLING WITH PASSIVE DIFFUSION TUBE

NO$_2$ (ppb) Range	Sampling Period					
	1 week		4 weeks		8 weeks	
	Data points	Av. Coeff. Var. (%)	Data points	Av. Coeff. Var. (%)	Data points	Av. Coeff. Var. (%)
0 - 4.9	7	15.5				
5.0 - 9.9	41	6.5	1	4.7	1	11.7
10.0 - 14.9	53	6.7	7	6.1	4	5.9
15.0 - 19.9	46	5.4	3	5.6	2	10.7
20.0 - 24.9	24	5.7	2	8.8	1	16.9
25.0 - 29.9	11	6.9	1	6.7		
30.0 - 34.9	10	5.3				
35.0 - 39.9	2	6.7				

TABLE 2

THE EFFECT OF EXPOSURE TIME

ON SAMPLER PERFORMANCE

SAMPLING PERIOD	CONCENTRATION IN SAMPLING PERIOD
	AVERAGE OF WEEKLY CONCENTRATIONS OVER THE SAME PERIOD
2 WEEKS	1.01 ± 0.05
4 WEEKS	0.97 ± 0.06
8 WEEKS	0.93 ± 0.07

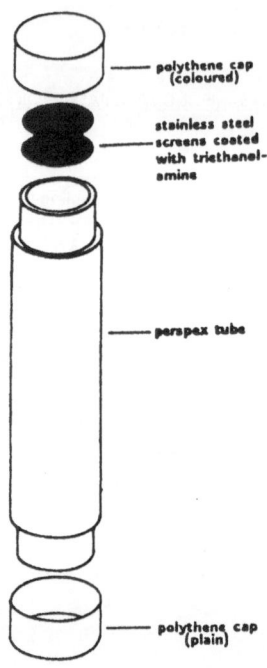

FIG. 1.   DIFFUSION TUBE SAMPLER FOR NO$_2$

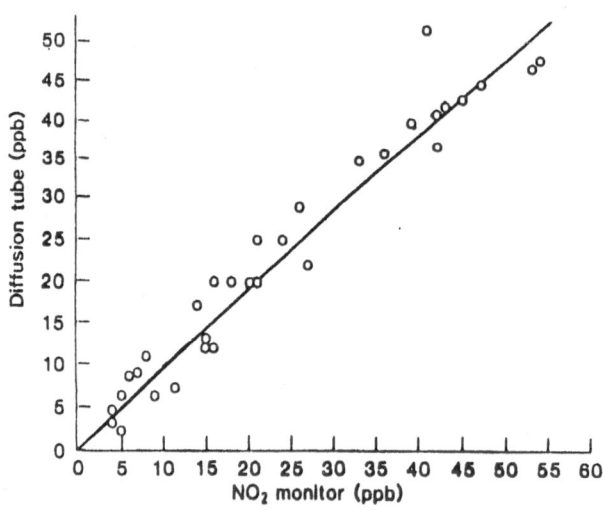

FIG. 2.   COMPARISON OF DIFFUSION TUBE SAMPLER FOR NO$_2$ WITH OTHER METHODS

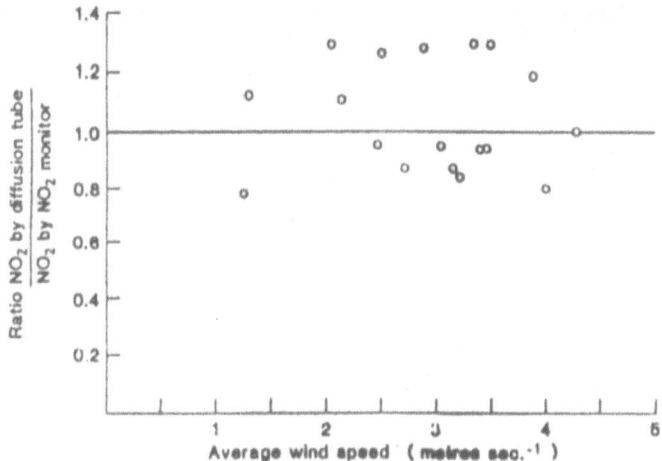

FIG. 3.   EFFECTS OF WIND SPEED ON THE PERFORMANCE OF THE NO₂ SAMPLER

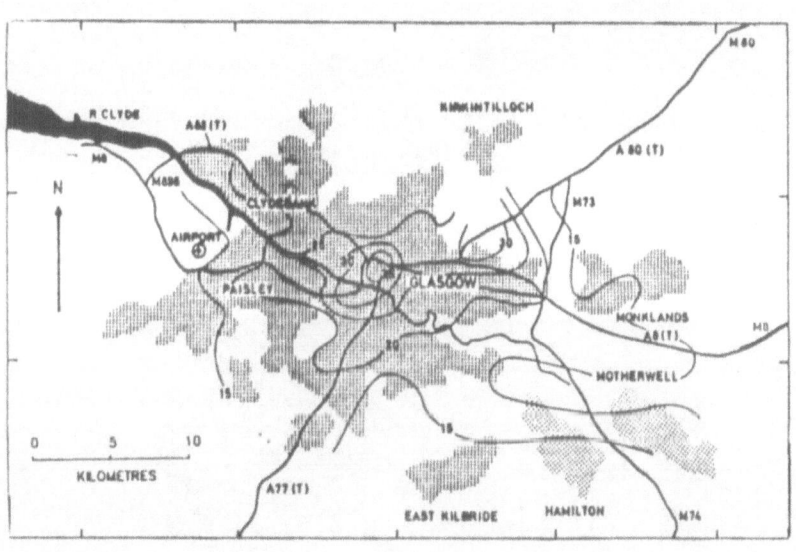

FIG. 4.   NO₂ DISTRIBUTION IN GLASGOW

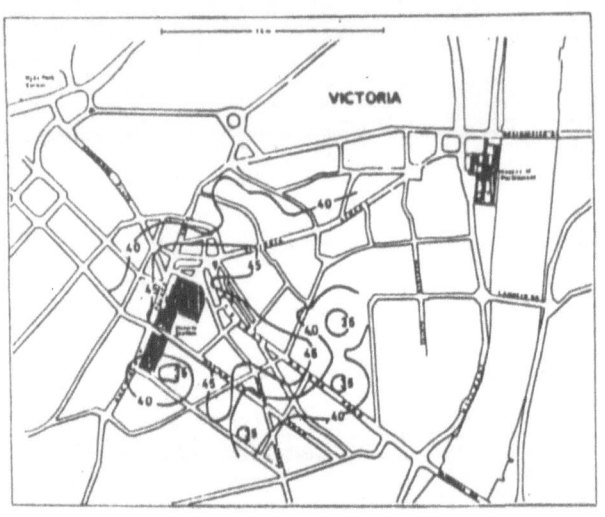

FIG. 5. NO₂ DISTRIBUTION IN VICTORIA

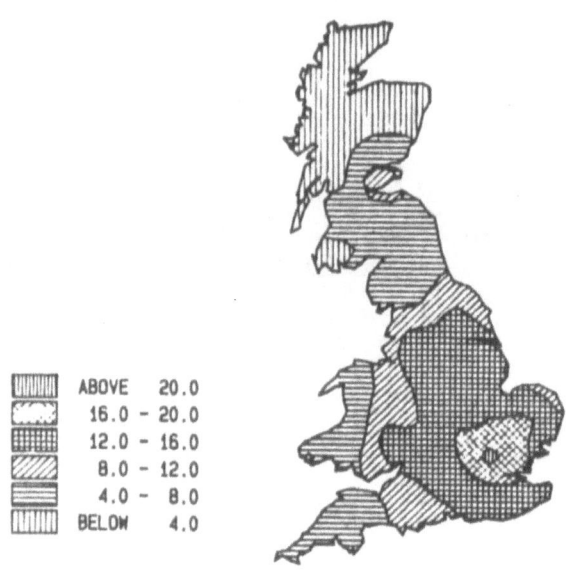

FIG. 6. NO₂ DISTRIBUTION IN GREAT BRITAIN

Studies on the spectroscopy and chemistry of some atmospheric hydroperoxides with a tunable diode laser

J. Bechara, K.H. Becker and K.J. Brockmann
Physikalische Chemie / Fachbereich 9
Bergische Universität – GH Wuppertal D–5600 Wuppertal 1, FRG

Summary

A Tunable Diode Laser Spectrometer (TDLS) has been developed and used to measure hydroperoxy compounds in atmospheric mixing ratios and to study their chemical behaviour under atmospheric conditions. The TDLS was coupled to a White cell which allows in combination with a 2f technique, the measurement of H_2O_2 mixing ratios below 0.2 ppb. The 2f signal was calibrated against H_2O_2 concentrations via UV absorption and checked by comparison with a peroxyoxalate chemiluminescence method. The reactions of some biogenic alkenes with ozone were studied under atmospheric conditions to determine the yield of H_2O_2 formed in these systems.
The absorption cross–sections of methylhydroperoxide near 1320 cm^{-1} were also measured in order to assess the possibility of its determination in the atmosphere.

1. Introduction

Hydroperoxides are important atmospheric constituents formed by photooxidation processes under conditions of low NO_x concentrations. The hydroperoxides play a special role in the aqueous phase oxidation of sulfur dioxide into sulfuric acid. The atmospheric mixing ratios of hydrogen peroxide have been found to lie in the sub–ppbV level for most of the time, reaching ppbV levels in clean dry air. The high solubility of H_2O_2 can lead to significant concentrations in cloud and rain water and thus, result in a significant contribution to the acidification process (1). Methylhydroperoxide on the other hand, is expected to have atmospheric mixing ratios comparable to H_2O_2 (2). In addition, it has been shown to oxidize SO_2 in solution nearly as efficiently as H_2O_2, though much less soluble (3).

Concentration profiles of the H_2O_2 content of rain and fog droplets inside forest areas show a morning maxima (4). This observation suggests an additional H_2O_2 source within forests.

Rapide and accurate in–situ measurements of the atmospheric mixing ratios of these two hydroperoxides under various atmospheric conditions are, thus,necessary in order to assess their role in the acidification process and generally, in the chemistry of the atmosphere.Tunable diode lasers with their narrow linewidth and high brightness provide a highly selective and sensitive tool for trace gas measurements. In the following, a TDLS system allowing the measurement of H_2O_2 mixing ratios in the sub–ppbV levels is described. IR absorption cross–sections of CH_3OOH were also measured and are reported. The possibility of atmospheric monitoring of CH_3OOH is also discussed.

2. Experimental

The laser beam from a Mesa–stripe diode (Spectra Physics) was passed through a chopper for amplitude modulation at 400 Hz and a monochromator for single mode selection. The emerging single mode beam was then splitted in three separate components which were directed, respectively, into a reference cell containing N_2O for absolute wavelength calibration, into an internally coupled Fabry–Perot etalon (FSR $\sim$ 0.01 cm^{-1}) providing the wavelength scale and into the infrared long–path absorption cell. Three separate Cu/Ge detectors were used for detection, the signals of which where then amplified and phase–sensitive detected at the chopper frequency. For high sensitivity measurements, the laser diode injection current was modulated at a frequency of 3 kHz and the signals demodulated at twice that frequency (second derivative technique) providing an enhancement factor of 100 for the signal to noise ratio. The wavelength range of interest was scanned by ramping the diode injection current. A micro–computer controlled the laser scanning, data collection, manipulation and storage.

The absolute calibration of the second derivative (2f) signal in terms of H_2O_2 concentration was carried out by simultaneously measuring the UV absorption of a H_2O_2/synthetic air mixture in a flow stream passing through a UV cell (25.2 cm) and the IR cell. The UV absorption cross–sections of H_2O_2 were taken from the literature (5). The concentration of H_2O_2 was adjusted by varying the temperature of a urea clathrate placed in the flow stream which provided a stable source of pure H_2O_2. An all PTFE–glass flow system and a fast flow rate were used to minimize the loss of H_2O_2 at the walls. However, a residual wall effect was still present and was corrected for by reversing the direction of flow. The linearity of the 2f signal with respect to the normal absorption signal, i.e. the concentration of H_2O_2, was checked by using successively weaker H_2O_2 lines in the range where the laser power stayed constant. The 2f signal was found to be linear down to the detection limit i.e. 10^{-5}. This corresponds to H_2O_2 ambient mixing ratios in the sub–ppb range when the system is coupled to a 136 m White cell. The detection limit for H_2O_2 is 0.2 ppb defined for a S/N of 1 and a time constant of 1 s. The H_2O_2 TDL system calibration was compared with a wet chemical method based on peroxyoxalate chemiluminescence (6). The agreement between the two methods was better than 4 %.

Methylhydroperoxide was prepared by hydrolysis of $(CH_3)_2SO_4$ with aqueous KOH and H_2O_2. After neutralisation with H_2SO_4, CH_3OOH was extracted with ether and dried with anhydrous Na_2SO_4 (7). The product was purified by vacuum distillation and its purity and identity were checked by means of low resolution FTIR spectra, by comparison with spectra reported previously (8).

For the absorption cross–section measurements, synthetic air containing CH_3OOH was slowly flowed through the PTFE–glass line by bubbling air through a glass bubbler containing pure methylhydroperoxide. The concentration of CH_3OOH was determined by UV absorption (9). The total pressure was measured with two calibrated capacitance manometers (MKS baratron). To minimize the relative error due to laser background fluctuations, the peak transmission was kept in the range of 0.2 to 0.75 using IR cells with different lengths (50, 15.5, 5.5 cm).To check for losses on the walls the flow of the gas mixture was reversed and the measurements repeated. The 100 % transmission level was determined before and after every measurement. The laser output was checked for single mode by saturating a N_2O line near the CH_3OOH absorption. The residual transmission was less than 0.5 % in each case.

To study the terpene – ozone reactions under atmospheric conditions a 140 l coolable photoreactor was used. All reactions were carried out under atmospheric pressure with varying terpene concentrations (10–60 ppm) in an excess of ozone (20–80 ppm). The initial ozone concentrations were measured with a 3 m UV–absorption cell and the terpene concentrations were determined by gas chromatography. The effect of water vapor on the reaction system was also investigated.

3. Results

High resolution FTIR spectra of CH_3OOH were first recorded by means of a Bruker Instrument (IFS 120 HR, resolution 3×10^{-3} cm^{-1}) at a total pressure of 3 torr in a 18.7 cm long cell. The whole IR region between 400 and 4000 cm^{-1} was thus surveyed in order to assess the general appearance of the absorption bands and locate strong isolated features suitable for atmospheric monitoring. The strongest absorption band near 2970 cm^{-1} (C–H stretch) was unresolved and did not show any narrow isolated features suitable for TDL monitoring. On the other hand the C–H deformation band near 1320 cm^{-1} showed strong isolated features and was subsequently used for absorption cross–section measurements (Fig. 1). However, even under near Doppler–limited conditions (Pressure = 150 mTorr) this band showed strong overlapping and blending of the lines and an unresolved continuous background contributing about 20% to the total absorption.

The peak absorption cross–section σ_P of the strongest lines was determined by plotting (ln 1/T) against CH_3OOH number density n according to the Beer–Lambert law:

$$\ln (1/T) = \sigma_P \, n \, l \qquad (1)$$

where T is the transmission and l is the cell length in cm. Peak absorption cross–sections were then obtained by dividing the slope of the plot by the cell length. Table I summarizes the peak cross–sections of the main lines observed. The line at 1320.5030 cm^{-1} was found to be the only strong line free from interference due to other atmospheric constituents absorbing in this region and could, therefore, be used for atmospheric monitoring. The dependence of the peak cross–section on total air pressure was then measured for total pressures between 2.5 and 90 torr and is shown in Fig. 2 for the line at $\nu = 1320.5030$ cm^{-1}. Due to the overlapping of the lines it was not possible to derive air broadening coefficients.

Since the unresolved background does not contribute to the 2f signal, an effective cross–section σ_{eff}, obtained by substracting the contribution of the background from σ_P, has been used in evaluating the system sensitivity for CH_3OOH monitoring. The minimum measurable CH_3OOH concentration is given by:

$$n_{min} = a_{min} / \sigma_{eff} \, l \qquad (2)$$

where a_{min} is the minimum detectable absorption i.e. 10^{-5}. The minimum detectable atmospheric mixing ratio m in ppbV is then calculated from the following equation:

$$m(ppbV) = n_{min} \, P_t/P_a \qquad (3)$$

where P_t and P_a are the total pressure in the White cell and atmospheric pressure respectively. This gives a sensitivity of 2–3 ppbV for CH_3OOH for the line at $\nu = 1320.5030$ cm^{-1}.

The gasphase reactions of ozone with isoprene and different terpenes leads to a hydrogen peroxide production. The observed H_2O_2 concentrations increased in the order isoprene < β–pinene < α–pinene < Δ^3carene < d–limonene. All observed hydrogen peroxide concentrations were in the order of a few parts per thousand compared to the reacted hydrocarbons. In presence of water vapor the H_2O_2 concentrations increased significantly. With 10 Torr partial pressure of water the measured H_2O_2 yield increases to about 2% of the reacted d–limonene.

4. Conclusion

A highly sensitive TDL system for atmospheric trace gas detection has been developed and calibrated for H_2O_2. By modulating the laser output wavelength and detecting at the second harmonic, absorptions as small as 10^{-5} can be measured with a time constant of 1 s. The TDL system, coupled with a 136 m White cell, allows H_2O_2 mixing ratios below 0.2 ppbV to be measured. However, the detection limit of the present system for methylhydroperoxide (2–3 ppbV) is still above the expected ambient mixing ratios. High frequency modulation of the diode laser in the MHZ–GHz frequency range would provide an alternative in the future. For the time being, wet chemical methods which can discriminate against H_2O_2 are the only available techniques for measuring CH_3OOH as the predominant component of the organic hydroperoxides.

Futher work is in progress to elucidate the mechanism of the formation of

H_2O_2 and of other hydroperoxides in the system ozone + biogenic alkenes.

Table I: Peak absorption cross–sections (σ_P) of the strongest CH_3OOH lines near 1320 cm^{-1} under near Doppler–limited conditions.

Wavenumber (cm^{-1})	σ_P (10^{-19} cm^{-1})
1319.8295	5.0
1319.8347	5.0
1320.0761	10.5
1320.2948	9.0
1320.5030	8.7
1320.8599	14.0
1321.0154	7.3
1321.1610	4.0
1321.2699	3.7

References

(1)   Lind, I.A., Lazrus, A.L. and Kok, G.L. (1987); Aqueous phase oxidation of S (IV) by hydrogen peroxide, methylhydroperoxide and peroxyacetic acid; J. Geophys. Res. Vol. 92, 4171 – 4177
(2)   Calvert, I.G. et al., (1985), Chemical mechanisms of acid generation in the troposphere; Nature Vol. 317, 27 – 35,
(3)   Lind, J.A. and Kok, G.L., (1986); Henry's law constants of hydrogen peroxide, methylhydroperoxide and peroxyacetic acid; J. Geophys Res. Vol. 91, 7889–7895,
(4)   Jakob, P. and Klockow, D.; privat communication
(5)   Molina, L.T., Schinke, D., Molina, M.J., (1977);UV absorption spectra of hydrogen peroxide vapor; Geophys. Res. Lett. Vol. 4, 580 – 582
(6)   Jakob, P., Tavares, T.T. and Klockow, D.,(1986); Metodology for the determination of gaseous hydrogen peroxide in ambient air; Z. Anal. Chemie Vol. 325, 359 – 365
(7)   Rieche, A. and Hitze, F. (1929), Über Methylhydroperoxid; Ber. Dtsch. Chem. Ges. B. Vol. 62, 2458 – 2474
(8)   Niki, H., Maker, P.D.and Savage, C.M. (1983), FTIR study of the kinetics and mechanism for the reaction HO + CH_3OOH; J. Phys. Chem. Vol. 87, 2190 – 2193
(9)   Molina, M.J. and Arguello, G. (1979); UV absorption spectra of methylhydroperoxide vapor; Geophys. Res. Lett. Vol 6, 953 – 955

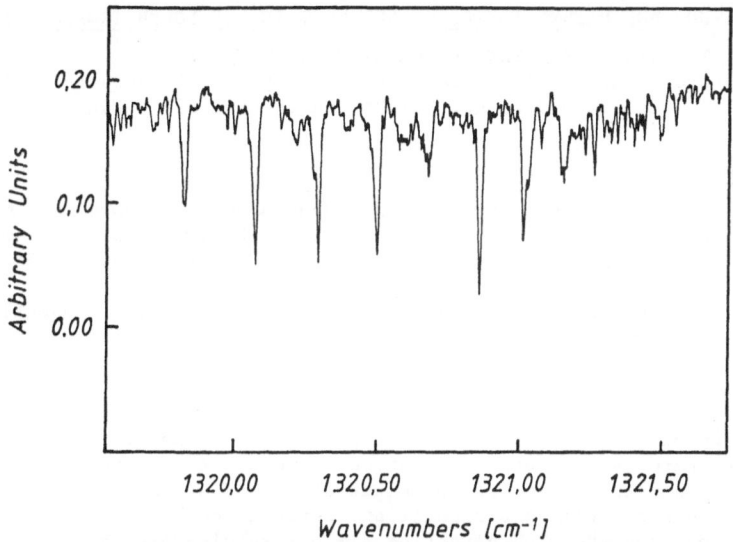

Fig. 1. High resolution FTIR spectrum of CH_3OOH near 1320 cm⁻¹ recorded with a Bruker Instrument (IFS 120 FR) at a total pressure of 3 torr and a resolution of 3×10^{-3} cm⁻¹.

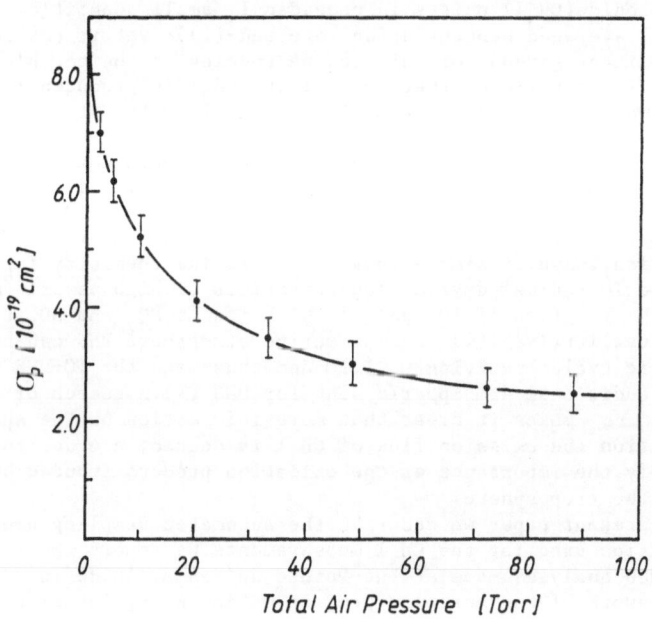

Fig. 2. Dependence of the peak absorption cross–section σ_p of CH_3OOH at $\tilde{\nu} = 1320.5030$ cm⁻¹ on total pressure in synthetic air.

A METHOD FOR METHYL IODIDE MEASUREMENT IN THE LOWER TROPOSPHERE

M. TSETSI, F. PETITET, P. CARLIER, G. MOUVIER
Laboratoire de Physico-Chimie de l'atmosphère, Université Paris 7,
2 Place Jussieu, F-75251 Paris Cedex 05, FRANCE

Summary

The levels of methyl iodide (CH_3I) have been determined over a coastal region of Western Europe,at the Pointe de Penmarc'h-Bretagne-France (53,10grN-7,40grW) during a preliminary campaign in July 1988 and a more systemetic study in June 1989.
We are using an automated sampling procedure by direct injection be-cause of the very aleatory characteristic of methyl iodide recovery yield for various pre-concentration techniques.Methyl iodide has been measured by electron capture(EC) gas chromatography(GC) and was deter-mined in the range of 0,1 to 2,5 parts per trillion(pptv).
The temporal variation in the tropospheric abundance of CH_3I suggests that there is a strong correlation between high concentrations of CH_3I in air and the occurrence of algae at the coastlines of the Pointe de Penmarc'h as well as a significant correlation between these concen-trations and the wind direction.All our results are based on more than 1000 measurements of CH_3I during June 1989;the observed concentrations are of the same order of magnitude with those cited in the literature.

1. INTRODUCTION

Methyl iodide(CH_3I) exists in exceedingly small quantities (0,8pptv is the global averaged concentration reported-(1)), yet it has been found to be the dominant gaseous organic iodine species in the earth's lower at-mosphere (1).It is believed that most of the CH_3I is produced in the oceans possibly by marine algae,and is mainly removed from the atmosphere by solar ultraviolet radiation (lifetime ~5 days).
Chameidis and Davis (1980)(2) and more recently Jenkin,Cox and Candeland (1985)(3),have shown that iodine species may play an important role in the tropospheric photochemical system: reactions involving I,IO,O_3 and NO_2 could be important for atmospheric processes in coastal areas with high CH_3I production.Computer simulations of the iodine chemistry for such re-gions yielded IO radicals daytime concentrations as high as $5x10^8 moles/cm^3$. Apart from the reaction of IO radicals with itself,NO_2 and NO,it has been proposed by Carlier(1985)(4) that in marine atmosphere DMS can be oxidized in a catalytic cycle involving O_3,I,IO and thus make the IO+DMS reaction a significant additional atmospheric sink for DMS (5).A search of the scien-tific literature makes it clear that more information on the spatio-tempo-ral distribution and emission flux of CH_3I is needed,in order to evaluate quantitatively the importance of the oxidation process induced by the IO radicals in the troposphere.

In the present paper we describe the automated sampling procedure by direct injection used for the CH_3I measurements.We report the results of more than 1000 analyses made at the Pointe de Penmarc'h during June 1989 -in the framework of the European project OCEANO NOX-, in terms of some daily profiles;we also discuss the influence of meteorological variables on our measurements.

2. EXPERIMENTAL

Two general approaches for sampling atmospheric halocarbons have been attempted by various scientists, involving air sampling with or without pre-concentration of the sample (6). However, the storage of a sample prior to analysis leads to difficulties for substances, like CH_3I, due to adsorption on the vessel walls or reactions with them. Thus in this paper, we report an automated sampling procedure, by direct injection of the air sample, without any pre-concentration technique.

The concentration of CH_3I is determined by a DELSI DI 700 gas chromatograph equiped with a frequency modulated electron capture detector with a [63]Ni source.

The essential components of the experimental set-up are shown schematically in figure 1.

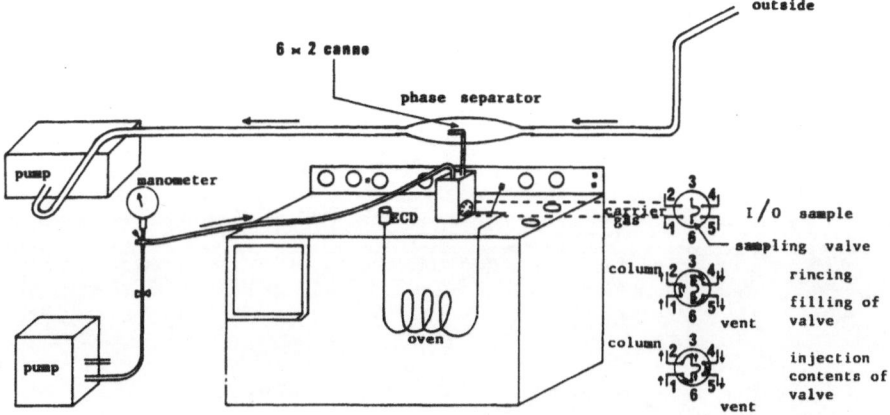

Figure 1. Schematic of the experimental arrangement used for CH_3I measurements.

Outside air was permanently pumped through a principal glass tube, at a flow rate of $2m^3/h$. Two minutes before the injection, a small portion of this flux was flushed by a lateral canalization onto the six-ports injection valve ($V=1,88\ cm^3$), which was automatically operated by a pneumatic jack. The total analysis cycle was of the order of 15 minutes.

Methyl iodide analysis was performed on a 4x1/8 inch stainless steel GC column packed with 10% OV 101 on 80/100 mesh Chromosorb. The temperature of the oven was maintained at 40°C. The detector was operated at 150°C. The carrier gas was Nitrogen and was flowed at a rate of 10ml/min.

Gaseous standards of CH_3I were generated at ppb levels from a permeation tube connected directly to the sampling valve; the repeatability of permeation standards was satisfactory and we have verified a good linearity of the method in the range of 0,5 to 100 ppb.

On the other hand, we have controlled that an identical quantity of CH_3I diluted in the air or in an organic solvent (THF), provided an equivalent response; using standard solutions of CH_3I in this solvent, we have reconfirmed the range of linearity of the EC detector, for methyl iodide's injections corresponding to the contents of $1,88\ cm^3$ of air, for concentrations of 0,1 to 1000pptv.

3. RESULTS AND DISCUSSION

The geographical location of the sampling site is shown in figure 2; the choice of the sampling site is reported elsewhere (7).For wind direction analysis the four sectors shown in fig.2,were used.

During the campaign of June 1989 most of the air masses encountered were advected with North-Westerly to North-Easterly winds (sectors 4 and 1) which can be classified as far atlantic and continental air masses respectively,thus did not favoured our study.We should also stress here the total lack of precipitation events during all this period and a quasi-optimal solar irradiation.These remarks have to be borne in mind in the interpretation of the concentration data.

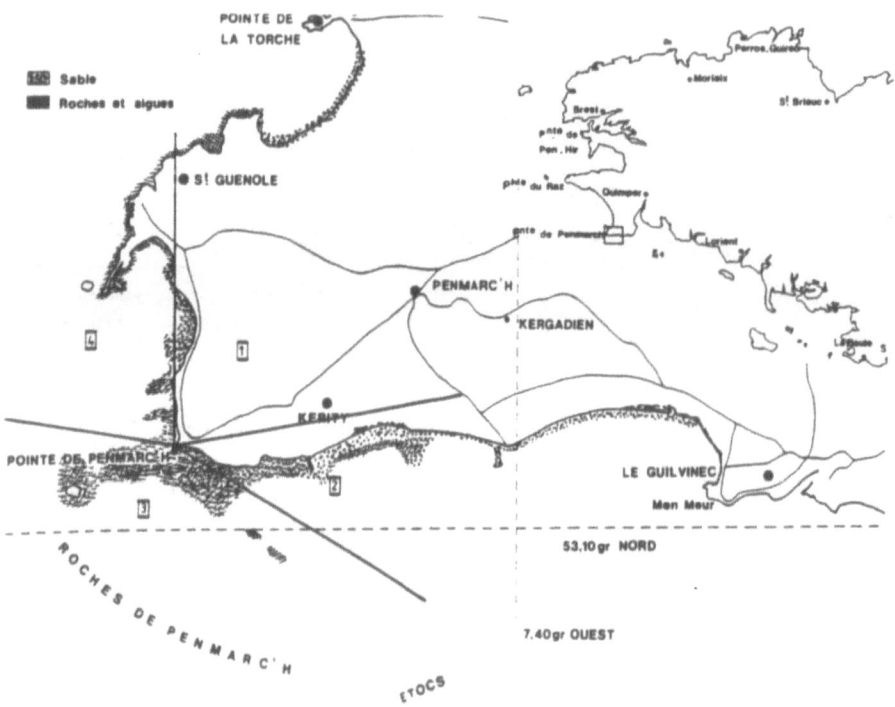

Figure 2. Geographical location of the "Pointe de Penmarc'h" and definition of the sectors used in wind direction analysis.

In figure 3,are reported some daily profiles of CH_3I and the corresponding data on wind direction.Each data point represents an average of 9 to 12 analyses.Low level concentrations occuring in the afternoon,may be ascribed to changes in wind direction from sector 4 (algae field influence) to sector 1 (continental air) which was the most unfavourable;in sector 1 the sampling point can receive methyl iodide produced by the algae field or the ocean,only by the local turbulence.

Figure 4 shows the dependance of CH_3I concentration on wind direction for the four sectors used.The highest CH_3I concentrations are associated with winds in sectors 2 and 3;these sectors include air streams coming from the Atlantic ocean and the local algae field.

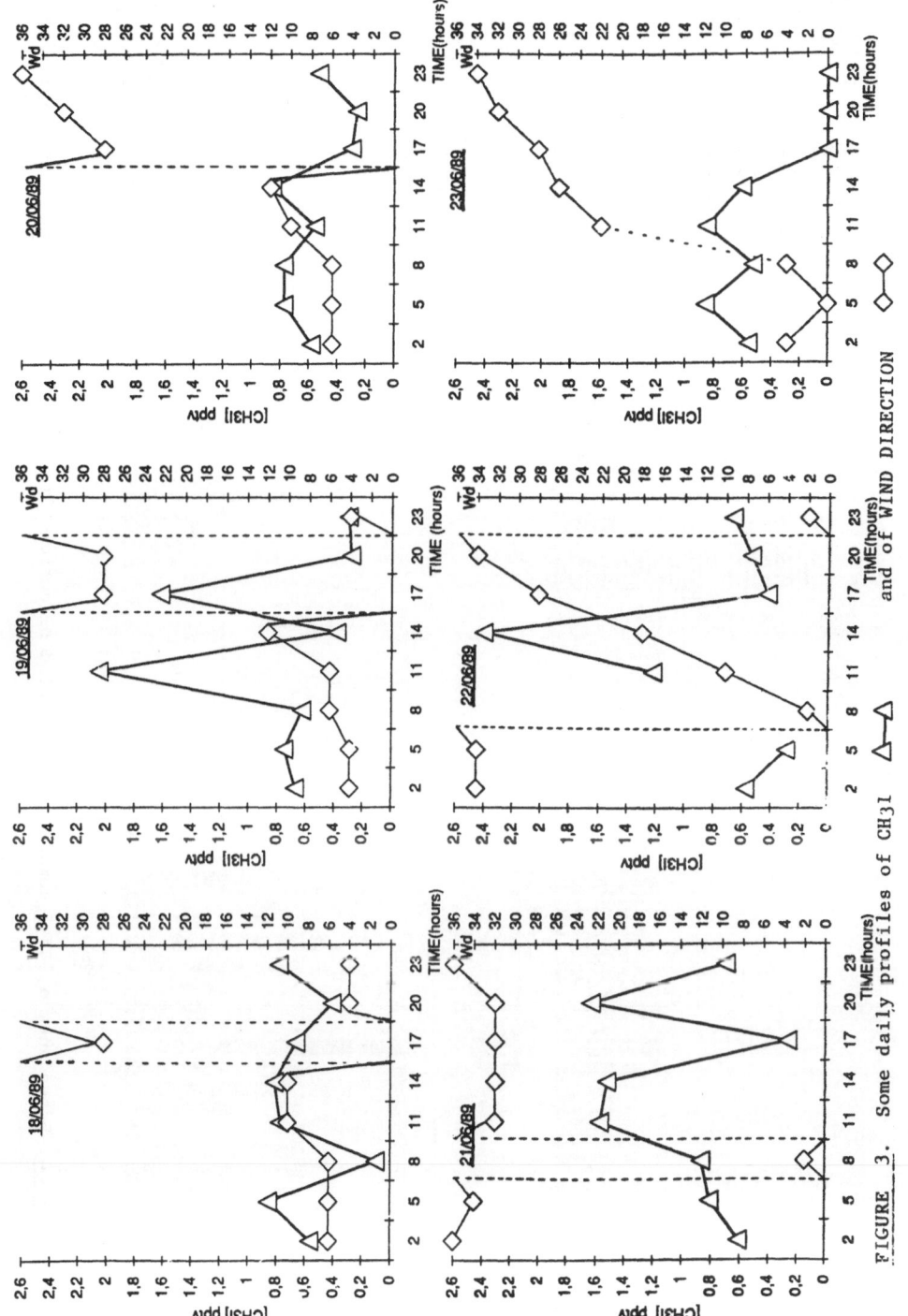

FIGURE 3. Some daily profiles of CH3I and of WIND DIRECTION

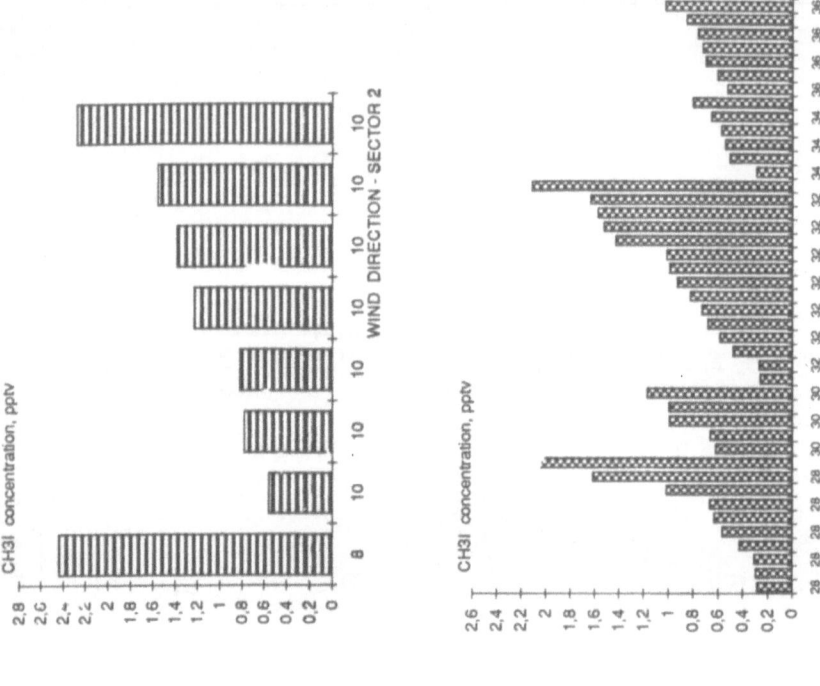

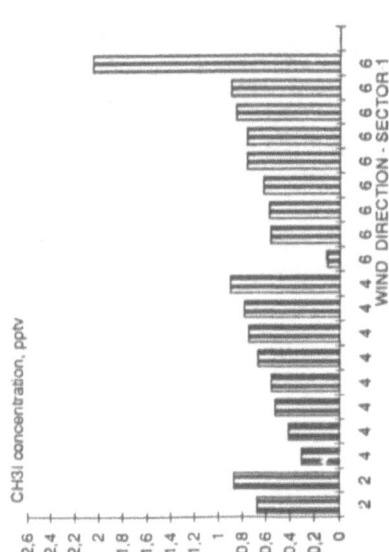

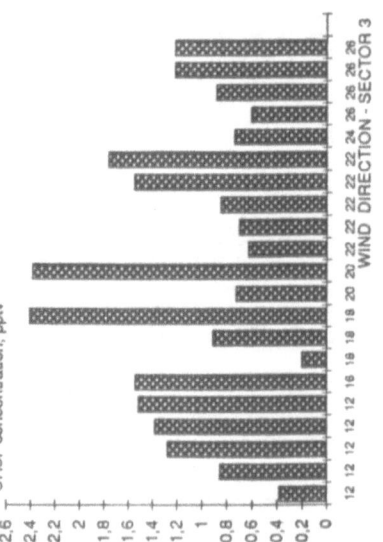

FIGURE 4. Dependance of CH3I concentration on Wind Direction- Cmax=2.44ppt- Cmin=0.10ppt- C̃=0.94±1.1 pptv

4. CONCLUSION

Methyl iodide concentrations in Penmarc'h are slightly lower than those cited in the literature.

However,on the average,the highest concentrations are observed when the wind was blowing from the quadrants E-S and S-W,i.e from the seaside and the ocean.This indicates that the local algae field plays a major role; this role is again consistent with high concentrations reported in high biomass productivity regions.Thus,taking into account,the exceptionally unfavourable meteorological conditions,the results reported in this paper are in reasonable agreement with similar studies and validate the method used.

Further interpretations on these results are actually in progress.

5. ACKNOWLEDGEMENTS

This work has been executed in the framework of the European project OCEANO-NOX which is supported by the Commission of the European Communities and,by a doctoral fellowship (Sectoral grant n° B/88000011, M.Tsetsi) of the Environmental Research Programme of the European Economic Communities.

The participants of the OCEANO-NOX project are grateful to the French Marine Nationale and the "Service des Phares et Balises" for the free access to the signal station at St Pierre and the old light house of Penmarc'h.We would also like to thank the staff of these establishments for their logistical support during the field experiments.

REFERENCES

(1) RASMUSSEN R.A., KHALIL M.A.K., GUNAWARDENA R., HOYT S.D., Atmospheric Methyl Iodide (CH3I).J.Geophys.Res.,87,No C4, 3086-3090, 1982.

(2) CHAMEIDIS W.L., DAVIS D.D., Iodine:Its Possible Role in the Tropospheric Photochemistry.J.Geophys.Res.,85,No C12, 7383-7398, 1980.

(3) JENKIN M.E., COX R.A., CANDELAND D.E., Photochemical Aspects of Tropospheric Iodine Behaviour.J.Atmos.Chem.,2, 359-375, 1985.

(4) CARLIER P., "L'ozone serait-il l'oxydant principal du Sulfure de Dimethyle en milieu océanique?", in Atmospheric Ozone, C.S.Zerefos and A.Ghazi,Eds., Reidel,Dordrecht, 815-819, 1985.

(5) BARNES I., BECKER K.H., CARLIER P., MOUVIER G., FTIR Study of the DMS/ $NO_2/I_2/N_2$ Photolysis System:The Reaction of IO Radicals with DMS.Int. J.Chem.Kin.,19, 489-501, 1987.

(6) RUSSEL J.W., SHADOFF L.A., The Sampling and Determination of Halocarbons in Ambient Air Using Concentration on Porous Polymer.J.Chromat., 134, 375-384, 1977.

(7) PASHALIDIS S., CARLIER P., MOUVIER G., Daily Profile Study of the Atmospheric DMS Concentration Near to an Intense Coastal Source at the "Pointe de Penmarc'h", ibidem.

LABORATORY INTERCOMPARISONS IN THE EMEP
(EUROPEAN MONITORING AND EVALUATION PROGRAMME)

J.E. HANSSEN
Norwegian Institute for Air Research
P.O. Box 64, 2001 Lillestrøm, Norway

Summary

In the air and precipitation quality monitoring programme (EMEP) under
the Convention on Long-range Transboundary Air Pollution, laboratory
intercomparisons of the chemical analyses of relevant samples have been
performed since 1977 in order to ensure data comparability and control
the participating laboratories performance. Although at the start of the
programme recommended methods were described, several different methods
have been in use since EMEP is based on national programmes in 24 different countries without a strict standardization. The present paper describes the organization of the intercomparisons and results from some of
the individual test as well as some trends in the results during the
eleven intercomparisons arranged up to now.

In order to ensure data comparability and to give the participating laboratories in the EMEP analysis programme a possibility to control their performance against common reference samples, interlaboratory tests are organized every year by the Chemical Co-ordinating Centre (CCC). In 1988 the
tenth intercomparison of this kind have been arranged.
The kind of samples distributed to the participants are given in Table
1.
29 laboratories reported 1.384 data to the CCC. 1,252 results were from
synthetic samples and 132 from real samples. More laboratories than before
now perform the complete analysis programme for the precipitation samples.
The range of the ratios between the arithmetic mean values and the theore-
thical values for all the parameters and samples are given in Table 2.
As seen from the table, only a few arithmetic mean values were more than
5% away from the theoretical value.
Table 3 shows the range of the relative standard deviations found for the
same samples and parameters. Most relative standard deviations are between
5 and 10%.
Examples of the graphical presentation of the results are given in Figure
1 and 2. Pairs of samples with almost similar concentrations are used.
The results from the laboratories, marked with a cross and a number are
plotted in a so-called Youden two-sample chart. For the fully drawn axes
the origin is the point for the two expected, theoretical values for the
concentration in the two samples. The origin of the dotted axes are the
two arithmetic mean values found from the reported results. The radii of
the circles in the figures are 10% of the mean value of the two expected
values for sulphate in precipitation and 0.10 pH-units for pH in precipitation.

Most of the points are located within the circles, and in the lower left
or upper right quadrant of the graph, which means that systematic errors
dominate over random errors (both results are either too high or too
low).
Frequency distributions of the results for sulphate and nitrate in precipitation are shown in Figure 3.
Of the 1,252 reported results for the synthetic samples with theoretical
values, 120 (9.6%) were outside $\pm$ 20% of the expected value. 2/3 of these
results were reported from six of the laboratories. Twelve laboratories
had no results more than 20% from the expected value. Three laboratories
had only one result more than 20% away.
 Some examples of the trend in the performance of the EMEP laboratories
during the intercomparisons arranged from 1977 to 1988 are shown in
Figures 4 and 5.
The relative standard deviations found for the results for sulphate and
nitrate in precipitation are shown in Figure 4. For each intercomparison
the relative standard deviation for each of the four samples are shown.
The samples are arranged in increasing concentration order from left to
right.
The overall trends are that the relative standard deviations have decreased during the years. However, two factors make it difficult to discuss
the trends: Every year one or more new laboratories have joined the programme and the relative standard deviations increase when the concentration in the samples approach the lower part of the calibration range for
the analytical methods. In the first intercomparisons most of the samples
had lower concentrations than in the last ones.
The relative difference between the arithmetic mean values and the theoretical values for the sulphate analyses in precipitation samples during the
intercomparisons is shown in Figure 5. The results for the four different
samples in each year are presented in the same way as in Figure 4. As for
the standard deviation, the overall trend is that the arithmetic mean values
are much nearer to the expected ones in the last intercomparisons than in
the first years.

SAMPLES DISTRIBUTED

SYNTHETIC SAMPLES :

4   SO_2 in absorbing solutions   (15 labs)
4   SO_2 on impregnated filters   (9 labs)
4   SO_4 on filters (wet chem.)   (19 labs)
4   SO_4, NO_3, NH_4, H, Na, Mg, Cl, Ca, K
      pH and conductivity in precipitation (30 labs)

REAL SAMPLES :

6   SO_4 on filters (XRF and wet chem.) (26 labs.)

Table 1

MEAN VALUES VS. EXPECTED VALUES

(all data, range for 4 samples).

SO_4	0.95 - 0.96
NO_3	0.96 - 0.97
NH_4	1.01 - 1.03
pH (as H)	0.97 - 1.01
H	1.02 - 1.06
Cl	0.96 - 0.97
Na	0.98 - 0.99
Mg	0.99 - 1.06
Ca	0.98 - 1.12
K	0.99 - 1.06
Cond.	0.96 - 0.98

SO_2	in abs. sol.	1.01 - 1.02
SO_2	on imp.filter	0.97 - 0.98
SO_4	on filter (low)	1.03 - 1.13
SO_4	on filter (high)	0.96 - 1.08

Table 2

RELATIVE STANDARD DEVIATION

(Outliers not included, range for 4 samples).

SO_4	5.2 -	8.9
NO_3	5.1 -	9.8
NH_4	9.0 -	11.1
pH (as H)	5.0 -	8.9
H	5.1 -	7.7
Cl	4.3 -	5.8
Na	4.2 -	5.7
Mg	5.9 -	12.8
Ca	5.0 -	20.2
K	5.9 -	14.3
Cond.	6.3 -	7.6

SO_2 in abs. sol.	8.1 -	10.0
SO_2 on imp.filter	2.8 -	5.2
SO_4 on filter (low)	13.0 -	28.4
SO_4 on filter (high)	5.2 -	13.8

Table 3

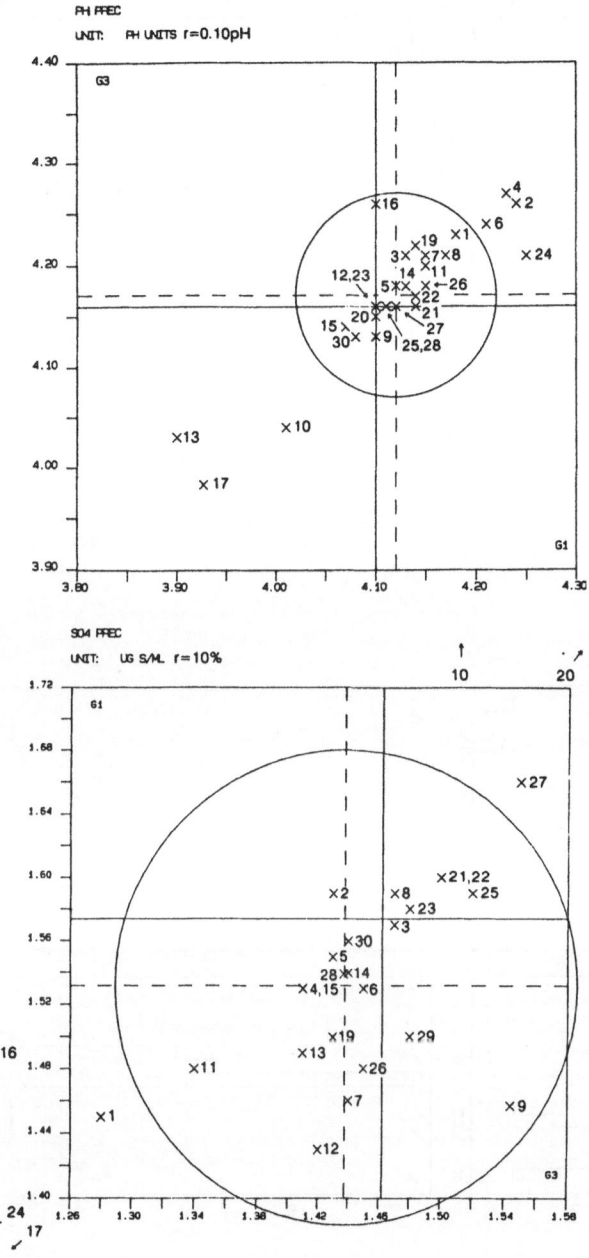

Figure 1 and 2: Youden two-sample chart for the results of two precipitation samples for pH and sulphate.
For explanation: see text.

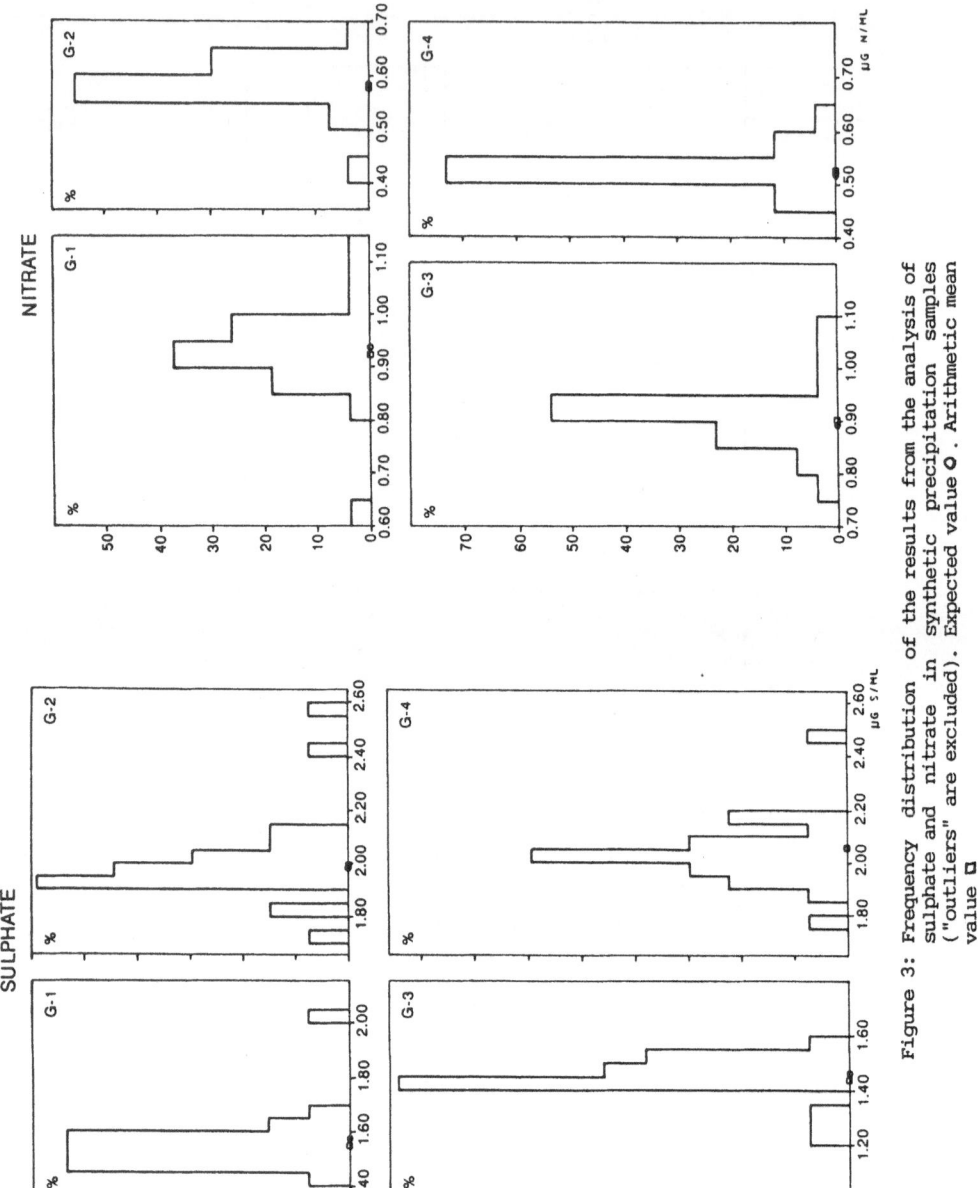

Figure 3: Frequency distribution of the results from the analysis of sulphate and nitrate in synthetic precipitation samples ("outliers" are excluded). Expected value ⬤. Arithmetic mean value ☐

- 42 -

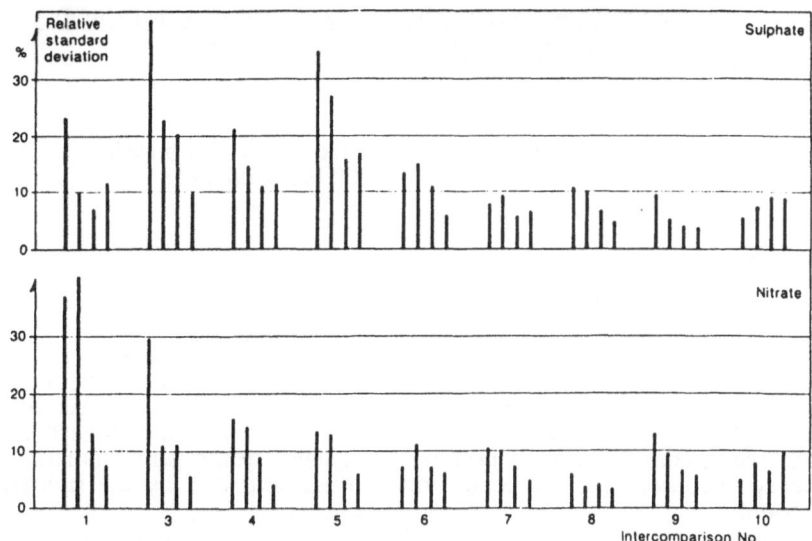

Figure 4: Relative standard deviations of the analyses of sulphate and nitrate in the EMEP intercomparisons frm 1977 to 1988. (After K. Nodop, Personal communication.) For explanation: see text.

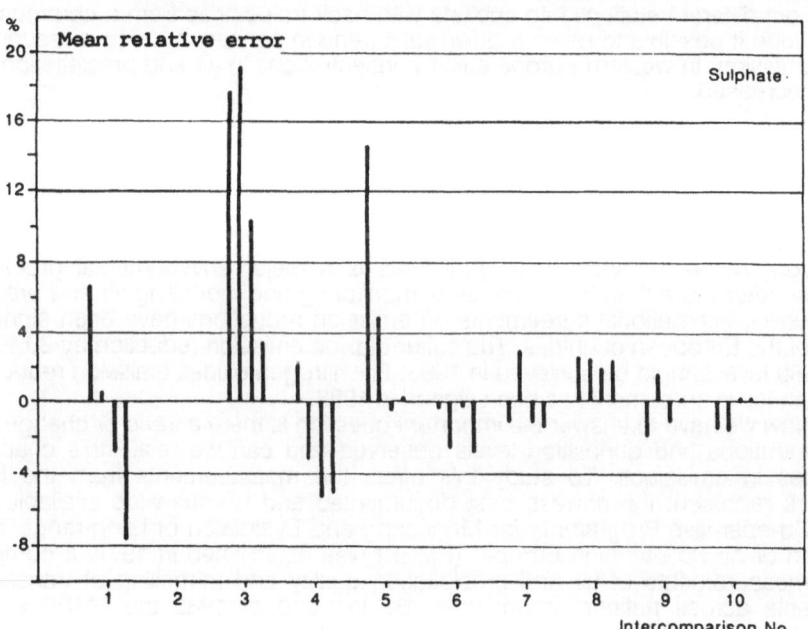

Figure 5: Trend in the mean relative error for the sulphate analyses in the EMEP intercomparisons from 1977 to 1988. For explanation: see text.

AIR AND PRECIPITATION QUALITY AND TRENDS IN EUROPE
RESULTS OF THE EMEP NETWORK

K. Nodop and H.-W. Georgii
Universitätsinstitut für Meteorologie und Geophysik
Feldbergstr. 47, D - 6000 Frankfurt/M 1

Summary

Results of the EMEP network consisting of 87 background stations in 24 countries are summarized in this presentation. The large-scale spatial distribution of S and N compounds is shown in maps for 1985. The areas of maximum concentrations of SO_2 and NO_2 compare well with major emission areas in Europe. Wet deposition patterns of non-marine sulfate, nitrate and ammonium are mainly governed by their concentration fields and less by the precipitation amount.

After international agreements on sulfur and oxidized nitrogen emission reduction have been signed by most European countries it is important to investigate how the changes in emissions affect concentration and deposition levels observed. By a single station statistic looking at averages or frequency distributions no significant trend could be observed. First a stratification of the data from different stations into subsets with back trajectories from a common area made it possible to relate a downward trend in concentration to a reduction in emission. In western Europe sulfur concentrations in air and precipitation have decreased.

1. INTRODUCTION

Long-range transport of air pollutants is a major environmental problem in Europe. After more than two decades of monitoring and modelling air and precipitation quality, international agreements on emission reductions have been signed by most of the European countries. The sulfur dioxide emission reduction by 30 % from the 1980 level should be achieved in 1993. For nitrogen oxides emission reduction a less restrictive agreement has been signed in 1988.

Now we have to answer the important question is there a trend or change in the concentrations and deposition levels observed and can we relate this change to changes in emissions. To study this effect the measurements from the EMEP network represent the newest, best documented and Europe-wide available data. The "Co-operative Programme for Monitoring and Evaluation of Long-range Transmission of Air Pollutants in Europe" (EMEP) was established in 1977. It comprises daily measurements of air and precipitation quality and modelling of transport of pollutants across national boundaries. By the end of 1985 the EMEP network consists of 87 background stations in 24 countries.

The relation between emissions and concentrations/depositions is very complex. The emissions are irregular spaced and vary not only from year to year but also within much shorter time periods. The concentration measured at a given site is the result of physical and chemical transformations and removal mechanisms during atmospheric transport. These processes are depending on meteorological conditions and exhibit strong variations in time. The result is a very noisy data set which makes trend detection difficult. A stratification of the data into subsets according to the most important influences might be helpful to detect a significant change.

This presentation shows that in a subset of data from different stations with back trajectories from a common area a trend in concentration could be related to a change in emissions. With this simple model it is possible to validate emission reductions by observations with only a minimum of additional information. As a matter of general information the large scale sulfur and nitrogen concentration and wet deposition fields in Europe are presented at first.

2. SPATIAL DISTRIBUTION OF CONCENTRATION AND WET DEPOSITION

The annual averages for 1985, as the last year of final data in EMEP, have been used to illustrate the large scale variability of the most important compounds (Fig. 1 and 2). Both the concentration and the deposition fields represent "average minimum fields", since the measurements are taken from background stations only and spike concentrations near emission sources are omitted.

The highest SO_2 concentrations (>11 μgSm^{-3}) are found in central Europe. For sulfate in aerosol (SO_4^A) the area of maximum concentrations is further east with values around 5 μgSm^{-3}. On contrast the area of highest NO_2 concentrations (>8 μgNm^{-3}) is found over the Netherlands. The different location of the SO_2 and NO_2 maximum corresponds very well with the location of the highest SO_2 and NO_x emissions (1).

The annual volume-weighted averages vary in Europe from 0.2 to 3.5 mgS/l for non-seasalt sulfate (nss SO_4^{2-}), from 0.1 to 1 mgN/l for NO_3^- and from 0.2 to 2 mgN/l for NH_4^+. The spatial distribution is similar but the location of the area with highest concentration is different - as could be seen for their precursors. nss SO_4^{2-} concentrations are highest in eastern Europe and NO_3^- concentrations are highest in Denmark and southern Sweden. Two separate maximum areas can be seen for NH_4^+, from northern France to Denmark and from Poland to Romania.

The wet depositon fields (Fig. 2) have been constructed by multiplying the concentration in 1985 at each grid point with the precipitation amount in 1985 at each grid point. The precipitation amount has been derived from the meteorological networks (2). The annual precipitation amount varies between 400 and 1900 mm. The mountain areas at the coast of Norway and Scotland and the alps receive most precipitation. The wet deposition of nss SO_4^{2-} varies from 0.6 to 1 gSm^{-2}. NO_3^- wet deposition lies between 0.3 to 0.7 gNm^{-2} and NH_4^+ wet deposition between 0.3 and 0.9 gNm^{-2}. The areas of maximum wet deposition have the same location as for the concentrations of the respective species. When comparing the concentration fields and the wet deposition fields it can be seen, that the concentrations dominate the large scale distribution of the wet deposition and not the precipitation amount.

3. TREND IN CONCENTRATION DATA

Emission changes are expected to influence downwind concentration levels. Most of the western European countries have reported sulfur emission reductions by 17 to 60 % between 1978 and 1985 (1). Therefore one would expect a downward trend in S concentration data as well.

At Langenbrügge the median concentrations of SO_2, SO_4^A and NO_2 are shown (Fig. 3). Langenbrügge is located in northern FRG directly at the german-german border. From 1978 to 1985 no significant trend can be observed. The changes in concentration over these eight years are within the sampling and analytical precision.

A more promising approach to detect a trend is to look at the changes in the data grouped into different transport sectors. The origin of an air mass was determined by 72 hours back trajectories at the 850 mb level (2). For each day the trajectories were grouped into one of the eight equal sized sectors (N, NE, E,..). When no sector could be allocated the value U was used. Daily data together with this daily sector values were used to calculate median concentrations for each sectors.

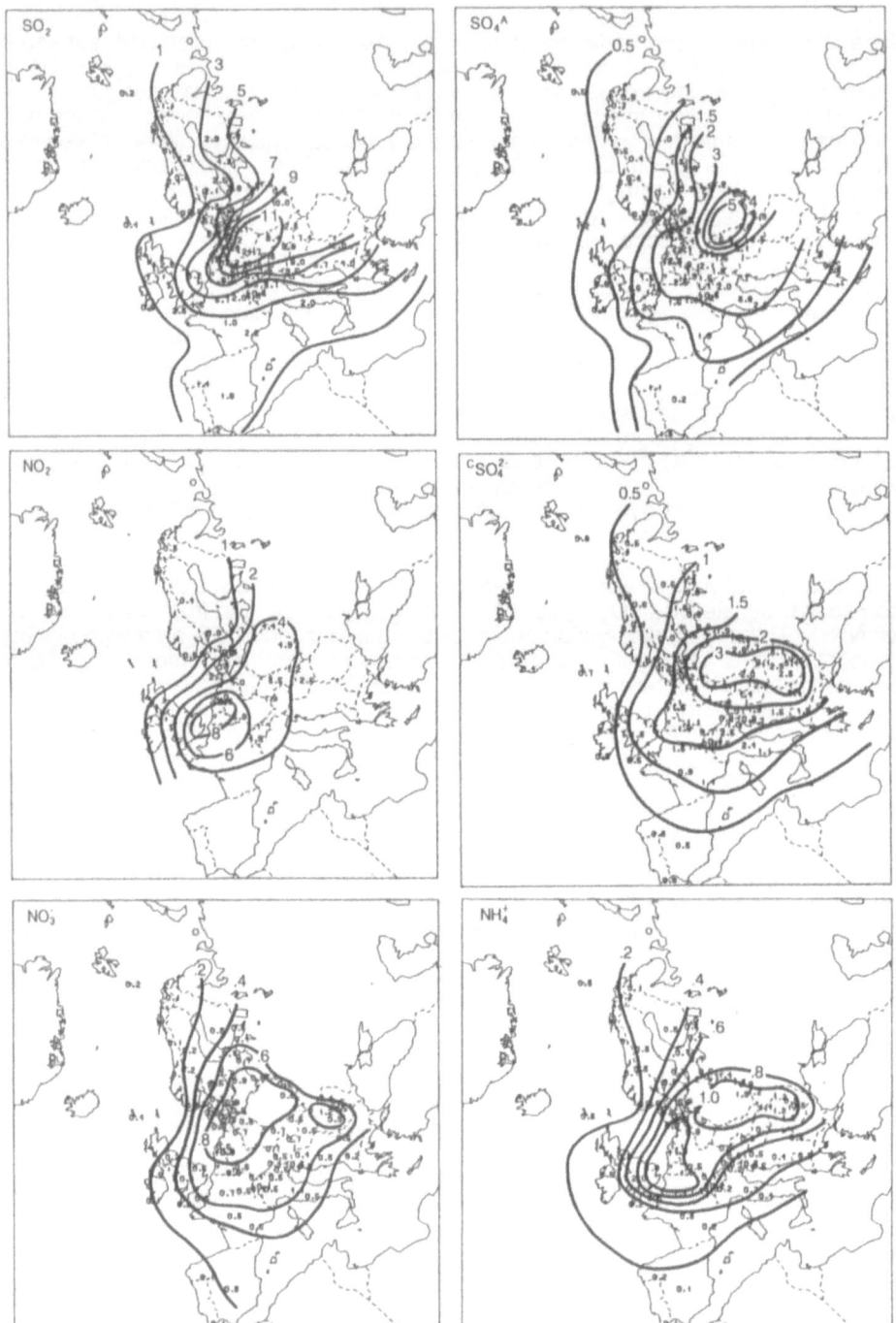

Fig. 1 Annual arithmetic average of sulfur dioxide, sulfate in aerosol and nitrogen dioxide in μgSm^{-3} and μgNm^{-3}; annual volume-weighted averages of non-marine sulfate, nitrate and ammonium in precipitation in mgS/l and mgN/l; 1985

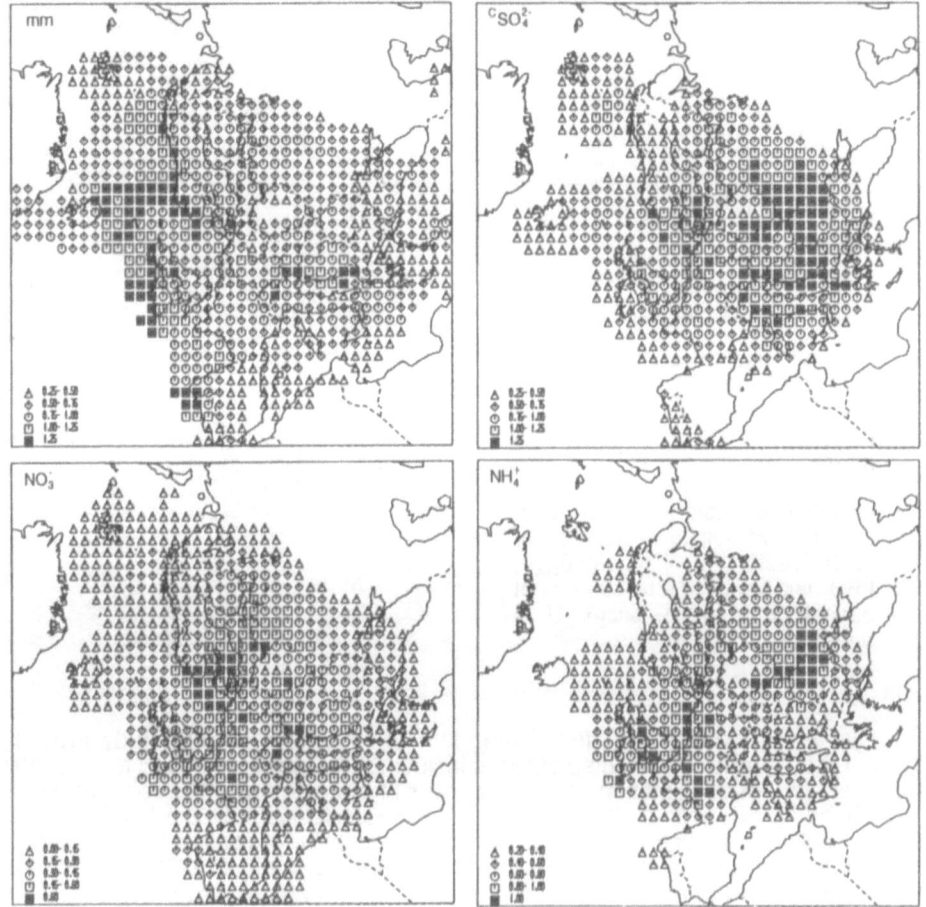

Fig. 2 Annual wet deposition of precipitation amount in metre, non-marine sulfate, nitrate
and ammonium in gSm⁻² and gNm⁻², 1985

In order to have sufficient number of data the median SO_2 and SO_4^A concentrations in the period 1978 to 1980 were compared to the median concentrations in the
period 1983 to 1985 for each sector (Fig.4). The highest S concentrations are found
in the NE, E, SE and S sector pointing to the highest sulfur emission areas in Europe.
This is true for both time periods investigated. But the change in concentration for
each sector is different. In the eastern and southern sectors the concentration of SO_2
and SO_4^A has increased. Especially in the east the SO_2 median went up from 7.9 to
17.2 μgSm^{-3}. Whereas in the other directions both the SO_2 and SO_4^A values have
decreased. This shows clearly that the influence of emission reduction in one area
can be compensated by an emission increase in another area. Therefore an annual
average or a median concentration may not be altered.

4. DETECTION OF CHANGE IN REGIONAL EMISSIONS

The previous chapter has already shown the usefulness of grouping the data
according to transport direction. It was a single-station statistic looking in different
directions. Now we want to focus on one area and see whether a change in
concentrations occurs at stations being located "around" our area of interest.

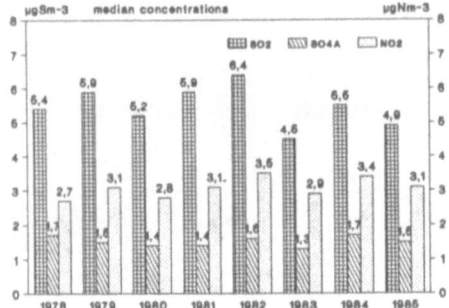

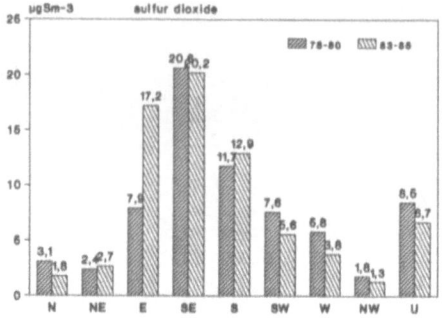

Fig. 3 Annual median concentrations at Langenbrügge, FRG, 1985, sulfur dioxide, sulfate in aerosol and nitrogen dioxide in μgSm^{-3} and μgNm^{-3}

Fig. 4 Median concentrations of sulfur dioxide and sulfate in aerosol for each sector at Langenbrügge, FRG, left bars 1978 to 1980, right bars 1983 to 1985, sector U = undefined, in μgSm^{-3}

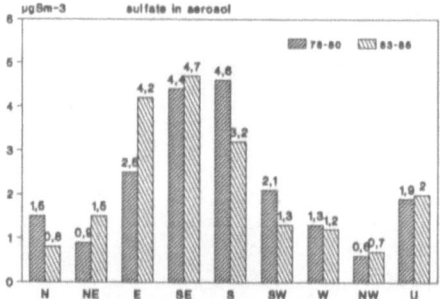

Due to its location the United Kingdom represents a perfect study area. In Figure 5 the UK and the selected stations including the sectors pointing towards the UK are shown. At each station separately it was investigated whether concentrations have changed, but only in air masses with back trajectories from the UK. For sulfur dioxide, sulfate in aerosol and sulfate and nitrate in precipitation the median concentrations have been calculated for the three-year periods 1978 to 1980 and 1983 to 1985. Due to the prevailing wind direction the number of daily values in the western sectors is much larger than in the eastern sectors. The results are summarized in Table I.

The results are very impressing. At nearly all stations and for all components there is a strong decrease in concentration. For SO_2 the decrease is between 34 and 82 % and for SO_4^A between 13 and 60 %. The concentrations in precipitation are in 1983 to 1985 by 29 to 73 % (nss SO_4^{2-}) and 17 to 75 % (NO_3^-) lower. Only for those stations having less than 10 values in the first period, the results are more uncertain explaining their extreme values.

The SO_2 concentration levels at the different stations can be explained by their distance to central England where the main sulfur emissions occur. The concentration decreases with increasing distance to the United Kingdom. This is less evident for sulfate and nitrate due to their much longer residence time in the atmosphere.

The fact, that the strong decrease in concentration occurs at all stations, gives confidence in that the trend is caused by emission reductions. The selected time periods are long enough to exclude meteorological variability as reason for the decrease. Also the sampling and analytical uncertainties are much lower than the observed trend. The United Kingdom has stated, that they have reduced their sulfur emissions by 29 % from 1978 to 1985. This compares very well with the change in concentrations at the surrounding stations. But from measurements alone the reduction can only be qualitatively validated.

sulfur dioxide							sulfate in aerosol				
Stat.	sect	N	78-80	N	83-85	%	N	78-80	N	83-85	%
DK 7	S	59	0.8	108	0.2	-75	59	2.4	107	2.1	-13
N 08	SW	121	1.1	167	0.2	-82	121	1.5	167	0.6	-60
N 01	SW	109	1.5	159	0.3	-80	108	2.0	159	0.8	-60
S 02	W	219	3.4	142	2.2	-35	229	2.5	142	1.2	-52
DK 3	W	225	4.7	163	2.1	-55	225	2.9	163	1.7	-41
D 01	W	253	3.3	184	1.5	-55	272	2.2	185	1.0	-55
NL 2	W	60	4.1	204	2.7	-34	38	1.2	201	1.2	0
F 05	N	68	7.4	106	4.1	-45	34	3.6	41	2.5	-31
UK 3	NE	5	7.7	55	3.8	-51	5	2.0	62	2.3	+15
IR 1	E	6	2.5	58	1.8	-28	6	1.1	58	2.0	+82
UK 2	S	61	8.0	101	1.8	-78	63	2.5	101	1.9	-24

sulfate in precipitation							nitrate in precipitation				
Stat.	sect	N	78-80	N	83-85	%	N	78-80	N	83-85	%
DK 7	S	39	1.1	47	0.3	-73	40	0.4	62	0.1	-75
N 08	SW	82	0.9	133	0.5	-44	81	0.4	131	0.2	-50
N 01	SW	75	1.4	104	0.8	-43	74	0.6	104	0.4	-33
S 02	W	131	1.1	57	0.7	-36	131	0.5	57	0.4	-20
DK 3	W	158	1.1	75	0.6	-45	158	0.4	75	0.3	-25
D 01	W	-	-	83	0.9	-	-	-	88	0.7	-
NL 2	W	67	1.4	133	1.0	-29	65	0.6	125	0.5	-17
F 05	N	5	0.8	128	1.5	+88	-	-	-	0.9	-
UK 3	NE	4	5.0	12	3.9	-22	4	1.8	12	2.5	+39
IR 1	E	2	2.6	11	2.3	-12	-	-	-	-	-
UK 2	S	41	1.3	63	0.6	-54	23	0.7	64	0.4	-43

Table I Station code and sector pointing towards the United Kingdom (Fig. 5), N number of daily values in this sector, median concentrations for 1978 to 1980 and 1983 to 1985, and difference in percent; sulfur dioxide and sulfate in aerosol in μgSm^{-3}, non-marine sulfate and nitrate in precipitation in mgS/l and mgN/l

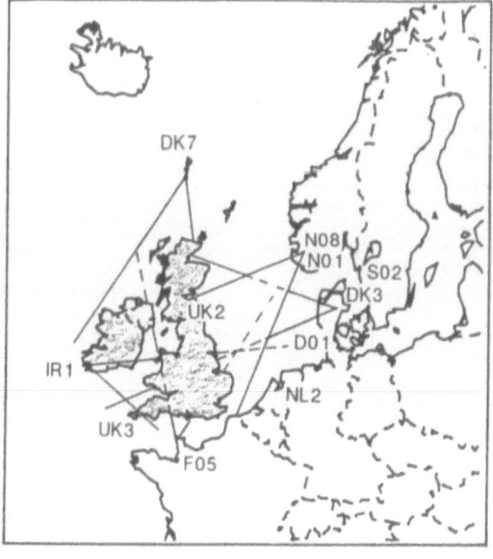

DK 7	Færøerne	1000
N 08	Skreådalen	800
N 01	Birkenes	900
S 02	Rörvik	900
DK 3	Tange	700
D 01	Westerland	600
NL 2	Witteveen	600
F 05	La Hague	400
UK 3	Goonhilly	500
IR 1	Valentia	700
UK 2	Eskdalemuir	200

Station code and name and distance in km to central England

Fig. 5 The selected stations with sectors pointing towards the United Kingdom

5. CONCLUSIONS

The existance of a trend in concentration or deposition data is difficult to detect due to the inherent noise in such time series. Very often annual averages or frequency distributions alone do not exhibit a trend. In this paper we could show, that by application of meteorological information, e.g. back trajectories or sector values, changes in concentrations could be related to changes in emissions. At this point we are not able to quantify the emission changes from the changes in concentrations observed. As a next step transport model calculations have to be conducted which can also address the non-linear relation between emission and deposition (3).

6. ACKNOWLEDGEMENTS

This work was supported by the Umweltbundesamt,FRG. Part of this study was conducted during the stay of Nodop at the Norwegian Institute for Air Research, Lillestrøm.

REFERENCES

(1)   UN (1987) National Strategies and Policies for Air Pollution Abatement, Convention on Long-Range Trans-Boundary Air Pollution. ECE/EB.Air/14 United Nations, New York

(2) Saltbones (1988) personal communications

(3) Clark P.A., Fisher B.E.A.,.Scriven R.A. (1987) The wet deposition of sulphate and its relationship to sulphur dioxide emissions. Atmos. Env. Vol 21 1125-1131

THE RELATIONSHIP BETWEEN OZONE, NITROGEN OXIDES AND VOLATILE ORGANIC COMPOUNDS IN BOUNDARY LAYER EPISODES

Øystein Hov

Department of meteorology, University of Bergen, Allegaten 70, N-5007 Bergen, Norway.

Summary
Model calculations of atmospheric boundary layer ozone formation to a remote European site (Langesund in South Norway) and a rural site not far from major emissions (Bottesford in the southern part of the United Kingdom) show that ozone and the emission of NO_x along the trajectories are correlated at Langesund and not correlated at Bottesford. This suggests that in central Europe, ozone formation is suppressed by high NO_x emissions, while 500-1000 km downwind of the sources, the NO_x emissions enhance the formation of ozone.

Introduction
The summertime episodic increase in ozone in the atmospheric boundary layer is well established over large parts of the industrialized world. Averaged over one month, ozone in the atmospheric boundary layer in Europe typically has a maximum in June or May-June. If parallel measurements are made of species which are good tracers of anthropogenic pollution, e.g. chlorofluorocarbons or submicron particles, and only the days where the tracers are at the background or northern hemisphere average level are taken into account when calculating the monthly average ozone concentration, a maximum ozone value is usually found in April in the northern hemisphere. In southern Scandinavia the monthly maximum concentration of ozone is typically 40-50 ppb in April for clean air, and about 60 ppb in May-June when all days are counted (Grennfelt and Schjoldager, 1984).

The April maximum for "clean" air is also influenced by pollution, since it turns out that the ozone concentrations over Europe some decades ago probably were only about one half of the present "clean" air values (see e.g. Bojkov, 1986). Free tropospheric measurements over e.g. Hohenpeissenberg in southern Germany indicate that the "clean" air growth in the ozone concentrations over the last decades, also has taken place there, indicating that the "clean" air growth in ozone is a phenomenon which probably has occurred over industrialized continents and extending up to the tropopause. It is not documented that this is a global phenomenon, however, and taking into account the limited lifetime of free tropospheric ozone (1-3 months) it is to be expected that the large-scale growth in concentration is confined to e.g. sizeable fractions of Europe and North America.

It is likely that the growth in tropospheric ozone has come about as a result of the increase in the global emissions of NO_x and CH_4, while in the atmospheric boundary layer the increase in nonmethane hydrocarbon emissions is important for the formation of the highest concentrations of ozone (Hov, 1988).

Model calculations
In this paper the relationship between precursor emissions of NO_x and volatile organic compounds (VOC) and the formation of ozone in the atmospheric boundary layer, will be discussed by looking at the behaviour of ozone at a rural site remote from pollution sources (Langesund on the south coast of Norway) and Bottesford not far from major emission areas in the

southern part of the United Kingdom. This is done through model calculations of ozone formation in the boundary layer, where a Lagrangian model is
used based on the formulation of the acid rain model in EMEP (European
Evaluation and Modelling Programme) (Eliassen et al., 1982). Receptor
oriented trajectories were calculated for 4 days prior to arrival based on
925 mb winds. This was done for every 6 h, and along the trajectory
meterorological information about the depth of the mixed layer (the mixing
height), precipitation and temperature, as well as about the emissions of
NO_x and VOC, specified on the EMEP-grid (covering Europe including the
European part of USSR, as well as the eastern part of the Atlantic Ocean,
with a grid where the mesh size is 150x150 km^2 at 60°N).

Calculations were done for a 20 d period in 1982, starting 20 May 1982.
Emissions were based on the annual emission figures for NO_x given by almost
all the European countries to ECE in Geneva, while the figures for VOC were
partly based on ECE-figures, partly on OECD information and partly estimated (Iversen et al., 1989).

In Figure 1 is shown calculated (stars) and measured ozone concentrations at Bottesford in the UK, and at Langesund in South Norway. The
measured values are hourly means close to the surface, while the calculated
values are representative of approx. 6 h around the receptor time, and are
boundary layer average values. Concentration variation with height is not
resolved in the model, and the mixing height at noon is taken to represent
the boundary layer depth over 24 h. The measured values at night are
strongly influenced by dry removal and possibly also by NO_x emissions
nearby, and can therefore not be compared directly with the calculated
value, while during the day when the boundary layer is better mixed,
measurements and calculated values can be compared. Both for Bottesford
and Langesund it is seen that the diurnal maximum in the ozone concentration is quite well calculated, except for one day (28 May) when the Langesund values are far off the measured values, indicating that an inadequate description of the meteorological situation has been used in the model.

In Figure 2 is shown the calculated concentration every 6 h of ozone,
emission of NO_x along the trajectory (in ppb) and the ratio RO_3NO_x which is
the total ozone formed along the trajectory divided by the total emission
of NO_x along the trajectory. In Figure 2 relative units are used on the
vertical axis, 1.0 corresponds to the maximum value given in the upper
right hand corner. At Langesund it can be seen that there is a good
correspondance between ozone peaks and the emission of NO_x along the
trajectories, except for a few cases where an anticorrelation is shown
(e.g. at midnight between day 3 and 4). The maximum ratio RO_3NO_x is seen
to occur for a few arrival times where very little NO_x was emitted along
the trajectory, for high O_3 cases the ratio typically is between 5 and 10.
At Bottesford, the situation is quite different because there are quite a
few cases of anticorrelation between the integrated NO_x emissions and the
O_3 concentration calculated at the receptor. This is particularly marked
for the days 19 and 20, when as much as 63.9 ppb of NO_x was emitted along
one of the trajectories, and it is also seen that the RO_3NO_x-ratio is
between 3 and 5 in the cases where high O_3 is calculated.

These differences are further elaborated upon in Figure 3, where the
calculated ozone concentration at the receptor point is plotted against the
total emission of NO_x along the trajectory, and where there is good correlation (r=0.84) at Langesund and no correlation at Bottesford (r=-0.20).
In Figure 4 is shown the ozone concentration calculated in the reference
case (horizontal axis) plotted against the ozone concentration calculated

when the NO_x emissions were reduced by 25%. The ozone concentration is scaled against the same maximum value in both cases. It can be seen that NO_x emission reduction in general gives rise to an ozone reduction at Langesund, while there is much more of a tendency to calculate higher ozone with decreased NO_x emissions at Bottesford (the number of + above the diagonal is much higher for Bottesford than for Langesund). For a 50% reduction in NO_x emissions (Figure 5) the same holds as for the 25% emission reduction case, but the plot for Bottesford now shows one set of cases where NO_x emission reduction reduces ozone, and another set of calculations where the NO_x emissions in the reference situation were so high that they had an inhibiting effect on the ozone formation.

A 25% reduction in the emissions of VOC was calculated to reduce the concentration of ozone at the receptor for all trajectories (Figure 6), and particularly at Langesund the reduction in VOC dramatically reduced the highest ozone concentrations. A combined 25% reduction of both NO_x and VOC is seen to have a limited effect on ozone both at Bottesford and Langesund (Figure 7).

Acknowledgement
The work reported here was funded by The Royal Norwegian Research Council for Science and Technology through contract MK.11.24232 to Norwegian Institute for Air Research.

References
Bojkov, R.D. (1986) Surface ozone during the second half of the nineteenth century. J. Climate Appl. Meteor., 25, 343-352.
Eliassen, A., Hov, Ø., Isaksen, I.S.A., Saltbones, J. and Stordal, F. (1982) A lagrangian long-range transport model with atmospheric boundary layer chemistry. J. Appl. Meteor., 21, 1645-1661.
Grennfelt, P. and Schjoldager, J. (1984) Photochemical oxidants in the troposphere: A mounting menace. Ambio, 13, 61-67.
Hov, Ø. (1988) Photochemical oxidant episodes, acid deposition and global atmospheric change. The relationships with emission changes of nitrogen oxides and volatile organic compounds. Lillestrøm (NILU OR 12/88).
Iversen, T., Saltbones, J., Sandnes, H., Eliassen, A. and Hov, Ø. (1989) Airborne transboundary transport of sulphur and nitrogen over Europe-model descriptions and calculations. Oslo (EMEP MSC-W R 2/89).

Figure 1 Calculated ozone concentrations every 6 h at Langesund and Bottesford for 20 d starting 20 May 1982 (stars). Measured hourly values are shown in full line.
Figure 2 Calculated ozone concentrations every 6 h at Langesund and Bottesford together with integrated NO_x emissions along each trajectory (RELNOX) and ratio of ozone formation to NO_x emission (RO_3NO_x). The abscissa is given in numbers relative to the maximum value which is given in the upper right hand corner for each parameter shown.
Figure 3 Calculated ozone concentration versus total emission of NO_x along the trajectory for Langesund and Bottesford; r is correlation coefficient.
Figure 4 Scatter plot of ozone calculated in the reference case (horizontal axis) versus a calculation where the NO_x emissions were reduced by 25%.
Figure 5 Same as Figure 4 but NO_x emissions reduced by 50%.
Figure 6 Same as Figure 4 but VOC emissions reduced by 25%.
Figure 7 Same as Figure 4 but both NO_x and VOC emissions reduced by 25%.

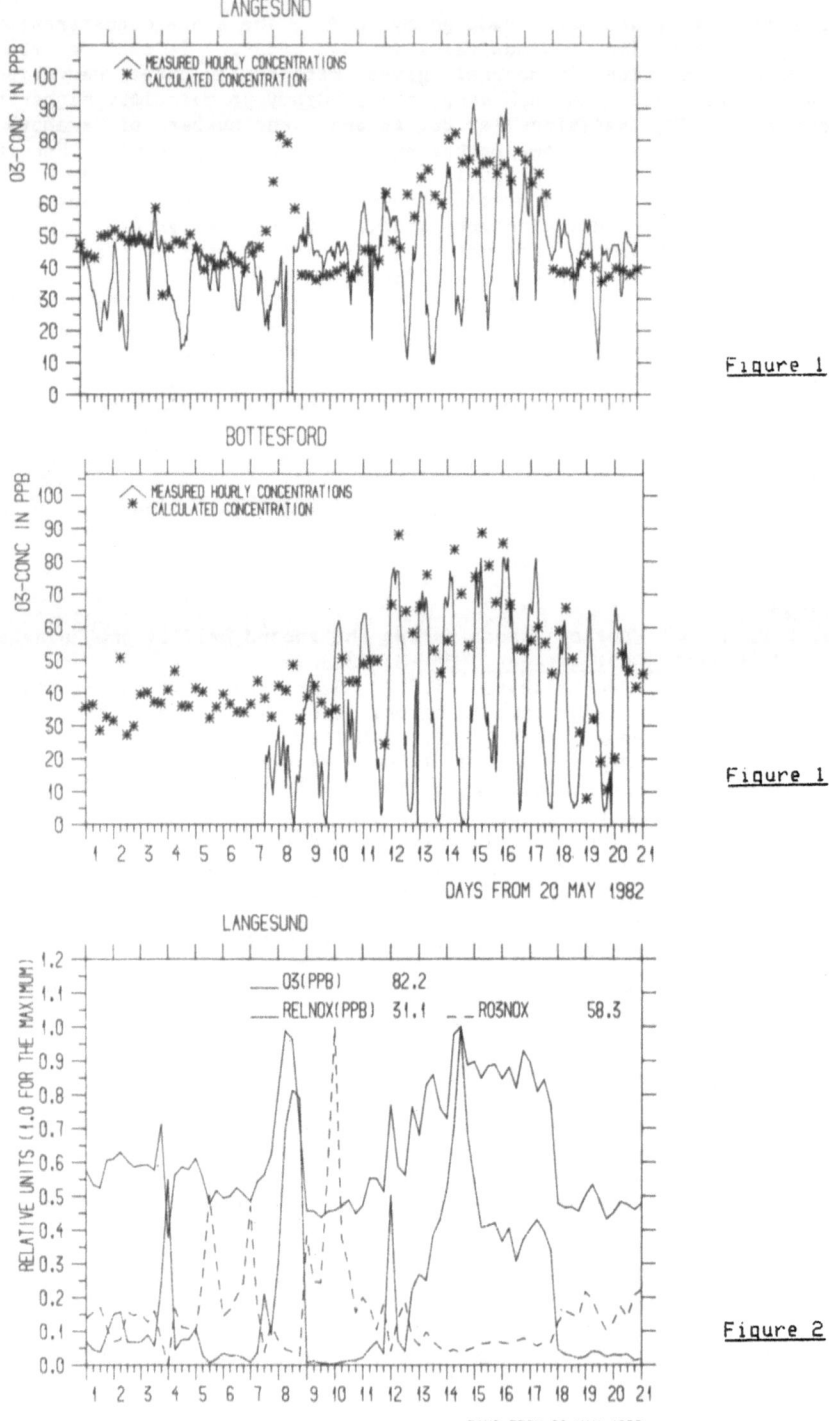

LANGESUND

Figure 1

BOTTESFORD

Figure 1

DAYS FROM 20 MAY 1982

LANGESUND

Figure 2

DAYS FROM 20 MAY 1982

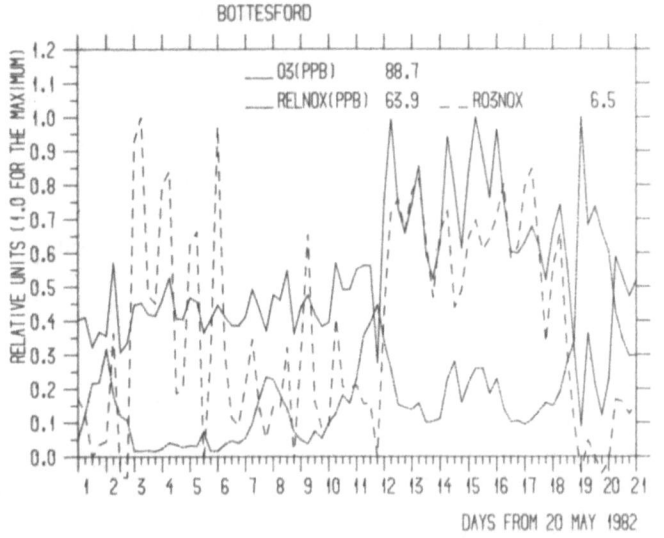

Figure 2

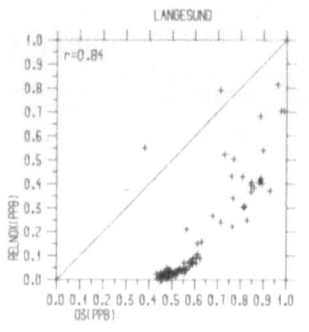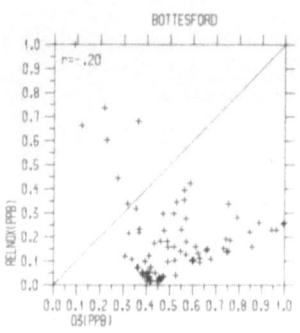

Figure 3

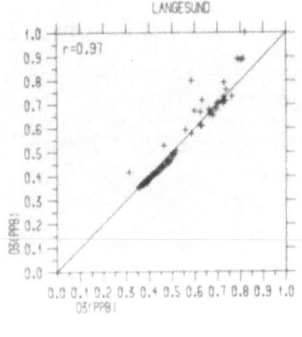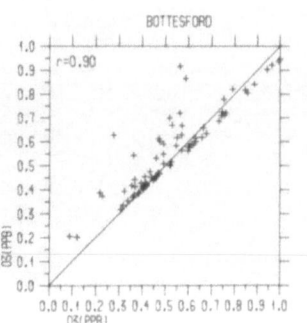

Figure 4

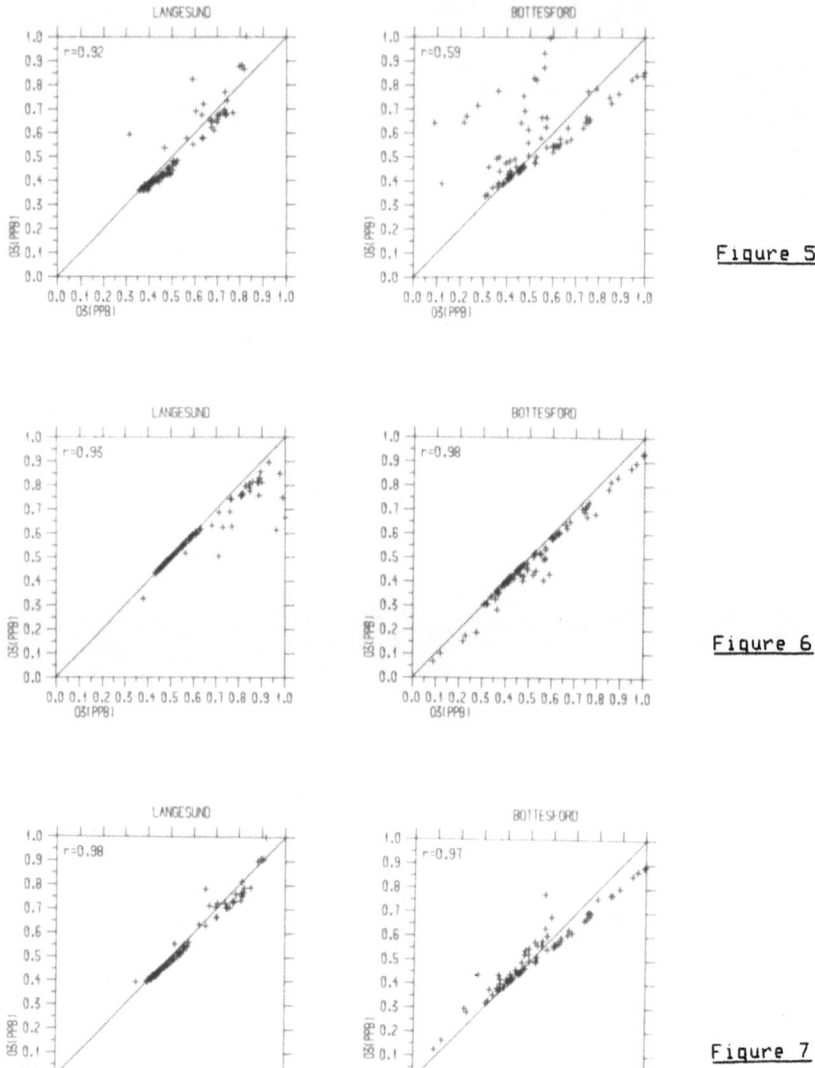

Figure 5

Figure 6

Figure 7

GROUND BASED CLOUD AND FOG EXPERIMENTS

W. JAESCKE and K.H. ENDERLE, Zentrum für Umweltforschung (ZUF),
J. W. Goethe-Universität Frankfurt am Main (F.R.G.)
S. FUZZI, G. ORSI and M.C. FACCHINI, Istituto FISBAT CNR,Bologna (Italy)
A. BERNER and G. REISCHL, Institut für Experimentalphysik (IEP),
Universität Wien (Austria)

Summary

Joined field campaigns were performed each year in November since 1985 at S. Pietro Capofiume in the Po-Valley. Main objectives were the investigation of microphysics and chemistry in radiation fog. The measured parameters include aerosol and droplet characteristics, concentrations of trace substances in the aerosol and liquid phase, gas phase concentrations and meteorological conditions near the surface. Also intercomparisons between different fog collection devices were done. The evolution of a typical observed fog event is described and discussed below. The comparison of meteorological data, liquid water content and ion concentrations of sulfate, nitrate and ammonium indicates that changes in the concentration of the trace substances are mainly due to vertical mass flux divergences in the surface layer. An important consequence of this finding is that future fog experiments must provide the resolved observation in and above the fog layer.

1. INTRODUCTION

Rapidly increasing interest in the effects of harmful substances possibly deposited by fog and captive clouds has led to extensive studies of the dynamics and chemistry of hydrometeors in recent years. Several groups attempted to understand the microphysics and chemical conversion processes occurring in clouds and fogs (1). Main prerequisite of these studies is the knowledge of the physical behavior of such systems and appropriate instruments for probing and collecting cloud droplets (fog and aerosol collectors, optical probes).

In order to investigate fog episodes in 1985-87 fog field campaigns were performed at S. Pietro Capofiume, Italy (FISBAT measuring site (2)) as a joint cooperation of ZUF (FRG), FISBAT (Italy) and IEP (Austria).

2. EXPERIMENTAL SETUP

Aerosol particles with diameter < 5 μm were sampled with a 0.45 μm pore size teflon filter (millipore) after passing a preseperator (impactor stage). All airborne particles > 5 μm in diameter are assumed to be droplets and were sampled by the Berner fog impactor (cut-off 5 μm). Due to its design this collector provides a predictable sharp collection efficiency and constant flowrate. It fulfills the requirements for fog sampling (3,4). A detailed description of the Berner fog collector is given in (5). It consists of a round nozzle for the formation of the jet and a collection disk and funnel assembly underneath the nozzle. The sample is collected in a vial.As this collector only provides a flowrate of 100 m^3/h ZUF developed on the same physical principles an improved fog impactor which is shown in Fig. 1. The collector provides a flowrate of 450 m^3/h and will allow a more detailed analysis of the collected water due to the resulting higher volume of the sample.

This type of fog collector was successfully tested in a field experiment in the Po-Valley in 1986 (6). Air-temperature and windspeed were monitored as 1-minute averages.

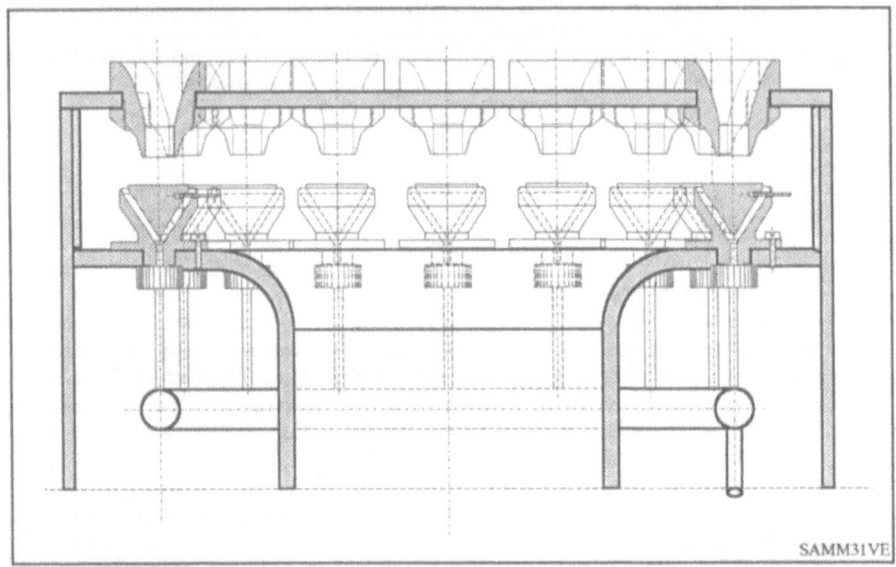

Fig.1: Fog impactor design

3. OBSERVATION AND DISCUSSION

The typical evolution of LWC and chemical species during a nocturnal radiation fog event on 17./18. Nov. 1987 will be described and discussed. The field site is situated in a homogeneous flat rural landscape. After a clear and sunny day, followed by haze at sunset, fog formed close to 9:00 p.m.. This formation happened instantaneously over a height of about 30 m. Fig. 2 represents the time evolution of the LWC observed with the fog impactor (5). This illustration characterizes typical fog events observed during the campaign showing a significant breakdown or dissipation after a first dense period (LWC > 0.1 g/m^3) of 2 - 3 hours. In a second period dense fog formed again at 2:00 a.m. and persisted till morning.

In order to illustrate the concentration of the chemical species during this fog night the sulfate-concentration is shown.The time evolution of the sulfate concentration in liquid and particulate phase (Fig. 3) is representative for the similar behavior of all major ions (ammonium, nitrate, sulfate). Fig. 3 displays the amount of sulfate in aerosol particles in 1 m^3 of cloudy air (particulate sulfate: solid line) as well as the amount of sulfate in the droplets in 1 m^3 of cloudy air (drop sulfate: hatched area).

The first fog period (9:00 p.m.- 1:00 a.m.) starts with a significant decrease in the particulate sulfate concentration which is obviously caused by the nucleation of the drops. This removal of the sulfate aerosol particles by the condensation process and the incorporation in the liquid phase is illustrated in Fig. 3 by the hatched area. After this 1-2 hour period of fog formation the drop sulfate also decreases - even stronger than the particulate sulfate.

However, with the beginning of the second fog period both sulfate concentrations increase. After this short increase the drop sulfate stays nearly constant until the fog dissipates at 9:00 a.m.. The particulate sulfate shows some more fluctuations in this period but also no dramatic changes can be observed.

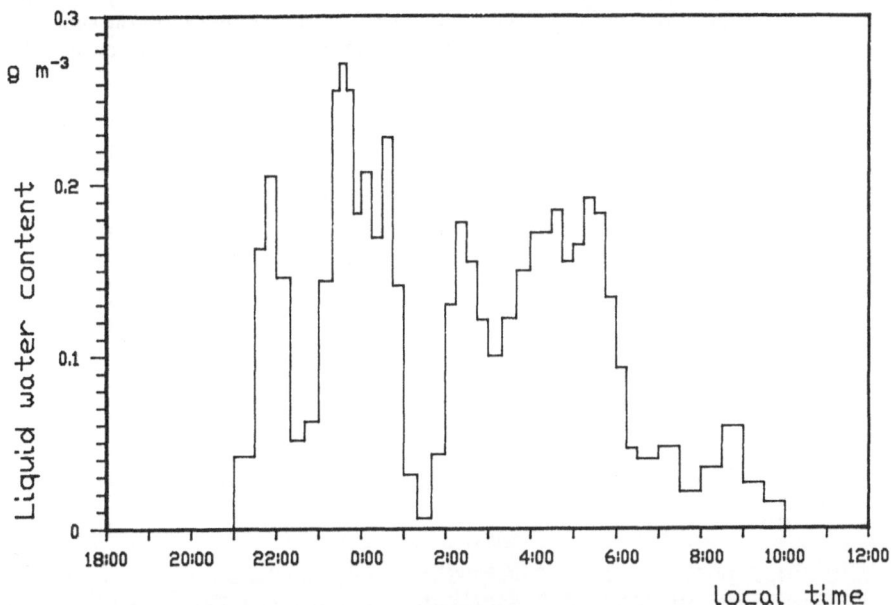

Fig.2: LWC determined from the sampled volumina of the Berner Fog Impactor, 17./18. Nov. 1987.

Thus, the first period of this fog is characterized by the decrease of both sulfate concentrations, while during the second period the concentrations are more or less constant. Only in the first period the sulfate concentration in the liquid phase exceeds the particulate concentration during strong condensation episodes (see 9:00 p.m.-10:00 p.m. and 11:00 p.m.-12:00 p.m.). The comparison of the time evolution of the LWC (Fig. 1) and the drop sulfate concentration also shows that new condensation intervals are always (in both periods) coupled with increases of the drop sulfate concentration.

This confirms that during the increases of LWC new aerosol particles must have been activated in supersaturated air (this holds for both fog periods). The first fog period allows the following interpretation:
With the formation of fog the major part of the sulfate containing aerosol particles gets activated. The amount of the drop sulfate is now higher than the particulate sulfate concentration as long as condensation is dominant. In periods of LWC decrease or dissipation a part of the drop sulfate becomes particulate due to drop evaporation (see time intervals 10:00 p.m.-11:00 p.m.and 0:00 a.m.-2:00 a.m.). The decrease of the total sulfate concentration (particulate + liquid) is obviously attributed to the deposition of big fog droplets onto the ground.

However, these interpretations do no longer hold for the second fog period. Here we see no decrease of the total sulfate concentration and no significant increase in one phase which is coupled with a decrease in the other. An explanation of this discrepancy is possible by considering the time evolution of the horizontal wind speed. The increase of the wind at 2:00 a.m. indicates a change of the meteorological conditions simultaneously with the beginning of the second fog period. Assuming that this increase in wind speed is coupled with an enhanced turbulent mixing from above the following interpretation is suggested: Vertical transport of water vapor and sulfate from higher altitudes into the cold surface layer causes the sudden fog formation, the increase in temperature (2 °C, not illustrated here) and sulfate concentration in liquid and particulate phase.

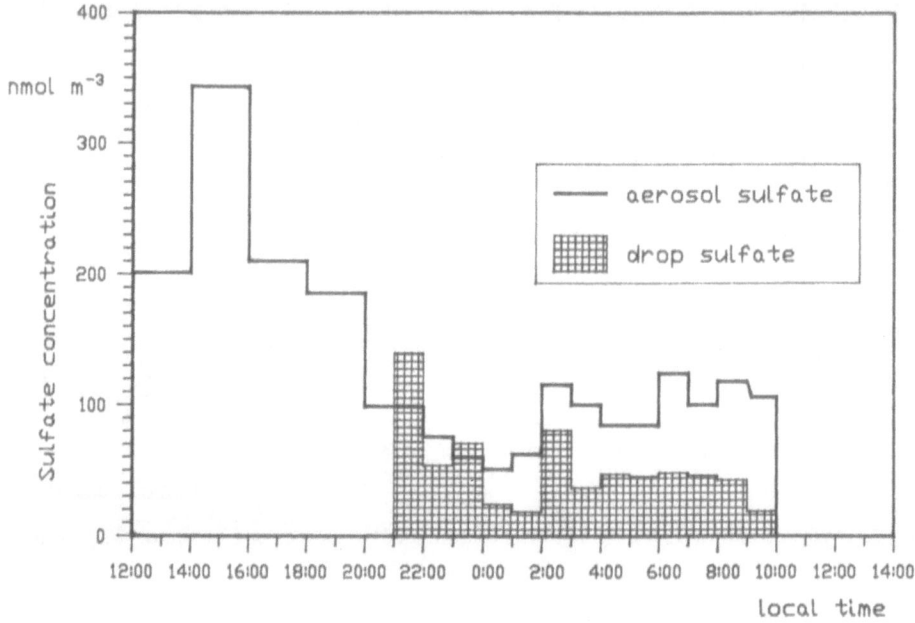

Fig.3: Sulfate concentration in aerosol particles and droplets, 17./18. Nov. 1987

In contrast to the first fog period the loss of sulfate mainly due to the deposition of large fog droplets becomes compensated by turbulent mixing of fresh aerosol particles from layers above the fog. This also explains the long persistence of this dense fog period of more than 4 hours. The entrainment of fresh airmasses from above provides the fog system with water vapor to maintain supersaturation and with new condensation nuclei to form new droplets, which compensates the loss of deposited water and sulfate mass. This implies a flux equilibrium for the total sulfate mass between the deposition onto the ground and turbulent transport into the fog system from above. This idea is not contradictory to the evolution during the first fog period. Here also mixing processes must have taken place to provide the system with new condensation nuclei and water vapor for reenhancement of LWC and drop sulfate (10:00-12:00 p.m.). Ceasing the turbulent vertical mass fluxes (indicated by the low wind at 0:00-2:00 a.m.) lead to the fast decrease of the LWC and the total sulfate mass.

4. CONCLUSION

This field experiment shows that a calm and stable looking atmospheric system like radiation fog contains a complex structure of dynamical and thermodynamical atmospheric processes. They control, beside the meteorological evolution of the fog, the pollutant concentration in the system. Thus, the interpretation of the observed time evolution of pollutant concentrations in order to investigate chemical reactions (especially in acid production) ongoing in the liquid phase cannot be done assuming that the fog layer is a closed system.

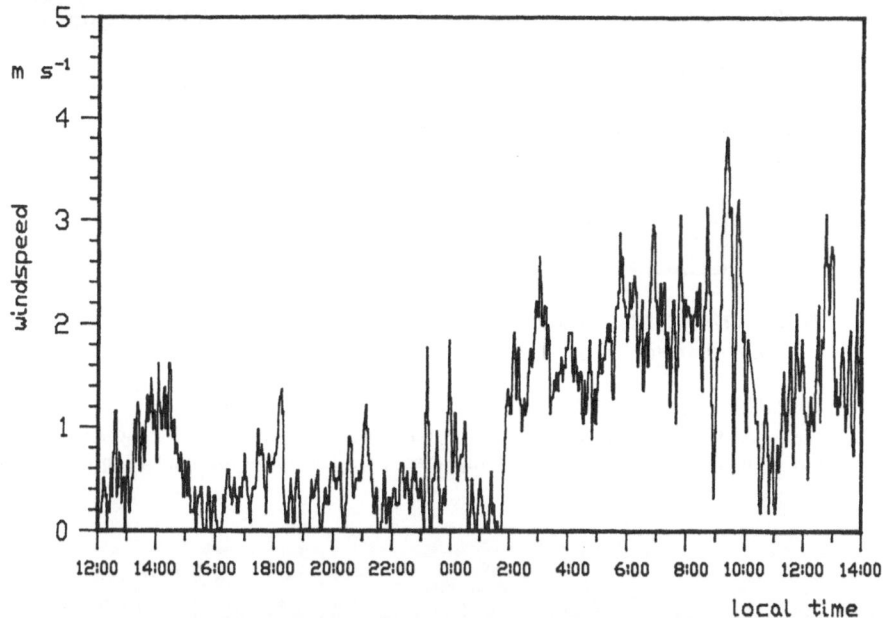

Fig.4: Horizontal windspeed obseved on 17./18. Nov. 1987 in 2 m height

Variations in chemical concentrations are mainly a consequence of a changing mass flux balance of the species. In order to detect chemical reactions these mass fluxes into and out of the fog layer have to be determined with a sufficient time resolution. Additionally the concentration change by a chemical reaction has to be significant with respect to the precision and accuracy of the used analytical method. Also the speed of a chemical reaction has to be fast compared with the turn over time of the fog system (ca. 1 hour) if atmospheric reaction constants should be derived. For future fog field experiments special emphasis must be layed on vertically resolved measurements within and above the fog layer to determine the chemical composition of the gas, aerosol and liquid phase as well as the meteorological data and the turbulent structure (exchange parameters) of the surface layer.

ACKNOWLEDGEMENT

The German part of this work was funded by the "Bundesministerium für Forschung und Technologie (BMFT) under project No. 07VND010. The Austrian part was carried out under contract P5693, "The Formation of Acidic Particles and Fog" of the "Fonds zur Förderung der wissenschaftlichen Forschung Österreich" and the Italian part was sponsored by the "Progetto Finalizzo Energetica 2, Sottoprogetto Ambiente e Salute" of the Italian National Research Council. The support of the respective authorities is gratefully acknowledged.

REFERENCES

(1)   JAESCHKE, W. (ed.) (1987). Chemistry of multiphase atmospheric systems. Heidelberg, Berlin New York, Springer Verlag.

(2)   FUZZI, S. (ed.) (1987). Heterogeneous Chemistry Project. FISBAT report HACP 2, Bologna, Italy.

(3)   ENDERLE, K.H. and W. JAESCHKE (1988). Sampling and chemical analysis of fog water with respect to microphysical processes. In: W. Jaeschke and K.H. Enderle (Eds.), Chemistry and Physics of Fog Water Collection. BPT Bericht 6/88 GSF München, pp.7-74.

(4)   ENDERLE, K.H., W. JAESCHKE (1988). Problems of fog sampling. Presented at the Annual GAeF Meeting in Hannover, F.R.G., 1987.

(5)   BERNER, A. (1988). The collection of fog droplets by a jet impaction stage. Sci. Total Environment 73, pp. 217-228.

(6)   BERNER, A., G. REISCHL, K.H. ENDERLE, W. JAESCHKE, S. FUZZI, G. ORSI, M.C. FACCHINI (1988). The liquid water content of a radiation fog measured by an FSSP 100 optical probe and a fog impactor. Sci. Total Environment 77, pp. 133-140.

INTERCOMPARISON OF WET-ONLY COLLECTORS
FOR MEASURING WET DEPOSITION

P. Winkler

Deutscher Wetterdienst, Meteorologisches Observatorium Hamburg
Frahmredder 95, D-2000 Hamburg 65

Summary

20 precipitation wet only collectors of 11 different designs have been investigated in a field intercomparison. For the bulk ions, deviations between the various types exceeded 20 %. For the trace metals, even larger differences occured. Three main reasons were made responsible:

1. Poor sensitivity of precipitation sensors.
2. Wetting losses in the funnels, causing a memory effect.
3. Interchange between rain water and materials of the collection device.

By rinsing experiments with diluted HNO_3 we found that appreciable amounts of trace metals accumulate at the funnel walls. Under high outside pollution concentrations gases penetrated into the samples with measurable changes in composition. Only rain events were evaluated.

1. Introduction

An intercomparison of 20 wet only collectors of 11 different designs has been performed at the Meteorological Observatory of Hamburg in cooperation with the Institute of Anorganic and Applied Chemistry of the University of Hamburg between 1986 and 1988. Major goals were

- to characterize the samplers not only by their deviation from the average, but to find out the reasons for the deviation.
- to create the fundamentals for defining a suitable threshold intensity at which the sensors should open the lid.
- to find a method how suitable reference values can be derived in order to judge the individual samplers.

The following parameters were studied: precipitation amount, electrical conductivity, pH-value, Cl^-, NO_3^-, SO_4^-, NH_4^+, Na^+, K^+, Ca^{++}, Mg^{++}, Fe, Pb, Cu, Cd. Additionally, the time open and the number of lid movements were monitored.

In this paper the general aspects of precipitation collection are presented. Details on the individual instruments can be taken from the final project report (Winkler et al., 1989).

2. Precipitation structure and threshold intensity

By means of a distrometer (Joss and Waldvogel, 1967) the relative cumulative amount of precipitation was derived as function of the intensity (fig. 1). It can be seen that in Hamburg 97 % of the total precipitation amount are falling at intensities above 0.1 mm/hr while only 3 %

are connected with smaller intensitities. A similar result has been obtained for the station Hohenpeissenberg near Munich (Winkler and Jobst, 1988). Therefore, this curve seems to be of more general importance. Consequently, the sensors should be activated at an intensity of about 0.05 mm/hr.

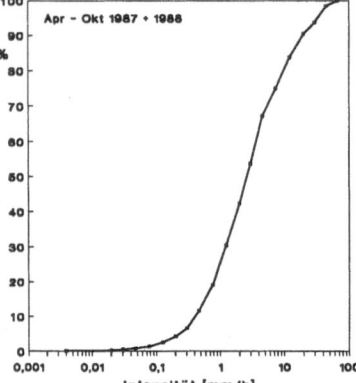

Fig.1: Relative cumulative distribution of precipitation amount as function of intensity, derived from data of the distrometer.

In fig. 2 the response curves of various sensors are given as function of the precipitation intensity. It can be seen, that sensors A, D and E met the recommended threshold intensity while others (C, I, F) need intensities of up to nearly 1 mm/hr in order to recognize all events.

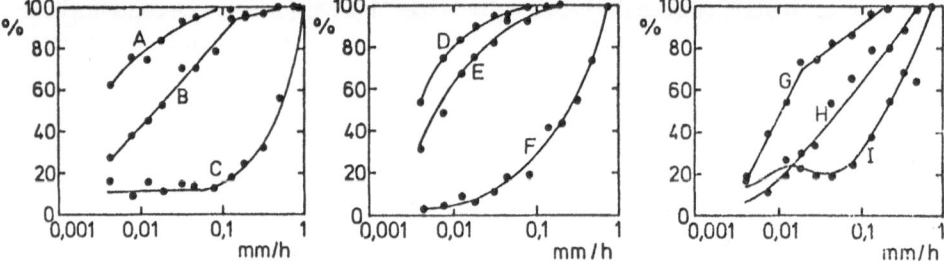

Fig.2 Response of various precipitation sensors at low precipitation intensities.
The following table gives an example of the opening time and the number of lid movements for a single day. The poorest instrument opened slightly above 3 hours while the longest time was nearly 11 hours. No clear correlation between the opening time and the number of lid movements can be seen.

Table 1: 08.08.1987, 8:30

	1	2	3	4	5	6	7	8	9	10
				Opening time (hours)						
Nr. 1-10	5.37	6.08	3.06	9.54	7.53	10.11	5.37	7.09		6.06
Nr.11-20	6.06	5.12	5.03	9.33	6.42	10.03	0.00	10.50		0.08
				Number of lid movements						
Nr. 1-10	35	10	19	11	27	17	28	20		27
Nr.11-20	32	12	19	14	33	14	1	14		6

There exist also large differences between identical instruments, which are standing in the same column. It must be emphasized, that some sensors did not reliably recognize the end of precipitation events. One sensor, for example, was still activated 20 min after the nominal end in about 50 % of all events. This does not mean that an instrument with the correct opening time collects the precipitation more completely because its sensor can respond poorly at the beginning as well as at the end of events. The response of some sensors proved to be temperature dependent.

3. Reasons for concentration deviations

The reference value was derived from those instruments, the sensors of which were not too bad and which were free of pollution. From some lids water could drip into the funnel during the closing movement so that these instruments collected more rain than had been collected with an open standard gauge.

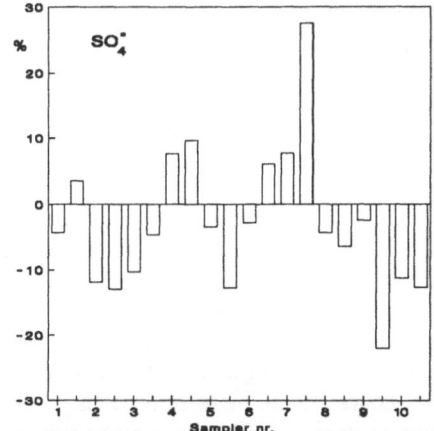

Fig.3 shows as an example the results for SO_4^-, neighboured columns belonging to the same type of instruments. We see differences of more than 20% between the highest and lowest deviations. It must be noted that the 0% line represents the reference but not the true value. The true value is most probably lying 4-5% above the reference. For other ions the deviation pattern from the 0% line was similar. Deviations from this pattern were usually a strong hint for a contamination problem. E.g. collector 2 was much above +50% for Cd while it was normally below -10% for the other ions.

This ordinary deviation pattern could be interpreted in terms of two competing effects:

1.)Due to poor sensitivity of the precipitation detector some instruments do not collect the first highly concentrated rain drops. The measured concentrations were found to be low by 15 - 20 % on the average. To demonstrate this we selected two groups of instruments, the first with low threshold intensity detectors and the second with high threshold intensity detectors (Fig.4). While nearly all ion concentrations of the first group varied between 94 and 106 %, the ion concentrations of the second group varied between 83 and 94 %. During a single event where at low intensities the electrical conductivity was monitored with a pre-

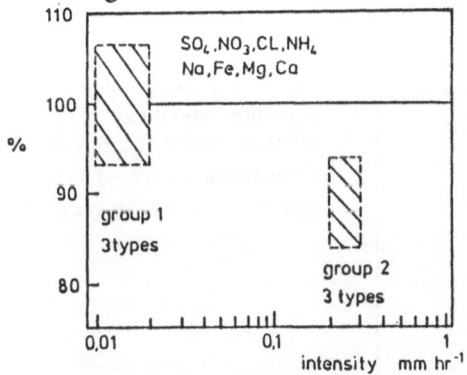

Fig.4: Comparison of two groups of instruments, the detectors of which show different sensitivity.

cipitation monitor we evaluated the electrical conductivity of the samples as function of the opening time. During this event a period with conductivities around 300 µS/cm was observed at low intensities. Later, as the intensity increased, the conductivity droped to values around 50 µS/cm. We see from fig. 5 that instruments with opening times below 6 hours lost the period with conductivities up to 300 µS/cm at low intensities and had only 70 % of the conductivity as compared with instruments which were open more than 12 hours.

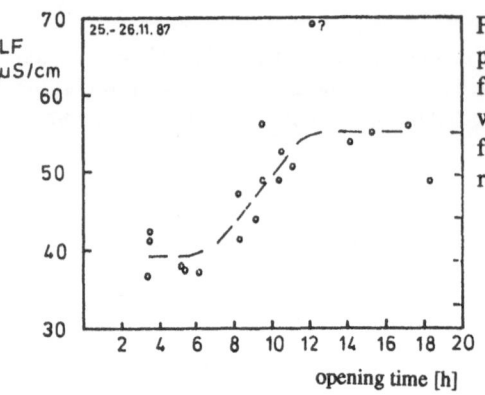

Fig.5: Electrical conductivity of precipitation of precipitation samples from 20 collectors as function of the opening time. The instruments with short opening times had poor sensors and failed to collect the first highly concentrated raindrops of the event.

2.)A certain fraction of the collected rain water sticks to the funnel walls, denoted as wetting loss. The amount of the wetting loss depends mainly on the total funnel surface, deep funnels with a large total surface having wetting losses which correpond up to 0.24 mm precipitation height. The water evaporates in the funnel releasing there the dissolved substances. This material is redissolved by the next event causing a memory effect, which increases the concentration in the sample. This concentration increase depends on the amount of the wetting loss, the trace substance content of the previous event and the concentration of the following eventand the relative productivity of both events.

In fig. 6 we try to assess the magnitude of this memory effect. The solid line shows the relative cumulative precipitation amount as function of the productivity of events. In a climate like in Hamburg where events of small productivity occur very frequently we find 5 % of the total precipitation amount being connected with productivities below 0.24 mm per event. This means that for the instrument with the highest wetting loss (0.24 mm) 5 % of the total precipitation amount never reaches the sampling bottle. Since the events with small productivity carry usually higher than normal concentrations of the trace substances, we see that the deposition of substances as function of productivity is still higher (dotted

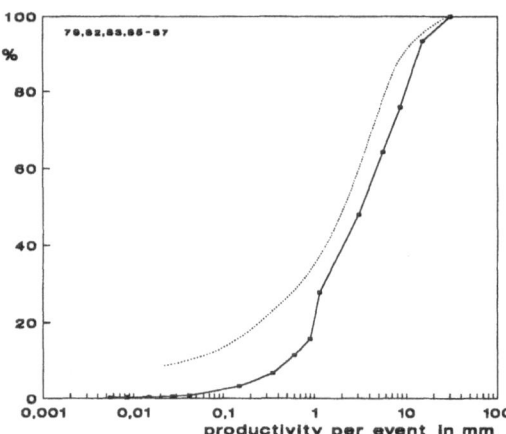

Fig.6: Relative cumulative precipitation amount (_____) and relative cumulative deposition of an arbitary ion (---) as function of the productivity per event.

line in fig. 6) and may reach 10-15 % at productivities as low as 0.1 mm per event. In other words: up to 20 or even 25 % of the material deposited remains in the funnel and is brought into the sample only by redissolving. The exact course of this curve has to be determined, because the curve presented in fig. 6 is derived from an instrument having a wetting loss itsself.

By a suitable combinition of the two effects an instrument can show ion concentrations near the reference value because it has a poor sensor and a high memory effect and these two effects compensate each other. Nevertheless the amount collected would be much to small. Because funnel instruments cannot avoid the memory effect, the funnel size and shape should be standardized so that this systematic error is of similar magnitude for different instrument types. Another recomendation is that the deposition is calculated more correctly if the measured concentration is combined with the precipitation amount from the same sampler and not with the amount of a standard gauge.

4. Trace metals

Not all trace metals are dissolved in rain water but a very large fraction is present in insoluble form. Usually, the samples are acidified with HNO_3 in order to include both fractions. The literature recommends also thorough acid washing of the sampling bottles in order to avoid high blancs (Ross, 1986). Samples are usually filtered before analysis, however, the facts during collection are not fully considered: rainwater comes into contact with the funnel or other parts of the sampling device and trace metals may be absorbed or adsorbed in the funnel. If the pH of a rain event is low enough, some of this absorbed material may be dissolved again. It is not known, whether the metals accumulate at the funnel walls or whether an absorption equilibrium is reached after some time.

In order to collect information on this problem we conducted rinsing experiments in which the funnels were washed with HNO_3(corresponding 1.4 mm precipitation height) after the funnels had been rinsed twice with double distilled water. The HNO_3 solutions as well as the distilled water still contained trace metal concentrations exceeding the average concentrations found in rain water. This shows that appreciable amounts of metals are absorbed and escape analysis respectively wet deposition determination unless special procedures are applied. From this experiences we can conclude that the trace metal concentration of individual, e.g. daily samples are very unreliable and must not be over interpreted. The determination of wet deposition of metals requires that

a)the dissolved and undissolved fraction of metals have to be determined routinely and

b)the funnels are washed with HNO_3 one or two times per year and the solutions are analyzed for metals absorbed in the funnel.

Not all problems in connection with trace metal deposition could be solved satisfactory during our research project.

5. Sample pollution and gas diffusion

In one experiment during a smog situation in February distilled water was exposed in the instruments under dry conditions. The amount corresponded to 1 mm precipitation height. After 24 hours the electrical conductivity in some samples exceeded 100 μS/cm and the SO_4^- concentration reached nearly 14 mg/l. This experiment shows that under high pollution conditions the samples may be contaminated due to diffusion of soluble gases. This is also an important result for siting criteria or the question up to how many days a sampling period can be tolerated. In some climates where areas exist in which soluble gases reach high concentrations by natural processes the sampling period should be short (1 day). Under clean conditions, sampling periods of a week may be tolerable. However, in warmer climates long sampling periods may also become problematic. De Pena et al. (1985) have found, contrary to Sisterson et al. (1985) that weekly samples showed lower concentrations than accumulated daily samples. The differences become larger during summer.

6. Conclusions and outlook

Some of the problems of the determination of the wet deposition reported here could only be found by the intercomparison of various instrument types. Such results cannot be obtained when samples of only one type are intercompared. The deviations of individual instruments from the reference value could be interpreted in terms of poor sensitivity of the sensor and of memory effects caused by the wetting loss. Several contamination problems have been overcome because the manufacterors have improved their instruments meanwhile. Few experiments with improved instruments have shown that, if the present state of the art is fully exploited, the wet deposition of the main ions can be determined with an accuracy of better than 10 % with various collectors. In the case of trace metals the measuring and analytical techniques have to be improved in order to achieve a comparable level of standardization.

It may be mentioned that for the investigation of precipitation sensors a precipitation chamber has been developed in which the response of sensors at low intensities can be studied under laboratory conditions.

One major problem still unsolved is the quantification of the wet deposition by events of low productivity. Some preliminary studies in that direction have been undertaken, however, further fundamental studies on precipitation statistics and instrumental development is necessary to solve this problem.

Acknowledgement:
This work was supported by the Ministery of Research and Technology under grant 07431073.

Literature

de Pena, R.G., Walter, K.C., Lebowitz, L., Wicka, J.G.:"Wet deposition monitoring. - Effect of sampling period." Atm. Environ. **19** (1985) 151-156.

Joss, J., Waldvogel, A.: "Ein Spektrograph für Niederschlagstropfen mit automatischer Auswertung". PAGEOPH **68** (1967) 240-246.

Ross, H.: "The importance of reducing sample contamination in routine monitoring of trace metals in atmospheric precipitation". Atm. Environ. **20** (1986) 401-405.

Sisterson, D.L., Würfel, B.E., Lesht, B.M.: "Chemical differences between event and weekly precipitation samples in northeastern Illinois". Atm. Environ. **19** (1985) 1453-1469.

Winkler, P., Jobst S.: Comparison of various precipitation gauges and sensors for the determination of wet deposition. In: K. Grefen, J. Löbel (eds.). Environmental Meteorology, (1988) 193-200.

Winkler, P., Jobst, S., Harder, C.: "Meteorologische Prüfung und Beurteilung von Sammelgeräten für die nasse Deposition". Schlußbericht Proj. 07431073, GSF-Berichte, München, 1989.

FIELD INTERCOMPARISON EXERCISE ON NITRIC ACID AND NITRATE MEASUREMENT: CRITICAL EVALUATION OF THE RESULTS

I. Allegrini, A. Febo, C. Perrino

C.N.R. - Istituto Inquinamento Atmosferico
Via Salaria Km 29,300 - C.P.10
00016 Monterotondo Stazione (Roma), Italy

Summary

A five days intercomparison of measurement techniques for nitric acid and particulate nitrate has been held in the Area della Ricerca di Roma during September 1988. The study design included 4-hours and 12-hours sampling periods, and, for two days, additional 24-hours samplings, so as to evaluate the self-consistence of each method. Also, two samples for each group were taken from a nitric acid permeation source, in order to check the accuracy of the methods in laboratory conditions. The evaluation of the results is not performed through a simple regression of the data, that is by assessing agreement or disagreement between pairs of methods, but according to the following steps: 1) the drawbacks of the different techniques are critically evaluated; 2) the self-consistence of each method is checked, also in the ligth of the presence of potential interfering species; 3) a study of the sum and ratio between reported nitric acid and reported nitrate concentrations is carried out; this represents an interesting tool for the data comprehension and interpretation. The application of these criteria to the data set gathered during the intercomparison shows that the diffusion techniques yield the most reliable results, while the teflon-nylon filter packs do not allow a correct discrimination between nitric acid and nitrate, particularly in the presence of high ammonium nitrate concentrations. Filter packs using a cellulose prefilter are only able to measure total nitrates.

1. INTRODUCTION

A Field Intercomparison Exercise on Nitric Acid and Nitrate Measurement, organized by the C.N.R. and the Commission of the European Communities, has been held in the Area della Ricerca di Roma in Montelibretti (Rome) during september 18-24, 1988. The objective of the study was to evaluate, under field conditions, the results yielded by different techniques normally used for measuring these two compounds.

Nitric acid and particulate nitrate are important reaction products of atmospheric nitrogen oxides, and are involved in both the problem of ground acidification and photochemical smog. As the concentration level of these species is too low to allow the use of continuous measurements techniques, accurate accumulation methods are strongly needed; thus, a great deal of attention has been devoted, during the last ten years, to development and intercomparison of specific sampling and analysis methods.

In principle, the purpose of an intercomparison should be the evaluation of the performances of different techniques in terms of sensitivity, precision, reproducibility, time resolution, simplicity and cost effectiveness. A field validation of the methods in terms of accuracy and selectivity is assumed to have been already carried out in the development phase of each technique. However, several laboratories and field intercomparisons on nitric acid and nitrate measurement, carried out between 1982 and 1988 (1-5), show many discrepancies among the tested methods which could not be attributed solely to experimental variability, and suggest the existence of interfering mechanisms which could not be identified. This state-of-art suggested to plan a study design which could be of help in a field evaluation of the methods and give information about the specific interfering mechanisms which might cause disagreement between the tested techniques.

2. STUDY DESIGN

A compromise between the time resolution of the less sensitive methods taking part in the intercomparison and the need for gaining a sufficient number of data during the 5-day exercise led to establish a four-per-day schedule, with three four-hour samplings during the day (8.00 to 12.00; 12.00 to 16.00; 16.00 to 20.00) and a twelwe-hour sampling during the nigth (20.00 to 8.00). In addition, the schedule included two 24-hour measurements, which started at 20.00 of the 20th and 21st September. These samplings, to be performed side-by-side to the main sampling line, are fundamental for the evaluation of the self-consistence of the method. That is, a reliable method is expected to yield a total agreement between the concentration measured by the 24-hour sampling line and the average concentration of the 4-hour and 12-hour samplings. The 24-hour period started in the evening in order to enhance the possible artifact due to evolution of ammonium nitrate and to nitrous acid collection (mostly during the nigth) and subsequent nitrite to nitrate oxidation (during the following day) (6-7).

Besides, two nitric acid determinations from a constant nitric acid permeation source, having a permeation rate of 1.43 ± 0.05 μg/min, have been scheduled for all the participating groups. This experiment allowed to check, in laboratory conditions, the collection efficiency for nitric acid of the whole sampling line (including the inlet) actually employed in the intercomparison.

As it was decided to focus on the evaluation of the accuracy of the methods, replicate measurements were not included in the sampling schedule. The data yielded by replicate measurements, in facts, allow an easy statistical evaluation of the precision of each sampling technique, but do not give information about the reliability of the data.

A number of ancillary measurements have been performed during the whole study, including HONO, SO_2, NH_3, PAN, H_2O_2, HCHO, NO_2, SO_4^{2-}, NH_4^+, ozone, TSP and routine metereological parameters. Besides yielding a detailed characterization of the air composition during the intercomparison, these measurements can also help in identifying the possible interfering species and in clarifying the interference mechanisms.

The intercomparison was run by 16 groups from 11 European countries, employing many different techniques. Unfortunately, only few participants followed the schedule completely. Among these, a group for each one of the main categories of techniques represented in the intercomparison: the teflon-nylon filter pack, the denuder technique and the cellulose-impregnated filter pack. The results yielded by these groups are used for a critical discussion of the data and for gathering useful information about the suitability of these techniques for accurate determination of nitric acid and particulate nitrate.

3. METHOD

To interpretate the intercomparison results, two fundamental pieces of information, which can be gained from laboratory experiments, are needed for each method: the first one is the predicted deposition pattern of nitrate on each stage of the sampling line; the second one is the link between the amount of nitrate determined on the collecting medium/a and the atmospheric concentration of HNO_3 and/or NO_3^-. Thus, for each stage of the sampling line and for each species to be determined, it was necessary to know: 1) the collection efficiency; 2) the collection efficiency of other co-collected species; 3) the nature of the possible link between the co-collected compounds and the species of interest (HNO_3 or NO_3^-).

A filter pack is usually constituted by an inlet and two or more filters set in series: the first filter is assumed to collect particulate nitrate, while the second or the following ones are assumed to measure gaseous nitric acid. For this kind of device, nitrate deposition pattern on the first two filters is described by the following relationships:

$$M_1 = A^p\ E^p_{IN}\ (1-E^p_1) + A^g\ E^g_{IN}\ (1-E^g_1) + \sum Q_1 - \alpha\ [A^p\ E^p_{IN}\ (1-E^p_1)] \qquad [1]$$

$$M_2 = A^g\ E^g_{IN}\ E^g_1\ (1-E^g_2)\ \beta + \sum Q_2 + \alpha\ [A^p\ E^p_{IN}\ (1-E^p_1)]\ \beta \qquad [2]$$

where superscripts p and g refer to particulate nitrate and gaseous HNO_3 respectively; subscripts 1, 2, IN refer to the first, second filter and inlet respectively; M is the amount collected; E is the penetration efficiency; A is the atmospheric amount entering the system; Q is the nitrate amount yielded by interferent species; α is a loss factor mainly due to evolution and/or displacement of HNO_3 from particulate NO_3^-; β is the nitric acid transmission factor between the 1st and the 2nd filter.

In the case of a teflon-nylon filter pack, the values assumed in operative conditions by some of the parameters of equations [1] and [2] are well known: $E^p_1 \cong 0$; $E^g_1 \cong 1$; $E^g_2 \cong 0$; $Q_1 \cong 0$, while the values assumed by the parameters E^p_{IN}, E^g_{IN} and β depend on the inlet geometry and costruction material of the inlet and filter holder; differently, the values assumed by the parameters Q_2 and α are not predictable "a priori": the value of Q_2, which is usually quite low, depends on the concentration of HONO, NO_2 and on the presence of atmospheric oxidants (9), while the value of α depends on the ratio between ammonium nitrate and total nitrate and on the termodynamic conditions during the sampling time. As the value of α can be neither known nor zeroed, a considerable uncertainty affects the possibility of determining the atmospheric concentration of HNO_3 and particulate nitrate from the values M_1 and M_2. From the above discussion it results that on teflon-nylon filter packs:

measured HNO_3 > true HNO_3;
measured NO_3^- < true NO_3^-;

but: measured $(HNO_3+NO_3^-) \equiv$ true$(HNO_3+NO_3^-)$
and, for high values of α,:
measured HNO_3/NO_3^- >> true HNO_3/NO_3^- .

In the case of a cellulose-impregnated filter pack, as the value of E^g_1 approaches zero we can conclude that this technique is substantially unable to discriminate between HNO_3 and particulate nitrate. Therefore, this method can be successfully used only for total nitrate determination; this, however, can also be performed simply by using a single impregnated filter.

A denuder system is usually constituted by one or more cylindrical or annular tubes, whose walls are coated with an appropriate chemical for selectively collecting the gaseous species to be determined, followed by one or more filters for particle collection. A detailed description of the deposition functions which are expected on denuder systems has been reported by Febo et al. (8).

As the gaseous species are removed before reaching the filter surface, denuder systems are free from the problem of gas-particle interaction; however, only an accurate choice of the coatings, denuder configuration and inlet materials can avoid or largely reduce the problem of interferent collection and gas adsorption at the inlet (7). For example, the denuder configuration of group IC (see Table I) was made of: 1) two NaCl coated annular denuders, the first for selectively collecting HNO_3 and the second for taking into account the little interference due to particle deposition on the denuder walls, by using the Absolute Differential Technique (7,8); ii) a poliethylene cyclone; iii) a Na_2CO_3/glycerine coated annular denuder for removing HONO, which could interfere with the particulate nitrate determination (9); iv) a nylon filter for nitrate collection. The possible sligth difference between the measured and the true values (measured HNO_3 > true HNO_3; measured NO_3^- < true NO_3^-) should be, in this case, only due to insufficient correction for particulate interference.

4. RESULTS AND DISCUSSION

The results of the 24-hour samplings (24h) and the average of the four-per day measurements (AVG) are compared in Table I. The ratio P between 24h and AVG data, that is, the self-consistence of the method, is also reported: in ideal conditions P = 1. If the 24h and the AVG results are not consistent, clear indications about the causes of deviation can be obtained from the study of the sum S and the ratio R between HNO_3 and NO_3^- concentrations. For example, R should be strongly influenced by an interconversion between nitrate and HNO_3 occurring on teflon-nylon filter packs, while S should remain constant (S is not a function of α).

The results of Table I show that for the three denuder systems P approaches 1 for all the examined parameters; thus, the self-consistence of this method is proved to be very good.

For cellulose-impregnated filters it clearly appears that HNO_3 values are largely underestimated and nitrate values are measured in excess, while S is consistent. This indicates that HNO_3 is retained on the prefilter (E_f^g<<1); the results obtained with this technique at the permeation source (0.42 and 0.43 μg/min) confirm these findings.

For teflon-nylon filter packs only $P(S) \cong 1$, while $P(HNO_3)>1$, $P(NO_3^-)<1$ and $P(R)>>1$. Good values of P are obtained, instead, for particulate sulfate, reported in the last column of Table I. The values of S obtained by group C are rather low; a possible explanation is that a loss of nitric acid occurred at the inlet; the permeation source results (0.78 μg/min) confirm this hypothesis.

For further discussion, groups A and IC will be used, as they yielded good results at the permeation source (1.24 and 1.41 μg/min, respectively) and constitute a complete data set.

The observation that $P(S) \cong 1$ and $P(R) >> 1$ is consistent only with the hypothesis of a nitrate conversion into nitric acid, a phenomenum which is particularly enhanced when the 24-hour samplings start in the evening. In fact, most of the ammonium nitrate is collected during the night and the early morning; the increase in temperature and decrease in relative humidity which occur during the following hours cause a remarkable release of nitric acid and ammonia from the ammonium nitrate collected on teflon filters. As a result, when the determination is carried out with a teflon-nylon filter pack, HNO_3 concentration is overestimated, nitrate concentration is underestimated and the value of R, which is important for atmospheric chemistry studies, results to be totally unreliable.

GROUP	DATE		HNO_3	NO_3^-	$HNO_3+NO_3^-$	HNO_3/NO_3^-	$SO_4^=$
A (TN)	9/20	24h	3.46	1.97	5.43	1.76	8.57
"	"	AVG	2.51	2.91	5.42	0.86	8.00
"	"	**24h/AVG**	**1.38**	**0.68**	**1.00**	**2.05**	**1.07**
"	9/21	24h	4.78	3.05	7.83	1.57	12.61
"	"	AVG	3.15	4.96	8.11	0.63	11.74
"	"	**24h/AVG**	**1.52**	**0.61**	**0.96**	**2.49**	**1.07**
C (TN)	9/20	24h	2.46	1.50	3.96	1.64	7.96
"	"	AVG	2.14	2.33	4.47	0.92	7.10
"	"	**24h/AVG**	**1.15**	**0.64**	**0.88**	**1.78**	**1.12**
"	9/21	24h	3.66	2.37	6.03	1.54	–
"	"	AVG	2.46	4.27	6.73	0.58	–
"	"	**24h/AVG**	**1.49**	**0.55**	**0.89**	**2.65**	**–**
IB (CI)	9/20	24h	0.15	4.91	5.06	0.03	8.64
"	"	AVG	0.76	5.13	5.89	0.15	8.54
"	"	**24h/AVG**	**0.18**	**0.96**	**0.86**	**0.20**	**1.01**
"	9/21	24h	0.26	7.16	7.42	0.04	12.80
"	"	AVG	0.74	7.09	7.83	0.10	12.34
"	"	**24h/AVG**	**0.35**	**1.01**	**0.95**	**0.40**	**1.04**
IC (AD)	9/20	24h	1.93	3.51	5.44	0.55	7.73
"	"	AVG	2.04	3.56	5.60	0.57	8.29
"	"	**24h/AVG**	**0.95**	**0.98**	**0.97**	**0.96**	**0.93**
"	9/21	24h	2.02	5.42	7.44	0.37	13.23
"	"	AVG	2.10	5.50	7.60	0.38	13.03
"	"	**24h/AVG**	**0.96**	**0.98**	**0.98**	**0.97**	**1.01**
F (CD)	9/21	24h	2.06	5.39	7.45	0.38	11.49
"	"	AVG	1.63	5.20	6.83	0.31	11.45
"	"	**24h/AVG**	**1.26**	**1.04**	**1.09**	**1.22**	**1.00**
M (CD)	9/20	24h	1.79	–	–	–	–
"	"	AVG	1.94	–	–	–	–
"	"	**24h/AVG**	**0.92**	–	–	–	–
"	9/21	24h	1.91	–	–	–	–
"	"	AVG	2.02	–	–	–	–
"	"	**24h/AVG**	**0.94**	–	–	–	–

AVG=average of the four-per-day determinations; TN=teflon-nylon filter pack; CI=cellulose-impregnated filter pack; AD=annular denuders; CD= cylindrical denuders. Concentrations are expressed in $\mu g/m^3$.

From the above description of filter pack characteristics and from the results of Table I, it results that:
$$HNO_3 \ (24h) > HNO_3 \ (AVG) > true \ HNO_3$$
thus, we can estimate the minimum value of the relative error (R.E.) occurring in the 24-hour determination of nitric acid, without knowing or estimating the true value of the HNO_3 air concentration and without using any result yielded by other techniques:

$$R.E. > \frac{M_2(24h) - M_2(AVG)}{M_2(AVG)}$$

In the case of group A, the error results to be greater than 40% and 52% for the samplings of 9/20 and 9/21 respectively.

From Table I it also results that:

$$M_2^{TN}(24h) >> M_2^{TN}(AVG) >> M_2^{AD}(24h) \cong M_2^{AD}(AVG) > true \ HNO_3$$

thus, a better estimate of the minimum value of the error associated to the 24h filter pack determination can be obtained by using the denuder results, as follows:

$$\text{R.E.} > \frac{M_2^{TN}(24h) - M_2^{AD}(24h)}{M_2^{AD}(24h)}$$

By using the denuder results of group IC, the value of R.E. of group A results to be greater than 79% and 137% respectively. By applying the same procedure, the errors associated to the average of the short period determinations can also be estimated, and result to be 25% and 52% respectively.

These observations constitute an evidence that during the intercomparison experiment a noticeable artifact, due to nitrate to nitric acid conversion, occurred on teflon-nylon filter packs; thus, this technique suffers from an unpredictable uncertainty, whose extent depends on the atmospheric conditions. The error in the determination of atmospheric HNO_3 and nitrate concentrations can be fairly large, as the evolved nitric acid amount may reach 100% of the ammonium nitrate collected on the first filter.

CONCLUSIONS

A critical analysis of the data of the Field Intercomparison Exercise on Nitric Acid and Nitrate Measurement yielded the following conclusive information about the performances of filter pack and denuder techniques:
- cellulose-impregnated filter packs can be reliably used only as total inorganic nitrate collectors;
- teflon-nylon filter packs can be seriously affected by the bias due to nitrate-nitric acid conversion, mostly caused by the volatilization of ammonium nitrate; thus a correct discrimination between the gaseous and the particulate phase is not assured in all environmental conditions;
- among the tested methods, diffusion techniques proved to be the most reliable ones, as they yielded very good results in both the permeation source and the self-consistence experiments.

A good reliability of denuder results also depends on the use of configurations appropriate for reducing any possible interference and for checking the performance of the device (that is more denuders placed in series); the compact dimensions of annular denuders satisfy this requirement.

REFERENCES

(1) SPICER, C.W. et al. Atmos. Environ. 16 (1982) 1487-1500.
(2) ANLAUF, K.G. et al. Atmos.Environ. 19 (1985) 325-333.
(3) HERING, S.V. et al. Atmos Environ. 22 (1988) 1519-1539.
(4) FOX, D.L. et al. Atmos. Environ. 22 (1988) 575-585.
(5) FERM, M. et al. Atmos. Environ. 22 (1988) 2275-2281.
(6) FEBO, A., DE SANTIS, F., PERRINO, C., Proceedings of the IVth European Symposium on Physico Chemical Behaviour of Atmospheric Pollutants; Stresa 23-25/9/1986 pp.121-125.
(7) PERRINO, C., DE SANTIS, F., FEBO, A. Atmos. Environ. in press.
(8) FEBO, A., DE SANTIS, F., PERRINO, C., GIUSTO, M., Atmos. Environ. 23 (1989) in press.
(9) PERRINO, C., DE SANTIS, F., FEBO, A., Atmos. Environ. 22 (1988) 1925-1930.

TRENDS OF PHOTOCHEMICAL OXIDANTS OBSERVED DURING THE INTERCOMPARISON FIELD EXERCISE ON NITRIC ACID AND NITRATE MEASUREMENTS HELD IN MONTELIBRETTI.

P. CICCIOLI, M. POSSANZINI, A. CECINATO, V. DI PALO, E. BRANCALEONI, A. BRACHETTI, S. MURA
Istituto sull'Inquinamento Atmosferico del CNR, Area della Ricerca di Roma, C.P.10, 00016, Monterotondo Scalo - Italy

D. PERNER, U. PARCHATKA, H. KARBACH, I.C. ESLICK
Max Planck Institut fur Chemie, Abteilung Luftchemie, Saarstrasse 23, 6500 Mainz, Federal Republic of Germany.

Summary
Data of Ozone, PAN, PPN, NO2, H2O2 and HCHO, collected during the field intercomparison exercise on nitric acid and nitrate measurement held at the Area della Ricerca di Roma (Montelibretti) from the 18th to the 24th of September 1988, are presented. Measurements carried out with different techniques are compared and differences critically discussed. Trends are consistent with the reactivity of the various species and the meteorological situation occurring over the site. A good correlation between PAN, PPN and HNO3 is found.

1. INTRODUCTION

Although the main task of the intercomparison held in Montelibretti the third week of September 1988 was to test, in the field, the various methodologies available for the collection and measurement of HNO3 and nitrate in particulate matter, it was decided that other species should have been measured during the same campaign so that a comprehensive view of what was happening in terms of transport of pollutants and their chemical reactivity could be achieved. Among them, photochemical pollutants were considered of key importance because their production occurs through the same chain of reactions that gives rise to the formation of acidic species. In addition to ozone and NO2, that are currently monitored at any control station, it was decided to measure PAN, PPN formaldehyde and hydrogen peroxide as these components can strongly affect the levels of nitric acid in the atmosphere (1) . In addition, some of them (i.e. PAN) can be used to assess the impact of photochemical pollution arising from anthropogenic sources as well as for tracking the movement of air masses containing oxidants and acidic species. The second task of our measurements was to compare some of the methods and instrumentation presently available for the monitoring of photochemical pollutants in view of their application in forest areas. This work was done as a part of project sponsored by the Commission of the European Communities. This paper presents the results obtained during the days when the intercomparison exercise was held. Data collected with various techniques and instrumentation are compared and the trends observed discussed on the light of the meteorological situation occurring during exercise and the information collected in previous years. The results fully confirmed the usefulness of monitoring photochemical oxidants for a better understanding the time evolution of acidic species in the atmosphere.

2. EXPERIMENTAL

The experimental setup for the monitoring of photochemical oxidants was comprised of several instruments. Ozone was detected with two commercially available analyzers based on UV absorption detection but provided by different sources (DASIBI and COLUMBIA). Monitoring of NO2 was carried out with an instrument based on chemiluminescence detection induced by Luminol (SCINTREX). Two instruments were used for the measure of PAN and PPN; both were based on gas chromatography combined with electron capture detection. The first instrument was a home made analyzer used for several years for the monitoring of PAN and PPN in our Institute (2). The other analyzer was a commercial one, recently made available by CARLO ERBA STRUMENTAZIONE. Together with the ozone monitors, they were placed in two different sites. One was in the laboratory and the other in the monitoring station located in the N-NE corner of the area. HCHO and H2O2 were collected in the laboratory with the Annular Denuder Method (ADM) and analyzed according to the procedures described in Ref. 3 and 4. Determinations of HCHO, NO2, HNO2 and ozone were also carried out by using Differential Optical Absorption Spectroscopy (DOAS) (5) with an instrument built in Germany and transported to Italy specifically for this exercise. It was capable to detect also SO2 and HNO2, whose results will be discussed by other authors. The deuterium lamp was placed inside the Tiber valley whereas the receiving unit was near to the laboratory. The pathlength was 3.8 Km. A comprehensive description of all the methods adopted can be found in the preliminary report prepared by our Institute (6) that has been printed and distributed to all the partecipants. A revised version of the report is in preparation and will be available soon.

3. RESULTS AND DISCUSSION

Before to present the results, it is important to say something about the meteorological situation that occurred during the intercomparison exercise. This will be extremely useful for the interpretation of the trends that were observed. The days before the exercise started, the site was under the influence of a low pressure area with a minimum established over Southern Italy. The sky was cloudy and some rain was falling the 17th and the 18th. Conditions were unfavorable to photochemical smog formation and the levels of ozone were those typically found in winter season (maximum daily value < 30 ppbv) with a flat diurnal profile. The maximum temperatures of these two days were not exceeding 22°C. This low pressure caused an air masses circulation seldom observed in our site. It was characterized by NE moderate winds prevailing over the sea-land breeze circulation typical of the Tiber valley during mild seasons. These effects were preventing transport of photochemical oxidants and their precursors from the city of Rome. After the first day of the intercomparison (19th), the low pressure area moved rapidly SE and sunny weather was observed. Higher maximum temperatures then the previous days were measured but they never exceeded 26°C. Conditions favorable to photochemical smog formation were established over the site and the city of Rome. However, the wind direction was variable so that transport could take place for a short period of time and usually in the late afternoon. The weather evolution is well reflected by the trends shown in Figure 1 where data of ozone and NO2, collected by UV absorption and chemiluminescence, are reported together with the levels of the same pollutants measured by DOAS. Ozone data collected by UV absorption show a constant increase in the mixing ratios along the week. Except for the first day, when a flat diurnal trend is observed, the ozone profiles are characterized by a maximum centered in the early afternoon. These profiles are typical of our site and have been often recorded in previous campaign from March to October (2,7). From the levels measured and trends found it is impossible to say, however, to

Ozone and NO2 Measured at Montelibretti

Ozone and NO2 Measured at Montelibretti (DOAS)

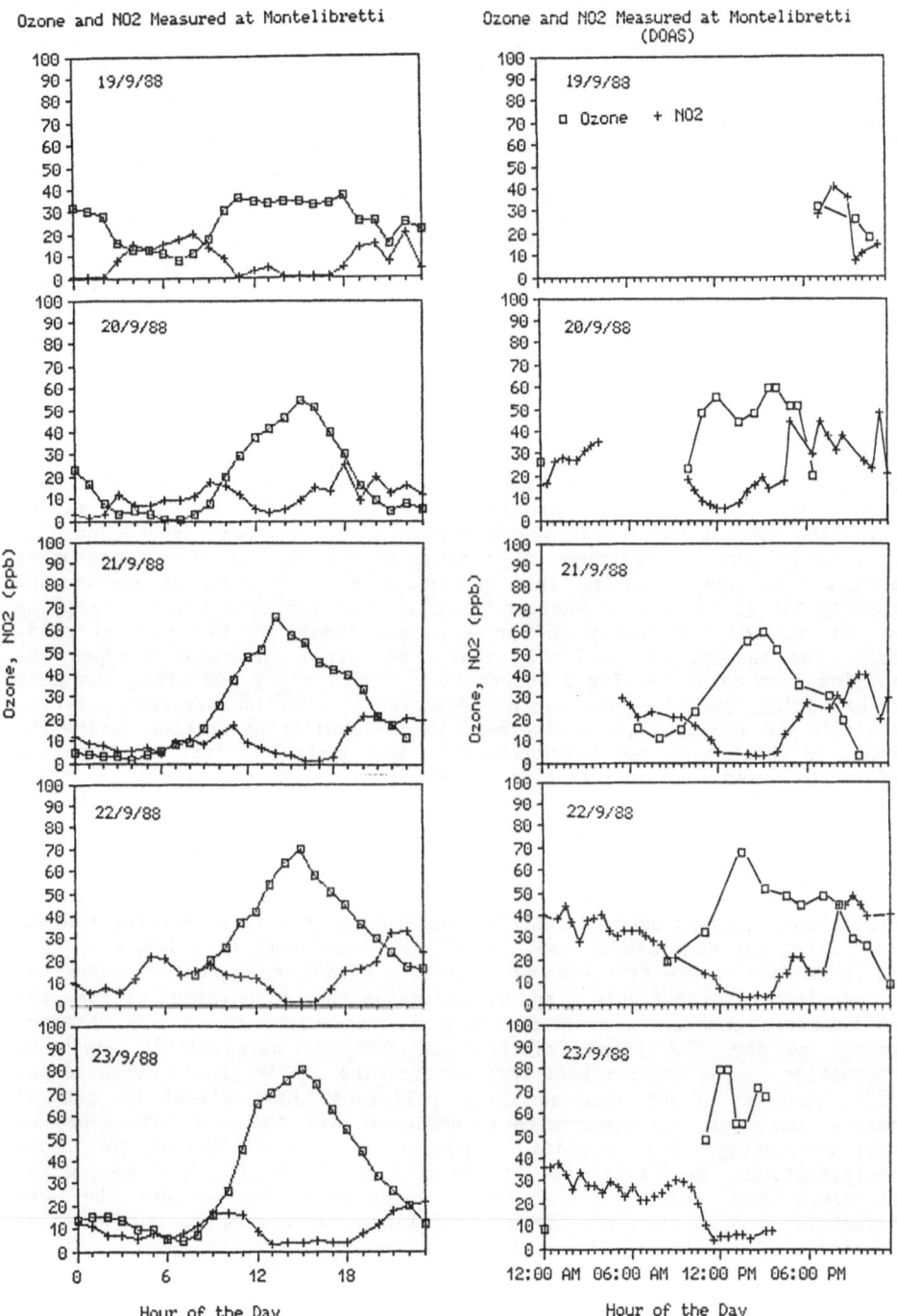

Figure 1. Trends of O_3 and NO_2 recorded during the monitoring campaign held in Montelibretti the 18th to the 23rd September 1988.

what extent photochemical pollution was determined by reactions occurring
near to the site, where moderate anthropogenic emission of photochemical
precursors is combined with natural background, or caused by transport from
the city. The small shoulders detected on some of the profiles shown in
Figure 1 are not high enough to be considered diagnostic for identifying
the city input. Ozone levels recorded by DOAS are close to those measured
by UV absorption. The profiles are similar but, in two instances, they
show a more complex trend suggesting transport from the city (see the
partial sets of data of the 20th and the 23rd). Unfortunately, the number
of samples analyzed during these two days is not sufficient to draw
definite conclusions. NO2 measured by chemiluminescence detection and DOAS
shows similar trends and close values were measured during sunny hours when
a drop in the mixing ratio was detected. These profiles are in good
agreement with the trends recorded in the same site in previous monitoring
campaign (2,7) where data were collected with a different instrument. The
drop occurring in the middle of the day is consistent with the reactivity
of NO2 toward OH and organic radicals to give PAN, PPN and nitric acid and
with photochemical dissociation that gives rise to NO and atomic oxygen
which, in turn, leads to the formation of OH radicals and ozone. Therefore,
diurnal trends of ozone and NO2 fit well with the theory of their
reactivity and formation and are typical of semirural areas (8). A marked
difference between the levels of NO2 measured by chemiluminescence and DOAS
was recorded during nighttime. A possible explanation of this discrepancy
is that DOAS was measuring this pollutant over an area of several Km
crossing the Tiber valley whereas chemiluminescence detector was sampling
the air outside the valley and in a higher position. As several roads
across the valley, it is likely that nocturnal inversion trapped the
emission from cars causing a substantial build up of NO2 where the DOAS
was measuring. Such increase was not detected by chemiluminescence detector
as little NO2 sources were active near to the monitoring station during the
night. In this case, the NO2 measured was just that left by diffusion
before the inversion layer developed. The drop in the concentration of NO2
measured at the monitoring station during the night can be explained by the
reactivity of NO2 toward organic species and ozone to give NO3 radicals,
N2O5 and acylnitrates. Figure 2 reports the data of PAN, PPN and HNO3
collected during the same days. A good agreement is observed between the
values of PAN recorded by the two different instruments used. A very nice
correlation is found between PAN, PPN and HNO3 with a ratio between PAN and
HNO3 mixing ratios close to one. Trends are consistent with the fact that
all these species are originated by reaction of NO2 with radicalic species.
A comparison between Figure 1 and 2, indicates that high values of PAN, PPN
and HNO3 are concurrent with the drop in concentration of NO2 observed
during the day. The trends of PAN and PPN are particularly reach of
information on the sources that were determining the levels of oxidants and
acidic species in our site as these pollutants have almost no natural
sources. Moreover, the experience accumulated over the last five years on
their monitoring has provided an amount of data sufficient for their
interpretation. Basically, the trends of PAN and PPN obtained during the
intercomparison exercise differ from the ozone trends because they are
characterized by marked day-by-day variations. According to our previous
experience (2,7), the mixing ratio of PAN and PPN remains rather low for
most part of the day and minimum values are measured during the night.
However, very sharp peaks, not exceeding one or two hour duration, are
very often recorded from March to October. They occur in the afternoon,
usually between 2 and 6 p.m. (2). This peak often corresponds to the
maximum value of vz, the vertical component of the wind speed. When

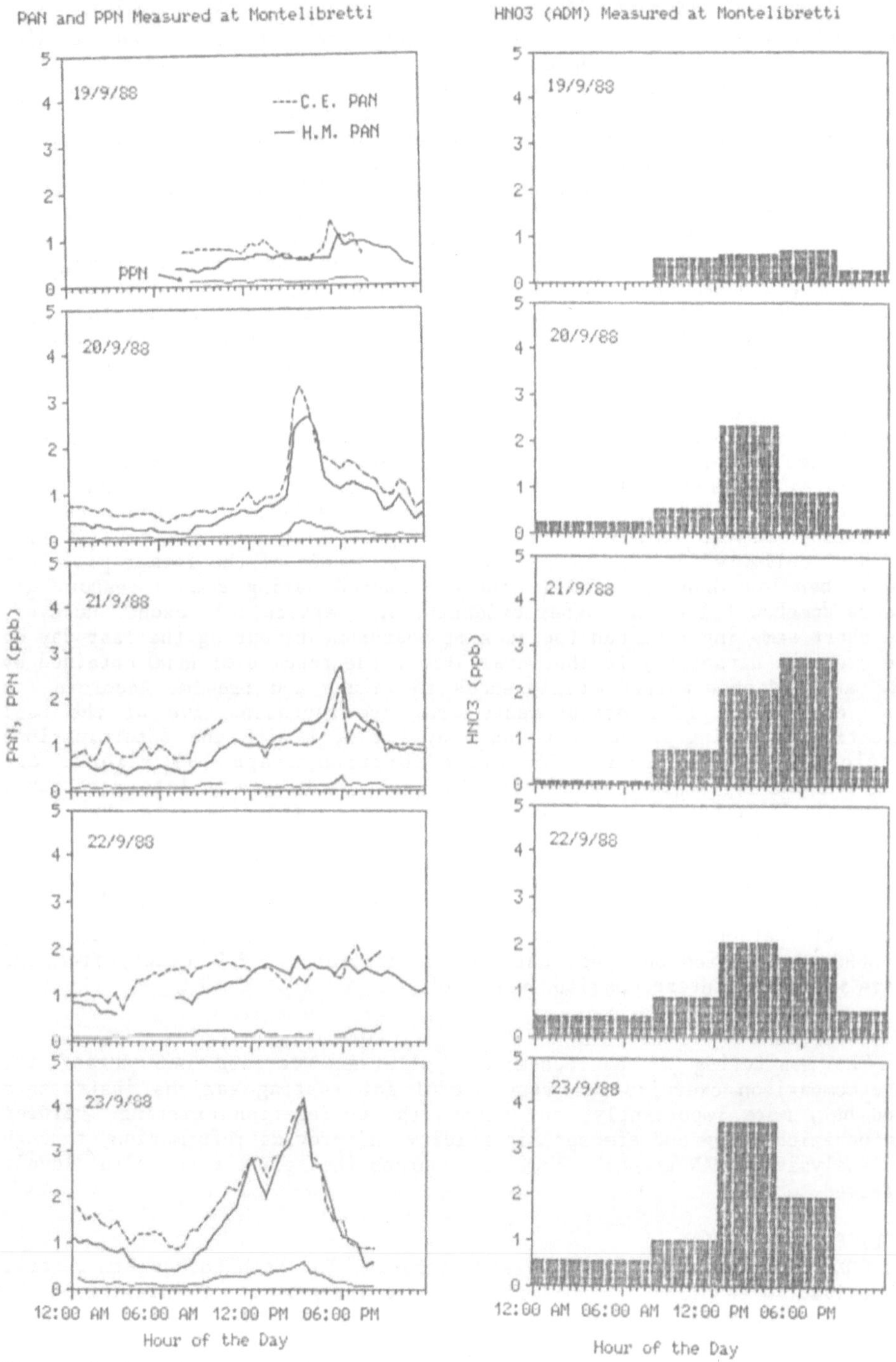

Figure 2. Trends of PAN, PPN and HNO$_3$ recorded the same days as in Figure1
C.E.= Carlo Erba PAN Analyzer, H.M.= Home made PAN Analyzer; data
of PPN were collected with the Carlo Erba PAN Analyzer.

photochemical smog episodes occur, a sharp peak of ozone can be also detected. It is always concurrent with the PAN and PPN peaks. These observations are made when the sea breeze blowing from W-SW reach the maximum intensity (2). The analysis of both meteorological and chemical parameters indicates clearly that such peaks of PAN and PPN are associated to transport of oxidants and their precursors from the city to the site. Therefore, the well detectable peaks of PAN and PPN in the graph of figure 2 can be attributed to the input coming from the city of Rome. They are not (or barely) seen the 18th and 22nd, whereas appear very clearly the last day of measurement. In the days when the maxima were not observed, the levels of photochemical oxidants and acidic species were almost exclusively arising from local sources or long range transport. The last day the input from the city accounted for most part of the levels of atmospheric acidity and photochemical smog pollution present in the sampling site. As can be seen from the graphs, the plume from the city reaches the site and moves away quite rapidly. Sometimes it is characterized by more than one peak. The profiles shown in figure 2 are extremely representative of our site as they summarize well what can happen during an entire year. It is interesting to notice that ozone data collected by DOAS somehow match those of PAN indicating that such technique might be more accurate than UV absorption for the monitoring of ozone. Figure 3 reports the data of H2O2 and HCHO collected during the same week. The levels of the former pollutant are rather low when compared to those measured during summer season, its trends somehow follow the other oxidants, and particularly ozone. However, the short sampling time and the lack of measurements during the last day of the exercise hinder any further evaluation. The results of HCHO obtained by DOAS and ADM show marked differences in values and trends. According to our experience, DOAS measurements are more representative of the real situation occurring in our site as they fit well with the determinations previously made by using ADM and adsorption traps (9). Trends are characterized by higher values in the early morning and late afternoon. During the day, photolysis of HCHO gives rise to a decrease in the mixing ratios that largely compensates for the increased rate of anthropogenic emission and hydrocarbons decomposition. Levels from DOAS are also in better agreement with the levels of PAN, PPN and ozone. According to our view, data from ADM were influenced by unidentified sources which affected our determinations in a unpredictable way. It should be noted that sampling of HCHO was carried out just outside the laboratory, 200 m away from the place where the intercomparison was held.

4. CONCLUSIONS

The monitoring of photochemical pollutants carried out during the intercomparison exercise was very useful for testing various instruments used but, more importantly, to confirm the correlation existing between photochemical smog and atmospheric acidity and provide information, through the analysis of PAN trends, about the sources that gives rise to the levels observed.

5. LITERATURE CITED

(1) COX R.A., "Rates, reactivity and mechanism for homogeneous atmospheric oxidation reactions" Proceedings of the 1st European Symposium on the Physico-Chemical Behaviour of Atmospheric Pollutants. B. Versino and H. Ott Edrs., EUR 6621, 1980, pp.91-107.
(2) CICCIOLI P., BRANCALEONI E., DI PALO C., LIBERTI M., DI PALO V., Acqua e Aria, 7, 675, (1986).
(3) POSSANZINI M., DI PALO V., LIBERTI A., Sci. of Total Environ., 77, 203,

HCHO Measured at Montelibretti

H2O2 (ADM) Measured at Montelibretti

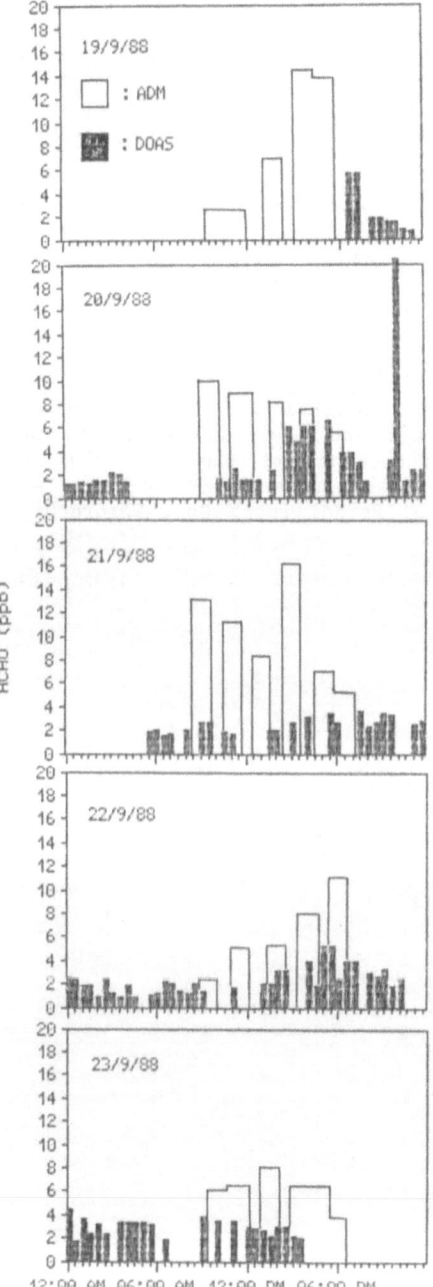

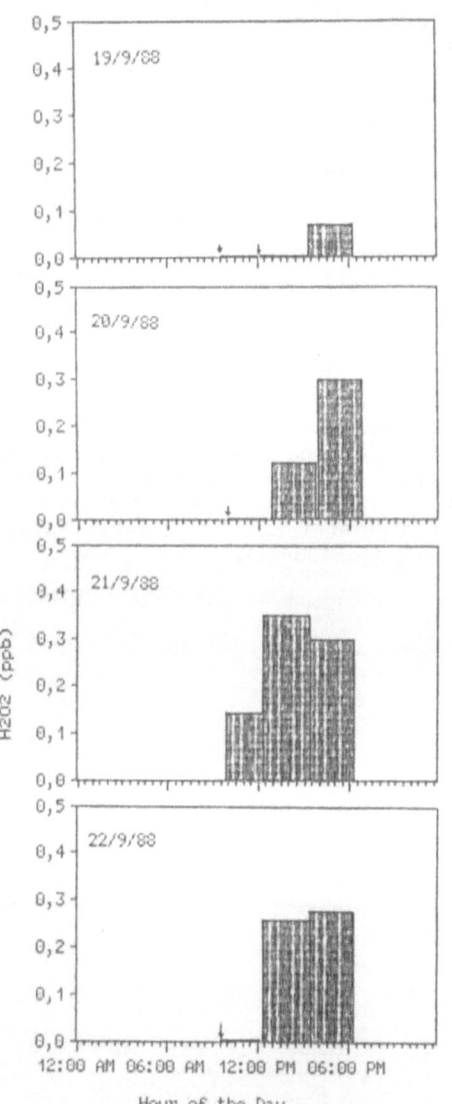

Figure 3. Trends of HCHO and H_2O_2 recorded the same days as in Figure 1. ADM = Annular Denuder Method; DOAS=Differential Optical Absorption Spectroscopy.

(1988).

(4) POSSANZINI M., CICCIOLI P., DI PALO V., DRAISCI R., Chromatographia, 23, 829, (1987).

(5) PLATT U., PERNER D., PATZ W., J. Geophys. Res., 84, 6329, 1979.

(6) Preliminary Report on the Field Intercomparison Exercise on Nitric Acid and Nitrate Measurements, Area della Ricerca Sep.18-24, 1988, C. Perrino Edr., Printed by CNR, Istituto Inquinamento Atmosferico with the support of the Commission of the European Communities, COST Project 611, December 1988.

(7) CICCIOLI P., BRANCALEONI E., DI PALO C., BRACHETTI A., CECINATO A., "Daily trends of photochemical oxidants and their precursors in a suburban forested area. A useful approach for evaluating the relative contributions of natural and anthropogenic hydrocarbons to the photochemical smog formation in rural areas in Italy" Proceedings of the 4th European Symposium on the Physico-Chemical Behaviour of Atmospheric Pollutants, G. Angeletti and G. Restelli Edrs., D. Reidel Publ. Co, Dordrecht, 1987, pp. 551-559.

(8) SPICER C.W., Sci. Total Environ, 24, 183, (1982).

(9) CICCIOLI P., DRAISCI R., CECINATO A., LIBERTI A., " Sampling of aldehydes and carbonyl compounds in air and their determination by liquid chromatographic techniques" Proceedings of the 4th European Symposium on the Physico-Chemical Behaviour of Atmospheric Pollutants, G. Angeletti and G. Restelli Edrs., D. Reidel Publ. Co., Dordrecht, 1987, pp. 133-141.

Gas phase measurements of NH_3 and $NH_4^{\pm}$ with Differential Optical Absorption Spectroscopy and Gas Stripping Scrubber in combination with Flow Injection Analysis.

A. Neftel and A. Blatter
Swiss federal Institute for Agricultural Chemistry and
Environmental Hygiene, CH-3097 Liebefeld

T. Staffelbach
Physics Institute University of Bern CH-3012 Bern

Summary

There is a growing need for NH_3 and NH_4^+ gas phase
measurements with high temporal resolution. We used a
commercially avaiable DOAS instrument for a continuous
determination of NH_3 concentration over path between 80m
and 250m in the UV region from 200 to 230nm. With an
integration time of 1 minute we reached a detection
limit of 5 $\mu g/m^3$, for a 75m path and 2 $\mu g/m^3$ for a 250m
path.
To test these results a two channel gas stripping system
combined with FIA was developed. NH_3 is fluorometrically
measured by the reaction of o-phthalaldehyde in the
presence of $NaHSO_3$ buffered at pH 11. One channel,
equipped with a glas spiral scrubber, measures NH_3 and
NH_4^+, whereas the second channel, equipped with a
diffusion scrubber measures only NH_3. Continuous
measurements can be made with a time resolution of 2
minutes. The detection limit is 1 $\mu g/m^3$. Both systems
are suitable for studying the processes that lead to
loss of NH_3 during different agricultural activities.

1.) Introduction

Asman and Diederen stated (1987) in a report about
ammonia and acidification "There is an urgent need for
continuous measurement techniques for NH_3 and NH_4^+ aerosol to
study emission, deposition, reaction and scavenging
processes" We started to develop an instrumentarium to
characterize the occurence of NH_3 and NH_4^+ (NH_x) in the air
during the different steps of the nitrogen cycle in
agricultural activities.
We used two different methods to measure NH_x. First for the
determination of gas phase NH_3 concentration we used a
Differential Optical Absorption Spectrometer (DOAS) from the
Swedish company OPSIS. DOAS systems became commercially
avaiable only recently.
Secondly we developed a point measuring system where NH_3 and
NH_4^+ has to be differentiated. The analysis is done by means
of a Flow Injection Analysis system. NH_3 is collected with a
diffusion scrubber (Dasgupta, 1986) and NH_x is collected with
a glass coil.

Experimental Setup

2.1) DOAS System

The use of DOAS systems for trace components measurements were first described in the literature by Platt and Perner 1979. The schematic outline of a DOAS system is shown in Figure 1 (Edner, 1986). The system uses a rotating disk with slits placed in between the monochromator and the photomultiplier to scan a specific spectral section of 40nm. Approximately 1000 scans are taken per minute.

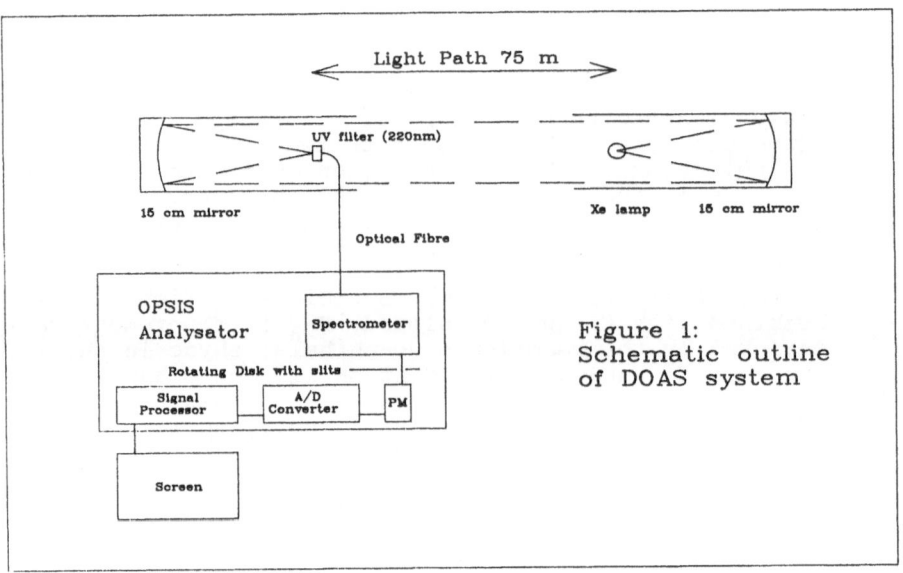

Figure 1:
Schematic outline
of DOAS system

Fig. 2 Absorption spectra of ammonia (700mg/m³ NH₃)
over a pathlength of 136mm

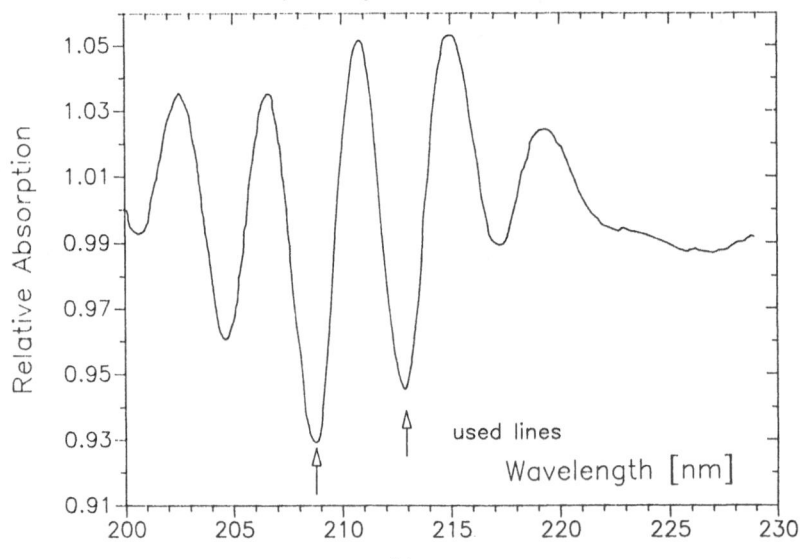

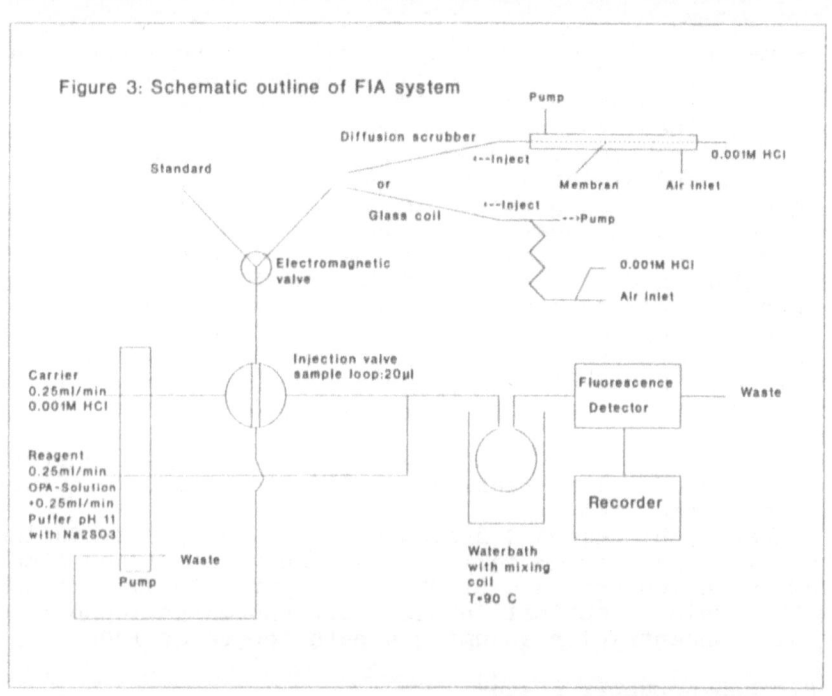

Figure 3: Schematic outline of FIA system

Figure 4: Recorder output

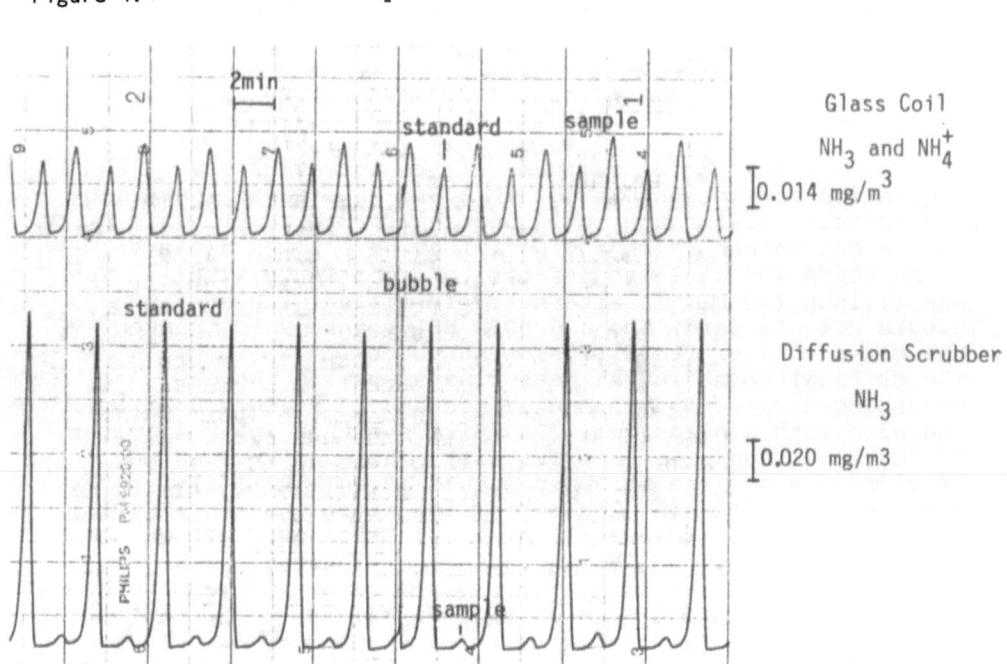

NH$_3$ is measured in the wavelength region 200 to 220nm, thus in the far UV. Figure 2 shows the differential NH$_3$ absorption spectrum as recorded with the OPSIS DOAS analyzer and the absorption lines used to determine NH$_3$ are indicated. This short wavelength requires the use of a high pressure Xenon lamp with fused silica glass as the light source. As a consequence ozone is produced in the lamp housing and omit therefore the ozone measurement on this light path.
The short wavelength also limits the usable distance range to approx. 250m, due to Raleigh and Mie scattering by the air molecule, as well as broad band absorption of ozone, oxygen and other compounds. The receiver is connected with an optical fibre to the spectrometer allowing for fast and uncomplicated setup of the system. Only recently have fibers with acceptable transmission characteristics in the far UV range become available.
In order to get rid of scattered light in the spectrometer a UV interference filter peaking at 220nm is placed immediatly in front of the fiber.
The linearity of the DOAS system was checked by introducing in the light path small cells (length between 1mm and 10mm) flushed with 1000ppm and 2% NH$_3$ in a synthetic air standard gas. The achieved optical density corresponds to 10ppb to 2ppm air concentration assuming a path length of 100m.

2.2 FIA gas scrubber system.

Basically we followed the ideas which are described in the literature by Lazrus et al (1986); Dasgupta et al (1988) and Genfa et al (1989). NH$_x$ is measured with a fluorometric system. The ternary reaction of o-phtaldialdehyde (OPA) with sulfite and NH$_3$ yielding intensely fluorescent product. (Details of the reaction are described by Genfa (1989)). We used the same chemical setup as Genfa with the exception of a 10 times higher sodium sulfite concentration in the phosphate buffer solution. Figure 3 is a schematic diagramm of the analytical system. A self made dual filter fluorimeter with a Cadmium light source was used in the detector. The excitation wavelength was set at 365nm (halfwidth 10nm) and the emission wavelength, at 450nm (halfwidth 10nm). This choice guarantees a large discrimination between different amino acids (by at least a factor of 100 (Genfa, 1989)). The sample loop (20 μl) is alternatively filled with NH$_3$ in 0.001M HCl standard, and a 0.001M HCl blank, which passed by the glass coil or the diffusion scrubber.
The critical point in the measuring system is the gas scrubbing device. One channel, intended to measure NH$_x$, is equipped with a glass coil (10 spirals with a inner diameter of 2mm and a radius of approx. 1cm). The gas flow was set to $2*10^{-3}$m^3/min and the scrubber liquid (0.001M HCl) flow rate was set to 0.25ml/min. The collection efficiency for NH$_3$ is 100% for gas flow below $3*10^{-3}$m^3/min. Previously, it was not possible to determine the collection efficiency for NH$_4^+$ aerosols. Theoretical considerations however, suggest that collection efficiency will be size dependent. Larger aerosols (with radii typically > 1μm) will be forced to the walls due

to centrifugal force in excess of frictional force (Stoke force); and very small aerosols are likely to diffuse to the wet walls. In the intermediate size range however, the various forces should be more or less in balance. Thus it is difficult to estimate a collection efficiency for aerosols of intermediate size.

The second channel, equipped with a diffusion scrubber (designed by Dasgupta) with a porous polypropylene membrane (Celgard X-20 inner diameter 400μm, wall thickness 25μm) measures only NH_3. The collection efficiency was determined to be 40%. The detection limit was determined to 0.5μg/m^3 NH_x Figure 4 shows a typical recorder output from the FIA gas scrubber system, when it measures background air.

3. Results and Discussion

3.1 Measurements of calibration gas

Known concentrations of gas phase NH_3 were made by dynamic dilution of NH_3 in synthetic air (commercially avaiable standards) with zero air using a Monitor Labs calibrator. Dilution factors up to 2000 can be achieved.

a) DOAS system

Either a 1000ppm or a 2% standard was fed directly through stainless steel tubing to the small cells switched between the light source and the analyzer. The recorded linearity was sufficient, but the slope was 60% too high (see Figure 5). Span and offset factors can be easily set in the OPSIS system, but they require a physical explanation. Our hypothesis that the absorption cross section used by the OPSIS machine in the reference spectrum is too low was later confirmed by the OPSIS manufacturer. All results reported below the DOAS value were adjusted with a span factor of 0.6.

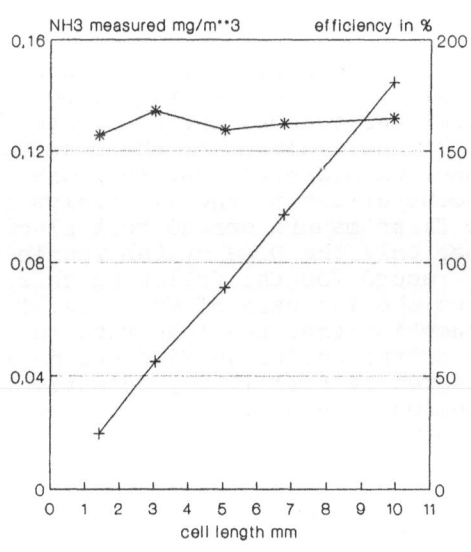

Figure 5: Calibration curve DOAS

b) FIA system
    The calibration of the FIA system using the glass coil
as gas scrubber gave on the other hand only 83% recovery. The
same efficiency was found when NH_3 was trapped in a wash
bottle with 0.001M HCl solution as absorbing liquid. We
conclude therefore that the indicated concentration in the
standard gas was effectively 17% lower than certified. Figure
6 shows the calibration curve for the diffusion scrubber and
the glass coil.

Figure 6: Calibration curves

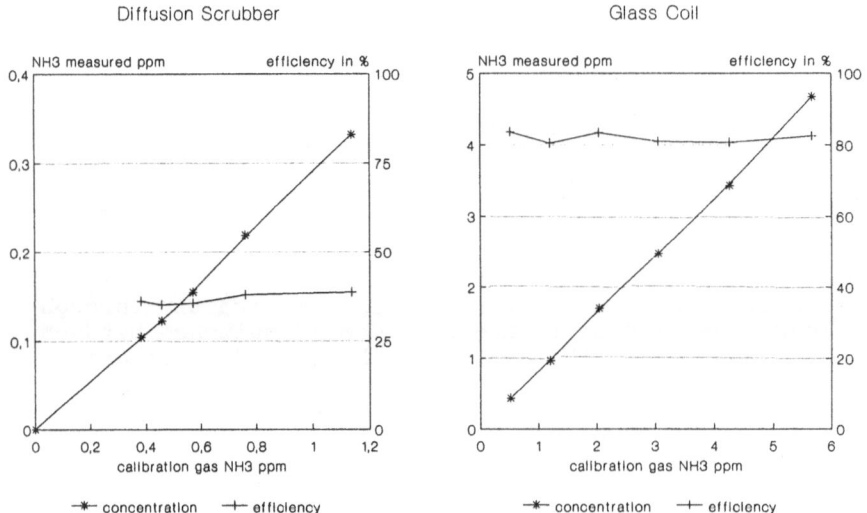

3.2 Measurements of NH_3 emission after cattle slurry
application

    For the DOAS, a 75m, path 2m above a grass field, was
installed within the area of our research station. For
logistical reasons the FIA system was installed at the east
end of the field, near the receiver (See Figure 7). Thus it
cannot be expected that the DOAS analyzer yields the same NH_3
concentrations,as the FIA analyzer.
The first manure spread took place the 31st of May. At that
time, only the DOAS system was installed. Figure 8 shows the
NH_3 record for the following three days. The integrated curve
shows the increase of NH_3 loss. This curve is based on the
assumption that the loss rate is proportional to the measured
concentration. We applied our data to the NH_3 profile
measured by Ferm (1987) to estimate the total NH_3 loss
assuming a mean wind velocity of 0.5m/sec. The loss can be
estimated to be about 20 to 40% of the total NH_3 content in
the cattle slurry.
The second manure spread took place the 22nd of August when
both systems were installed. In addition, a wind and
temperature sensor was mounted on the OPSIS system. Figure 9
shows the NH_3 concentration recorded with both systems. At

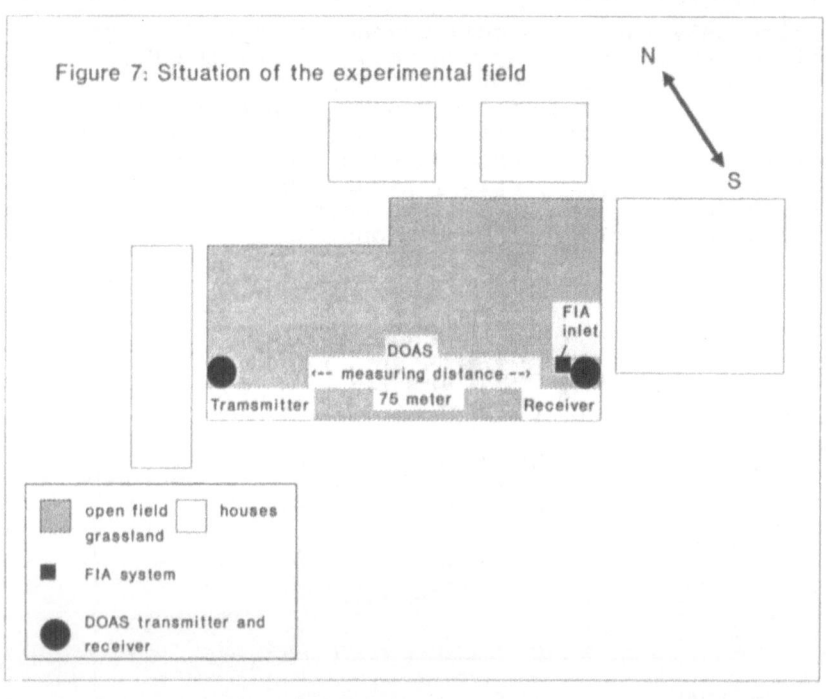

Figure 7: Situation of the experimental field

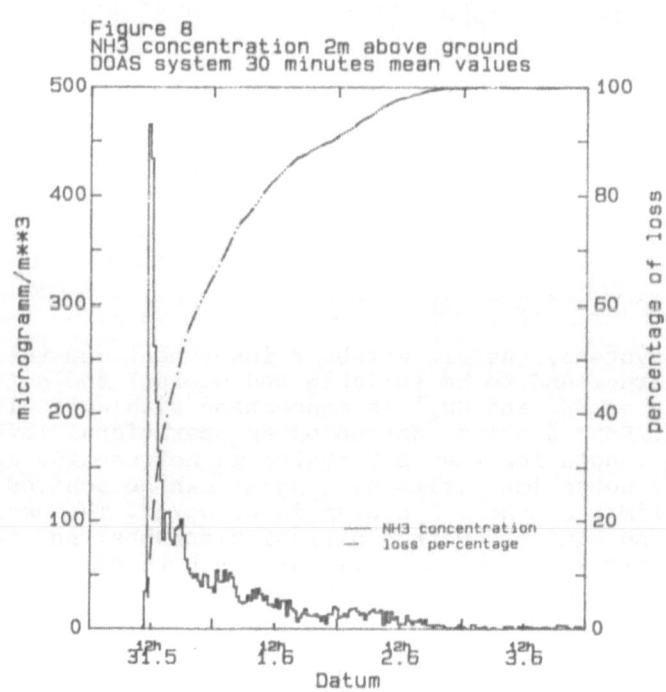

Figure 8
NH3 concentration 2m above ground
DOAS system 30 minutes mean values

that time only the glass coil channel of the FIA system was installed, thus the FIA system measures NH_3 and NH_4^+. After manure spread the contribution of NH_4^+ to NH_x will be small. Two hours after spread of the manure a thunderstorm began, causing a sharp temperature decrease, rainfall and high wind velocities. As a consequence of this the recorded NH_x concentrations dropped rapidly resulting in a smaller loss of NH_3 to the atmosphere. NH_3 losses were not estimated due to the "wild" meteorological conditions.

Figure 9:
NH3 concentration 2m above ground
DOAS and FIA system
10 minutes mean values

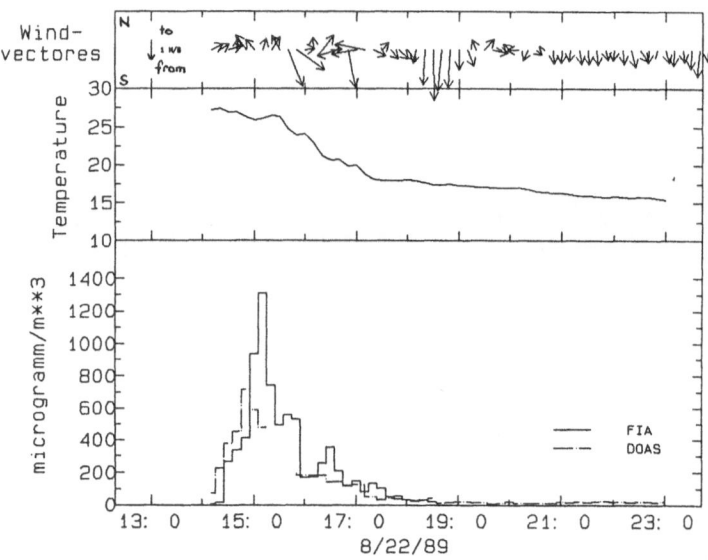

4. Conclusions and Outlook

Both systems, the FIA scrubber instrument and the DOAS analyzer were found to be reliable and usefull tools for the measurement of NH_3 and NH_4^+ in connection with agricultural activities. Both systems can run on an operational level. Ideal path length for the DOAS system is between 150 and 200m, where detection limits of 2 $\mu g/m^3$ can be achived. The detection limit of the FIA system is 0.5$\mu g/m^3$. The use of two channels, one equipped with a diffusion scrubber and the other equipped with a glass coil, allows both NH_3 and NH_4^+ to be measured and distinguished. Improvements can be made by replacing the glass coil with a continuous nebulizer which should be 100% efficient for both NH_3 and NH_4^+. Instead of injecting the sample alternating with the standard, we plan to modulate the gas phase by switching between ambient air,

zero air and calibration gases, while running 0.001M HCl continuously through the liquid system .
In addition, we have begun testing a passive sampler that would provide a cheap tool for monitoring the spatial distribution of NH_3 e.g. at a forest border close to intensively used agricultural land.

Acknowledgements:
This work was supported by the Swiss National Science Foundation within the NFP 14+ program (contract nr. 4.094.088.14). Prof. P. Dasgupta was so kind to introduce us to the world of manufactering diffusion scrubbers. We thank Dr. Anne Arquit for the editorial assistance. We thank Dr. F.X. Stadelmann and M. Bongard for their support. We are grateful to the personnel of the field group of our federal institute for setting up the DOAS system.

References:

Asman W.A.H. and Diederen H.S. (1987)
Ammonia and Acidification
Proceedings of a Symposium of the EURASAP held at the National Institute of Public Health and Environmental Hygiene, Bilthoven, The Netherlands, 13.-15. April 1987 8-10

Dasgupta P.K., McDowell W.L. and Rhee J.S. (1986)
Porous Membrane-based Diffusion Scrubber for the Sampling of Atmospheric Gases. Analyst **111** 87-90

Dasgupta P.K.,Dong S.,Hwang H.,Yang H.C. and Genfa Z. (1988)
Continuous liquid-phase fluorometry coupled to a diffusion scrubber for the real-time determination of atmospheric formaldehyde, hydrogen peroxide and sulfur dioxid
Atmospheric Environment **22** 949-963

Edner H.,Sunesson A.,Svanberg S.,Unéus L. and Wallin S.(1986)
Differential optical absorption spectroscopy system used for atmospheric mercury monitoring. Applied Optics **25** 403-409

Ferm M. and Christensen B.T. (1987)
Determination of NH_3 volatilization from surface-applied cattle slurry using passive flux samplers.
In Proceedings of a Symposium of the EURASAP held at the National Institute of Public Health and Environmental Hygiene, Bilthoven, The Netherlands, 13.-15. April 1987 28-41

Genfa Zhang and Dasgupta P.K. (1989)
Fluorometric measurement of aqueous ammonium ion in a flow injection system
Anal.Chem. **61** 408-412

Platt U., Perner D. and Pätz H.W. (1979)
Simultanneous measurements of atmospheric CH_2O, O_3 and NO_2 by Differential Optical Absorption. J. Geophys. Res. **84** 6329-6335

SIMULTANEOUSLY SAMPLING OF NH_3, HNO_3, HNO_2, HCl, SO_2 AND H_2O_2 IN AMBIENT AIR BY A WET ANNULAR DENUDER SYSTEM

M.P. KEUKEN, R.P. OTJES AND J. SLANINA
Netherlands Energy Research Foundation (ECN),
P.O.Box 1, 1755 ZG Petten, The Netherlands

SUMMARY

The automated wet denuder developed at ECN, is based on air sampling by an aqueous solution present in the annulus of a rotating annular denuder. After a sampling time of 35 minutes at a flow rate of 30 $1.min^{-1}$, automatically the absorption solution is pumped into a sample tube and the denuder is recoated. The obtained solutions are analyzed off-line. The limits of detection for NH_3, HNO_3, HNO_2, HCl and SO_2 are 0.1 $\mu g.m^{-3}$, and for H_2O_2 0.01 $\mu g.m^{-3}$.
During an intercomparison experiment for the measurement of HNO_3 and nitrate-containing aerosols in ambient air at Rome from 19 to 23 September 1988, ECN participated with a parallel wet annular denuder system. Advantages of the wet annular denuder are demonstrated by the detection of a strong diurnal variation of HNO_3 and NH_3 concentrations at Rome. The reproducibility better than 5 % between the parallel denuders for NH_3 and HNO_3 show the potential use of wet denuders for gradient measurements.
Results of experiments in The Netherlands for the simultaneously measurement of NH_3, HNO_3, HNO_2, HCl, SO_2 and H_2O_2 over a year demonstrate the application of the wet denuder as a field instrument.

1. INTRODUCTION

To evaluate the acidification of the atmosphere, sampling techniques are required for the determination of gases (e.g., SO_2, NH_3 and HNO_3), and NH_4^+-, NO_3^-- and SO_4^{2-}-containing aerosols in ambient air.
By using denuder/filter pack combinations sampling problems encountered with filter techniques, such as the interaction between gases and aerosols, are reduced and the gases are stabilized by chemical fixation (1). In addition information is obtained on the chemical speciation of air pollutants.
Advantages of denuder/filter pack systems are the relative simple and low cost equipment. However, disadvantages are the labour-intensive coating and extraction procedures involved, and the time delay between sampling and analysis.
The use of thermodenuders, based on thermal desorption of the collected species and on-line detection by a sensitive monitor of the liberated gases, eliminates these shortcomings. In an accompanying paper in this volume thermodenuder systems for the measurement of NH_3, HNO_3 and NH_4NO_3 are described. A disadvantage of the thermodenuder technique is the rather expensive equipment (as it includes a NO_x monitor), especially, if several components are to be measured simultaneously.

ECN has developed a so-called "wet annular denuder" for the simultaneous sampling of gases such as NH_3, HNO_3, HNO_2, HCl, SO_2 and H_2O_2 (2). Advantages of this apparatus are on-line recoating, a high sampling flow of 30 $l.min^{-1}$ and a compact instrument for field monitoring. The main disadvantage is the time delay between sampling and analysis of the absorption solutions.

During a field intercomparison experiment from 19 to 23 September 1988 at Rome, Italy, two wet annular denuders were used for the simultaneous determination of HNO_3 and NH_3 in ambient air. In The Netherlands for more than a year already, the wet annular denuder is located at 18 m height 1 m above the canopy of a forest to measure simultaneously air concentrations of NH_3, HNO_3, HNO_2, HCl, SO_2 and H_2O_2.

2. INSTRUMENTATION

In Figure 1 the wet annular denuder system[*] is shown.

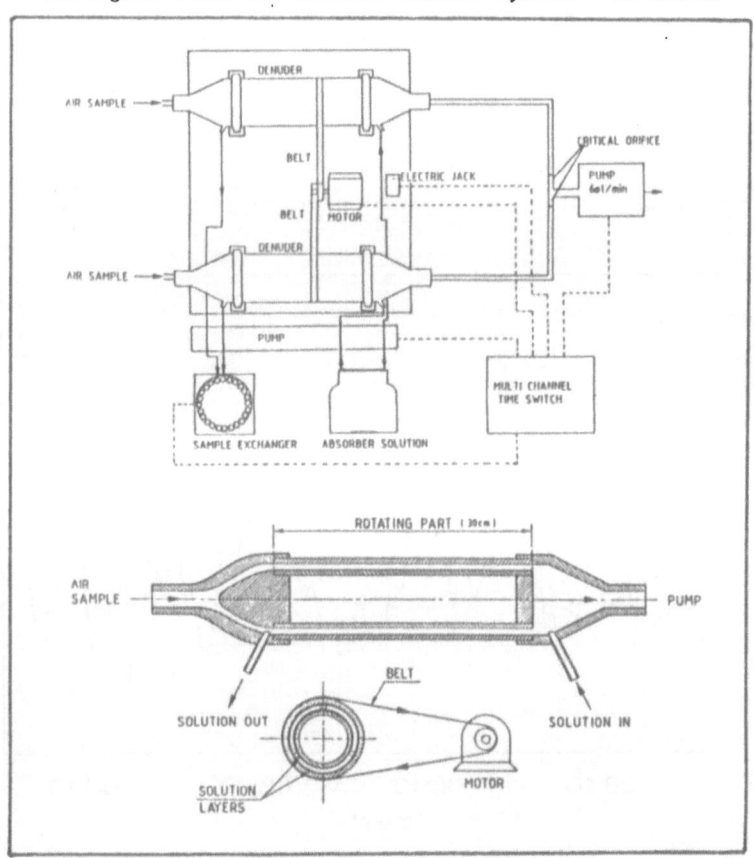

Figure 1 : The wet annular denuder setup for parallel sampling of NH_3 and HNO_3.

(*) The wet annular denuder is commercially available from: van Essen Instruments (Delft, The Netherlands).

The wet annular denuder consists of two concentric glass tubes with a length of 30 cm. The inner tube has an outer diameter of 42 mm, the outer tube an inner diameter of 45 mm, leaving an annulus of 1.5 mm. An amount of solution (about 20 ml) is present in the annulus of the denuder tube. During sampling the denuder is rotated around its axis at a speed of 40 r.p.m. and the solution film so obtained on the walls of the annulus acts as the coating of a conventional denuder. After a sampling time of 35 minutes at a flow rate of 30 $l.min^{-1}$, automatically the absorption solution is pumped into a sample tube and the denuder is recoated. The solutions so obtained are analyzed off-line for NH_4^+, NO_3^-, NO_2^-, Cl^-, SO_3^{2-} and H_2O_2 (2). Up to a time delay of two weeks between sampling and analysis, no significant change in the concentrations of the collected species was observed at ambient temperatures below 20 °C.

3. FIELD EXPERIMENTS

In Figure 2 the results are shown of NH_3 measurements by two parallel wet annular denuder systems at Rome (19 -23 September 1989).

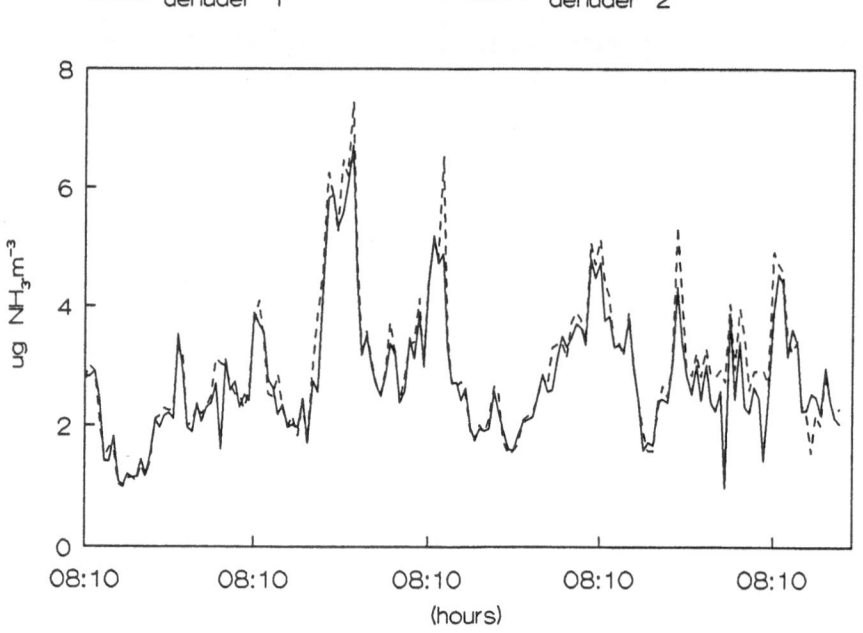

Figure 2 : Ammonia concentration in air at Rome in the period 19-23 September 1988 measured by two parallel wet annular denuders.

For NH$_3$ and HNO$_3$ the coefficients of correlation are 0.978 and 0.985, respectively, between 168 parallel measurements in the range of 0.1 - 8 μg.m^{-3} NH$_3$ and 0.1 - 7 μg.m^{-3} HNO$_3$. The average of the differences in NH$_3$ and HNO$_3$ concentration are 0.3 μg.m^{-3} NH$_3$ and 0.2 μg.m^{-3} HNO$_3$ with a standard deviation of 0.3 μg.m^{-3} NH$_3$ and 0.3 μg.m^{-3} HNO$_3$ (68 % level of confidence).

At Rome a cyclone was tested to eliminate interferences of NH$_4^+$- and NO$_3^-$-containing aerosols on the determination of NH$_3$ and HNO$_3$ air concentrations (2). Parallel measurements were performed with and without the cyclone in the sample inlet of a denuder. In Figure 3 the results are shown of the differences in NH$_3$ and HNO$_3$ air concentrations.

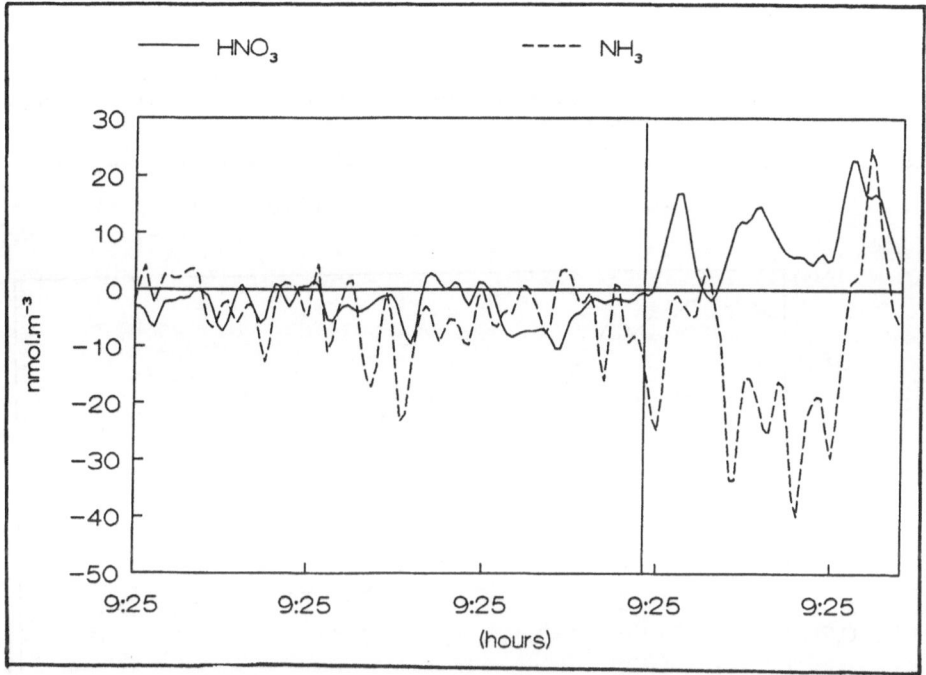

Figure 3 : Differences in air concentrations of NH$_3$ and HNO$_3$ sampled by two wet denuders with and without a cyclone in the sample inlet of one denuder.

During the first three days both denuders sample without a cyclone in the sample inlet and the reproducibility between both denuders is better than 5%, which was illustrated in Figure 2. However, the last two days differences between a denuder without and with a cyclone in the sample inlet demonstrate that HNO$_3$ concentrations are lower with a cyclone due to sorption into the cyclone and NH$_3$ concentrations are higher with a cyclone due to evaporation of NH$_4$NO$_3$. The latter effect also enhances HNO$_3$ concentrations but sorption of HNO$_3$ is apparently more important. It

is concluded that no preseparator (even a PTFE coated cylone as used in
this experiment) should be present in the sample inlet when measuring NH_3
and HNO_3, as the cure seems worse than the disease.

Within the framework of a Dutch national programme on acid
deposition a wet denuder is located about 1 m above the canopy of a
forest at 18 m height. From April to October 1988 simultaneously the air
concentrations of NH_3, HNO_3 and HCl were daily measured. Since January
1989 these measurements have been extended with HNO_2, SO_2 and H_2O_2. Up to
June 1989 average concentrations are: NH_3 5 $\mu g.m^{-3}$, SO_2 8 $\mu g.m^{-3}$, HNO_3 1
$\mu g.m^{-3}$, HNO_2 0.5 $\mu g.m^{-3}$, HCl 1 $\mu g.m^{-3}$ and H_2O_2 0.05 $\mu g.m^{-3}$. The NH_3
concentrations
are higher by a factor two than results obtained at Rome, which is not
surprising as The Netherlands are notorious for their high NH_3 emissions
due to intensive agricultural activities (3) and average HNO_3
concentrations in The Netherlands are low compared to Rome as a
consequence of the lower photochemical activity.

In Figure 4 the results are shown for the HNO_3 and HNO_2 air
concentrations. These are average diurnal concentrations over 20 days
from January to June (1989).

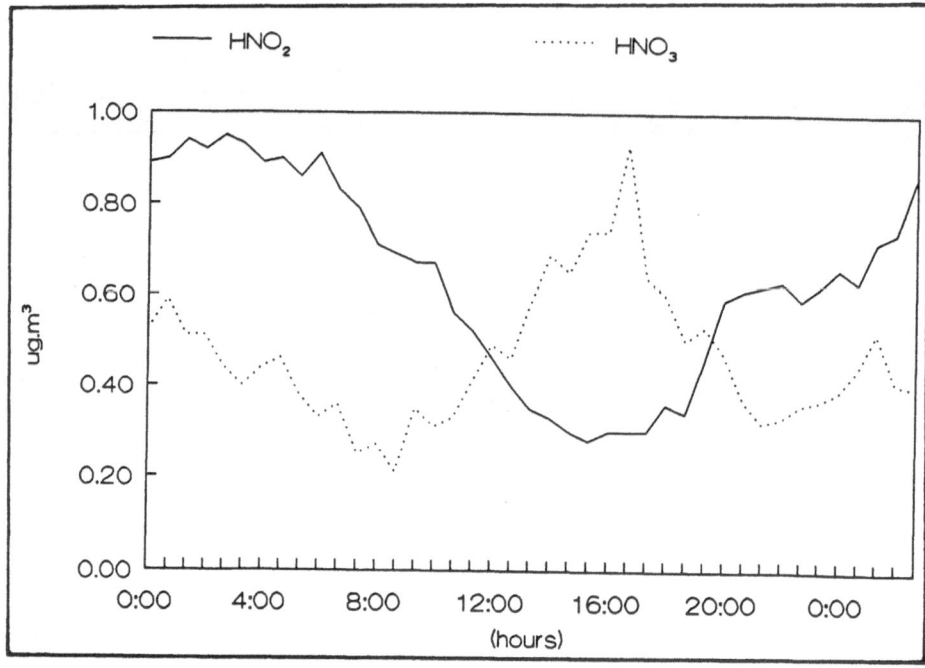

Figure 4: Air concentrations of HNO_3 and HNO_2 at 18 m height 1 m above
the canopy of a forest in The Netherlands (January-June 1989).

It clearly reflects the instability of HNO_2 over the day due to
photochemical induced oxidation to HNO_3. These are the first systematic
measurements of HNO_2 in The Netherlands; more data are needed to evaluate
the effects of HNO_2 deposition. The "dip" in the HNO_3 concentration in the

morning is probably caused by increased dry deposition due to formation of dew on vegetation.

In Figure 5 the results of HNO$_3$ concentrations are differentiated for the period January-March and April-May 1989, which illustrates the relation between higher nitric acid air concentrations and increased photochemical acitivity in May.

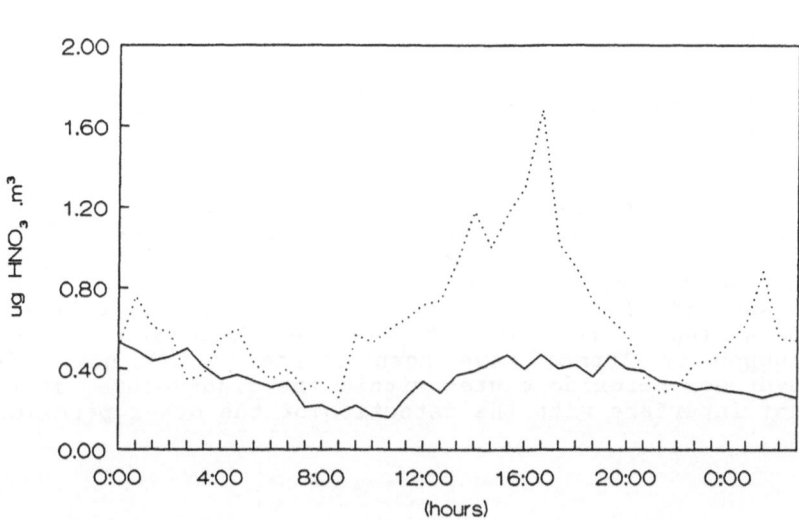

Figure 5: Air concentrations of HNO$_3$ at 20 m height within the canopy of a forest in The Netherlands in the period January -March and April-May 1989.

Both the results in Rome and The Netherlands show the potential of the automated wet annular denuder as a monitoring device. Another application could be dry deposition flux measurements by means of the gradient method.

Acknowledgements-Research on the development of the thermodenuder and the wet annular denuder was funded by the Dutch Ministry of Economic Affairs and the European Community (Contract N°: EV4V-00883-C).

REFERENCES

(1) Keuken M.P. (1989) Determination of acid-deposition-related compounds in the lower atmosphere. ECN-218.

(2) Keuken M.P., Schoonebeek C.A.M., van Wensveen-Louter A. and Slanina J. (1988) Simultaneous sampling of NH$_3$, HNO$_3$, HCl, SO$_2$ and H$_2$O$_2$ in ambient air by a wet annular denuder system. Atmos. Environ. 22, 2541-2548.

(3) Buisman E., Maas H.F.M. and Asman W.A.H. (1987) Anthropogenic NH$_3$ emissions in Europe. Atmos. Environ. 21, 1009-1022.

DETERMINATION OF ORGANIC PEROXIDES

K. Bächmann and J. Hauptmann
Institut für Anorganische und Kernchemie
D-6100 Darmstadt, Hochschulstr. 4, F.R.G

Summary

The major chemical pathway for the degradation of hydrocarbons is their photochemical oxidation by hydroxy radicals. The products vary depending on the concentration of nitrogen oxides. At low NO_x-levels, the production of organic hydroperoxides is favoured. Until now the determination of these compounds was limited to analysis of the total amount of peroxide and the analysis of hydrogen peroxide. This paper describes a method to separate and determine organic peroxides by means of HPLC and post-column reaction detection. Depending on the kind of peroxide, detection limits of 100ppb or 10pmol have been reached until now. Since hydrogen peroxide elutes within the dead-volume, it does not interfere with the detection of the other peroxides.

1. INTRODUCTION

For two reasons, a detailed interpretation of the oxidation mechanism of biogenic hydrocarbons is of importance. Firstly, the role of the degradation products in forest decline and secondly their role in the balance of ozone. A comparison of oxidation in clean (low NO_x-levels) and polluted (high NO_x-levels) air is of considerable interest for atmospheric chemistry. Although most of the reaction pathways are known, especially for methane, the mechanisms for the higher hydrocarbons have not yet been cleared up entirely because of many competitive reactions on the one hand and the lack of efficient quantitative and qualitative detection methods on the other hand.

One of the non-radical species formed during the oxidation of hydrocarbons are organic hydroperoxides. During the daytime, the hydrocarbons emitted by vegetation, mostly terpenes like isoprene, cumene or pinene, are oxidized either by ozone or by hydroxy radicals. Abstraction of one hydrogen-radical leads to the reactive alcoperoxy radicals (1). The reactions of these radicals depend on the concentration of nitrogen oxides:

1. In urban, polluted regions with high NO_x-levels, alcoxyradicals, nitrates and peroxynitrates are formed, the former yielding aldehydes and ketones (2).

2. On the other hand, when hydrocarbons are accumulated in comparison to NO_x, in regions of clean air at elevated sites, the alcoperoxyradicals tend to react with perhydroxyradicals to give primary, secondary and tertiary hydroperoxides (3) [1].

(1) $RH + OH + O_2 \longrightarrow ROO + H_2O$

(2) $ROO + NO \longrightarrow RO + NO_2$

$ RO \longrightarrow$ aldehydes + ketones

$ ROO + NO \longrightarrow RONO_2$

$ RO_2 + NO_2 \longrightarrow RO_2NO_2$

(3) $RO_2 + HO_2 \longrightarrow ROOH + O_2$

Aldehydes and ketones are detected by fluorescence after derivatisation with dansyl hydrazine and separation by HPLC [2]. Peroxynitrates are determined directly after separation by gas chromatography. The measurement of peroxides in the atmosphere was successful only for the determination of hydrogen peroxide, a few measurements of the total amount of non-H_2O_2 peroxides and, recently, the determination of hydroxyalkyl hydroperoxides which are formed in the ozonolysis of alkenes [3]. The latter method can not be applied to alkyl hydroperoxides because it is based on the easy decay of the hydroxy compounds to hydrogen peroxide.

2. EXPERIMENTAL

The separation of the peroxides gave rise to considerable problems during the development of our method. Due to the low thermal stability of these compounds a separation by gas chromatography is not possible. The separation must be carried out by liquid chromatography, where the expected excess of hydrogen peroxide raises difficulties. The separations of hydroperoxides described in literature are restricted to those of biological interest [4,5]. The detection methods usually used here are : direct UV/vis detection [6], polarographic detection [7] and the detection of a post-column derivatisation product. Since UV/vis detection yields insufficient detection limits, a post-column derivatisation had to be applied.

The derivatisations in question are: reaction with iodide [8] and detection via UV/vis-absorption, reaction with luminol [9,10] and registration of the chemiluminescence and, finally, reaction with p-hydroxyphenylacetic acid (POHPA) [11] and detection via fluorescence. Since the limits of detection of the two former methods proved to be insufficient, the reaction with POHPA was investigated.

The peroxides are reduced by catalytical amounts of peroxidase, the p-hydroxyphenylacetic acid giving a fluorescent dimer (figure 1).

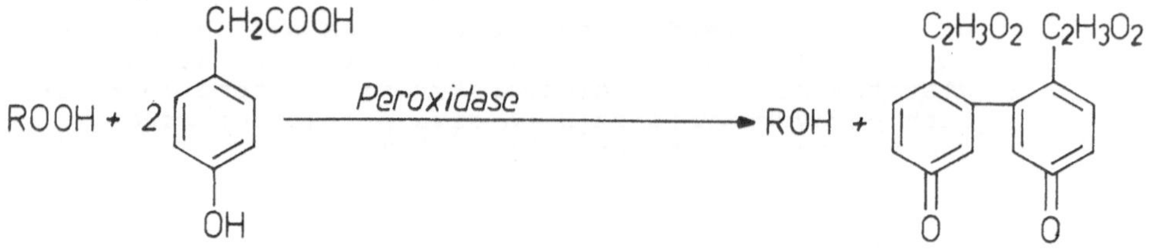

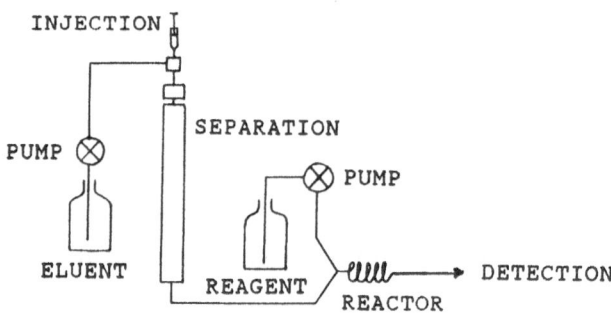

Figure 1. Mechanism of the reaction of POHPA with peroxides

Due to steric effects the reaction of peroxidase with alkyl hydroperoxides is very slow [12], and a different catalyst had to be found. Cytochrome c proved to be an efficient and stable catalyst for this reaction. The 0.25m tris buffer solution was deaerated by applying a vacuum source for 20 min followed by addition of 10mg POHPA, 3mg cytochrome c and 5g guanidinium chloride per 100ml. Since the catalyst showed considerable loss of activity during a period of a few days, the solution had to be freshly prepared every day. The reaction coil, a knitted open tube reactor with a total volume of 2ml, was heated to 60°C to accelerate the reaction. Further increase of the temperature resulted in decomposition of some peroxides. The fluorescence was registered at an excitation wavelength of 315nm and an emission wavelength of 425nm.

To achieve a sufficient separation of the short-chained hydroperoxides, neither a polar silica gel nor an unpolar RP-18 material was useful. Complete separation was achieved on mid-polar materials like RP-2 and SAS Hypersil. A methanol/water mixture (50/50 v/v) was used as an eluent. Flow rates were optimized to 0.2 ml/min for the eluent and 0.1 ml/min for the reagent for best results. Increasing the flow rates raised the limits of detection due to the shorter reaction times. A scheme of the system is shown in figure 2.

Figure 2. Scheme of the post-column derivatisation system

The system was tested and optimized with a mixture of three commercially available hydroperoxides (t-butyl, t-amyl and cumole hydroperoxide). The peroxide content of the stock solutions was determined by iodometry.

Figure 3 shows a typical chromatogram for the separation of the three alkyl hydroperoxides.

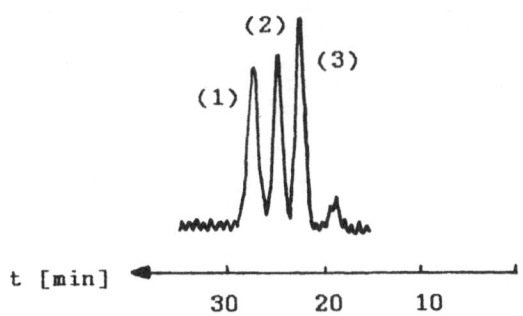

Figure 3. Chromatogram of the peroxide test mixture. Column: SAS Hypersil (5μm, 250*4mm), eluent: water/methanol (50/50 v/v), flow rate : 0.2ml/min; flow rate of reagent: 0.1ml/min; temperature: 60°C; detector at 315/425nm, attenuation 4; (1) 1ppm cumole hydroperoxide; (2) 1ppm t-amyl hydroperoxide; (3) 1ppm t-butyl hydroperoxide.

3. RESULTS AND DISCUSSION

The developed method is convenient for separation and detection of short-chained hydroperoxides in the upper ppb-levels. Detection limits of 100 ppb or 10pmol can be reached.

Since the peroxides are separated, an excess of hydrogen peroxide does not interfere with the detection of the other peroxides as it was the case with the previously existing methods.

Furthermore, the determination of peroxy compounds produced during laboratory experiments, simulating the oxidation of natural hydrcarbons, is to be tried. In this way, the separation and detection method can be a valuable instrument for the clarification of some atmospheric processes.

REFERENCES

[1]   J. G. CALVERT, S. MADRONICH (1987) "Theoretical study
       of the initial products of the atmospheric oxidation of
       hydrocarbons", Journal of Geophysical Research 92(D2),
       2211-2220.
[2]   W. SCHMIED, M. PRZEWOSNIK and K. BÄCHMANN (1988)
       "Spurenbestimmung von Aldehyden und Ketonen in der
       Troposphäre durch Festphasenderivatisierung mit DNSH",
       submitted to Zeitschrift für Analytische Chemie
       (Fresenius)
[3]   E. HELLPOINTER and S. GÄB (1989) " Detection of methyl,
       hydroxymethyl and hydroxyethyl hydroperoxides in air
       and precipitation", Nature 337(6208), 631-632.

[4]   T. MIYAZAWA, K. YASUDO, K. FUJIMOTO and T. KANEDA
        (1988) "Determination of phosphatidylcholine hydro-
        peroxide in human plasma by chemiluminescence high
        performance liquid chromatography", Analytical Letters
        21(6), 1033-1044.
[5]   K. AKASAKA, H. OHURI and H. MEGURO (1988) "An aromatic
        phosphine reagent for the high performance liquid
        chromatography fluorescence determination of hydroper-
        oxides - determination of phosphatidylcholine hydro-
        peroxides in human plasma", Analytical Letters 21(6),
        965-975.
[6]   R. K. JENSEN, M. ZINBOAND and S. KORCEK (1983) "High
        performance liquid chromatography determination of hy-
        droperoxidic product formed in the autooxidation of
        n-hexadecane at elavated temperatures", Journal of
        Chromatographic Science 21, 394-397.
[7]   M. O. FUNK and W. J. BAKER (1985) "Determination of
        organic peroxides by high performance liquid chroma-
        tography with electrochemical detection", Journal of
        Liquid Chromatography 8(4), 663-675.
[8]   R. S. DEELDER, M. G. F. KROLL and J. H. M. VAN DEN
        BERG (1976)"Determination of trace amount of hydrogen-
        peroxide by column liquid chromatography and colori-
        metric detection", Journal of Chromatography 125,
        307-314.
[9]   Y. YAMAMOTO, M. H. BRODSKY, J. C. BAKER and B. N. AMES
        (1987) "Detection and characterization of lipid hydro-
        peroxides at picomole levels by high performance liquid
        chromatography", Analytical Biochemistry 160(1), 7-13.
[10]   T. MIYAZAWA, K. YASUDO and K. FUJIMOTO (1987)
        "Chemiluminescence - high performance liquid chroma-
        tography of phosphatidylcholine hydroperoxide", Analy-
        tical Letters 20(6), 915-925.
[11]   G. L. KOK, K. THOMPSON and A. L. LAZRUS (1986)
        "Derivatization technique for the determination of
        peroxides in precipitation", Analytical Chemistry
        58(6), 1192-1194.
[12]   K. G. PAUL, P. I. OHLSSON and S.WOLD (1979) "Formation
        of horseradish peroxidase compound I with alkyl hydro-
        peroxides", Acta Chemica Scandinavica B33, 747-754.

TROPOSPHERIC OH RADICAL MEASUREMENT TECHNIQUES: RECENT DEVELOPMENTS

A. HOFZUMAHAUS, H.-P. DORN and U. PLATT*

Kernforschungsanlage Jülich GmbH,
Institut für Chemie 3: Atmospärische Chemie,
Postfach 1913, D-5170 Jülich, F.R. Germany

*present address:
Institut f. Umweltphysik, Univ. Heidelberg,
Im Neuenheimer Feld 366,
D-6900 Heidelberg, F.R. Germany

Summary

Two techniques to measure tropospheric OH radicals in situ are currently being developed in our laboratory. First, the already established laser long-path absorption spectroscopy is modified by folding a ligth path of about 4 km length in a multireflection system of 20 m base length. The detection limit will be about $1-2 \cdot 10^6$ OH cm^{-3}. Second, a laser-induced fluorescence instrument has been build , which uses the 308 nm excitation of the OH radicals at a reduced pressure and has an expected detection limit of $5 \cdot 10^5$ cm^{-3}. In both developments special care is taken to avoid the self-generation of OH from ozone UV-photolysis by the probing laser beam.

1. INTRODUCTION

The reaction of hydoxyl radicals (OH) with atmospheric trace gases is one of the most important chemical processes in the troposphere. In many cases that reaction provides the first and rate limiting step in the chain of oxidation reactions which transform many atmospheric pollutants like NO_2, SO_2, CO or CH_2O into acidic, water soluble end-products. The measurement of ambient OH radicals allows to test the underlying photochemical reaction theory by comparison with model calculations (1). However, the detection of atmospheric hydroxyl is a difficult task because of its very low concentration in the range of 10^6 molecules cm^{-3}. In addition, a number of possibly interfering trace gases are present in the ambient atmosphere. A very sensitive and highly specific detection method is, therefore, necessary to measure tropospheric OH radicals.

2. OH MEASUREMENT TECHNIQUES: RECENT DEVELOPMENTS

In the past we have developed the laser long-path absorption technique (laser LPA) (2-4) and demonstrated its usefulness by observation of distinct diurnal concentration profiles of OH (1,3-6). The main advantage of this method is its inherent calibration: based on Lambert-Beer's law, absolute OH concentrations can be evaluated directly from observed atmospheric absorption spectra. Because of the very low abundance of ambient hydroxyl a long optical light path is required ranging typically from 5.8 to 15 km. For good visibility a detection limit of $5 \cdot 10^5$ molecules cm^{-3} is accessible at 10 km path length.

An experimental arrangement of this kind provides spatially averaged data, is difficult to move and places a number of constraints on the measurement site. In order to extend our experimental OH investigations to maritime sites and the upper troposphere, we need mobile in situ instrumentation, which can be placed on board ships or air-planes. Two recent developments are in progress in our laboratory: first, the laser-LPA OH-spectrometer will be combined with a multiple reflection system of 20 m base length, providing an optical path-length of about 4 km at 200 reflections. This experimental set-up was operated successfully 1988 on board the ship 'Polarstern'. Here, instead of the laser-system, a Xe-arc lamp was used to measure IO and NO_3 radicals. The mechanical design turned out to be sufficient stable, i.e. the detected light signal was hardly disturbed by oceanic wind, vibrations and ship movements. At present we are investigating theoretically and experimentally the extent of the ozone/water interference in the multi-reflection system. This interference leads to self-generation of OH radicals by the probing laser beam. At the laser wavelength of 308 nm ambient ozone is photolyzed and the thus produced electronically excited $O(^1D)$-atoms react readily with water vapour to form OH.

In our usual LPA experiment the laser intensity is reduced by beam expansion which keeps the interference well below the detection limit. However, in the multi-reflection system the folded laser beams overlap partially and are focused at one side of the mirror-system. Therefore, the energy flux distribution of the laser is much more complicated and a careful analysis is necessary to determine the applicability for tropospheric OH measurements.

Our second development is a laser-induced fluorescence (LIF) experiment, which represents another direct spectroscopic means to measure ambient OH. Is is far more sensitive than the absorption method. Thus ony a small detection volume of about 1 cm length is necessary to reach a similar detection limit as the LPA-technique for several km of path-length. Hence, the LIF experiment is ideally suited for mobile in situ observations.

In the past, the O_3/H_2O interference proved to be a severe drawback for LIF-techniques (7-11) and is even nowadays controversely discussed (11,12) for the experiment of Hard et al. (13,14). Our LIF instrument differs from this and other current LIF experiments (13,15-18) by suppressing this interference far below the detection limit. A brief outline of the experiment is given in the following text.

3. LIF TECHNIQUE : A NEW APPROACH
3.1 DETECTION METHOD

For the LIF-detection of tropospheric OH radicals usually a pulsed, tunable UV-laser is employed to excite the radicals at one of the rotational lines of the $OH(A^2\Sigma^+ \leftarrow X^2\Pi)$ electronic transition. This excitation can be performed on the (0,0) vibrational band at 308 nm (Fig. 1a) or on the (1,0) band at 282 nm (Fig. 1b). The excited molecules return to their electronic ground state $(X^2\Pi, v''= 0)$ either by photon emission or relaxation processes induced by collisions with air molecules M. In both excitation schemes (Fig. 1a, 1b) most of the emitted fluorescence is observed at about 308 nm. The detected intensity of this light is a measure for the absolute OH concentration.

Up to now, all current LIF-techniques for tropospheric OH measurements (11, 13, 15-18) have used the excitation wavelengths at 282 nm (Fig. 1b). That way, the fluorescence signal at 308 nm is easily seperated from laser stray-light by using a narrow-band interference filter in front of the detector.

We have developed an LIF experiment which, for the first time, uses the 308 nm excitation scheme (Fig. 1a) for the detection of atmospheric OH. One reason is the higher detection sensitivity: the OH absorption cross sections of the (0,0) band at 308 nm are 4 times larger than for the (1,0) band at

282 nm. Furthermore, the fluorescence is a factor of 1.4 more efficient. This results in an overall increase in sensitivity by a factor of 5.6. The second reason for our approach is the much smaller O_3/H_2O interference as discussed below (see 3.3).

A disadvantage of the method is the spectral coincidence of the laser stray-light and the OH fluorescence, which is weaker by several orders of magnitude. To seperate the two signals the pressure in the detection cell (Fig.2) is reduced to 1–2 mbar. As a consequence the lifetime of the excited OH is increased from 7 ns at 1000 mbar, where it is limited by collisional quenching, to about 300 ns, the radiative lifetime being about 700 ns. By gated photon counting the fluorescence can easily be detected after the laser pulse whose duration is only 10 ns. In addition the photomultiplier is switched off during the laser pulse by a gating electronic, to prevent saturation or damage while the intense flash of laser stray-light hits the photocathode. In this way we were able to record OH excitation spectra in the laboratory with the apparatus schematically shown in Fig.2.

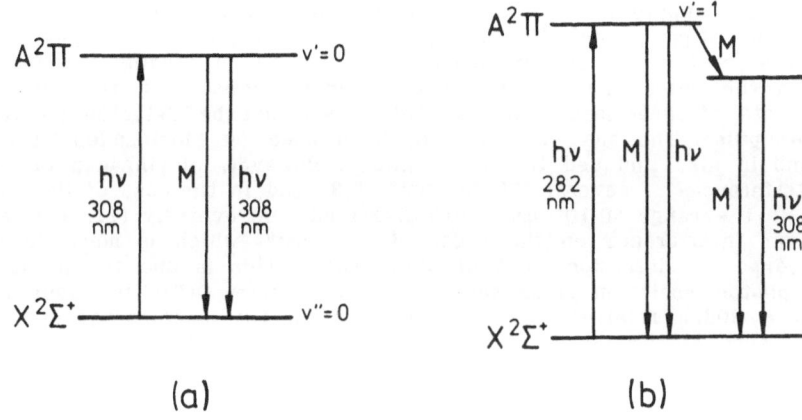

Fig.1: Simplified excitation- and fluorescence scheme for the LIF-detection of OH radicals.

3.2 EXPERIMENTAL DETAILS

A tunable, pulsed dye laser is pumped by a frequency doubled, Q-switched Nd-YAG laser. The resulting dye laser output is frequency doubled to give pulses at 308 nm with a spectral band-width of typical 0.0015 nm, a duration of about 10 ns and a maximum repetition rate of 50 Hz. The laser fluence is kept low (15 μJ cm^{-2}) to avoid saturation of the OH absorption on the $Q_1(2)$-line of the $A^2\Sigma^+$, v'= 0 <-- $X^2\Pi$, v''=0 transition. The laser beam crosses a steady flow of ambient air in the fluorescence cell which enters through an inlet nozzle. The pressure in the system is maintained at 1–2 mbar by a vacuum pump, not shown in Fig.2. At right angles to the gas flow and laser beam, respectively, a photomultiplier/opt. filter combination is fitted to detect the OH fluorescence by gated photon counting. The laser pulse energy is measured by calibrated photodiodes whose outputs are analogue integrated and digitized by an AD-converter. This and the photon counter are read out by a microcomputer which normalizes the OH-signal relative to the laser pulse energy for each laser shot. The normalized signals are then successively averaged. The computer also controls the laser wavelength. The OH signal plus background is measured on-resonance, whereas the background is determined seperately by off-resonance measurements.

The sensitivity of the present LIF-instrument is estimated to 0.004 detected fluorescence photons per 10^6 OH cm^{-3} in ambient air. This number is calculated from excitation and fluorescence efficiencies of OH and several parameters specific for the used apparatus. For the same conditions, background signals were measured of 0.008 noise photons for synthetic air from a tank and 0.013 photons for laboratory air. Using the latter number, a 1σ-detection limit of $5 \cdot 10^5$ OH cm^{-3} is to be expected for an overall integration time of approximately 6 min. However, the absolute sensitivity of the method has still to be validated by calibration and the influence of polluted or aerosol-loaded air on the background signal must be studied.

3.3 O_3/H_2O INTERFERENCE

The O_3/H_2O interference, which has been described above (see 2.), is very efficiently suppressed by our LIF-technique compared to the methods used so far by other groups (13, 15-18). Up to now, the most promising way to avoid the problem was the approach by Hard et al. (13) using a reduced pressure in the detection chamber. This procedure slows down the kinetic formation rate of laser made OH, so that it will not be detected by the short laser pulse. The same concept has been used by Shirinzadeh et al. (11,18) and is also included in our technique. However, Shirinzadeh et al. report interferences of several 10^5 OH cm^{-3} (18) and in the case of Hard et al. levels in the range of 10^6 cm^{-3} are discussed controversely (11, 12). We estimate our interference on the order of 10^3 cm^{-3} which is much lower than the expected detection limit of $5 \cdot 10^5$ cm^{-3}. This is due to the fact that the photodissociation cross section of O_3 to form $O(^1D)$ is about 30 times less at 308nm relative to 282 nm. In addition, the laser energy can be lowered because of the higher detection sensitivity, thereby reducing the interference further by a factor of 5-6. This results in an interference suppression by a total factor of about 150 due to the improved excitation method at 308 nm.

4. CONCLUSIONS

Two techniques to measure tropospheric OH radicals in situ are currently being developed in our laboratory. The laser long-path absorption spectrometer will be combined with a folded light path of about 4 km in a 20 m multi-reflection system. The extent of a possible O_3/H_2O interference is presently explored. Second, a prototype of a laser-induced fluorescence instrument has been build which uses the 308 nm excitation at a reduced pressure. The O_3/H_2O interference is suppressed by two orders of magnitude below the expected detection limit of about $5 \cdot 10^5$ cm^{-3}. The system will now be calibrated and tested in the field.

REEFERENCES

(1) Perner, D., Platt, U., Trainer, M., Hübler, G., Drummond, J., Junkermann, W., Rudolph, J., Schubert, B., Volz, A., Ehhalt, D.H., Rumpel, K.J. and Helas, G. (1987). Measurements of tropospheric OH concentrations: a comparison of field data with model predictions. J. Atm. Chem. 5, 185-216.

(2) Hübler, G., Perner, D., Platt, U., Tönnissen, A. and Ehhalt, D.H. (1984). Groundlevel OH radical concentration: New measurements by optical absorption. J. Geophys. Res. 89D, 1309-1319.
(3) Dorn, H.-P., Callies, J., Platt, U. and Ehhalt, D.H. (1988). Measurements of tropospheric OH concentrations by laser long-path absorption spectroscopy. Tellus 40B, 437-445.
(4) Hofzumahaus, A., Dorn, H.-P., Callies, J., Platt, U. and Ehhalt, D.H. (1990). Tropospheric OH concentration measurements by laser long-path absorption spectroscopy. To be published.
(5) Platt, U., Rateike, M., Junkermann, W., Hofzumahaus, A. and Ehhalt, D.H. (1987). Detection of atmospheric OH radicals. Free Rad. Res. Comms. 3, 165-172.
(6) Platt, U., Rateike, M., Junkermann, W., Rudolph, J. and Ehhalt, D.H. (1988). New tropospheric OH measurements. J. Geophys. Res. 93D, 5159-5166.
(7) Wang, C.C., Davis, L.I., Wu, C.H. and Japar, S. (1976). Laser-induced dissociation of ozone and resonance fluorescence of OH in ambient air. Appl. Phys. Lett. 28, 14-16.
(8) Hanabusa, M. and Wang, C.C. (1977). Pulsewidth dependence of ozone interference in the laser fluorescence measurement of OH in the atmosphere. J. Chem. Phys. 66, 2118-2120.
(9) Ortgies, G., Gericke, K.-H. and Comes, F.J. (1980). Is UV laser induced fluorescence a method to monitor tropospheric OH ? Geophys. Res. Lett. 7, 905-908.
(10) Davis, D.D., Rodgers, M.O., Fischer, S.D. and Asai, K. (1981). An experimental assessment of the O3/H2O interference problem in the detection of natural levels of OH via laser induced fluorescence. Geophys. Res. Lett. 8, 69-72.
(11) Shirinzadeh, B., Wang, C.C. and Deng, D.Q. (1987). Pressure dependence of ozone interference in the laser fluorescence measurements of OH in the atmosphere. Appl. Opt. 26, 2102-2105.
(12) Hard, T.M., Chan, C.Y. Mehrabzadeh, A.A. and O'Brien , R.J. (1989). Pressure dependence of ozone interference in the laser fluorescence measurement of OH in the atmosphere: comment. Appl. Opt. 28, 26-27.
(13) Hard, T.M., O'Brien, R.J., Chan, C.Y. and Mehrabzadeh, A.A. (1984). Tropospheric free radical determination by FAGE. Environ. Sci. Technol. 18, 768-777.
(14) Hard, T.M., Chan, C.Y., Mehrabzadeh, A.A., Pan, W.H. and O'Brien, R.J. (1986). Nature 322, 617-620.
(15) Bakalyar, D.M., Davis, L.I., Chuan Guo, James, J.V., Spiros Kakos, Morris, P.T. and Wang, C.C. (1984). Shot noise limited detection of OH using the technique of laser-induced fluorescence. Appl. Opt. 23, 4076-4082.
(16) Davis, L.I., Chuan Guo, James, J.V., Morris, T.M., Postiff, R. and Wang, C.C. (1985). An airborne Lidar instrument for detection of OH using the technique of laser-induced fluorescence. J. Geophys. Res. 90, 12835-12842.
(17) Rodgers, M.O., Bradshaw, J.D., Sandholm, S.T., KeSheng, S. and Davis, D.D. (1985). A 2-λ laser-induced fluorescence field instrument for ground-based and airborne measurements of atmospheric OH. J. Geophys. Res. 90, 12819-12834.
(18) Shirinzadeh, B., Wang, C.C. and Deng, D.Q. (1987). Diurnal variation of the OH concentration in ambient air. Geophys. Res. Lett. 14, 123-126.

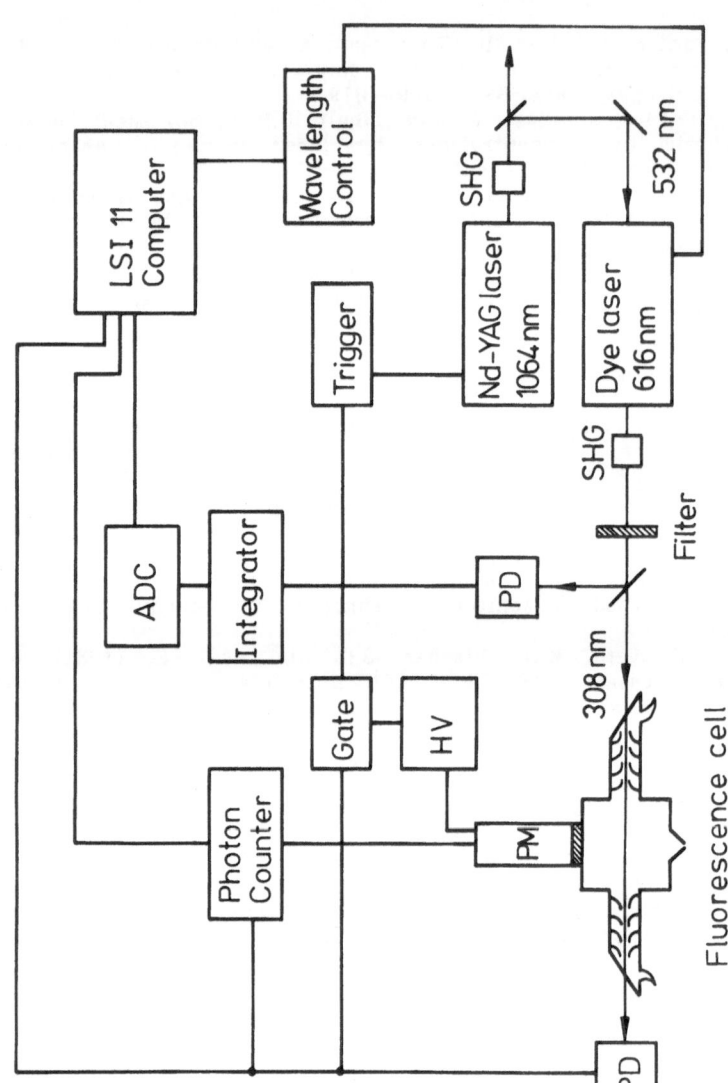

Fig.2: Schematic of the OH detection system by LIF using 308 nm excitation at reduced pressure. Abbreviations: SHG second harmonic generation crystal, PD photodiode, PM photomultiplier, HV high voltage supply, ADC analog-digital converter.

THREE METHODS OF HYDROGEN PEROXIDE DETERMINATION IN RAINWATER

Philippe LAGRANGE and Janine LAGRANGE

Ecole Européenne des Hautes Etudes des Industries Chimiques
de Strasbourg, URA 405 au CNRS, 1 rue Blaise Pascal,
67000 STRASBOURG (France)

Summary

Various analytical methods are presented for the determination of
hydrogen peroxide in rainwater :
- a polarographic determination using the H_2O_2 reduction on the
  dropping mercury electrode in five different media ;
- a voltamperometric determination using the H_2O_2 oxidation on a
  rotating disk electrode ;
- a spectrophotometric determination using the peroxidase enzyme to
  catalyse the oxidation reaction of the indicator ABTS.
For the three methods, detection limits are estimated.
The voltamperometric method is the most sensitive with a detection
limit : 5.10^{-9} M . This method could be adapted to an inexpensive
portable amperometric titrator which could measure an electrical
current at constant potential. The current is proportional to
hydrogen peroxide concentration. This method allows to perform one
determination each 3 minutes.

H_2O_2 is believed to be a major contributor to oxidation processes
that convert dissolved S(IV) compounds to sulphuric acid in rain, cloud,
or fog waters at pH between 2 and 7. Produce by re-combination of
hydroperoxyl radicals, H_2O_2 is extremely soluble in water (Henry's law
constant = 10^5 M atm^{-1}) leading to significant aqueous concentration in
rainwater (about 10^{-8} - 10^{-4} mol dm^{-3} ; the highest concentration may be
obtained in summer). Accurate determination of H_2O_2 concentration in
precipitation is very important to understand its involvement in
acidification of cloud. Various analytical techniques are available for
H_2O_2 determinations including principally chemiluminescence, photometry
and fluorometry (1-13). These methods are the most sensitive and allow
achievement of sub-micromolar detection limits. In these methods, H_2O_2 is
often determined after reaction involving peroxidase enzyme to catalyse a
reaction between H_2O_2 and a chromophore or a fluorophore, or after
oxidation of an indicator substrate by H_2O_2 promoted by a transition
metal. All these methods are not independent and are sometimes prone to
interference.
      Then, we have chosen to develop two voltamperometric methods, and to
compare the obtained results with a spectrophotometric method involving
the peroxidase enzyme to catalyse the oxidation reaction of an indicator
by H_2O_2.

1. DETERMINATION OF TRACES OF H_2O_2 BY VOLTAMPEROMETRIC METHOD
      1.1. Polarographic reduction of H_2O_2
      Polarographic method with a dropping mercury electrode is performed
in five different media with a H_2O_2 concentration ranging from 10^{-4} to
10^{-9} M.

- [phenobarbital] = 0.1 M ; pH = 8.8,
  phenobarbital : sodium 5-phenyl barbiturate ;
- [OT] = 4 x 10^{-4} M ; [NaClO$_4$] = 0.1 M ; pH = 5.1,
  OT : sodium dioctylsulfosuccinate ;
- [HClO$_4$] = 0.01 M ; [NaClO$_4$] = 0.09 M ;
- [borate] = 0.01 M ; [NaClO$_4$] = 0.05 M ; pH = 9.0 ;
- [phosphate] = 0.1 M ; pH = 7.5.

For the five media, a diffusion current, due to the reduction of H$_2$O$_2$:

$$H_2O_2 + 2 H^+ + 2 e^- \longrightarrow 2 H_2O$$

is clearly developed and can be used for determination of H$_2$O$_2$ concentration.

Experimental conditions for each medium (pH and concentration of the buffer, or OT, or phenobarbital) have been chosen to obtain a reduction wave the highest as possible.

Solutions may be carefully deaerated before titration, because the first reduction wave of O$_2$ gives H$_2$O$_2$ leading to an additional amount of H$_2$O$_2$ in solution.

Three polarographic techniques have been performed : tast polarography, polarography with linearly increasing pulses and polarography with constant amplitude superimposed pulses (polarographic unit : Tacussel PRG5). In acidic, phosphate and borate media, the reduction wave (or peak) is very extended, but quantitative measurements may be done at - 1.3 V/SCE : tast polarography, or polarography with increasing pulses, or at - 1.0 V/SCE : polarography with constant pulses. In phenobarbital or OT, a sharply defined wave (or peak) has been observed (as it was showed by Shaw (14) and Semerano (15)). In each case, the diffusion current is proportional to H$_2$O$_2$ concentration (from 10^{-4} M to the detection limit). The detection limits determined for the five media and the three techniques are given table I.

medium	polarography		
	"tast"	"constant pulses"	"increasing pulses"
phenobarbital	1.2 x 10^{-6}	1.2 x 10^{-6}	6 x 10^{-7}
OT	3 x 10^{-6}	3 x 10^{-6}	3 x 10^{-6}
HClO$_4$	8 x 10^{-7}	6 x 10^{-7}	4 x 10^{-7}
borate	4 x 10^{-7}	4 x 10^{-7}	5 x 10^{-7}
phosphate	4 x 10^{-7}	2 x 10^{-7}	8 x 10^{-7}

Table I : H$_2$O$_2$ determination : detection limits (M)

In the first part of our studies, the most sensitive method, for the determination of H$_2$O$_2$ at very low concentrations, agrees with measurements in phosphate medium by polarography with increasing pulses.

1.2. Voltamperometric oxidation of H$_2$O$_2$

A voltamperometric method with a rotating platinum or gold electrode (16) was carrying out. We have measured the diffusion current corresponding to the following indicating reaction :

$$H_2O_2 \longrightarrow O_2 + 2\,H^+ + 2\,e^-$$

Various experimental conditions have been tested : medium (perchlorate, nitrate, chloride, K^+ or Na^+), pH and buffer, nature and rotation speed of the electrode. Finally, we have fixed our experimental conditions to obtain the best sensitivity to determine H_2O_2 concentration at a low level as follows :

- medium : $KNO_3 = 1$ M, pH = 7.5, phosphate buffer ;
- Pt rotating electrode (Tacussel EDI), rotation speed of the disk = 1000 rpm, diameter of the disk = 2mm . A "clean" electrode was obtained by a pretreatment : polarization at the potential of +0.21 V/SCE during 4 minutes.

In this case, it is not necessary to deaerated the titrated sample before measurement, O_2 gives O_3 at a higher potential than the H_2O_2 oxidation potential.

Measurements of diffusion current were performed at + 0.4 V/SCE leading to a linear variation of this current in function of H_2O_2 concentration from 10^{-4} M to the detection limit :

Detection limit	5×10^{-9} M

1.3. Measurements of H_2O_2 in rainwater

Between the different voltamperometric titrations of H_2O_2 describe in this paper, the last one (oxidation of H_2O_2 on a Pt rotating electrode) is the most suitable method to determine H_2O_2 in rainwater. This method could be adapted to an inexpensive portable amperometric titrator which could measure an electrical current at constant potential. Furthermore desoxygenation is not necessary and it is possible to perform one determination each 3 minutes.

The procedure may be describe as follows in quantitative determination of H_2O_2 in a precipitation sample (10 ml).

- Measurement of the diffusion current at +0.40 V/SCE (i_A) after addition of KNO_3 and phosphate buffer solution (1 ml).

- Calibration of the method by addition to the sample in the titration cell of a measured and negligible volume of standard solution of H_2O_2 (y mole) : measurement of i_C at + 0.40 V/SCE.

- Preparation of an analytical blank : measurement of i_{ref} at +0,40V/SCE. Interferences caused by others compounds giving undesirable reactions of oxidation at + 0.40 V/SCE have been eliminated. For this purpose, after the measurement of i_A and i_C, a selective elimination of H_2O_2 has been carried out "in situ" by addition of catalase enzyme (2.2 units/ml). Then, i_{ref} has been measured one minute after this addition to ensure the complete destruction of hydrogen peroxide. In this condition, no reaction occurs between catalase and the various organic peroxides which should be present in the analysed rainwater sample. Catalase decomposes H_2O_2 more readily than methyl hydroperoxide but catalase does not react with the majority of the organic peroxides.

Then, H_2O_2 concentration in the rain water sample can be written as :

$$[H_2O_2] = y \; \frac{i_A - i_{ref}}{i_C - i_A} \; \times 10^2 / 1.1$$

We added to some samples of rainwater, sulphite, sulphate, nitrate, chloride, Na^+, K^+, NH_4^+, Fe^{3+} and Mn^{2+}, species which seem to be relevant for aqueous phase atmospheric chemistry. No interference of these species was detected.

2. DETERMINATION OF TRACES OF H_2O_2 BY SPECTROPHOTOMETRIC MEASUREMENTS

H_2O_2 has been determined by spectrophotometric measurements using ABTS (2,2' -azinobis(3-ethyl benzothiazoline 6-sulfonic) acid : H_2L) as reagent (17). The peroxidase enzyme (POD) catalyses reaction between H_2O_2 and ABTS. The reaction products are water and the oxidised form of ABTS (L). L has a characteristic colour caused by a very intense absorption band in the visible spectrum ($\epsilon = 34,000$; $\lambda = 417$ nm). For calibration purpose, a wide range of H_2O_2 concentrations was investigated and spectrophotometric measurements were performed in the following conditions.
- Composition of the solution in the measurement cell :
  [ABTS] = 4.4 x 10^{-4} M ; [POD] = 1 unit/ml ;
  [phosphate buffer] = 0.1 M ; pH = 7.5 ; [H_2O_2] = y M
  (y : from 10^{-4} to 2 x 10^{-8} M),
- Composition of the solution in the reference cell :
  [phosphate buffer] = 0.1 M ; pH = 7.5 ; [H_2O_2] = y M
  (y : from 10^{-4} to 2 x 10^{-8} M) ; [catalase] : 0.2 sigma units/ml,
  one minute's wait, then addition of [ABTS] = 4.4 x 10^{-4} M +
  [POD] = 1 unit/ml.
Spectrophotometric measurements have been immediately done at 417 nm. We observe a linear variation of absorbance in function of H_2O_2 concentration (from 10^{-4} M to the detection limit).

Detection limit	2 x 10^{-8} M

3. COMPARISON OF THE VOLTAMPEROMETRIC AND SPECTROPHOTOMETRIC METHODS
Rainwater samples were collected on the roof of the Institute of Chemistry on the campus of the University Louis Pasteur in Strasbourg (May 1988). They were analysed simultaneously by the voltamperometric and the spectrophotometric methods. Some results are given table II.

sample n°	Rotating Pt voltamperometry (M)	Spectrophotometry (M)
1	2.5 x 10^{-7}	2.1 x 10^{-7}
2	1.8 x 10^{-7}	1.5 x 10^{-7}
3	0.7 x 10^{-7}	0.6 x 10^{-7}
4	8.8 x 10^{-7}	8.3 x 10^{-7}
5	11.3 x 10^{-7}	11.2 x 10^{-7}
6	7.8 x 10^{-7}	7.4 x 10^{-7}

Table II

The results found by the voltamperometric method are generally higher than those obtained by the spectrophotometric method (from 2 to 5 %). If the concentration of H_2O_2 is higher than 10^{-6} M, no variation between the two methods are observed.

REFERENCES

(1) HEIKES B.G., LAZRUS A.L., KOK G.L., KUNEN S.M., GANDRUD B.W., GITLIN S.N., SPERRY P.D. - J. Geophysical Research, 1982, 87, 3045.

(2) KOK G.L., HOLLER T.P., LOPEZ M.B., NACHTRIEB H.A., YUAN M. - Environ. Sci. Technol., 1978, 12, 1072.

(3) IBUSUKI T. - Atmos. Environment, 1983, 17, 393.

(4) SHAW F. - The Analyst, 1980, 105, 11.

(5) KLOCKOW D., JACOB P. - Nato Asi Series, Vol. G6, Chemistry of Multiphase Atmospheric Systems, Jaeschke ed., Berlin, 1986.

(6) SCOTT G. SEITZ W.R., AMBROSE J. - Anal. Chim. Acta, 1980, 115, 221.

(7) WILLIAMS III D.C., HUFF G.F., SEITZ W.R. - Anal. Chem., 1976, 48, 1003.

(8) WILLIAMS III D.C.SEITZ W.R. - Anal. Chem., 1976, 48, 1478.

(9) FREW J.E., JONES P., SCHOLES G. - Anal. Chim. Acta, 1983, 155, 139.

(10) TAMAOKU K., MURAO Y., AKIURA K., OHKURA Y. - Anal. Chim. Acta, 1982, 136, 121.

(11) LAZRUS A.L., KOK G.L., GITLIN S.N., LIND J.A., Mc LAREN S.E. - Anal. Chem., 1985, 57, 917.

(12) LAZRUS A.L., KOK G.L., LIND J.A., GITLIN (S.N.), HEIKES B.G., SHETTER R.E. - Anal. Chem., 1986, 58, 594.

(13) PEINATO J., TORIBIO F., PEREZ-BENDITO D. - Anal. Chem., 1986, 58, 1725.

(14) SHAW E.H. - Jr. Proc. S. Dakota Acad. Sci., 1948, 27, 92.

(15) SEMERANO G. - Rend. Accad. Nazl. Lincci, 1949, 6,721.

(16) KUSU F. NISHIKAWA Y., TAKAMURA K. - Tokyo Yakka Daigaku Kenkyu Nempo, 1980, 30, 1363.

(17) GAWEHN K., WIELINGER H., WERNER W. - Z. Anal. Chem.,1970, 252, 222.

MEASUREMENTS OF THE NON FILTERABLE SHARE OF
HEAVY METALS AND METALLOIDS IN AMBIENT AIR

P. Bruckmann, P. Bonitz, U. Düwel and Th. Reich
Umweltbehörde Hamburg, Dep. Luftuntersuchungen
Hamburg, FRG

Summary

Based on hints in the literature that some heavy metals
like Cd and As may not be completely deposited on filters
during ambient air measurements of suspended particulate
matter (S.P.M.) we tested and developped a method for the
complete collection of air borne metallic or metalloid
compounds. Briefly the method consists in a series of
cooled impingers filled with HNO_3, which are integrated
by glass pipes into a conventional apparatus for the measurement of S.P.M. The results of two series of measurements in ambient air are presented. It can be shown that
arsenic (15 - 28 %), lead (15 - 17 %) and copper
(42 - 55 %) compounds have significant shares which are
non filterable, whereas cadmium (72 - 73 %), nickel and
chromium compounds even predominantly pass through the
filters. The experiments, however, revealed also some
analytical shortcomings of our method, which are discussed. Further work is in progress to improve the method
and to validate further the preliminary results.

1. INTRODUCTION
     The usual method for the measurement of the heavy metal
content in ambient air consists in the collection of suspended
particulate matter (S.P.M.) on filters, e. g. by means of high
volume samplers, followed by the analysis of the particulates
by physico-chemical methods, e. g. atomic absorption spectroscopy (AAS). This usual method depends critically on the assumption that heavy metals and metalloids can be completely deposited on filters. Because of the low vapour pressures at ambient temperatures of most metallic compounds except Hg, this
condition seemed to be self-evident.
     There are, however, some indications in the literature
that this assumption may not be valid. Walsh (1) found
2 - 22 %% of arsenic on specially prepared back up filters during ambient air measurements. Vogg et al. (2) and Ecker (3)
analyzed high amounts of Cd (up to 150 %) and smaller shares
of Pb (up to 30 %) within gas scrubbers placed behind cascade
impactors (3) and high volume samplers (2), respectively.
     Iron and manganese compounds did not show this behaviour
(3). Finally, Boecker et al. (4) compared the results of simultaneous measurements of S.P.M. with the conventional filter

technique and with an Andersen impactor. Whereas the concentrations of S.P.M. and lead were comparable with both methods, 380 % more Cd was retained by the impactor than by the high volume sampler.

These indications would have great environmental impact if they can be assured. Apparently, the exposition of the public to metallic and metalloid compounds would have been considerably underestimated so far. Some of these compounds are toxic (e. g. Pb, Cd), others even have carcinogenic properties (e. g. As, Cr VI, Ni, Cd)   , so that valid exposition data for risk assessment are urgently needed. In addition, budget estimations of these highly persistent compounds in the environment based on ambient air concentrations would have to be revaluated.

We report here on two series of ambient air measurements with the aim to determine the non filterable share of the elements Pb, Cd, As, Ni, Cu, Cr, Zn and their compounds.

2. EXPERIMENTAL
2.1 Sampling

In order to preserve the comparability of our data with published heavy metal concentrations determined by high volume sampling, the sampling train to collect the metal compounds eventually passing through the filters was integrated into a conventional set up for dust measurements according to VDI 2463 (5) (figure 1).

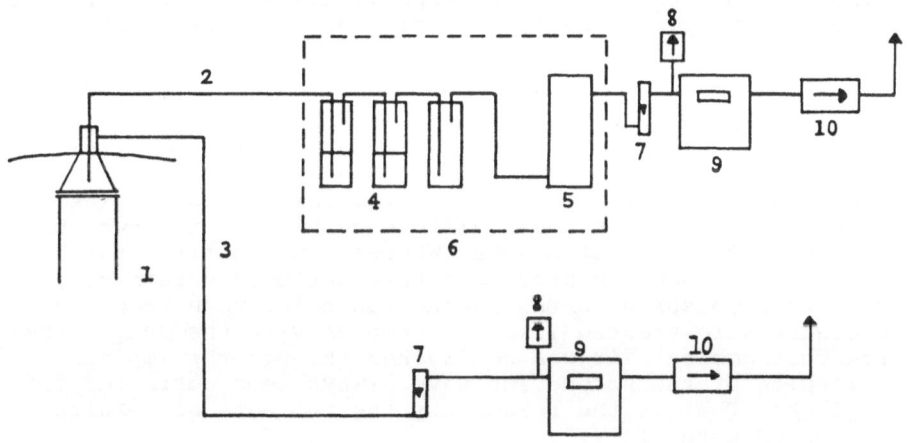

Figure 1: Sampling unit for the total collection of metallic compounds in ambient air. 1: filter head according to VDI 2463 (5); 2: glass tubing for partial air stream; 3: main stream with flow meter 7, pressure gauge 8, gas meter 9 and pump 10; 4: impinger and condensation trap; 5: air dryer; 6: cooling box

Throughout the experiments, membran filters with 120 mm diameter were used (Sartorius SM 11 302 with 3.0 µm pore width in
a first series of sampling, SM 11 303 with 1.2 µm pore width
in the second one). The air stream of 15 m³/h was divided into a main stream, which remained untreated, and a partial air
stream of 1.5 - 2 m³/h, which was transferred by glass line
tubing starting 1 cm above the filter into two chilled impingers filled with 25 % HNO_3 as absorption fluid. The impingers
and an additional condensation trap were kept at - 12 °C.
The whole sampling apparatus except the filter head is an all
glas system to avoid contaminations. The sampling period was
24 h in all cases. The sampling site was an industrial region
about 5 km south east of downtown Hamburg.

2.2 Analysis
    The filters were digested with a mixture of HNO_3, $HClO_4$
and HF with a temperature program rising from 100 °C to
200 °C (6). The elements were determined by graphite furnace
atomic absorption spectrometry (Perkin-Elmer Z 3030 with
HGA 600), using the standard addition method. For higher concentrations of Pb, Cu and Cd, flame atomization with a
Hitachi 180 - 80 was used. After sampling, the glass lines
and the emptied impingers were rinsed twice with 25 % HNO_3,
the rinsing fluids were combined with the condensate and the
absorption fluids, and the combined solution was condensed
to a few milliliters. The elements were measured by graphite
furnace AAS. Great care had to be taken during the whole procedure to avoid contaminations. Quality assurance consisted
in frequent blanks, additional experiments with a filter head
made entirely from glass and plastic and recovery experiments
(vide infra).

3. RESULTS
    Two sampling runs have been performed with 20 and 18
samples, respectively. Before the second run the sampling
apparatus has been improved in some minor details. Membran
filters of 3.0 µm (first series) and 1.2 µm (second series)
have been employed. The main difference between the two runs
is the consideration of blanks. Whereas in the first run
blanks of the complete procedure have not been determined,
blanks were measured in the second run after each sampling.
The blanks were treated like the samples with the only difference that ambient air was not sucked through the impingers.
The results of the second run have always been corrected for
the blanks. That is the reason why the two sets of results
are treated separately.
    The blanks of the non filterable elements (e. g. Pb
10 ng/m³, Cd 0.5 ng/m³; Ni 16 ng/m³) are considerably higher
than blanks from the used reagents (65 % HNO_3, Merck Supra-
pur). This shows that the sampling method has still to be improved to avoid even the smallest contaminations. On the other
hand, the results of most samples were much higher than the
blanks, so that the main conclusions are not put into question
by this fact.
    In figures 2 and 3 the main results of the two sampling
runs are presented by comparing the average concentrations

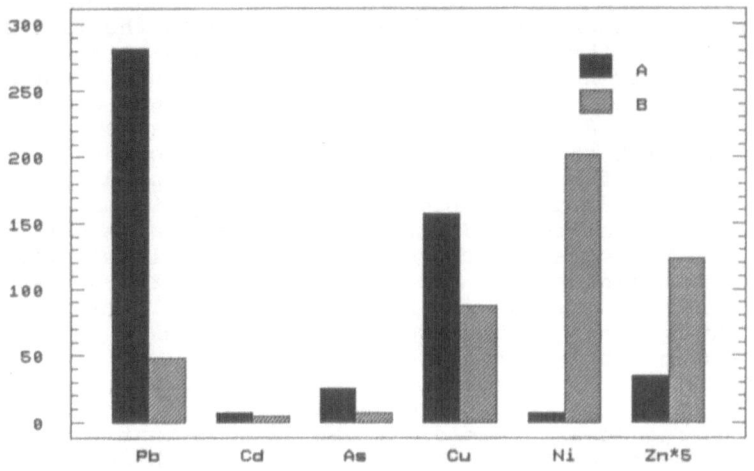

Figure 2: Average concentrations (ng/m³) of various heavy
metals and As deposited on the filter (A) and
passing through (B) (20 measurements, first run)

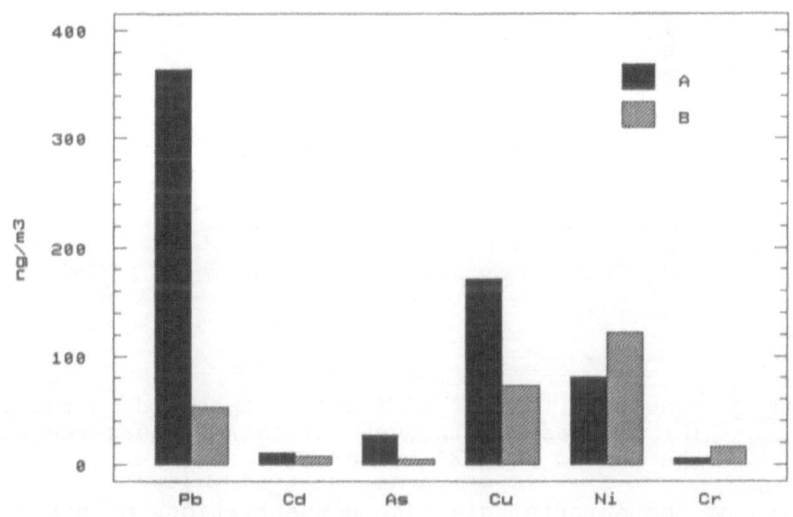

Figure 3: Average concentrations (ng/m³) of various heavy
metals and As deposited on the filter (A) and
passing through (B) (18 measurements, second run,
corrected for blanks)

of various elements having been deposited on the filters and
having passed through. It can be established that all measured elements have significant shares passing through the
filters. Whereas the non filterable percentages of Pb (17 and
15 %) and As (28 and 15 %) are not exceptionally high, surprisingly high shares of non filterable compounds > 100 % have
been obtained for Ni, Cr and Zn. The elements Cu (55 % and
42 %) and Cd (73 and 72%) show an intermediate behaviour. As
has already been pointed out, the non filterable concentrations are on average well above the blank values. Turning to
absolute concentrations and summing both shares up, more than
one order of magnitude (> 100 ng/m³) higher levels of Ni have
been found in the atmosphere as expected on the basis of
previous ambient air measurements.

In figure 4 the results of single determinations
(24 h-samples) of the second run are considered in more detail, taking Cd as example. It can be seen that the shares of
filterable to non filterable Cd amounts are highly variable
from experiment to experiment, the reasons for this are unknown. This holds more or less also for the other elements.

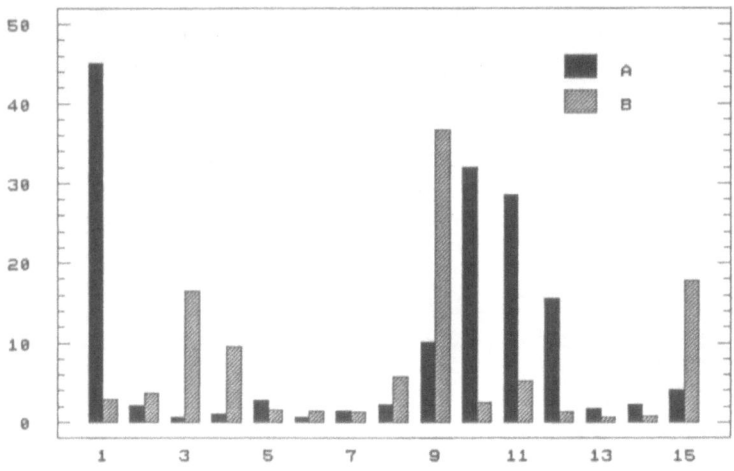

Figure 4: Concentrations of Cd (ng/m³) deposited on the filter
(A) and passing through (B). Single measurements of
the second run, 24 h-averages.

In view of the surprisingly high concentrations of non filterable heavy metal compounds, it seems self-evident that extensive quality assurance of the results is mandatory. The
frequent determination of blanks has already been mentioned.
In addition, several experiments have been performed to check
the method further. Recovery rates of the whole method were
determined by filling solutions with known amounts of metals
and metalloids in the sampling device and treating them like

ordinary samples. The recoveries were quite satisfactory with percentages between 90 and 110 %. In order to cherk possible contaminations arising from the conventional filter head (5) containing teflonized metal parts (filter holder from stainless steel, tubus from aluminium), 7 parallel measurements were undertaken with a second sampling train and a modified filter holder according to VDI 2463 Bl. 7 (7), where all metal parts had been removed. Whereas dust concentrations and element concentrations were comparable within the uncertainties of the methods, the element concentrations behind the filters did not agree. Systematically smaller element concentrations behind the filter head entirely made from plastic, however, were not observed. This shows that the reproducibility of the method is still unsatisfactory, but no systematic contaminations could be detected.

Finally, a second sampling train with two additional impingers was placed behind the first train, in order to check the precipitation efficiency of the impingers. Three such experiments have been performed. Quantities of Ni well above the detection limit were analyzed in the second sampling train in all three experiments, for Cu and Cd in one out of the three experiments. Consequently the collection efficiency has also to be improved.

4. DISCUSSION

In table I the non filterable shares of metal and metalloid compounds having been measured so far are collected.

Table I: Shares (%) of non filterable elements compared with concentrations on the first filter. a) 2nd run with 1.2 μm pore width; results from the 1st run (3 μm) in brackets

	Walsh (1)	Ecker (3)	Vogg (2)	Boecker (4)	this work
As	2 - 22				15 (28)
Pb		≤ 30	8	0	15 (17)
Cd		≤ 150	120	380	72 (73)
Fe, Mn		0			
Cu					42 (55)
Ni					152 (3015)
Cr					339 (-)

Although different shares have been obtained e. g. for Cd, agreement seems to be quite satisfactory for As andPb, taking into account the different methods, measuring sites and periods. More important, the arising general pattern (Cd >> As ~ Pb) is the same. It is therefore highly improbable that contaminations alone count for the observed effects in different laboratories, although there are still imperfections of the measuring methods to overcome. But it seems clear now

that conventional methods to measure the content of elements
on filters underestimate their concentrations in ambient air,
in part considerably. It seems surprising that so little
attention has been paid to this fact up to now.

The mechanisms which prevent metal compounds to be deposited completely on filters are still open for question.
Boecker et al. (4) assume water solubility of e. g. Cd compounds to be the main mechanism which causes Cd compounds to
pass through the filter in soluble form during humid weather
conditions. Vogg et al. (2) assume more generally chemical
reactions on the filters. In agreement with these assumptions
is the observation that the non filterable shares do not depend on the pore size of the filters (2); compare also the
last row in table I. In addition, comparable amounts of elements as deposited in the impingers have been found already
in the transfer glass lime connecting filter head and impingers (cf figure 1) during our experiments.

More work is in progress to improve our method and to
validate further our results.

REFERENCES

(1) WALSH, P. A., DUCE, R. A. and FASCHING, J. L., Impregnated
    filter sampling system for collection of volatile arsenic
    in the atmosphere. Environ. Sci. Technol. 11 (1977), 163
(2) BRAUN, H., METZGER, M. and VOGG, H. Neue Erkenntnisse
    über metallische Schadstoffe in der Luft. Z. Anal. Chem.
    317 (1984), 304 - 308
(3) ECKER, F. J., Experimentelle Untersuchungen zur Impaktion
    einiger Schwermetallaerosole in den Waldgebieten des
    Taunus; Diplomarbeit 1985, Institut für Meteorologie und
    Geophysik, Universität Frankfurt
(4) BOECKER, W. and HÄNSEL, Ch., Der Einfluß der Probenahme-
    technik auf die Erfassung des Kadmiumgehaltes im Aerosol
    und seine Ursachen. Z. gesamte Hyg. 34 (1988), 75 - 78
(5) VDI 2463, Bl. 9, Messungen der Massenkonzentration
    (Immission) Filterverfahren. LIS/P-Filtergerät, Febr. 1987
(6) NAUMANN, K., BERGMANN, J. and DANNECKER, W., in:
    Fortschritte in der atomspektrometrischen Spurenanalytik
    Vol. 1 (1984), 543 - 555, Verlag Chemie, Weinheim, 1984
(7) VDI 2463, Bl. 7. Messen der Massenkonzentration von Par-
    tikeln in der Außenluft. Filterverfahren, Kleinfiltergerät
    GS 050/3

STATISTICAL METHODS IN INTERCALIBRATION STUDIES OF SO_2 AND NO_x MONITORS

Jari A. Walden
Finnish Meteorological Institute, Air Quality Department
Sahaajankatu 22 E, SF-00810 Helsinki, Finland

Summary

Intercomparison studies of sulphur dioxide and nitrogen oxides monitors were performed using results from field and laboratory experiments. The results were evaluated by a linear multivariate method. Regression analysis and principal component analysis (PCA) were used for calculating the operational parameters (lower detection limits, precision) from the laboratory and field measurements.

1. INTRODUCTION

Determination of the concentrations of gaseous compounds in ambient air can be quite problematic, even in the case of sulphur dioxide, which has been routinely measured for a long time. Problems are focused in areas with very low pollutant levels, where the detection limit of the measuring technique may not be exceeded. In addition to low concentrations, also interferences from other gaseous compounds may deteriorate the accuracy of air quality measurements.

In order to obtain reliable air quality data, it is essential to know the performance characteristics of the measuring instrument or method. In case of a controlled or reference situation, the measurement parameters can be calculated, but unexpected behaviour may occur (electrical disturbances, operational malfunctioning and interferences) when the instruments are functioning in the real atmosphere.

Classical regression analysis has been used to evaluate the laboratory results when the calibration concentrations are known. Performance characteristics, such as the lower detection limit (LDL) and precision have been calculated from these results. Principal component analysis (PCA) and orthogonal linear relation (OLR) have been used in field measurements when no reference method is available.

The tested sulphur dioxide monitors were: Kimoto model 318 (Kimoto Electric CO, Japan), Kimoto model 365, Monitorlabs model 8850 (Monitorlabs Inc., USA),

Environnement AF20M (Environnement SA, France) and Apsa 300E
(Horiba Ltd, Japan). The first monitor is based on the
conductometric method, and the others on the UV-fluorescence
method. A detailed description of these experiments is given
in Walden et al.(1).

The tested nitrogen oxides monitors were Monitorlabs
model 8840 (Monitorlabs Inc., USA), Environnement AC30m
(Environnement SA, France) and Teco 14BE (Thermo Electron
Instruments, USA). These monitors are based on the
chemiluminescence method.

2. MATERIALS AND METHODS

2.1. Classical regression analysis

In regression analysis our model is the following:

(1) $Y = \alpha + \beta x + \epsilon$

Here Y is the measurement of a stochastic variable, α is the
fixed bias of the analyzing method, β is the proportional
bias of the analyzing method (or monitor), x is the value
for the reference method, and the random error term is ϵ.
We assume that the error term is normally distributed, with
zero mean and a standard deviation σ. This assumption means
that the mean of Y is also normally distributed with $N(\alpha +
\beta x, \sigma^2)$. To estimate the parameters α and β we can form the
maximum likelihood function of the measurements and solve
the quadratic equation with the least squares methods, i.e.

(2) $\underline{G} = \Sigma_i (x_i - \mu_i(\phi))^2 = \Sigma_i (y_i - \alpha - \beta x_i)^2$; $i = (1 \ldots n)$

The function is then minimized with respect to the parameter
ϕ (in this case α and β), and we can calculate the unbiased
estimate for parameter σ^2:

(3) $\sigma^2 = s^2 = \Sigma_i (x_i - \mu_i(\theta))^2 / (n-2)$; $i = (1 \ldots n)$

The degree of freedom (n-2) ensures that the estimate of s^2
is unbiased. S is often called the precision of the method.
Equations for parameters α and β, including their confidence
limits are presented in many textbooks (e.g. Morrison (2)).

2.2. Orthogonal linear relation

When both variables X and Y are subject to error i.e.
both variables are stochastic we have:

(4) $\underline{Z}_i = \underline{\mu} + \underline{\delta}_i + \underline{E}$

where $\underline{Z}_i$ includes the stochastic variables x and y, $\underline{\mu}$ is a
matrix for the mean of μ_x and μ_y, and x_{oi} and y_{oi} (= $\underline{\delta}$) are
the measurements. We assume that the measurement errors, e
and f are normally distributed, with zero means and standard
deviations σ_e and σ_f, respectively. We also assume that the
linear relation $y_i = \alpha + \beta x_i$ holds, which means that $\sigma_f = \tau \sigma_e$.

We now make a transformation to a new set of coordinates, by solving the equation (Morrison (2)):

(5) $\underline{\Sigma}' = \underline{P}(\theta)^{-1}\underline{\Sigma}\ \underline{P}(\theta)$

where $\underline{\Sigma}$ is the covariance matrix of measurements in (x, y)-coordinates, and $\underline{P}(\theta)$ is the rotational matrix formed by vectors $\underline{p}_1$ and $\underline{p}_2$. $\underline{P}(\theta)^{-1}$ is the inverse matrix, and $\underline{\Sigma}'$ is the sample dispersion matrix in (u,v)-coordinates. The geometrical situation is shown in Fig.1.

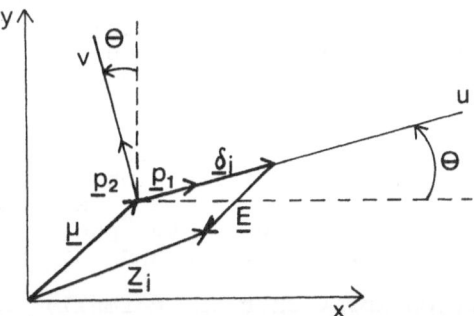

Figure 1. Rotation of the coordinate axes in OLR analysis.

We have to solve the eigenvalue problem $\underline{P}\underline{\Sigma}' = \underline{\Sigma}\underline{P}$ to get the diagonal matrix $\underline{\Sigma}'$, with the diagonal (defined as σ^2_1 and σ^2_2) elements being the eigenvalues of matrix $\underline{\Sigma}$. We also require that the largest eigenvalue is σ^2_1 with eigenvector $\underline{p}_1$, and the minimun eigenvalue (minimizes orthogonal distance between data points and the principal axis $\underline{p}_1$) σ^2_2 with eigenvector $\underline{p}_2$. The characteristic roots of matrix $\underline{\Sigma}$ are determined by equation $\det(\underline{\Sigma}- \kappa I) = 0$. We also use the fact that $\mathrm{Tr}(\underline{\Sigma}) = \mathrm{Tr}(\underline{\Sigma}')$ and $\mathrm{Det}(\underline{\Sigma}) = \mathrm{Det}(\underline{\Sigma}')$ (variables are invariant under orthogonal transformation). If we have an estimate for the coefficient τ, we can have explicit equations for the eigenvalues and with the help of eq.(3), we have estimates for variances $S_{1\Phi}^2$ and $S_{2\Phi}^2$. Omitting the details of this calculation (see Heidam (3)), we give estimates for the error variances

(6a) $S_{e\Phi}^2 = S_x^2[(1-\beta/c_{o\Phi})(n-1)/(n-2)]$

(6b) $S_{f\Phi}^2 = (\tau S_{e\Phi})^2 = S_{2\Phi}^2 = S_y^2[(1-r^2 c_{o\Phi}/\beta)(n-1)/(n-2)]$

Here S_x and S_y are the unbiased estimates of the measurements x and y, r is the correlation between the variables x and y, β is the slope of the regression line and $c_{o\Phi}$ is the slope of the orthogonal regression line. This can be written in the form $c_{o\Phi} = \tan\theta_{1\Phi} = (\tan\theta_1)^\Phi$, where the angle θ is shown in Fig.1 and the index $\Phi = \pm1$, depending on $\tau S_x \gtrless S_y$, ensures that the orthogonal distance of the observations to the OLR line is minimum.

OLR analysis can give mutually exact information about the dependence structure of the variables, but in a case of a number of stochastic variables, a special kind of factor

analysis, called the principal component analyasis (PCA), is
needed. The mathematical treatment of the PCA is similar to
the OLR, and we solve eq.(5) for the sample covariance
matrix to get the eigenvalues to form the principal axes. In
general only a few principal axes explain most of the total
variation of the sample. A more detailed description about
the PCA is presented in Morrison (2).

3. Results

The experimental set-up and data acquisition system
used in these monitor intercomparison studies are presented
in Walden et al.(1). The laboratory concentrations ranged
from 0 to 1000 ppb(SO_2), from 0 to 500 ppb(NO_2), and from 0
to 100 ppb(NO). During the field experiments in Helsinki the
hourly SO_2 concentrations were between 0 to 60 ppb (0 to 30
ppb in Imatra), and the arithmetic mean, calculated as an
average value from three monitors, was 22 ppb (8 ppb in
Imatra). In the case of the NO_x measurements the
concentration ranges were from 0 to 60 ppb(NO), and from 0
to 90 ppb(NO_2), and the arithmetic means of hourly values
were 2 ppb(NO) and 15 ppb(NO_2).
The laboratory calibration runs normally contained 10
different concentration levels, and regression analysis was
used in estimating the characteristic parameters of the
monitors. The precision of a monitor in certain
concentration range was calculated according to equation
$S_p = S/\beta$, where S is the square root from eq.3 and β is the
slope of the regression line. For SO_2 monitors the precision
was between 1 and 6 ppb in the concentration range 0 to 250
ppb. One of the SO_2 monitors had a continuous zero level
drift and this affected the results. For the NO_x monitors,
the precision was calculated in the NO_2 concentration range,
and the values were about 1 ppb(NO_2).
The lower detection limit (LDL), defined as the
concentration which is distinguishable from zero with a
certain probability (95%), was calculated from equation:

(7) $S_{LDL} = t_{gd}[\ S_o{}^2 + (S_{yo}/\beta)^2 \ ]^{1/2}$

Here $S_o{}^2$ is the variance from the clean air measurements,
S_{yo} is given by eq.3 when the concentration approaches
zero, t_{gd} is the Student t-factor, and β is the slope of the
regression line. The LDL values varied between 2 and 5 ppb
with the SO_2 monitors, and between 1 and 2 ppb with the NO_x
monitors. Omitting the drifting SO_2 monitor, we get LDL
values around 2 ppb, i.e. almost the same as for the NO_x
monitors. If we use criteria, $C_c = 3*LDL$ for accurate
concentration measurements, we get the critical
concentration values between 6 and 15 ppb for the tested
SO_2-monitors and 4 to 6 ppb for the NO_x-monitors.
When the field data were analyzed with the OLR
analysis, the assumption $\tau = 1$ (i.e. $\sigma_e = \sigma_f$) was made for
any pair of monitors. The operating precision, calculated
from eq.6b, is presented in Table 1a for the SO_2 monitors
and in Table 1b for the NO_x monitors.

Table 1a. The operating precision (S_p (ppb)) with respect to other SO_2-monitors according to the OLR analysis (eq.6b). Figures without parenthesis are from Helsinki and those in parenthesis are from Imatra.

```
Kimoto365   5.7 (4.3)
Ml.8850     0.8 (0.7)    4.9 (9.2)
Env.af20m   1.8 (0.9)    4.3 (4.2)    1.1 (.9)
Apsa300e    8.0 (3.5)   11.0 (5.6)    8.0 (3.4)    8.3 (3.0)
            Kimoto318   Kimoto365    Ml.8850      Env.af20m
```

Table 1b. Operating precision (S_p (ppb, in NO_2-mode)) of NO_x-monitors.

```
Env. ac30m              0.9
Teco 14B/E              1.0                      1.0
                        Ml.8840            Env. ac30
```

The low operating precision values related to monitors Kimoto 318, Monitorlabs 8850 and Environnement af20m indicate good agreement between the measuring signals of these monitors. The same conclusion can be drawn with NO_x-monitors in NO_2-mode.

Sample covariance matrices were used in the PCA analysis and in the Appendix we present the covariance matrices and the eigenvalues for SO_2 and NO_x field data. The first principal components are given by:

(8a) P_{1SO_2} = .281(Ki.318) + .580(Ki.365) + .287(Ml.8850) + .310(Env.af20) + .637(Apsa300E) : Helsinki

(8b) P_{1SO_2} = .139(Ki.318) - .303(Ki.365) + .182(Ml.8850) + .151(Env.af20) + .912(Apsa300E) : Imatra

(8c) P_{1NO} = .085(Ml.8840) + .175(Env.ac30m) + .981(Teco14be)

(8d) P_{1NO_2}= .620(Ml.8840) + .595(Env.ac30m) + .511(Teco14be)

The P_1 (8a) explains 89 % of the variability of the field data in Helsinki, and 91 % (8b) in Imatra. The first principal axis (8c) for the NO field data explains 68 % of the total variability of the results, and for NO_2 data (8d) this percentage is 98 %. We found that the three SO_2-monitors mentioned in connection with Table 1a give also now almost the same contribution to the principal axes, and that there are differences in results between Helsinki and Imatra. With NO_x-monitors there are differences between the monitors in the NO-mode, but the NO_2-mode is very similar with all monitors.

4. Conclusions

Statistical methods offer a valuable tool for analyzing monitored data, especially when no reference methods exist. In OLR analysis, the assumption $\tau = 1$ does not necessarily do justice to all monitors, and exact values can be used if the impact of interferences is known . Assumption about the normal distribution of the errors is reasonable, and provides a convenient way of defining the maximum likelihood function in estimating the parameters (eq.(2)). In this type of monitor intercomparison studies, the sample covariance matrix is a proper way of doing the PCA.

According to the laboratory experiments, the calculated LDL-values are rather high for measurements at low pollutant levels, as can be seen from the SO_2 field experiments in Imatra. The calculated average concentration value (8 ppb) means that there are periods when the concentrations have been below the LDL-values. Some interferences or malfunctioning may be responsible for the rather exceptional results of the Kimoto 365 during the field experiments in Imatra (eq.8b and Appendix).

The LDL-values of the NO_x-monitors, were almost as high as of the SO_2-monitors. During the field period the situation with the NO concentration is even worse than with the SO_2 concentration in Imatra. In this case statistical methods do not give accurate results in NO mode. With NO_2 concentrations the situation was much better and good agreement between the results is apparent.

ACKNOWLEDGEMENTS

I express my thanks to Mr T. Säynätkari for reading the manuscript and to Mrs V. Lindfors for revising the language of the manuscript.

REFERENCES

(1) WALDEN, J.A., LÄTTILÄ, H.O., HYPPÖNEN, M., PLATHAN, P. J., VIRTANEN, T.O. (1987). Intercalibration of sulphur dioxide monitors. Finnish Meteorological Institute, Publications on air quality 2. Helsinki, Finland.
(2) MORRISON, D.F. (1978). Multivariate statistical methods. McGraw-Hill, Tokyo.
(3) HEIDAM, N.Z. (1980). Regressionanalyser for to variable. MST LUFT-A31. Riso, Denmark.

NOTE: Mention of trade name of the monitors does not constitute recommendation for use.

APPENDIX

Covariance matrix for SO_2 monitors in Helsinki, (Imatra).

```
Ki.318      94 (24)
Ki.365     183(-17)     411(92)
ML.8850     94 (28)     189(-26)    96(33)
af20m      102 (23)     207(-20)   103(27)    112(24)
Ap.300e    182 (83)     355(-182)  185(110)   198(92)    510(562)
         Kimoto 318  Kimoto365   ML 8850    Env.af20m  Apsa300e
```

Covariance matrix for NO_x monitors: NO and NO_2 in parenthesis

```
ML 8840      9.78 (48.3)
Env.ac30m    9.35 (45.9)    10.7  (44.9)
Teco 14b/e   3.63 (39.2)     9.98 (37.2)    69.0 (33.9)
                ML 8840     Env.ac30m      Teco 14b/e
```

Eigenvalues of the sample covariance matrices above.

SO_2-monitors		NO_x-monitors	
Helsinki	Imatra	NO	NO_2
1093	673	71.1	125
114	48.8	17.8	1.85
15.2	13.3	0.606	0.534
0.20	0.181		
0.0	0.682		

DETERMINATION OF SULFUR COMPOUNDS IN ATMOSPHERIC
MATRICES AFTER DERIVATIZATION USING A GC-FPD METHOD -
A LABORATORY STUDY WITH RESPECT TO POSSIBLE APPLICATIONS

G. LAMMEL and E. KRANZ
Kernforschungszentrum Karlsruhe GmbH
Laboratorium für Aerosolphysik und Filtertechnik 1
P.O. Box 3640, D - 7500 Karlsruhe

Summary

An analysis method for sulfur species using a GC-FPD technique to
determine the methyl derivatives (PENZHORN and FILBY, 1976) was
tested in a feasibility study for its applicability to samples of
atmospheric origin. A series of experiments was performed in our
laboratory. Solutions of sulfates, sulfites and hydroxymethane
sulfonates, as well as the salts thereof have been analysed.
Due to the so achieved detection limits of 0.4 ng S with respect to
sulfates, and 0.8 ng S with respect to the aforementioned reduced
sulfur species, respectively, the method seems to be applicable to
the determination of sulfate, and, under certain conditions, to the
reduced sulfur species in samples of atmospheric aerosol, but not,
however, in samples of cloud- and fogwater.
Further more, the analysis was found to be incapable to determine
specifically the amount of not neutralized sulfate as suggested by
PENZHORN and FILBY, 1976.
The applicability of the method for the determination of sulfur
species in samples of atmospheric origin is discussed.

1. Introduction

In the atmosphere, sulfur dioxide, SO_2, undergoes chemical transformation
in the gas-phase and even faster in the heterogeneous phase, if present, to
sulfuric acid and sulfate, respectively. Sulfate, SO_4^{2-}, is removed from the
atmosphere through wet and dry deposition.
The partitioning of SO_2 between gas-phase and aqueous phase is in such a way
that most of this species is found in the gas-phase. The effective solubility
of SO_2, however, can be greatly enhanced, if formaldehyde, HCHO, is present,
which forms the adduct hydroxymethanesulfonate, the anion of hydroxymethanesulfonic acid (HMSA), in the aqueous phase (RICHARDS et al. 1983, MUNGER et
al. 1984).
The gas-phase mixing ratio of sulfuric acid is restricted to low levels under
ambient conditions due to the effective nucleation with water vapor or ammonia
molecules and the heterogeneous condensation onto preexisting aerosol
particles.
Atmospheric sulfuric acid is therefore mainly found in the heterogeneous phase
where it is completely dissociated, forming sulfate and bisulfate ions.
Sulfate or bisulfate, respectively, is the main anion of the atmospheric
aerosol's soluble fraction, except of the marine aerosol. Concentrations
observed are in the order of several $\mu g/m^3$ at continental sites and up to
several tens of $\mu g/m^3$ at urban sites. In the free troposphere and strato-

sphere, a maximum of sulfuric acid vapor pressure corresponding to the order of 1 pptv is observed at about 35 km of height (FERGUSON and ARNOLD, 1981). The determination of particulate sulfate concentration usually is performed using filter or thermal denuder techniques (SLANINA et al., 1985) and subsequent ion chromatographic analysis. Thermal denuder techniques also allow the determination of sulfuric acid present in the gas-phase, in clusters and in the Aitken particle mode. Also measurement of the bisulfate ion concentration has been used as an indirect means for sulfuric acid determination in the atmosphere (FERGUSON and ARNOLD 1981, EISELE 1989).

The determination of the HMSA adduct in samples of environmental origin has been reported by means of indirect methods only insofar as using the reduced stability of the complex at high pH releasing sulfite (CHAPMAN 1986, JAESCHKE 1986, GROSCH and RUMPEL 1987).

PENZHORN and FILBY, 1976, suggested to determine sulfate, methyl sulfonate, and other sulfur species sensitively by derivatization forming the methyl esters of the corresponding acids with diazomethane, CH_2N_2, and subsequent gas chromatographic separation using a flame photometric detector specifically for sulfur. Microgram amounts of sulfate and less have been determined in samples of atmospheric aerosol particles (PENZHORN and FILBY 1976, PANTER and PENZHORN 1980). The authors found the method to be sensitive to acid, i.e. not neutralized sulfate, while neutralized sulfate was not found to show a similar effect. Thus the determination of methylated sulfates through GC-FPD was used to discriminate between neutralized and not neutralized sulfate (POSS et al. 1986, DLUGI et al. 1987, METZIG and POSS 1988, DLUGI 1989).

In a feasibility study of the method with respect to the determination of sulfur species relevant in atmospheric matrices, a series of experiments has been performed in our laboratory.

2. Experiments

Solutions of sulfates, sulfites, and of hydroxymethane sulfonates with varying pH, as well as the salts thereof have been analysed using the unmodified method of PENZHORN and FILBY: The methyl esters have been synthesized by adding a concentrated solution of diazomethane in ethyl ether containing an over-stoichiometric amount of the derivatizing reagent.

Inorganic salts (Merck, p.a. grade) have been used without any further purification. Diazomethane, CH_2N_2, was synthesized from p-toluylsulfonylmethyl-nitrosamide ('Diazald'). CH_2N_2 was kept at -18°C in ethyl ether solution.

For the derivatization reaction the mixture was kept cool for a certain time. The reaction was stopped by means of dilution with ethyl ether.

The analysis was performed by gas chromatographic separation of the methyl derivatives and flame photometric detection.

Several microliters of the methyl derivative solution containing 0.3 to 100 ng of sulfur was injected (split 1:10, 150°C) onto a packed column (1.5 m, 2 mm i.d., 10% Ucon LB550X on chromosorb WHP 80 - 100 mesh, carrier: nitrogen, chromatograph Pye Unicam 4550). The separation was performed isothermally (125°C). The FPD was kept at 250°C and was equipped with an optical filter selective for sulfur species.

3. Results

The reproducibility of the method, given by the variation of the relative and absolute peak areas, was better than 80 %. Reproducibility may have also been affected by a possible deterioration of the CH_2N_2 agent with ageing.

3.1. Sulfur-VI

Only one peak was detected when derivatives of solid ammonium sulfate and ammonium bisulfate, or solutions of the salts sodium sulfate, ammonium

sulfate, ammonium bisulfate, or diluted sulfuric acid were analyzed.
By comparison with direct gas chromatographic separation of the compound, the peak was identified to represent the elution of sulfuric acid dimethyl ether, $SO_2(OCH_3)_2$.
The reaction product was found to be the same, independently of the pH's variation between pH = 1 and pH = 6, i.e. independently of the degree of dissociation of sulfuric acid in this pH range.
The results are shown in table I.
For the determination of sulfates, a detection limit of 0.4 ng with respect to sulfur was achieved.

Table I: Detected peaks of S-VI compounds.
Retention time is given in minutes.

Sulfur compound		rel. peak area (time)
		0.8 min
$(NH_4)_2SO_4$	Salt	100 %
NH_4HSO_4	Salt	100 %
H_2SO_4	pH = 1	100 %
$(NH_4)_2SO_4$	Solution, pH = 1	100 %
NH_4HSO_4	Solution, ph = 1.5	100 %
Na_2SO_4	Solution, pH = 4	100 %
$(NH_4)_2SO_4$	Solution, pH = 5.5	100 %

The peak area detected was dependent on reaction time. An influence of the sample's acidity on the total peak area, however, was not observed varying the solutions' pH between 1 and 6.
No effect on the kind of product formed was observed, when the reaction time was varied between several minutes and one day.

3.2. Sulfur-VI

Solid Na_2SO_3 and solutions thereof at various pH, as well as solutions of $(NH_4)_2SO_3$ and $Na_2S_2O_5$ at various pH have been treated with CH_2N_2 and analysed by GC-FPD as described above. In aqueous solution $Na_2S_2O_5$ dissociates immediately forming sulfite and bisulfite.
Chromatograms of methyl derivatives of reduced sulfur compounds, whether solutions or solids, always showed several peaks. The resolution of the chromatographic separation was satisfactory.
The relative peak areas and the retention times of the peaks of solutions and salts studied are given in table II.
An identification of the eluted methyl derivatives was not achieved. However the retention time of peak No. 3 was identical to that of dimethyl sulfate, $SO_2(OCH_3)_2$. Dimethly sulfate might be the product of the methylation of oxidized sulfite or of oxidation of dimethyl sulfite, $SO(OCH_3)_2$, formed.
For the determination of the reduced sulfur species, sulfurous acid and hydroxymethanesulfonic acid (HMSA), detection limits of 0.8 ng with respect to sulfur were achieved.
The proportionality of the FPD's signal to the squared concentration of reduced sulfur species was verified within the range of 5 to 50 ng S (absolute

amount injected).

The total peak area, as well as the relative peak areas have been found to be sensitive to variation of the reaction time: For instance, peak No. 3 was found to decrease in area with increasing reaction time of sulfite solution under weakly acid conditions: While chromatograms of reaction mixtures after very short reaction time (several minutes) showed this peak dominating the total peak area with 95 % of its value, after 21 h of reaction time it held only for remaining 57 % of the total peak area.

Further, the total yield of derivatives of S-IV compounds as detected by GC-FPD was found to be pH-dependent: With decreasing acidity moving from pH = 1 to pH ≥ 4, the total peak area decreased about 90 % in case of Na_2SO_3.

Derivatives of sodium sulfite solutions and solid sodium sulfite showed different peak patterns. The peak pattern of solid sodium sulfite resembled very much the pattern of weakly acid ammonium sulfite solution, which was significantly different from those of weakly acid sodium sulfite solutions.

Table II: Detected peaks of S-IV compounds.
Retention times are given in minutes, relative peak areas in %.

Sulfur compound		rel. peak area (No., time)				
		1 0.33	2 0.40	3 0.80	4 0.93	5 2.0
Na_2SO_3	Salt	8	0	80	12	1
$HOCH_2SO_3Na$	Salt	0	2	0	78	20
Na_2SO_3	Solution, pH = 1	< 2	88	2	8	0
Na_2SO_3	Solution, pH = 2	< 2	87	6	6	0
$(NH_4)_2SO_3$	Sol., pH = 3	< 2	5	57	37	0
Na_2SO_3	Sol., pH = 4	23	34	20	22	0
$Na_2S_2O_5$	Sol., pH = 5	23	23	4	50	0
Na_2SO_3	Sol., pH = 12	65	15	12	9	0
$HOCH_2SO_3Na$	Sol., pH = 1	< 2	30	61	09	0
$HOCH_2SO_3Na$	Sol., pH = 4	< 3	33	50	17	0
$HOCH_2SO_3Na$	Sol., pH = 12	27	9	55	8	0

Chromatograms of derivatives of solid hydroxymethanesulfonic acid sodium salt (HMSANa) and the solution thereof showed peaks eluting at the same retention time as the derivatives of sulfite compounds did. The total peak area, as well as the relative peak areas, however, have been different. The total peak area detected indicated a reduced yield of methyl derivatives compared to the derivatization of sulfite, even in the pH range of maximum stability of the HMSA complex (DEISTER et al., 1986).

An additional peak, which might have indicated the elution of a derivative specific for the presence of HMSA was not detected.

4. Discussion

With detection limits of 0.2 ng of S-VI sulfur and 0.8 ng of S-IV sulfur, respectively, the study showed that the method is to be sufficiently sensitive for the determination of sulfur compounds in samples of atmospheric origin.

Based on a time resolution of 1 h for 'high-volume'-sampling of atmospheric aerosol and fluxes of air commonly used, the amount of particle sulfate sampled will exceed the critical value given by the detection limit in most cases at least by a factor of 10. If the volume of ethyl ether used for stopping the derivatization reaction was reduced by a factor of 5, which may be feasible, this holds true even in the case of low particle sulfate concentration ('continental' aerosol) and of size fractionation of aerosol by means of a multi-stage impactor, as well as for sulfate concentration typical for fog- and cloud-water samples (see for instance: OKITA 1968, MUNGER et al. 1983).

Assuming S-IV concentrations in samples of atmospheric origin to be at least a factor of 5 to 10 lower than those of S-VI (see for instance the case studies on polluted air masses by MUNGER et al. 1983, BELTZ et al. 1986, CHAPMAN 1986), the detection of S-IV compounds by means of the here described method will in most cases not be feasible, at the present state of the art. The analysis method was found not to be capable to determine specifically the amount of not neutralized sulfate in sulfate solutions, neither to discriminate between solid ammonium sulfate and ammonium bisulfate. Our experiments indicate that total-sulfate concentration can be determined with this method, whereas not neutralized sulfate, the so-called 'sulfuric acid', as suggested by PENZHORN and FILBY, 1976, cannot.

The effect of pH variation on the total detected peak area might be the result of a change in reactivity and stability of derivatives formed. The total peak area detected, however, does not necessarily reflect the yield of the derivatization reaction, as some other derivatives formed might be undetectable under the chosen separation conditions.

Maximum peak areas from solutions were observed with lower pH. This is in accordance with an enhanced formation of the protonated species, the methyldiazonium ion, $CH_3N_2^+$, which is the reactive electrophile of the derivatization reaction (EISTERT, 1941).

At the present state of knowledge, a differentiation between sulfite and HMSA with the described method is not possible.

Further investigation would be necessary in order to elucidate the reaction mechanisms of the reaction of diazomethane and the studied solid S-IV compounds and aqueous solutions thereof. The reaction mechanism might explain observed differences of peak patterns.

As a result of hygroscopicity of inorganic salts present, atmospheric aerosol particles undergo water uptake with increasing humidity (WINKLER and JUNGE, 1972). Under ambient conditions, the atmospheric aerosol therefore contains liquid water.

In case of the analysis of undried atmospheric aerosol particles and aqueous solutions of sulfur species, the derivatization reaction takes place at the interface of the two inmiscible solvents, ethyl ether and water. As water promotes decomposition of diazomethane, the yield and therefore also the reproducibility of the derivatization reaction is greatly influenced by the amount of water present.

An application of the method to the determination of sulfur species in environmental samples is suggested only if exclusion of humidity in the sample was guaranteed, or, at least, if fixed volumina of aqueous phase with stabilized pH were analysed.

As acidity of the sample would enhance the yield of derivatization reaction, further investigation is recommended with the aim to evaluate the effect of acids present other than sulfuric acid on the reaction of CH_2N_2 and sulfur species.

5. References

BELTZ, N., K.-H.ENDERLE, W.JAESCHKE, and H.OBENLAND, 1986: Measurement of
  sulfur species during the 1986 field experiment in S.Pietro Capofiume,
  in S.FUZZI (Ed.): Heterogeneous chemistry project, Report Ist. FISBAT,
  Bologna, Italy, pp. 16 - 20

CHAPMAN, E.G., 1986: Evidence for S(IV) compounds other than dissolved SO_2 in
  precipitation, Geophys. Res. Lett. 13, 1411 - 1414

DEISTER, U., R.NEEB, G.HELAS, and P.WARNECK, 1986: Temperature dependence of
  the equilibrium $CH_2(OH)_2 + HSO_3^- = CH_2(OH)SO_3^- + H_2O$ in aqueous solution,
  J. Phys. Chem. 90, 3213 - 3217

DLUGI, R., 1989: Chemistry and deposition of soot particles in moist air and
  fog, Aerosol Sci. Technol. 10, 93 - 105

DLUGI, R., S.JORDAN, S.MANEGOLD and H.MÄTZING, 1987: Die Entstehung sulfat-
  und nitrathaltiger Partikeln und Tropfen in der Atmosphäre, Report of
  the Nuclear Research Centre Karlsruhe, Rep. No. KfK-PEF 29

EISELE, F.L., 1989: Natural and anthropogenic negative ions in the troposphere, J. Geophys. Res. 94, 2183 - 2196

EISTERT, B., 1941: Neuere Methoden der präparativen organischen Chemie.
  Synthesen mit Diazomethan, Angew. Chem. 54, 99 - 105

FERGUSON, E.E., and F.ARNOLD, 1981: Ion chemistry of the stratosphere,
  Acc. Chem. Res. 14, 327 - 334

GROSCH, W., and K.J.RUMPEL, 1987: Grundzüge und Anwendungsmöglichkeiten eines
  Verfahrens zur kontinuierlichen Mikroanalytik von Regen-, Nebel- und
  Wolkenwasser, Verein Deutscher Ingenieure,
  VDI Report No. 608, pp. 421 - 433

JAESCHKE, W., 1986: Multiphase atmospheric chemistry, in: W.JAESCHKE (Ed.):
  Chemistry of multiphase atmospheric systems, NATO ASI Series Vol. G6,
  pp. 3 - 40, Springer Verlag Berlin

METZIG, G., and G.POSS, 1988: Optical and physico-chemical properties of
  tropospheric aerosols in the lee of the city of Karlsruhe,
  in K.GREFEN and J.LÖBEL (Eds.), Environmental Meteorology,
  Kluwer Acad. Publ., Amsterdam, pp. 23 - 37

MUNGER, J.W., D.J.JACOB, J.M.WALDMAN, and M.R.HOFMANN, 1983: Fogwater
chemistry in an urban atmosphere, J. Geophys. Res. 88, 5109 - 5121

MUNGER, J.W., D.J.JACOB, and M.R.HOFFMANN, 1984: The occurrence of bisulfite-
  aldehyde addition products in fog- and cloudwater,
  J. Atmos. Chem. 1, 335 - 350

OKITA, T., 1968: Concentration of sulfate and other inorganic materials in fog
  and cloud water and in aerosol, J. Met. Soc. Japan 46, 120 -127

PANTER, R., and R.-D.PENZHORN, 1980: Alkyl sulfonic acids in the atmosphere,
  Atmosph. Environ. 14, 149 - 151

PENZHORN, R.-D., and W.G.FILBY, 1976: Eine Methode zur spezifischen Bestimmung
  von schwefelhaltigen Säuren im atmosphärischen Aerosol,
  Staub - Reinhaltung der Luft 36, 205 - 207

POSS, G., C.WEILAND, and G.METZIG, 1986: First experiences with a set of
  chemical analyses of tropospheric aerosol within the POETA program,
  J. Aerosol Sci. 17, 264 - 267

RICHARDS, L.W., J.ANDERSON, D.L.BLUMENTHAL, J.A.McDONALD, G.L.KOK, and
  L.A.LAZRUS, 1983: Hydrogen peroxide and sulphur-IV in Los Angeles cloud
  water, Atmosph. Environ. 17, 911 -914

SLANINA, J., C.A.M.SCHOONEBECK, D.KLOCKOW, and R.NIESSNER, 1985: Determination
  of sulfuric acid and ammonium sulfate by means of a computer controlled
  thermodenuder system, Anal. Chem. 57, 1955 - 1960

WINKLER, P., and C.JUNGE, 1972: Growth of atmospheric particles as a function
  of the relative humidity - Part I: Method and measurements at different
  locations, J. Rech. Atmos. 4, 617 - 638

THE PHYSICO-CHEMICAL PROPERTIES OF AEROSOLS
IN RELATION TO THEIR EXTINCTION COEFFICIENT

G. METZIG

Kernforschungszentrum Karlsruhe GmbH
Laboratorium für Aerosolphysik und Filtertechnik I
Postfach 36 40, D-7500 Karlsruhe 1

Summary

A set of analytics is described to measure the physico-chemical and optical properties of tropospheric aerosols. With the measured data set and by the help of straight-forward Mie calculations it is tried to find a correlation between the physico-chemical aerosol properties and their optical properties. The evaluation scheme is presented. The agreement between calculations and measurements of the aerosol extinction coefficient was in most cases better than a factor of 2. The uncertainty in this procedure is the unknown particle bulk density, which may range between 1.4 and 2.0 g/cm^3.

1. INTRODUCTION

Within the framework of the POETA program (a German acronym for physico-chemical and optical properties of tropospheric aerosols) the aerosol is chemically described by several analytics. The particles are collected on bulkfilter and a high volume sampler in connection with a five-stage impactor (Andersen, particle size ranging from 0.5 to 15 μm) is also used. The particle composition is analyzed as well as the element distribution on the particles' surface, and the element composition of single particles. A set of chemical analyses is applied to determine the portion of free acid, elemental carbon, and some important anions and cations. The spectral extinction coefficient (450, 500, 550, 600, 650, 692.5, and 450 to 700 nm) of ambient aerosol particles is measured using a White Cell with an effective path length of 100m. Reliable data of the extinction coefficient is obtained, even under clean air conditions, i.e. after rainfall at about 10 μg/m^3 aerosol mass concentration or less. The samples were taken at ground and at the top of the meteorological tower at KfK-site at a height of 200m.

Extinction coefficient calculations are performed using straightforward Mie calculations. The results of the chemical analyses are used to calculate a start value of the effective index of refraction and the number size distribution is derived from mass size distribution measurements with a 10-stage QCM impactor (fig. 1).

2. MEASUREMENTS

2.1 CHEMICAL PROPERTIES

Qualified filter material is used for each chemical analysis. We selected six different filter materials on the basis of underground value and its variation: silver, borosilicate, quartzglass fibre, teflon, nuclepore, and cellulose acetate.

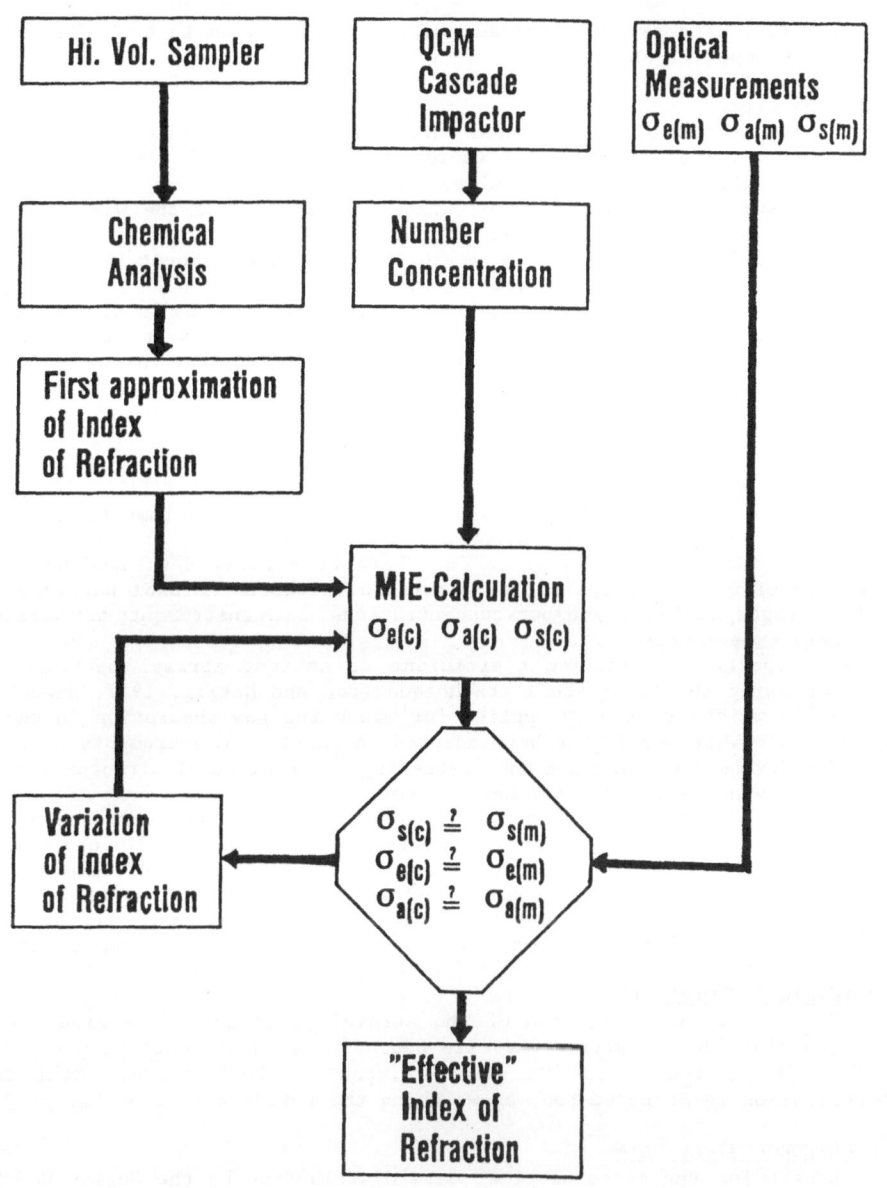

Fig. 1: EVALUATION SCHEME IN THE POETA PROGRAM

The following set of chemical analytics is applied to bulk filters as well as to impactor samples:
- Measurements of the water soluble and insoluble portion of the aerosol particles.
- Measurement of the pH-value and the specific conductivity of the water-soluble part. Using these two values: calculation of the free acid portion's (after Winkler, 1980).
- Measurement of SO_4^{2-}, NO_3^-, Cl^-, Na^+, K^+, and NH_4^+ of the water-soluble part by means of ionchromatography.
- Determination of soot by measuring the total amount of elemental carbon by means of thermal decomposition analysis.
- Bulk analysis of elements of the aerosol particles by total reflection X-ray fluorescence analysis (TRFA) for all elements heavier than P.
- Analysis of compounds on the particles' surface using electron spectroscopy (ESCA).
- Determination of elements of single particles by REM using energy-dispersive X-rays analysis (EDAX).

2.2 OPTICAL PROPERTIES
The second part of the POETA program includes the measurements of volume extinction, scattering, and absorption coefficient. The instruments for the extinction and scattering measurements have been newly developed with the objective of measuring optical properties of airborne aerosol particles at typical tropospheric mass/number-concentrations. Both instruments are suitable for field experiments.

The spectral extinction coefficient of ambient aerosol particles is measured using the 'White-Cell' technique (Poß and Metzig, 1987; Riedel et al., 1988) which is usually applied for measuring gas absorption in the IR region. The 'White Cell' has been adapted to particle measurements.

The device for measuring the scattering coefficient of airborne aerosol particles is not yet ready for measurements.

The absorption coefficient of aerosol particles is measured using the Chin-I Lin method (Chin-I Lin et al., 1973). The aerosol particles are sampled on nuclepore filters and the light transmission is measured before and after sampling. The difference in transmitted light through the filter is a measure for the absorption due to aerosols on the filter. Corrections are necessary for the scattered light fraction scattered into the backward hemisphere.

2.3 PHYSICAL PROPERTIES
Physical characterization of the aerosol particles is carried out by electron microscope analysis and size dependent mass concentration measurements using a ten stage QCM cascade impactor. Furthermore, total mass concentration is measured too, by weighing the teflon bulk filters.

2.4 METEOROLOGICAL DATA
Except for the meteorological data distribution by the German Weather Service, the following data is measured and recorded by the KfK institute for meteorology and climatology: wind velocity and direction, temperature, humidity, pressure, short-wave and global radiation. These data items are determined in ten minutes' intervals and for different heights of the KfK meteorological tower. The measurment positions are: ground, 20m, 30m, 40m, 50m, 60m, 80m, 100m, 130m, 160m, and 200m height.

3. RESULTS
The results of the physico-chemical analysis of the aerosol particles measured in the POETA program are discussed elsewhere (Metzig and Poß, 1988).

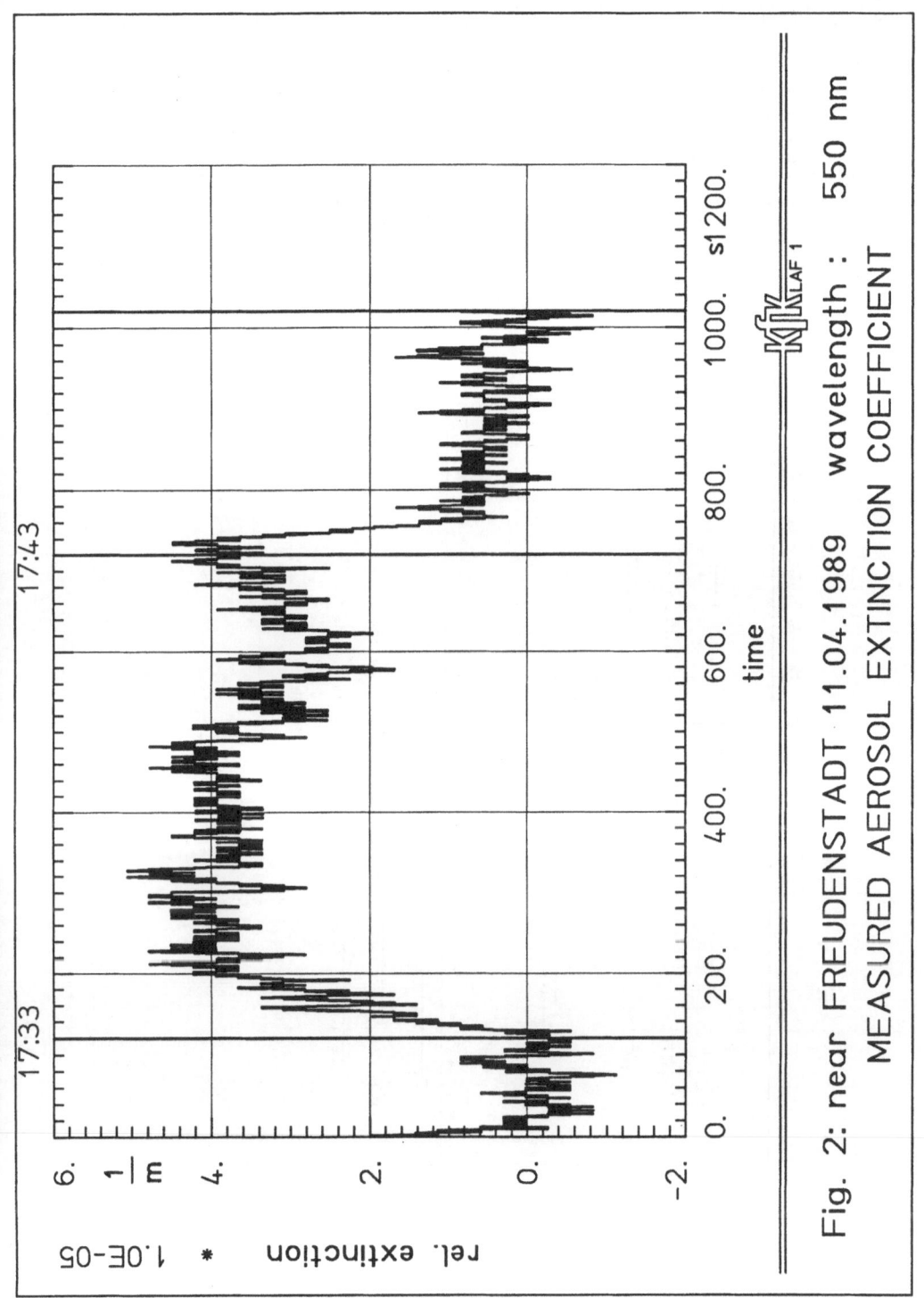

Fig. 2: near FREUDENSTADT 11.04.1989   wavelength :   550 nm
MEASURED AEROSOL EXTINCTION COEFFICIENT

Figure 2 shows as an example the measured aerosol extinction coefficient during a campaign during April this year near Freudenstadt (Black Forest). The air flow rate through the cell is $1m^3/h$. The volume of the cell is ten liter. During the first two minutes of sampling and at the end of the measuring cycle particles free air is measured. This air can contain absorbing gases, i.e. ozone etc. The average of the measured light extinction values of the first two minutes is set zero and the deviations are plotted only. Changing to particles loaden air, it takes about one minutes to fill the cell and about 2.5 minutes to clean the cell from airborn particles. In the example (fig. 2) an aerosol extinction coefficient at 550nm is measured between $3.*10^{-05}/m$ and $4.*10^{-05}/m$. The corresponding measured absorption coefficient is $1.4*10^{-06}/m$. During the time interval 14.15 - 20.32h the average mass concentration was

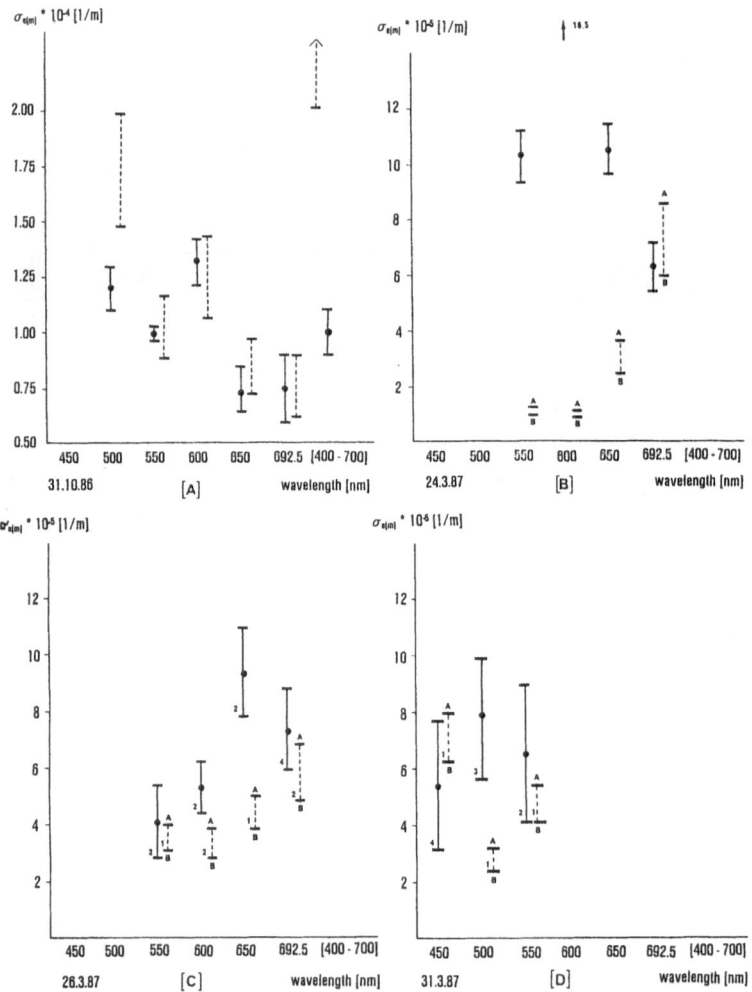

Fig. 3: Comparison between measured (solid) and calculated (dashed lines) extinction coefficients of airborne aerosol particles (see text).

19.5 $\mu g/m^3$. The following major chemical components of the particles are found: 8.5 $\mu g/(m^3$sampled air) elemental carbon and a soluble portion of 85%. The major anion and kation are chlorine and sodium, respectivly.

The experience made during several measuring campaigns is that agreement between calculations and measurements of the extinction coefficient is in most cases better than a factor of two (fig. 3). When reproducing the measured extinction coefficient by straight-forward Mie calculations, particle number distribution and complex index of refraction of the aerosol must be known. This means, that the particle bulk density must be known.

Usually an overlapping of the two bandwidths within a variation of calculated (dashed lines in fig. 3a-d) and measured extinction coefficients (solid lines in fig. 3a-d) can be observed. The bandwidths of the measurements represent the standard deviation of the extinction coefficient including measuring errors. The corresponding bandwidth of the theoretical values results from calculations with different number distributions which were obtained by assuming a particle bulk density of 1.4 and $2g/cm^3$. The left number of the solid lines' minimum indicates the quantity of extinction coefficient measurements at the respective wavelength. The left number of the dashed lines' minimum represents the number of distributions which are used to obtain a theoretical mean extinction coefficient.

Besides the cases of good agreement, there are some greater deviations at a wavelenght of 500-650nm (fig. 3b). It is striking that these deviations occur in connection with bi-modal distributions. However, compared with all the other measured mass concentrations during the same period no significant change in the total mass concentration can be noted. Although the second mode occurs at larger particle sizes at the expense of a drastic reduction of the number of fine particles, the measured extinction is unexpectedly high. This leads to the assumption that the second mode contains high-absorber material.

4. CONCLUSIONS

The accordance of the extinction coefficient measured and calculated within a factor of two can be regarded as the maximum achieveable result for average values. Taking into account the various influencing parameters such as mean size and broadness of the used particle distribution, uncertainties of the assumed particle bulk density and complex index of refraction, the margin of the variations cannot be further confined. Nevertheless, the results obtained using the White Cell technique are sufficient to describe the optical behaviour of the ambient aerosol in reliable limits.

5. REFERENCES

Chin-I Lin, M. Baker and R.J. Charlson, 1973: Absorption Coefficient of
    Atmospheric Aerosol - A Method for Measurement. Appl. Optics, Vol. 12,
    No. 6, 1356-1363
Metzig, G. and G. Poß, 1988: Optical and Physico-Chemical Properties
    of Tropospheric Aerosols in the Lee of the City of Karlsruhe. Ed.:
    K.Grefen and J. Löbel, Environmental Meteorology, Kluwer Academic
    Publisher, 23-37
Poß, G. and G. Metzig, 1987: Extinction Coefficient Measurements of
    Ambient Aerosol Particles Using a White-Cell. J. Aerosol Sci., Vol.
    18, No. 6, 895-898
Riedel, W.J., M. Knothe, W. Kohn and R. Grisar, 1988: An Anastigmatic
    White Cell for IR Diode Laser Spectroscopy. Fourth International
    Conference on Infrared Physics, Zürich, August 22 - 26, 504-505
Winkler, P., 1980: Observations on Acidity in Continental and in
    Marine Atmospheric Aerosols and in Precipitation. J. of Geophysical
    Research, 85, 4481-4486

GENERATION OF STANDARD ATMOSPHERES OF NITROUS ACID

I. Allegrini, M. Cortiello, A. Febo, C. Perrino

C.N.R. - Istituto Inquinamento Atmosferico
Via Salaria Km 29,300 - C.P.10
00016 Monterotondo Stazione (Roma), Italy

Summary
A method for generating standard atmospheres of nitrous acid in a wide
range of concentrations is described. The system is based on the
reaction of hydrogen chloride, produced by means of a low pressure
permeation device, and a sodium nitrite fluidized bed. The amount of
HONO generated in the unit time is independent of the flow rate and
equimolecular to the mass flow rate of HCl entering the system.
The analysis of the gaseous mixture has been performed by means of both
a Na_2CO_3-glycerol diffusion denuder and an instrumental method which
makes use of a chemiluminescence detector and allows the continuous
monitoring of NO, NO_2 and HONO concentrations.
An appropriate choice of the flow rate conditions of the system allows
to mantain NO_x impurities below the detection limit of the detection
technique (0.06 nmol/min). No trace of nitric acid has been observed.
The system is easy to built and shows a very good reproducibility and
stability (2% standard deviation over a period of about three months).

1. INTRODUCTION
The importance of nitrous acid as a key precursor to photochemical air
pollution and its role in the chemistry of nitrogen-air systems has been
widely recognized in the last few years (1-3). Besides, the occurring of
high concentrations of this compound in indoor environments has very
recently been shown, suggesting important public health implications (4,5).
For these reasons, the availability of a calibrated source of nitrous acid
constitutes a primary target. In addition, the possibility of obtaining
known concentrations of pure nitrous acid would allow to evaluate the
interference of this species in the measurement of other nitrogen
compounds, i.e. nitric acid and nitrate (6,7).
    Before 1986, only two methods for the generation of nitrous acid had
been reported in the literature: the first one involves the reaction of NO,
NO_2 and water vapour (8); the other one relies on the reaction of nitrites
with sulphuric acid (9). Both these methods lead to the production of a
gaseous mixture containing large amounts of nitrogen oxides.
    In 1986 Braman and de la Cantera reported a method (10), based on the
sublimation of oxalic acid into sodium nitrite, which allows to produce a
low, reasonably constant vapour concentration of nitrous acid in air,
containing low amounts of NO, NO_2 and HNO_3 (about 10% in the best
conditions). Although this method represent the first important step in
obtaining a nitrous acid calibration source, it presents some limitations,
concerning its poor reproducibility and moderate stability, the difficulty
of obtaining a wide range of concentrations and the need to operate in a

very limited range of relative humidity.

We describe here a new simple method to obtain constant amounts of very pure nitrous acid in a nitrogen stream, in a wide range of concentrations. The system is based on the constant production of hydrogen chloride, obtained by means of a low pressure permeation device; the hydrogen chloride flows into a glass trap containing sodium nitrite, producing a quasi-quantitative displacement of gaseous nitrous acid. Nitrous acid has been quantified by both a diffusion technique and an instrumental method.

2. EXPERIMENTAL

The nitrous acid generating system, comprised of a source of HCl, as described by Scarano et al. (11), a glass tube containing $NaNO_2$ and two different measuring systems, is depicted in figure 1.

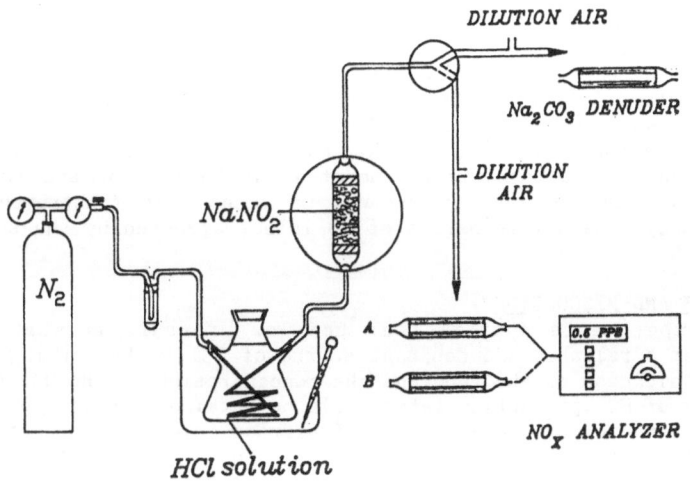

Figure 1: Schematic diagram of the HONO generation system. A is a NaCl coated denuder; B is a Na_2CO_3-glycerol denuder.

The hearth of the system for producing HCl, based on the permeation principle, is comprised of a gas-tight glass flask filled with a concentrated (9M or 12M) aqueous solution of HCl and containing a Teflon tube about 1.5 m in length having an outside diameter of 2 mm (wall thickness: 0.2 mm). The flask is placed into a thermostatic bath. An humidified inert carrier gas (nitrogen) is allowed to flow into the Teflon tube. Due to the difference in the partial pressure of HCl between the inside and outside of the tube, a constant amount of HCl in N_2, which only depends on the temperature and flow-dynamic conditions of the system, is produced in the time unit at the outlet of the Teflon tube.

HCl mass flow rates were measured by inserting a bubbler containing 20 ml of the Dionex eluent solution at the outlet of the Teflon tube; the analysis of the chloride content could be performed by ion chromatography, as the purposes of this work did not require high analytical precision.

A glass trap, partially filled with 0.8 g $NaNO_2$, is placed at the outlet of the Teflon tube in a vertical position, so as to avoid an excessive packing of the compound. Glass wool is inserted at both end of the trap. At the outlet of the system the flow is diluted up to 1.5 l/min with purified air and checked for its nitrous acid concentration by means of either an annular denuder or a chemiluminescence analyzer (Environment AC-30M), which makes use of a molibdenum converter operating at 350°C.

The annular denuder is coated with 1% Na_2CO_3 + 1% glycerol (w/w) in a water-methanol (1:1) solution, as described by Allegrini et al. (12). The collection efficiency of the device (80 mm in length, 10 and 13 mm inside and outside diameters, respectively) is better than 99% at the flow rate of 1.5 l/min. After sampling, the denuder is leached with 5 ml of the Dionex eluent and analyzed by IC for its chloride, nitrite and (possibly) nitrate content. Nitrous acid can be unequivocally identified with this technique, by studing the nitrite distribution along the denuder in function of the sampling flow rate (13).

The instrumental determination has been performed by placing a NaCl coated and a Na_2CO_3-glycerol coated denuders upstream of a chemiluminescence detector, as shown in figure 1. When line A is operating, nitric acid, if present, is selectively removed from the air stream by the NaCl coated denuder; when line B is operating, both nitric and nitrous acids are removed by the Na_2CO_3-gly coated denuder. Nitrous acid concentration is measured automatically on the NO_x channel of the analyzer, by subtracting the instrument response yielded when operating line A from the response obtained when operating line B. NO_2 concentration is measured by line B. The two lines are alternatively operated for 90 seconds and the data are collected during the last 30 seconds of each cycle (instrument response time: 60 sec). The determination of NO is not affected by the selection of line A or B.

3. RESULTS AND DISCUSSION

Our experiments confirm that the low pressure permeation device constitutes a reliable and constant source of HCl. By varying the length and wall thickness of the tube and the concentration of the HCl solution, a wide range of HCl production rates could be obtained, which were reflected in a wide range of production rates of nitrous acid. The production of HCl resulted to be fairly independent of flow rate variations, while changes in the temperature of the cell yielded variations in the HCl mass flow rate of approximately 10% per degree, as reported by Scarano et al. (11).

Preliminary tests showed that the humidification of the nitrogen flow was necessary to the production of nitrous acid. It was also found that placing the trap in a vertical position avoided excessive packing of the sodium nitrite, which resulted in poor efficiency in the generation of nitrous acid. The use of a hollow tube coated with sodium nitrite is unconvenient for the production of pure nitrous acid, since noticeable amounts of NO and NO_2, due to the dissociation of HONO, are observed in these conditions.

The composition of the effluent from the system is reported in Table I. The results show that no species other than nitrous acid were present in the nitrogen flow exiting the trap. Particularly, NO and NO_2 concentrations were below the detection limit of the chemiluminescence analyzer (0.06 nmol/min) and the analysis of the independent Na_2CO_3-gly denuder showed no trace of nitric acid. Nitrous acid production rate resulted to be quite constant; after three months, the standard deviation in the mass flow rate was still below 2%.

The results of Table I also indicate that there is a very good agreement between the data yielded by the two detection techniques and that the chemiluminescence analyzer exhibits an unitary response to nitrous acid, showing that the described system can be reliably used for HONO determinations.

Date 1989	NO	NO_2	HNO_2 AUTOM. ANAL.	DENUDER
*22-03	<0.06	<0.06	12.1 ± 0.3	12.4 ± 0.2
23-03	"	"	12.8	13.1
24-03	"	"	13.0	12.8
31-03	"	"	12.2	12.9
07-04	"	"	13.1	12.8
21-04	"	"	12.3	12.9
05-05	"	"	13.3	12.7
19-05	"	"	13.4	12.8
02-06	"	"	12.8	12.9
12-06	"	"	12.0	12.7
*22-06	"	"	12.9 ± 0.2	12.8 ± 0.3

* Average of 10 measurements.

Temperature of the HCl generation system: 30°C;   HCl concentration: 9 M;
HCl mass flow rate: 13.4 nmol/min;   temperature of the $NaNO_2$ trap: 25°C;
flow rate: 800 ml/min;   sampling time: 15 min. Concentrations are expressed in nmol/min.

The identification of nitrous acid was confirmed by studing the nitrite
deposition pattern on the Na_2CO_3+gly denuder. This analysis showed that a
species having the same diffusion coefficient as nitrous acid and an high
reactivity on sodium carbonate was present in the effluent stream.
    The effect of a variation in the temperature and flow rate is shown in
Figure 2. The results of Figure 2a show that the production of HONO follows
the increase with the temperature which is exhibited by the HCl production
rate and   that,   within the experimental errors, equimolecular amounts of
HCl and HONO are respectively removed and produced in the unit time. In
these conditions, no influence of the flow rate could be observed.

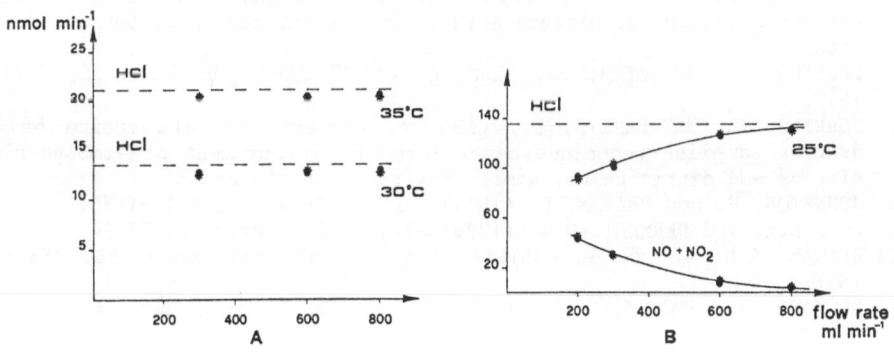

Figure 2: HONO production rate as a function of the flow rate.
          A: concentration of the HCl solution: 9 M;
          B: concentration of the HCl solution: 12 M.

When the HCl concentration is increased, by placing a 12M HCl acqueous
solution in the generating system, noticeable and equimolecular amounts of
NO and NO_2 are observed. Fig. 2b shows that the mass flow rate of
interferent NO_x strongly decreases when the flow rate increases and that
in any case the molar amount of HCl equals the molar amount of the sum of
the three nitrogen compounds. These findings suggest that NO and NO_2 are
yielded by the dissociation of HONO, in agreement with the following
reaction scheme:

$$HCl + NaNO_2 \longrightarrow HNO_2$$
$$2HNO_2 \longrightarrow NO + NO_2 + H_2O$$

The strong decrease of NO_x with the increase in the flow rate at high
HONO concentration and the absence of NO_x at low HONO concentration indicate that the dissociation of nitrous acid is strongly dependent on its
concentration in the system. Thus, even at very high values of HONO mass
flow rate, the composition of the gaseous mixture can be optimized by
increasing the flow rate. For practical use, this system is able to supply
concentrations up to 2.0 ppm of nitrous acid without yielding appreciable
amounts of NO_x.

4. CONCLUSION
A nitrous acid flow generation system, based on the reaction of HCl on
a sodium nitrite fluidized bed, has been described.   The system allows to
produce high-purity nitrous acid in a wide range of concentrations (up to
2.0 ppm) and exhibits a very good stability and reproducibility (standard
deviation below 2% over a three months period).   The method constitutes a
valuable aid for studing air chemistry and for testing sampling and
analytical techniques.

5. REFERENCES

(1) PITTS, J.N. Jr. and FINLAYSON-PITTS, B.J. (1977) Adv. Envir. Sci.
     Technol. 7, 75-162.
(2) PITTS, J.N. Jr. (1983) Envir. Hlth Perspec. 47, 115-140.
(3) COX, R.A. (1984) J. Photochem. 25, 43-48.
(4) PITTS, J.N. Jr., WALLINGTON, T.J., BIERMANN, H.W. , WINER, A.M. (1985)
     Atmos. Environ. 19, 763-767.
(5) FEBO, A. and PERRINO, C. Prediction and experimental evidence of high
     air concentration of nitrous acid in indoor environments. Submitted to
     Nature.
(6) PERRINO, C., DE SANTIS F., FEBO, A. (1988) Atmos. Environ. 22, 1925-
     1930.
(7) PERRINO, C., DE SANTIS, F., FEBO, A. Criteria for the choice of a
     denuder sampling technique devoted to the measurement of atmospheric
     nitrous and nitric acids. Atmos. Environ., in the press.
(8) JOHNSTON, H. and GRAHAM, R. (1974) Can. J. Chem. 52, 1415-1423.
(9) COX, R.A. and DERMENT, R.G. (1976/1977) J. Photochem. 6, 23-24.
(10) BRAMAN, R.S. and de la CANTERA, M.A. (1986) Anal. Chem. 58, 1533-
     1537.
(11) SCARANO, E., CALCAGNO, C., CIGNOLI, L. (1979) Anal. Chem. Acta 110,
     95-106.
(12) ALLEGRINI, I., DE SANTIS, F., DI PALO, V., FEBO, A., PERRINO, C.,
     POSSANZINI, M., LIBERTI, A. (1987) Sci. Total Envir. 67, 1-16.
(13) FEBO, A., DE SANTIS, F., PERRINO, C., GIUSTO, M. (1989) Atmos.
     Environ. 23, 1517-1530.

A NEW ANALYZER FOR SULFURIC ACID AND SULFUR AEROSOLS IN AMBIENT AIR: DEVELOPMENT AND MEASUREMENTS

R. Böhm, G. W. Israël
Fachgebiet Luftreinhaltung, Sekr. KF 2,
Technische Universität Berlin, D-1000 Berlin 12

Summary

A semi-continuous analyzer has been developed for simultaneous monitoring of ambient sulfuric acid and sulfur aerosols. Sulfur aerosol concentration is measured directly with a flame photometric detector. Sulfuric acid is first adsorbed and accumulated in a heated diffusion denuder. Subsequently it is released by increasing the temperature to 700°C and analyzed by the same detector. With a time resolution of 30 min. the detection limit of the analyzer amounts to 0.1 μg m^{-3} for sulfuric acid and 1.0 μg m^{-3} for sulfur aerosols (calculated as SO_4^{2-}). The automated and computer controlled analyzer was applied to a clean and highly polluted air region (Edelmannshof; Berlin West)) and a smog chamber. Sulfuric acid was detected only in winter, where it reached up to 2.7 μg m^{-3} at Edelmannshof and up to 0.74 μg m^{-3} in Berlin (West). Corresponding SO_2 concentrations were much lower at Edelmannshof. In summer the ratio of SO_4^{2-}/SO_2 was found to be 10 times higher at Edelmannshof than in Berlin (West). The photochemical experiments, which were carried out with α-pinene, SO_2, NO_x, O_3, H_2O and light in Teflon bags, showed good reproducibility. The relative humidity influenced greatly the composition of the formed sulfur aerosols. More than 90% of these aerosols were identified as SO_4^{2-} using ion chromatographic analysis. Up to 50% of the converted SO_2 were recovered as suspended aerosols in the bag.

1. Introduction

The only commercially available method for the determination of ambient H_2SO_4-levels (Ströhlein, 1986) has been reported by Niessner and Klockow (1980). Its time resolution is about one day, which is not sufficient to study short term episodes. Moreover, it requires laborious handling of a specially coated tube and subsequent analysis by ion chromatography for each measurement. In order to be able to estimate the potential acute harmful effects of ambient sulfuric acid and to conduct field studies of conversion products of SO_2 in the atmosphere, we developed a new analyzer that allows simultaneous monitoring of ambient sulfuric acid and sulfur aerosols with high sensitivity and time resolution.

2. Apparatus

Figure 1 shows a schematic diagram of the analyzer. Ambient air enters the instrument with a flow rate of 18 l min^{-1}. The inlet is designed to ensure quantitative collection of particles smaller than 10 μm, followed by a diffusor to extract a flow rate of 1.4 l min^{-1} isokinetically for analysis. By passing the gas through a diffusion scrubber all sulfurous gases are removed. Subsequently, the sample stream is routed either through a particle filter or through a H_2SO_4 diffusion denuder into the flame photometric detector (FPD) for sulfur content analysis. In the first case purified air reaches the detector to establish the baseline. In the other case H_2SO_4 is removed by the diffusion denuder and the detector registers sulfur aerosols only. The diffusion denuder for H_2SO_4 consists of a thin-walled uncoated stainless steel tube, which is heated directly by a welding transformer controlled by a Pt 100 temperature sensor. H_2SO_4 is vaporized at a mean gas temperature of 135°C and diffuses to the inner tube wall, where it is adsorbed and retained. Higher boiling sulfur aerosols pass the denuder unaffected and are measured continuously by the detector. After the H_2SO_4 has been accumulated for 15 minutes, the diffusion denuder is heated within 10 s up to 700°C, while being rinsed with purified air. This releases the H_2SO_4 quantitatively and carries it into the FPD for analysis. With a time resolution of 30 min. the detection limit of the analyzer amounts to 0.1 μg m^{-3} for sulfuric acid and 1.0 μg m^{-3} for sulfur aerosols (calculated as SO_4^{2-}). With an interface and a decoder the whole system is operated by a personal computer and works automatically. Measurement campaigns of 4 weeks are possible without any maintenance (Israël and Böhm, 1987).

3. Measurement campaigns at Edelmannshof and Berlin (West)

In 1988 during each season measurement campaings were carried out in a clean and highly polluted air region (Edelmannshof; Berlin (West)) to compare the concentration of sulfuric acid and sulfur aerosols with other atmospheric compounds like SO_2, NO_x, O_3, suspended particles and meteorological parameters. Excepted sulfuric acid all measured concentrations were normalized to the mean value and begin at midnight.

The Edelmannshof was located in a forest 500 m above sea level and 30 km northeast from the city of Stuttgart (Southern Germany). Supporting data were measured at the same location by the LfU of Baden Württemberg. **Table1** gives an survey to the results for all campaigns. It is obvious that sulfuric acid was detected only in autumn and winter, whereas it was below the detection limit of the analyzer of 0.1 μg m^{-3} in spring and summer. Due to the occurrence of H_2SO_4 only in the winter time this campaign will be discussed in more detail.

During the winter campaign (25.2. – 6.3.1988) temperatures varied between -5 and 4°C with wind speeds of about 1 m s^{-1} coming from the south west. In the first days it was often cloudy and snow fell, whereas later the sky was more sunny resulting in varying conditions. According to the meteorological situation a mean O_3 concentration of 48 μg m^{-3} (Max. = 101 μg m^{-3}) was oberserved indicating photochemical conversions in the atmosphere (**figure 2**). SO_2 concentrations averaged 13.2 μg m^{-3} (Max. = 29 μg m^{-3}), whereas concentrations of sulfuric acid ranged from <0.1 to 2.7 μg m^{-3} (mean value = 0.22 μg m^{-3}). The mean ratio of H_2SO_4/SO_2 was about 0.017 (Max. 0.080). A linear correlation between H_2SO_4 and SO_2 did not exist (r = 0.30; n = 526). This may be due to the presence of ammonia in the air neutralizing a part of the free acid. H_2SO_4 occurred very often in the form of short term peaks with a strong dynamic. As indicated by the wind direction of SW the measuring location was prevailingly influenced by polluted air from Stuttgart. Accordingly in the same air mass the course of sulfur dioxide and sulfate aerosols corresponded very well, but with a lower dynamic than the sulfuric acid. The mean ratio of SO_4^{2-}/SO_2 was about 0.50.

The measurement campaigns in Berlin (West) were carried out in down town near a major traffic artory. Supporting data were provided by the "Senator für Stadtentwicklung und Umweltschutz" of Berlin (West). **Table 2** summarizes the results of all campaigns. Like in the clean air region sulfuric acid was found to be detectable only in winter (short term peaks from <0.1 to 0.74 μg m^{-3}).

During the winter campaign (16.1. – 26.1.1988) the SO_2-concentration reached an average of 177.8 μg m^{-3} (Max. = 480 μg m^{-3}) far exceeding the SO_2-level in the clean air region (average = 13.2 μg m^{-3}). However, the mean concentration of sulfuric acid hardly differed (0.24 μg m^{-3} in Berlin (West); 0.22 μg m^{-3} at Edelmannshof). The mean ratio of H_2SO_4/SO_2 was about 0.003 (Max. = 0.010). A linear correlation between H_2SO_4 and SO_2 was found to be very weak (r = 0.50; n = 526). The mean ratio of SO_4^{2-}/SO_2 was 0.15, which is much lower than at Edelmannshof (0.50). 21 % of the suspended particles consisted as SO_4^{2-} (26,1 μg m^{-3}) verifying a result of Israël et al. (1983), that the contribution of the sulfate aerosols were about 19 % in winter 1981/82 in Berlin (West). Additionally it was confirmed by the high time resolution of the analyzer that sulfate aerosol concentration peaked especially at a wind direction coming from south east.

4. Smog chamber experiments

Photochemical experiments have been performed in a smog chamber of the Fraunhofer Institute for Toxicology and Aerosol Science. The experiments were carried out in 400-450 l Teflon film bags (Du Pont, FEP 200 A) inserted in a smog chamber described elsewhere (Nolting et al., 1988). Reactants were α-pinene, NO_2, SO_2, H_2O and light. As a result of the photochemical experiments **figure 3** shows a typical time profile of the irradiation of 100 ppb SO_2, 100 ppb NO_2 and 300 ppb α-pinene at a relative humidity (r.h.) of <2%. Gas concentrations and aerosol data are separated into two plots for better clearer presentation. Replicate measurements, indicated by the dashed lines, showed good reproducibility. After the beginning of irradiation (bolt) ozone rose up to 78 ppb from the reaction of α-pinene and NO_2 forming up to 10^6 particles per cm^3. Due to coagulation and condensation the particle diameter increased from a few nm up to 150 nm after 2 hours. The geometric standard deviation σ_g of a log normal distribution was about 1.2 representing nearly monodisperse aerosols. H_2SO_4 and SO_4^{2-}-concentrations increased at the same time and reached 9.4 and 3.7 μg m^{-3} respectively. According to the course of the inert dilution standard hexaflourben-zol all compounds were diluted to a third of the starting concentration (F.E. = area units of the GC-analysis).

A repetition of the above-mentioned experiment at a r.h. of 60% showed a good reproducibility as well. The important difference in the results lies in the fact that the relative humidity had a great effect on the formation of sulfuric acid (1.0 μg m^{-3}) and sulfate aerosols (24.6 μg m^{-3}). No change could be observed in the courses of α-pinene, NO_2 and ozone. In the dry experiment the ratio of H_2SO_4/SO_4^{2-} was about 2.5 indicating more sulfuric acid than sulfate. In the wet one most of the formed aerosols appeared as higher boiling sulfates resulting in a ratio of 0.04. At the same time the total sulfate concentration was about 60 % higher than in the dry experiment.

In the past parallel measurements of the analyzer and a filter sampling method with IC-analysis revealed that more than 98% of ambient sulfur aerosols existed in the form of SO_4^{2-} (Israël and Böhm, 1987). Therefore sulfur aerosol data of the analyzer are calculated and expressed as SO_4^{2-}. To check whether this assumption was valid for the aerosols formed in the smog chamber aerosol samples were also analyzed by IC at the end of each experiment. In comparison with the results of the analyzer more than 90% of these aerosols were identified as SO_4^{2-}.

Between 30 and 50% of the converted SO_2 were recovered as sulfate in the aerosol phase during the experiments. According to a study of McMurry and Grosjean (1985) gas-to-particle conversions in Teflon film smog chambers were strongly influenced by aerosol deposition due to diffusion and electrostatic forces. They reported that a large fraction (up to 83%, typical values 33 - 70%) of the formed aerosols was deposited on the smog chamber walls. Thus our recovery rates appear to be plausible.

5. Conclusions

The new analyzer has been applied succesfully for simultaneous monitoring of sulfuric acid and sulfur aerosols in ambient air and smog chamber studies. The measurement campaigns in the clean air region (Edelmannshof) and Berlin (West) in 1987/88 revealed that sulfuric acid appears very often in the form of short term peaks with a strong dynamic behavior. It occurred only in winter and varied between <0.1 and 2.7 μg m^{-3}. The smog chamber experiments with α-pinene, NO_2, SO_2, H_2O and light showed sulfuric acid formation only at dry conditions (r.h. <2%), whereas mainly other sulfur aerosols were detected under atmospheric conditions (r.h. = 60%).

Acknowledgements – This work was financially supported by the Projekt Europäisches Forschungszentrum (PEF) of the Kernforschungszentrum Karlsruhe (KfK).

References
Israël G.W., Heits B., Wengenroth K. and Bauer H.-W., **1983**: Staubbelastung und Staubeigenschaften während Smogsituationen, Bericht des Fachgebiets Luftrein-haltung, TU Berlin, 95-97
Israël G.W. and Böhm R., **1987**: Verbesserung und Prüfung der Anwendbarkeit eines Meßverfahrens zur quasikontinuierlichen Messung von Schwefelsäure und Gesamtsul-fat in Reinluftgebieten, Forschungsbericht KfK 30, Kernforschungszentrum Karlsruhe
McMurry P.H. and Grosjean D., **1985**: Gas and Aerosol Wall Losses in Teflon Film Chambers, Environ. Sci. Technol., 19, 1176-1182
Niessner R. and Klockow D., **1980**: A Thermoanalytical Approach to Speciation of Atmospheric Strong Acids, Intern. J. Environ. Anal. Chem., 8, 163-175
Nolting F., Behnke W. and Zetzsch C., **1988**: A smog chamber for studies of the reactions of terpenes and alkenes with ozone and OH, J. Atmos. Chem., 6, 47-59
Ströhlein Instruments, **1986**: Diffusionsabscheiderprobenahmesystem für die Erfassung starker Säuren im atmosphärischen Aerosol, Ströhlein Instruments GmbH&Co, D-4044 Kaarst 1

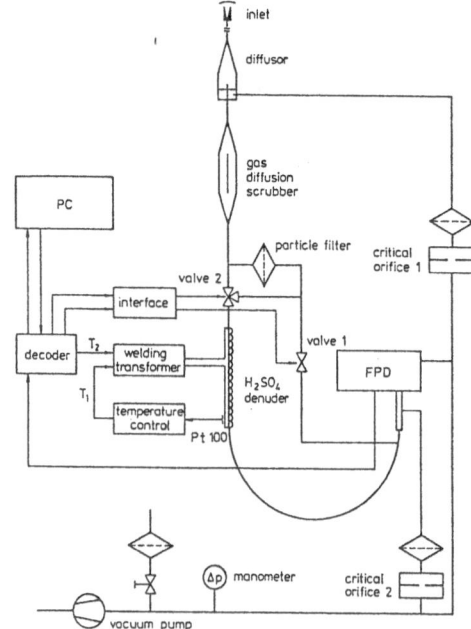

Figure 1:
Diagram of the analyzer

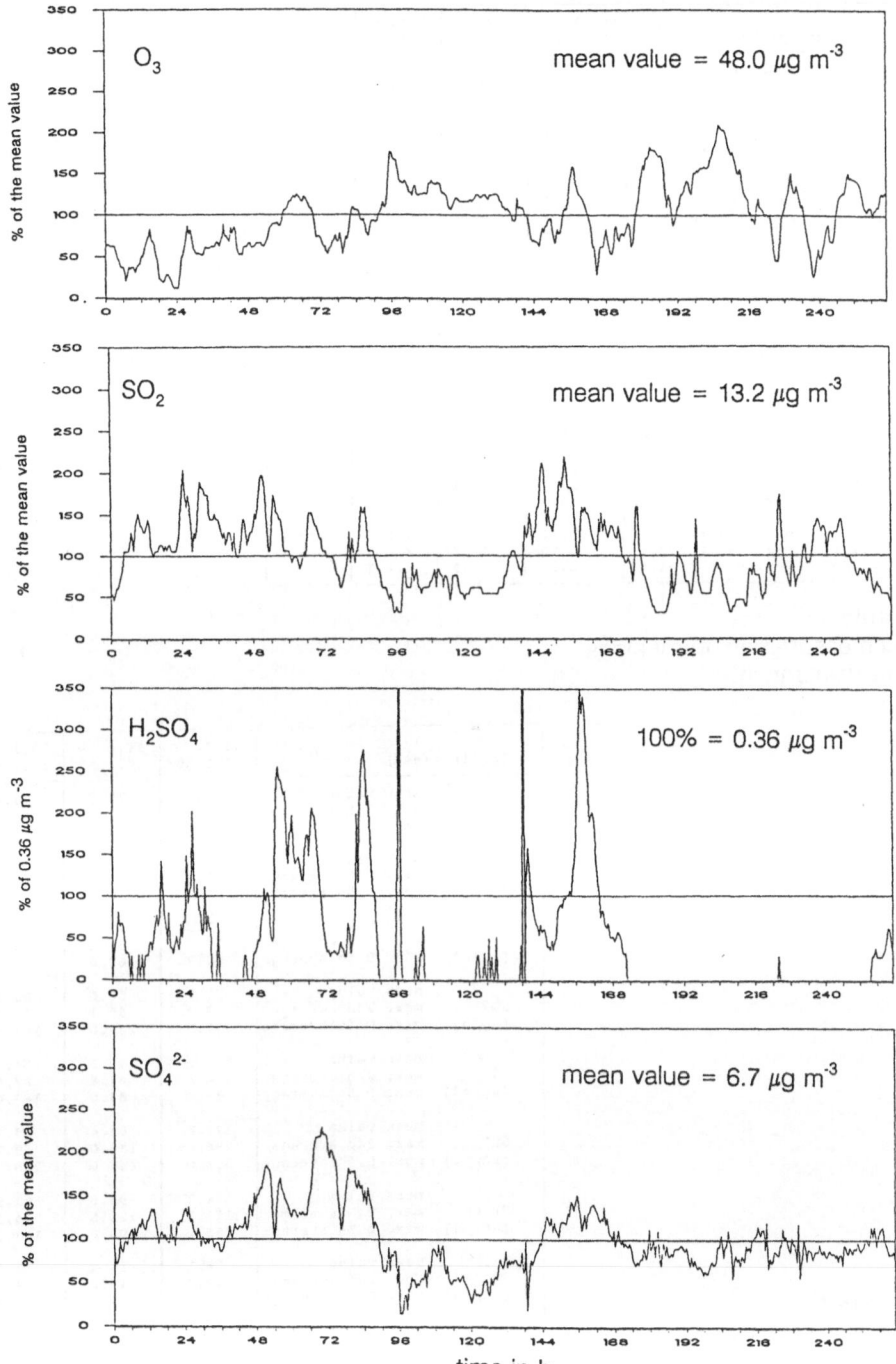

Figure 2:
Concentrations during the measurement campaign in winter 1988 at Edelmannshof

Edelmannshof		24.11. – 4.12.87	25.2. – 6.3.88	8.5. – 21.5.88	8.8. – 19.8.88
Temp. [°C]	mean value	1.4	-1.1	15.0	19.6
	max. 24h average	3.5	0.8	17.3	21.2
	max. 0.5h average	5.2	4.1	22.0	24.2
Wind-V [m/s]	mean value	2.3	1.2	–	–
	max. 24h average	3.1	2.1	–	–
	max. 0.5h average	4.4	3.6	–	–
NO [µg/m3]	mean value	2.0	1.1	2.4	–
	max. 24h average	14.7	3.9	3.9	–
	max. 0.5h average	24.0	10.0	12.0	–
NO2 [µg/m3]	mean value	22.0	22.0	6.2	9.4
	max. 24h average	36.2	34.6	13.7	19.1
	max. 0.5h average	49.0	69.0	66.0	40.0
O3 [µg/m3]	mean value	10.3	48.0	56.6	117.2
	max. 24h average	54.4	75.0	87.0	184.0
	max. 0.5h average	70.0	101.0	109.0	223.0
SO2 [µg/m3]	mean value	32.8	13.2	5.9	3.7
	max. 24h average	90.4	20.3	10.2	6.4
	max. 0.5h average	116.0	29.0	16.0	25.0
SO4-- [µg/m3]	mean value	11.6	6.7	9.0	9.3
	max. 24h average	19.3	10.9	10.6	14.0
	max. 0.5h average	22.7	15.5	17.0	21.1
H2SO4 [µg/m3]	mean value	≤0.12	≤0.22	< 0.1	< 0.1
	max. 24h average	0.20	0.50	< 0.1	< 0.1
	max. 0.5h average	0.38	2.73	< 0.1	< 0.1

Table 1:
All measurement campaigns
at Edelmannshof

Berlin (West)		16.1. – 26.1.88	14.4. – 19.4.88	12.7. – 18.7.88
Temp. [°C]	mean value	2.3	12.5	17.6
	max. 24h average	4.9	15.5	20.7
	max. 0.5h average	9.6	22.4	24.2
Wind-V [m/s]	mean value	2.6	2.8	2.5
	max. 24h average	3.4	3.8	3.3
	max. 0.5h average	5.7	5.9	5.2
NO [µg/m3]	mean value	7.5	17.7	15.6
	max. 24h average	17.7	65.1	19.1
	max. 0.5h average	53.2	169.0	85.0
NO2 [µg/m3]	mean value	5.3	54.6	35.2
	max. 24h average	7.2	65.1	45.4
	max. 0.5h average	11.3	124.0	102.0
O3 [µg/m3]	mean value	4.5	42.2	54.1
	max. 24h average	6.4	47.8	82.0
	max. 0.5h average	12.0	83.0	143.0
SO2 [µg/m3]	mean value	177.8	81.4	43.7
	max. 24h average	296.2	136.0	59.7
	max. 0.5h average	558.0	443.0	250.0
Staub [µg/m3]	mean value	124.7	98.6	53.7
	max. 24h average	195.5	125.5	74.3
	max. 0.5h average	325.0	304.0	183.0
SO4-- [µg/m3]	mean value	26.1	15.6	11.5
	max. 24h average	42.4	18.3	14.1
	max. 0.5h average	75.3	45.4	44.2
H2SO4 [µg/m3]	mean value	≤0.24	≤0.11	< 0.10
	max. 24h average	0.42	0.12	< 0.10
	max. 0.5h average	0.74	0.24	0.13

Table 2:
All measurement campaigns
in Berlin (West)

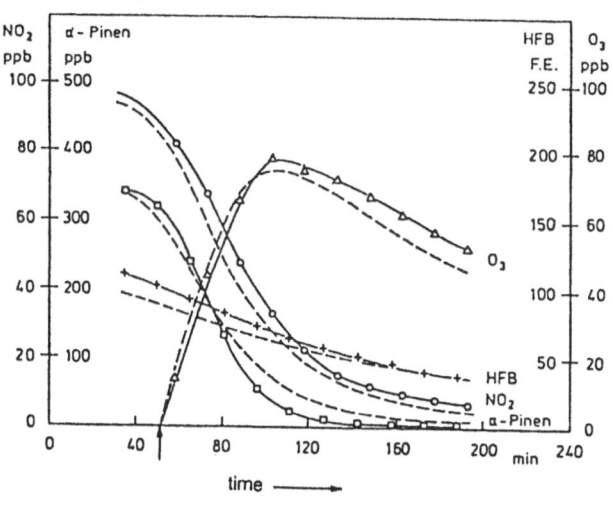

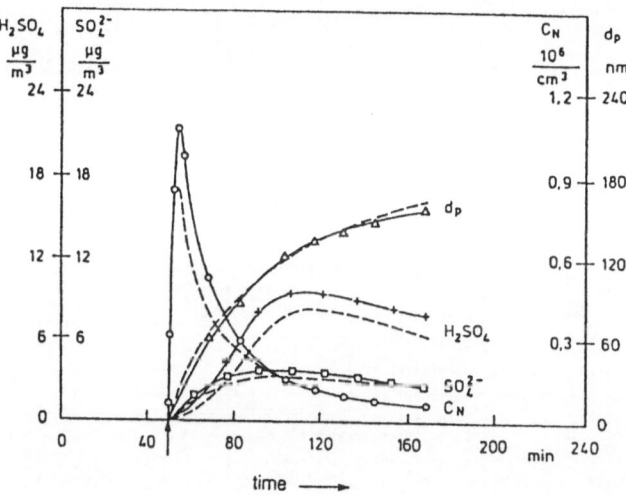

Figure 3:
Photochemical experiment with 100 ppb SO$_2$, 100 ppb NO$_x$, 300 ppb α-pinene, r.h. < 2%

SESSION II/A

ATMOSPHERIC CHEMICAL

AND

PHOTOCHEMICAL PROCESSES

R.A.COX
Harwell Laboratory
Didcot, UK

The scientific work related to Working Party 2 was presented at the Symposium in three separate sessions. The first two sessions were devoted to reports of work in the two EUROTRAC subprojects, HALIPP (Heterogeneous and Liquid Phase Processes) and LACTOZ (Laboratory Chemistry related to Tropospheric Ozone), which are currently co-ordinated within the framework of COST611. The third session covered other activities within the Working Party including the CEC programme OCEANO-NO$_X$, concerned with atmospheric chemistry in the coastal boundary layer, and also for the first time, laboratory stratospheric chemistry which is now incorporated in Working Party 2. A total of 17 papers were presented in the two EUROTRAC sessions and a further 8 papers in the third session, of which three were devoted to stratospheric ozone. In addition 13 poster presentations on laboratory studies were displayed, providing overall a unique and exciting forum for exchanging research results in this discipline.

HALIPP Workshop

The heterogeneous and liquid phase studies fall into two areas: a) surface and phase change phenomena and b) liquid phase chemical reaction kinetics and mechanism.

The presentations at the Symposium relating to the first area included laboratory experimental evidence for increased radical production in a smog chamber in the presence of aerosols. Specifically photochemical decomposition and oxidation of NaCl aerosol is proposed, leading to the production of Cl_2 and atomic chlorine (Behnke and Zetsch). Other studies demonstrated photochemical oxidation process of aromatics on solid aerosols.

Excellent progress has been made in providing data for a realistic picture of the detailed cloudwater processes leading to scavenging and oxidation of sulphur dioxide and nitrogen compounds. New measurements of the trace gas exchange kinetics at the air/water interface provide values for the mass accommodation coefficients of nitrogen compounds: HNO_3, N_2O_5, $CH_3COO_2NO_2$ (PAN) and HONO. The accommodation coefficient values were all >5x10^{-3} and consequently uptake into atmospheric cloud water particles is likely to be diffusion controlled. These measurements were made using a novel experimental/modelling approach developed at the University of Bonn by Dr. U. Schurath and his co-workers.

The kinetics and mechanism of the aqueous phase oxidation of bisulphite ions, HSO_3^-, has been investigated by the elementary reaction approach. Both the uncatalysed and FE^{2+} catalysed systems have been studied in the programme by two groups from the UK (Salmon et al, University of Leeds and McElroy and Waygood, CERL Leatherhead), the group from France (Langrange and Lagrange) and Dr. P. Warneck of the Max Planck Institute for Chemistry in Mainz. The rate coefficients for the sequential oxidation reactions of the radical ions SO_3^-, SO_4^- and SO_5^- have been determined by kinetic spectroscopy and various mechanistic approaches have been made.

The main conclusion from this work is that the rates of liquid phase oxidation reactions are likely to cover a wide range due to the critical dependence of the rate on oxidising agents such as H_2O_2 or catalysts such as FE^{2+}, the atmospheric abundance of which is very variable. There are particular problems remaining in the rate and mechanism of the H_2O_2 mediated reaction and, concerning the FE^{2+} oxidation, the study of McElroy et al has shown that current cloud chemistry models are based on an incorrect mechanism which overpredicts the rate of Fe(II) oxidation. In addition, quantum yield measurements on Fe(II) - hydroxy complexes show that photolysis in clouds probably contributes significantly to OH production in the aqueous phase.

LACTOZ Workshop

The primary aim of the EUROTRAC sub-project LACTOZ is to provide the necessary chemical and photochemical data for a proper description of the formation of tropospheric ozone. This involves the chemistry which controls the degradation of volatile organics and the life cycles of nitrogen oxides in the troposphere. Laboratory results on a number of different aspects of this problem were presented at the Symposium.

One of the most important species in the nitrogen oxides chemistry is peroxyacetyl nitrate (PAN) and new information on both the formation and loss of this compound was presented. The thermal decomposition of PAN was investigated by E. Ljungstrom and co-workers (Goteborg) with a view to establishing the occurence of a second channel leading to methyl nitrate formation, which has been suggested as an alternative atmospheric loss process for PAN. It seems that CH_3ONO_2 formation is a very minor and probably heterogeneous pathway. The atmospheric behaviour of PAN is best described by the well established, homogeneous reaction forming acetylperoxy radicals which leads to the pseudo-equilibrium:

$$CH_3COO_2NO_2(+M) <-> CH_3COO_2 + NO_2$$

New measurements of the reverse reaction of CH_3COO_2 with NO_2 by Bridier et al. indicate that the rate coefficient for atmospheric conditions is higher by about 50% than the value currently used in models. More results will be obtained so that a full evaluation of the kinetics and equilibrium constant for this very important reaction can be established with high accuracy.

Other reactions involving nitrogen-containing molecules have also been investigated and results were reported which enable the lifetimes of the organic nitrates and the scavenging of organics by reaction with NO_3 to be quantitatively described. An interesting mechanistic outcome of these studies is that the conversion of NO_x to HNO_3 via the H-abstraction reaction of NO_3 with organics is rather slow. The more rapid reactions of NO_3 with unsaturated molecules, e.g. olefins, gives rise to formation of organic nitrates, which are removed rather slowly from the atmosphere. Data was also presented on the reaction of NO_3 with other radicals, which has more interest for stratospheric chemistry, (see below).

Good progress has also been reported in the understanding of peroxy radical chemistry. In particular a new technique for laboratory detection and measurement of the CH_3O_2, involving laser photofragment emission spectroscopy has been reported from the University of Gottingen (Hartmann et. al.). This enabled a measurement of the important kinetic parameters for the reaction of CH_3O_2 with HO_2:

$$CH_3O_2 + HO_2 -> CH_3OOH + O_2$$

$$\rightarrow HCHO + H_2O + O_2$$

These results agree fairly well with those obtained using absorption spectroscopy for measurement of the radical concentrations: a matrix isolation product study of the CH_3O_2 self reaction, conducted by J. Crowley and co-workers (Max Planck Institute of Chemistry, Mainz) has provided more information on the branching ratios for the various reactions of the CH_3O_2 radical that occur in methyl radical oxidation. The data base required for modelling the oxidation of methane in the troposphere is now relatively well defined, the major outstanding uncertainty being in the branching ratio for the two channels in the $CH_3O_2 + HO_2$ reaction, and its temperature dependence.

Kinetic data has also been reported for the more complex peroxy radicals formed in the atmospheric oxidation of volatile organics. Jenkin and Cox presented a poster describing studies of the $HOCH_2CH_2O_2$ radical, which is formed in the atmospheric oxidation of ethylene and a collaborative study between Patras University and the University of Bonn has provided information on the reaction of accetylperoxy with sulphur compounds. The data base necessary to establish the effect of difference organic structures on the reactivity of peroxy radicals is starting to be consolidated.

Mechanistic and kinetic information on the formation and reaction of the adduct formed in the reaction of OH with benzene and toluene was reported by two groups (Dr. Devolder, University of Lille and Dr. Zetsch, Fraunhofer Institute). Both studies show that a very fast reaction occurs between the adduct and NO_2, but the rate of reaction of the adduct with O_2 is slow and the NO_2 reaction (which presumably forms nitroarenes) may compete under atmospheric conditions. However it is clear that other pathways exist in the oxidation mechanism of the simple aromatics, since a range of oxygenated products are found under pseudo atmospheric conditions. Sorting out these mechanisms continues to present a major challenge to laboratory chemists.

A comprehensive set of rate data, including temperature dependences for reactions of ozone with simple alkenes was presented by Dr. Tracy (Dublin). This is the first data on these important "non-photochemical" reactions within LACTOZ.

Kinetics studies related to the Marine Boundary Layer Chemistry - OCEANOX PROJECT.

Several papers were presented concerning the chemistry of iodine and sulphur compounds relevant to the marine boundary layer. Data for the reaction of IO radicals with dimethyl sulphide (DMS) and other sulphur compounds was presented. New work in the US at Georgia Institute of Technology was referred to, which indicated that the reaction of DMS with IO was very much slower than indicated by the earlier experiments at Wuppertal and Orleans. This discrepancy needs to be resolved. The data for kinetics of the reactions of I and IO with NO_x and HO_x species is now starting to consolidate and it should be possible to describe the Iodine radical chemistry in the marine boundary layer with much greater certainty by the end of this project.

Tropospheric halogen Chemistry

There is renewed interest in the tropospheric chemistry of halogenated organic compounds at the present time, in view of the foreseen introduction of replacement compounds for the fully halogenated CFCs. The potential replacements contain H atoms as well as fluorine and chlorine (in some cases) in the molecule. These molecules undergo attack by OH in the troposphere which results in a reduced atmospheric lifetime and consequently a lower potential for depletion of stratospheric ozone than the fully halogenated CFCs. In addition to the need to establish the rate constants for reaction with OH for these compounds there is a requirement to establish the degradation mechanisms and products, so that the full environmental impact of these molecules can be assessed.

Papers were presented at the Symposium on the atmospheric behaviour of the chloroacetaldehydes, which are model compounds for the expected products of the hydrochlorofluorocarbon degradation. Very little information has been available on this topic and interesting and novel data were reported.

Stratospheric Chemistry

Chemistry relating to stratospheric ozone is included under the new terms of reference of COST611 Working Party 2. There is particular interest at the present time in the chemistry related to polar ozone depletion in the lower stratosphere. This involves novel chemistry in which the chlorine reservoir molecules, e.g. HCl, $ClONO_2$, react on the surface of the polar stratospheric clouds formed during wintertime in the arctic and antarctic regions, to form photochemically labile chlorine reservoirs such as Cl_2, Cl_2O or $ClONO$. which can readily form Cl and ClO by photolysis in visible light. The radicals produced thus are then involved in several catalytic cycles which destroy ozone: e.g.

$$ClO + ClO + M \rightarrow Cl_2O_2 + M$$

$$Cl_2O_2 + h\nu \rightarrow ClOO + Cl$$

$$ClOO + M \rightarrow Cl + O_2 + M$$

$$2(Cl + O_3 \rightarrow Cl + O_2)$$

$$net: \quad 2O_3 \rightarrow 3O_2$$

Bromine oxide radicals are also involved in this chemistry through their coupling with ClO reactions:

$$BrO + ClO \rightarrow Br + Cl + O_2$$

$$Br + O_3 \rightarrow BrO + O_2$$

$$Cl + O_3 \rightarrow ClO + O_2$$

$$net: \quad 2O_3 \rightarrow 3O_2$$

The reaction of BrO with ClO has two other channels, one producing OClO, which does not lead to ozone depletion because of the rapid photolysis of OClO to produce O (and hence O_3), and one producing BrCl which acts as a temporary reservoir species for bromine and chlorine atoms during darkness (BrCl is also rapidly photolysed).

At the conference papers were presented on some of the latest experimental data pertaining to the BrO + ClO reaction (G. Poulet et. al.) and the formation and loss of the chlorine oxide dimer, Cl_2O_2 (Hayman and

Cox). These allow ozone depletion by these catalytic cycles to be modelled and a comparison to be made with observed ozone depletion in the polar springtime (the "Antarctic Ozone Hole").

Conclusions

Overall the standard of laboratory work reported at the Symposium was very high. The presentations and the discussions were lively and informative. There is no doubt that the work carried out under Working Party 2 is providing a valuable insight into the budgets and pathways of pollutant molecules in the atmosphere and forms a basis for a quantitative understanding of the atmospheric processes controlling the impact of pollutants current and future in the environment.

LABORATORY KINETIC STUDIES OF REACTIONS OF IODINE
SPECIES OCCURRING IN MARINE ATMOSPHERE

J.L. JOURDAIN, G. LAVERDET, G. LE BRAS, H. MAC LEOD,
A. MELLOUKI and G. POULET
CRCCHT/CNRS 45071 ORLEANS-Cedex 2 France

Summary

The reactions I + HO$_2$ ---> HI + O$_2$ (1) and OH + HI ---> H$_2$O + I (2) have been studied at 298 K using the Discharge Flow- EPR method. Reaction (2) was also studied by the Laser Photolysis-Resonance Fluorescence technique. The rate constants were found to be k_1 = (3.1 ± 1.2) x 10^{-13} and k_2 = (3.0 ± 0.5) x 10^{-11} (units are cm^3 molecule^{-1}s^{-1}). These data allow for assessing the importance of HI as a reservoir of active iodine in atmospheres such as the marine boundary layer.

1.INTRODUCTION

The potential importance of iodine tropospheric chemistry has been discussed these recent years (1,2). In the marine boundary layer CH$_3$I is known to be emitted (3) and to produce I atoms from photolysis. I atoms are rapidly converted into IO radicals by reaction with O$_3$. The IO radicals can be converted back to I atoms by photodissociation and by reaction with NO and also with DMS (dimethylsulfide) which is emitted in large amounts in marine coastal atmospheres (4). The reaction of IO with DMS may also lead to removal of O$_3$ (5,6) :

$$IO + DMS ---> I + DMSO$$
$$I + O_3 ---> IO + O_2$$
$$\text{-----------------------}$$
$$DMS + O_3 ---> DMSO + O_2$$

The rate of such a process will depend on the concentrations of the active species I and IO in the atmosphere. These concentrations are in turn dependent on the removal rates of these species. IO radicals are considered to be removed by reaction with itself, NO$_2$ and HO$_2$, whereas I atoms would be mainly removed by reaction with HO$_2$:

$$I + HO_2 ---> HI + O_2 \qquad (1)$$

The HI reservoir formed can regenerate I atoms through the reaction:

$$OH + HI ---> H_2O + I \qquad (2)$$

These two reactions have been studied at 298 K using the discharge flow-EPR method for reactions 1 and 2 and also the laser photolysis-resonance fluorescence method for reaction 2.

2.REACTION I + HO2

2.1- Experimental :

The experimental arrangement is schematically represented in Figure 1. The reactor was coated with halocarbon wax to reduce wall reactions. Iodine atoms and HO_2 radicals were generated separately in the reactor which crossed the EPR cavity. The central tube could be moved along the reactor axis. Iodine atoms were analysed directly by EPR, whereas HO_2 radicals were monitored indirectly by converting HO_2 to OH by the rapid reaction $NO + HO_2 \longrightarrow NO_2 + OH$. Excess NO was added 10 cm upstream of the EPR cavity.

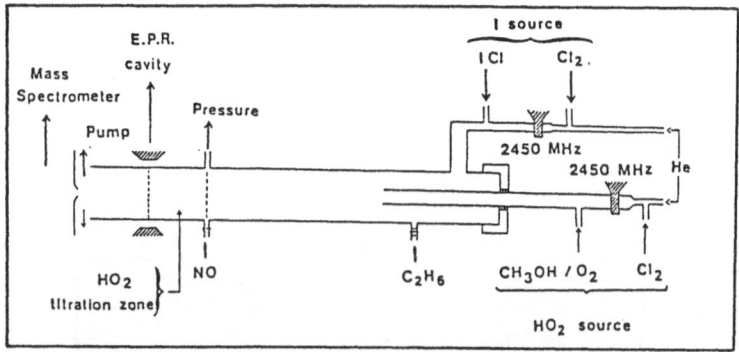

Fig 1 : Scheme of the discharge flow apparatus

Iodine atoms were produced by the fast reaction : $Cl + ICl \longrightarrow I + Cl_2$, with excess of ICl or Cl atoms. With excess of ICl, the reaction $OH + ICl$ occurring in the HO_2/OH conversion zone, was taken into account by computer simulation. Its rate constant has been determined for the purpose of the present study: $k = (1.1 \pm 0.1) \times 10^{-11}$ cm^3 $molecule^{-1}s^{-1}$ at 298 K (7). With excess of Cl, this excess was trapped by adding C_2H_6, in order to prevent the fast reaction of Cl with HO_2.

HO_2 radicals were produced by the sequence :

$$Cl + CH_3OH \longrightarrow CH_2OH (+ HCl) \cdots O_2 \longrightarrow HO_2 (+ HCHO).$$

The linear velocities of gases in the reactor were in the range 15 - 23 m s^{-1} and total pressures in the range 1.0 -1.5 Torr.

2.2- Results and discussion :

The reaction of iodine atoms with HO_2 radicals was studied under pseudo-first order conditions using as large as possible excess of iodine atoms over HO_2. Consequently the rate constant k_1 was derived from the expression : $-d\ln [HO_2] /dt = k_1 [I]$. The absolute concentrations of iodine atoms were measured before and after each experiment.

The indirect monitoring of HO_2 by OH was shown to be valid from computer simulations of the secondary chemistry occurring in the $[HO_2] / [OH]$ conversion zone. The $[HO_2]/[OH]$ ratio was found to be effectively independent of the HO_2 concentration at the NO introduction position, for the different concentrations used for the other species which reacted with OH (i.e. CH_3OH, ICl or C_2H_6).

The following ranges of initial concentrations of the reactants were used: $[I]_o = (0.53 - 2.22) \times 10^{14}$ cm^{-3} and $[HO_2]_o = (1 - 3) \times 10^{11}$ cm^{-3}. The $[I]_o$ range is rather narrow, but it could not be extended more : at lower $[I]_o$, the reaction rates became too low to be measured precisely, whereas at higher $[I]_o$, ICl (excess of ICl over Cl) or C_2H_6 (excess of Cl over ICl) reacted too quickly with OH in the conversion zone, due to the large concentrations required.

The pseudo-first order data obtained are reported in Table 1. The rate constant derived from 14 experiments at 298 K is :

$$k_1 = (3.1 \pm 1.2) \times 10^{13} \text{ cm}^3 \text{ molecule}^{-1} \text{s}^{-1}$$

The error represents one standard deviation to which an error of 10% has been added to account for systematic errors. The error due to the wall recombination of both iodine atoms and HO$_2$ radicals was found to be negligible from separate measurements of their recombination rates.

$(I)_o$ $(10^{14} \text{ cm}^{-3})$	k_1 $(10^{-13} \text{ cm}^3 \text{ s}^{-1})$
0.53	3.6
0.69	3.9
0.88	4.3
0.94	2.9
1.09	4.8
1.11	1.9
1.17	2.0
1.27	2.0
1.29	4.6
1.62	1.5
1.88	1.8
1.92	2.5
1.99	4.9
2.22	2.3
Mean Value	$k_1 = (3.1 \pm 1.2) \times 10^{-13}$

Table 1

The present determination of k_1 is in good agreement with the value recently obtained by Jenkin and Cox (8) : $k_1 = 4.5 \times 10^{-13}$ at 298 K. These authors have also studied the temperature dependence of k_1 in the range 283 - 353 K : $k_1 = (1.76 \pm 0.75) \times 10^{-11}$ exp (- 1090 ($\pm$ 130)/T). The technique used was the modulated photolysis of $Cl_2/CH_3OH/I_2/O_2/N_2$ mixtures at 50 Torr.

With these new data, the rate constants are now available for the complete series of reactions of HO$_2$ with the halogen atoms (Table 2). It can be seen that the reactivity is decreasing from F to I and this decrease is correlated with the exothermicity for the channel X + HO$_2$ ---> HX + O$_2$. A factors are comparable and indicate a direct H atom transfer mechanism as recently discussed (9).

Reaction	ΔH kcal mol^{-1}	$k_{(298K)}$ 10^{-11} cm^3 molecule^{-1} s^{-1}	A	E/R
F + HO$_2$ → HF + O$_2$	-87	~ 5		
→ OH + FO	+13.3			
Cl + HO$_2$ → HCl + O$_2$	-54	3.2	1.8	(-170 ± 200)
→ OH + ClO	+1	0.91		
Br + HO$_2$ → HBr + O$_2$	-39	0.2	1.5	(600 ± 140)
→ OH + BrO	+10.8			
I + HO$_2$ → HI + O$_2$	-22.2	0.038	1.47	(1090 ± 130)
→ OH + IO	+21.9			

Table 2

3.REACTION OH + HI
3.1- Experimental :
This reaction has been studied using both the discharge flow- EPR and the laser photolysis-resonance fluoresence methods.

With the discharge flow method, the OH radicals were produced by the reaction H + NO$_2$ ---> OH + NO. The OH kinetics were monitored by EPR in the presence of high excess of HI which was flowed through the central tube. The experimental conditions were typically :

$$[OH]_o = (2 - 5) \times 10^{11} \text{ cm}^{-3}, \quad [HI]_o = (2.5 - 18.8) \times 10^{12} \text{ cm}^{-3}$$
$$v = 24 - 26 \text{ m/s and } P = 1 \text{ Torr.}$$

With the laser photolysis method, the OH radicals were produced by photolysing HNO$_3$ at 248 nm using an excimer laser. OH concentrations were monitored as a function of time by resonance fluorescence. The fluorescence was excited by a H$_2$O/Ar microwave discharge lamp and detected by a photomultiplier (Hammamatsu R 292) fitted with a narrow-band interference filter centered at 309 nm. The signal from the photomultiplier was amplified and fed into a multichannel averager (ATNE). OH fluorescence signals were accumulated and summed typically for 256 laser shots. Experiments were carried out under slow flow conditions in Argon (P = 27 Torr). The initial concentrations of reactants were :

$$[OH]_o \simeq 2 \times 10^{11} \text{ cm}^{-3} \text{ and } [HI]_o = (0.5 - 4.5) \times 10^{13} \text{ cm}^{-3}$$

In both types of experiments, special care was taken to eliminate I$_2$ which reacts very rapidly with OH.

3.2- Results and discussion :
The value of the rate constant was derived from the expression - d ln $[OH]$ / dt = k_2 $[HI]$ in both series of experiments. The pseudo-first order plots are represented in figures 2 and 3.

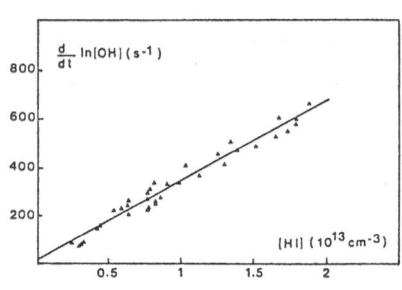

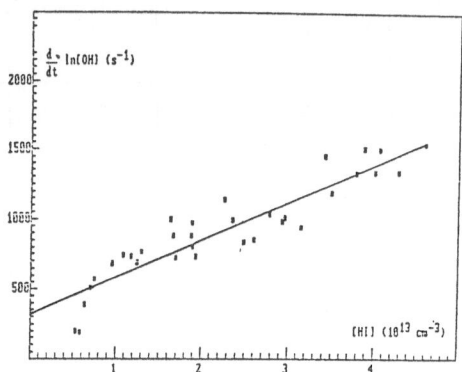

Fig 2 : Discharge flow
experiments

Fig 3 : Laser photolysis
experiments

The rate constants obtained at 298 K by the discharge flow and the laser photolysis methods are, respectively :

$$k_2 = (3.3 \pm 0.2) \times 10^{-11} \ cm^3 \ molecule^{-1}s^{-1} \ (DF)$$
$$k_2 = (2.7 \pm 0.2) \times 10^{-11} \ cm^3 \ molecule^{-1}s^{-1} \ (LP)$$

The low intercept of figure 2 corresponds to a low wall loss rate of OH, whereas the intercept of figure 3 is essentially the sum of the reaction rate of OH with HNO_3 and the OH diffusion rate. The DF and LP determinations of k_2 are in good agreement and the following mean value can be considered from these two experiments : $k_2 = (3.0 \pm 0.5) \times 10^{-11}$ $cm^3 \ molecule^{-1}s^{-1}$, where the error includes 10 % systematic error. This result is approx. 2.5 times higher than the previous determination of Takacs and Glass (10). They found $k_2 = (1.3 \pm 0.5) \times 10^{-11}$ at 298 K using the discharge flow-EPR method. Their determination might have been underestimated as a result of a poor time resolution in their experiments.

4. ATMOSPHERIC APPLICATIONS

Reactions 1 and 2 determine the potential role of HI as a reservoir of active iodine, in addition to the other reservoirs $IONO_2$, HOI and iodine oxides (2) formed by the reactions :

$$IO + NO_2 + M \ ---> \ IONO_2 + M$$
$$IO + HO_2 \ ---> \ HOI + O_2$$
$$IO + IO \ ---> \ I_2O_2 \ --->... \ ---> \ (IO)_x$$

A recent model calculation by Jenkin and Cox (8), based on the present value of k_2 and the mean value of k_1 obtained from the present work and from their own determination, has shown that HI would contribute to 5 - 10 % to the storage of active iodine under the typical conditions of the marine boundary layer. This conclusion is however not definitive, since the rate constants of the reactions of formation and removal of the reservoirs involving IO are uncertain and remain to be determined or precised.

REFERENCES

(1) Chameides, W.L. and Davis, D.D. (1980). J. Geophys. Res., 85, 7383.

(2) Jenkin, M.E., Cox, R.A. and Candeland, D.E. (1985). J. Atmos. Chem., 2, 359.

(3) Lovelock, J.E. , Maggs, R.J. and Wade, R.J. (1973). Nature, 241, 194.

(4) Lovelock, J.E., Maggs, R.J. and Rasmussen, R.A. (1972). Nature, 237, 452.

(5) Barnes, I., Becker, K.H., Carlier, P. and Mouvier, G. (1987). Int. J. Chem. Kinetics, 19, 489.

(6) Martin, D., Jourdain, J.L., Laverdet, G. and Le Bras, G. (1987). Int. J. Chem. Kinetics, 19, 503.

(7) Jourdain, J.L., Mellouki, A., Poulet, G., Laverdet, G. and Le Bras, G. (1988). Air Pollution Research Report 17 (CEC). COST 611/Lactoz - Halipp meeting, Norwich, 20-22 sept 1988. p. 21.

(8) Submitted for publication.

(9) Toohey, D.W. and Anderson, J.G. (1989). J. Phys. Chem., 93, 1049.

(10) Takacs, G.A. and Glass, G.P. (1973). J. Phys. Chem., 77, 1948.

REACTIONS OF IO RADICALS WITH SULPHUR CONTAINING COMPOUNDS

I. Barnes, V. Bastian and K.H. Becker

Physikalische Chemie / Fachbereich 9, Bergische Universität-GH Wuppertal

5600 Wuppertal 1, FRG

Summary

The reactions of IO radicals with the organic sulphur compounds CH_3SH, CH_3SCH_3, CH_3SOCH_3 and CH_3SSCH_3 and the inorganic sulphur compounds SO_2, CS_2 and H_2S have been investigated in a 420 l glass chamber at 1 bar total pressure in synthetic air by 298 K.

The reaction of IO with CH_3SCH_3 was found to lead solely to the formation of CH_3SOCH_3 which reacts further with IO forming $CH_3SO_2CH_3$. SO_2 and HCHO are products of the reactions of IO with CH_3SH and CH_3SSCH_3. However, for IO + CH_3SSCH_3 two other products are also present one of which has been positively identified as $(CH_3SO_2)_2O$ and the other tentively assigned to CH_3SO_2OI. Analogous to the reaction of OH with CS_2 the reaction of IO with CS_2 leads to the formation of SO_2 and COS. The products of IO +SO_2 have not yet been identified.

MSA was not found as a product of the reaction of IO with the organic sulphur compounds in contrast to their OH initiated oxidation in the presence of NO_x where MSA is a major product.

Preliminary computer simulations of the experimental results indicate that the rate constants for the reactions of IO with H_2S, CH_3SH and CH_3SSCH_3 are of the same order of magnitude as that for IO + CH_3SCH_3. IO reacts moderately fast with CH_3SOCH_3 and only very slowly with SO_2 and CS_2.

1. INTRODUCTION

It is estimated that between 97 and 216 Tg of gaseous sulphur is cycled annually through the atmosphere whereby approximately half of the sulphur is of biogenic origin /1,2/. Volatile sulphur compounds of biogenic origin include hydrogen sulphide (H_2S), carbonyl sulphide (COS), methylmercaptan (CH_3SH) and also dimethyl sulphide (CH_3SCH_3) which is the most abundant of these compounds and is emitted with a flux of approximately 40 ± 20 Tg $year^{-1}$ from the world oceans /3/.

Presently, the major degradation pathway for sulphur compounds in the atmosphere is considered to be reaction with OH radicals. Experimental studies have shown that the OH initiated oxidation of CH_3SH, CH_3SCH_3 and CH_3SSCH_3 in the presence of NO_x leads to the formation of SO_2 and substantial amounts of methanesulphonic acid (MSA) /4-6/. Recent field measurements show that MSA is an important constituent of marine aerosols /7,8/. Other oxidation products of CH_3SCH_3 detected in marine aerosols include dimethyl sulphoxide (CH_3SOCH_3) and dimethyl sulphone ($CH_3SO_2CH_3$).

Recent laboratory studies have shown that the reaction of IO radicals with CH_3SCH_3 is fast ($k=3\times10^{-11}$ cm^3 s^{-1}). This reaction leads to the formation of CH_3SOCH_3 as major product and may constitute an important sink for CH_3SCH_3 in the marine atmosphere /9-11/. To date, very little is known about the reactions of IO with other atmospheric trace gases. Our knowledge of marine chemistry does not rule out the possibility that halogen oxides, in general, may also contribute to the oxidation of marine sulphur compounds. In order to better assess the role of halogen oxides compared to OH in the chemistry of the marine atmosphere, studies have been performed on the reactions of IO with the organic sulphur compounds CH_3SH, CH_3SCH_3, CH_3SOCH_3 and CH_3SSCH_3 and the inorganic sulphur compounds SO_2, CS_2 and H_2S.

2. EXPERIMENTAL

The reactions of IO with the various sulphur compounds were investigated in a 420 l Duran glass chamber at 1 bar total pressure of synthetic air at 298 K. Experimental details have been given elsewhere /12/.

The IO radicals were produced by the photolysis of CH_2I_2 (~20 ppmv) in the presence of O_3 (~30 ppmv) using 2-6 Philips TLA 40W/05 fluoresence lamps (320 nm $< \lambda < $ 480 nm). Concentrations of the sulphur compounds were in the range 15-30 ppmv. Reactants and products were monitored using *in situ* long-path FTIR spectroscopy. Infrared absorption spectra of the reaction mixtures were recorded at 1 cm^{-1} resolution by co-adding 30 interferograms over a period of 1 min and collecting 20 such spectra over a period of 20 min.

3 RESULTS AND DISCUSSION

3.1 CH_2I_2/O_3/air photolysis system: the IO radical source

Since this is the first study employing the photolysis of CH_2I_2 in the presence of O_3 as an IO source and also because CH_2I_2 has been detected in the marine atmosphere /13/ some aspects of the reaction system will be discussed. The primary step in the photolysis of CH_2I_2 is the formation of I-atoms which under the present experimental conditions react rapidly with O_3 to form IO radicals. It is found that for every molecule of CH_2I_2 photolysed two molecules of O_3 are consumed. The observed products are CO, HCHO, HCOOH and $(HCO)_2O$. The results are consistent with the following simplified reaction mechanism,

$$CH_2I_2 + h\upsilon \longrightarrow CH_2I + I$$
$$I + O_3 \longrightarrow IO + O_2$$
$$CH_2I + O_2 \longrightarrow CH_2O_2 + I$$
$$CH_2O_2 \dashrightarrow products (CO, HCHO, HCOOH, (HCO)_2O)$$

In the above mechanism both carbon-iodine bonds are broken and explains the loss of two molecules of O_3 for every photolysed CH_2I_2 molecule. Support for the fission of both carbon-iodine bonds can be found in the studies of Schmitt and Comes /14/ and Gregory and Style /15/ on the photolysis of CH_2I_2.

In the IR product spectra of the CH_2I_2/O_3/air reaction mixtures two broad bands were observed at 795 and 605 cm^{-1} and are tentively assigned to an iodine oxide. The molecular formula of the iodine oxide is presently unclear, however, similar bands were also formed in the dark reaction between I_2 with O_3 which is known to lead to the formation of I_4O_9 /16/.

Since CH_2I_2 is known to be emitted to the atmosphere its reactions with OH and NO_3 are currently being investigated. A preliminary rate constant of $k=(5.7\pm0.4)\times10^{-12}\,cm^3\,s^{-1}$ has been obtained for the reaction of OH with CH_2I_2. The k-value was obtained relative to the reaction of OH with n-butane ($k=2.53\times10^{-12}\,cm^3\,s^{-1}$) at 1 bar total pressure by 298 K using the thermal decomposition of pernitric acid in the presence of NO as the OH source. An upper limit of $<1\times10^{-16}\,cm^3\,s^{-1}$ can be put on the reaction of NO_3 with CH_2I_2. The reaction was studied relative to NO_3 + propene using the thermal decomposition of N_2O_5 as the NO_3 radical source.

3.2 Reactions of IO with (a) CH_3SCH_3, (b) CH_3SOCH_3 and (c) CH_3SSCH_3

(a) Although the products of the reaction of IO with CH_3SCH_3 are already known /9,10/ the reaction was reinvestigated to test the photolysis of CH_2I_2 in the presence of O_3 as a IO radical source. During the photolysis of $CH_3SCH_3/CH_2I_2/O_3$/air reaction mixtures CH_3SCH_3 was observed to decay with the formation of equivalent amounts of CH_3SOCH_3. The result leads to the conclusion that CH_3SCH_3 is reacting solely with IO radicals since its reaction with O_3 and radicals such as O and OH which might be present in the system is known to lead to the formation of SO_2. In the previous two studies on the reaction /9,10/ it was suggested that the reaction of IO with CH_3SCH_3 also results in the formation of I-atoms,

$$IO + CH_3SCH_3 \longrightarrow CH_3SOCH_3 + I$$

In the present work it is found that the photolysis of one molecule of CH_2I_2 leads to the loss of between 7 to 9 molecules each of CH_3SCH_3 and O_3. This results strongly suggests the presence of a chain reaction in the system and gives support to the above proposed mechanism for the IO + CH_3SCH_3 reaction.

(b) Preliminary computer simulations of $CH_3SOCH_3/CH_2I_2/O_3$/air photolysis mixtures using the experimental data indicate that the reaction of IO with CH_3SOCH_3 is approximately a factor of ten slower than the corresponding reaction with CH_3SCH_3. The only observable product of the reaction was $CH_3SO_2CH_3$ and for every molecule of CH_2I_2 photolysed between 2 to 3 molecules each of CH_3SOCH_3 and O_3 were consumed. Because of difficulties in calibrating $CH_3SO_2CH_3$ it not possible to say whether the reacted CH_3SOCH_3 results in the formation of equivalent amounts of $CH_3SO_2CH_3$. However, the lack of other products would seem to support that this is probably the case. The stoichiometry of the reacted CH_2I_2, $CH_3SO_2CH_3$ and O_3 indicates that I-atoms are probably being formed in the reaction of IO with CH_3SOCH_3. The evidence is not as conclusive as for IO + CH_3SCH_3 since the reaction in this case is much slower and the fate of the IO radicals is mainly controlled by their fast self-reaction. The following mechanism is suggested,

$$IO + CH_3SOCH_3 \longrightarrow CH_3SO_2CH_3 + I$$

(c) IO was found to react with CH_3SSCH_3 at a rate comparable to that for CH_3SCH_3. Every photolysed CH_2I_2 molecule led to the loss of 2 molecules of CH_3SSCH_3 and 4 molecules of O_3. SO_2 and HCHO were formed as products each with a yield of ~50% on a molar basis.

The IR product spectrum shows the presence of at least two other sulphur containing products each containing the sulphone functional group ($-SO_2$). One of these products has been positively identified as methanesulphonic anhydride (($CH_3SO_2)_2O$) and the other is tentively assigned to CH_3SO_2OI. Due to calibration difficulties the yield of $(CH_3SO_2)_2O$ is not known. The results suggest that I-atoms are not being recycled in the reaction system. The stoichiometry of the reacted CH_3SSCH_3, CH_2I_2 and O_3 can be explained by the following reaction mechanism,

$$CH_2I_2 + O_2 + h\upsilon \longrightarrow CH_2O_2 + 2I$$

$$I + O_3 \longrightarrow IO + O_2$$
$$CH_3SSCH_3 + IO \longrightarrow CH_3SOI + CH_3S$$
$$CH_3S + O_3 \longrightarrow CH_3SO + O_2$$

---

net reaction: $CH_2I_2 + 2CH_3SSCH_3 + 4O_3 \longrightarrow$ products

The reactions of CH_3SO in air are known to lead to the formation of SO_2 and HCHO and it is thought that the further reactions of CH_3SOI are responsible for the formation of the sulphone products.

3.3 Reactions of IO with H_2S and CH_3SH

Work on the reactions of IO with H_2S and CH_3SH is still very preliminary. Initial indications are that the rate constants for both reactions are of a similar order of magnitude as that for $IO + CH_3SCH_3$.

GC and FTIR analyses show that the reaction of IO with H_2S leads to the formation of SO_2 with ~100% yield. Analogous to the reaction $OH + H_2S$ /17/ the dominant reaction pathway at room temperature is expected to be abstraction with the formation of hypoiodous acid (HOI). No evidence for the formation of HOI could be found in the FTIR product spectrum. It is probably unstable under the present experimental conditions and may also be consumed by fast secondary reactions /18/. The stoichiometry of the reacted H_2S, CH_2I_2 and O_3 in $H_2S/CH_2I_2/O_3$/air photolysis systems is consistent with the following mechanism,

$$CH_2I_2 + O_2 + h\upsilon \longrightarrow CH_2O_2 + 2I$$

$$I + O_3 \longrightarrow IO + O_2$$
$$IO + H_2S \longrightarrow HS + HOI$$
$$HS + O_3 \longrightarrow HSO + O_2$$

---

net reaction: $CH_2I_2 + 2H_2S + 4O_3 \longrightarrow$ products

The reaction of IO with CH_3SH in synthetic air leads to the formatiom of SO_2 and CH_3SSCH_3. Here the reaction is also expected to proceed via an abstraction mechanism /17/,

$$CH_3SH + IO \longrightarrow CH_3S + HOI$$

The yields of SO_2 and CH_3SSCH_3 are variable and depend on the initial concentration of O_3 due to the competing reactions,

$$CH_3S + O_3 \longrightarrow CH_3SO + O_2$$
$$CH_3S + CH_3S + M \longrightarrow CH_3SSCH_3 + M$$

SO_2 can be formed by the further reactions of CH_3S with O_2 and O_3 and CH_3SSCH_3 by the recombination of CH_3S radicals as shown above. As in the studies on $IO + H_2S$ no evidence has yet been found for the formation of HOI in this reaction.

Further work is continuing on both reactions to better characterise the products and elucidate the reaction mechanism.

3.4 Reactions of IO with SO_2 and CS_2

On the basis of the experimental data preliminary computer simulations of $SO_2/CH_2I_2/O_3$/air and $CS_2/CH_2I_2/O_3$/air photolysis systems indicate that the rate constants for the reactions of IO with SO_2 and CS_2 are at least 3 orders of magnitude slower than that for $IO + CH_3SCH_3$.

Analogous to the reaction of OH with CS_2 in air /19/ the reaction with IO is found to lead to the formation of SO_2 and COS as products with one molecule of each being formed for every reacted CS_2. In the reaction of IO with SO_2 the formation of a broad featureless absorption in the region 1400 to 850 cm^{-1} with a maximum at 1200 cm^{-1} is observed. The compound giving rise to this absorption has not yet been identified.

4. CONCLUSIONS

IO radicals have been found to react rapidly with H_2S, CH_3SH, CH_3SCH_3 and CH_3SSCH_3 and moderately fast with CH_3SOCH_3, and only very slowly with SO_2 and CS_2. Preliminary computer simulations of the reaction systems indicate that the rate constants for the reactions of IO with H_2S, CH_3SH and CH_3SSCH_3 are of the same order of magnitude as that for IO with CH_3SCH_3. Therefore, as has been suggested for CH_3SCH_3 reaction with IO may also be a sink for these compounds in the marine atmosphere. The reactions of IO with SO_2, CS_2 and CH_3SOCH_3 are too slow to be of any significance in the atmosphere.

The reaction of IO with the organic sulphur compounds does not lead to the formation of MSA as has been observed for the corresponding reactions of OH with these compounds in the presence of NO_x. However, the reactions of IO with CH_3SH and CH_3SSCH_3 probably lead to the formation of CH_3S radicals which under atmospheric conditions where NO_x is present could react further to form MSA. The reaction of IO with CH_3SCH_3 and CH_3SSCH_3 leads to the formation of CH_3SOCH_3 and $CH_3SO_2CH_3$, respectively, both of which have been found in marine aerosols.

Only in the case of CH_3SCH_3 was conclusive evidence found for the conversion of IO to I which could lead to the catalytic destruction of O_3 /11/.

REFERENCES

(1)   MOLLER, D. (1984). Atmos. Environ. 18, 19-27.

(2)   BROWN, K. and BELL, J.N.B. (1986). Atmos. Environ. 20, 537-540.

(3)   ANDREAE, M.O. (1986). The Ocean as a Source of Atmospheric Sulphur Com
       pounds. In, The Role of Air-Sea-Exchange in Geochemical Cycling, P. Baut-Menard
       (editor). NATO ASI Series C185, D. Reidel, Dordrecht, pp 331-342.

(4)   HATAKEYAMA, S., IZUMI, K. and AKIMOTO, H. (1985).
       Atmos. Eniviron. 19, 135-141.

(5)   BARNES, I., BASTIAN, V., BECKER, K.H., FINK, E.H. and NELSEN, W. (1986).
       J. Atmos. Chem. 4, 445-466.

(6)   BARNES, I., BASTIAN, V. and BECKER, K.H. (1988).
       Int. J. Chem. Kinetics 20, 415-431.

(7)   KOLAITIS, L.N., BRUYNSEELS, F.J., VAN GRIEKEN, R.E. and ANDREAE, M.O.
       (1989). Environ. Sci. Technol. 23, 236-240.

(8)   WATTS, S.F., WATSON, A. and BRIMBLECOMBE, P. (1987).
       Atmos. Environ. 21, 2667-2672.

(9)   BARNES, I., BECKER, K.H., CARLIER, P. and MOUVIER, G. (1987). Int. J. Chem.
       Kinetics 19, 489-501.

(10)   MARTIN, D., JOURDAIN, J.L., LAVERDET, G. and LE BRAS, G. (1987). Int. J.
       Chem. Kinetics 19, 503-512.

(11)   BARNES, I., BECKER, K.H., MARTIN, D., CARLIER, P., MOUVIER, G.,
       JOURDAIN, J.L., LAVERDET, G. and LE BRAS, G. (1989). The impact of halogen
       oxides on the DMS oxidation in the marine atmosphere. In, ACS Symposium Series
       No 393: Biogenic Sulfur in the Environment, E.S. Saltzman and W.J. Cooper (eds),
       Washington, pp 464-476.

(12)   BARNES, I., BASTIAN, V., BECKER, K.H., FINK, E.H., KLEIN, Th., NELSEN, W.,
       REIMER, A. and ZABEL, F. (1984). Untersuchung der Reaktionssyteme
       $NO_x/ClO_x/HO_x$ unter troposphärischen Bedingungen. BPT-Bericht 11/84, ISSN
       01761077, Gesellschaft für Strahlen- und Umweltforschung mbH., München.

(13)   CLASS, Th. and BALLSCHMITER, K. (1988). J. Atmos. Chem. 6, 35-46.

(14)   SCHMITT, G. and COMES, F.J. (1980). J. Photochem. 14, 107-123.

(15)   GREGORY, R.A. and STYLE, D.W.G. (1936). Trans. Faraday Soc. 32, 724-736.

(16)   JENKIN, M.E. and COX, R.A. (1985). J. Phys. Chem. 89, 192-199.

(17)   HYNES, A.J. and WINE, P.H. (1987). J. Phys. Chem. 91, 3672-3676.

(18)   COX, R.A., JENKIN, M.E., STOCKER, D.W. and CLEMITSHAW, K.C. (1984).
       Laboratory Investigations of Gaseous Hypoiodous Acid. Report: AERE-R 11290,
       ISBN 0-7058-0978-1, H.M. Stationery Office, England.

(19)   BARNES, I., BECKER, K.H., FINK, E.H., REIMER, A., ZABEL, F. and NIKI, H.
       (1983). Int. J. Chem. Kinetics 15, 631-645.

ATMOSPHERIC BEHAVIOUR OF CHLOROACETALDEHYDES

J. Starcke, F. Zabel, L. Elsen, W. Nelsen, I. Barnes, and K. H. Becker

Bergische Universität-Gesamthochschule Wuppertal, Physikalische Chemie / FB 9,
Gaußstr. 20, 5600 Wuppertal 1, FRG

Summary

Chloroacetaldehydes are intermediates in the atmospheric degradation of several important chlorinated alkanes and alkenes. The most probable loss processes which determine their atmospheric lifetimes are photolysis and reactions with OH radicals and Cl atoms. In the present work, UV absorption spectra between 235 and 345 nm and the rate constants for the reactions with OH and Cl have been determined at room temperature for CCl_3CHO, $CHCl_2CHO$, and CH_2ClCHO. The experiments were performed in large reaction chambers from Duran glass (400-500 l volume). UV absorption coefficients were measured using a built-in White mirror system (optical pathlengths ≤ 150 m) and a diode array spectrometer. Rate constants were determined by relative rate techniques with CH_3ONO and Cl_2 photolysis as stationary photolytic OH and Cl sources, respectively. Reaction mixtures were analyzed by in-situ long-path IR absorption using an FTIR spectrometer and by gas chromatography. From the results it is suggested that the tropospheric degradation of chloroacetaldehydes proceeds by photolysis and reaction with OH, both being of comparable importance.

1. INTRODUCTION

Chloroacetaldehydes are intermediates in the tropospheric degradation of several common chlorinated alkanes ($\geq C_2$) and alkenes ($\geq C_3$). For example, chloral (CCl_3CHO) and monochloroacetaldehyde have been detected as intermediates in the oxidative degradation of methyl chloroforme (1) and 1,3-dichloropropene (2), respectively. Similarly, monochloroacetaldehyde is expected to be an intermediate in the atmospheric oxidation of 1,2-dichloroethane which is used in large amounts for the vinyl chloride production. As was stated earlier (2), the atmospheric lifetimes of chloroacetaldehydes could be larger than those of some of their parent compounds. Kinetic data of atmospheric relevance for chloroacetaldehydes are very sparse. The rate constant for the reaction of CCl_3CHO with OH radicals was determined at room temperature (1,9) and found to be lower than the corresponding OH rate constant of acetaldehyde by a factor of 8. Absorption spectra are published for solutions, showing a marked dependence of the absorption coefficients on the solvent (3).

In the atmosphere, the degradation of chloroacetaldehydes will probably be initiated by photolysis and reaction with OH radicals, possibly also by reaction with Cl atoms. In the present work, the gas phase absorption spectra and the rate constants for reactions with OH and Cl have been determined in large reaction chambers (ca. 400 l) for CCl_3CHO, $CHCl_2CHO$, and CH_2ClCHO.

2. EXPERIMENTAL

The experiments were performed in three reaction chambers from Duran glass:

(a) A temperature controlled 420 l chamber which is equipped with a White mirror system for long-path IR absorption measurements (typical optical pathlength: 50.4 m). A Globar light source and a Fourier transform spectrometer (Nicolet 7199) were used for spectral analysis.

(b) A 420 l chamber identical to (a) above except that temperature is not controlled. Much higher OH concentrations are available in this reactor as compared to (a).

(c) A 480 l chamber equipped with a White mirror system for long-path absorption measurements in the UV (optical pathlengths between 6 and 150 m).

UV spectra were measured in chamber (c) with a spectral resolution of 0.7 nm, using a 25 W deuterium lamp, a 22 cm Spex monochromator and a diode array spectrometer (PAR 1412). The wavelength scale was calibrated with several Hg lines from a low pressure mercury arc lamp. UV spectra of CCl_3CHO, CCl_2HCHO, CH_2ClCHO and, for comparison, of CH_3CHO were measured with an optical pathlength of 51.6 m and integration times of several seconds.

For the kinetic experiments, relative rate techniques were applied. OH radicals and Cl atoms were generated by stationary photolysis of $CH_3ONO+NO$ mixtures and Cl_2, respectively. The relative concentrations of the chloroacetaldehyde R and the reference compound Ref were measured as a function of time, and the rate constants were determined according to the equation,

$$\text{(I)} \qquad \frac{d \ln ([R]_o / [R]_t)}{d \ln ([Ref]_o / [Ref]_t)} = \frac{k_R}{k_{Ref}}$$

Relative concentrations of R ($R = CH_2ClCHO$, $CHCl_2CHO$, CCl_3CHO) and Ref (Ref = n-butane, ethane) were measured either in-situ by long-path IR absorption using absorption bands close to 1060 cm^{-1} (for the chloroacetaldehydes) or by sampling with gas tight syringes and gas chromatographic analysis (for the reference substances).

Research grade CH_3CHO and CCl_3CHO were used without further purification. CH_2ClCHO was prepared from a commercial aqueous solution by ether extraction and fractionated destillation. $CHCl_2CHO$ was obtained in the same way after treatment of the diethylacetal with H_2SO_4. The aldehydes were stored at -10 °C. $CHCl_2CHO$ and CH_2ClCHO polymerized within 2 or 3 weeks.

In a typical experiment, 0.025 mbar aldehyde, 0.025 mbar n-butane, 0.05 mbar CH_3ONO, and 0.05 mbar NO were mixed with 1000 mbar synthetic air. Then [R] and [Ref] were measured for some minutes in the dark and for 20 min with the photolysis lights on. There was no measurable wall loss of the aldehydes during the reaction time.

3. RESULTS AND DISCUSSION

The UV spectra of CCl_3CHO, $CHCl_2CHO$, CH_2ClCHO and, for comparison, of CH_3CHO were measured with the diode array spectrometer between 235 and 345 nm and are presented in fig. 1. Absorption cross sections were determined according to $\ln (I_o /I)= \sigma \times N \times d$. The shape of the absorption spectra is similar for all of the aldehydes with a maximum at ca. 295 nm. Absolute cross sections are comparable as well, except for chloral where the absorption is stronger by a factor of 2. It appears that the absorption maximum is shifted to the visible by 10 nm for the partly halogenated aldehydes as compared to CH_3CHO and CCl_3CHO. The gas phase absorption cross sections from this work agree with the absorption coefficients obtained in cyclohexane solutions (3) within a factor of 2. The structure of the absorption as it appears at a spectral resolution of 0.7 nm becomes weaker with increasing chlorine content.

The rate constants for the reactions

$$CCl_3CHO \quad + \quad OH ----> \quad products \quad (k_1)$$
$$CHCl_2CHO \quad + \quad OH ----> \quad products \quad (k_2)$$
$$CH_2ClCHO \quad + \quad OH ----> \quad products \quad (k_3)$$

have been determined from plots of $\ln ([R]_o / [R]_t)$ as a function of $\ln ([Ref]_o / [Ref]_t)$. This is shown for $R = CH_2ClCHO$ in fig. 2. According to eq. I, the slope of the corresponding straight line is equal to k_R / k_{Ref}. With $k_{Ref} = 2.56 \times 10^{-12}$ cm^3/s (4) for Ref = n-butane at 300 K, the values of table I are obtained for k_1, k_2, and k_3.

It is well known that in the OH - initiated degradation of chlorine rich gaseous compounds Cl atoms can be released which then may react with the reactant and / or reference compound and thus lead to erraneous k_R / k_{Ref} ratios. Eventhough k_R / k_{Ref} did not vary when different initial concentrations of n-butane and NO were used, the values of k_1 and k_2 are considered to be preliminary and lower limits to the true rate constants. The use of a more favorable reference substance is planned.

The rate constants for the reactions

$$(4) \quad CCl_3CHO \quad + \quad Cl -----> \quad products \quad (k_4)$$
$$(5) \quad CHCl_2CHO \quad + \quad Cl -----> \quad products \quad (k_5)$$
$$(6) \quad CH_2ClCHO \quad + \quad Cl -----> \quad products \quad (k_6)$$

have been determined in the same way as k_1, k_2, and k_3, with ethane as the reference compound. A plot of $\ln ([R]_o / [R]_t)$ as a function of $\ln ([Ref]_o / [Ref]_t)$ is shown in fig.3 for $R = CH_2ClCHO$. The room temperature values thus obtained for k_4, k_5 and k_6 with $k_{Ref} = 5.70 \times 10^{-12}$ cm^3/s at 300 K are included in table I.

In table I, the rate constants of the reactions with OH (k_{OH}) and Cl(k_{Cl}) are compared for various compounds RCHO containing the formyl group, including the chloroacetaldehydes. The rate constants decrease with increasing electron drawing effect of the substituent R (except k_5 and k_6 being exchanged), with k_{Cl} being roughly a factor of 5 faster than k_{OH}.

4. ATMOSPHERIC IMPLICATIONS

The absorption cross reactions and the rate constants from this work may be used to estimate the atmospheric lifetimes of chloroacetaldehydes.

From the absorption cross sections of fig. 1, upper limits may be derived for the photolysis rates of chloroacetaldehydes in the atmosphere, using the actinic fluxes from ref. 6 and assuming a quantum yield of 1 as an upper limit. The figures thus obtained for certain conditions are included in table II. For CH_3CHO, it may be deduced from measured quantum yields (6) that the actual photolysis lifetime is about ten times the values calculated in table II, i. e. between 3 and 4 days at the given conditions. If the quantum yields are similar for the chloroacetaldehydes, their photolytic lifetimes at the given conditions are between 1 and 2 days.

With an average tropospheric OH concentration of 5×10^5 cm^{-3} (7) and an average Cl concentration of 3×10^3 cm^{-3} (8), lifetimes with respect to degradation by OH and Cl may be estimated from the rate constants of table I. The corresponding figures are included in table II. Considering that the photolysis rates under the conditions 40° N, noontime are higher than the average, it may be concluded that the most important tropospheric loss processes for chloroacetaldehydes are photolysis and OH reaction, both being of comparable importance.

ACKNOWLEDGEMENT
Financial support by the "Umweltbundesamt" is gratefully acknowledged.

REFERENCES
(1) L. Nelson, I. Shanahan, H. W. Sidebottom, J. Tracey, and O. J. Nielsen, Int. J. Chem. Kinet., submitted for publication (1989)
(2) E. C. Tuazon, R. Atkinson, A. M. Winer, and J. N. Pitts, Jr., Arch. Environ. Contam. Toxicol. 13 (1984) 691
(3) A. Kirrmann and J. Cantacuzène, Comptes Rendus 248 (1959) 1968
(4) R. Atkinson, D. L. Baulch, R. A. Cox, R. F. Hampson, Jr., J. A. Kerr, and J. Troe, Int. J. Chem. Kinet. 21 (1989) 115
(5) H. G. Libuda, F. Zabel, and K. H. Becker, manuscript in preparation
(6) B. J. Finlayson-Pitts and J. N. Pitts, Jr. : "Atmospheric Chemistry", Wiley, New York 1986
(7) K. M. Jeong and F. Kaufman, Geopys. Res. Lett. 6 (1979) 757
(8) H.B. Singh and J. F. Kasting, J. Atmos. Chem. 7 (1988) 262
(9) S. Dóbé, L. A. Khachatryan, and T. Bérces, Ber. Bunsenges. Physik. Chem. 93 (1989) 847

R-CHO	k_{OH} [10^{-12}cm^3/s]	Ref.	k_{Cl} [10^{-12}cm^3/s]	Ref.	$\dfrac{k_{Cl}}{k_{OH}}$
CH$_3$CHO	14	4	76	4	5.3
HCHO	10	4	73	4	7.3
CH$_2$ClCO	3.2	this work	10.8	this work	3.4
CHCl$_2$CO	(2.8)	this work	12.7	this work	(4.5)
CCl$_3$CHO	(1.2)	this work	9.0	this work	(7.5)
	1.6 ± 0.2	9			5.6
	1.8 ± 0.3	1			5.0
ClCHO	< 0.32	5	0.75	5	> 2.3

Table I. Rate constants for the reactions of chloroacetaldehydes and related aldehydes OH and Cl at 300 K (in parentheses: preliminary values, probably lower limits)

R-CHO	J^{-1} [hours] at 40 °N, noon, July (lower limit, for Φ=1)	(k_{OH}x[OH])$^{-1}$[hours] with [OH]=5x10^{5}cm^{-3} (ref.7)	(k_{Cl}x[Cl])$^{-1}$[hours] with [Cl]=3x10^{3}cm^{-3} (ref.8)
CH$_3$CHO	8.0	40	8600
CH$_2$ClCHO	4.4	174	1200
CHCl$_2$CHO	3.9	< 198	7300
CCl$_3$CHO	3.3	< 463	10300

Table II. Tropospheric lifetimes of chloroacetaldehydes with respect to photolysis and reactions with OH and Cl.

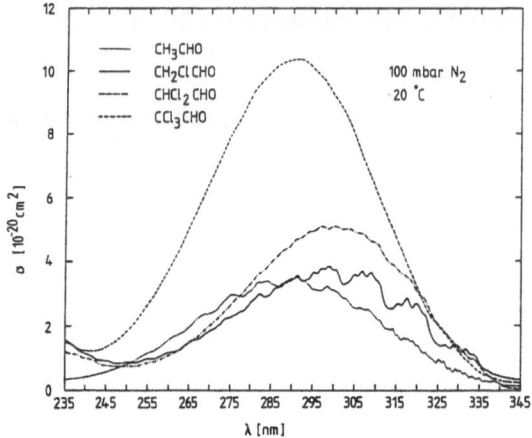
Fig.1: UV spectra of chloroacetaldehydes

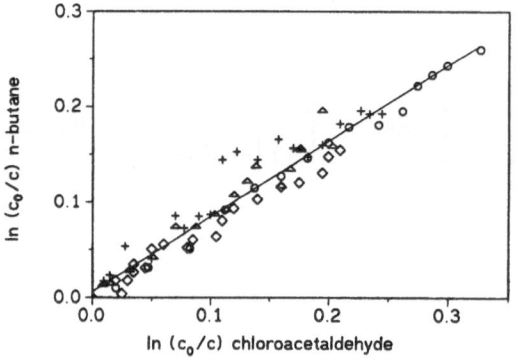
Fig.2: Determination of the rate constant for the reaction $CH_2ClCHO + OH$ ----> products by the relative rate method with n-butane as reference; different symbols correspond to individual runs

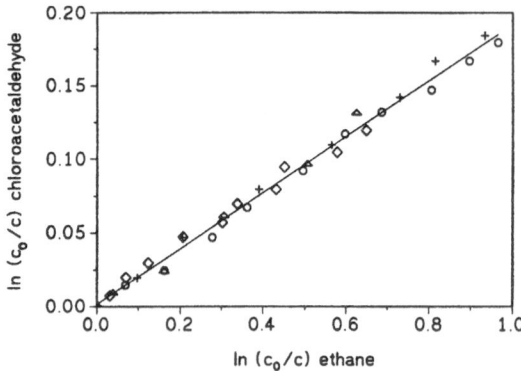
Fig. 3: Determination of the rate constant of the reaction $CH_2ClCHO + Cl$ ----> products by the relative rate method with ethane as reference; different symbols correspond to individual runs

PRODUCTS AND MECHANISM OF THE REACTION BETWEEN NO₃ AND DIMETHYLSULPHIDE IN AIR

J.Hjorth, N.Jensen, C.Lohse, G.Restelli and H.Skov

Commission of the European Communities, Joint Research Centre,
Ispra Establishment, Environment Institute, 21020 Ispra (VA), Italy.

Summary

The reaction between NO_3 and DMS in air has been studied in a 480 L reaction chamber. Intermediates and end products were identified by FT-IR and ion chromatography. HNO_3, CH_2O, SO_2 and methanesulphonic acid were found to be main products of the reaction, with the latter as the most abundant sulphur containing product. Organic peroxynitrates and CH_3SNO_2 have been identified as intermediates from their characteristic spectral features. Based on the results obtained a mechanism for the reaction DMS + NO_3 is proposed, which has hydrogen abstraction as the first step. In the marine troposphere this reaction with formation of HNO_3 which is removed by deposition may represent an efficient nighttime sink for NO_x.

1. INTRODUCTION

The most important natural sources of reduced sulphur in the atmosphere are considered to be dimethylsulphide CH_3SCH_3 (DMS), hydrogensulphide H_2S, methylmercaptan CH_3SH, ethylmercaptan CH_3CH_2SH, dimethyldisulphide CH_3SSCH_3, carbon disulphide CS_2 and carbonyl sulphide COS. DMS alone, produced by marine phytoplankton, is estimated to account for approximately 25% of all naturally and anthropogenically emitted gaseous sulphur[1-3].

In the daytime the main sink for DMS is its reaction with the hydroxyl radical, OH, while in the nighttime the reaction with the nitrate radical, NO_3, is important[4]. In fact the DMS + NO_3 reaction rate constant has been measured by several groups and is found to be fast: $(1.0 \pm 0.3) \times 10^{-12}$ $cm^{+3}molecule^{-1}s^{-1}$ at 298 K and atmospheric pressure.

Little is known about mechanism and products of the NO_3 reaction with DMS. The products have been studied[5], with FT-IR and gas chromatography as analytical techniques. HNO_3, NO_2, CH_2O and small amounts of SO_2 were found; neither dimethylsulphoxide ($CH_3S(O)CH_3$) nor methanesulphonic acid (CH_3SO_3H) were detected and no reaction mechanism was proposed.

In the present laboratory study long path FT-IR spectroscopy and ion chromatography have been used for measuring intermediates and end products of the DMS + NO_3 reaction in air.

In this study evidence has been obtained for hydrogen abstraction as the first reaction step:

$$(1) \quad CH_3SCH_3 + NO_3 \longrightarrow CH_3SCH_2 + HNO_3$$

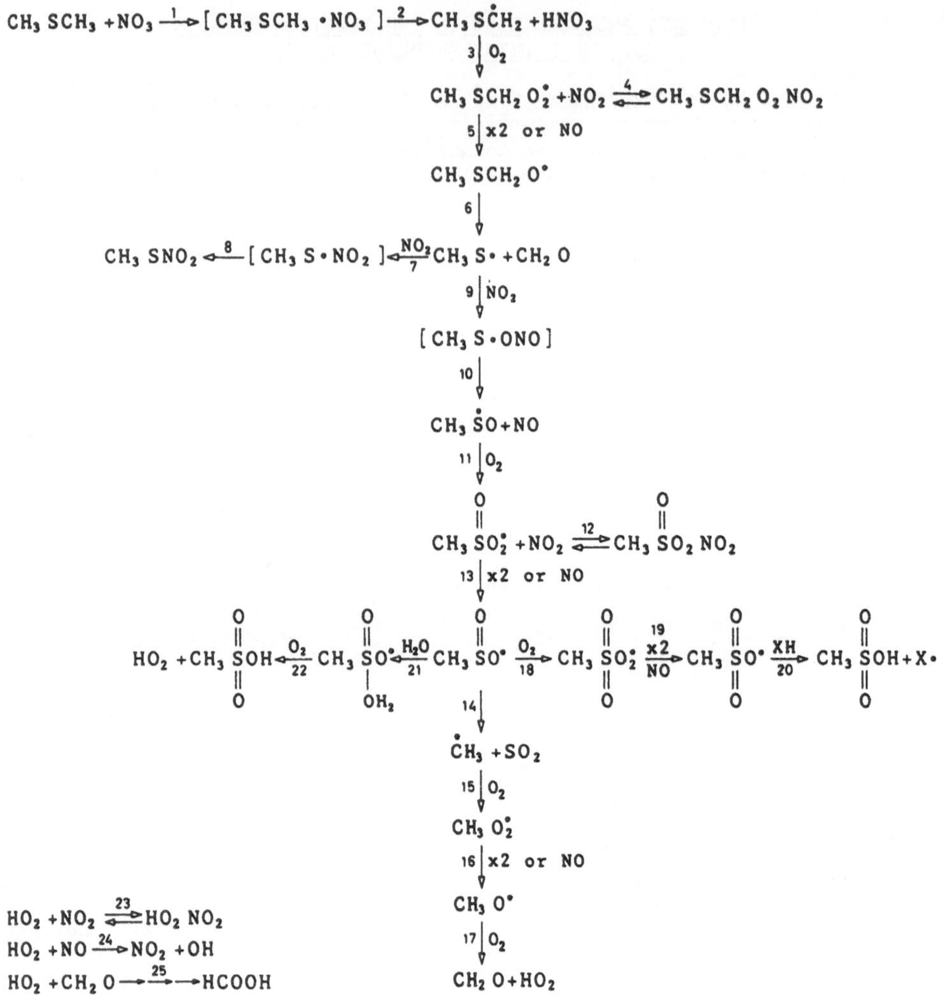

$$CH_3SCH_3 + NO_3 \xrightarrow{1} [CH_3SCH_3 \cdot NO_3] \xrightarrow{2} CH_3S\overset{\bullet}{C}H_2 + HNO_3$$

$$\downarrow 3 \;\; O_2$$

$$CH_3SCH_2O_2^{\bullet} + NO_2 \underset{4}{\rightleftharpoons} CH_3SCH_2O_2NO_2$$

$$\downarrow 5 \;\; \text{x2 or NO}$$

$$CH_3SCH_2O^{\bullet}$$

$$\downarrow 6$$

$$CH_3SNO_2 \xleftarrow{8} [CH_3S \cdot NO_2] \xleftarrow[7]{NO_2} CH_3S^{\bullet} + CH_2O$$

$$\downarrow 9 \;\; \overset{\bullet}{N}O_2$$

$$[CH_3S \cdot ONO]$$

$$\downarrow 10$$

$$CH_3\overset{\bullet}{S}O + NO$$

$$\downarrow 11 \;\; O_2$$

$$\overset{O}{\underset{||}{C}}H_3SO_2^{\bullet} + NO_2 \underset{12}{\rightleftharpoons} \overset{O}{\underset{||}{C}}H_3SO_2NO_2$$

$$\downarrow 13 \;\; \text{x2 or NO}$$

$$HO_2 + CH_3\overset{O}{\underset{||}{S}}OH \xleftarrow[22]{O_2} CH_3\overset{O}{\underset{||}{S}}O^{\bullet} \xleftarrow[21]{H_2O} CH_3\overset{O}{\underset{||}{S}}O^{\bullet} \xrightarrow[18]{O_2} CH_3\overset{O}{\underset{||}{S}}O_2^{\bullet} \xrightarrow[NO]{19 \; x2} CH_3\overset{O}{\underset{||}{S}}O^{\bullet} \xrightarrow[20]{XH} CH_3\overset{O}{\underset{||}{S}}OH + X^{\bullet}$$

(with OH_2 on the $CH_3SO^{\bullet}$ at step 21)

$$\downarrow 14$$

$$\overset{\bullet}{C}H_3 + SO_2$$

$$\downarrow 15 \;\; O_2$$

$$CH_3O_2^{\bullet}$$

$$\downarrow 16 \;\; \text{x2 or NO}$$

$$CH_3O^{\bullet}$$

$$\downarrow 17 \;\; O_2$$

$$CH_2O + HO_2$$

$$HO_2 + NO_2 \underset{23}{\rightleftharpoons} HO_2NO_2$$

$$HO_2 + NO \xrightarrow{24} NO_2 + OH$$

$$HO_2 + CH_2O \xrightarrow{25} HCOOH$$

Figure 1.
Mechanism proposed for the reaction between DMS and NO_3, based on measurements of products and intermediates.

and no evidence for the alternative addition reaction:

(2) $CH_3SCH_3 + NO_3 \longrightarrow CH_3S(ONO_2)CH_3 \longrightarrow CH_3S(O)CH_3 + NO_2$

Methanesulphonic acid has been found to be the most abundant sulphur containing product. In addition HNO_3, CH_2O and SO_2 have been found to be major products while minor quantities of pernitric acid (HO_2NO_2), HCOOH, CO and NO have been positively identified. Unstable intermediates such as organic peroxynitrates and thionitric acid-S-methyl ester (CH_3SNO_2) have been tentatively identified from their characteristic infrared spectral features. H_2SO_4 was found in the liquid samples collected for ionchromatography analysis.

The principal products appear to be the same as for the reaction between DMS + OH, with the exception of HNO_3, which comes from the hydrogen abstraction by NO_3.

Based on the results obtained in this and previous investigations a mechanism for the DMS reaction with NO_3 is proposed.

2. EXPERIMENTAL

The experiments were performed in a 480 L volume gas reactor made of a 1.5 m long and 0.6 m diameter glass tube coated with a Teflon film. A multiple reflection White type mirror system was included in the reactor, adjusted to give a total optical path length of 81 m and coupled to a Bruker IFS 113 V FT spectrometer. Spectra were obtained by coadding 5-50 scans recorded at 1 cm^{-1} instrumental resolution.

Reactants were mixed in purified air at 295 $\pm$ 2 K and 740 Torr total pressure. NO_3 was obtained in equilibrium with N_2O_5 and NO_2.

Time-dependent concentrations of DMS, N_2O_5, HNO_3, SO_2, NO_2, CH_2O, CO, HCOOH, NO and HO_2NO_2 were followed from their infared spectral absorptions. Methanesulphonic acid and H_2SO_4 were measured by bubbling known volumes (20-40 L) of the reaction mixture through deionized water. The liquid was then analyzed by a Dionex ion chromatograph with suppression techniques.

Table 1 and Figs. 1-4 summarize the results obtained.

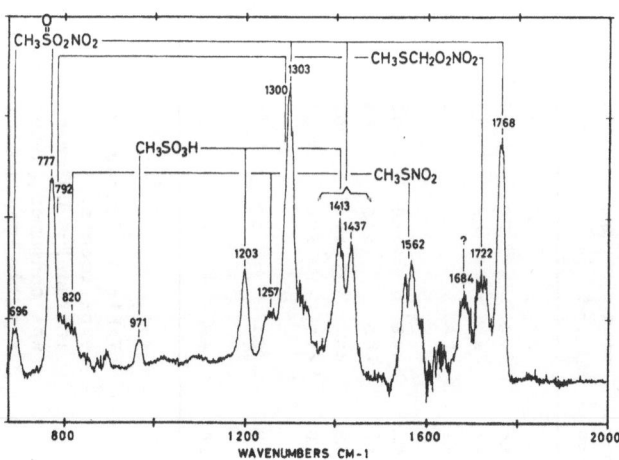

Figure 2.
IR-spectrum of the reaction mixture, recorded just after N_2O_5 has been consumed; DMS, NO_2, HNO_3, CH_2O, SO_2 and H_2O were subtracted from the original spectrum.

TABLE 1

Run No.	N_2O_5 Initial conc. ppm	DMS Initial conc. ppm	DMS Conc. (ppm) when $[N_2O_5]=0$	DMS Consumed ppm	NO_2 Initial conc. ppm	NO_2 Conc. (ppm) when $[N_2O_5]=0$	HNO_3 Initial conc. ppm	HNO_3 Conc. (ppm) when $[N_2O_5]=0$	CH_2O $\Delta[CH_2O]$ ppm	SO_2 $\Delta[SO_2]$ ppm	CH_3SO_3H ppm	CH_3SO_3H % of consumed DMS	H_2SO_4 ppm	H_2SO_4 % of consumed DMS
1	11.7	17.2	8.7	8.5	6.4	12.1	6.3	17.7	6.6	1.0	4.1	48	2.0	24
2	10.6	16.4	9.2	7.2	6.0	11.0	3.8	15.3	5.9	1.0	3.1	43	0.3	4
3	6.8	8.4	4.0	4.4	1.3	4.8	3.4	10.4	4.2	0.6	1.7	39	0.7	17
4	5.8	8.3	2.4	5.9	5.1	7.0	4.3	10.0	3.7	0.9	-	-	-	-
5	9.0	5.3	<0.5	>4.8	10.3	14.1	8.4	17.2	3.7	0.7	1.9	40	0.2	5
6	>12.0	5.4	<0.5	>4.9	6.5	12.1	26.1	>50.0	4.3	1.1	3.0	61	0.5	11
7	15.7	4.9	<0.1	>4.8	18.2	23.3	8.6	31.5	4.6	0.3	3.7	77	0.9	19
8	10.4	6.6	<1.0	>5.6	5.1	9.5	4.1	16.6	4.3	0.7	-	-	-	-
9	7.2	12.6	6.1	6.5	8.1	14.1	3.2	11.0	5.7	0.9	1.1	17	0.3	4
10	6.5	15.0	6.6	8.4	2.0	7.7	3.6	8.6	6.1	0.8	0.2	3	0.3	4

Table 1. Initial and final concentration of reactants and products in 10 experiments.
In run 1-4 DMS was in excess and the measurements were made just after N_2O_5 was consumed (after 3-15 min. reaction time). In run 5-8 N_2O_5 was in excess and the measurements were made just after N_2O_5 was consumed (after 6-20 min. reaction time). In run 9 NO was added and the measurements were made after [peroxynitrate] $\cong$ 0 (after $\approx$ 18 min. reaction time). In run 10 the measurements were made after [peroxynitrate] $\cong$ 0 (after 220 min reaction time).

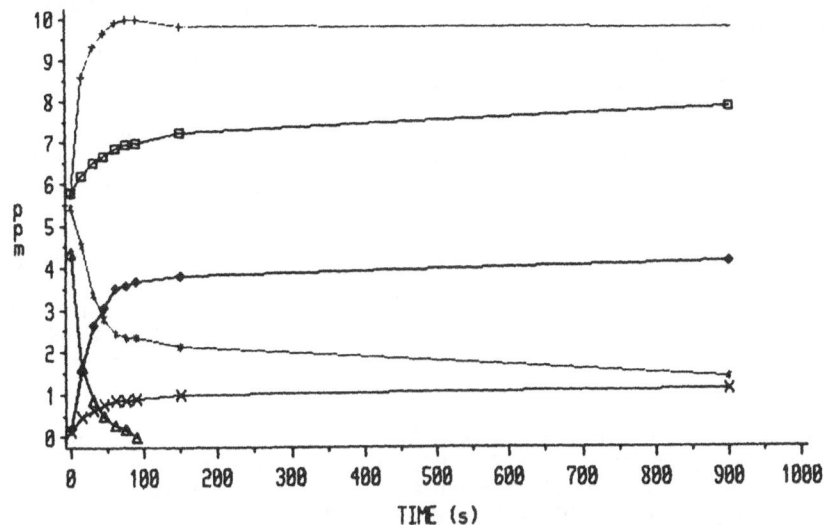

Initial conc.: DMS = 8.3 ppm, NO$_2$ = 5.1 ppm, HNO$_3$ = 4.3 ppm
and N$_2$O$_5$ = 5.8 ppm
DMS = star, NO$_2$ = square, HNO$_3$ = plus,
N$_2$O$_5$ = triangle, CH$_2$O = diamond and SO$_2$ = x

Figure 3.
Typical concentration vs time plot of reactants and products in an experiment where excess DMS was used (run No. 4 in Table 1).

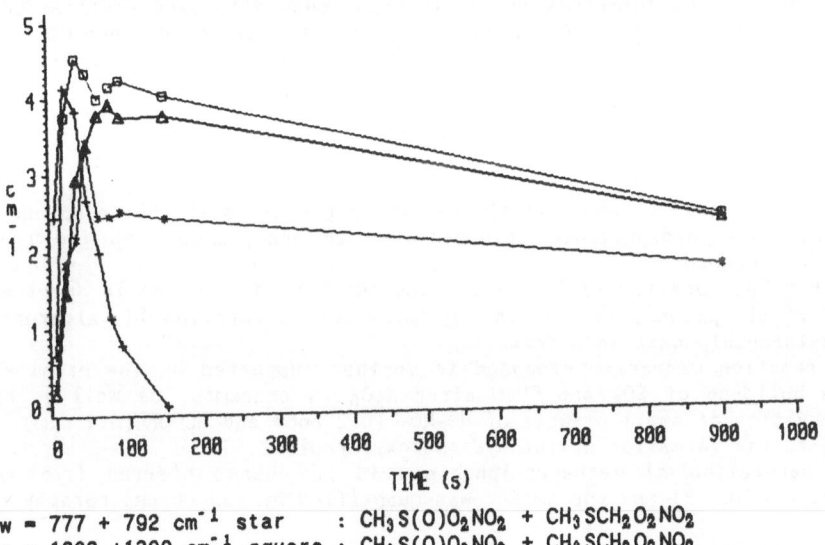

w = 777 + 792 cm^{-1} star   : CH$_3$S(O)O$_2$NO$_2$ + CH$_3$SCH$_2$O$_2$NO$_2$
w = 1303 +1300 cm^{-1} square : CH$_3$S(O)O$_2$NO$_2$ + CH$_3$SCH$_2$O$_2$NO$_2$
w = 1722 cm^{-1} plus   : CH$_3$SCH$_2$O$_2$NO$_2$
w = 1768 cm^{-1} triangle   : CH$_3$S(O)O$_2$NO$_2$

Figure 4.
Build-up and decay of the two peroxynitrate intermediates in accordance with the reaction scheme proposed.

3. DISCUSSION

Hydrogen abstraction appears to be the first step of the reaction between DMS and NO_3. This is supported by the immediate formation of HNO_3 with the stoichiometry expected (see e.g. run no. 4 in Table 1 and Fig. 3: initial N_2O_5 = 5.8 ppm $\simeq$ consumed DMS = 5.9 ppm $\simeq$ ΔHNO_3 = 5.7 ppm) and by the lack of observation of dimethylsulphoxide, which would produce a clearly visible spectral feature at 1100 cm^{-1}. However the reported observation of a negative or very small temperature dependence for the reaction rate constant[6-7] indicates an initial addition step, as shown in Fig. 1.

The observation of a fast and large build up of CH_2O (Fig. 3) indicates, that the radical formed by reaction 2, reacts with O_2 to produce CH_2O (see reactions 3-6 in Fig.1). In the presence of NO_2 the equilibrium 4 should be established and the bands at 792, 1300 and 1722 cm^{-1} in Fig. 2 have been assigned to the peroxynitrate compound. The fast build-up and subsequent decay of the peroxynitrate compound (Fig. 4), indicate that the mechanism then proceeds through reaction 3-6.

Balla et al.[8] have carried out kinetic studies on the reaction of the methylthiyl radical $CH_3S\bullet$ with O_2, NO_2 and NO. These studies suggest that under the experimental conditions used in this study the reaction with NO_2 is the main fate for the $CH_3S\bullet$ formed by reaction 6.

Two possible reaction routes are envisaged and shown as reactions 7 and 8 or reactions 9 and 10, but in both cases an adduct would be initially formed (a negative temperature dependence for the rate constant was in fact observed[8]). Bands at 820, 1257 and 1562 cm^{-1} have been assigned to the thionitric acid-S-methyl ester (CH_3SNO_2) by Niki et al.[9], these bands can also be seen in this experiment (Fig. 2).Reactions 9 and 10 have been suggested to be the main route in agreement with the results obtained by photolytic cleavage of CH_3SSCH_3 in N_2 + NO_2 [10], where a nearly stoichiometric relationship was observed between the consumption of CH_3SSCH_3 and the formation of NO. In this study only very small amounts of NO were detected (about 0.2 ppm), which is not unexpected since NO is known to react very fast with peroxy-radicals and the hydroperoxyl-radical (HO_2) which both are present in this chemical system (Fig. 1).

Reaction 10 is then followed by reaction 11 and equilibrium 12 is established when NO_2 is present. Spectral features at 696, 777, 1303, 1413, 1437 and 1768 cm^{-1} (Fig. 2) have been tentatively assigned to the methyl sulfinyl peroxynitrate $CH_3S(O)O_2NO_2$ [10]; these bands show a parallel fast decrease when NO is added to the system. Fig 4 shows the build up and decay of the two peroxynitrate intermediates in accordance with the reaction scheme proposed.

SO_2 has been positively identified and quantified from the IR spectra. The most likely pathway leading to SO_2 formation is reaction 14, alternatively to methanesulphonic acid formation.

The reaction mechanism proposed is further supported by the parallel and slow build-up of SO_2 and CH_2O after N_2O_5 is consumed, as well as by the observation of small amounts of HO_2NO_2 (0.2 ppm) and HCOOH (0.1 ppm), which support the formation of the hydroperoxyl radical.

The generation of methanesulphonic acid has been inferred from the IR spectra (Fig. 2) and the amount was quantified by ion chromatography (Table 1).

After the formation of $CH_3S(O)O\bullet$, two possible pathways to methanesulphonic acid are proposed, reactions 18-20 or 21-22 ; where XH in reaction 20 could be DMS, CH_2O or HO_2. No evidence in favor of one or the other pathway has been obtained.

H_2SO_4 has only been detected by ion chromatography. Its formation is attributed to oxidation of SO_2 or methanesulphonic acid either in the gasphase or in the liquid phase.

4. CONCLUSION

Products have been identified and a mechanism has been proposed for the reaction between NO_3 and DMS in air.
The principal products appear to be the same for the two oxidation reactions NO_3 + DMS and OH + DMS, with the exception of HNO_3, which comes from the hydrogen abstraction by NO_3. Also the ratio SO_2 : CH_3SO_3H (approximately 1:4) in the products of the two oxidation reactions, appears to be the similar[11].
In the marine troposphere the reaction between NO_3 and DMS, with the formation of HNO_3, which is removed by deposition, may represent an efficient nighttime sink for NO_x.

5. ACKNOWLEDGEMENT

The authors gratefully acknowledge the help of Mr. G. Ottobrini in the experimental work and of Mrs. H. Geiss and Mrs. G. Serrini in the ion chromatography measurements.
C. L. as visiting scientist and N. J. and H. S. as grantholders acknowledge financial support by the E.E.C. Commission to carry out this work.

REFERENCES

(1) Andreae M. O. and Andreae T. W., J. Geophys. Res. 1988, 93, 1487.
(2) Andreae M. O. and Raemdonck H., Science 1983, 221, 744.
(3) Barnes I., Bastian V., Becker K. H. and Wirtz K., Dechema-Monographs 1987, 104, 59.
(4) Winer A. M., Atkinson R. and Pitts Jr. J. N., Science 1984, 224, 156.
(5) Tyndall G. S., Burrows J. P., Schneider W., Bingemer H. and Moortgat G. K., Paper presented at Workshop on "Chemistry related to tropospheric ozone", Cost 611 (Working Party 2), Cologne, 12-13 Nov. 1985.
     And
     Burrows J. P., Tyndall G. S., Schneider W., Bingemer H., Moortgat G. K. and Griffith D. W. T., NBS Spec. Publ. (US) 1986, 716, 137.
(6) Dlugokencky E. J. and Howard C. J., J. Phys. Chem. 1988, 92, 1188.
(7) Wallington T. J., Atkinson R., Winer A. M. and Pitts Jr. J. N., J. Phys. Chem. 1986, 90, 5393.
(8) Balla R. J., Nelson H. H. and Mcdonald J. R., Chem. Phys. 1986, 109, 101.
(9) Niki H., Maker P. D., Savage C. M. and Breitenbach L. P., J. Phys. Chem. 1983, 87, 7.
(10) Barnes I., Bastian V., Becker K. H. and Niki H., Chem. Phys. Lett. 1987, 140, 451.
(11) Hatakeyama S., Izumi K. and Akimoto H., Atmospheric Environment 1985, 19, 135.

REACTIONS OF PEROXYACETYL RADICALS WITH SULFUR COMPOUNDS

G. MINESHOS[1], S. GLAVAS[1] and U. SCHURATH[2]

[1]University of Patras, Department of Chemistry GR-26110 PATRAS.

[2]Institut für Physikalische Chemie der Universität Bonn. Wegelerstr.12 D-5300 BONN.

Summary.
Reaction [4] of the peroxyacetyl radical with sulfur compounds was studied at 55°C in N_2 at 1000 mbar total pressure.The radical was generated in equilibrium with peroxyacetyl nitrate and NO_2 in large excess and the pseudo first order decay of PAN was measured by GC/ECD in the absence and presence of several 100 ppm CH_3SH, C_2H_5SH. n-C_4H_9SH and $(CH_3)_2S$. Rate constants of 1.9, 1.4, 5.0 and $0.4 \cdot 10^{-16}$ cm^3 s^{-1} were obtained, relative to the reported rate constant of reaction [-1] at room tmperature. An electron capturing reaction product of NO_2 with the thiols is tentatively assigned to R-SNO (R=CH_3, C_2H_5, C_4H_9).An acceleration of the PAN decay is attributed to NO_2 removal by secondary reactions.

1. INTRODUCTION

NO_2 is a unique photochemical precursor of ozone in the troposphere. The partitioning of total odd nitrogen between NO_2 and photochemically inactivated transient reservoir species such as PAN is therefore of great importance. The lifetime of PAN in the atmosphere is determined (a) by the strongly temperature-dependent reversible dissociation of PAN into NO_2+CH_3CO_3, and (b) by secondary reactions of the peroxyacetyl radicals with NO and other possible substrates.Extremely slow reactions of PAN with various trace gases including SO_2 as the only sulfur compound have been superficially studied by Pate et al (1) before the radical mechanism of these reactions was definitely established (2). A study of CH_3CO_3 radical reactions with various olefins. which took advantage of the correct mechanism. has been reported (3).Here we present an analogous study of peroxyacetyl radical reactions with reduced organic sulfur compounds which are known to be emitted by natural sources.

2.EXPERIMENTAL

All experiments were carried out in a 4.5 L cylindrical glass vessel equipped with two septum ports for sampling and a teflon stopcock for its connection to a vacuum line which in turn was provided with teflon stopcocks. The glass vessel was immersed in an oil bath maintained at a constant temperature of 55 ± 0.5°C. Nitrogen (99.99%) at 1000 mbar was always used as the matrix atmosphere. in order to avoid interferences from oxygen on the ECD used for the monitoring of PAN and other electron capturing compounds.Samples were withdrawn with a gas-tight syringe, and injected into a Shimadzu GC equipped with an ECD operating in the constant current-variable frequency mode.Separation was achieved at 30°C on a glass column, 60 cm long, 3mm i.d., packed with 4.8% QF-1 + 0.18% diglycerol on Chromosorb G-AW-DMCS (80-100 mesh).The PAN peak appeared 3.3 min after injection.

NO_2 was prepared daily from NO (99.85%) and oxygen.A solution of pure PAN in tridecane was obtained by the method of Gaffney et al. (4). The appropriate amount of the NO_2 was first introduced into the evacuated reaction vessel.Subsequently PAN in tridecane was injected into the gas handling system, and its vapour flushed into the vessel as it was brought to 1000 mbar total pressure with pure nitrogen.The final mixing ratios were 2.2 ppm NO_2 and typi- cally 350 ppb PAN.The PAN concentration in the reaction ves- sel was analyzed every 15 minutes for approximately 90 minutes, in order to establish its slow decay rate.After this time between 2 and 10% of CH_3SH (99.5%), C_2H_5SH (99%), C_4H_9SH (99%), and $(CH_3)_2S$ (99%) in N_2 were injected with a syringe, yielding mixing ratios from 50 to 2500 ppm.Mixing in the glass vessel was achieved by repeated pumping with the syringe.Analysis of PAN and other detectable products was continued until more than 90% of PAN was consumed.

3. RESULTS

To check our system control experiments were first carried out with PAN in the presence of excess NO at 28°C in 1000 mbar N_2. The first order plot of its decay was linear throughout yielding a slope of $7 \cdot 10^{-4} s^{-1}$. This value agrees with the reported (5) rate constant of reaction [1],

$$CH_3COO_2NO_2 \longrightarrow CH_3COO_2 + NO_2 \qquad [1]$$

which is the rate limiting in the presence of excess NO.

PAN in nitrogen gas containing 2.2 ppm NO_2 at 55°C decayed at a rate of only $(3.5 \pm 2) \cdot 10^{-5} s^{-1}$, although the first order rate constant of reaction [1] is known to be $2.4 \cdot 10^{-2} s^{-1}$ at this temperature.The decay rate is reduced be- cause reaction [1] is no longer rate determining in the presence of NO_2. The following mechanism is assumed:

$$PAN \rightleftharpoons CH_3CO_3 + NO_2 \qquad [1,-1]$$
$$PAN + wall \longrightarrow products \qquad [2]$$
$$CH_3CO_3 + wall \longrightarrow products \qquad [3]$$

Under steady state conditions this mechanism yields

$$-\frac{d\ln[PAN]}{dt} = k_2 + \frac{k_1 \cdot k_3}{k_{-1}[NO_2]} \qquad [A]$$

This agrees with the experimental results shown in Figure 1. From the intercept of the graph we obtain $k_2 = 2 \cdot 10^{-5} s^{-1}$,

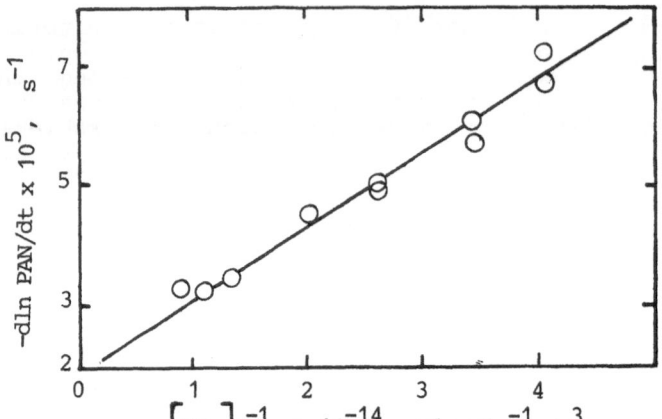

Figure 1. Plot of first order decay constant of PAN vs
1/[NO$_2$] present,cf.equation [A].Measurements at 28°C.

while the slope yields a rate constant $k_3=0.26$ s^{-1} for
removal of CH$_3$CO$_3$ by surface collisions.This latter value is
based on $k_1=0.024$ s^{-1} (5) and $k_{-1}=4.5 \cdot 10^{-12}$ cm^3 s^{-1} (6).The
surface-to-volume ratio of the vessel yields a sticking coef-
ficient α of the order 10^{-4} if k_3 is surface collision rate
limited.However, because k_3 is more likely to be diffusion
limited at ambient pressure, a larger sticking coefficient
would be consistent with our measurements.

When the sulfur compound was added after 90–100 minutes,
PAN started to decay significantly faster.This must be taken
into account by including a reaction of the CH$_3$CO$_3$ radical
with the sulfur compound X in the above mechanism.

$$CH_3CO_3 + X \longrightarrow products \qquad\qquad [4]$$

which leads to the following first order decay rate of PAN
under steady state conditions:

$$-\frac{d\ln[PAN]}{dt} = \{k_2 + \frac{k_1 \cdot k_3}{k_{-1} \, [NO_2]}\} + \frac{k_1 \cdot k_4 \, [X]}{k_{-1} \, [NO_2]} \qquad [B]$$

Figure 2 shows, however, that the decay rate was not
constant in time, but increased as the reaction
proceeded.This is indicative of a more complex reaction
mechanism, as will be discussed in the next SECTION.

It is assumed that the secondary reactions which are
responsible for the rate acceleration at extended reaction
times are unimportant at early reaction times. Rate constants
k_4 can then be obtained by plotting the <u>initial</u> slopes of
semilog plots, like the one displayed in Figure 2, versus the
sulfur compound concentration [X].These plots are shown in
Figure 3.They are indeed linear within experimental accuracy,
yielding intercepts between 0 and $3 \cdot 10^{-5}$ s^{-1}, in slight disagreement with the intercept of $4 \cdot 10^{-5}$ s^{-1} calculated from
equation [B].Combining the slopes in Figure 3 with $k_1=2.4 \cdot 10^{-2}$ s^{-1}, $k_{-1}=4.5 \cdot 10^{-12}$ cm^3 s^{-1} (assuming that this third order
reaction is essentially temperature independent), and the ex

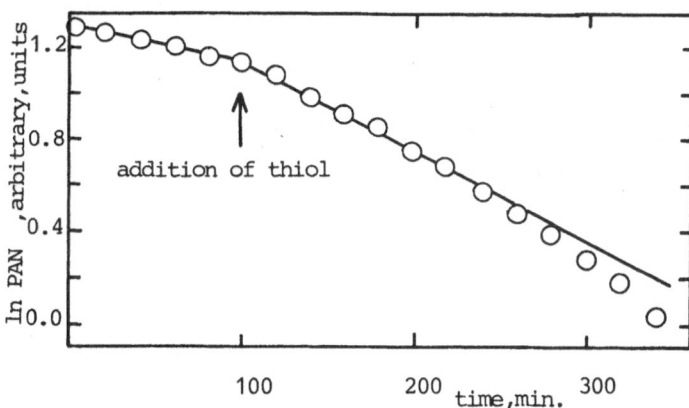

Figure 2. Semilog plot of the PAN concentration vs time, at 55°C, before and after the addition of 150 ppm CH₃SH.

perimental NO_2 mixing ratio of 2.2 ppm, yields the following rate constants k_4:

$$X = CH_3SH \qquad k_4 = 1.9 \cdot 10^{-16} \ cm^3 \ s^{-1}$$
$$X = C_2H_5SH \qquad k_4 = 1.4 \cdot 10^{-16} \ cm^3 \ s^{-1}$$
$$X = C_4H_9SH \qquad k_4 = 5.0 \cdot 10^{-16} \ cm^3 \ s^{-1}$$
$$X = (CH_3)_2S \qquad k_4 = 0.4 \cdot 10^{-16} \ cm^3 \ s^{-1}$$

These rate constants refer to a temperature of 55°C.

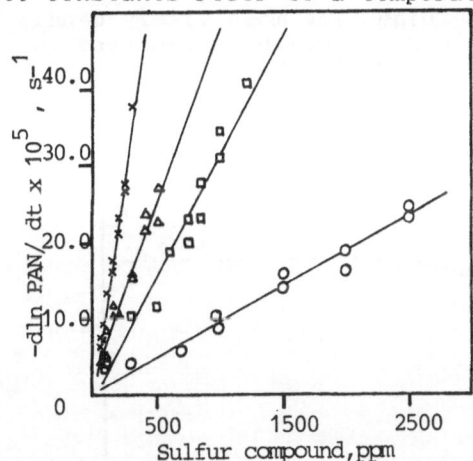

Figure 3. Initial first order rate constants of PAN decay, as function of sulfur compound concentration, △ CH₃SH,□ C₂H₅SH ✕ C₄H₉SH, ○ (CH₃)₂S.

4. DISCUSSION

The obtained rate constants k_4 are dependent on the assumption that the simple reaction mechanism [1] – [4] is adequate, at least during the first 30 minutes of the reaction while the initial slope is determined. The fact that the decay rate of PAN accelerates at longer reaction times can have one or both of the following causes:

1.   NO₂ as the inhibitor of PAN decay (cf. denominator of equation [B]) is removed by the reaction.
2.   NO₂ is converted to NO, which reacts rapidly with CH_3CO_3 radicals.
It is reported in the literature (7,8) that CH_3SH reacts with NO₂ to form NO and other products.However the rate constant given in ref.(8), log $k(cm^3 \ s^{-1}) = -12.78 - 2747/T$, is much too low to account for the acceleration of the dacay rate of PAN.It is thus likely that secondary reactions are operative which enhance the rate of NO₂ removal and/or its conversion to NO.
     The products of the reaction between CH_3CO_3 and the thiols are unknown.The peak of an electron capturing product with retention times of 1.07, 1.90 and 11.80 minutes was observed to grow while the decay of PAN proceeded in the presence of CH_3SH,C_2H_5SH and C_4H_9SH respectively.It was intially thought that these peaks were due either to primary or secondary products of reaction [4]. However,the same peaks grew in at essentially the same rate in the absence of PAN, showing that the unknown compounds must be due to a reaction of NO₂ with the thiols.Small amounts of the unknown compound were also formed in the reaction of 300 ppb PAN with 100 ppm CH_3SH at 23°C in the absence of added NO₂.Because only a very small amount of NO₂ is evolved in this system from the decomposition of PAN, the yield of the unidentified product must also be very small, and hence the sensitivity for its detection by the ECD must be high. The most likely candidate is R-SNO (R= CH_3, C_2H_5 and C_4H_9), which is reported to be formed in the reaction of NO₂ and CH_3SH (7).
     An interesting result was obtained when the formation of the unknown electron capturing compound during the reaction of 2 ppm NO₂ with 85 ppm CH_3SH in N₂ was measured at 23°C.

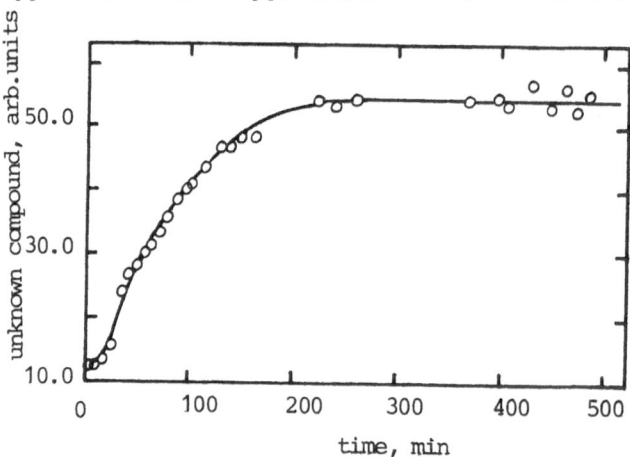

Figure 4.Formation of an unidentified electron capturing compound,in the reaction of 2.2 ppm NO₂ with 85 ppm CH_3SH,at 23°C.

As shown in Figure 4 the product concentration reached a plateau after 200 minutes reaction time,although according to the rate constant given in ref.(8) only a few % of the NO_2 should have reacted until this time. We also observed an induction period preceding the maximum rate of product formation. We conclude that NO_2 is depleated in the presence of the sulfur compounds by secondary reactions, which is consistent with the acceleration of the PAN decay rate in our study of reaction [4].

The rate constants of the sulfur compounds with peroxyacetyl radicals are comparable in magnitude with those of the most reactive olefins (3), which are known to yield epoxides (9). Although the products of reaction [4] remain to be identified, abstraction of the sulfur bound hydrogen is more likely than sulfoxide formation, besause $(CH_3)_2S$ is sigificantly less reactive than the thiols.In conclusion, the reactions of CH_3CO_3 with the sulfur compounds studied are too slow to shorten their lifetimes under atmospheric conditions.

5.ACKNOWLEDGEMENTS

This work was supported by the Commission of the European Communities.A NATO grant is also gratefully acknowledged.

6.REFERENCES

(1) Pate,C.T.,Atkinson, R., and Pitts,J.N.Jr., J.Environ.Sci.Health, AII(1), 19, (1976).
(2) Cox,R.A. and Roffey, M.J., Environ.Sci.Technol.11,900, (1977).
(3) Schurath,U. and Vipprecht,V., in "Physico-Chemical Behaviour of Atmospheric Pollutants", B.Versino and H.Ott (Eds.), D.Reidel,Dordrecht,1980,pp.157.
(4) Gaffney, J.S., Fajer, R. and Senum,G.I., Atmos.Environ. 18,215, (1984).
(5) Schurath, U.,Kortmann, U. and Glavas, S., in "Physico-Chemical Behaviour of Atmospheric Pollutants", B.Versino and G.Angeletti (Eds.), D.Reidel, Dordrecht,1984,pp.27.
(6) Basco,N.and Parmar,S.S., Int.J.Chem.Kinet. 19,115,(1987).
(7) Balla,R.J. and Heicklen,J., J.Phys.Chem. 88,5314,(1984).
(8) Balla,R.J. and Heicklen,J., J.Phys.Chem. 89,4596,(1985).
(9) Diaz, R.R., Selby,K. and Waddington,D.J., J.C.S. Perkin II,360,(1977) and references therein.

THE BrO + ClO REACTION AND THE POLAR STRATOSPHERIC OZONE

G. POULET, I.T. LANCAR ,G. LAVERDET and G. LE BRAS
CRCCHT/C.N.R.S. 45071 ORLEANS cedex2 France

Summary

The rate constant of the title reaction has been measured at 298 K using the discharge flow - mass spectrometry method: $k_1 =$ $(1.13 \pm 0.15) \times 10^{-11}$ cm^3 molecule^{-1}s^{-1}. The measurement of the products has led to the determination of the branching ratios: (0.43 ± 0.10) for the OClO channel and (0.12 ± 0.05) for the BrCl one at 298 K.

1.INTRODUCTION

The role of halogen compounds in the photochemistry involved in the large ozone variations observed in the antarctic stratosphere during springtime is now recognized. If chlorine chemistry is clearly implied, bromine compounds may also contribute to the ozone depletion, as suggested earlier by modelling (1), via the synergistic effect induced by the first channel of reaction 1 :

$$BrO + ClO \longrightarrow Br + Cl + O_2 \quad (1a)$$

Other recent modelling (2) has shown that the observed diurnal variations of OClO in the antarctic stratosphere could be explained if the OClO channel (1b), which is the only one known source of OClO, is associated with a third channel forming BrCl (1c):

$$BrO + ClO \longrightarrow Br + OClO \quad (1b)$$
$$BrO + ClO \longrightarrow BrCl + O_2 \quad (1c)$$

Previous kinetic data have been obtained for reaction 1 (3-7) with k_1 ranging from 0.25 to 1.4 x 10^{-11} cm^3 molecule^{-1} s^{-1} at 298 K. Discrepancies were also observed for the branching ratios since BrCl was not detected previously, except in the recent study of Sander and Friedl (7) who measured a branching ratio of 8% for channel (1c).

2.EXPERIMENTAL

The discharge flow method was used with a pyrex reactor, coated with halocarbon wax in order to reduce the catalytic recombination of active species. The detection of all the species involved in the study of reaction (1) was made by means of a molecular beam mass spectrometer (Balzers QMG 420). The configuration used for the introduction of reactants is schematically shown in Figure 1.

ClO was produced in excess over BrO, with the following sources of radicals :

$$Cl + OClO \longrightarrow 2ClO$$
$$Br + O_3 \longrightarrow BrO + O_2$$

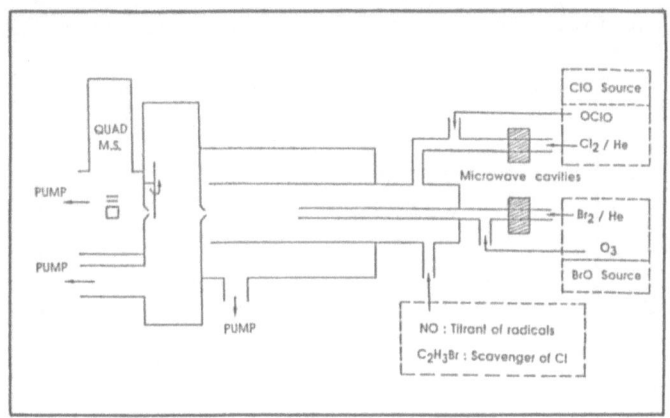

Figure 1 : Schematic diagram of the discharge flow apparatus.

Cl and Br, produced by microwave discharge in Cl_2 and Br_2 respectively, were used in excess over OClO and O_3, in order to keep negligible any secondary reaction involving OClO and O_3. The excess Cl was scavenged by C_2H_3Br following :

$$Cl + C_2H_3Br \longrightarrow Br + C_2H_3Cl$$

The introduction of C_2H_3Br, upstream from the region where the reaction between BrO and ClO took place, made impossible the fast reaction of BrO with Cl. C_2H_3Br was preferred to C_3H_8 or C_4H_{10} (isobutane) as a scavenger for Cl since, when these molecules were used, fast secondary reactions were observed between ClO and the alkyl radicals formed in the scavenging reaction.

Titration of XO (BrO and ClO) radicals was made using NO :

$$XO + NO \longrightarrow X + NO_2$$

O_3 was prepared from an ozonizer and OClO from the reaction H_2SO_4 + $KClO_3$ whereas BrCl was produced in situ for the calibration experiments from the fast reaction :

$$Cl + Br_2 \longrightarrow BrCl + Br$$

3. RESULTS
3.1- Overall rate constant measurements:
At T = 298 K and P = (0.60 - 1) Torr, with (ClO) = (10-30) x (BrO), ranging from 5×10^{12} to 5×10^{13} molecule cm^{-3}, the decay rates of BrO (measured at m/e = 95 or 97) as a function of ClO concentration were strictly linear (Fig.2), leading to the following value for k_1 :

$$k_1 = (1.13 \pm 0.15) \times 10^{-11} \text{ cm}^3 \text{ molecule}^{-1}\text{s}^{-1}$$

The error includes one standard deviation and an estimated 10% of systematic error. The zero intercept (15 ± 13) s^{-1}, which represents the heterogeneous loss rate of BrO, is in acceptable agreement with the upper limit of 3 s^{-1} measured in the absence of ClO.

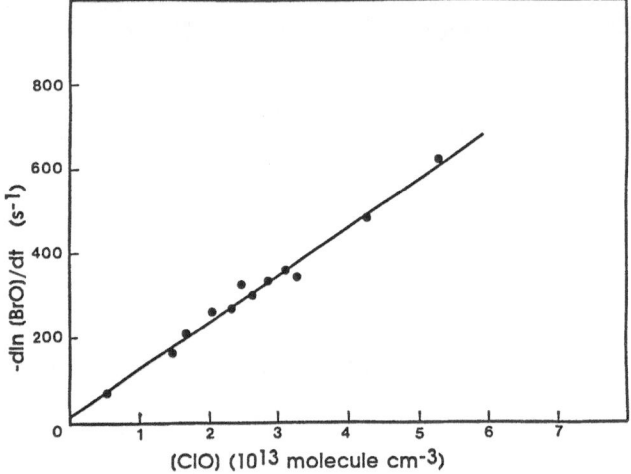

Figure 2 : Pseudo-first order rate constant measurements.

3.2- Branching ratio measurements:

In a first series of experiments, OClO produced in channel (1b) and BrCl produced in channel (1c) were detected for a reaction time corresponding to a total consumption of BrO. After appropriate calibrations, OClO and BrCl concentrations were measured as a function of BrO concentration,as shown in Figure 3.

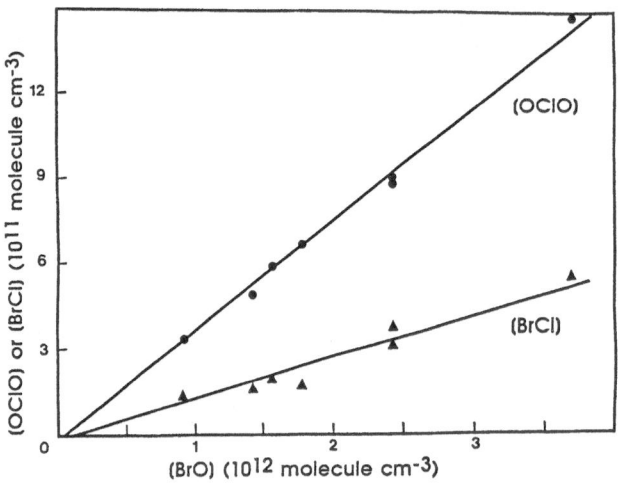

Figure 3 : Absolute measurements of OClO and BrCl channels.

A computer simulation of the whole mechanism was necessary to account for possible secondary reactions involving OClO and BrCl. After these corrections, the branching ratio for channels (1b) and (1c) were found to be, at 298 K :

$$k_{1b}/k_1 = (0.43 \pm 0.10)$$
$$k_{1c}/k_1 = (0.12 \pm 0.04)$$

The errors include one standard deviation and 15% systematic error.

In a second series of experiments, the same ClO source was used but BrO was produced differently, from the conversion of Br atoms produced by Cl + C_2H_3Br into BrO, via the addition of an excess of O_3. One advantage of this procedure was the absence of Br_2 in the reactor, which suppressed any additional formation of BrCl coming from Cl + Br_2 ---> BrCl + Br . The measurement of BrCl as a function of ClO allowed for the direct determination of the ratio k_{1c}/k_{1b}, which was given by the slope of the straight line shown in Figure 4:

$$k_{1c}/k_{1b} = (0.28 \pm 0.05)$$

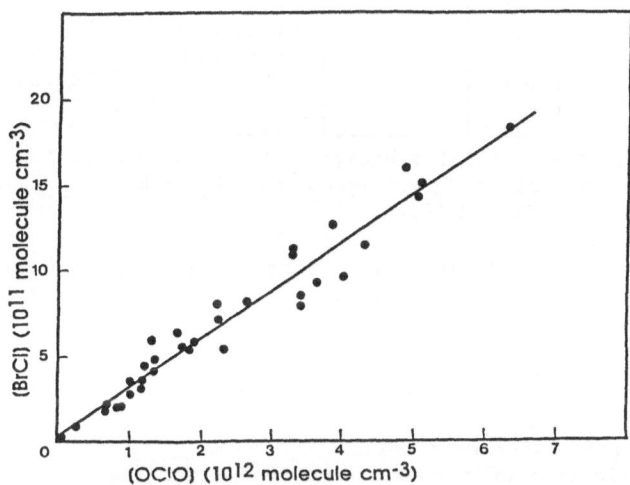

Figure 4 : Relative measurements of BrCl channel.

Under the conditions used, no correction was necessary and, using the value obtained for k_{1b}/k_1 in this study, the resulting branching ratio for the BrCl formation was :

$$k_{1c}/k_1 = (0.12 \pm 0.05)$$

This result is in excellent agreement with the first determination.

4.DISCUSSION AND CONCLUSION

The present results can be compared with the previous ones which are summarized in Table I. In this table, the very recent data of Sander and Friedl (8,9) have been included. The values of k_1 and of k_{1b}/k_1 obtained at 298 K in this work are in good agreement with the other recent determinations, considering the experimental difficulties of such radical-radical reaction studies.

TABLE I : KINETIC DATA FOR REACTION (1) AT 298 K :

$$
\begin{array}{lll}
 & ---> & Br + Cl + O_2 \quad (1a) \\
BrO + ClO & ---> & Br + OClO \quad (1b) \\
 & ---> & BrCl + O_2 \quad (1c)
\end{array}
$$

k_1 (10^{-11} cm^3 molecule^{-1}s^{-1})	Method (a)	k_{1a}/k_1	k_{1b}/k_1	k_{1c}/k_1	ref
0.25±0.04	F.P.				3
1.34±0.34	D.F.	0.50	0.50		4
0.82±0.10	D.F.	0.45±0.10	0.55±0.10	< 0.02	5
1.40±0.20	D.F.		0.45		6
1.29±0.16	F.P.		0.59±0.10		8
1.29±0.19	D.F.	(b)	0.48±0.07	0.08±0.03	9
1.13±0.15	D.F.		0.43±0.10	0.12±0.05	this work

Notes: (a) F.P. = Flash Photolysis; D.F. = Discharge Flow
       (b) Not directly measured, but $(k_{1a} + k_{1c})/k_1$ found to be (0.48±0.10).

The value obtained for k_{1c}/k_1 correctly agrees with that of ref 9 (0.08 ± 0.03) ,but it disagrees with ref 5 where no evidence was found for BrCl production.

The mechanism of reaction (1) can be compared with the general mechanism for the low pressure recombination of XO radicals (X = halogen atom) which proceeds via a metastable intermediate (X_2O_2) possessing a non unique structure. Here, the BrCl production would result from the formation of a four-center complex.

The implication of the existence of a significant BrCl channel in reaction 1 may be important in the stratospheric ozone chemistry: first, reaction 1 is the only recognized source of OClO and, as previously shown (2), the inclusion of the new channel 1c in a photochemical model of the antarctic stratosphere leads to a better agreement between the calculated and observed OClO diurnal variations during the formation of the "ozone hole"; secondly, during the polar night, BrO can be rapidly converted into the inactive form BrCl. This conversion can occur within less than one hour if the typical stratospheric concentrations of BrO and ClO are considered. Thus, channel 1c of the BrO + ClO reaction would make BrCl an efficient nighttime reservoir of bromine in the antarctic stratosphere. A similar behaviour can be anticipated in the arctic stratosphere from the preliminary analysis of a recent campaign where no BrO was observed during the night whereas daytime mixing ratios of a few pptv were measured.

REFERENCES

(1) Yung, Y.L., Pinto, J.P., Watson, R.T. and Sander, S.P. (1980). J. Atmos. Sci., 37, 339.

(2) Salawitch, R.J., Wofsy, S.C. and McElroy, M.B.(1988). Planet Space Sci., 36 , 213.

(3) Basco, N.and Dogra, S.K.(1971). Proc. Roy. Soc. Lond. A, 323, 417.

(4) Clyne, M.A.A. and Watson, R.T.(1977). J. Chem. Soc. Far. Trans.I, 73, 1169.

(5) Hills, A.J., Cicerone, R.J., Calvert, J.G. and Birks, J.W. (1988). J. Phys. Chem., 92 , 1853.

(6) Toohey, D.W. and Anderson, J.G. (1988). J. Phys. Chem., 92 , 1705.

(7) Sander, S.P. and Friedl, R.R. (1988). Geophys. Res. Lett., 15, 887.

(8) Sander, S.P. and Friedl, R.R. (1989). J. Phys. Chem., 93, 4764.

(9) Friedl, R.R. and Sander, S.P. (1989). J. Phys. Chem., 93, 4756.

THE KINETICS OF THE REACTION ClO + ClO = Cl$_2$O$_2$ AND ITS IMPLICATIONS FOR POLAR OZONE DEPLETIONS.

G.D.HAYMAN and R.A.COX
Chemical Physics Group, B. 551, Harwell Laboratory
Didcot, Oxon, OX11 0RA, UK.

Summary

The Molecular Modulation technique has been applied to study the kinetics of the association reaction of ClO radicals to form a dimer, Cl$_2$O$_2$.

$$ClO + ClO + M = Cl_2O_2 + M \qquad (1)$$

The ClO radical was generated by the low intensity photolysis at 350 nm of mixtures of Cl$_2$/Cl$_2$O or Cl$_2$/O$_3$. The kinetics of the ClO radical were followed by UV absorption spectroscopy. Modulation waveforms were recorded as functions of the temperature (203-243K) and the pressure (10-200 Torr (233,243K)/10-600 Torr (203-223K)). The observed rate constants showed a strong pressure dependence at each temperature and were analysed in terms of the Troe parameterisation. The temperature dependence of the limiting low pressure rate constant was determined to be

$$k_o(T) = (6.1\pm0.6) \times 10^{-32} (T/300)^{-1.1\pm0.7} \text{ cm}^6 \text{ molecule}^{-2} \text{ s}^{-1}$$

1. Introduction

The major depletions of stratospheric ozone observed over Antarctica each spring [1,2] and the measurements of elevated concentrations of the ClO radical [3] and the related species, OClO [4] has once again focussed attention on the chemistry which influences the abundance of stratospheric ozone. The chemistry believed to control ozone concentrations at mid-latitudes is perturbed in the low temperature regions of the polar stratosphere. Heterogeneous reactions at the surface of polar stratospheric clouds which are formed at these low temperatures, release photolytically labile chlorine-containing species from inactive reservoirs and sinks [5,6]. At the same time, nitrogen oxides are sequestered into the condensed phases as HNO$_3$. Photolysis in the near UV liberates active chlorine which catalyses the ozone depletion. Three catalytic cycles have been recognised as causing the ozone loss; these are a ClO dimer cycle [7], given below, a coupled BrO-ClO cycle [8] and a cycle coupling HO$_2$ with ClO [9].

$$
\begin{aligned}
ClO + ClO + M &= Cl_2O_2 + M \\
Cl_2O_2 + h\nu &= ClOO + Cl \\
ClOO + M &= Cl + O_2 + M \\
2(Cl + O_3 &= ClO + O_2) \\
\text{net} \qquad 2 O_3 &= 3 O_2
\end{aligned}
$$

Recent studies have focussed on aspects of this cycle;

    (a)   determination of the rate constant at room temperature [10].
    (b)   the number and structure of the products formed in the ClO self-reaction [7,11].
    (c)   the stability of the ClO dimer [11].
    (d)   the quantum yield and UV absorption spectrum, particularly in the long-wavelength region, leading to the ClO dimer photolysis rate [11].
    (e)   the products of the ClO dimer photolysis [11].

These studies have revealed that the rate determining step of the cycle is the rate of formation of the dimer, that the dimer does absorb significantly in the long wavelength region and appears to proceed with unit quantum efficiency to give the products as shown.

The rate constant for the dimerisation reaction is still the subject of considerable uncertainty. The present study was undertaken to extend the measurements of the rate constant to the temperatures and pressures appropriate to the polar stratosphere. At the same time as the work presented in this paper was in progress, an independent study was also undertaken by Sander and co-workers using the flash photolysis technique [12]. The results derived from the two studies are compared.

2. Experimental

The molecular modulation technique has been described in greater detail elsewhere [13] but involves the production of transient species by low-intensity intermittent photolysis. The concentration of the radicals builds up and reaches a steady-state when production and loss are balanced. When photolysis stops, the radical concentration decays. The deviation of the radical absorption profile from the square wave driving the photolysis gives the kinetic data of interest.

The ClO radicals were produced by photolysis of Cl_2 in the presence of O_3 or Cl_2O in a jacketted silica reaction vessel, 120 cm in length and 3 cm in diameter, which could be stabilised at temperatures between 200 and 330 K. The silica cell was surrounded by six photolysis lamps (Phillips, Blacklight). Cl_2O and O_3 were generated in-situ by passing a Cl_2/N_2 mixture through a column of yellow HgO or by electric discharge on O_2. The gas mixtures, $Cl_2/Cl_2O/N_2$ and $Cl_2/O_3/O_2/N_2$ were flowed continuously through the cell with a typical residence time of 10 s. The residence time was chosen as a compromise between dimer lifetime, number of photolysis cycles per residence time and signal strength.

The concentrations of ClO, initial reagents (Cl_2, Cl_2O and O_3) and the product Cl_2O_2 could be followed by UV absorption spectroscopy. The emission of a deuterium lamp was passed longitudinally through the cell. A monochromator (SPEX 1700, 1 m focal length) was used to disperse the incident radiation before detection by a photomultiplier tube (EMI, 9783B). The reagent conentrations were determined at 360 nm ($\sigma[Cl_2] = 1.30 \times 10^{-19}$ cm^2 molecule^{-1} [14]) or 260 nm ($\sigma[Cl_2O] = 1.83 \times 10^{-20}$ cm^2 molecule^{-1} [15], ($\sigma[O_3] = 1.09 \times 10^{-17}$ cm^2 molecule^{-1} [16]). ClO was monitored at 277.2 nm which corresponds to the 11-0 vibrational band of the A($^2\Pi$)-X($^2\Pi$) electronic transition. The output of the photomultiplier was then digitised and converted into absorption-time profiles which could be stored on a microcomputer for storage and subsequent analysis.

Typical reagent concentrations were Cl_2, $0.5-2 \times 10^{16}$ molecule cm^{-3}, Cl_2O/O_3 $4-8 \times 10^{14}$ molecule cm^{-3}. N_2 was added as the bath gas and its concentration was adjusted to bring the total pressure to the desired value. For temperatures below 223K, pressures up to atmospheric pressure could be used but above this temperature, the accessible pressure range was restricted because of the effect of the thermal decomposition of the dimer on the ClO kinetics.

3. Results

Absorption waveforms were recorded at 10 K intervals from 203 to 243 K and at pressures up to 600 Torr. A typical absorption profile recorded is shown in figure 1a. Each waveform recorded contained not only absorptions due to ClO but also components from the precursors (Cl_2O or O_3) and the product Cl_2O_2. To overcome the difficulty of extracting a ClO profile from the composite absorption, computer modelling was used to analyse the waveform and extract the rate constants of interest. A simplified model containing the following reactions was used

$$
\begin{array}{rcll}
Cl_2 + h\nu & = & Cl + Cl & (2) \\
Cl_2O + h\nu & = & ClO + Cl & (3) \\
Cl + Cl_2O & = & Cl_2 + ClO & (4) \\
\\
ClO + ClO + M & = & Cl_2O_2 + M & (1) \\
Cl + Cl_2O_2 & = & Cl_2 + ClOO & (5) \\
\text{flow in/out} & & &
\end{array}
$$

For the Cl_2/O_3 system, Cl_2O can be replaced by O_3 and reaction (3) was removed.

As the experiments were conducted on flowing systems, the simulations were taken out to at least four residence times to ensure reproducibilty of the modulation profiles.

One of the key parameters in the above analysis is the ClO cross-section for the 11-0 band at 277.2 nm which is highly temperature dependent. For the current study, the temperature dependence was based on a recent study of Sander and Friedl [17] for the 12-0 band and scaled so that the value of the 11-0 band at 298 K was 7.26×10^{-18} cm^2 molecule^{-1} [18].

$$\sigma(T, 277.2nm) = 4.77 \times 10^{-18} \exp(125/T) \text{ cm}^2 \text{ molecule}^{-1}$$

The cross-sections of ClO were then fixed at the values given by the above expression. The simulations were carried out using the Harwell FACSIMILE numerical integration package [19]. The programme has a facility to optimise given parameters to obtain the best fit of the simulated profiles with the observed data. The rate constant for the formation of the dimer (k_1) and the photolysis rate constant of Cl_2 were optimised using a non-linear least squares routine. The photolysis rate constant showed little variation for experiments conducted with the same number of lamps but compensated for inaccuracies or variations in the measured Cl_2 concentration. Figure 1a compares the observed data with the simulated waveform obtained after optimisation. The excellent agreement provides strong support for the model interpretation. Figure 1b presents the contributions to the total absorption by the other species present in the system. The observed rate constants determined at 223 K are plotted as a function of pressure in figure 2 and the averaged rate constants obtained as a function of pressure for the 5 temperatures between 203 and 243 K are summarised in Table 1. The strong pressure dependence observed is clear evidence for an association reaction.

The rate constants obtained as a function of pressure at each temperature were then interpreted in terms of the Troe formalism [20]. The pressure range of the data precluded a simultaneous determination of the three parameters used in the parameterisation. Therefore, F_c was fixed at 0.6, a value adopted in the latest kinetics data evaluations [18], whilst the low pressure slope and the high pressure limiting value were varied. Table 2 shows the result of this analysis together with the results for the low pressure rate constant determined in an earlier study [10].

A weighted linear regression ($w_i = 1/\sigma_i^2$) gave the following expressions

$$k_{1,0}(T) = (6.1\pm0.6) \times 10^{-32} (T/300)^{-1.1\pm0.7} \text{ cm}^6 \text{ molecule}^{-2} \text{ s}^{-1}$$
$$\text{for } 203 < T/K < 243$$

$$k_{1,0}(T) = (5.7\pm0.5) \times 10^{-32} (T/300)^{-1.3\pm0.3} \text{ cm}^6 \text{ molecule}^{-2} \text{ s}^{-1}$$
$$\text{for } 203 < T/K < 338$$

4. Discussion

The low pressure rate constants derived in the present study clearly show the negative temperature dependence expected for an association reaction of two radicals. The results appear in good agreement with the rate constants obtained using the Cl_2/O_2 photolysis system at temperatures near room temperature (268-338 K) [10] but with a smaller temperature coefficient than our earlier work suggested. A recent and independent study of the title reaction by Sander et al. [12] using the flash photolysis technique has indicated that the rate constant has a much stronger temperature dependence.

$$k_{1,0}(T) = 1.8 \times 10^{-32} (T/300)^{-3.6} \text{ cm}^6 \text{ molecule}^{-2} \text{ s}^{-1}$$
$$\text{for } 194 < T/K < 247$$

It appears from figure 3 that there is good agreement at the lowest temperatures used, not only with the low pressure and limiting high pressure rate constants but also with the observed bimolecular rate parameters. There is however a large difference between the two data sets at the higher temperatures (230-247). The reason for this discrepancy is not clear but may reflect inadequate data, an inappropiate data treatment, an incorrect choice of cross-section or perhaps a subtle difference in the chemistry occurring on the different timescales of the two techniques. The existence of an intercept indicating a pressure-independent component to the rate constant is a further possibilty. The data however suggest that this is not the case. The origin of this difference at the higher temperatures needs to be rationalised given the central rôle of the ClO dimer in the polar ozone depletions.

5. Stratospheric Implications

The rate constants determined in this work and from the recent flash photolysis studies [12] are in good agreement for conditions appropiate to the polar stratospheres (north and south). They are both significantly slower (about 40%) than the extrapolation of the earlier room temperature results would indicate [10] but are however within the extrapolation errors suggested by these earlier data. The title reaction is the rate determining step in the ClO dimer catalytic cycle and its efficiency for ozone depletion scales with this rate constant. The smaller value of this rate constants promotes the rôle of the other two cycles (BrO-ClO and ClO-HO$_2$). Additional loss mechanisms may also be required before a full quantitative description of this phenomenon can be given.

Acknowledgements

The support of the Fluorocarbon Panel of the Chemical Manufacturers' Association and the UK Department of the Environment is gratefully acknowledged.

References

(1)   Farman, J.C., Gardiner, B.G. and Shanklin, J.D., (1985) Nature, **315**, 207.

(2)   Stolarski, R.S., Krueger, A.J., Schoeberl, M.R., McPeters, R.D., Newman, P.A. and Alpert, J.C., (1986) Nature, **322**, 808.

(3)   da Zafra, R.L., Jaramillo, M., Parrish, A., Solomon, P.M., Connor, B. and Barrett, J., (1987) Nature, **328**, 408.

        Solomon, P.M., Connor, B., da Zafra, R.L., Parrish, A., Barrett, J. and Jaramillo, M., (1987) Nature, **328**, 411.

(4)   Solomon, S., Mount, G.H., Sanders, R.W. and Schmeltekopf, A.L., (1987) J.Geophys.Res., **92D**, 8329.

(5)   Molina, M.J., Tso, T.-L., Molina, L.T. and Wang, F.C.-Y., (1987) Science, **238**, 1253.

(6)   Tolbert, M.A., Rossi, M.J., Malhorta, M. and Golden, D.M., (1987) Science, **238**, 1258.

(7)   Molina, L.T. and Molina, M.J., (1987) J.Phys.Chem., **91**, 433.

(8)   McElroy, M.B., Salawitch, R.J., Wofsy, S.C. and Logan, J.A., (1986) Nature **321**, 759.

(9)   Solomon, S., Garcia, R.R., Rowland, F.S. and Wuebbles, D.J., (1986) Nature, **321**, 755.

(10)   Hayman, G.D., Davies, J.M. and Cox, R.A., (1986) Geophys.Res.Lett., **13**, 1347.

(11)   Cox, R.A. and Hayman, G.D., (1988) Nature, **332**, 796.

(12)   Sander, S.P., Friedl, R.R. and Yung, Y.L., (1989) Science, **245**, 1095.

(13)   Cox, R.A., Sheppard, D.W. and Stevens, M.P., (1982) J.Photochem, **19**, 189.

        Jenkin, M.E. and Cox, R.A., (1985) J.Phys.Chem., **89**, 192.

(14)   Schneider, W. (MPI,Mainz), personal communication.

(15)   Watson, R.T., (1977) J.Phys. and Chem.Ref.Data, **6**, 871.

(16)   Molina, L.T. and Molina, M.J., (1986) J.Geophys.Res., **91D**, 14501.

(17)   Sander, S.P. and Friedl, R.R., (1989) J.Phys.Chem., **93**, 4764.

(18)   NASA Panel for Data Evaluation, 'Chemical Kinetic and Photochemical Data for Use in Stratospheric Modelling', (1987) Evaluation number 8, JPL Publication 87-41.

(19)   Curtis, A.R. and Sweetenham, W.P., 'FACSIMILE/CHECKMAT user manual', (1987) AERE report R-12805, (HMSO, London).

(20)   Troe, J., (1979) J.Phys.Chem., **83**, 114.

Table 1 : Averaged rate constants for ClO + ClO = Cl$_2$O$_2$ as a function of temperature and Pressure

Temperature in K	Pressure in Torr	Number of Observations	Rate Constant, k_1 in 10^{-13} cm^3 molecule^{-1} s^{-1}
203	10.4(0.1)[1]	3	0.62 (5)[2]
	20.2(0.1)	2	0.84 (2)
	31.3(0.1)	3	1.10 (3)
	41.1(0.1)	3	1.36(14)
	49.9(0.1)	2	1.63 (5)
	76.1(0.3)	2	2.60(20)
	104.8(0.4)	3	2.85(15)
	201.8(0.7)	3	5.12(15)
	378.0(3.0)	3	7.95(40)
213	10.3(0.5)	5	0.59 (3)
	20.3(0.2)	3	0.93 (4)
	30.3(0.1)	2	1.31 (3)
	43.1(0.4)	1	1.43 (4)
	50.1(0.6)	5	1.62 (4)
	75.8(0.3)	3	2.22 (4)
	102.9(0.2)	3	2.83 (7)
	198.4(0.2)	2	4.78 (16)
	389.6(1.8)	2	10.27 (53)
223	20.5(0.2)	4	0.69 (4)
	40.3(0.9)	9	1.19(15)
	60.0(0.4)	2	1.84 (4)
	81.7(0.2)	2	2.21(12)
	105.3(0.4)	4	2.50(17)
	148.3(0.2)	2	3.40(17)
	207.4(0.1)	5	5.15(47)
	409.6(6.5)	2	6.93(40)
	583.3(1.6)	2	9.32(52)
233	20.1(0.1)	3	0.68 (2)
	39.8(0.1)	2	1.13 (3)
	59.0(0.1)	2	1.71 (6)
	80.5(0.6)	3	2.18(14)
	106.0(1.9)	4	2.40(11)
	156.1(0.6)	3	3.73(21)
	204.2(0.7)	3	4.64 (4)
243	10.1(0.1)	4	0.21 (-)
	20.0(0.1)	3	0.55 (1)
	36.4(0.2)	2	0.81 (9)
	51.1(0.4)	5	1.33 (3)
	65.6(0.2)	2	1.66 (1)
	102.2(0.2)	2	2.40(40)

Notes (1) Parentheses represent range of pressures of measurements
(2) Parentheses represent one standard deviation in units of the last digit

**Table 2 : Low and high pressure limiting rate constants
for the association reaction ClO + ClO = Cl$_2$O$_2$**

Temperature in K	$k_o(T)$ in 10^{-32} cm^6 molecule^{-2} s^{-1}	$k_\infty(T)$ in 10^{-12} cm^3 molecule^{-1} s^{-1}
203	8.5±0.4[a]	5.4±0.9
213	10.0±0.4	3.0±0.7
223	8.7±0.4	4.3±0.9
233	7.8±0.4	7.2 (+5.8,−3.2)[b]
243	7.3±0.4	9.0 (+18.0,−6.0)
268	6.8±0.6[c]	−
288	6.7±1.6[c]	−
308	5.6±0.7[c]	−
338	3.6±0.7[c]	−

Notes (a) One standard deviation in units of the last digit
(b) Confidence limits (5% and 95%)
(c) Modified data from earlier work [10]

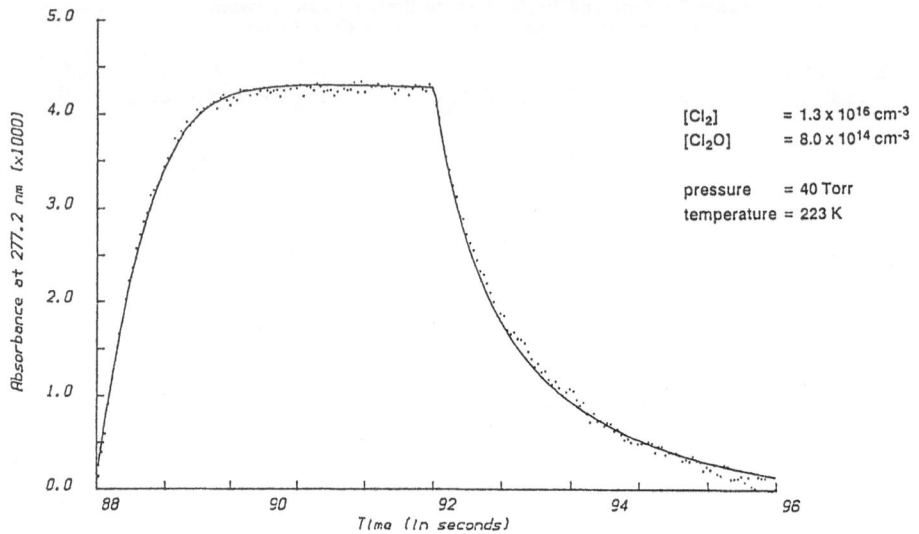

Figure 1a – Observed and optimised waveforms obtained when Cl_2 was photolysed at 350 nm in the presence of Cl_2O at 223 K and a pressure of 20 Torr.

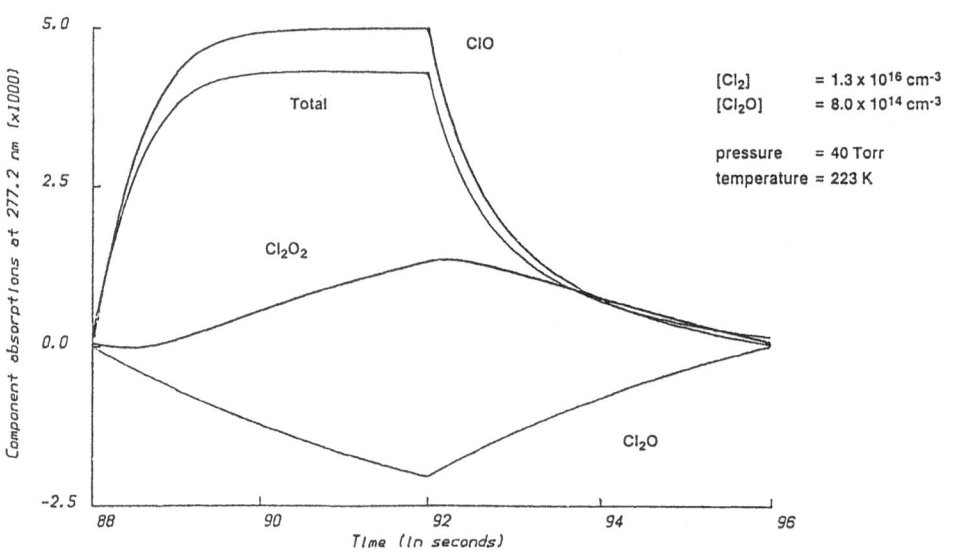

Figure 1b – Contribution of the individual absorbing species to the simulated waveform shown in figure 1a.

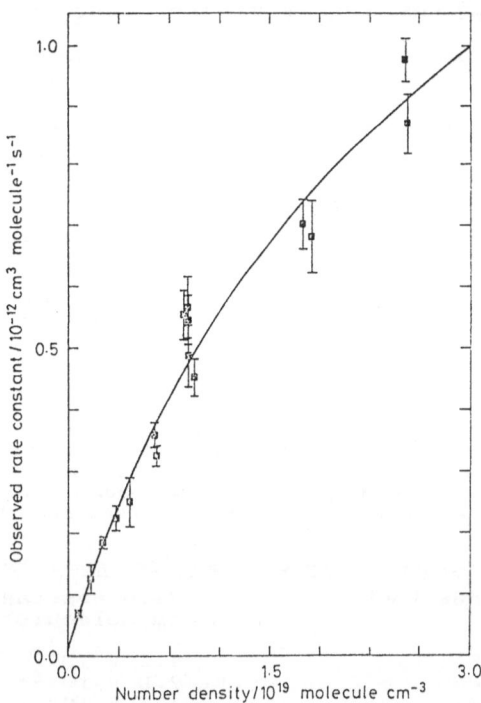

Figure 2 – Pressure dependence of the observed bimolecular rate constant at 223K.

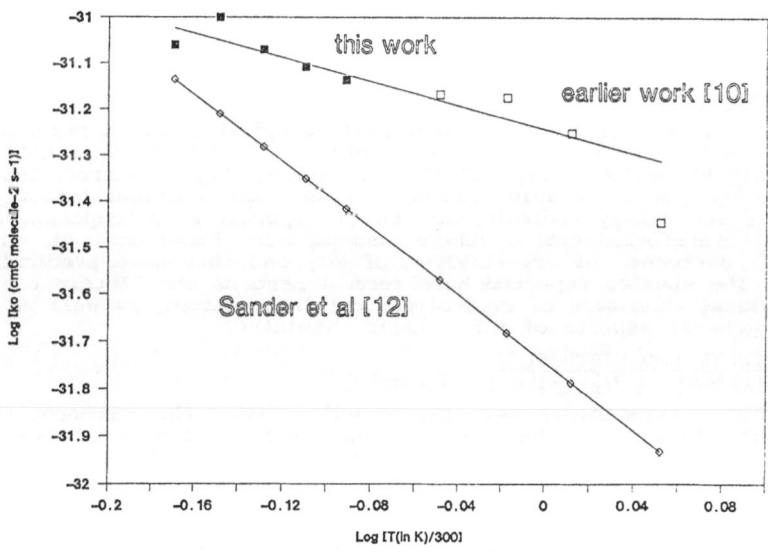

Figure 3 – Comparison of the temperature dependence for the low pressure slope obtained in this work and the work of Sander et al.[12].

THE REACTIONS OF NO$_3$ WITH ATOMS AND RADICALS

Peter Biggs, Anne C. Brown, Carlos Canosa-Mas, Peter Carpenter,
Paul S. Monks, and Richard P. Wayne.

Physical Chemistry Laboratory, University of Oxford,
South Parks Road, Oxford, OX1 3QZ, UK

SUMMARY

This paper reports direct investigations of interactions of the NO$_3$ radical with atomic and radical species. Reaction partners now studied quantitatively or qualitatively by 'direct' techniques include F, Cℓ, H, O, N, OH, HO$_2$, CℓO, and NO$_3$ itself. The reactions of Br and BrO with NO$_3$ have been invoked in 'indirect' measurements. Discharge-flow methods have been used in general, although work on Cℓ, CℓO, and HO$_2$ was conducted in collaboration with Dr R.A. Cox at AERE, Harwell, using the modulated-photolysis technique. For studies of the self-reaction of NO$_3$, we employed two techniques: laser flash photolysis (using F$_2$ photolysis in the presence of HNO$_3$ as the source of NO$_3$), and a newly-developed stopped-flow method. The rate constants obtained at room temperature are tabulated below

Room-temperature rate constants for some reactions of NO$_3$

Reactant	Rate constant cm^{3}molecule^{-1}s^{-1}	Reference
H	1.1×10^{-10}	1
OH	2.0×10^{-11}	1
O	1.7×10^{-11}	2
HO$_2$	3.5×10^{-12}	3
Cℓ	5.5×10^{-11}	4
CℓO	4.0×10^{-13}	4
N	$\geqslant 3 \times 10^{-11}$	–
Br	1.2×10^{-11}	5
BrO	7×10^{-12}	5
NO$_3$	3.3×10^{-16}	6

1. INTRODUCTION

The nitrate radical is now well-established as an important oxidant in tropospheric chemistry at night. Although apparently relatively unreactive, the radical does undergo rapid interactions with other unpaired-spin species. Some such reactions (notably with HO$_2$ and alkoxy radicals) may be of importance in tropospheric chemical transformations; others enhance our knowledge of the general patterns of reactivity of NO$_3$ and thus have predictive value. The studies reported here form a part of our interest in the physical chemistry of radical-radical interactions, as well as in the general aspects of atmospheric chemistry.

2. RESULTS AND DISCUSSION
2.1 Reactions of NO$_3$ with H, OH, and O

These experiments were performed[1,2] using the discharge-flow technique, the reaction between F and HNO$_3$ being employed to generate NO$_3$, and a discharge through the molecular parent gas being used to produce H or O atoms. Hydroxyl radicals were generated either by the reaction between H and NO$_3$ itself (see below), or by the reaction of H with NO$_2$. Multipass absorption at λ = 662 nm was used to obtain absolute concentrations of the NO$_3$ radical. Concentrations of hydroxyl radicals and of atomic oxygen were determined by resonance fluorescence. Since preliminary experiments showed that OH was a product of the interaction of H with NO_3

$$H + NO_3 \longrightarrow OH + NO_2 \tag{1}$$

it was possible to employ the OH-resonance fluorescence intensity to monitor [H] in these experiments. Various analytical and numerical methods[1,2] of obtaining multi-parameter least-squares fits to the data were adopted, and all gave essentially the same results. The criteria for determinations of absolute concentrations of H and O were considered, and a new method devised[7] for application to reactions with species such as NO_2 under conditions where conversion is incomplete on the time scale of the experiments.

The reaction

$$OH + NO_3 \longrightarrow HO_2 + NO_2 \tag{2}$$

competes with formation of OH, so that [OH] varies with contact time in a complex manner, building up to a maximum at times around 5 ms in our system. Our data were analysed by numerical integration of the complete kinetic scheme, although we have shown[1] that a simple treatment based on formation of OH in reaction (1), and losses described by pseudo first-order processes including reaction (2) gives almost identical results. Direct measurements of k_2 using OH prepared by the reaction between H and NO_2 upstream from the point of NO_3 addition gave essentially identical results.

Figure 1 shows some results from the experiments on the reaction

$$O + NO_3 \longrightarrow NO_2 + O_2 \tag{3}$$

between O and NO_3. It demonstrates, for four experimental runs,

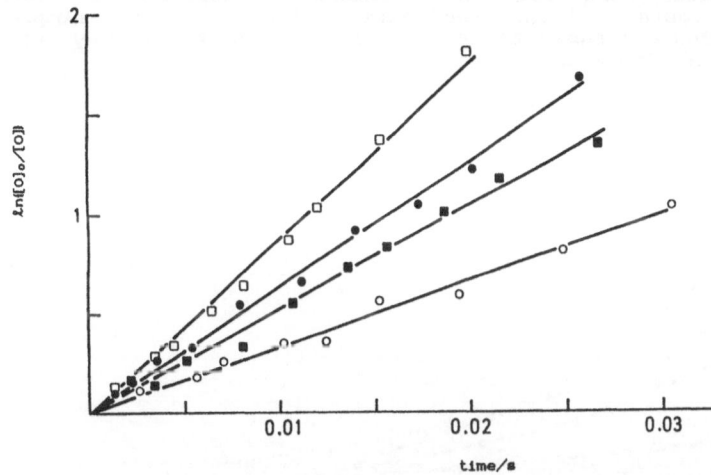

Figure 1. Time-dependent loss of O in the presence of NO_3 plotted as a first-order process □ , $[NO_3]_0 = 5.4 \times 10^{12}$ molecule cm^{-3}; ● , $[NO_3]_0 = 3.6 \times 10^{12}$ molecule cm^{-3}; ■ , $[NO_3]_0 = 4.1 \times 10^{12}$ molecule cm^{-3}; ○ , $[NO_3]_0 = 2.3 \times 10^{12}$ molecule cm^{-3}.

that the time-dependent loss of O is almost first order. The pseudo first-order rate constants obtained in all the experiments are displayed as a function of [NO_3] in figure 2.

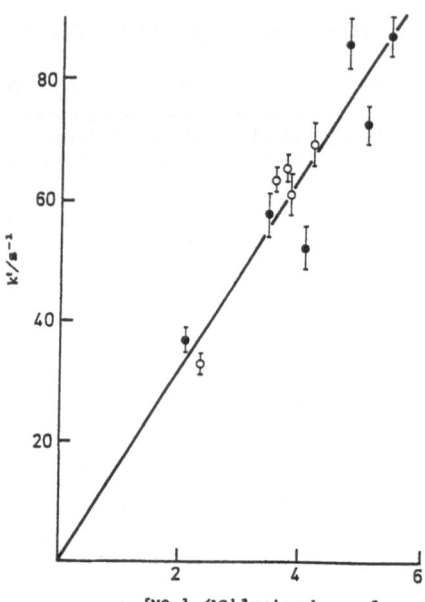

Figure 2. Pseudo-first-order rate constants for the loss of O in the reaction with NO_3. ○ , P = 2.1 mmHg; ● , P = 3.7 mmHg.

Numerical integration of the full kinetic scheme, with k_3 as fitting parameter, and with the measured losses of NO_3 incorporated, gave almost the same results as the simple treatment. Figure 3 shows the decays of O fitted in this way for the four runs of figure 1.

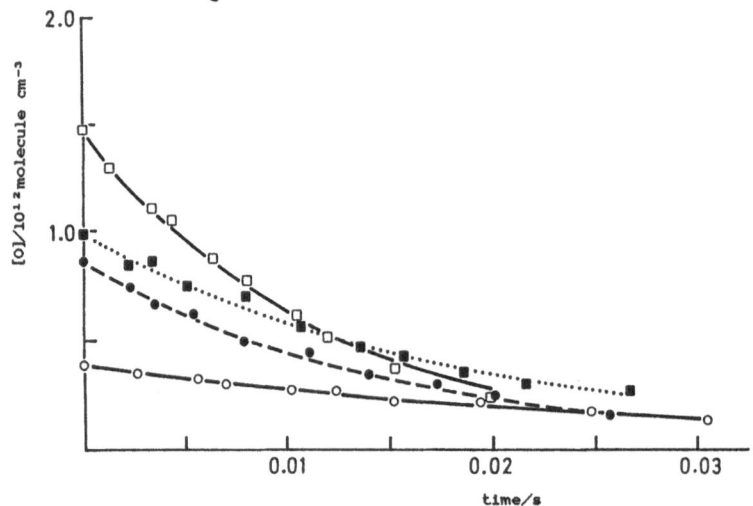

Figure 3. Decay of O in the presence of NO_3 shown as a direct function of time. Symbols are as for figure 1.

Reaction (1)–(3), with H, OH, and O all possessed rate constants that were large at room temperature, and no attempt was made to obtain the temperature dependences of the rate constants. The derived rate constants are reported in the summary.

2.2   Reaction with Br and BrO

In our discharge-flow investigations[5] of the reactions of NO_3 with $HC\ell$ and with HBr, we observed an acceleration of rate with contact time that we attributed to secondary reactions of halogen atoms and haloxy radicals with NO_3. The rate constants needed to model the behaviour with $HC\ell$ were entirely consistent with the known rate constants for the $C\ell$ and $C\ell O$ reactions. For the Br and BrO reactions, we find that rate constants of 1.2×10^{-11} and 7×10^{-12} $cm^3 molecule^{-1}s^{-1}$ provide a good fit to the data. These rate constants are ones for which the Orléans group[6] have evidence.

2.3   Reaction between NO_3 and N

We have carried out a preliminary qualitative study of the reaction of N atoms with NO_3. Oxygen free nitrogen was diluted with He and passed through a microwave discharge before meeting the main flow of helium. NO_3 was added to the flow via a sliding injector. Evidence for any possible interaction of N with NO_3 was gathered by monitoring the loss of NO_3 only. The contact times were in the range 0.02 to 0.04s. Under these conditions, the observed loss of NO_3 was independent of the contact time so that all the N atoms were consumed within 0.02s. A rather simplistic calculation, assuming that one atom of N consumes one NO_3 molecule in a single, second order process, leads to a lower limit for the apparent rate constant of 3×10^{-11} $cm^3 molecule^{-1}s^{-1}$.

2.4   Self-reaction of NO_3

The self-reaction between two NO_3 radicals

$$NO_3 \quad + \quad NO_3 \quad \longrightarrow \quad 2NO_2 \quad + \quad O_2 \qquad (4)$$

was investigated in two ways. In the first, excimer laser flash photolysis of F_2 (at $\lambda = 308$ nm) was used to generate F that reacted with HNO_3 to produce NO_3 radicals. The laser repetition rate was adjusted (2-28 Hz) to allow build-up of the desired $[NO_3]$ (in the range 10^{14}-10^{15} molecule cm^{-3}), and the decay of NO_3 followed, by CW dye laser absorption at $\lambda = 623$ nm, when the laser was turned off.

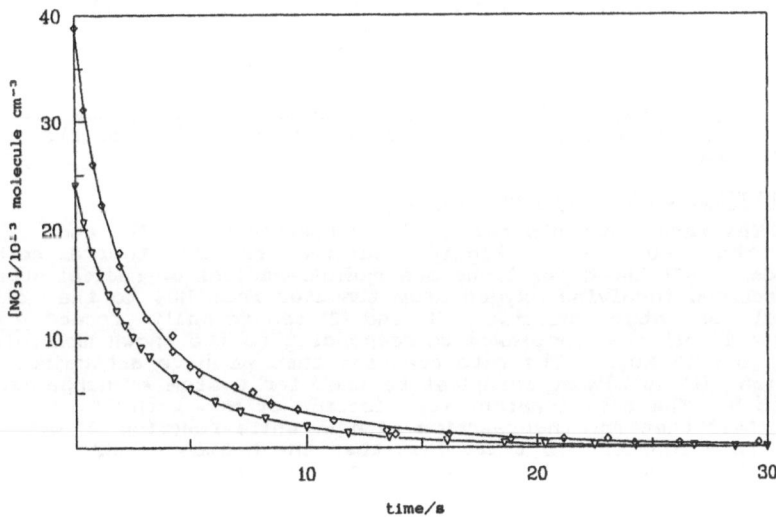

Figure 4. Experimental and calculated NO_3 decays in the flash photolysis of F + HNO_3 mixtures.

Figure 4 shows decay curves for two starting values of $[NO_3]$. The curves were simulated by numerical integration of the relevant kinetic equations, including reversible reaction with NO_2 and a possible first-order wall loss. The mean value for k_4 in five preliminary experiments was $(3.3\pm0.6) \times 10^{-16}$ cm^{3}molecule^{-1}s^{-1}, a value that agrees closely with the value reported a decade ago by Graham and Johnston[9] of $(2.3\pm0.8) \times 10^{-16}$ cm^{3}molecule^{-1}s^{-1}. The first-order loss of NO_3 had a rate constant of about 0.06 s^{-1}, confirming that wall loss of NO_3 is slow in a glass system.

Our second method for investigating reaction (4) employed a solenoid-controlled stopped-flow system with F atoms produced conventionally by a microwave discharge, followed by reaction with HNO_3 to generate NO_3. Optical absorption at λ = 662 nm was used to monitor $[NO_3]$ after the flow through the absorption cell was by-passed. Signal averaging over many shots was employed to yield decay curves that were analysed for the rates of first- and second-order components. The second-order rate constant thus obtained was $(2.5\pm0.2) \times 10^{-16}$ cm^{3}molecule^{-1}s^{-1}, and is thus consistent with the value obtained in the laser photolysis experiments.

2.5 Reaction between NO_3 and alkyl radicals

The purpose of these experiments was to provide information for future work, when we shall study the reactions of NO_3 with organic radicals. The reactions of H atoms with ethylene and of F atoms with ethane were used to produce ethyl radicals. We have evidence of loss of NO_3 that is likely to be the result of the interaction of NO_3 with ethyl and/or methyl radicals, the latter being produced by the fast reaction of H atoms with ethyl radicals. A series of similar experiments with isobutene was also carried out, but here isobutene is regenerated by the very fast disproportionation reaction between H and t-C_4H_9. Isobutene reacts with NO_3 on the time scale of our experiments, so that it is not possible to increase its concentration sufficiently to ensure complete depletion of the H atoms before the injection of NO_3. However, model calculations showed that a fraction of the NO_3 loss could be due to the reaction with t-butyl radicals. In all the experiments described in this section, the contact time between the NO_3 and the reaction partner was varied between 0.02 and 0.04s. The final concentrations of NO_3 do not depend on the contact time, so that the processes leading to loss of NO_3 are completed in less than 0.02s. Although these experiments did not yield any quantitative information, they have been useful in showing that a fast reaction of NO_3 with alkyl radicals does take place and, furthermore, that the reaction of F with alkanes is a clean and convenient source of alkyl radicals for flow tube experiments.

3. DISCUSSION OF RATE MEASUREMENTS

The rate constants measured for reactions (1), (2), and (3) show that NO_3 is a highly reactive radical towards certain species. All three reactions are radical-radical or radical-atom interactions involving oxygen atom transfer from NO_3 to the other radical or atom. Reactions (1) and (2) can formally proceed via energized collision complexes corresponding to the known molecules $HONO_2$ and HO_2NO_2. The rate constant that we have determined for reaction (1) is almost identical to that for O-atom exchange from NO_2 to H. The rate constant for reaction of NO_3 with Cl is less than half that for the reaction with H, while reaction (3) with O has a rate constant 6.5 times less than the H-atom reaction. We note that the total electronic degeneracies of H, OH, Cl, and O are 2, 4, 4, and 9. There must be a geometrical penalty for the reaction with OH, so that the measured reactivities appear to show[1][2] the influence of the number of available potential energy surfaces, only one of which can be effective in promoting reaction. The rate constants for the reactions of NO_3 with HO_2, ClO, and NO_3 are smaller by factors of 30, nearly 300, and more than 300000 than that for the H-atom interaction, but the major channels cannot involve only O-atom abstraction from NO_3.

4. ATMOSPHERIC PROCESSES

It seems unlikely that either reaction (1) or reaction (2) is of great importance in atmospheric chemistry. The NO_3 radical is rapidly photolysed by visible radiation, so that reaction with any photochemically-generated radical has to compete with this photochemical loss. Atomic oxygen concentrations in the stratosphere are relatively high during the day, and a simple calculation[2] shows that the rate of the reaction could reach roughly 10% of the photolysis rate. The reaction could, therefore, make a minor, but significant, contribution to the destruction of NO_3. That conclusion does not argue for any major change in the interpretation of daytime stratospheric chemistry, since NO_3 already plays so small a role during the hours of sunlight. Atomic oxygen concentrations in the stratosphere drop sharply at night. However, observed stratospheric concentrations of NO_3 are often smaller than those calculated so that there may be some hitherto-unknown scavenging process that removes NO_3. Reaction between O and NO_3 might conceivably be of importance just after sunset at altitudes high enough for there to be a time lag between the cessation of O_3 photolysis and the recombination of virtually all O with O_2. However, the most important conclusion to be drawn for atmospheric chemistry from the present experiments is the confirmation that radical-radical reactions of NO_3 can be fast, and that transformations involving other radicals that persist at night because of their otherwise low reactivity remain plausible contributors to night-time atmospheric chemistry.

REFERENCES

1. BOODAGHIANS, R.B., CANOSA-MAS, C.E., CARPENTER, P.J. and WAYNE, R.P. (1988) The reactions of NO_3 with OH and H. J. Chem. Soc., Faraday Trans. 2, 84, 931.

2. CANOSA-MAS, C.E., CARPENTER, P.J. and WAYNE, R.P. (1989) The reaction of NO_3 with atomic oxygen J. Chem. Soc., Faraday Trans. 2, 85, 697.

3. HALL, I.W., WAYNE, R.P., COX, R.A., JENKIN, M.E. and HAYMAN, G.D. (1988) Kinetics of the reaction of NO_3 with HO_2. J. Phys. Chem., 92, 5049.

4. COX, R.A., FOWLES, M., MOULTON, D and WAYNE, R.P. (1987) Kinetics of the reactions of NO_3 radicals with Cℓ and CℓO. J. Phys. Chem., 91, 3361.

5. CANOSA-MAS, C.E., SMITH, S.J., TOBY, S. and WAYNE, R.P. (1989) Laboratory studies of the reactions of the nitrate radical with chloroform, methanol, hydrogen chloride and hydrogen bromide. J. Chem. Soc., Faraday Trans. 2, 85, 709.

6. BIGGS, P., CANOSA-MAS, C.E., and WAYNE, R.P. (1989) The self-reaction of NO_3 radicals. Second International Conference on Gas Kinetics, NIST Gaithersburg, Md.

7. CANOSA-MAS, C.E., and WAYNE, R.P. (1989) Determination of the absolute concentration of O atoms and OH radicals in laboratory studies. In preparation.

8. POULET, G., MELLOUKI, A and LE BRAS, G. (1988) Discharge-Flow study of NO_3 reactions with free radicals and inorganic molecules. in 'Tropospheric NOx chemistry - gas phase and multiphase aspects', O.J. Nielsen and R.A. Cox, eds., Air Pollution Report 9, EVR 11440, CEC, Brussels, pp 90 - 95.

9. GRAHAM, R and JOHNSTON, H.S. (1978) The photochemistry of NO_3 and the kinetics of the N_2O_5—O_3 system. J. Phys. Chem., 82, 254.

POSSIBLE HETEROGENEOUS REACTION PROCESSES INDUCING DEPLETION
OF THE OZONE LAYER IN THE STRATOSPHERIC ANTARCTICA

Jürgen Müller
Umweltbundesamt, Pilotstation Frankfurt
Frankfurter Str. 135, D-6050 Offenbach

Summary

Increasing methane increases the stratospheric water vapour which during
the Antarctic winter time is frozen in stratospheric particles. The
particles are built up according to the specific vapour pressure of
the constituting substances: sulfuric acid in the core of the particles
and other less volatile substances like HNO_3 and HCl in surrounding
shell together with H_2O-ligands. Induced by the low temperatures in the
Antarctic winter ($- 80°$ C) in addition to HNO_3 and HCl also NO_2 is
frozen on the surface of the ice particles. Due to reactions with HCl
located on the surface CL_2 and other ozone attacking species are set
free.
It is postulated that during the Antarctic sunrise in spring-time by
charge transfer processes in the particles hydrated electrons are
induced which react at the surface with H^+-ions, N_2O and chlorinated
halocarbons. By these processes the ozone attacking radicals H, OH and Cl
are generated.

1. INTRODUCTION

In 1974 by Molina and Rowland (1) the reaction of halocarbons with
stratospheric ozone was discovered. Since some years a hole in the ozone
layer of the lower stratosphere above the Antarctica is observed during
spring time.

By Crutzen and Arnold (2) the ozone hole mainly was explained by reaction
of $ClONO_2$ with HCl generated on the surfaces of the stratospheric cloud
particles. The reaction product Cl_2 is released into the gas phase during
winter time. By sunrise in spring Cl_2 is split into ozone attacking
radicals. O_3 and Cl react into ClO and O_2. ClO is assumed to react with
ClO forming CL_2O_2 which ends up in O_2 and two Cl-radicals keeping up the
destruction chain.

Inspite of the convincing explanations the overall understanding of the
ozone hole is still incomplete. Many processes may happen which up to date
are not taken into account. Also, the closure of the hole at the end of
the Antarctic spring time – in the presence of activated halocarbons – is
not thouroughly handled.

The man-made emissions into the stratosphere contribute to a growing complexity of the stratospheric chemistry destroying at least temporarily the steady-state of ozone production and decay.

2. CHEMICAL STRUCTURE OF STRATOSPHERIC AEROSOLS

The stratospheric particles which are generated in the Antarctic winter at temperatures down to - 80° C beside condensed sulfuric acid, nitric acid and hydrochloric acid mainly consist of ice. Also $HNO_3 \cdot 3H_2O$ was found in sampled "stratospheric haze particles" (3).

H_2O reaches the stratosphere by vertical transport through the tropical tropopause but half of the stratospheric H_2O-concentration is assumed to be generated by oxidized methane (4). The world-wide methane production is increasing, as observed p.e. by trend measurements at the base-station Izaña on the Can. Islands (Fig. 1). Increasing water vapour in the stratophere contributes to higher concentrations of stratospheric particles.

The stratospheric aerosols are assumed to be chemically structured in a similar way as observed for tropospheric aerosols. However, the quantity of involved different chemical substances is some order of magnitudes less.

Impactor measurements and surface analyses revealed that aerosols are built up in a macroscopic shell structure with heavy-volatile substances in the core of the particles surrounded by substances with rising volatility (5). At the surface of the particles volatile compounds and adsorbed gases are located.

This model applied to the chemical structure of stratospheric aerosols implies H_2SO_4 to be in the core of the particles surrounded by $(HNO_3 . 3H_2O)$ and HNO_3. HCl together with adsorbed stratospheric gases is located at the surface (Fig. 2).

Adsorbed nitrous oxides like $ClONO_2$, NO_2 or molecular HNO_3 push away HCl and release Cl_2 into the gas-phase.

3. RADICALS INDUCED BY HYDRATED ELECTRONS

By reaction of O_3 and O with radicals ozone as destructed (1, 2):

(1) $x_1O + O_3 \longrightarrow x_1O + O_2$

(2) $x_1O + O \longrightarrow x_1 + O_2$

$O + O_3 \longrightarrow 2O_2$

$x_1 = NO, OH, H, Cl, Br$

O-atoms generated by photolytic split (hv < 240 nm) of O_2 predominantly combine with O_2 to form stratospheric ozone (Chapman 1930, (6)).

However, at the begin of Antarctic spring time due to weak light-intensity no O-atoms are available. Therefore, ozone destruction predominantly is based on equation (1).

It is assumed that during the Antarctic sunrise by charge transfer processes in the surfaces of the particles hydrated electrons are created. These species are adequate to react with ions, gases or radicals in the water-film of the conducting surfaces of the particles. The following reactions of e^-aq are known (7):

(3) $\quad e^-aq \; + \; H_3O^+ \; \text{-----}> \; H^{\bullet} + H_2O$

(4) $\quad e^-aq \; + \; e^-aq \; \text{-----}> \; H_2 + 2 \; OH^-$

(5) $\quad e^-aq \; + \; H \qquad \text{-----}> \; H_2 + OH^-$

Chlorocarbons, N_2O and O_3 are known as scavengers for hydrated electrons:

(6) $\quad e^-aq \; + \; R \; CH_2Cl \; \text{-----}> \; Cl^- + R \; CH_2^{\bullet}$

(7) $\quad e^-aq \; + \; N_2O \qquad \text{-----}> \; N_2 \; + \; O^-$

$\qquad \quad O^- \quad + \; H_2O \qquad \text{-----}> \; OH \; + \; OH^-$

(8) $\quad e^-aq \; + \; O_3 \qquad \text{-----}> \; O_2 \; + \; O^-$

$\qquad \quad O^- \quad + \; H_2O \qquad \text{-----}> \; OH \; + \; OH^-$

(9) $\quad e^-aq \; + \; O_2 \qquad \text{-----}> \; O_2^-$

$\qquad \quad H^+ \quad + \; O_2^- \qquad \text{-----}> \; HO_2$

The intermediate products of reaction (7) formerly were clarified (8).

If the reactions (3) to (9) take place during the Antarctic sunrise several additional radicals are produced which directly or indirectly contribute to the ozone destruction. In sequence of reaction (1) several further combinations may occur keeping up the radical generation:

$$x_iO \; + \; x_jO \; \xrightarrow{M} \; \begin{array}{l} x_i \; + \; x_j \; + \; O_2 \\[4pt] x_i \; + \; Ox_jO \end{array}$$

$$M \; = \; O_2, \; N_2$$

At the end of the Antarctic spring time by rising temperatures the stratospheric ice particles are dissolved. The equilibrium of ozone decay and ozone production is reestablished.

Due to sufficient radiation intensity O_2-molecules are split which gives rise to ozone production. The evaporating ice particles additionally inject NO_2 into the gas phase which acts as another ozone source. Cl-radicals set free from halocarbons are bound in HCl-molecules and NO and NO_3 in HNO_3-molecules.

REFERENCES

1. Molina, M. J.; Rowland, F. S. (1974)
   Nature 249, pp. 810 - 814

2. Crutzen, P. J.; Arnold, F. (1986)
   Nature 324, pp. 651 - 665

3. Pueschel, R. F. (1989) c/o Eur. Aerosol Conf. (Vienna), Proc. c/o
   J. Aerosol Sc.

4. Blake, D. R.; Rowland, F. S. (1988)
   Science, 239, pp. 1129 - 1131

5. Müller, J. (1988), J. Aerosol Sc. 19,7, pp. 1161 - 1164

6. Chapman, S. (1930), Mem. Roy. Meteorol. Soc., 3, pp. 103 - 125

7. Müller, J. (1973), Naturwissenschaften, 5, pp. 256 - 257

8. Buxton, G.V. and G.A. Salmon (1989): Univers. of Leeds, UK,
   pers. comm.

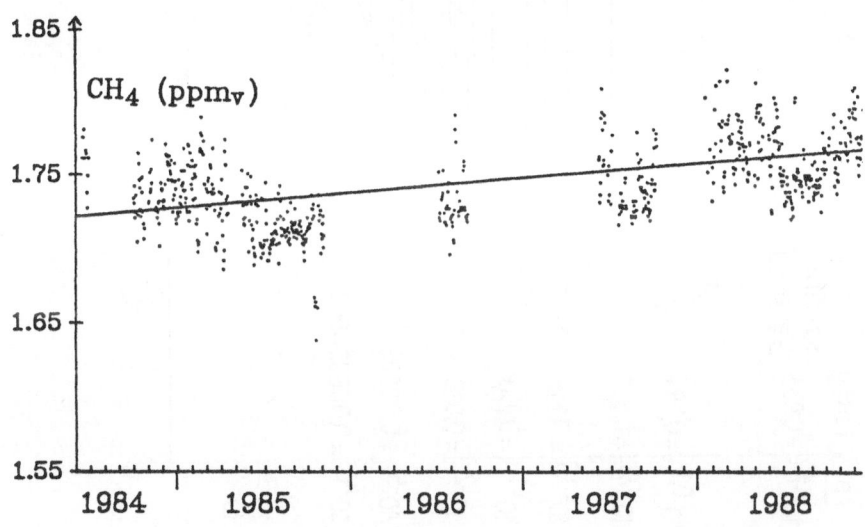

Figure 1 : Trend of methane at the base-station Izana
            (Can. Islands)
            (Meteorologie Consult, Glashütten)

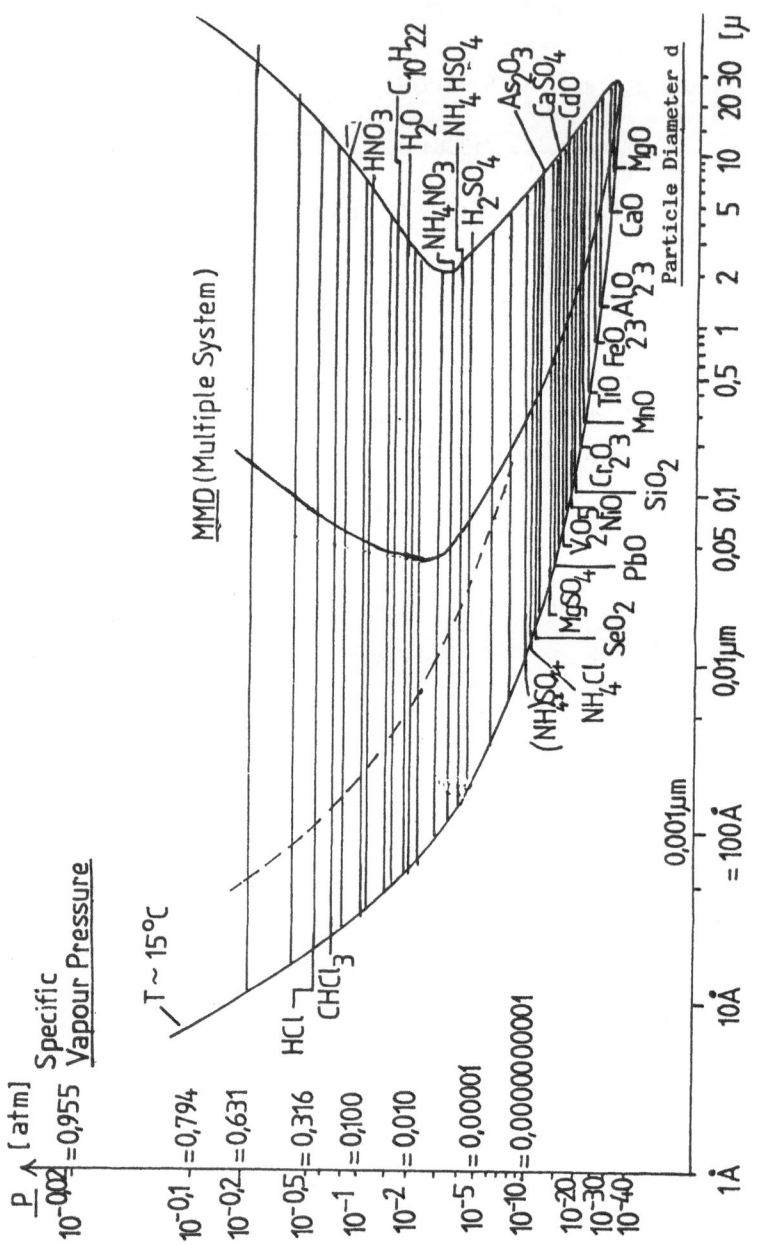

Fig. 2: Specific vapour pressures of aerosol substances versus particle diameter ("Macroscopic shell structure")

DEGRADATION PRODUCTS OF CHLORINATED HYDROCARBONS

K. Baechmann and J. Polzer
Institut fuer Anorganische Chemie und Kernchemie
D - 6100 Darmstadt, Hochschulstr. 4, F.R.G.

SUMMARY

Since chlorinated hydrocarbons are known to be responsible for the destruction of the stratospheric ozone
layer and contribute to the `greenhouse effect` it is
necessary to investigate the chemical reaction cycles in
which the halocarbons are involved.
Among the tropospheric degradation products of chlorinated hydrocarbons, toxic compounds such as phosgene or
chlorinated acetic acids were identified.
The following paper describes a sensitive analytical
method for the determination of phosgene and other
volatile halocarbons in ambient air using capillary
GC/ECD. Quantitative detection of concentrations as low
as 7 pptV phosgene is possible, a reproducibility of 8 %
can be achieved.
Results of measuring phosgene in the troposphere and
diurnal variation of its concentration are presented.

1. INTRODUCTION

Large amounts of halogenated hydrocarbons have been
released into the atmosphere in recent years. Most of them
are of anthropogenic origin, only few are released from
natural sources.
Table 1 presents some of these volatile chlorinated
hydrocarbons, their mean concentration derived from long term
measurements in Kolmbach (Odenwald, F.R.G.) /1/ and their
expected tropospheric lifetime /2,3/. The reactive compounds
i.e. the hydrogen-containing species and unsaturated compounds decompose in the troposphere whereas the inert species
reach the stratosphere.

Table 1: Chlorinated hydrocarbons in the troposphere

compound	concentration (pptV)	lifetime (a)
CCl_4	137	60 - 100
$CH_3 CCl_3$	/	6 - 10
$CH_2 Cl_2$	319	0,5
$CHCl_3$	228	0,3 - 0,6
$CCl_2 = CCl_2$	508	0,4
$CHCl = CCl_2$	643	0,02

In the troposphere the hydrogen-containing compounds are degradated by attack of hydroxyl radicals and hydrogen abstraction, in the case of unsaturated compounds attack of O_3, O_2, and O radicals is also discussed.

Phosgene has been identified as an atmospheric decomposition product of several chlorinated hydrocarbons, for example di- and trichloromethane, 1,1,1-trichloroethane, tri- and tetrachloroethene.

Fig. 1 shows the reaction mechanism for the formation of phosgene from $CHCl_3$ and CCl_2CCl_2 under tropospheric conditions /4,5,6/.

$CHCl_3$	+	OH	------>	CCl_3	+	H_2O	
CCl_3	+	O_2	------>	CCl_3O_2			
CCl_3O_2	+	NO	------>	CCl_3O	+	NO_2	
CCl_3O	+	O_2	------>	$COCl_2$	+	ClO_2	

$$CCl_2 =\!=CHCl \quad + \quad O_3 \quad ------> \quad \overset{\displaystyle O}{\underset{\displaystyle Cl_2C---CHCl}{\overset{O\quad O}{}}}$$

$$\overset{\displaystyle O}{\underset{\displaystyle Cl_2C---CHCl}{\overset{O\quad O}{}}} \quad --^{h\,v}---> \quad COCl_2 \quad + \quad HCl \quad + \quad CO_2$$

Figure 1 : Formation of phosgene in the troposphere

Although the sources for phosgene are well known there is a lack of information about the sinks. Model calculations expect hydrolysis in clouds, hydrometeors or maritime regions to be an effective sink for phosgene /5/. Another possible sink process could be heterogeneously catalysed decomposition on surfaces. Photolytic processes or gaseous phase hydrolysis are known to be very slow in the lower troposphere /5,7/.

Some authors therefore predict an accumulation of phosgene in the troposphere in the future /5,8/.

2. EXPERIMENTAL
The main problems of the analysis are the low detection limit which has to be achieved for determination of phosgene under tropospheric conditions and the high reactivity of this compound in contact with solid phases (i.e. column packings).

The method described avoids significant losses of phosgene by use of chemically inert tubes, a capillary column and a careful enrichment step and analytical procedure. The high resolution of the capillary column allows the separation and simultaneous detection of other ECD sensitive compounds. The method can be easily adapted for collecting samples in the field.

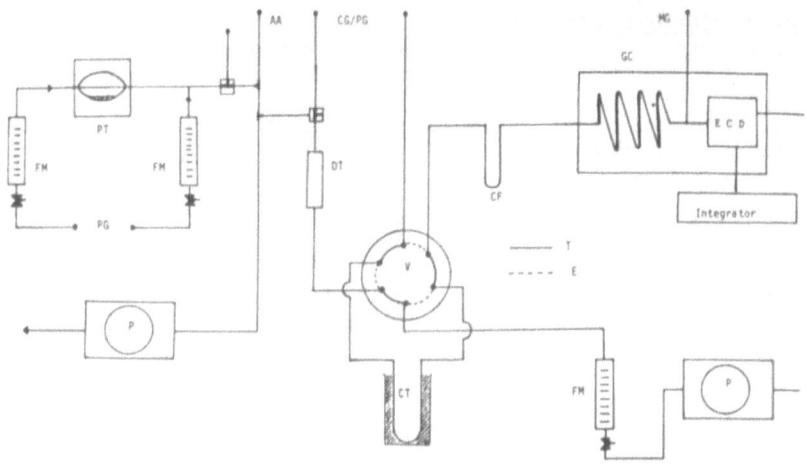

Figure 2 : Analytical system: FM: flow-meter; PG: purge gas
(N₂); PT: permeation tube; AA: ambient air; CG/PG: carrier
gas/purge gas (He); DT: drying tube; V: six-port valve; P:
pump; CT: cold trap; CF: cold focussing; T: transfer; E:
enrichment; GC: gas chromatograph; MG: make-up gas

Air samples were collected by a system of two pumps as
shown in Fig. 2. Ambient air was sucked through a stainless
steel covered teflon loop kept under liquid nitrogen (-196
°C) where cryogenic enrichment of the trace compounds takes
place. Volumes of air sampled during the enrichment procedure
varied between 0,3 and 1 l.

Analysis was started by switching the six-port valve
which conducted carrier gas through the teflon loop. The
analytes were vaporized by resistant heating of the stainless
steel cover of the teflon loop, and cryogenically focused
again in a deactivated capillary column to provide sharper
peaks, followed by separation on a non polar capillary column
(50 m x 0,32 mm ID, 0,53 µm SE 30 CB) and ECD.

Calibration was carried out using a dynamic dilution
system, the phosgene standard was generated with a teflon
permeation tube kept at constant temperature. The purge gas
from the permeation tube was diluted with phosgene-free
ambient air down to concentrations expected in the
troposphere. Enrichment and analysis were carried out
analogous to the analysis of air samples in order to
compensate possible losses of phosgene during the analytical
procedure.

Sample collection in the field was realized in a teflon
loop at -196 °C. Then the loop was closed by two teflon stopcocks and kept under liquid nitrogen until it was coupled to
the GC and analysed in the usual way.

3. RESULTS AND DISCUSSION

Quantitative detection of concentrations as low as 7 pptV phosgene is possible, a reproducibility of 8 % can be achieved.

Fig. 3 presents a chromatogram containing characteristic compounds of air samples.

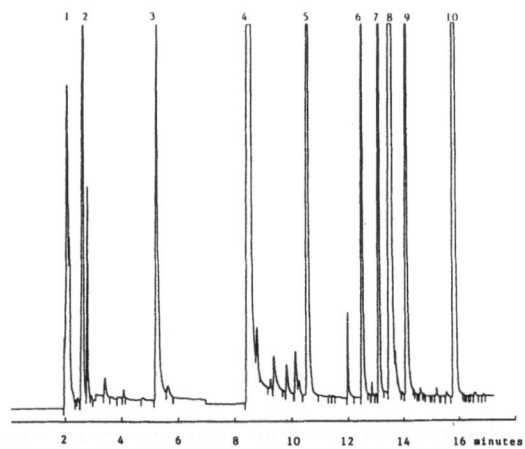

Figure 3 : Gaschromatogram of an airsample, 0,61, Darmstadt, capillary column, temperature program: -30 °C 7 min; 25 °C/min to 200 °C. 1 O_2/CHClF$_2$; 2 CCl_2F_2; 3 $COCl_2$; 4 CCl_3F; 5 $CCl_2F-CClF_2$; 6 $CHCl_3$; 7 CCl_3-CH_3; 8 CCl_4; 9 $CCl_2=CHCl$; 10 $CCl_2=CCl_2$.

During our investigations concentrations of phosgene which we determined in airsamples varied between 8 and 87 pptV during daytime, whereas nighttime concentrations up to 143 pptV were determined.

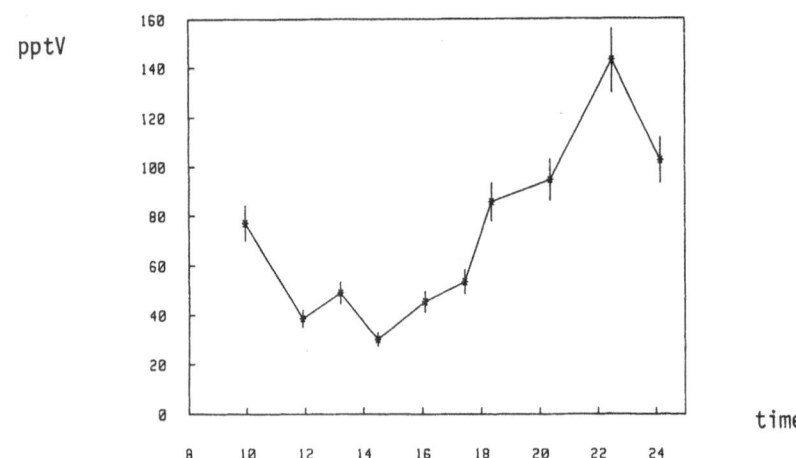

Figure 4 : Diurnal variation of phosgene

– 218 –

Resulting from the varying phosgene concentrations, we measured over several days/nights, a tropospheric lifetime of a few months has to be assumed. Until now lifetimes of a few years have been stated /5,8/.

In future, continuous monitoring of phosgene and its source compounds especially trichloroethene is desirable in order to confirm the expected source and sink processes.

REFERENCES

/1/ Baechmann, K. and Kessel, M. unpublished results
/2/ Noy, T. et al. (1987) "Trace analysis of halogenated hydrocarbons in gaseous samples by online enrichment in an adsorption trap, on column cold trapping and capillary gas chromatography", J. Chromatogr. 393, 343-356.
/3/ Fabian, P. (1986) Handbook of Environmental Chemistry 4A, Hutzinger, Springerverlag Berlin.
/4/ Gay, B. et al. (1976) "Atmospheric oxidation of chlorinated ethylenes", Environ. Sci. Technol. 10, 58-67.
/5/ Helas, G. and Wilson, S.R. (1987) "Considerations on sources and sinks of phosgene in the troposphere", submitted to Tellus.
/6/ Atkinson, R. (1985) "Kinetics and mechanisms of the gas phase reactions of hydroxyl radical with organic compounds under atmospheric conditions", Chem. Rev. 85, 69-201.
/7/ Butler, R. and Snelson, A. (1979) "Kinetics of the homogeneous gas phase hydrolysis of CCl_3COCl, CCl_2HCOCl, $CH_2ClCOCl$ and $COCl_2$", J. Air Pollut. Control. Assoc. 29, 833-837.
/8/ Singh, H.B. (1976) "Phosgene in ambient air", Nature 264, 428-429.

ATMOSPHERIC REMOVAL PROCESSES FOR CHLORINE-CONTAINING COMPOUNDS

H.W.SIDEBOTTOM and O.RATTIGAN
Department of Chemistry, University College Dublin, Dublin, Ireland

J.J.TREACY
Department of Chemistry, Dublin Institute of Technology, Dublin, Ireland

O.J.NIELSEN
Chemistry Department, Risø National Laboratory, DK-4000 Roskilde, Denmark

SUMMARY

The chlorine atom initiated oxidation of 1,1,1-trichloroethane has been investigated at $298 \pm 3K$ in air at 1 atmosphere total pressure. Product analysis data suggest phosgene is the major reaction product while smaller amounts of chloral were also detected. Rate constants were determined for the reaction of atomic chlorine with CH_3CCl_3 and CCl_3CHO using a relative rate technique. Attempts to model the product concentration profiles for the oxidation of CH_3CCl_3 indicated that the major reaction channel for CCl_3CH_2O radicals is reaction with O_2 to give CCl_3CHO and that CCl_2O is a secondary product arising from the rapid reaction of Cl atoms with CCl_3CHO.

1. INTRODUCTION

Chlorocarbons which are reactive with respect to oxidation in the troposphere may eventually provide a flow of chlorine into the stratosphere. Hence an understanding of the oxidation mechanisms for chlorocarbons is of importance in order to establish the eventual sink for chlorine in these molecules. A number of mechanistic investigations have been reported in which the chlorine atom initiated oxidation of chlorinated hydrocarbons has been studied [1].

$$Cl_2 + h\nu \quad \rightarrow \quad 2Cl \qquad (1)$$

$$Cl + RH \quad \rightarrow \quad HCl + R \qquad (2)$$

$$R + O_2 + M \quad \rightarrow \quad RO_2 + M \qquad (3)$$

However, interpretation of the oxidation mechanism for chloroalkyl peroxy radicals produced in these systems can be complicated by secondary reactions of chlorine atoms with the primary oxidation products

In this work the oxidation of CCl_3CH_2 radicals produced by the photolysis of chlorine in the presence of 1,1,1-trichloroethane was investigated . In order to ascertain the importance of reactions of chlorine atoms with the primary oxidation products a relative rate technique was used to obtain rate constants for the reactions of Cl atoms with CH_3CCl_3 and CCl_3CHO.

2. EXPERIMENTAL

Product analysis studies were carried out in a conventional static system at $298 \pm 3K$ using both gas chromatography and infrared analysis to determine product concentrations. A Gow-Mac series 550 gas chromatograph, incorporating a gas density balance detector system was used for all quantitative chromatographic analyses. Calibration curves relating the pressure of the various reactants and products to optical density of a given infrared band were prepared using authentic samples of the reagents. Product identification was confirmed by means of gas chromatography coupled mass spectrometry. Radiation of wavelengths centered around 360 nm was isolated from a Hanovia

500W medium pressure mercury source by means of the appropriate Corning glass filters. The T-shaped reaction vessel of volume 135 cm^3 was housed in the sample compartment of a Perkin-Elmer model 257 infrared spectrometer, with sodium chloride windows in the infrared beam and a pathlength of 10 cm. Along the perpendicular axis, with a pathlength of 11.5 cm, was the mercury source and its associated filter system.

Relative rate experiments were carried out in an approximately 70 liter FEP Teflon reaction chamber. The chamber was surrounded by 10 blacklamps (Philips TL 20W/08) and the temperature maintained at $298 \pm 3K$ by forced air circulation. Chlorine atoms were generated by the photolysis of chlorine. Mixtures of Cl_2, reactant and the reference compound, all in the range 1-20 ppm, were flushed from a Pyrex bulb into the reactant chamber by a stream of ultra-zero air. The chamber was then filled with air up to atmospheric pressure. During the course of the reaction samples were removed from the reaction chamber and analysed on a Gow-Mac gas chromatograph equipped with a flame ionization detector.

3.   RESULTS AND DISCUSSION

Bertrand et al. [1] have proposed a general mechanism for the chlorine atom photosensitized oxidation of chloroethyl radicals which provides a useful basis on which to discuss the results from this work

$$Cl_2 + h\nu \quad \rightarrow \quad 2Cl \tag{1}$$

$$Cl + CH_3CCl_3 \quad \rightarrow \quad HCl + CCl_3CH_2 \tag{2}$$

$$CCl_3CH_2 + O_2 + M \quad \rightarrow \quad CCl_3CH_2O_2 + M \tag{3}$$

$$2CCl_3CH_2O_2 \quad \rightarrow \quad CCl_3CH_2O_2CH_2CCl_3 + O_2 \tag{4}$$

$$\rightarrow \quad 2CCl_3CH_2O + O_2 \tag{5}$$

$$CCl_3CH_2O \quad \rightarrow \quad oxidation\ products \tag{6}$$

Under the experimental conditions employed, $p(Cl_2)$ = 2-20 torr, $p(CH_3CCl_3)$ = 2-100 torr, phosgene was the major product resulting from photolysis of chlorine / 1,1,1-trichloroethene mixtures in air at $298 \pm 3K$. Trace amounts of 1,1,1-trichloroacetaldehyde (chloral) were also detected, however, as the photolysis time increased the relative yield of chloral decreased. The primary products of CCl_3CH_2 radical oxidation are dependent upon the relative importance of the various possible reaction channels for loss of CCl_3CH_2O radicals, reaction (6). Decomposition may occur via C-H or C-C bond rupture:

$$CCl_3CH_2O \rightarrow CCl_3CHO + H \quad : \quad \Delta H° = 19 kcal\ mol^{-1} \tag{6a}$$

$$CCl_3CH_2O \rightarrow CCl_3 + CH_2O \quad : \quad \Delta H° = 5 kcal\ mol^{-1} \tag{6b}$$

Both these processes are endothermic and it is expected that analogous to the reactions of methoxy and ethoxy radicals with O_2, H atom abstraction will occur:

$$CCl_3CH_2O + O_2 \rightarrow CCl_3CHO + HO_2 \quad : \quad \Delta H° = -30 kcal\ mol^{-1} \tag{6c}$$

Chloral was observed as a minor product from CCl_3CH_2 radical oxidation and presumably arises from the reaction of trichloroethoxy radicals with O_2 in reaction (6c). It is possible that the major product of the oxidation phosgene, is formed via the lessthermodynamically favoured C-C bond dissociation reaction (6b). Trichloromethyl radicals produced in this process would be quantitatively converted to CCl_2 in the presence of O_2 [2]

$$CCl_3 + O_2 + M \quad \rightarrow \quad CCl_3O_2 + M \qquad (7)$$

$$2CCl_3O_2 \quad \rightarrow \quad 2CCl_3O + O_2 \qquad (8)$$

$$CCl_3O \quad \rightarrow \quad CCl_2O + Cl \qquad (9)$$

A more plausible explanation for CCl_2O formation is the Cl atom initiated oxidation of chloral formed in reaction (6c)[3]

$$Cl + CCl_3CHO \quad \rightarrow \quad HCl + CCl_3CO \qquad (10)$$

$$CCl_3CO + O_2 + M \quad \rightarrow \quad CCl_3COO_2 + M \qquad (11)$$

$$2CCl_3COO_2 \quad \rightarrow \quad 2CCl_3CO_2 + O_2 \qquad (12)$$

$$CCl_3CO_2 \quad \rightarrow \quad CCl_3 + CO_2 \qquad (13)$$

In order to test this hypothesis rate constants for the reaction of Cl atoms with CH_3CCl_3 and CCl_3CHO were determined using a relative rate technique. In the presence of Cl atoms the reactant and reference compound decay via the reactions

$$Cl + CH_3CCl_3 \quad \rightarrow \quad HCl + CCl_3CH_2 \qquad (2)$$

$$Cl + CCl_3CHO \quad \rightarrow \quad HCl + CCl_3CO \qquad (10)$$

$$Cl + Reference \quad \rightarrow \quad products \qquad (14)$$

Assuming that reaction with Cl is the only significant loss process for both reactant and reference compound then:

$$\ln([CH_3CCl_3]_o/[CH_3CCl_3]_t) = k_2/k_{14}\ln([ref]_o/[ref]_t) \qquad (15)$$

and

$$\ln([CCl_3CHO]_o/[CCl_3CHO]_t) = k_{10}/k_{14}\ln([ref]_o/[ref]_t) \qquad (16)$$

The rate constant ratios k_2/k_{14} and k_{10}/k_{14} were determined from plots of equations (15) and (16) respectively, Figures 1 and 2. The rate constant ratios were independent of reaction time, relative reactant concentrations and light intensity in agreement with the proposed mechanism. The derived rate constants for the reaction of Cl with CH_3CCl_3 and CCl_3CHO are given in Table I together with the corresponding data for C_2H_6 and CH_3CHO. The reference compounds used were $CHCl_3$ for CH_3CCl_3, $k_{14}(CHCl_3) = 1.23 \times 10^{-13}$ [4] and CH_3Cl for CCl_3CHO, $k_{14}(CH_3Cl) = 4.9 \times 10^{-13}$ [4]. The errors quoted are twice the standard deviation from the fit and do not include an estimate of the error in the reference rate constant.

TABLE I: Rate constants for the reaction of Cl atoms with various compounds at $298 \pm 3K$

$$Cl + RH \rightarrow HCl + R$$

RH	k_{Cl}, cm^3 molecule^{-1} s^{-1}	Reference
CH_3CCl_3	$(1.1 \pm 0.05) \times 10^{-14}$	This work
CCl_3CHO	$< 3.7 \times 10^{-14}$ $(4.6 \pm 0.4) \times 10^{-12}$	5 This work
CH_3CH_3	5.7×10^{-11}	6
CH_3CHO	8.0×10^{-11}	6

Comparisons of the above data show that chlorine substitution decreases the reactivity of a compound with respect to attack by Cl atoms. Thus in going from ethane to 1,1,1-trichloroethane the rate constant per available H atom is reduced by three orders of magnitude. Similarly the reactivity of CCl_3CHO is considerably reduced relative to CH_3CHO. Since chlorine substitution of these compounds is unlikely to significantly affect the C-H bond dissociation energies the reduction in reactivity must be polar rather then enthalpy related. It is proposed that chlorine substitution increases the repulsive polar forces in the transition state for reaction of the electrophilic Cl atom.

The results from this work show that CCl_3CHO is approximately 400 times more reactive towards attack by atomic chlorine than CH_3CCl_3. Thus chloral produced in reaction (6c) as a primary product in the photolysis of Cl_2/CH_3CCl_3/air mixtures will be rapidly removed from the system by reaction with Cl atoms. Computer modelling of the Cl atom initiated oxidation of CH_3CCl_3 under the experimental conditions employed shows quite clearly that a mechanism involving CCl_3CHO formation by H atom abstraction from CCl_3CH_2O by O_2 followed by rapid loss of chloral by reaction with Cl atoms, leading to CCl_2O formation, is consistent with the observed product concentration profiles. Hence the present results support the conclusion that CCl_2O is a secondary reaction product and does not arise from the thermodynamically unfavourable C-C bond cleavage of the CCl_3CH_2O radical.

REFERENCES
1. BERTRAND, L., EXSTEEN-MEYERS, L., FRANKLIN, J.A., HUYBRECHTS, G. and OLBREGTS, J., Int.J.Chem.Kinet., 3, 89 (1971)
2. LESCLAUX, R., DOGNON, A.M. and CARALP, F., J.Photochem., A41, 1 (1987)
3. OHTA, T. and MIZOGUCHI, I., Int.J.Chem.Kinet., 12, 717 (1980)
4. WATSON, R.T., J.Phys.Chem.Ref. Data, 6, 871 (1977)
5. WINE, P.H., SEMMES, D.H. and RAVISHANKARA, A.R., Chem.Phys.Lett., 90,128 (1982)
6. WALLINGTON, T.J., SKEWES, L.M., SIEGL, W.O., CHING-HSONG, W. and JAPAR, S.M., Int.J.Chem.Kinet., 20, 867 (1988)

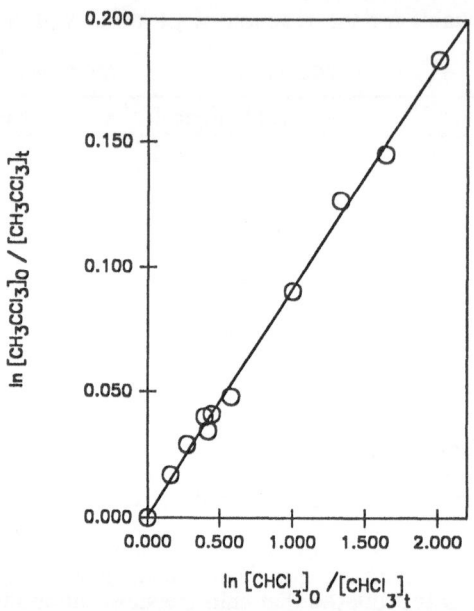

Fig.1:   Plot of $\ln([CH_3CCl_3]_0/[CH_3CCl_3]_t)$ against $\ln([CHCl_3]_0/[CHCl_3]_t)$ for the $CH_3CCl_3/CHCl_3$ system at 298K

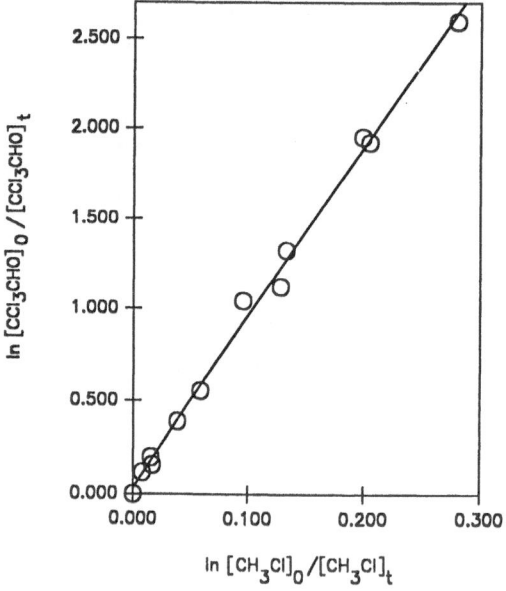

Fig.2:   Plot of $\ln([CCl_3CHO]_0/[CCl_3CHO]_t)$ against $\ln([CH_3Cl]_0/[CH_3Cl]_t)$ for the CCl_3CHO/CH_3Cl system at 298K

UV SPECTRUM AND KINETICS OF THE TRICHLOROMETHYL RADICAL.

T. ELLERMANN, J. MUNK, O.J. NIELSEN and P. PAGSBERG.
Department of Chemistry, Risø National Laboratory,
DK-4000 Roskilde, Denmark.

SUMMARY

The trichloromethyl radical (CCl_3) was produced by pulse radiolysis of CCl_4/Ar or $CHCl_3/SF_6$ mixtures through the following reactions: $Ar^* + CCl_4 \rightarrow Ar + CCl_3 + Cl$ or $F + CHCl_3 \rightarrow HF + CCl_3$. An absorption band with a maximum at 210 nm was assigned to the trichloromethyl radical. The absorption cross section of the trichloromethyl radical was determined by comparison of the amplitude of the transient absorption signal of CCl_3 to that of CH_3 and FO_2 produced by $F + CH_4 \rightarrow HF + CH_3$ and $F + O_2 + M \rightarrow FO_2 + M$. The maximum absorption cross section at 210 nm was determined to be $2.2 \pm 0.3 \times 10^{-17}$ cm^2 molecules^{-1}. Based on the initial yield and the observed second order kinetics of CCl_3 we have determined an apparent bimolecular rate constant of $k_7 = 7.8 \pm 1.2 \times 10^{-12}$ cm^3 molecules^{-1} sec^{-1} for the dimerisation $2 CCl_3 + M \rightarrow C_2Cl_6 + M$ at 295 K and total pressure of 1 atm. A value of the rate constant for the reaction $CCl_3 + Cl \rightarrow CCl_4$ has been determined to $86 \pm 1.3 \times 10^{-11}$ cm^3 molecules^{-1} sec^{-1}. Preliminary work with the reaction between CCl_3 and O_2 has given the rate constant for the addition of O_2 to CCl_3 to be $1.7 \pm 0.4 \times 10^{-12}$ cm^3 molecules^{-1} sec^{-1}.

1. INTRODUCTION

The trichloromethyl radical is produced in the atmosphere as an intermediate in the degradation of CCl_4, $CHCl_3$ and CH_3CCl_3. The UV absorption spectrum from 230 to 280 nm of the trichloromethyl radical in aqueous solution has previously been reported to be a featureless spectrum falling of steeply from 230 to 280 nm with a maximum of $\sigma(230 \text{ nm}) = 8.8 \pm 1.7 \times 10^{-18}$ cm^2 molecules^{-1} (1). The trichloromethyl radical is known to produce phosgene in the presence of oxygen (2). Phosgene has recently been detected both in the troposphere and the stratosphere (3). Although the end product of the reaction between the trichloromethyl radical and oxygen has long been known, there are still uncertainties regarding the complete reaction mechanism.

In this paper the gas phase UV spectrum of the trichloromethyl radical is presented together with kinetic data for the dimerisation reaction of CCl_3 and the between Cl and CCl_3. Besides this preliminary work on the oxidation of CCl_3 is presented.

2. EXPERIMENTAL

The trichloromethyl radical was produced by pulse radiolysis of gas mixtures of CCl_4/Ar or $CHCl_3/SF_6$ and the formation and decay kinetics were followed by transient UV absorption spectrophotometry. The details about the experimental setup have been described previously (4). With both types of gas mixtures a 30 nsec pulse of 2 MeV electrons from a field emission accelerator were used to initiate the radical reactions. With gas mixtures of 1 - 10 mbar CCl and 1 atm Ar excited Ar^* is formed primarily after irradiation. Secondly Ar^* react with CCl_4, which split in CCl_3 and Cl:

1. $Ar^* + CCl_4$ $\rightarrow$ $Ar + CCl_3 + Cl.$

Irradiation of gas mixtures of 1 - 50 mbar $CHCl_3$ and 1 atm SF_6 leads to the formation of flourine atoms, which produces CCl_3 by hydrogen abstraction:

2. SF_6 -> $SF_5 + F$,
3. $F + CHCl_3$ -> $HF + CCl_3$.

The initial yield of flourine atoms produced by pulse radiolysis of 1 atm SF_6 was obtained by titration of flourine with oxygen and methane:

4. $F + CH_4$ -> $HF + CH_3$,
5. $F + O_2 + M$ -> $FO_2 + M$.

The concentration of CH_3 and FO_2 could be obtained from their well known absorption cross sections (5,6). All kinetic experiments were carried out at room temperature and a total pressure of 1 atm. Simulation of the reaction mechanism were carried out with CHEMSIMUL (7).

3. RESULTS AND DISCUSSION

The trichloromethyl spectrum was recorded in the 210-270 nm region with a spectral band pass of 0.8 nm. Because the two spectra produced with $CHCl_3$ and CCl_4 as precursers to CCl_3 show the same general features (figure 1), we believe that our assignment of the spectrum to the trichloromethyl radical is correct. The shape of the spectrum is in good accordance with the spectrum in aqueous solution (1) and the gas phase spectrum recently produced by Lesclaux and coworkers (8) except for the two small features at 225 and 240 nm.

The yield of CCl_3 in the $CHCl_3/SF_6$ system depends on the initial yield of flourine atoms produced by the pulse and the balance between the formation and decay of CCl_3:

3. $F + CHCl_3$ -> $HF + CCl_3$,
6. $F + CCl_3 + M$ -> Products + M,
7. $2 CCl_3 + M$ -> $C_2Cl_6 + M$.

By increasing the partial pressure of $CHCl_3$ a situation can be reached, where practically all flourine atoms are converted into CCl_3 (figure 2). A minimum value of the absorption cross section of CCl_3 can therefore be calculated from the initial yield of flourine $(F)_o$:

$$(F)_o = 8.0 \pm 1.2 \times 10^{14} \text{ molecules cm}^{-3},$$
$$\sigma(230 \text{ nm}) \geq 8.4 \pm 1.3 \times 10^{-18} \text{ cm}^2 \text{ molecules}^{-1}.$$

A more exact value of the absorption cross section could be determined by fitting simulated yields of CCl_3 to the experimental absorption signal of CCl_3 for increasing partial pressures of chloroform. The reaction mechanism consist of reaction 3,6,7 and 8:

8. $F + SF_5 + M$ -> $SF_6 + M$.

k_3 and k_7 were obtained from the observed formation and decay (figure 3) of CCl_3 in the $CHCl_3/SF_6$ system (listed in table I). k_8 was obtained from the literature (9) and k_6 was adjusted to give proportionality between calculated CCl_3 concentrations and those derived from experimental traces. This procedure gave a sligthly higher absorption cross signal:

$$\sigma(230 \text{ nm}) = 8.6 \pm 1.3 \times 10^{-18} \text{ cm}^2 \text{ molecules}^{-1}.$$

In the CCl_4/Ar system a nearly instantaneous formation of CCl_3 was observed. The decay of CCl_3 was considerably faster than in the $CHCl_3/SF_6$ system (figure 3). This was due to the production of both CCl_3 and Cl (reaction 1) and the very fast reaction between CCl_3 and Cl:

9. $CCl_3 + Cl \quad \rightarrow \quad CCl_4$.

From the decay of CCl_3 in the CCl_4/Ar system, the rate constant k_9 can be extracted, when it was assumed that the high pressure limit of k_7 has been reached at 1 atm total pressure.

When oxygen was added to a gas mixture of $CHCl_3$ and SF_6, the decay of the transient absorption changes from the second order decay of CCl_3 to a biphasic curve. The first part was pseudo first order with respect to oxygen and must therefore represent the reaction between CCl_3 and O_2:

10. $CCl_3 + O_2 + M \rightarrow \quad CCl_3O_2 + M$.

From the variation of the halflife with partial pressure of oxygen the rate constant was determined to be $1.7 \pm 0.4 \times 10^{-12}$ cm^3 molecules^{-1} sec^{-1}. The second part shows second order kinetics and is believed to represent the dimerisation of the CCl_3O_2 radical:

11. $2\ CCl_3O_2 \quad \rightarrow \quad 2\ CCl_3O + O_2$.

Work is in progress to establish the rate constant for reaction 11 and the UV spectrum of CCl_3O_2.

All rate constants determined in this work are listed in the table, and as can be seen there is generally good agreement with literature data. With respect to reaction 6, Wörsdörfer and Heydtmann (10) have identified two reaction channels at low pressure (3 mbar):

6a. $F + CCl_3 \rightarrow FCCl_3^* \rightarrow FCCl_2 + Cl$,
6b. $F + CCl_3 \rightarrow FCl + CCl_2$.

At 1 atm total pressure we assume, that the excited intermediate will be stabilized very fast, and that the main product therefore will be $CFCl_3$. Product studies need to be conducted to verify this assumption.

The literature data for the recombination of CCl_3 differs by a factor of 100, but three values are concentrated in the range from $6.6 - 12.1 \times 10^{-12}$ cm^3 molecules^{-1} sec^{-1} (11-13). These three determinations have been carried out at temperatures above 340 K, but as the activation energy is near to zero (11), they should be compareble to our value. Various pressures from about 200 mbar to 1 atm have been used in the three determinations and our work. It can therefore be concluded that the high pressure limit has been reached at 1 atm total pressure.

There exist only a few earlier measurements of k_{10}. Cooper and coworkers (14) found a value three times higher than our value. Their value is believed to be in error, because they have used ethane as chlorine scavenger and the products are therefore a mixture of CCl_3O_2 and $C_3H_5O_2$. Extrapolation of the low pressure work done by Ryan and Plumb (15) to 1 atm total pressure gives a value in good agreement with our work.

ACKNOWLEDGEMENT
     Thanks are due to Dr. R. Lesclaux and coworkers (University of Bordeaux, France) for making their results known to us prior to publication.

REFERENCES

(1) Lesigne, B., L. Gilles and R.J. Woods, Can. J. Chem., 52, 1135, 1974.
(2) Alfassi, Z.B. and S. Mosseri, J. Phys. Chem., 88, 3296, 1984.
(3) Wilson, S.R., P.J. Crutzen, G. Schuster, D.W.T. Griffith and G. Helas, Nature, 334, 689, 1988.
(4) Hansen, K.B., R. Wilbrandt and P. Pagsberg, Rev. Sci. Instrum., 50, 1532, 1979.
    Pagsberg, P., J. Eriksen and H.C. Christensen, J. Phys. Chem., 83, 35, 1979.
(5) Macpherson, M.T., M.J. Pilling and M.J.C. Smith, J. Phys. Chem., 89, 2268, 1985.
(6) Pagsberb, P., E. Ratajczak, A. Sillesen and J.T. Jodkowski, Chem. Phys. Letters., 141, 88, 1987.
(7) Rasmussen, O.L. and E. Bjergbakke, Risø-R-395, Risø National Laboratory, Denmark, 1984.
(8) Lesclaux, R., private communication, 1988.
(9) Herron, J.T., Int. J. Chem. Kinetics, 19, 129, 1987.
(10) Wörsdörfer, U. and H. Heydtmann, 10. Int. Symp. on Gas Kinetics, Swansea, 1988.
(11) DeMaré, G.R. and G. Huybrechts, Trans. Faraday Soc., 64, 1311, 1973.
(12) White, M.L. and R. Kuntz, Int. J. Chem. Kinetics, 5, 187, 1973.
(13) Matheson, I.A., H.W. Sidebottom and J.M. Tedder, Int. J. Chem. Kinetics, 6, 493, 1974.
(14) Cooper, R., J.B. Cumming, S. Gordon and W.A. Mulac, Radiat. Phys. Chem., 16, 169, 1980.
(15) Ryan, K.R. and I.C. Plumb, Int. J. Chem. Kinetics, 16, 591, 1984.
(16) Foon, R. and N.A McAskill, Trans. Faraday Soc., 65, 3005, 1969.

		This work 10^{12} x k_{295}	Literatur data		
			10^{12} x k	Techn.[a]	Ref.
3.	F + $CHCl_3$	3.5 ± 0.9	4.2 ± 0.3	DF-MS	(10)
			0.9	RR	(16)
6.	F + CCl_3	110			
6a.			60 ± 10	DF-MS	(10)
6b.			50 ± 10	DF-MS	(10)
7.	CCl_3 + CCl_3	7.8 ± 1.2	6.6	RR	(11)
			7.8	P-GC	(12)
			12.1	RR	(13)
8.	F + SF_5		17		(9)
9.	CCl_3 + Cl	86 ± 13	105	RR	(11)
10.	CCl_3 + O_2	1.7 ± 0.4	2.5	DF-MS[b]	(15)
			5.2 ± 0.5	UV-PR	(14)

Table I: Rate constants determined in this work and literature data. All rate constants are in units of cm^3 molecules^{-1} sec^{-1}. Termolecular rate constants are given as apparent second order rate constants at a total pressure of 1 atm.
   a. DF-MS: Discharge flow - mass spectrometry, RR: Relative rate method, P-GC: Pyrolysis - gas chromatography, UV-PR : UV spectroscopy - pulse radiolysis.
   b. Extrapolated rate constant ar 1 atm total pressure.

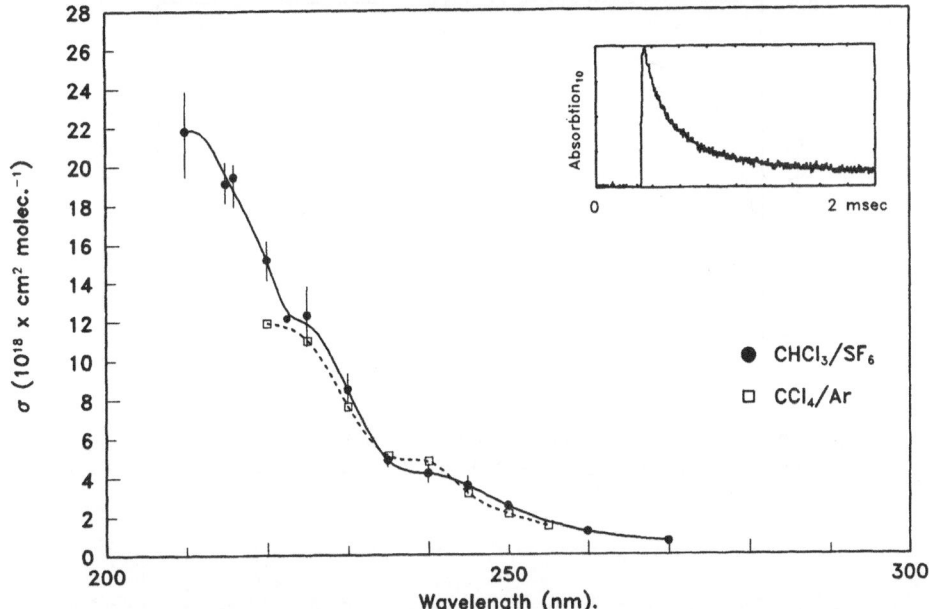

<u>Figure 1</u>   The CCl_3 spectrum recorded in the range from 210 - 280 nm with
           $CHCl_3$ or CCl_4 as precurser. Spectral band pass = 0.8 nm. Temp. =
           295 K. Gas mixtures: 2 mbar $CHCl_3$ and SF_6 to 1 atm. 1 mbar CCl_4
           and Ar to 1 atm. The insert shows transient CCl_3 absorption in
           the $CHCl_3/SF_6$ system.

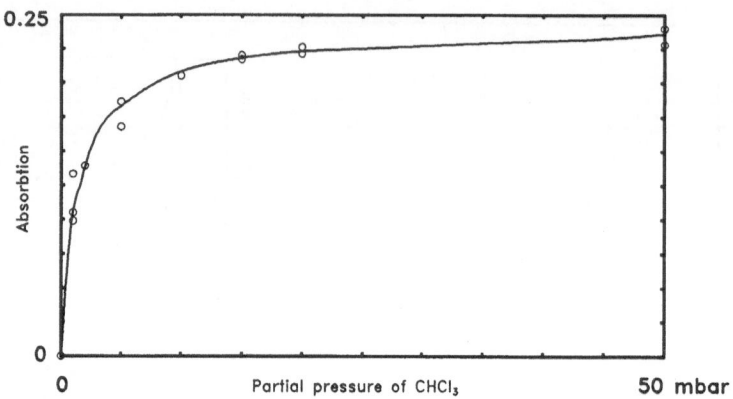

Figure 2 The yield of CCl$_3$ monitored at 230 nm as a function of partial pressure of CHCl$_3$. Gas mixture: 0-50 mbar CHCl$_3$ and SF$_6$ to 1 atm. Decadic absorption.

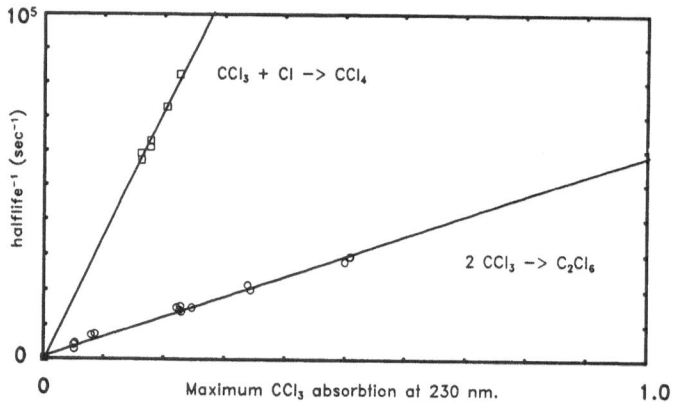

Figure 3 Determination of k$_7$ and k$_9$ at 295 K and 1 atm total pressure. Reciprocal halflife as a function of initial yield of CCl$_3$.

FTIR STUDIES OF THE REACTION BETWEEN THE NITRATE RADICAL AND HALOETHENES.   Part I. Vinylchloride.

J. HJORTH [*], C.J. NIELSEN [#], I.M.W. OLSEN [#] and G. OTTOBRINI [*]
[*] Commission of the European Communities, JRC - Ispra Establishment, I-21020 Ispra (VA) Italy.
[#] Department of Chemistry, University of Oslo P.O. Box 1033 Blindern, N-0315 Oslo 3, Norway.

The vapour phase reaction at 740 mmHg and 296 K between the nitrate radical and vinyl-chloride has been studied by long path FTIR spectroscopy. The reaction mechanism for the NO_3 initiated degradation of vinylchloride appears to follow the pattern for unsaturated compounds: the nitrate radical adds to the more electron rich sp^2 hybridized carbon followed by the formation of peroxy- and oxy radicals, of alcohols and carbonyl compounds.
The following products have been identified unambiguously by their IR spectra: CO, CH_2O, CHOCl and HCl. In addition, the spectra indicate the existence of α–nitroxy-acetylchloride, O_2NO-CH_2-COCl, as an oxidation product as well as a nitrate-pernitrate intermediate.

INTRODUCTION

The antropogenic emissions of haloethenes are insignificant relative to other organic compounds. However, in local areas near petrochemical industries this type of chemicals are a major concern. The mechanisms of the gas phase degradation of haloethenes are not yet fully known. Particularly the night time chemistry of these compounds needs to be studied further.

Prior studies [1-4] have shown the nitrate radical to react slowly with the haloethenes: typical rate constants are of the order 10^{-16} cm^3 molecule^{-1} s^{-1}. In the present communication we submit our preliminary results from product studies by long path FTIR spectroscopy of the reaction between the nitrate radical and vinylchloride, $CH_2=CHCl$.

EXPERIMENTAL

Infrared spectra were recorded of $CH_2=CHCl$ / NO_3 mixtures (i.e. NO_2 and N_2O_5) in a 480 l reactor with a Bruker 113 V FTIR instrument employing an 1 cm^{-1} spectral resolution. The reactor, constructed from a 150 cm long, 60 cm diameter Duran glass tube with flanges of aluminium, was coated with Teflon and included a 80 m White type multireflection mirror system.

NO_3 was provided by the thermal equilibrium:

$$N_2O_5 \quad \overset{M}{\rightleftharpoons} \quad NO_2 + NO_3$$

N_2O_5, in turn, was produced in the reaction chamber by mixing NO_2 and O_3 prior to introducing the vinylchloride. Typical start concentrations in the experiments were: 25 ppm vinylchloride, 2 ppm NO_2 and 20 ppm N_2O_5. The reactions were terminated after 5 to 8 hours by adding an excess of NO.

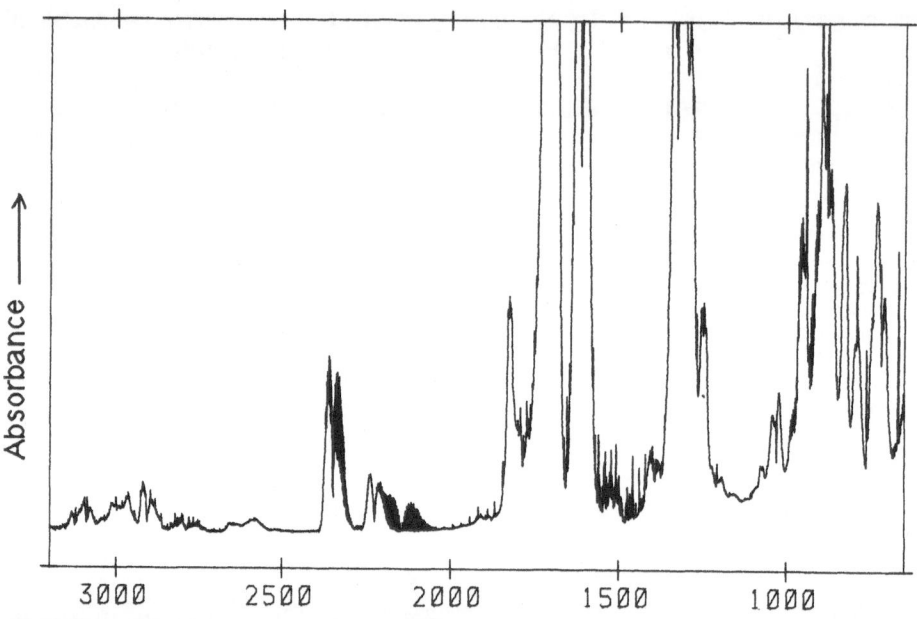

Figure 1. IR spectrum of the reaction pot-pourri $CH_2=CHCl$, NO_2 and N_2O_5 in purified air immediately after mixing.

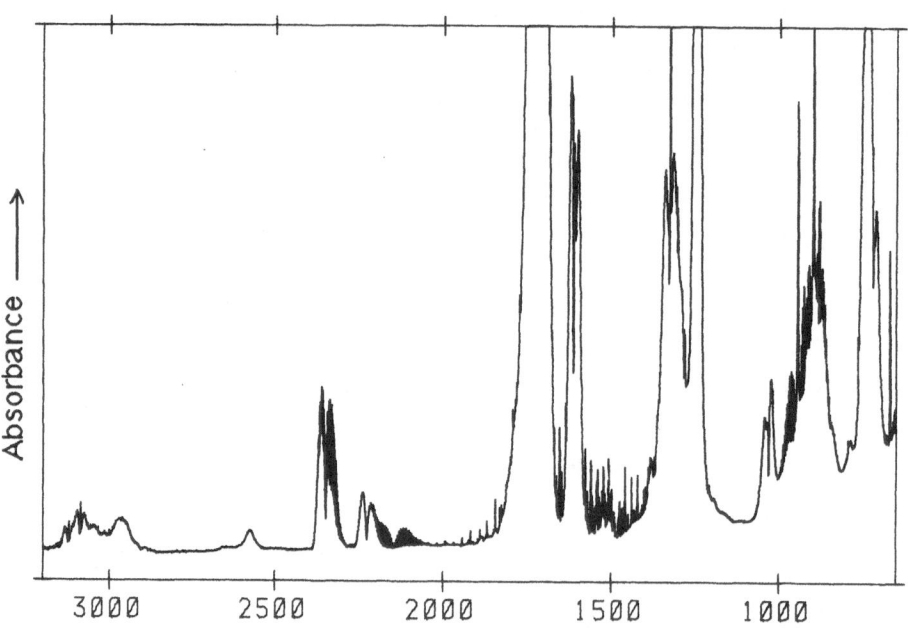

Figure 2. IR spectrum of the reaction hotchpotch $CH_2=CHCl$, NO_2 and N_2O_5 in purified air 6 hours after mixing.

RESULTS

Vibrational spectra. Survey spectra of the mixture $CH_2=CHCl$, NO_2 and N_2O_5 in purified air immediately after mixing and after six hours of reaction are shown in Figs. 1 and 2, respectively. From a quick perusal one can identify CO, CH_2O and HCl as oxidation products (the CO_2 concentration remains constant throughout the experiment). In addition to the bands originating from HNO_3, NO_2, N_2O_5 and vinylchloride, absorption bands at ca. 790 and ca. 1305 cm^{-1} indicate the presence of some organic pernitrate ($R-O_2NO_2$) and bands at ca. 835 and ca. 1290 cm^{-1} some organic nitrate ($R-ONO_2$). The production of HNO_3 seemed to be larger than in similar experiments with non-halogenated, unsaturated hydrocarbons and no traces of HO_2NO_2 or HONO were found in the spectra.

A closer inspection of the spectra reveals **Q** branches at 1783 and 738.6 cm^{-1} and associated **PR** branches with rotational structures identical to those of the v_2 and v_4 bands in formylchloride, CHOCl [5]. However, the most striking feature in the IR spectrum is the band at 1824 cm^{-1}. Considering the band intensity (Fig. 2) it is difficult, if not impossible, to explain this band except by a carbonyl stretching vibration. The frequency, 1824 cm^{-1}, is unusually high and the only type of carbonyl compound that would fit this wavenumber is an acidochloride with a strong electronegative substituent in α-position. In accordance with the proposed reaction mechanism (see later) we tentatively attribute the 1824 cm^{-1} band to α-nitroxy-acetylchloride, O_2NO-CH_2-COCl.

The relative intensity of the 1824 cm^{-1} band varies with the experimental conditions. It seems that the water vapour content and NO_2 concentration are critical factors in this connection. This is perhaps not surprising considering the nature of the (suggested) origin to the absorption band.

Adding excess NO to the reaction mixture forces a break down of the peroxy radicals and thereby accelerates the formation of the subsequent oxidation products. In one experiment it was found that the amount of formylchloride increased by ca. 50 % after the NO was added. At the same time the nitrate band at 835 cm^{-1} and the pernitrate band at 790 cm^{-1} decreased by ca. 20 and 50 %, respectively. Apparently the experimental conditions favours the formation of relatively stable nitrate compounds. It further seems that the band at 790 cm^{-1} is not entirely due to an organic pernitrate intermediate. It is also informative that the 1824 cm^{-1} band, interpreted as due to α-nitroxy-acetylchloride, does not change significantly in intensity after the addition of NO.

In Fig. 3 are shown the time profiles of the "pernitrate", the "nitrate" and of the supposed "α-nitroxy-acetylchloride" bands at 790, 835 and 1824 cm^{-1}, respectively. Due to the lack of reference spectra the band intensities have been scaled relative to the last spectrum in the series of measurements. There is obviously some correlation between the 835 and 1824 cm^{-1} bands. On the other hand, if the 1824 cm^{-1} band as postulated originates from α-nitroxy-acetylchloride then it can be concluded that the reaction mixture contains several organic nitrates.

An analogous plot showing the time profiles of the oxidation products, CO, HCl, CH_2O and CHOCl, is given in Fig. 4. Again the measured IR absorptions have been scaled according to the last

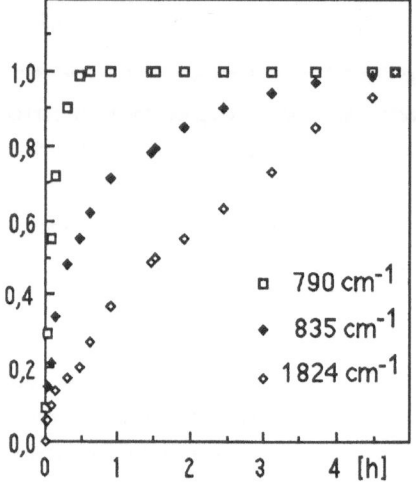

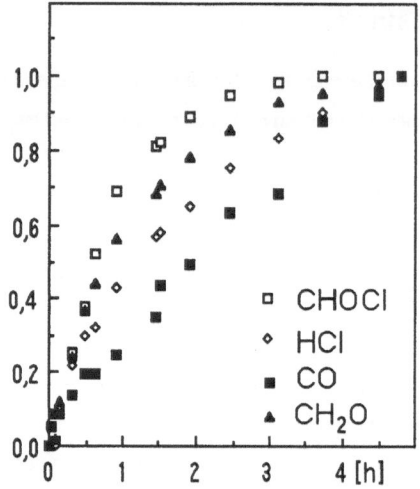

Figure 3. Scaled time profiles of the bands at 790, 835 and 1824 cm^{-1} in the mixture: CH$_2$=CHCl, NO$_2$ and N$_2$O$_5$ in purified air.

Figure 4. Scaled time profiles of the CHOCl, HCl, CO and CH$_2$O bands in the reaction mixture: CH$_2$=CHCl, NO$_2$ and N$_2$O$_5$ in purified air.

spectrum recorded before NO was added. As can be seen, there is clearly some correlation between the concentrations of the oxidation products. However, it is equally obvious that there must exist several reaction paths to these products. For example, it is evident that CO and HCl cannot result from a thermal break down of CHOCl alone.

Reaction mechanism. The electronic distribution in vinylchloride is often visualized in terms of the two mesomeric forms:

$$\begin{array}{c} H \\ \\ H \end{array} C = C \begin{array}{c} H \\ \\ Cl \end{array} \rightleftharpoons \begin{array}{c} H \\ \\ H \end{array} C^- - C \begin{array}{c} H \\ \\ Cl \end{array}^+ $$

Hence, the nitrate radical is expected to add to the double bond primarily at the unsubstituted carbon atom. The subsequent reactions are then expected to follow the proposed mechanism for the NO$_3$ initiated degradation of unsaturated hydrocarbons [6] (see also references therein). In the reaction mechanism suggested below the unambiguously identified products have been typed in bold italics.

(1) $\qquad$ CH$_2$=CHCl + NO$_3$ $\quad \rightarrow \quad$ O$_2$NO-CH$_2$-ĊHCl

(2) $\qquad$ O$_2$NO-CH$_2$-ĊHCl + O$_2$ $\quad \rightarrow \quad$ O$_2$NO-CH$_2$-CHCl-Ȯ$_2$

(3) $O_2NO\text{-}CH_2\text{-}CHCl\text{-}\dot{O}_2 + NO_2 \rightleftarrows O_2NO\text{-}CH_2\text{-}CHCl\text{-}O_2NO_2$

(4) $2\ O_2NO\text{-}CH_2\text{-}CHCl\text{-}\dot{O}_2 \xrightarrow{a} 2\ O_2NO\text{-}CH_2\text{-}CHCl\text{-}\dot{O} + O_2$

                              $\xrightarrow{b} O_2NO\text{-}CH_2\text{-}CHCl\text{-}OH + O_2$

                                   $+ O_2NO\text{-}CH_2\text{-}COCl$

(5) $O_2NO\text{-}CH_2\text{-}CHCl\text{-}\dot{O} \rightarrow NO_2 + \mathit{CH_2O} + \mathit{CHOCl}$

(6) $O_2NO\text{-}CH_2\text{-}CHCl\text{-}\dot{O} + O_2 \rightarrow H\dot{O}_2 + O_2NO\text{-}CH_2\text{-}COCl$

(7) $O_2NO\text{-}CH_2\text{-}CHCl\text{-}\dot{O} + NO_2 \rightarrow O_2NO\text{-}CH_2\text{-}CHCl\text{-}ONO_2$

The somewhat "unusual" molecule 1-chloro-2-nitroxy-ethanol, formed in (4b), is likely to decompose rapidly:

(8) $O_2NO\text{-}CH_2\text{-}CHCl\text{-}OH \rightarrow O_2NO\text{-}CH_2\text{-}CHO + \mathit{HCl}$

Formylchloride, formed in reaction (5), is known to be unstable and the reported half-life is ca. 10 min [5]. However, under our experimental conditions the half-life of formylchloride appeared to be more than one hour.

(9) $\mathit{CHOCl} \rightarrow \mathit{CO} + \mathit{HCl}$

Both formaldehyde and nitroxy ethanal are known to react with NO_3. These reactions are also quite slow, of the order 10^{-15} - 10^{-16} cm^3 molecule^{-1} s^{-1} [7-9], and have not been included in this discussion. The break down of the alkoxy radicals formed in (4a) may also lead to the formation of new alkyl radicals which then may undergo steps similar to those in (2-7). Instead of a formaldehyde production this will lead to the formation of new small nitroxy compounds such as nitroxy methanol, nitroxy methanal and dinitroxy methane.

We realize that the results so far presented are poor evidence for the proposed reaction mechanism for the degradation of vinylchloride in an NO_2 / N_2O_5 atmosphere. The alternative reaction mechanism given below (I-VII), where the nitrate radical adds to the less electron rich sp^2 hybridized carbon atom, also predicts the formation of CO, HCl, CH_2O and CHOCl as oxidation products. Some of the nitrate containing products will be different. More important, CO and HCl will be formed by the thermal decomposition of CHOCl alone and there is no reaction path that will lead to a compound with a carbonyl stretching vibration above 1800 cm^{-1}. This does, however, not exclude the possibility that the initial NO_3 attack may take place at both carbon atoms.

(I) $CH_2{=}CHCl + NO_3 \rightarrow \dot{C}H_2\text{-}CHCl\text{-}ONO_2$

(II) $\dot{C}H_2\text{-}CHCl\text{-}ONO_2 + O_2 \rightarrow \dot{O}_2\text{-}CH_2\text{-}CHCl\text{-}ONO_2$

(III) $\dot{O}_2\text{-}CH_2\text{-}CHCl\text{-}ONO_2 + NO_2 \rightleftarrows O_2NO_2\text{-}CH_2\text{-}CHCl\text{-}ONO_2$

(IV) $2 \ \dot{O}_2\text{-}CH_2\text{-}CHCl\text{-}ONO_2 \xrightarrow{a} 2 \ \dot{O}\text{-}CH_2\text{-}CHCl\text{-}ONO_2 + O_2$

                      $\xrightarrow{b} CH_2OH\text{-}CHCl\text{-}ONO_2 + O_>$

                      $+ CHO\text{-}CHCl\text{-}ONO_2$

(V) $\dot{O}\text{-}CH_2\text{-}CHCl\text{-}ONO_2 \rightarrow CH_2O + CHOCl + NO_2$

(VI) $\dot{O}\text{-}CH_2\text{-}CHCl\text{-}ONO_2 + O_2 \rightarrow H\dot{O}_2 + CHO\text{-}CHCl\text{-}ONO_2$

(VII) $\dot{O}\text{-}CH_2\text{-}CHCl\text{-}ONO_2 + NO_2 \rightarrow O_2NO\text{-}CH_2\text{-}CHCl\text{-}ONO_2$

We hope to be able to distinguish between the two proposed reaction mechanisms from a comparison with IR spectra of the different nitroxy compounds which are currently being synthesized. We also expect to elucidate the reaction mechanism by on-going simulations with the FACSIMILE program [10] and by experiments employing isotope-labelled compounds.

As stated previously, we did not detect HO_2NO_2 or HONO in the IR spectra. This may either be explained by our experimental conditions where either NO_3 or NO will be present when HO_2 is formed or, alternatively, by the rate constant for reaction (6) being at least an order of magnitude smaller than the rate constant k_5 for the thermal decomposition of the oxy radicals.

Acknowledgement. I.M.W.O. appreciates financial support from the Royal Norwegian Council for Scientific and Industrial Research, NTNF.

REFERENCES

[1] Atkinson, R., Aschmann, R. and Goodman, M.A., *Int. J. Chem. Kin.* **19**, 299 (1985).

[2] Andersson, Y. and Ljungström, E., *Atm. Environ, in press.*

[3] Andersson, Y. and Ljungström, E., *Int. J. Chem. Kin., in press.*

[4] Andersson, Y., Wängberg, I., Ljungström, E., Hjorth, J. and Ottobrini, G., *CEC Air Poll. Res. Report,* **17**, 95 (1988).

[5] Hisatsune, I.C. and Heicklen, J., *Can. J. Spectrosc.,* **18**, 77 (1973).

[6] Hjorth, J., Lohse, C., Nielsen, C.J. , Restelli, G. and , H., *J. Phys. Chem., submitted.*

[7] Hjorth, J., Ottobrini, G. and Restelli, G., *J. Phys. Chem.,* **92**, 2669 (1988).

[8] Atkinson, R., Plum, C.N., Carter, W.P.L., Winer, A.M. and Pitts, J.N. Jr., *J. Phys. Chem.,* **88**, 1210 (1974).

[9] Morris, E.D. and Nikki, H., *J. Phys. Chem.,* **78**, 1337 (1974).

[10] Chance, E.M., Curtis, A.R., Jones, I.P. and Kirby, C.R., *Report R 8775, United Kingdom Atom Energy Authority, Harwell (1977).*

SPECTROSCOPIC MEASUREMENTS OF NO_2, O_3, SO_2, IO AND NO_3 IN MARITIME AIR

T. Brauers[*], H.-P. Dorn[*] and U. Platt[x]
[*]Institut für Atmosphärische Chemie
Kernforschungsanlage Jülich
[x]Institut für Umweltphysik
Universität Heidelberg

Summary

During July 1988 measurements of NO_2, NO_3, O_3, SO_2, and IO were made at Pointe de Penmarc'h, a coastal site in western Britanny. NO_2 and NO_3 concentrations range from the detection limit up to 1.6 ppb and 2.6 ppt, respectively. It appears that dimethyl sulfide is an important sink to NO_3 radicals under those conditions. An upper limit of 0.4 ppt can be given for the iodine oxide concentration.

1. INTRODUCTION

Oceans, in particular coastal regions, are known to be sources of biogenic compounds like hydrocarbons, halogen containing organic species, and reduced sulphur species. Dimethyl sulfide (DMS), for instance, is the dominant component of the natural sulfur cycle in the oceanic atmosphere (1).

The removal of the above trace species from the atmosphere is initiated by oxidation reactions with free radicals. The daytime removal is controlled by hydroxyl radicals, OH. The night-time oxidation of reduced trace gases, on the other hand, mainly proceeds through reactions with nitrate radicals, NO_3 and also O_3.

The reaction of DMS with OH radicals has been proposed as the major loss mechanism of DMS in troposphere. Experimentally determined atmospheric residence times of DMS are ranging between 10 minutes and more than 10 hours (2). Calculations indicate that the OH reaction may be too slow to explain the shortest lifetimes of DMS observed. However, the reaction of DMS with halogene oxide radicals as speculated by Carlier (3) is sufficiently fast to offer a possible explanation for the observed loss of DMS at least during daytime.

Another possible, nighttime sink for DMS is the oxidation by nitrogen trioxide radicals. However, the sink strength of this reaction is not well known because of the lack of measurements of the NO_3 concentrations over the oceans.

Nitrogen trioxide is almost exclusively formed by the oxidation of NO_2 by O_3:

(1) $NO_2 + O_3 \longrightarrow NO_3 + O_2$

In a subsequent reaction N_2O_5 quickly reaches an equilibrium between formation and thermal decay (at tropospheric temperatures):

(2) $NO_2 + NO_3 \; <\text{-M-}> \; N_2O_5$

The NO_3 lifetime is potentially determined by homogeneous reactions with biogenic species like unsaturated hydrocarbons (see for example (4)). In addition either NO_3 or N_2O_5 may react heterogenously on aerosol surfaces. Finally dry deposition represents a possible loss.

Atmospheric IO has been proposed to derive from methyl iodide, for which concentrations up to 20 ppt in coastal areas with a mean of 1.2 ppt have been reported (5). Several laboratory studies (2,6,7) and model calculations (7,8) were carried out to investigate the reactions of of methyl iodide and its products in the atmosphere.

Photolysis of CH_3I leads to iodine atoms

(3) $CH_3I + h\nu ---> CH_3 + I$

The main reaction of the I atoms under tropospheric conditions is:

(4) $I + O_3 \ --> \ IO + O_2$ $k_5 = 9.6 \cdot 10^{-13} \ cm^3/s$

The IO radical reacts with NO, NO_2, and with itself. Model calculations suggest IO concentrations up to $5 \cdot 10^7 cm^{-3}$ (7).

Barnes et al. (3) proposed following reaction of DMS:

(5) $IO + DMS ---> DMSO + I$ $k_6 = (3 \pm 1.5) \cdot 10^{-11} \ cm^3/s$

From the rate constant and a typical estimated IO concentration ($10^7 cm^{-3}$) a DMS lifetime of around one hour can be expected during daytime.

2.EXPERIMENTAL

The measurements were carried out at the west coast of Brittany, France, Pointe de Penmarc'h. This location is characterized by large algae fields extending several hundred meter off coast. These fields are responsible for the emission of volatile organic compounds. In particular they are thought to lead to the emissions of DMS and CH_3I.

Measurements of the stable trace gases NO_2, O_3, SO_2, and the free radicals NO_3 and IO were carried out by long path Differential Optical Absorption Spectroscopy (DOAS) (9). The light path between a 500 W Xenon arc lamp on the top of a lighthouse at the Pointe de Penmarc'h 43 m above sea-level (Fig.1,A), and the analyzing spectrometer located near Le Guilvinec approximately 5 m above sea-level (Fig.1,B) had a length of 5.93 km and extended mainly over the coastal algae fields (Fig.1).

In addition to the long path measurements, in situ data of ^{222}Rn and aerosol concentration, photolysis frequencies of NO_2 and O_3, ozone concentration, and metereological parameters were recorded at the top of the lighthouse. DMS, HC, and CH_3I have also been measured and will be presented seperately (10,11,12).

3.RESULTS

The ozone concentrations measured in situ on the top of the lighthouse (Fig.2b) ranged from 20 to 70 ppb with a mean of 30 ppb, which is typical for clean oceanic air. The short spikes in ozone concentration at the beginning of the observation period were strongly correlated with corresponding changes in wind direction (Fig.2a).

Simultanously recorded ozone data by the DOAS system (Fig.2c) show comparable values during the whole campaign. The DOAS data, however, do not show those sharp peaks. This is probably due to the spatial integration by the DOAS technique (6 km horizontal distance, 40 m altitude).

The concentration profiles of SO_2 and NO_2 show similarities (Fig.2d, Fig.2e) in the following sense: The observed values were low - often below the detection limit of 100 ppt and 200 ppt, respectively - with maximum values around 300 ppt for SO_2 and 600 ppt for NO_2 during the first nine days of the observation period. During the last three days, July 14 - 16, the concentrations of SO_2 and NO_2 increased with peak values of 1 ppb and 1.6 ppb, respectively.

Measurements of NO_3 radicals could be made during 5 nights (Fig.2f). The NO_3 concentrations were highest (2.6 ppt) in the night July 14/15, when NO_2 concentration were highest.

The other radical species, IO, could not be detected. Averaging over 5 sunny days leads to an upper limit of 10^7 molecules/cm^3.

4.DISCUSSION

Since the reaction of NO_2 with O_3 is the only source of NO_3, one can calculate the lifetime using measured quantities

$$(6) \quad \tau_{tot}(NO_3) = [NO_3]_{stat} / ([NO_2] [O_3] k_1)$$

assuming steady state conditions. The NO_3 lifetimes thus obtained lie between 5 s and 550 s as shown in table 1 and are integral values over all loss mechanisms. In the following we try to estimate the individual contributions to the total removal (average $\tau_{tot}(NO_3) = 125$ s) for the night July 14/15:

Dimethylsulfide (DMS) concentrations (10,12) range from 200 ppt at groundlevel to 20 ppt at 40 m altitude. An exponential mean along the altidudinal interval covered by the DOAS measurement was calculated to be 70 ppt. The NO_3 lifetime against the reaction with DMS can be determined by

$$(7) \quad NO_3 + DMS \to products \qquad k_7 = 8.6 \cdot 10^{-13} cm^3 s^{-1}$$
$$(8) \quad \tau_{DMS}(NO_3) = 1 / [DMS] k_7$$

The thus calculated value is around 600 s.

Similar considerations apply for the reactions of NO_3 with unsaturated hydrocarbons (HC) like propene and butenes

$$(9) \quad NO_3 + HC \to products$$

HC concentrations from air samples taken at sea level and on the top of the lighthouse during the same night (10) lead to $\tau_{HC}(NO_3) > 3000s$. The measured hydrocarbons have, therefore, a negligible effect on the steady state NO_3 concentration.

In addition we have to consider heterogeneous removal of NO_3 by reactions on the surfaces of aerosols, either of NO_3 itself or of N_2O_5. As a crude estimate of the total aerosol surface the mean value of the nephelometer data $3 \cdot 10^{-5} m^{-1}$ can be taken to calculate the residence time of NO_3 against loss on aerosols. With the thermal velocity of NO_3 molecules of 290 ms^{-1} at 293 K the collision frequency is in the order of 0.01 s^{-1}. Assuming a accommodation coefficient of A = 0.1 as an upper limit, NO_3 lifetime would be of the order of or more than 1000 s.

In order to consider the heterogenous loss of N_2O_5 we calculate the equilibrium N_2O_5 concentration (eq.2) from

$$(10) \quad [N_2O_5] = k_{eq} [NO_2] [NO_3] \qquad k_{eq} = 4.4 \cdot 10^{-11} cm^3 \text{ at 293 K.}$$

Calculated values for four night are listed in table 2, showing that the N_2O_5 concentration can range from 3% to 95% of NO_3 concentration. A mean value during the night July 14/15 is around 70%. The time to reach equlibrium is approximately 50 s at 293 K. The accommodation coefficient of N_2O_5 is reported to be 0.05. These values lead to a lower limit of NO_3 lifetime due to heterogenous N_2O_5 loss of 2500s.

From these estimates of the identificable NO_3 sinks it appears, that the reaction of NO_3 with DMS is the strongest contributing about 20%. Nevertheless unidentified loss processes of NO_3 have to be considered which could contribute up to 50 %.

The chemical lifetime of DMS has been proposed to be the result of reactions with OH, NO_3, and IO. The observed NO_3 concentrations correspond to a DMS lifetime of about 4.5 h for the maximum NO_3 of 2.7 ppt,and 37 h for our NO_3 detection limit of 0.3 ppt. The upper limit of the IO concentration ($10^7 cm^{-3}$) corresponds to lower limit of the DMS lifetime of about one hour. In comparision the DMS lifetime against the reaction with OH should be around 30 h assuming a daytime average OH concentration of $10^6 cm^{-3}$.

5.CONCLUSION

The field measurements have confirmed for the first time the predicted interactions between free radicals and dimethyl sulfide in marine air. It has been shown that DMS is an important scavenger for NO_3 under those conditions. However, other loss processes for NO_3, either heterogenous or due to hydrocarbons not measured during our study, also exist. Moreover DMS lifetimes due to NO_3 reactions could be calculated. For iodine oxide it was possible to obtain an upper limit of the concentration.

REFERENCES

(1) M.O.Andreae et al., JGR 90 (1985) 12891
(2) B.C.Nguyen et al., in N. Brutsaert, G.H.Jirka eds. Gas transfer at water surfaces (1984) 539
(3) I.Barnes et al., Int J Chem Kin 19 (1986) 489
(4) A.Winer et al., Science 224 (1984) 156
(5) R.D.Rasmussen et al., JGR 87 (1982) 3086
(6) D.Martin et al. 1986, Int J Chem Kin 19 (1987) 503
(7) M.E.Jenkin et al., J Atm Chem 2 (1985) 359
(8) W.L.Chameides and D.D.Davis, JGR 85 (1980) 7383
(9) U.Platt and D.Perner, Fresenius Z Anal Chem 317 (1984) 309
(10) B.C.Nguyen et al. private communication
(11) B.Bonsang et al. private communication
(12) P.Carlier et al. private communication

Table 1:

time		O_3	NO_2	NO_3	$t(NO_3)$	N_2O_5
day	h	ppb	ppb	ppt	s	NO_3
07.07.88	23	33.53	0.28	0.14	18.96	0.30
08.07.88	22	29.05	0.58	0.54	40.07	0.63
09.07.88	1	30.66	0.71	0.08	4.83	0.78
09.07.88	2	32.76	0.49	0.15	11.21	0.54
09.07.88	3	29.43	0.58	0.13	9.66	0.64
09.07.88	4	32.36	0.34	0.71	80.88	0.37
09.07.88	5	31.53	0.20	0.12	24.10	0.22
09.07.88	23	30.60	0.07	0.51	293.05	0.08
10.07.88	0	31.18	0.15	1.43	375.37	0.17
10.07.88	1	30.60	0.10	0.71	279.27	0.11
10.07.88	2	30.02	0.06	0.73	546.39	0.06
12.07.88	20	32.14	0.03	0.27	379.32	0.03
14.07.88	21	22.54	0.66	1.26	106.76	0.72
14.07.88	22	21.88	0.68	1.06	88.37	0.75
14.07.88	23	21.22	0.87	1.50	101.46	0.96
15.07.88	1	23.55	0.45	0.67	79.00	0.49
15.07.88	2	22.89	0.23	1.12	261.05	0.26
15.07.88	3	22.62	0.82	1.56	105.54	0.90

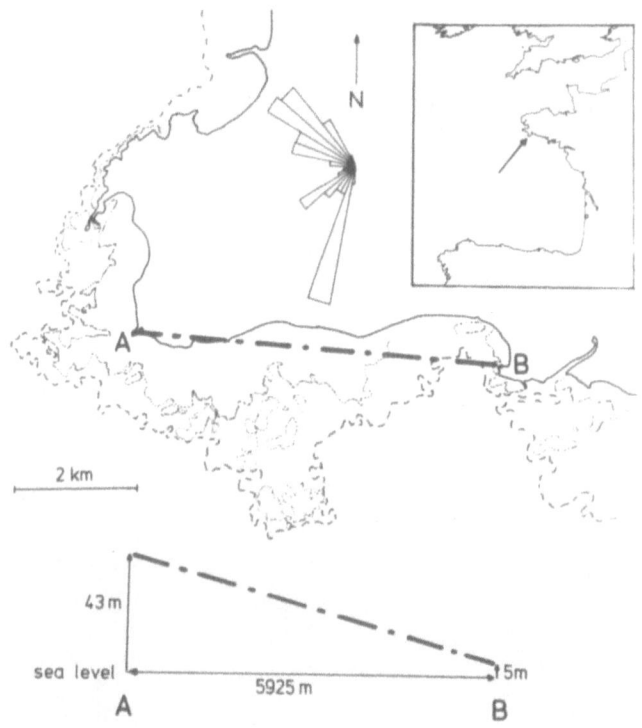

Figure 1:
- Insert: coastal line of France, arrow indicates measurement site
- Upper part: Measurement site
  A: lighthouse at Pointe de Penmarc'h
  B: DOAS spectrometer near Le Guilvinec
  solid line: coast line at high tide
  dashed line: coast line at low tide
  dotted line: areas with strong algae vegetation
  dash-dotted line: light path of the DOAS experiment
  histogram: distributions of the wind directions during July 5th and 16th
- Lower part: elevation of the optical light path

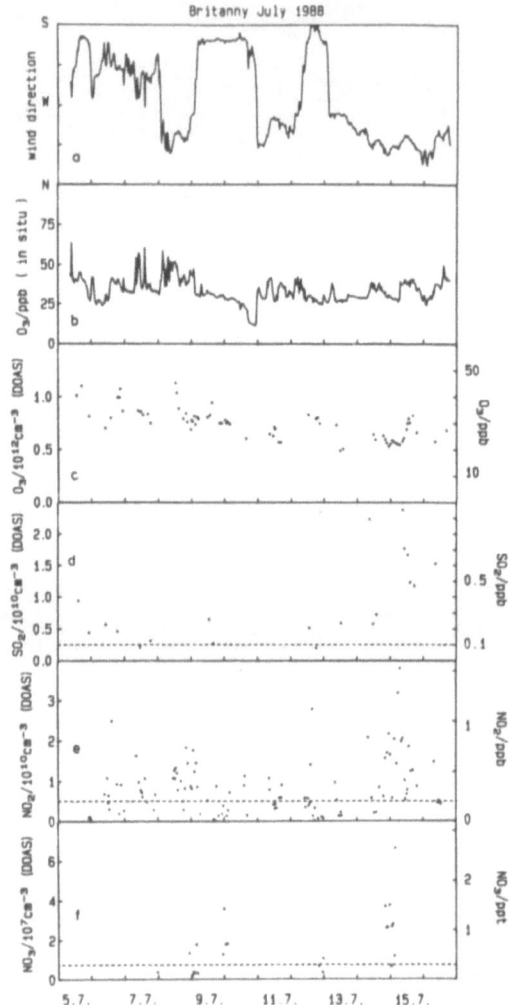

Figure 2:

a: wind direction
b: ozone concentration in situ
c: Ozone concentration measured by DOAS system
d: SO_2 measured by DOAS, the dashed line indicates the detection limit of 100 ppt
e: NO_2 measured by DOAS, the dashed line indicates the detection limit of 200ppt
f: NO_3 measured by DOAS, the dashed line indicates the detection limit of 0.3 ppt

SESSION II/B

ATMOSPHERIC CHEMICAL

AND

PHOTOCHEMICAL PROCESSES

JOINT EUROTRAC – COST 611

PROJECT HALIPP

A PULSE RADIOLYSIS STUDY OF THE CHEMISTRY OF OXYSULPHUR RADICALS IN AQUEOUS SOLUTION

GEORGE V. BUXTON, G. ARTHUR SALMON AND NICHOLAS D. WOOD

Cookridge Radiation Research Centre, The University of Leeds, Cookridge Hospital, Leeds LS16 6QB, UK

Summary

Pulse radiolysis was used to study the spectra and some reactions of the radicals $SO_4^{.-}$, $SO_3^{.-}$ and $SO_5^{.-}$. The use of CO to generate $SO_4^{.-}$ in the pulse radiolysis system without generating $\cdot OH$ was investigated, but the reaction between $CO_2^{.-}$ and $S_2O_8^{2-}$ ($k_7 \sim 1 \times 10^5$ dm^3 mol^{-1} s^{-1}) was too slow for this purpose. The reaction between $\cdot OH$ and $S_2O_8^{2-}$ was shown to proceed with $k_8 = (1.4 \pm 0.4) \times 10^7$ dm^3 mol^{-1} s^{-1}. The reaction of $SO_4^{.-}$ with formaldehyde was studied and k_{12} determined to be $(1.26 \pm 0.05) \times 10^7$ dm^3 mol^{-1} s^{-1} at $20 \pm 1°C$. The temperature dependence of the reaction was studied and gave $E_a = (10.8 \pm 1.3)$ kJ mol^{-1} and lg (A/dm^3 mol^{-1} s^{-1}) = 9.0 ± 0.2. The spectra of $SO_3^{.-}$ and $SO_5^{.-}$ were observed in deaerated and aerated solutions, respectively, of SO_3^{2-} and HSO_3^- and their molar absorptivities were determined. In deaerated solutions $SO_3^{.-}$ decayed by second-order kinetics with $2k_{13} = (7.6 \pm 0.1) \times 10^8$ dm^3 mol^{-1} s^{-1}, independent of pH. The reaction of $SO_3^{.-}$ with O_2 proceeds with $k_{14} = (2.25 \pm 0.1) \times 10^9$ dm^3 mol^{-1} s^{-1}. In oxygenated solutions $SO_5^{.-}$ decayed by second-order kinetics with $2k_{18} = (1.2 \pm 0.12) \times 10^8$ dm^3 mol^{-1} s^{-1}. Also in oxygenated solutions the chain oxidation of SO_3^{2-}/HSO_3^- occurred, which in unbuffered solutions at pH 4 led to the generation of SO_2. Some conclusions are drawn concerning the chain propagation steps.

1. INTRODUCTION

In 1982 Chameides and Davis (1) proposed that the chain oxidation of HSO_3^- in the aqueous phase of cloud droplets may represent a major pathway for the oxidation of atmospheric SO_2 to sulphuric acid. This suggestion has been incorporated in a number of models of cloud and raindrop chemistry. The chain oxidation of aqueous sulphite has been extensively studied since the early work of Backstrom and Haber, but the full details of the mechanism have not yet been elucidated. Hayon et al. (2) have proposed a mechanism which invokes the involvement of $\cdot OH$, $SO_3^{.-}$, $SO_5^{.-}$ and $SO_4^{.-}$ radicals. The current status of the mechanism has been extensively reviewed by Deister and Warneck (3) and its relationship to cloudwater chemistry has been critically reviewed by McElroy (4). The work described here is intended to clarify some features of the spectra and chemistry of the pertinent oxysulphur radicals. The radiolysis of water and aqueous solutions (reaction 1) results in the generation of hydrated electrons, e_{aq}^-, hydroxyl radicals, a small yield ($\sim 10\%$) of hydrogen atoms and small amounts of hydrogen peroxide and hydrogen (5).

$$H_2O \xrightarrow{\text{\Large$\sim\!\!\sim\!\!\sim$}} e_{aq}^-, \cdot H, \cdot OH, H_2, H_2O_2, H^+ \tag{1}$$

Saturation of the solutions with N_2O results in the conversion of e_{aq}^- to $\cdot OH$ by reaction 2 ($k_2 = 9.1 \times 10^9$ dm^3 mol^{-1} s^{-1}, (6)).

$$N_2O + e_{aq}^- \xrightarrow{H^+} N_2 + \cdot OH \tag{2}$$

$SO_4^{.-}$ is generated by reaction 3 ($k_3 = 1.2 \times 10^{10}$ dm^3 mol^{-1} s^{-1}, (6))

$$e_{aq}^- + S_2O_8^{2-} \rightarrow SO_4^{.-} + SO_4^{2-} \tag{3}$$

and $SO_3^{.-}$ by reactions 4 ($k_4 = 4.5 \times 10^9$ dm^3 mol^{-1} s^{-1} (6)) and 5 ($k_5 = 5.5 \times 10^9$ dm^3 mol^{-1} s^{-1} (6)).

$$\cdot OH + HSO_3^- \rightarrow H_2O + SO_3^{.-} \tag{4}$$
$$\cdot OH + SO_3^{2-} \rightarrow OH^- + SO_3^{.-} \tag{5}$$

Thus, pulse radiolysis provides a means of generating, and observing the fast reactions of, the radicals implicated in the chain oxidation.

2. EXPERIMENTAL

Details of the techniques of pulse radiolysis used in this laboratory have been reported elsewhere (7). The thiocyanate dosimeter was used to measure the dose taking $G\epsilon_{475} = 2.38 \times 10^{-4}$ m^2 J^{-1}. Cells of 10 or 25 mm optical path length were used as appropriate. Solutions of SO_3^{2-} and HSO_3^- were prepared using triply distilled water which was deaerated before the addition of Analar sodium sulphite (BDH). HSO_3^- solutions were adjusted to pH 4 by the addition of Aristar sulphuric acid. Some experiments were carried out at pH 5 using solutions buffered with phosphate (4 x 10^{-2} mol dm^{-3} KH_2PO_4, 4 x 10^{-4} mol dm^{-3} K_2HPO_4). Solutions were saturated with argon, nitrous oxide or oxygen as appropriate by bubbling with the gas. To ensure that autoxidation was negligible in the solution containing both oxygen and SO_3^{2-}/HSO_3^-, separate solutions of O_2 and SO_3^{2-}/HSO_3^- were mixed using a rapid-mix apparatus less than 1s before the radiation pulse was delivered.

Formaldehyde solutions were prepared by heating paraformaldehyde in a stream of argon which was bubbled into water to dissolve the formaldehyde. The concentration of the stock solution was determined by oxidising it to formic acid with hydrogen peroxide in sodium hydroxide and back titration of the excess base.

3. RESULTS AND DISCUSSION

SO_4^-. The absorption spectrum of SO_4^- has been characterised in a number of earlier studies (2, 8, 9). The absorption extends from 280 to 570 nm with a maximum at 450 nm. Values of the extinction coefficient range from 450 to 1600 dm^3 mol^{-1} cm^{-1}, but McElroy (10) has recently determined the value to be 1600 ± 100 dm^3 mol^{-1}cm^{-1}. To further check this value we carried out the pulse radiolysis of argon saturated solutions of 1 mmol dm^{-3} peroxydisulphate and 2 × 10^{-2} mol dm^{-3} t-butanol using 0.6 μs, 20 Gy pulses. The absorption spectrum of SO_4^- generated in reaction (3) had the general shape previously observed and assuming $G_{e_{aq}^-} = 270$ nmol J^{-1} (11) at the scavenging power involved gave $\epsilon_{450} = 1500 \pm 50$ dm^3 mol^{-1} cm^{-1}, in good agreement with the values obtained by McElroy (10) and Chawla and Fessenden (12).

It would be advantageous in the radiolysis system to be able to generate the radical in the absence of the ·OH which is generated in large yields when aqueous systems are irradiated. An approach to this problem, which we have investigated, is to saturate the solutions with carbon monoxide so that ·OH will react to form CO_2^-, reaction 6.

$$CO + \cdot OH \rightarrow CO_2^- + H^+ \tag{6}$$

In view of the large negative reduction potential of $CO_2(-2.1$ V), it seemed plausible that reaction 7 would occur.

$$CO_2^- + S_2O_8^{2-} \rightarrow CO_2 + SO_4^{2-} + SO_4^- \tag{7}$$

The decay of CO_2^- was followed at 290 nm in solutions irradiated with 0.6μs, 6–8 Gy pulses and containing 0.012 mol dm^{-3} N_2O, 5 x 10^{-3} mol dm^{-3} CO and 0.01 to 0.05 mol dm^{-3} $S_2O_8^{2-}$. Mixed-order kinetics were observed due to the occurence of reaction 7 and the self-reaction of CO_2^-. A simplified kinetic treatment of the data indicated $k_7 \sim 1 \times 10^5$ dm^3 mol^{-1} s^{-1}, which makes reaction 7 too slow to be a viable route to SO_4^- under pulse radiolysis conditions.

In the course of this investigation we observed that the yield of CO_2^- in reaction 6 was considerably less than G_{OH}, the radiolytic yield of ·OH. An explanation of this behaviour is that reaction 8 occurs.

$$\cdot OH + S_2O_8^{2-} \rightarrow S_2O_8^- + OH^- \tag{8}$$

To test this possibility, the yield of CO_2^- was determined using 200 ns, 6–8 Gy pulses for solutions with [CO] = 5 x 10^{-4} mol dm^{-3}, [N_2O] = 0.012 mol dm^{-3} and [$S_2O_8^{2-}$] in the range 0.1 to 0.05 mol dm^{-3}. A mechanism involving reactions 2, 3, 6 and 8 predicts that $G(CO_2^-)$ should conform to equation 9,

$$\left[G_{OH} + f_{N_2O} G_{e_{aq}^-} \right] \Big/ G(CO_2^-) - 1 = \frac{k_8[S_2O_8^{2-}]}{k_6[CO]} \tag{9}$$

where $f_{N_2O} = k_2[N_2O]/(k_2[N_2O] + k_3[S_2O_8^{2-}])$. Since k_2 and k_3 are known (6), the left hand side of equation 9 was evaluated and plotted against $[S_2O_8^{2-}]$. A linear-least-squares-regression treatment of the data gave $k_8/k_6 = (9.3 \pm 2.5) \times 10^{-3}$ and taking $k_6 = (1.5 \pm 0.3) \times 10^9$ dm^3 mol^{-1} s^{-1} (6) gives $k_8 = (1.4 \pm 0.4) \times 10^7$ dm^3 mol^{-1} s^{-1}. This value is not in agreement with an earlier estimate (13).

On pulse radiolysis of an argon saturated solution containing 0.05 mol dm^{-3} S$_2$O$_8^{2-}$ with 6–8 Gy pulses, no evidence was found for the formation after the pulse of the absorption with $\lambda_{max} = 310$ nm reported by von Sonntag et al. (14), which they ascribed to S$_2$O$_8^{3-}$ formed in reaction 10.

$$SO_4^{-} + S_2O_8^{2-} \rightarrow SO_4^{2-} + S_2O_8^{-} \tag{10}$$

These authors have recently confirmed the absence of this absorption (15).

The importance of the presence of formaldehyde in cloudwater on the rate of SO$_2$ oxidation has been exphasised particularly with regard to the effect of reaction 11 for which $k_{11} \sim 10^9$ dm^3 mol^{-1} s^{-1} (6).

$$\cdot OH + CH_2(OH)_2 \rightarrow H_2O + \cdot CH(OH)_2 \tag{11}$$

However, reaction 12 is of possible major significance as this would serve to inhibit the chain oxidation.

$$SO_4^{-} + CH_2(OH)_2 \rightarrow HSO_4^{-} + \cdot CH(OH)_2 \tag{12}$$

Consequently we considered it timely to investigate the rate of this process.

Argon saturated solutions of formaldehyde containing 0.01 mol dm^{-3} S$_2$O$_8^{2-}$ were irradiated with 0.6μs, 6 Gy pulses in a 25 mm path length cell. The reaction was monitored by following the absorption of SO$_4^{-}$ at 450 nm. For solutions with [HCHO] = 0.03 to 0.2 mol dm^{-3}, SO$_4^{-}$ decayed by first-order kinetics to leave a small long-lived absorption. A plot of k_{obs} versus [HCHO] was linear and yielded $k_{12} = (1.26 \pm 0.05) \times 10^7$ dm^3 mol^{-1} s^{-1} at 20 ± 1°C which is considerably less than k_{11}.

The temperature dependence of k_{12} was studied from 5 to 60°C. An Arrhenius plot of the data gave a good straight line with $E_a = (10.8 \pm 1.3)$ kJ mol^{-1} and $\log_{10}(A/dm^3 mol^{-1} s^{-1}) = 9.0 \pm 0.2$. Thus the relative slowness of reaction 12 seems to be determined mainly by entropic factors.

On the assumption that the long-lived product referred to above is formed in the same yield as SO$_4^{-}$, we estimate it has $\epsilon_{450} \sim 200$ dm^3 mol^{-1} cm^{-1}. Pulse radiolysis of an N$_2$O saturated formaldehyde solution, which is expected to yield the hydrated formyl radical in reaction 11, gave an absorption which declined monotonically from 230 nm towards the red with $\epsilon_{230} = 830$ dm^3 mol^{-1} cm^{-1} and $\epsilon_{450} \sim 1\%$ of this. Thus these data cast doubt on the products of reactions 11 and 12 being the same species. Further study is needed to resolve this question.

SO$_3^{-}$ The radical was generated and its spectrum measured by pulse radiolysis of N$_2$O saturated 2 mmol dm^{-3} solutions of SO$_3^{2-}$ and HSO$_3^{-}$ at pH 9 and 4 respectively, using 100 Gy, 0.6μs pulses. The signals were fully developed at the end of the pulse as expected in view of the known rates of reactions 4 and 5. The spectrum was independent of pH (see fig. 1) but in the SO$_3^{2-}$ solutions it was not possible to measure the spectrum accurately at $\lambda < 250$ nm due to the strong absorption by SO$_3^{2-}$ ion.

The values of ϵ given assume that $G(\cdot OH) = 580$ nmol J^{-1} for the $\cdot$OH scavenging power involved here, i.e. $k[S] = 10^7$ s^{-1} (11). The shape of the spectrum is very similar to that reported previously (2,16), but the molar absorptivity at λ_{max} of 900 ± 15 dm^3 mol^{-1} cm^{-1} determined in this study is considerably less than that (1380 dm^3 mol^{-1} cm^{-1}) obtained by Huie et al. (16).

The decay of SO$_3^{-}$ obeyed second-order kinetics at both pHs. Rate constants determined at 250, 280 and 340 nm were independent of wavelength and pH with $2k_{13} = (7.6 \pm 0.1) \times 10^8$ dm^3 mol^{-1} s^{-1}. If allowance is made for the difference in extinction coefficients obtained in this work and in that by Huie et al. (16), it seems that similar values of $2k/\epsilon$ were found in the two studies.

$$SO_3^{-} + SO_3^{-} \rightarrow S_2O_6^{2-} \tag{13}$$

The rate of reaction 14

$$SO_3^{-} + O_2 \rightarrow SO_5^{-} \tag{14}$$

was measured at pH 4 and 9 by irradiation of solutions containing 2 mmol dm^{-3} SO$_3^{2-}$/HSO$_3^-$ and 1.4 x 10^{-4} to 6.6 x 10^{-4} mol dm^{-3} O$_2$ with 0.2µs, 25 Gy pulses. The reaction was followed at 340 nm where the difference between ϵ for SO$_3^-$ and SO$_5^-$ is a maximum (see below). The decay of absorption obeyed first-order kinetics at both pHs and yielded linear plots of k_{obs} versus [O$_2$] from which we found k_{14} = (2.25 $\pm$ 0.1) x 10^9 dm^3 mol^{-1} s^{-1}, independent of pH. This is somewhat larger than the value obtained previously (16) (k_{14} = 1.1 x 10^9 dm^3 mol^{-1} s^{-1}).

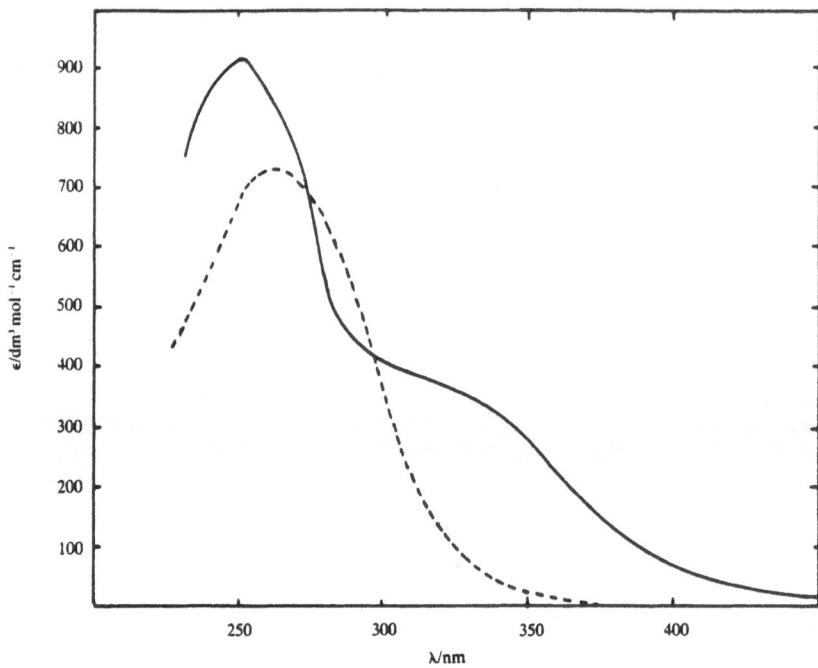

Figure 1.   Spectra of SO$_3^-$(———) and SO$_5^-$(- - - -), see text.

SO$_5^-$. The spectrum of the radical was measured by pulse radiolysis of 2mmol dm^{-3} solutions of SO$_3^{2-}$ and HSO$_3^-$ (pH 9 and 4 respectively) containing 0.67 mmol dm^{-3} O$_2$ using 0.6µs, 100 Gy pulses and observing the signal 9µs after the pulse. At pH 4 no absorption was observed for λ>360nm, but at pH 9 there was a small absorption in this wavelength region with λ_{max} = 450nm (~6% of that at 250nm) which we attribute to SO$_4^-$. It is also assumed that the yield of H· is converted to HO$_2^-$ /O$_2^-$ according to the pH. Therefore, the experimentally observed spectra were corrected for the presence of these radicals to obtain the spectrum of SO$_5^-$ shown in figure 1.

The observation of SO$_4^-$ at pH 9 lends support for reaction 15 being a chain propagation step at this pH.

$$SO_5^- + SO_3^{2-} \rightarrow SO_4^- + SO_4^{2-} \tag{15}$$

The corresponding reaction at pH 4 (reaction 16) seems not to occur, probably because the presence of the proton on the sulphur of the bisulphite ion prevents the O-atom transfer.

$$SO_5^- + HSO_3^- \rightarrow SO_4^- + H^+ + SO_4^{2-} \tag{16}$$

Thus, the likely propagation reaction at pH 4 is reaction 17.

$$SO_5^- + HSO_3^- \rightarrow HSO_5^- + SO_3^- \tag{17}$$

In unbuffered solutions at pH 4, the absorption at 280nm decayed by second-order kinetics leaving a permanent absorption due to SO_2 which is formed as a result of the solution becoming acid. On delivering a second pulse to the sample, the behaviour indicated that most of the O_2 was depleted. In buffered solutions at pH 5, the long-lived absorption did not form and the absorption at 280nm decayed by second-order kinetics. At pH 9 the decays were second-order leaving a small long-lived absorption which was consistent with it being O_2^-. In the oxygenated solutions, the radiation-induced chain reaction is occurring and the pulse radiolysis technique observes the concentration of the chain propagating radicals. Thus, we identify the second-order decay with the removal of SO_5^- in reaction 18.

$$SO_5^- + SO_5^- \leftarrow S_2O_8^{2-} + O_2 \tag{18}$$

At both pH we found $2k_{18} = (1.2 \pm 0.12) \times 10^8 \; dm^3 \; mol^{-1} \; s^{-1}$, but at the lowest $[O_2]$, the observed second-order rate constant was consistently larger. This effect requires further study, but is probably due to the occurrence of reaction 13 following the depletion of O_2 by the radiation-induced chain reaction.

4. ACKNOWLEDGEMENTS

This work is part of the HALIPP subproject of EUROTRAC and was supported by the Commission of the European Communities.

5. REFERENCES

(1) CHAMEIDES, W.L. and DAVIES, D.D. (1982). The free radical chemistry of cloud droplets and its influence on the composition of rain. J. Geophys. Res. 87, 4863-4877.

(2) HAYON, E., TREININ, A. and WILF, J. (1972). Electronic spectra, photochemistry and autoxidation mechanism of the sulfite-bisulfite- pyrosulfite systems. The SO_2^-, SO_3^-, SO_4^- and SO_5^- radicals. J. Am. Chem. Soc. 94, 47-57.

(3) DEISTER, U. and WARNECK, P. (1989). On the photoxidation of SO_3^{2-} in aqueous solution. In press.

(4) McELROY, W.J. (1986). The aqueous oxidation of SO_2 by OH radicals. Atmospheric Environment 20, 323-330.

(5) BUXTON, G.V. (1987). Radiation chemistry of the liquid state: (1) Water and homogeneous aqueous solutions. In Radiation Chemistry, Principles and Applications. Edited by Farhataziz and M.J. Rodgers. VCH Publishers, New York, 321-349.

(6) BUXTON, G.V., GREENSTOCK, C.L., HELMAN, W.P. and ROSS, A.B. (1988). Critical review of rate constants for reactions of hydrated electrons, hydrogen atoms and hydroxyl radicals in aqueous solution. J. Phys. Chem. Ref. Data 17, 513-886.

(7) DAINTON, F.S., ROBINSON, E.A. and SALMON, G.A. (1972). Pulse radiolysis of solutions of stilbene. I. Evidence for triplet and singlet excited state formation. J. Phys. Chem. 76, 3897-3904.

(8) LESIGNE, B., FERRADINI, C. and PUCHEAULT, J. (1973). Pulse radiolysis study of the direct effect on sulfuric acid. J. Phys. Chem., 77, 2156–8.

(9) DOGLIOTTI, L. and HAYON, E. (1967). Flash photolysis of persulfate ions in aqueous solutions. Study of the sulfate and ozonide radical anions. J. Phys. Chem., 71, 2511–2516.

(10) McELROY, W.J. (1988). Kinetic and spectroscopic studies of the SO_4^- and Cl_2^- radicals in aqueous solution. In Mechanisms of Gas Phase and Liquid Phase Chemical Transformations inTropospheric Chemistry. Edited by R. A. Cox, Commission of the European Communities, Brussels, 129–134.

(11) SCHULER, R.H., HARTZELL, A.L. and BEHAR, B. (1981). Track effects in radiation chemistry. Concentration dependence for scavenging of OH by ferrocyanide in N_2O-saturated aqueous solutions. J. Phys. Chem. 85, 192-199.

(12) CHAWLA, O.P. and FESSENDEN, R.W. (1975). Electron spin resonance and pulse radiolysis studies of some reactions of SO_4^-. J. Phys. Chem., 79, 2693–2700.

(13) ROEBKE, W., RENZ, M. and HENGLEIN, A. (1969). Pulsradiolyse der Anionen $S_2O_8^{2-}$ und HSO_5^- in wässriger Lösung. Int. J. Radiat. Phys. Chem. 1, 39-44.

(14) SCHUCHMANN, H.-P., DEEBLE, D.J., OLBRICH, G. and von SONNTAG, C. (1987). The SO_4^- induced chain reaction of 1,3-dimethyluracil with peroxydisulphate. Int. J. Radiat. Biol. 51, 441-453.

(15) von SONNTAG, C. (1989). Private communication.

(16) HUIE, R.E., CLIFTON, C.L. and ALSTEIN, N. (1989). A pulse radiolysis and flash photolysis study of the radicals SO_2^-, SO_3^-, SO_4^- and SO_5^-. Radiat. Phys. Chem., 33, 293-297.

KINETICS OF THE REACTIONS OF THE SO_4^-

RADICAL WITH SO_4^-, $S_2O_8^{2-}$, H_2O and Fe^{2+}

W..J. McElroy and S.J. Waygood
National Power, Technology and Environmental Centre,
Kelvin Avenue, Leatherhead, KT22 7SE, UK.

Summary

The sulphate radical anion, SO_4^-, has been identified as a key intermediate in the aqueous aerobic oxidation of SO_2 initiated by free radicals. The absorption spectrum of SO_4^- has been obtained following photolysis of $K_2S_2O_8$ solutions at 248 nm. Kinetic spectrophotometric techniques have been employed to investigate the decay of SO_4^- in the presence of $S_2O_8^{2-}$ and Fe(II). Several rate constants have been determined at 293 K: $2k(SO_4^- + SO_4^-) = (8.8 \pm 0.3) \times 10^8$ $M^{-1}s^{-1}$ (at infinite dilution), $k(SO_4^- + S_2O_8^{2-}) = (6.1 \pm 0.6) \times 10^5$ $M^{-1}s^{-1}$ (at infinite dilution) and $k(SO_4^- + H_2O) = (500 \pm 60)$ s^{-1}. The reaction of SO_4^- with Fe(II) proceeds via a rapid equilibration step followed by electron transfer, which is rate-determining.

$$SO_4^- + Fe^{2+} \underset{k_{-1}}{\overset{k_1}{\rightleftarrows}} [Fe(II)(SO_4)]^+ \overset{k_2}{\rightarrow} [Fe(III)(SO_4)]^+ \rightleftarrows Fe^{3+} + SO_4^{2-}$$

Values of k_1, k_{-1} and k_2 at 293K are $(3.0 \pm 1.9) \times 10^8$ $M^{-1}s^{-1}$, $(4.8 \pm 3.0) \times 10^4$ s^{-1} and $(6.5 \pm 2.1) \times 10^3$ s^{-1} respectively. The implications of these results for current cloud and precipitation chemistry models are discussed.

1 Introduction

The oxy-sulphur radicals SO_3^-, SO_4^- and SO_5^- have been proposed as species involved in the aerobic oxidation of SO_2 in aqueous solution (1). However a critical assessment of available literature data (2) has shown that many uncertainties remain with regard to the detailed kinetics and mechanism of this reaction. Recent model calculations (3,4) predict that during daytime the majority of soluble iron in polluted boundary layer cloud is present as ferrous iron, Fe(II). It has been proposed (4) that reaction of Fe^{2+} with SO_4^- is the major route by which Fe(II) is re-oxidised to Fe(III) under the conditions simulated. These predictions are based on a single determination of the rate coefficient for this reaction (5). It is important to establish the validity of this conclusion, particularly as Fe(III) is itself a catalyst for the oxidation of dissolved SO_2. In the present study the optical absorption spectrum of SO_4^- has been investigated. Kinetic spectrophotometric techniques have been used to study the decay of the sulphate radical in the presence of $S_2O_8^{2-}$ and Fe(II).

2 Experimental

The sulphate radical was produced by photolysis of aqueous $K_2S_2O_8$ solutions at 248 nm. The apparatus has been described in detail previously (6) and consisted of an excimer laser, thermostatted quartz photolysis cell and xenon arc lamp monitoring light source. For kinetic measurements transient absorptions were monitored using a monochromator, photomultiplier and a 150 MHz digital oscilloscope, with a signal averaging capability, interfaced to an IBM PC-AT microcomputer. Time constants for the detection circuit were variable in the range 0.3 - 3 μs. Absorption spectra were obtained using a polychromator and 512 element diode array with an optimum resolution of 0.488 nm per diode. The gate on the array was synchronised with the laser pulse using a digital delay generator. Gatewidths were variable in the range 180 ns to 6 ms.

$K_2S_2O_8$ (Fluka puriss. p.a., > 99.5%), $FeSO_4$, $HClO_4$ and $NaClO_4$ (BDH Analar) were used without further purification. Argon (Air Products, high purity grade) was passed through a rare gas purifier before use. Solutions were prepared using water with resistivity greater than 18 MΩ cm .

3 Results and Discussion

The optical absorption spectrum of SO_4^- obtained by photolysis of a 4 x 10^{-2} M $K_2S_2O_8$ solution (Figure 1a) shows two overlapping bands, the one of greater intensity having a maximum at 450 nm. This is in agreement with spectra reported previously (7,8,9,10).

It has been suggested in a recent study (11) that the intensity of the band around 350 nm increases with time. This was attributed to the formation of $S_2O_8^-$ [1] and a rate constant of 1.2 x 10^6 M^{-1}s^{-1} was ascribed to the reaction. No evidence of an increase in absorption was found in the present investigation for similar experimental conditions. The most likely explanation of the observations of Schuchmann et al. (11). is the presence of chloride ion as an impurity in their source of $S_2O_8^{2-}$. This would lead to the formation of Cl_2^- (6) which has an absorption maximum at 340 nm. Calculation shows that contamination at a level of 0.5% could explain the results.

$$SO_4^- + S_2O_8^{2-} \rightarrow SO_4^{2-} + S_2O_8^- \qquad [1]$$

The decay of SO_4^- in aqueous solution at 293K following photolysis of 1 x 10^{-3} M $K_2S_2O_8$ was monitored at 450 nm and a typical decay is shown in Figure 1b. Absorption signals were identical in Ar, N_2O and O_2 saturated solutions, confirming the absence of hydrated electrons, and demonstrating that photolysis of $S_2O_8^{2-}$ at 248nm is a clean source of SO_4^-. Analysis of the decays over more than three half-lives showed that these were mixed order with respect to $[SO_4^-]$. The second-order component has previously been ascribed to self-reaction of SO_4^- (1,8,9,10,12,13) while the first-order component has been assigned to pseudo-first -order processes (1,4,10).

$$SO_4^- + SO_4^- \rightarrow S_2O_8^{2-} \qquad [2]$$

$$SO_4^- + X \rightarrow \text{decay by first} - \text{order process(es)} \qquad [3]$$

The influence of ionic strength was investigated by the addition of $NaClO_4$. Decays were analysed using a combined first- and second-order non-linear curve fitting procedure. It is clear (Table I) that $2k_2$ shows a strong positive correlation with I, the ionic strength. For a reaction between two ions in dilute solution, $\log_{10} k = \log_{10} k_0 + 1.02 z_a z_b I^{0.5}$, where z_a and z_b are the ionic charges and k_0 is the rate coefficient at infinite dilution (6). Least squares analysis of of log $2k_2$ versus $I^{0.5}$ gives a value for $2k_2$ at infinite dilution of (8.8 ± 0.3) x 10^8 M^{-1}s^{-1} with a slope of (0.92 ± 0.12) M$^{-0.5}$.

I (/M)	$[SO_4^-]_i$ ($/10^{-6}M^{-1}$)	k_3 ($/10^3s^{-1}$)	$2k_2$ ($/10^9M^{-1}s^{-1}$)
3.0×10^{-3}	5.5 ± 0.1	1.3 ± 0.1	1.0 ± 0.1
1.1×10^{-2}	5.4 ± 0.2	1.4 ± 0.2	1.2 ± 0.1
1.9×10^{-2}	5.7 ± 0.2	1.4 ± 0.3	1.2 ± 0.1
2.7×10^{-2}	5.6 ± 0.2	1.4 ± 0.1	1.3 ± 0.1
3.5×10^{-2}	5.6 ± 0.2	1.5 ± 0.2	1.3 ± 0.1
5.1×10^{-2}	5.5 ± 0.3	1.7 ± 0.3	1.4 ± 0.1
6.7×10^{-2}	5.5 ± 0.1	1.4 ± 0.2	1.5 ± 0.1
8.3×10^{-2}	5.5 ± 0.3	1.7 ± 0.1	1.7 ± 0.1

Table I. Rate constants for the decay of SO_4^- in aqueous solution. (All errors represent $\pm 2\sigma$, $[SO_4^-]_i$ is the initial radical concentration)

Earlier broad-band flash photolysis studies (1,8) were carried out at ionic strengths comparable to the highest used in the the present study and are essentially in agreement with values given in Table I when expressed as $2k_2/\varepsilon$. The higher rate coefficients reported at very low pH (12,13) are most likely attributable to enhanced reactivity due the protonation of SO_4^- to form HSO_4. Tang et al. (10) obtained a value of $2k_2/\varepsilon = (3.9 \pm 0.6) \times 10^5$ $M^{-1}s^{-1}$ at infinite dilution by considering 11 experiments at low ionic strength (I < 0.04 M), although there was considerable scatter in the data. The lowest reported rate coefficient (9) was obtained by photolysis of $S_2O_8^{2-}$ solutions at 193 nm, where the decay at 450 nm exhibited a fast initial first-order component which was ascribed to the formation of an excited state of SO_4^-. This complicated the analysis of the experimental data since only linear least squares fitting procedures were employed.

$[S_2O_8^{2-}]$ (/M)	n	$[SO_4^-]_i$ ($/10^{-6}M$)	k_3 ($/10^2s^{-1}$)
1×10^{-3}	29	4.2- 6.0	13 ± 2
8×10^{-4}	25	3.5-4.0	12 ± 3
6×10^{-4}	25	2.5-2.8	8.9 ± 1.6
5×10^{-4}	7	2.6-2.8	9.0 ± 1.5
4×10^{-4}	22	1.3-1.8	7.2 ± 1.3
2×10^{-4}	12	0.8-1.0	7.8 ± 2.3

Table II. Variation of k_3 with precursor concentration. (n represents the number of experiments)

The first-order component, k_3, shows an apparent dependence on precursor concentration (Table II) although there is also a significant residual first-order component which is most likely attributable to reaction of SO_4^- with water [4]. Negligible variation in k_3 was observed when alternative sources of demineralised water were used.

$$SO_4^- + H_2O \rightarrow OH + HSO_4^- \qquad [4]$$

Analysis of a 0.1 M $K_2S_2O_8$ solution by both ICP M/S and ion chromatography confirmed the absence of impurities. The dependence of k_3 on precursor concentration was therefore due to reaction of SO_4^- with $S_2O_8^{2-}$ [1]. Reaction with an undetected product of $S_2O_8^{2-}$ photolysis can also be excluded as k_3 exhibited no dependence on the laser pulse energy. Analysis of the data, taking into account the likely ionic strength dependence of reaction [1], gives values for k_1 and k_4 of $(6.1 \pm 0.6) \times 10^5$ $M^{-1}s^{-1}$ and (500 ± 60) s^{-1} respectively.

In previous investigations of the observed first-order component the quality of the experimental data was insufficient to allow reliable interpretation of the measurements and the influence of precursor concentration was not investigated. Pennington et al. (4) concluded that k_4 was of the order of 10^3-10^4 s^{-1} while Hayon et al. (1) placed an upper limit on k_3 of 3×10^3 s^{-1}. Clearly, the results quoted here are not inconsistent with these findings. Tang et al. (10) ascribed the first-order component solely to reaction [4] to which a rate coefficient of (360 ± 90) s^{-1} was attributed. An arbitrary correction of (50 ± 50) s^{-1} was made to allow for reactions of SO_4^- other than [4].

The decay of SO_4^- in the presence of Fe^{2+} was monitored at 450 nm and a typical trace is shown in Figure 1c. The reaction conditions ($[S_2O_8^{2-}] = 1 \times 10^{-3}$ M and $[Fe^{2+}] = 7.5 \times 10^{-5}$ M) were chosen to preclude interference from reaction [5] for which a rate constant of 27 $M^{-1}s^{-1}$ has been reported (15). Reagent solutions were mixed immediately before entry into the photolysis cell such that less than 5% depletion of Fe^{2+} should have occured before the laser pulse.

The data suggest that SO_4^- reacts with Fe^{2+} via a rapid equilibration step [6] to form a complex which subsequently decays to yield products [7]. Such a reaction scheme was adopted for analysis of the experimental data using the FACSIMILE (AERE, Harwell) curve fitting program. Reactions [2] and [3] were also included for completeness.

$$Fe^{2+} + S_2O_8^{2-} \rightarrow Fe^{3+} + SO_4^- + SO_4^{2-} \qquad [5]$$

$$SO_4^- + Fe^{2+} \rightleftarrows X \rightarrow Products \qquad [6,7]$$

From a set of twelve experiments the following average rate constants were obtained; $k_6 = (3.0 \pm 1.9) \times 10^8$ $M^{-1}s^{-1}$, $\&kM6. = (4.8 \pm 3.0) \times 10^4$ s^{-1} and $k_7 = (6.5 \pm 2.1) \times 10^3$ s^{-1}. Figure 1c shows a fitted decay curve superimposed on typical experimental data. $\&K6.$ has the value $(6.4 \pm 2.3) \times 10^3$ M^{-1} and equilibrium is established in approximately 50 μs. The rate constants also gave a good fit to the experimental data obtained for initial SO_4^- concentrations in the range 3-7 μM. In the only previous investigation (15) it was concluded that the reaction was a simple second-order electron transfer process with a rate constant of 9.9×10^8 $M^{-1}s^{-1}$. This value is higher than that for k_6 reported here as no allowance was made for the contribution of reaction [2].

Several mechanisms can be postulated for the reaction of Fe^{2+} with SO_4^- based on the above reaction scheme. Comparison of the present results with those for the reaction of OH with Fe^{2+} (16,17) enables conclusions to be drawn concerning the most likely mechanism. Stuglik et al. (16) proposed that oxidation of Fe^{2+} by OH radicals proceeds via hydrogen atom abstraction followed by inner-sphere charge transfer. However, in the case of SO_4^-, the rate constant for the initial abstraction reaction would have to be greater than the diffusion controlled limit to account for the observed decay. Jayson et al. (17) concluded that the reaction of Fe^{2+} with OH, proceeds via simple electron transfer in which the first step is the formation of a loosely-bound precursor complex. Since removal of OH was faster than the known rate of solvent exchange within the inner coordination sphere of Fe^{2+} (18), this suggested that the subsequent step was outer-sphere electron transfer to form a successor complex which dissociates to give the reaction products. The reaction of the SO_4^- radical with Fe^{2+} is much slower than the analogous reaction of OH (17) and no clear distinction on the basis of experimental data alone can be made between an outer- or inner-sphere electron transfer process. However theoretical calculations based on consideration of the electrostatic interactions (19) between Fe^{2+} and SO_4^- suggest that the derived value of &K6. is a factor of approximately 1000 larger than would be predicted for an outer-sphere process. It is, therefore, likely that the reaction proceeds by inner-sphere electron transfer, which is slower than ligand exchange.

While the role of iron and manganese as catalysts in the aerobic oxidation of dissolved SO_2 has been appreciated for many years, the possible impact of transition metal ions on the concentration of oxidising species such as OH, &HO2. and H_2O_2 in atmospheric droplets has only recently received consideration. Several modelling studies have included a detailed assessment of the impact of transition metal chemistry on the composition of passive and precipitating cloud systems (3,4). Examination of the reaction schemes shows that, without exception, all transition metal-free radical reactions have been treated as simple second-order processes. It is clear from the present study that the rate-determining step is not necessarily outer- or inner-sphere complex formation, as has been assumed in these models. This has also been demonstrated in an experimental study of the interactions of Mn(II) with the &HO2. and &O2M. radicals (20). Kinetic and mechanistic data are not firmly established for many of the transition metal-free radical reactions incorporated in current cloud chemistry models. It is apparent that further detailed studies of these reactions is required if the predictions of such models are to have any validity.

4 Acknowledgements

This work was carried out at the Technology and Environmental Centre as part of CEC contract number EN4V/0080-C(AM) and is published by permission of National Power.

5 References

(1) Hayon, E., Treinen, A. and Wilf, J., 1972, J.A.C.S., 94 , 47
(2) McElroy, W. J., 1986, Atmos. Environment, 20 , 323
(3) Graedel, T.E., Mandlich, M.L. and Weschler, C.J., 1986, J.G.R., 91 , 5205
(4) Jacob, D. ,Gottlieb, E.W. and Prather, M.J., 1989, personal communication
(5) Heckel, E., Henglein, A. and Beck, G., 1966, Ber. Bun. phys. Chem., 70 , 149
(6) McElroy, W.J., 1989, J.Phys. Chem., in press
(7) Hayon, E. and McGarvey, J.J., 1967, J. Phys. Chem., 71 , 1472
(8) Dogliotti, L. and Hayon, E., 1967, J. Phys. Chem., 71 , 2511
(9) Huie, R.E., Clifton, C.L. and Altstein, N., 1989, Radiat. Phys. Chem., 33 , 361
(10) Tang, Y., Thorn, R.P., Mauldin, R.L. and Wine, P.H., 1988, J. Photochem. Photobiol. A., 44 , 243
(11) Schuchmann, H., Deeble, D.J., Olbrich, G. and von Sonntag, C., 1987, Int. J. Radiat. Biol, 51 , 441
(12) Dogliotti, L. and Hayon, E., 1967, J. Phys. Chem., 71 , 3802
(13) Lesigne, B., Ferradini, C. and Pucheault, J., 1973, J. Phys. Chem., 77 , 2156
(14) Pennington, D.E. and Haim, A., 1968, J.A.C.S., 90 , 3700
(15) Woods, R., Kolthoff, I.M. and Meehan, E.J., 1963, J.A.C.S., 85 , 2385
(16) Stuglik, Z. and Zagorski, Z.P., 1981, Radiat. Phys. Chem., 17 , 229
(17) Jayson, G.G., Parsons, B.J. and Swallow, A.J., 1972, J.C.S. Faraday I, 68 , 2053
(18) Swift, T.J. and Connick, R.E., 1962, J. Chem. Phys., 37 , 307
(19) Sutin, N., 1983, in 'Progress in Inorganic Chemistry', ed. Lippard, S. J., Wiley, New York, 30 , 441
(20) Cabelli, D.E. and Bielski, H.J., 1984, J. Phys. Chem., 88 , 6291

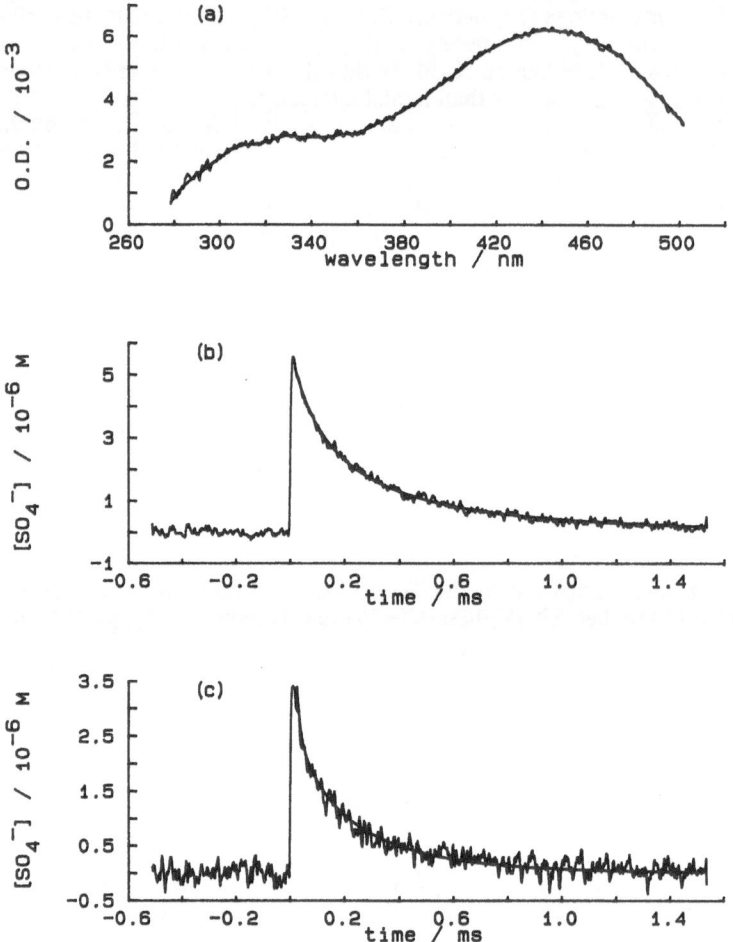

Figure 1. (a) Absorption spectrum of the sulphate radical anion, SO_4^-. (b) Decay of the sulphate radical at 450 nm. The line indicates the fit to the experimental data predicted by the rate constants $2k_2$ and k_3. (c) Decay of the sulphate radical at 450 nm in the presence of Fe^{2+}. The line represents the fit to the experimental data predicted by the rate constants k_6, k_{-6} and k_7.

MECHANISM OF THE HOMOGENEOUS LIQUID OXIDATION

OF SULPHUR(IV) TO SULPHUR(VI) BY HYDROGEN PEROXIDE

Janine LAGRANGE and Philippe LAGRANGE

Ecole Européenne des Hautes Etudes des Industries Chimiques
de Strasbourg, URA 405 au CNRS, 1 rue Blaise Pascal,
F-67000 STRASBOURG

Summary

This work is a contribution to elucidate the mechanism of the S(IV) oxidation by hydrogen peroxide in atmospheric water droplet. This laboratory study has been done by stopped-flow spectrophotometry. In order to determine the factors which should give interfering reactions, the kinetic study was performed, at first, in sodium perchlorate medium. This medium does not give interference, but stabilizes only the activity factors of the reacting species.
Between pH 3 and 7, the reaction rate appears proportional to the H^+, HSO_3^- and H_2O_2 concentrations. The influence of the buffers used was studied.
After this fundamental study, the roles of anions as SO_4^{2-} and Cl^- and of cations as Cu^{2+}, Mn^{2+}, Cr^{3+} and Fe^{3+} were determined in order to extrapolate the results to cloud or rain conditions.

1. INTRODUCTION

The oxidation of SO_2 in atmospheric water (rain, cloud, fog...) is principally due to the action of hydrogen peroxide, between pH 3 and 6. Many kinetic studies of the overall reaction :

$$HSO_3^- + H_2O_2 \longrightarrow H_3O^+ + SO_4^{2-} \qquad [1]$$

have been done, between pH 3 and 7, in a number of laboratories (1-7). On the basis of these studies, a significant portion of the sulphate and of the strong acidity found in atmospheric precipitation derives from this reaction [1]. These studies show consistent results concerning the observed rate law. But, the influence of different species (organic of inorganic trace components present in real atmospheric water) and also, the medium allowing the laboratory study (concentrated supporting electrolyte to keep constant the ionic strength and buffering system to control pH), is not systematically analysed. Some kinetic effects of these species have been observed. A new possible approach would be to examine, at first, the reaction [1] in a low complexing medium and to study the potential interferences by systematic additions of different species. We choose to study the kinetics of reaction [1] in $NaClO_4$ at constant ionic strength $I = 1$ M. This supporting electrolyte is known to give very low interactions with ions and molecules. Kinetic measurements were carried out by stopped flow spectrophotometry.

2. EXPERIMENTAL SECTION

2.1. Reagents

All chemical were analytical reagent grade. Sodium perchlorate, sodium sulphate, sodium chloride, sodium acetate, sodium monohydrogenophosphate, perchloric acid were obtained from Merck. Sodium sulphite, hydrogen peroxide solution, sodium hydroxide, hydrochloric acid and sulphuric acid were purchased from Prolabo.

The test solutions were prepared at constant ionic strength in a given supporting electrolyte : $NaClO_4$ or $NaCl$ or Na_2SO_4.

The analytical concentration of sulphite was determined by iodometry (8) : for each solution used in kinetic measurement, a given sample was simultaneously collected and added to a given solution of iodine in excess. Titrations by standardized thiosulphate solution allow to calculate the sulphite concentration of a given kinetic experiment. The concentration of hydrogen peroxide was determined by potentiometry using a standardized solution of ferrous ion in sulphuric medium (8).

The solutions have been freed from oxygen by bubbling argon before analysis and kinetic measurements.

2.2 pH measurements

pH measurements were carried out with a TB/HA glass electrode and an $Ag/AgCl/0.1$ M Cl^- reference electrode. The pH-meter was a Tacussel ISIS 20000 . Our standard solutions, in $NaClO_4$ or in $NaCl$, contained 10^{-2} M of a strong acid ($HClO_4$ in $NaClO_4$, HCl in $NaCl$), at a given constant ionic strength with a given supporting electrolyte. For these standard solutions, we wrote $-\log[H^+]$ = pH = 2.000. In Na_2SO_4 (I = 0.1 M), we used a solution 0.01 M $NaOH$ + 0.03 M Na_2SO_4 as pH standard for which we wrote pH = 11.60. Measurements were carried out at 25 ± 0.1 °C. The test solutions were generally buffered by addition of sodium acetate (0.04 M) or sodium phosphate (0.04 M) and the pH have been adjusted before kinetic measurements by addition of sodium hydroxide or of perchloric (or hydrochloric, or sulphuric) acid. The pH of the mixtures were checked after the kinetic measurements.

2.3. Kinetic measurements

The stopped-flow spectrophotometer, type Durrum Gibson equipped with a transcient recorder Datalab DL 905 was interfaced to an Apple II microcomputer. The on-line data acquisition and treatment system have been previously described (9). In all kinetic studies, the pH was kept constant by addition of a buffer and the sulphite ions were in sufficient excess to ensure that the observed step of the sulphite oxidation was always pseudo first order with respect to hydrogen peroxide. The measurements were carried out at 25 ± 0.1 °C and at 245 nm. At this wavelength, the absorbance of H_2O_2 is predominant. Each value of the pseudo first order rate constant (k_{obs}) was the average of at least three determinations.

3. RESULTS AND DISCUSSION

3.1. Measurements in 1 M $NaClO_4$

In a presence of an excess of sulphite ions and at constant pH, the following rate law has been checked :

$$-\frac{d[H_2O_2]}{dt} = k_{obs}\,[H_2O_2] = k\,[H_2O_2]\,[H^+]\,[\,HSO_3{}^-]$$ [2]

The experimental study has been done for different concentrations of H^+ and of monohydrogenosulphite ion, in presence of an acetate or a phosphate buffer (analytical concentration = 0.04 M). The results give the same rate law in the acetate and in phosphate buffer in the experimental conditions given in table I, but the rate constants are significantly different. One measurement has been done without buffer and the result agrees well with those caculated in acetate buffer. The k values are given in table I.

Buffer	pH range studied	$[HSO_3^-]$ range studied mol l^{-1}	k l^2 mol^{-2} s^{-1}
none *	3	0.005	$(2.65\pm0.10)\ 10^7$
0.04 M CH$_3$COOH/ CH$_3$COO$^-$	3 - 6.5	0.004 - 0.012	$(2.80\pm0.10)\ 10^7$
0.04 M H$_2$PO$_4^-$/ HPO$_4^{2-}$	3 - 7.1	0.004 - 0.012	$(7.10\pm0.30)\ 10^7$

Table I.- *Rate constant of the sulphite oxidation by hydrogen peroxide in NaClO$_4$ (I = 1 M) at 25 °C. (* measurement without buffer, and determination of k by the measurement of the initial rate). The given concentration of the buffers is the sum of the concentrations of the two forms in equilibrium.*

3.2. Measurements in NaCl

Two experimental series have been done. In the first one, the ionic strength is maintained constant (1 M) by addition of sodium perchlorate. In the second one, no addition of NaClO$_4$ has been done to the NaCl supporting electrolyte : the ionic strength is variable. The test solutions have been buffered by addition of 0.04 M acetic acid/acetate, because this buffer has a small effect on the value of k. Tables II and III give the results.

In our experimental conditions, the presence of "concentrated" Cl$^-$ ion seems increase the rate constant by a factor of 10.

$[Cl^-]$ mol l^{-1}	Ionic strength mol l^{-1}	acetic acid/ acetate buffer mol l^{-1}	k l^2 mol^{-2} s^{-1}
0	1	0.04	$(2.80\pm0.10)\ 10^7$
0.1	1	0.04	$(1.50\pm0.25)\ 10^8$
0.2	1	0.04	$(2.10\pm0.20)\ 10^8$
0.4	1	0.04	$(2.05\pm0.35)\ 10^8$
0.6	1	0.04	$(2.05\pm0.45)\ 10^8$

Table II.- *Rate constant of the sulphite oxidation by hydrogen peroxide in NaCl + NaClO$_4$ (I = 1 M) at 25 °C. The given concentration of the buffers is the sum of the concentrations of the two forms in equilibrium.*

[Cl⁻] mol l⁻¹	Ionic strength mol l⁻¹	acetic acid/ acetate buffer mol l⁻¹	k l^2 mol⁻² s⁻¹
0.1	0.1	0.04	$(1.00\pm0.25)\ 10^8$
0.2	0.2	0.04	$(1.10\pm0.25)\ 10^8$
0.4	0.4	0.04	$(1.05\pm0.25)\ 10^8$
0.6	0.6	0.04	$(1.20\pm0.35)\ 10^8$

Table III.- *Rate constant of the sulphite oxidation by hydrogen peroxide in NaCl at 25 °C. The given concentration of the buffers is the sum of the concentrations of the two forms in equilibrium.*

3.3. Measurements in Na₂SO₄

In this medium, the experimental measurements must be done taking into account the HSO_4^- species of the supporting electrolyte for pH less than 3. For this reason our kinetic measurements have been done between pH 4 and 7. The rate law given in [2] has been checked in presence of acetate and phosphate buffers. Table IV gives the kinetic results.

Buffer	pH range studied	[HSO₃⁻] range studied mol l⁻¹	k l^2 mol⁻² s⁻¹
0.04 M CH_3COOH/ CH_3COO^-	4 - 6.5	0.004 - 0.010	$(7.20\pm0.20)\ 10^7$
0.04 M $H_2PO_4^-$/ HPO_4^{2-}	5 - 7.0	0.004 - 0.010	$(2.80\pm0.25)\ 10^8$

Table IV.- *Rate constant of the sulphite oxidation by hydrogen peroxide in Na₂SO₄ (I = 0.1 M) at 25 °C. The given concentration of the buffers is the sum of the concentrations of the two forms in equilibrium.*

The difference between the rate constants determined in the acetate buffer and in the phosphate buffer seems of the same order of magnitude in NaClO₄ (I = 1 M) and in Na₂SO₄ (I = 0.1 M).

3.4. Influence of metallic traces
The data have been determined in NaClO₄ (I = 1 M) and in NaCl (I = 0.1 M). The rate constants corresponding to equation [2] are given in tables V and VI.
The metallic cations have a small effect on the kinetics studied here, in NaClO₄ the rate constant increases by a factor less than 3 and no effect was observed in NaCl 0.1 M.

3.5. Discussion
All the obtained data fit with the experimental law [2] and the influence of the additionnal species (buffer : HA/A⁻, Cl⁻, metal traces) is included in the calculated third order rate constant k. The observed effects may be summarized, as follows :
        - Results are practilly identical without buffer or with acetate

Metallic trace	Concentration of the metal mol l^{-1}	acetic acid/ acetate buffer mol l^{-1}	k l^2 mol^{-2} s^{-1}
none	-	0.04	$(2.80\pm0.10)\ 10^7$
Fe(III)	10^{-4}	0.04	$(7.10\pm0.25)\ 10^7$
Fe(III)	10^{-5}	0.04	$(7.40\pm0.25)\ 10^7$
Mn(II)	10^{-4}	0.04	$(6.05\pm0.25)\ 10^7$
Mn(II)	10^{-5}	0.04	$(6.45\pm0.25)\ 10^7$
Cu(II)	10^{-4}	0.04	$(6.90\pm0.25)\ 10^7$
Cr(III)	10^{-4}	0.04	$(6.45\pm0.25)\ 10^7$

Table V.- *Influence of metallic traces on the rate constant of the sulphite oxidation by hydrogen peroxide in $NaClO_4$ (I= 1 M), at 25 °C.*

Metallic trace	Concentration of the metal mol l^{-1}	acetic acid/ acetate buffer mol l^{-1}	k l^2 mol^{-2} s^{-1}
none	-	0.04	$(1.00\pm0.25)\ 10^8$
Fe(III)	10^{-5}	0.04	$(1.20\pm0.25)\ 10^8$
Mn(II)	10^{-5}	0.04	$(1.05\pm0.25)\ 10^8$
Cu(II)	10^{-5}	0.04	$(1.15\pm0.15)\ 10^8$
Cr(III)	10^{-5}	0.04	$(1.10\pm0.15)\ 10^8$

Table VI.- *Influence of metallic traces on the rate constant of the sulphite oxidation by hydrogen peroxide in NaCl (I= 0.1 M), at 25 °C.*

buffer. But, when the phosphate buffer replaces the acetate buffer (at the same concentration) in the same medium ($NaClO_4$ or Na_2SO_4), the rate constant k is approximately three times higher. k remains constant, for a given buffer, when the ratio HA/A^- changes with pH.

- The substitution of ClO_4^- by Cl^-, at constant ionic strength, increases the rate constant k (ten times higher).

- The values of k do not seem affected by metal traces in NaCl, but in $NaClO_4$, Cu(II), Mn(II), Cr(III) and Fe(III) (10^{-4} - 10^{-5} M) lead to an increament of k. In NaCl, the metal ions are probably complexed by Cl^-.

The kinetic results and the rate law agree with a part of the mechanism proposed in the literature : a reaction between H_2O_2 and HSO_3^- gives a peroxomonosulphite intermediate (fast equilibrium) ; an H^+ assisted rearrangement of this intermediate is the rate determining step. But, our results do not agree with a generalized hydrolysis of this intermediate, as it is often proposed. Presently, an other fast preequilibrium between H_2O_2 and the interfering species must be taken into account before the formation of the peroxosulphite.

REFERENCES

(1) HALPERIN J., TAUBE H., J. Amer. Chem. Soc. (1952) 74, 380.
(2) MADER P.M., J. Amer. Chem. Soc. (1958) 80, 2634.
(3) HOFFMANN M.R., EDWARDS J.O., J. Phys. Chem. (1975) 20, 2096.
(4) PENKETT S.A., JONES B.M.R., BRICE K.A., EGGLETON A.E.J., Atmos. Environ. (1979) 13, 123.
(5) MARTIN L.R., DAMSCHEN D.E., Atmos. Environ. (1981), 15, 1615.

(6) LEE Y.N., SHEN J., KLOTZ P.J., SCHWARTZ S.E., NEWMAN L., J. Geophys. Res. (1986) 91, 13.264.

(7) LIND J.A., LAZRUS A.L., KOK G.L., J. Geophys. Res. (1987) 92, 4171.

(8) CHARLOT G., (1974) Chimie Analytique Quantitative, Masson, Paris.

(9) LAGRANGE J., LAGRANGE P., J. Chim. Phys. (1984) 81, 425.

OH QUANTUM YIELDS FOR THE PHOTODECOMPOSITION OF FE(III) HYDROXO COMPLEXES
IN AQUEOUS SOLUTION AND THE REACTION OF OH WITH HYDROXYMETHANESULFONATE

H.J. Benkelberg, U. Deister and P. Warneck
Max-Planck-Institut für Chemie, Mainz, Federal Republic of Germany

Summary

Aqueous solutions of iron(III) perchlorate at pH=3 were irradiated
with light in the 300-400 nm wavelength region, and the quantum yields
for OH radicals resulting from the photodecomposition of $Fe(H_2O)_5OH^{2+}$
and $Fe(H_2O)_4(OH)^+_2$ complexes were determined as a function of
wavelength from the amounts of acetone produced when 2-propanol was
added to the solutions as a radical scavenger. The OH quantum yield
found was 0.182±0.041 at 300 nm, and it declined with increasing
wavelength toward 1/3 of that value at 360 nm. With these data we
estimate a photodissociation coefficient of $j=10^{-3}s^{-1}$ for summer
conditions at 45° N latitude. The reaction of OH with
hydroxymethanesulfonate in aqueous solution was studied at pH 4-6 by
competition kinetics. The source of OH was photodecomposition of the
NO^-_3 ion at wavelengths above 300 nm. Benzene was used as the competitor reagent. A comparison of the yields for the products sulfate
and phenol showed the reaction to be a chain process with an apparent
chain length of 5.8. It is suggested that sulfate radicals act as
chain carriers.

1. INTRODUCTION

Following the initial study of Chameides and Davis (1), several investigators have sought to delineate the chemistry occurring in the aqueous
phase of clouds and fogs by means of computer models. Pandis and Seinfeld
(2) have recently summarized this work in conjunction with a sensitivity
analysis, and McElroy (3) has further reviewed reactions that are specific
to the problem of sulfur(IV) oxidation. In these models, reactions of OH
radicals are an essential element. The major source of OH was found to be
the dissociation of the HO_2 radical, $HO_2 = H^+ + O^-_2$, followed by the reac-
tions $O^-_2 + O_3 \longrightarrow O^-_3 + O_2$ and $H^+ + O^-_3 \longrightarrow OH + O_2$. The HO_2 radical, in
turn, enters cloud drops by dissolution from the gas phase, where it is
produced in the sun-lit atmosphere by hydrocarbon oxidation reactions. The
formation of HO_2 in the aqueous phase itself is a minor source. The primary
production of OH in aqueous solution by the photodissociation of either
H_2O_2 or the NO^-_3 ion is fairly inefficient at solar wavelengths available
in the troposphere. However, Weschler et al. (4) have pointed out that the
photodissociation of complexes of the iron(III) ion with OH^- may be a
viable source of OH, if the associated quantum yields were favorable. Very
little information in this regard exists from previous studies. In particular, the wavelength dependence of the quantum yield is highly uncertain. We have therefore embarked on a reinvestigation of the photochemistry
of Fe(III) complexes in aqueous solution and report our first results here.

We have also studied the reaction of OH with hydroxymethanesulfonate
(HMS) by competition kinetics. HMS is formed by the interaction of formaldehyde with sulfite in aqueous solution, and it is believed to be an

important S(IV) species in fogs and clouds (5).

2. EXPERIMENTAL

The photolysis set-up used for studying iron(III) solutions combined a 150 W xenon arc lamp with an f/2 grating monochromator, whose slits were adjusted so that a spectral resolution of 4 nm at half-width was obtained. The light emerging from the exit slit was refocussed to pass in axial direction through a 2 cm diameter quartz cuvette of 1 cm depth, and then further onto a calibrated thermopile. The photon flux calculated from the thermopile output signal was about 1×10^{15} photons/s. Solutions contained in the cell were mostly irradiated for 10 min, although both shorter and longer irradiation times were sometimes used.

Milli Q, "organic free" water was used to prepare the solutions. Since the Fe^{3+} ion is prone to form complexes with many anions except with perchlorate, $Fe(ClO_4)_3$ was used to prepare iron(III) solutions, and the pH was likewise adjusted with perchloride acid. To keep such solutions stable, the pH values were confined to the region of pH<4. The iron(III) concentration was $2 \times 10^{-4}M$ or less in order to establish optically this conditions and to avoid inner filter effects. The pH dependent ion equilibria that one must consider under such conditions are

$$Fe^{3+} + OH^- = FeOH^{2+} \qquad\qquad \log_{10} K_1 = 11.8$$

$$Fe^{3+} + 2\ OH^- = Fe(OH)_2^+ \qquad\qquad \log_{10} K_2 = 22.3$$

$$2\ Fe^{3+} + 2\ OH^- = Fe_2(OH)_2^{4+} \qquad\qquad \log_{10} K_{22} = 25.1$$

In this representation, water molecules in the hydration sphere are not shown. Values for the equilibrium constants were taken from Smith and Martell (6) (for zero ionic strength).

With these data one calculates for pH=3 and total iron(III)=2×10^4M a composition of 10% Fe^{3+}, 68% $FeOH^{2+}$, 21% $Fe^+(OH)_2$, and 0.6% $Fe_2(OH)_2^{4+}$. The dimer is clearly negligible, and the dominant ion is $FeOH^{2+}$. There is some doubt about the validity of K_2, which might be smaller than shown, and this would further reduce the contribution of $Fe(OH)^+_2$. Since the absorption coefficients of the main species $FeOH^{2+}$ and $Fe(OH)^+_2$ are similar in magnitude, it will be clear that $FeOH^{2+}$ is the major photoactive ion under these conditions. To these solutions 2-propanol was added as a radical scavenger. Its reaction with OH leads to the formation of acetone with at least 87% yield. Acetone was determined by reaction with 2,4 dinitrophenylhydrazine to form the corresponding hydrazone, which was then isolated and assayed by high pressure liquid chromatography. Initially we attempted to analyze for acetone directly by the use of gas chromatography. Although good separation from 2-propanol and adequate sensitivity was achieved, the method failed because 2-propanol reacted with perchlorate in the heated inlet port of the gas chromatograph. This produced additional acetone and gave rise to large errors.

Quantum yields were calculated from the total product yield and the radiation dose delivered into the cell

$$I_o(1-10^{-\varepsilon cd})t/VN_A \rightleftharpoons I_o\ \varepsilon\ (\ln 10)cdt/VN_A$$

where the right hand side gives the approximation for optically thin conditions. I_o is the photon flux and ε the decadic absorption coefficient (both depend on wavelength); c is the total Fe(III) concentration, d and V, respectively, are length and volume of the cuvette, t is the photolysis time; N_A = 6.02 x 10^{23}molecules/mol.

The reaction of OH with HMS was studied in competition with that of OH with benzene, which leads mainly to the formation of phenol. OH radicals were produced by photodecomposition of NO^-_3 ions at wavelengths ≥ 300 nm. For this purpose the 150 W Xenon lamp was combined with a 295 nm cut-off filter (7). The solution contained constant concentrations of 10 mM NO^-_3 and 1 mM of benzene, whereas the concentration of HMS was varied between 2 and 100mM. The pH was initially between 5.6 and 5.8, but fell during the experiments to values of 4.1-4.6. The photolysis time was 2 h. The yields of phenol and sulfate were measured by methods of high pressure liquid chromatography.

3. <u>RESULTS AND DISCUSSION</u> (a) **Photolysis of Fe(III) complexes**

The absorption coefficients of Fe(III) solutions at pH=1 and 3 are shown in Figure 1a as a function of wavelength. The strong increase of ε at short wavelengths is due to Fe^{3+}. Its contribution to the total absorption at > 280 nM is negligible, however. The weak maximum at = 295 nM is mainly associated with $FeOH^{2+}$. A small contribution of $Fe(OH)^+_2$ must be tolerated.

The yield of acetone observed with 300 mm radiation rose with increasing concentration of 2-propanol from 0.07 at 50 μM to about 0.17 at 1 mM of 2-propanol and then attained a plateau value in the concentration range 1-10 mM. This type of saturation curve indicates that 2-propanol competes with another scavenger for OH. The competitor cannot be Fe^{2+} because on one hand not enough Fe^{2+} is produced and, on the other hand, the rate coefficient for OH + Fe^{2+} --> Fe^{3+} + OH^- is only about one tenth of that for the reaction with 2-propanol. Instead, the data suggest the presence of an impurity in the water used, presumably one or more organic compounds. In this case, the consumption rate of OH by the competitor substances should be constant. Steady state kinetics then leads to a reciprocal acetone quantum yield, which should rise lineally with the reciprocal concentration of 2-propanol

$$1/\phi_{acetone} = (1/\phi_o) (1 + k_y[Y]/k_p [\text{2-Propanol}])$$

where [Y] is the concentration of the impurity. Figure 2 shows that such a relation is indeed observed. From the intercept with the ordinate we calculate the limiting acetone quantum yield for large concentrations of 2-propanol as 0.17±0.09 at 300 nm wavelength.

In addition to acetone, the reaction of OH with 2-propanol produces some formaldehyde and acetaldehyde, which result from the attack of OH on the methyl groups. Measurements with OH generated by photodecomposition of H_2O_2 or NO^-_3 ion at 254 nm wavelength indicate relative product yields of 90% acetone and 5% each of formaldehyde and acetaldehyde. Since the last two compounds presumably are formed simultaneously, the yield of acetone would be 0.9/0.95 = 0.95%. From pulse radiolysis studies, Asmus et al. (8) have found relative yields for H atom abstraction by OH from 2-propanol of 85.5% at the α-position and 1.2% at the OH group. Since both pathways produce acetone, the yield of acetone should be at least 87% . At present it is not clear whether H abstraction at the ß-position would also produce some acetone, and this uncertainty must be tolerated. Accordingly, we use a yield of 91±4% as a correction factor to derive OH quantum yields from the measured acetone quantum yields. Figure 1b presents OH quantum yields as a function of wavelength in the 300 - 400 nm spectral region. The quantum yield is seen to decrease with increasing wavelength from a value of 0.18±0.01 at 300 nm toward about 1/3 that value at 360 nm. The effect is in

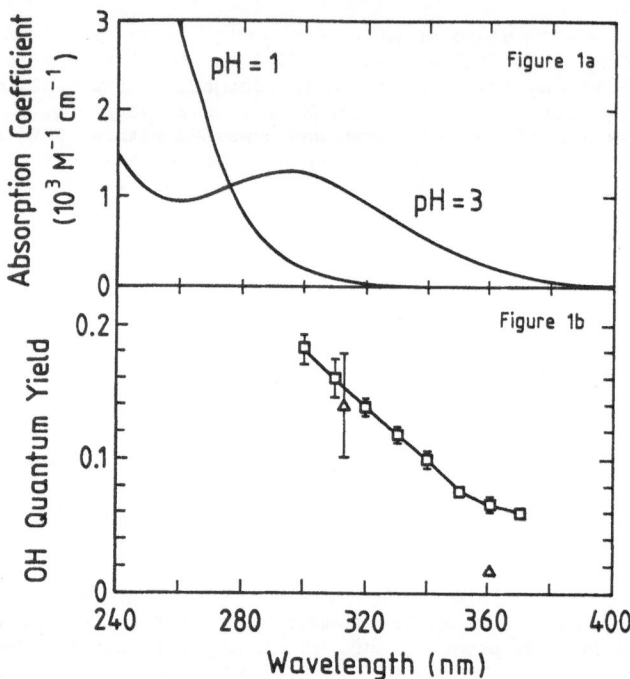

Figure 1: (a)   Absorption spectra   and (b) OH quantum yields at pH = 3
for Fe(III) solutions as a function of wavelength. Recent data by Faust and
Hoigné (9) are shwon by the triangles.

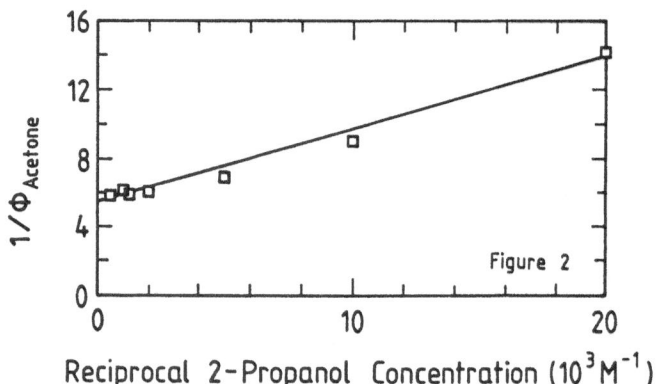

Figure 2: Relation between acetone quantum yields and scavenger concentration.

accord with the notion that the ejection of OH from the solvent cage into solution requires kinetic energy, which must be provided by excitation of the ion in excess of the dissociation energy. Since the release of OH into solution is in competition with in-cavity recombination of OH with Fe^{2+}, an increase of the excitation energy also increases the probability for the escape of OH from the solvent cage.

Figure 1a includes at wavelengths of 313 and 360 nm OH quantum yields recently reported by Faust and Hoigné (9). These data were obtained from the conversion of Fe(III) to Fe(II) in the presence of tert-butanol as scavenger. Although in their case the solutions were adjusted to pH = 4, the data are in good agreement with the present measurements.

With the help of our quantum yield data (extrapolated out to 400 nm) we have calculated the photodissociation coefficient of $FeOH^{2+}$ for (clear sky) summer conditions at 45 degree northern latitude. Ground level solar fluxes were calculated by the two stream procedure of Brühl and Crutzen (10). Absorption coefficients were taken either from our own data for pH = 3 or from Faust and Hoigné (9) at pH = 4. Their values are by a factor of almost two higher. The photodissociation coefficients obtained are

$$ j = \int q(\lambda)\ \sigma(\lambda)\ I(\lambda)\ d\lambda = 0.67 \times 10^{-3} \text{ and } 1.09 \times 10^{-3} s^{-1}. $$

The second value agrees well with results of similar calculations by Faust and Hoigné (9), whereas their in-sunlight measurement gave a lower value more similar to that calculated here for pH = 3. In order to assess the significance of these results for the aqueous chemistry of clouds, one must compare the rate of OH formation from the photodecomposition of Fe(III) complexes with that for the reaction $O^-_2 + O_3 \longrightarrow O^-_3 + O_2$. For the present purpose, we make use of recent model calculations by Lilieveld and Crutzen (11), who found that at pH = 5 the rate of OH production is $3 \times 10^{-9} Ms^{-1}$. This rate would decrease upon lowering the pH because the equilibrium $O^-_2 + H^+ = HO_2$ then shifts toward the right. Chameides and Davis (1) had previously estimated OH production rates in the range $10^{-12} - 10^{-9}$ M s^{-1}. The concentration of iron in fog, cloud and rain waters varies considerably (4), but values of about 2 μM seem typical for continental background clouds. If 50% of this Fe(III) is made available in ionic form, the OH production rate will be 1×10^{-9} M s^{-1}. Both rates are comparable in magnitude, and this demonstrates the importance of the process, which we studied, for the chemistry of clouds.

(b) Reaction of OH with hydroxymethanesulfonate

When HMS was added to aqueous solutions containing NO^-_3 and benzene and the solutions were irradiated at $\lambda \geq 300$ nm, the yield of phenol decreased only slightly from that observed in the absence of HMS, and then only in the presence of high HMS concentrations. On the basis of the competition kinetics involved we expect the relation

$$ \frac{[phenol]_o}{[phenol]} = 1 + \frac{k_x[X]}{k_{bz}[BZ]} $$

to apply, where $[phenol]_o/[phenol]$ is the ratio of the yield of phenol in the absence and presence of the added reactant X, and k_x/k_{bz} is the ratio of the two rate coefficients involved. The equation was tested for a number of reactants other than HMS, among them resorcin, acetone and acetonitrile. Good agreement of the k_x with literature data was obtained for k_{bz} = 7.8×10^9 $M^{-1}s^{-1}$. Figure 3a shows our data for HMS plotted in accordance with the above equation. Although the data fall on a straight line, the intercept with the ordinate occurs somewhat below unity. Nevertheless, from the slope, which is 0.0069 ± 0.0003, we obtained k_{HMS} = 5.4×10^7 $M^{-1}s^{-1}$. This value is two orders of magnitude smaller than the rate coefficients for the reactions of OH with bisulfite and formaldehyde, which are also present in small concentrations in equilibrium with HMS. The results suggest, therefore, that OH radicals react primarily with bisulfite and formaldehyde rather than with HMS, so that the value obtained should represent only an upper limit to the time rate coefficient. On the other hand, the rate coefficient value would be an apparent one, if the consumption of OH radicals were compensated by the production of another type of radical, which can also react with benzene to produce phenol. Accordingly, we have also measured the yield of sulfate and compared it with the loss of OH as calculated form the difference in the phenol yields. Up to 18 times more sulfate was formed than can be accounted for by the decrease in phenol production. This result indicates the occurrence of a chain reaction and it invalidates the above procedure for determining k_{HMS}.

In order to account for the excess of SO^-_4 production, we suggest the following reaction scheme with SO^-_4 radicals acting as chain carriers

$$OH + CH_2(OH)SO_3^- \quad\quad ---> H_2O + CH(OH)SO_3^-$$

$$CH(OH)SO_3^- + O_2 \quad\quad ---> HCOOH + SO_4^-$$

$$SO_4^- + CH_2(OH)SO_3^- \quad\quad ---> SO_4^= + H^+ + CH(OH)SO_3^-$$

$$SO_4^- + C_6H_6 \,(+H_2O, \,O_2) \quad\quad ---> C_6H_5OH + SO_4^= + H^+ + HO_2$$

$$HO_2 + HO_2 \quad\quad ---> H_2O_2 + O_2$$

$$H_2O_2 + HSO_3^- \quad\quad ---> SO_4^= + H^+ + H_2O$$

If for simplicity one assumes a constant chain length f, the rate of SO^-_4 formation under steady state conditions for OH will be given by:

$$\frac{[SO_4^=] - [phenol]/2}{[phenol]_0} = f/(1 + \frac{k_{bz}[BZ]}{k_{HMS}[HMS]})$$

Figure 3b shows the reciprocal expression on the left hand side of this equation plotted versus the concentration ratio [BZ]/[HMS]. The data are seen to fall on a straight line, whose intercept with the ordinate suggests $1/f$ = 0.17 ± 0.06 or f = 5.8. The slope of the line is 0.55 ± 0.05, which combined with the values for f and k_{bz} leads to k_{HMS} = 2.45×10^9 $M^{-1}s^{-1}$, a value 50 times greater than that suggested by the data in Figure 3a. Martin et al. (12) have studied the reaction of OH with HMS in competition with pinacol, using Fenton's reaction as a source of OH, and reported k_{HMS} = $1.25 \times 10^9 M^{-1}s^{-1}$. Although both values agree in magnitude, we cannot recommend either one. Martin et al. (12) were unaware of the chain character of the reaction, whereas our own evaluation depends on the questionable assumption of a constant chain length. Evidently, in order to determine the absolute rate coefficient k_{HMS} the reaction will have to be reinvestigated with improved experimental techniques.

Acknowledgements
    The work on the photodecomposition of iron(III) complexes was carried
out under contract EV4V-0080-C of the Commission of European Communities;
the study of the reaction of OH with hydroxymethanesulfonate was part of
the programme of Sonderforschungsbereich 233 (Dynamics and Chemistry of
Hydrometeors), which is supported by the Deutsche Forschungsgemeinschaft.

5. REFERENCES:

1) W.L. CHAMEIDES and D.D. DAVIS, J. Geophys. Res. 87, 4863-4877 (1982)
2) S.N. PANDIS and J.H. SEINFELD, J. Geophys. Res. 94, 1105-1126 (1989)
3) W.J. McELROY, Atmos. Environment 20, 323-330 (1986)
4) C.J. WESCHLER, M.L. MANDICH and T.E. GRAEDEL, J. Geophys. Res. 91,
    5189-5204 (1986)
5) P. WARNECK, J. Atmos. Chem. 8, 99-117 (1989)
6) R.M. SMITH and A.E. MARTELL, Critical Stability Constants, Vol. 4,
    Inorganic Complexes, Plenum, New York, (1976)
7) P. WARNECK and C. WURZINGER, J. Phys. Chem. 92, 6278-6283 (1988)
8) K.D. ASMUS, H. MÖCKEL and A. HENGLEIN, J. Phys. Chem. 77, 1218-1221
    (1973)
9) B.C. FAUST and J. HOIGNE, Atmos. Environment, to be published (1989)
10) C. BRÜHL and P.J. CRUTZEN, Geophys. Res. Lett. 16, 703-706 (1989)
11) J. LELIEVELD and P.J. CRUTZEN, to be published (1989)
12) L.R. MARTIN, M.P. EASTON, J.W. FOSTER and M.W. HILL, Atmos. Environ-
    ment 21, 563-568 (1989)

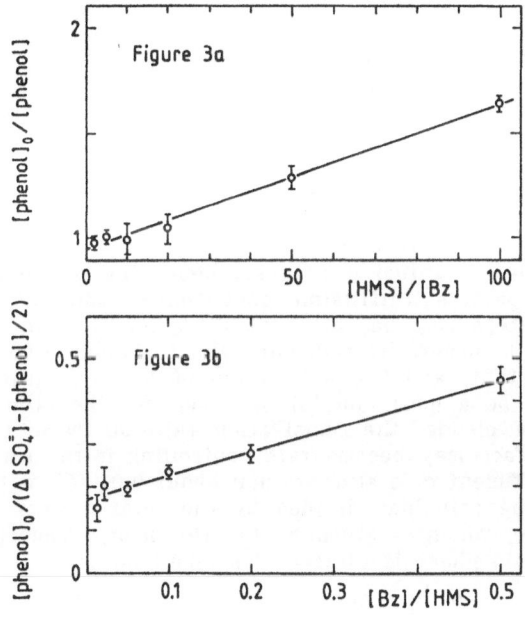

Figure 3: Competition for OH radicals between benzene and hydroxy-
methane- sulfonate (a) relative yields of phenol with increasing HMS
concentration; (b) reciprocal ratio of sulfate to phenol yields versus the
ratio of reactant concentrations.

TRACE GAS EXCHANGE KINETICS AT THE AIR/WATER INTERFACE

W. KIRCHNER, F. WELTER and U. SCHURATH
Institut für Physikalische Chemie der Universität Bonn
D-5300 Bonn 1, Wegelerstr. 12

Summary

The transfer rate of various trace gases from air at atmospheric pressure into a thin water jet of 45 µm radius, jet speed typically 600 cm s^{-1}, was measured in a cylindrical flow tube under isokinetic conditions. The air/water contact time was varied between 40 µs and about 1 ms by changing the jet lenth. The jet was trapped in a receiver capillary of 75 µm radius, and the water analyzed by ion chromatography to determine the solute concentration as function of contact time. It was shown by numerical calculations that under the experimental conditions the uptake rate is dependent on the mass accommodation coefficient α of the trace gas, which can thus be determined. Mass accommodation coefficients α of the order 0.005 to about 0.01 were obtained for HCl, HNO$_3$, HONO, N$_2$O$_5$, PAN, and N$_2$O$_4$. It is concluded that the absorption rate of these compounds by very small cloud and fog droplets is diffusion limited in either the gas or liquid phase, and does not depend on α.

1. INTRODUCTION

The presence of liquid water (cloud and fog droplets) in the atmosphere has a significant impact on the fate of soluble trace gases, and may also shorten the lifetimes of less soluble species if their reactions in water are fast. This has been discussed in greatest depth for the prototype species SO$_2$. Several authors (1-4) have singled out the rate limiting steps which may occur in the transition of a trace gas from interstitial air into small cloud droplets. The situation is confusing because, depending on their individual physico-chemical properties (diffusion coefficients, solubilities, chemical reactivity), different steps may be rate controlling for different trace gases. For larger rain drops, transport by molecular and turbulent diffusion in both phases are most important, and the assumption of Henry equilibrium at the drop surface seems to be a good one (5). However, for the other extreme of very small droplets in clouds, the penetration rate of trace gas molecules through the water surface may become rate controlling if the mass accommodation or sticking coefficient α is smaller than about 0.01 (3). α is defined as the ratio of penetrating collisions divided by the total number of collisions with the liquid surface, which is given by kinetic theory. The flux from the gas phase into the liquid phase is directly dependent on α:

$$F_{air \to water} = -\alpha \cdot \frac{\langle v \rangle}{4} \cdot c_g \qquad [1]$$

$\langle v \rangle$ denotes the mean molecular speed of the trace gas, c_g is the gas phase concentration at the surface of the droplet. Equation [1] must be modified if the flux $F_{water \to air}$ is non-negligible, see equation [2] below. Conditions under which F, and thus α, has a significant effect on the overall removal rate of a trace gas have been discussed by Schwartz (1).

Laboratory measurements of α are very difficult, and have long been restricted to the condensation and evaporation of pure compounds, particularly pure water, for which α is likely to be unity. Very small sticking coefficients of a few trace gases on concentrated sulfuric acid have been reported by Baldwin and Golden (6). Very recently two groups in the US have published measurements of α on water for SO_2, H_2O_2, NO_2, and ozone (7–9). We have also recently reported some preliminary sticking coefficients of NO_x compounds (10). While our measurements were carried out in air at atmospheric pressure on a fresh surface of very clean water, all other workers use low total pressures to reduce the rate limiting effect of molecular diffusion in the gas phase, and have to add chemical quenchers to the liquid phase to eliminate saturation effects. We believe that the advantage of our system is its simple geometry which alleviates full numerical calculations of transport and chemical rate processes in the phase boundary region.

2. EXPERIMENTAL

According to Schwartz (1) the reduction of uptake rates into liquid water by gas phase diffusion is less severe if the surface radius is small. Our studies are therefore conducted at the surface of an extremely thin jet of liquid water. Very short contact times are chosen to minimize surface saturation effects.

The jet reactor (Figure 1) consists of a glass tube of 3 mm i.d. mounted vertically in a thermostatted box, and of a very thin quartz tube of 100 μm i.d. and 200 μm o.d. which ejects the liquid jet. The quartz capillary forms the tip of a glass capillary of slightly larger diameter. The capillaries can be moved smoothly up and down the z axis of the flow tube by means of a micro motor drive. Pure water (Milli-Q) is supplied to the glass capillary via a calibrated flow meter (Rota) from a pyrex flask which is pressurized with He to produce a jet speed of ca. 600 cm s⁻¹. The emerging jet contracts rapidly, developing a constant radius r of 45 μm within less than 3 jet diameters from the orifice of the quartz tube. This implies (11) that plug flow is rapidly established in the free jet, and that its surface speed v_s equals the average jet speed $\langle v \rangle = F_w/\sigma$ (F_w = water flow rate, σ = cross section of jet). The jet is neatly trapped in a receiver capillary of 150 μm i.d. which can be fine adjusted in the x,y plane, and the water is collected for analysis via a teflon tube and valve. Entrainment of gas bubbles into the receiver capillary is prevented by adjusting the valve resistance properly.

Trace gases and synthetic air were blended in an evacuabale glass chamber of 408 liter volume in a thermostatted room. Mixtures of the labile compounds N_2O_5, HONO, and peroxyacetyl nitrate (PAN) in air were prepared and stored at a room temperature of about 5 °C to increase stability. Mixing ratios were either established by standard volumetric techniques, or measured in situ: NO_2 by HeCd laser-induced fluorescence inside the chamber; N_2O_5, HNO_3 and HONO by UV absorption spectrometry with a deuterium lamp and a Jobin-Yvon HRS2 monochromator. The folded optical path in the chamber was 304 cm long. A computer was used for scan control, data acquisition and spectrum evaluation on the basis of recommended UV spectral data.

Sample air from the glass chamber was pumped through the flow tube under laminar flow conditions. The required air flow rate for establishing isokinetic conditions of air and water was calculated assuming a Hagen-Poiseuille velocity profile of the air in the glass tube. The calculated dilution rate due to pumping was always in good agreement with the recorded time-concentration-profile of the trace gas in the glass chamber, which was measured optically during the experiment.

The water collected by the receiver capillary was analyzed for chloride, nitrate and nitrite by ion chromatography. Gas concentrations of about 100 ppm in air were typically needed to yield solute concentrations between 20 and 100 μM at a maximum jet length of 6 mm. A detection limit of 0.1 μM Cl⁻ in 100 μL water could be achieved. PAN was detected after hydrolysis as the sum of NO_2^- and NO_3^- in water. Since the compound hydrolyses very slowly in pure water (12), the jet was run with 0.1 mM LiOH, and a strongly alkaline eluent was used.

3. THEORETICAL BASIS OF THE METHOD

The net flux of a soluble trace gas (Henry coefficient H [M atm⁻¹]) through a water surface is given by gas kinetic theory, cf. equation [1]:

$$F = -\alpha \cdot \frac{\langle v \rangle}{4} \cdot (c_g - c_w/H) \qquad [2]$$

c_g [atm⁻¹] and c_w [M] are the surface concentrations of the trace gas in air and in water. When a jet of liquid water is exposed for some time t = z/v (z = jet length) to a trace gas in air in the flow tube, Figure 1, the concentration $\langle c \rangle$ of the trace gas dissolved in the trapped water is obtained by integrating the specific surface flux $F_{(z)}$ times surface area 2πr·dz, divided by volume $r^2 \cdot \pi \cdot dz$, along the jet:

$$\langle c \rangle = \frac{2}{r} \int_0^{t(z)} F_{(z)} \cdot dt \qquad [3]$$

$F_{(z)}$ varies as a function of z because c_g is depleted, while c_w increases with time.

The concentrations c_g and c_w needed to calculated $F_{(z)}$ can be obtained by solving the general equation of convective diffusion:

$$\left(\frac{\delta c}{\delta t}\right) = D \nabla^2 c - \vec{v} \nabla c \qquad [4]$$

The second term for convective transport may be neglected, because the measurements were carried out under isokinetic conditions. Furthermore we introduce the radial coordinate r and neglect axial diffusion. This yields a much simpler equation which can be solved numerically to yield c_r as a function of time:

$$\left(\frac{\delta c}{\delta t}\right) = D \cdot \frac{1}{r} \cdot \frac{\delta}{\delta r}\left(r \cdot \frac{\delta c}{\delta r}\right) \qquad [5]$$

This differential equation holds for both phases if equation [2] and the following boundary condition are taken into account at the jet surface, r = r_0 (D_g, D_w = diffusion coefficients of the trace gas in air and in the liquid phase):

$$- D_g \left(\frac{\delta c_g}{\delta r}\right)_{r_0} = - D_w \left(\frac{\delta c_g}{\delta r}\right)_{r_0} \qquad [6]$$

The differential equations for the coupled diffusion in air and in the water jet were resolved into finite differences and numerically integrated as function of time, using the FACSIMILE program on the University computer. α was treated as a parameter, and concentrations $\langle c \rangle$ obtained from equation [3] were compared with the measurements to determine α.

Rates of chemical reactions (e.g. hydrolysis) in the liquid phase are not yet (but will be) explicitly included in our model. Neglecting secondary reactions in the liquid phase may be dangerous for trace gases whose effective Henry coefficients are pH dependent. However, the effective Henry coefficient of HCl and HNO_3 in the surface water of the jet remains large on our experimental time scale, and outgassing ($\hat{=}$ second term in equation [2]) can be safely neglected. The same holds for N_2O_5 which was assumed to hydrolyse "instantaneously" because the "physical" Henry constant of the compound is unknown. The situation is more complex for HONO which has a "physical" Henry coefficient of 49 M atm⁻¹ at 25 °C (13), but hydrolyses to form H⁺ + NO_2^- (equlilibrium constant K = $5.9 \cdot 10^{-4}$ M, (13)). We calculate

$H_{eff} = H \cdot (1 + K/[H^+])$ at the surface of our jet to be ≥ 107 M atm^{-1} for jet lengths ≤ 6 mm. It could be shown by numerical simulations that the effect of outgassing on our measurements is small if $H_{eff}/\alpha > 1000$. This is confirmed by the relatively low value of α obtained for HONO and for all other gases. More reliable calculations require a model which includes chemical reaction rates in the liquid phase explicitly. Such a model will be available in the near future.

4. RESULTS

The general procedure for obtaining values of the sticking coefficient α consisted of four steps:

1. A dilute sample of a (preferably pure) trace gas in air was prepared and analysed in the glass chamber.

2. Sample air was pumped through the jet flow tube at a constant flow rate, and the jet length was varied several times in incremets of ca. 250 µm, to a maximum length of 6 mm, each time collecting a sample of water for later analysis.

3. The water samples were analysed by ion chromatography, and the results corrected for dilution of sample air in the glass chamber. Averages of the corrected solute concentrations for each jet length were plotted versus jet length $\triangleq$ contact time.

4. Numerical integrations were carried out as described in SECTION 3 for a range of assumed sticking coefficients, using evaluated literature data for the diffusion coefficients D_g and D_w, and for the Henry constant H. The results were compared with the plotted experimental data to encircle the "true" α. The procedure was repeated with an improved estimate of α until a good fit of the experimental data was achieved.

Step 4 is illustrated in Figure 2 which shows both experimental data and theoretical profiles for the interaction of gaseous HNO$_3$ with the jet. The data imply a mass accommodation coefficient of $0.9 \cdot 10^{-2}$. Relatively large error limits of +65 %, −30 % must be attachted to this value, as discussed in greater detail elsewhere (14). Concentrations of HNO$_3$ in the gas phase are based on absorption cross sections of Molina and Molina (15).

An alternative method is less dependent on absolute concentration measurements in the gas phase: Assuming that our absolute value of $\alpha_{(HCl)} = 1.0 \cdot 10^{-2}$ is correct (because analytical errors are minimized for this stable compound), measurements of relative uptake rates into the water jet have been carried out with binary mixtures of HCl and another trace gas. This method is only useful for trace gases which do not react with HCl in the gas phase, and whose uptake rates are unaffected by changes in surface pH due to dissolved HCl. Rapid reactions of HCl were found to occur in the gas phase with HONO and N$_2$O$_4$. Figure 3 shows normalized ratios [Cl$^-$]/[NO$_3$$^-$] for a binary mixture of $3 \cdot 10^{15}$ cm^{-3} HCl and $3 \cdot 10^{15}$ cm^{-3} N$_2$O$_5$ containing 10 % HNO$_3$ as an impurity. The ratios were normalized by dividing through the solute ratio at 6 mm jet length. Experimental data are again compared with numerical calculations. The method yields α(N$_2$O$_5$) = 0.01, which is compatible with the absolute result given above.

Measurements of α(HCl) and α(HNO$_3$) are comparatively simple, but considerably larger difficulties were encountered in our studies of N$_2$O$_5$, PAN, HONO, and N$_2$O$_4$, either because the gas phase concentrations of the compounds were based on ambiguous uv absorption cross sections (e.g. for HONO), or because Henry coefficients and/or kinetic data in the liquid phase were unavailable or very uncertain (e.g. for N$_2$O$_4$). A more detailed discussion (which we defer to a future publication) may be found in a recent thesis (14).

Table I summarizes the sticking coefficients which have been obtained by the liquid jet method. The value for N$_2$O$_4$ is based on the assumption that the nitrite formed in the jet when contacted with about 1 % NO$_2$ in air is actually due to dissolution and hydrolysis of N$_2$O$_4$, which is supported by the extremely low Henry coefficient of NO$_2$ (H = 10^{-2} M atm^{-1} (16)), and by our observation that the amount of nitrite formed was proportional to (p$_{NO2}$)^2.

Table I: Mass accommodation coefficients of various trace gases and their estimated uncertainties due to analytical errors only.

trace gas	mass accommodation coefficients:	
	absolute values	relative to α(HCl) = 0.01
HCl	0.01 (+20 %/-20 %)	reference gas
HNO$_3$	0.009 (+65 %/-30 %)	0.012 (± 20 %)
N$_2$O$_5$	0.005 (+50 %/-30 %)	0.01 (+25 %/-15 %)
HONO	0.006 (error uncertain)	reacts with HCl gas
PAN	0.001 (error uncertain)	0.01 (± 30 %)
N$_2$O$_4$	≥ 0.02	reacts with HCl gas

5. CONCLUSION

While the solubilities of the trace gases studied vary considerably, from (effective) Henry constants of order 10^6 M atm^{-1} for HCl and HNO$_3$ to H = 7,5 and 1.3 M atm^{-1} for PAN (12) and N$_2$O$_4$ (17), the sticking coefficients are surprisingly similar, consistently in the order 0.01 (some of the smaller values obtained by the absolute method, 2nd column in Table I, are uncertain, and the relative values, 3rd column, should be preferred where available). Possible systematic errors of our method are difficult to assess, particularly because no literature data are available for the same trace gases. Our method is not very sensitive to variations in the range 1 ≥ α ≥ 0.05, as can be seen in Figures 2 and 3, but lower values should be readily resolved under favourable conditions. A sticking coefficient of 0.11 has recently been determined for SO$_2$ (7), and it will be interesting to measure α(SO$_2$) with our method.

It should be noted that, for atmospheric chemistry applications, it is sufficient to ascertain a lower limit of α ≥ 0.01, because only in the case of significantly smaller sticking coefficients will there be an effect on the removal rate by cloud water (1).

6. ACKNOWLEDGEMENTS

This work was supported by Minister für Umwelt des Landes NRW, and by the Commission of the European Communities. A travel grant by NATO for our cooperation with the University of Patras, Greece, is also gratefully acknowledged.

7. REFERENCES

(1) S.E. Schwartz, Mass-Transport Considerations Pertinent to Aqueous Phase Reactions of Gases in Liquid-Water Clouds. In "Chemistry of Multiphase Systems", W. Jaeschke (Ed.), NATO ASI Series G: Ecological Series, Vol. 6, Springer, Berlin 1986, p. 415
(2) S.E. Schwartz, Mass-Transport Limitation to the Rate of In-Cloud Oxidation of SO$_2$: Re-Examination in the Light of New Data. Atmos. Environ. 22, 2491-2499 (1988)
(3) B.G. Heikes and A.M. Thompson, Effects of heterogeneous processes on NO$_3$, HONO and HNO$_3$. J. Geophys. Res. 88, 10883 (1983)
(4) W.L. Chameides, Possible Role of NO$_3$ in the Nighttime Chemistry of a

Cloud. J. Geophys. Res. 91, 5331–5337 (1986)

(5)   J.H. Topalian and D.C. Montague, A Theoretical Study of Chemical Kinetic Control in the Scavenging of Pollutant Gases by Cloud and Rain Drops. J. Atmos. Chem. 8, 1–18 (1989)

(6)   A.C. Baldwin and D.M. Golden, Heterogeneous Atmospheric Reactions: Sulfuric Acid Aerosols as Tropospheric Sinks. Science 206, 562–563 (1979)

(7)   D.R. Worsnop, M.S. Zahniser, Ch.E. Kolb, J.A. Gardner, L.R. Watson, J.M. Van Doren, J.T. Yayne, and P. Davidovits, Temperature Dependence of Mass Accommodation of SO_2 nad H_2O_2 on Aqueous Surfaces. J. Phys. Chem. 93, 1159–1172 (1989)

(8)   J.A. Gardner, L.R. Watson, Y.G. Adewuyi, P. Davidovits, M.S. Zahniser, D.R. Worsnop, and Ch.E. Kolb, Measurement of the mass accommodation coefficient of $SO_2(g)$ on water droplets. J. Geophys. Res. 92, 10887 (1987)

(9)   J.H. Lee and I.N. Tang, Accommodation coefficient of gaseous NO_2 on water surfaces. Atmos. Environ. 22, 1147 (1988)

(10) U. Schurath, W. Kirchner, F. Welter and J. Kames, Transport Rates of Atmospheric Trace Gases Across the Gas/Liquid Interface, pp. 125–128 in Air Poll. Res. Rep. 17, R.A. Cox (ed.), EUR 12035 (1988)

(11) S. Middleman, Profile Relaxation in Newtonian Jets. Ind. Eng. Chem. Fundam. 3, 118–122 (1964)

(12) M.W. Holdren, C.W. Spicer and J.M. Hales, Peroxyacetyl nitrate solubility and decomposition rate in acidic water. Atmos. Environ. 18, 1171 (1984), and results (to be published) from our laboratory

(13) W.L. Chameides, The Photochemistry of a Remote Marine Stratiform Cloud. J. Geophys. Res. 89, 4739–4755 (1984)

(14) W. Kirchner, Massen–Akkommodationskoeffizienten atmosphärischer Spurengase für die Grenzfläche Luft/Wasser, Dissertation, Bonn 1989

(15) L.T. Molina and M.J. Molina, UV–Absorption Cross Section of HO_2NO_2 Vapor. J. Photochem. 15, 97–108 (1981)

(16) H. Komiyama and H. Inoue, Absorption of Nitrogen Oxides into Water. Chem. Eng. Sci. 35, 154–161 (1980)

(17) H. Kramers, M.P.P. Blind and E. Snoek, Absorption of Nitrogen Tetroxide by Water Jets. Chem. Eng. Sci. 14, 115–125 (1961)

8. FIGURES

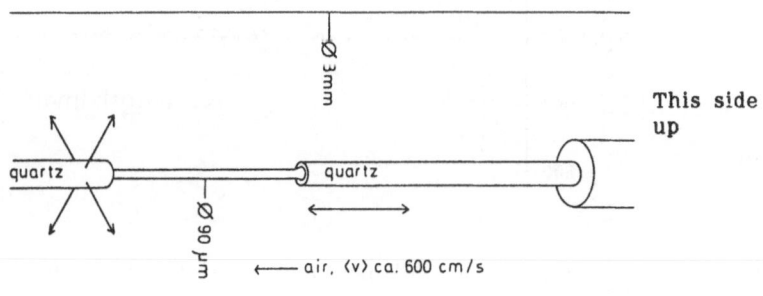

Figure 1:   Schematics of the jet flow tube: quartz tube diameter, flow tube diameter, and jet diameter approximately to scale. The quartz tube which ejects the jet is a factor of 2.5 longer than shown.

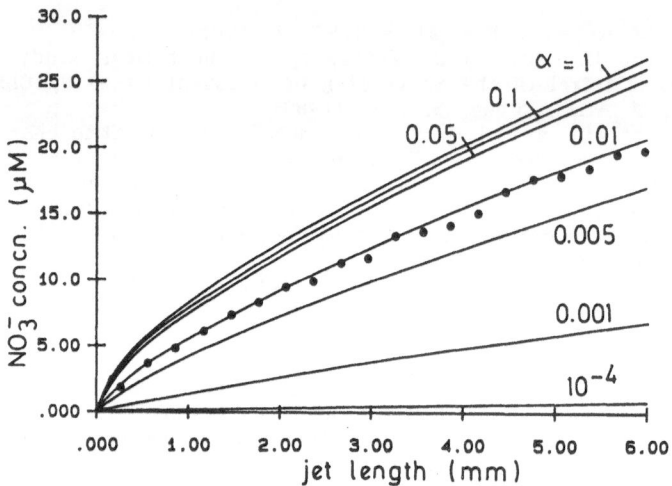

Figure 2: Nitrate concentration in water as function of jet lengt. The HNO3 concentration in the gas phase was $1.56 \cdot 10^{15}$ molecule cm^{-3}. The experimental results are compared with numerical calculations assuming different values of the sticking coefficient α.

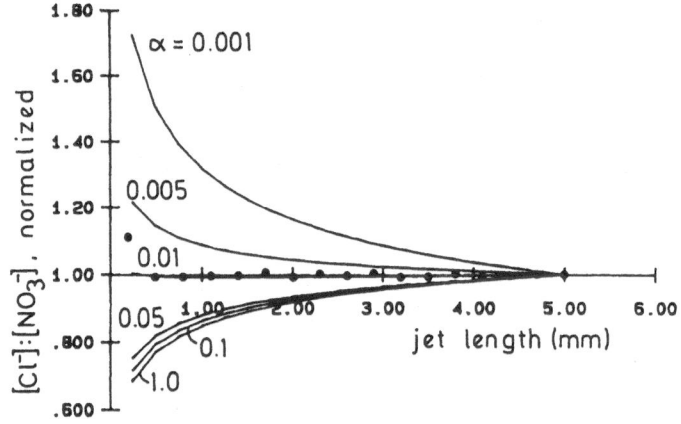

Figure 3: Co-absorption of HCl gas and N2O5 gas into the jet: the ratio [Cl$^-$]:[NO3$^-$], normalized to the ratio at 6 mm jet length, is plotted versus jet length. The experimental results are compared with a series of numerical calculations assuming α(HCl) = 0.01 and various values of α(N2O5).

HETEROGENEOUS PRODUCTION OF Cl ATOMS UNDER SIMULATED
TROPOSPHERIC CONDITIONS IN A SMOG CHAMBER

Wolfgang Behnke and Cornelius Zetzsch

Fraunhofer-Institut für Toxikologie und Aerosolforschung
D 3000 Hannover 61, FRG

Summary

The behaviour of sea spray in tropospheric chemistry is investigated
by exposing airborne NaCl aerosol to a photochemical smog in a chamber. In the presence of NO_x, an intrusion of NO_3^- into the aerosol is
observed, leading to a corresponding release of HCl, that is mainly
adsorbed on the glass walls of the chamber. A photochemical formation
of atomic Cl is observed, exceeding the expected production from the
well-known reaction of OH with HCl in the gas-phase by far, even if
all HCl released would be assumed to remain in the gas-phase. In
accompanying dark experiments, molecular Cl_2 (formed by a heterogeneous oxidation of HCl on the glass walls in the presence of O_3 and
humidity) is identified to be a major precursor of the atomic Cl.
Inserting a teflon bag into the chamber in the presence of HCl gas and
O_3 at 50% relative humidity, the aerosols SiO_2 and Fe_2O_3 are observed
to enhance the formation of Cl by factors of 2-10 as compared to the
production by OH + HCl; TiO_2 aerosol acts as a strong photocatalyst
and is effective by a factor of up to 1000, even in the absence of O_3.
Heterogeneous reactions occuring on the surface of the aerosols and on
the glass surface convert a considerable fraction of the HCl and a
portion of the hydrocarbons to chlorinated products like phosgene,
chloroacetone and 1,1-dichloroacetone.

1. INTRODUCTION

NaCl is an abundant constituent of the tropospheric aerosol, and about
10^9 t/a are emitted in the form of sea spray worldwide (1). The loss of Cl^-
from this aerosol is a well-known phenomenon in coastal regions (2-4). It
is mainly caused by the intrusion of acids (HNO_3 and H_2SO_4) and leads to a
release of HCl gas. Further sources of HCl are mostly anthropogenic, e.g.
the emissions from coal power plants (5) or from incinerators (5) or from
the photochemical degradation of the chloroalkenes (6) and the (natural)
methyl chloride (1). Average mixing ratios around 1 ppb are observed in the
lower northern hemispheric troposphere (7-8). The well-known reaction of
tropospheric OH with HCl (9) is then expected to produce Cl atoms. Furthermore, the formation of nitrosyl chloride and nitryl chloride from NaCl in
the presence of NO_2 and N_2O_5 has been observed in the laboratory (10, 11)
and has been proposed as a photolytic source of atomic Cl in the troposphere.

Atomic Cl may compete with OH in the degradation of trace gases;
considerable contributions are expected to the tropospheric consumption of
the lower non-methane alkanes (ethane and propane) and especially branched-
chain alkanes, like neopentane and tetramethylbutane because of their high
ratio of k_{Cl}/k_{OH} (12, 13).

Aerosol smog chamber experiments on the behaviour of NaCl (14) showed a strong formation of atomic Cl, much stronger than expected from the chemistry of NO_x and catalyzed by O_3 even in the absence of NO_x and NaCl, just in the presence of HCl, O_3 and humidity in a heterogeneous process occuring on the borosilicate glass surface of the smog chamber and on the surface of aerosols (15). In corresponding dark experiments, molecular chlorine was identified to be the main precursor of the atomic Cl (15-16). The efficiency of various aerosols (SiO_2, Fe_2O_3 and TiO_2) is compared and investigated in the present study.

2. EXPERIMENTAL

One part of the experiments is performed in a 2.3 m^3 smog chamber made of borosilicate glass (Duran 50, Schott), the other part in 0.8 m^3 teflon bags (FEP 200 A, DuPont). Details of the equipment (the solar simulator, the generation of aerosols, the analyzers for O_3 and NO_x, the automatic preconcentration of hydrocarbons combined with capillary gas chromatography and the GC/MS for the identification of reaction products) have been described previously (17-19). Most experiments were performed under irradiation with a UV-cutoff filter at 360 nm (Plexiglas). Gaseous HCl and HNO_3 are collected in NaF coated denuders and analyzed by ion chromatography (Sykam). Cl_2 is determined by the bleaching of methyl orange (20). The aerosol materials investigated are SiO_2 (Aerosil 200, DEGUSSA, 200 m^2/g), Fe_2O_3 (Sicotram orange, BASF, 100 m^2/g) and TiO_2 (PKP anatase, Bayer, 120 m^2/g). The initial concentrations of the aerosol range from 3-8 mg/m^3 in the glass chamber. In the teflon bag, about 1 mg/m^3 is attained, corresponding to aerosol surface to gas volume ratios of 0.1-0.2 m^{-1}.

3. EVALUATION OF THE OBSERVED CONCENTRATION TIME PROFILES

The concentration time profiles of Cl atoms and OH radicals are computed from the degradation of various hydrocarbons, chosen to cover a wide range of the ratio k_{Cl}/k_{OH} (21). Since the steady state approximation is well fulfilled for the active species OH and Cl in smog chamber experiments, the source and sink terms for Cl atoms must be equal. A minimum source term for Cl follows from the presence of OH by its reaction with HCl:

$$d[Cl]/dt = k_{OH}[OH][HCl]. \qquad (1)$$

The total production of atomic Cl must be equal to the total consumption of Cl atoms, that can be computed with reasonable precision from the time profiles of the injected hydrocarbons and their known rate constants for the reactions with Cl, until the hydrocarbons are degraded to a level of about 50%, where the additional consumption by the reaction products begins to become important (16):

$$Cl-c. = \sum_i k_i[KW_i(t)][Cl(t)]. \qquad (2)$$

In the presence of Cl_2, the photolysis of Cl_2 yields a further production term for atomic Cl:

$$d[Cl]/dt = 2 k_{ph}[Cl_2(t)]. \qquad (3)$$

The ratio of the sink term (Cl-c.) to the source term $d[Cl]/dt$ should be unity in the steady state, hence further sources of Cl not yet considered should show up in increased values for this ratio (Cl-c./ $d[Cl]/dt$).

4. RESULTS AND DISCUSSION
4.1 PRODUCTION OF Cl ATOMS

As shown above, the minimum production of atomic Cl under irradiation in the smog chamber can be computed from observed gas-phase concentration of HCl and its rate of reaction with OH. Measurements of the recovery of HCl (300 ppb injected at r.h. = 50% and T = 28°C) revealed that only 45% of the injected amount is remaining in the gas phase and that the remainder is adsorbed on the wall, even in the teflon bag. In the presence of all three aerosols investigated in the teflon bag (SiO$_2$, Fe$_2$O$_3$ and TiO$_2$), the concentration of gaseous HCl was about the same as in the absence of aerosol.

In the presence of TiO$_2$ aerosol and under irradiation at λ>360 nm, the hydrocarbons (32 ppb of pentafluorobenzene, 41 ppb of toluene, 151 ppb of n-hexane, 200 ppb of n-heptane, 165 ppb of 2,2,4-trimethylpentane and 165 ppb of 2,2,3,3-tetramethylbutane) are consumed by reactions with Cl atoms and OH radicals in the homogeneous gas phase alone. This result is in contrast to earlier experiments in the absence of HCl (19) where an additional, heterogeneous degradation, inversely dependent on the vapor pressure of the hydrocarbons, was observed (the lack of a heterogeneous degradation in the present experiments was observed to be dependent on the HCl/ hydrocarbon ratio and on the UV-cutoff of the irradiation of the solar simulator).

Figure 1 shows the ratio of the observed consumption of Cl atoms according to eq. (2) and the expected production of atomic Cl according to eq. (1). The ratio is small at the beginning, but after two hours it is about a factor of 1000 higher than can be explained from the reaction of OH with HCl in the gas-phase alone. This may again be caused by a heterogeneous process, similar to the heterogeneous photodegradation of hydrocarbons observed in previous experiments in the presence of TiO$_2$ aerosol (19). This high production rate of atomic Cl decreases with time, indicating that some precursor of the additional Cl atoms is vanishing.

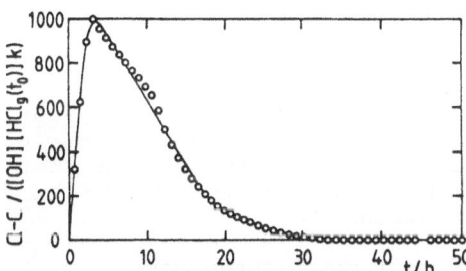 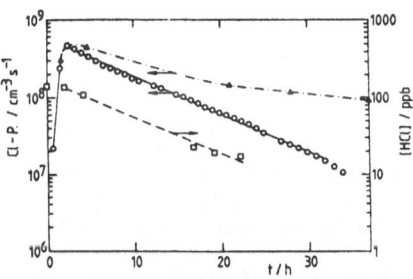

Fig. 1: The ratio of the consumption of atomic Cl by the hydrocarbons to the expected production by the reaction of OH with HCl in the presence of TiO$_2$ aerosol (λ>360 nm), [HCl]$_{injected}$ = 300 ppb. No O$_3$ is added in this experiment.

Fig. 2: A comparison of the production of atomic Cl by photolysis of the observed Cl$_2$ $\triangle$ (by the methyl orange technique) with the consumption of atomic Cl by the hydrocarbons (left hand scale) and with the time profile of gaseous HCl (right hand scale).

Figure 2 shows a comparison of the total source strength of atomic Cl (Cl.-c. from eq. 2) with the potential precursors, gaseous HCl and Cl$_2$, in a semilogarithmic diagram. The concentration time profile of Cl_2 was determined by the methyl orange technique (20), and the concentrations were converted to the photolytic production rate of atomic Cl (using a separately determined value of $k_{ph} = 4\times10^{-4}s^{-1}$)($\triangle$). During the initial phase of such experiments, the consumption of Cl atoms can be explained by the photolysis of Cl_2 alone. At a later stage, the observed (true) consumption becomes smaller than is expected if the total signal of the methyl orange technique would be due to Cl_2. Obviously, interferences by other compounds (which are able to bleach methyl orange but are less efficient in the production of atomic Cl) must be responsible for this discrepancy. The real production of atomic Cl is observed to be proportional to HCl in the gas-phase.

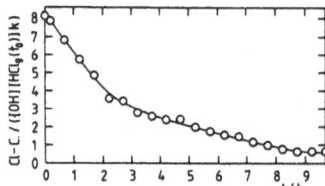

Fig. 3: The ratio of the consumption of atomic Cl by the hydrocarbons to the expected production of atomic Cl by the reaction of OH with HCl in the presence of SiO_2 aerosol.

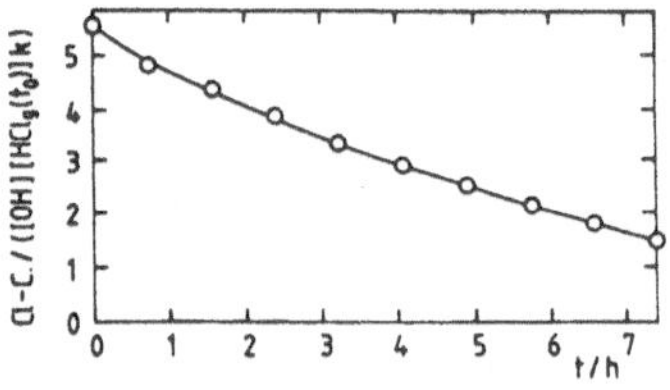

Fig. 4: Similar experiment as in fig. 3, but in the presence of Fe_2O_3 aerosol instead of SiO_2 (exp. conditions for both experiments: $\lambda > 300$nm, $[O_3]_0 = 1$ ppm, $[HCl]_{inj} = 300$ ppb).

4.2 HETEROGENEOUS FORMATION OF PRODUCTS IN THE PRESENCE OF HCl

The reaction pathways of atomic Cl are expected to be similar to those of OH in that abstraction of an H atom from a C-H-bond is expected as the first step of the reaction with paraffinic hydrocarbons (leadig to the same organic radical as with OH attack and HCl). On the other hand, a rapid decrease of HCl is observed in the present experiments (see e.g. fig. 2), and a part of the chlorine is recovered in organic compounds, which have been identified by GC/MS. At the end of the experiment of fig. 1 in the presence of TiO_2 aerosol after 24h of irradiation, about 15 ppb of phosgene and 1,1-dichloroacetone each, and 2 ppb of chloracetone were observed as chlorinated products from heterogeneous reactions. With SiO_2 and Fe_2O_3 (and O_3 and NO_x at $\lambda > 300$nm irradiation) only phosgene was observed at levels up to 5 ppb, but neither 1,1-dichloroacetone nor chloroacetone. Further experiments are required to understand the reaction pathways leading to these highly toxic, chlorinated compounds.

4.3 RADICAL PRODUCTION MECHANISMS IN THE PRESENCE OF SEMICONDUCTORS

The production of molecular chlorine in the dark via oxidation of Cl^- by O_3 (and hence the photolytic precursor of Cl) was observed to occur on the glass walls of the vessel and on the surface of aerosols, and it may have similar explanations as the formation of H_2O_2 from O_3 in impingers (22). Furthermore, the electrophilic properties of O_3 may be enhanced strongly in the adsorbed state by interaction with Lewis acid centers (23).

The tremendous production of atomic Cl in the presence of TiO_2 (even in the absence of O_3) may be explained from a similar phctocatalysis as has been observed previously in aerosol smog chamber experiments in the presence of TiO_2 aerosol (17,19):

$$TiO_2 + hv \rightarrow TiO_2 (h^+ + e^-) \qquad (4)$$

Since the positive sites formed from the excitons are known to oxidize the OH^- of the water layer by the reaction:

$$h^+ + OH^- \rightarrow OH \text{ (ads)}, \qquad (5)$$

a similar oxidation may occur with Cl^-, according to:

$$h^+ + Cl^- \rightarrow Cl \text{ (ads)}. \qquad (6)$$

Furthermore, the OH radicals on the surface of the photocatalyst may oxidize Cl^- as known from liquid-phase (24):

$$OH(ads) + H^+ + Cl^- \rightarrow Cl \text{ (ads)} + H_2O \qquad (7)$$

The atomic Cl(ads) is then converted to Cl_2 and HOCl in the liquid phase by a number of elementary steps, involving Cl_2^- (25), Cl_3^- (26) and ClO^- (27).

ACKNOWLEDGEMENT This work was supported by the Bundesminister für Forschung und Technologie (grant 0744105 8). The experimental contributions of M. Elend, H.U. Krüger, Petra Kühn, Gisela Pfahler, V. Scheer and J. Zorn and discussions with Dr. Friedrich Nolting are gratefully acknowledged.

REFERENCES
1. WARNECK, P. (1988). Chemistry of the natural atmosphere. Internat. Geophys. Ser. Vol. 41, Academic Press, San Diego, CA.
2. CADLE, R.D. and ROBBINS, R.C. (1960). III. Physical and chemical properties: Kinetics of atmospheric reactions involving aerosols. Disc. Faraday Soc. 30, 155-161.
3. MARTENS, C.S., WESOLOWSKI, J.J., HARRIS R.C. and KAIFER, R. (1973). Chlorine loss from Puerto Rican and San Francisco bay area marine aerosol. J. Geophys. Res. 78, 8788-8792.
4. HITCHCOCK, D.R., SPILLER, L.L. and WILSON, W.E. (1980). Sulfuric acid aerosol and HCl release in coastal atmospheres: Evidence of rapid formation of sulfuric acid particulates. Atmos. Environ. 14, 165-182.
5. LIGHTOWLERS, P.J. and CAPE, N.J. (1988). Sources and fate of atmospheric HCl in the U.K. and Western Europe. Atmos. Environ. 22, 7-15.
6. BECKER, K.H. and ZETZSCH, C. (1989). Das luftchemische Verhalten von flüchtigen Organohalogenverbindungen. In: Halogenierte organische Verbindungen in der Umwelt, VDI-Berichte 745, VDI-Verlag, Düsseldorf (in press).
7. VIERKORN-RUDOLPH, B., BÄCHMANN, K., SCHWARZ, B. and MEIXNER, F.X. (1984). Vertical profiles of hydrogen chloride in the troposphere. J. Atmos. Chem. 2, 47-63.
8. BÄCHMANN, K. and FUCHS, G. (1987). Chlorinated air pollutants: sources and distribution. In: Formation, Distribution and Chemical Transformation of Air Pollutants (ed. R. Zellner), DECHEMA Monogr. 104, pp. 79-89, VCH-Verlagsgesellschaft, Weinheim.
9. PARASKEVOPOULOS, G. and SINGLETON, D.L. (1988). Reactions of OH radicals with inorganic compounds in the gas phase. Rev. Chem. Intermediates 10, 139-218.
10. FINLAYSON-PITTS, B.J. (1983). Reaction of NO_2 with NaCl and atmospheric implication of NOCl formation. Nature 306, 676-677.
11. FINLAYSON-PITTS, B.J., EZELL, M.J. and PITTS, Jr., J.N. (1989). Formation of chemically active chlorine compounds by reactions of atmospheric NaCl particles with gaseous N_2O_5 and $ClONO_2$. Nature 337, 241-244.

12. ZETZSCH, C. (1987). Simulation of atmospheric photochemistry in the presence of solid airborne aerosols. Formation, Distribution and Chemical Transformation of Air Pollutants (ed. R. Zellner), DECHEMA Monograph 1C4, pp. 187-212, VCH-Verlagsgesellschaft, Weinheim.

13. SINGH, H.B. and KASTING, J.F. (1988). Chlorine-hydrocarbon photochemistry in the marine troposphere and lower stratosphere. J. Atmos. Chem. 7, 262-285.

14. BEHNKE, W., NOLTING, F. and ZETZSCH, C. (1984). Rate constants of chlorine atoms with hydrocarbons determined by an aerosol smog chamber technique. Abstr. 16th Inf. Conf. on Photochemistry, Harvard, MA.

15. BEHNKE, W. and ZETZSCH, C. (1989). Smog chamber investigations of the influence of NaCl aerosol on the concentration of O_3 in a photosmog system. Proc. Int. Ozone Sympos. 1988 (eds. R. Bojkov and P. Fabian), Deepak, Hampton, VA (in press); ZETZSCH, C., PFAHLER, G., and BEHNKE, W. (1988). Heterogeneous formation of chlorine atoms from NaCl in a photosmog system, J. Aerosol Sci. 19, 1203-1206.

16. BEHNKE, W. and ZETZSCH, C. (1989). Photochemischer Abbau von chlorierten Kohlenwasserstofffen in einer Smogkammer. In: Halogenierte organische Verbindungen in der Umwelt, VDI-Berichte 745, VDI-Verlag, Düsseldorf (in press); BEHNKE, W. and ZETZSCH, C. (1988) Simulation der troposphärischen Produktion von Chloratomen unter Berücksichtigung eines Aerosoleinflusses. Final Report BMFT-PTU 325-4007-0744158.

17. BEHNKE, W., HOLLÄNDER, W., KOCH, W., NOLTING, F. and ZETZSCH, C. (1988). A smog chamber for studies of the photochemical degradation of chemicals in the presence of aerosols. Atmos. Environ. 22, 1113-1120.

18. NOLTING, F., BEHNKE, W. and ZETZSCH, C. (1988). A smog chamber for studies of the reactions of terpenes and alkanes with ozone and OH. J. Atmos. Chem. 6, 47-59.

19. BEHNKE, W., NOLTING, F. and ZETZSCH, C. (1987). A smog chamber study on the impact of aerosols on the photodegradation of chemicals in the troposphere. J. Aerosol Sci. 18, 65-71.

20. KETTNER, K. and FORWERG, W. (1969). Bestimmung geringer Chlorkonzentrationen nach der Methylorange-Methode. Atmos. Environ. 3, 215-220.

21. BEHNKE, W., and ZETZSCH, C. (1987). The simultaneous determination of two active species in smog chambers: Evaluation of rate constants for Cl atoms in the presence of OH. Proceedings of the 3rd French German Workshop on Tropospheric Chemistry by Laboratory Studies, 6.-8. October 1987, Leinsweiler Hof, Rep. No. 15 of FB 8, Physikal. Chemie, pp. 59-60, Universität Gesamthochschule Wuppertal.

22. HEIKES, B.G. (1984). Aqueous H_2O_2 production from O_3 in impingers. Atmos. Environ. 18, 1433-1445.

23. BAILEY, P. (1982). Ozonation in Organic Chemistry II, Wiley, New York.

24. JAYSON, C.G., PARSONS, B.J. and SWALLOW, A.J. (1973). Some simple, highly reactive chloride derivates in aqueous solution. J. Chem. Soc. Faraday Trans. I, 69, 1597-1607.

25. WAGNER, I., KARTHÄUSER, J. and STREHLOW, H. (1986). On the decay of the dichloride anion, Cl_2^-, in aqueous solution. Ber. Bunsenges. Phys. Chem. 90, 861-867.

26. McELROY, M.B. (1988). A laser photolysis study of the reaction of SO_4^- with Cl^- and the subsequent decay of Cl_2^- in aqueous solution. Report TPRD/L/3274/R88, Central Electricity Generating Board, Leatherhead, UK.

27. EIGEN, M. and KUSTIN, K. (1962). The kinetics of halogen hydrolysis. J. Amer. Chem. Soc. 84, 1355-1360.

PHOTOTRANSFORMATION OF O-XYLENE
OVER ATMOSPHERIC SOLID AEROSOLS
IN THE PRESENCE OF O_2 AND H_2O

Juan CASADO, Jean-Marie HERRMANN and Pierre PICHAT
URA au CNRS "Photocatalyse, Catalyse et Environnement",
Ecole Centrale de Lyon,B.P.163, 69131 ECULLY Cedex, France

Summary

The photodegradation (300 < λ/nm < 400) of o-xylene over four oxides
which are present as aerosols in the atmosphere has been studied.
Ortho-xylene in the gas phase is stable over the SiO_2 and Fe_2O_3
samples used and is oxidised to o-tolualdehyde over TiO_2 and, to a
much lesser extent, over Al_2O_3. Poisoning of TiO_2 is observed. No
reaction occurs when o-tolualdehyde is used as a reactant instead
of o-xylene over TiO_2. In the presence of liquid water,
2-methyl-benzylalcohol, 2,3-dimethylphenol and 3,4-dimethylphenol
are produced (probably because of the formation of OH· radicals on
TiO_2) in addition to o-tolualdehyde, and all the aromatic compounds
progressively disappear , which suggests the mineralization of
o-xylene. No product of o-xylene is found when Al_2O_3 is used
instead of TiO_2 in liquid water.

1. INTRODUCTION

After toluene, xylenes are the most common aromatic pollutants in
urban atmospheres (1,2). They are known as highly toxic and potential
carcinogens(3). Therefore, the knowledge of the fate of these compounds
and in particular investigations of possible reactions in the adsorbed
phase, are deeply needed. Ortho-xylene photodegradation in the gas phase
has been studied under simulated atmospheric conditions by Atkinson et al.
(4-6 and refs. therein) who determined absolute rate constants for the
reaction of this pollutant with OH· radicals and found the occurrence of
both OH addition to the aromatic ring and H atom abstraction, mainly from
the CH_3 group, at room temperature (4). In the presence of NO_x (5-9),
tolualdehyde, dimethylphenols, dimethylnitrobenzene, maleic anhydride,
biacetyl, glyoxal, methylglyoxal, acetaldehyde, formaldehyde and
peroxyacetylnitrate, have been found and mechanistic schemes have been
suggested for their formation.
    The catalytic oxidation of o-xylene over metal oxides, in particular
V_2O_5, between 583 and 773 K leads eventually to phtalic anhydride (10,11).
Anatase-containing V_2O_5 shows the highest activity and selectivity to this
product (12).
    A relative enhancement in the photodegradation rate of o-xylene
and other atmospheric contaminants at room temperature in the presence of
TiO_2, especially at low relative humidities, has recently been reported
(13).
    However, to our knowledge, no extensive study has been carried out to
assess the role of various atmospheric oxides in the phototransformation
of o-xylene, which prompted us to undertake this research.

2. EXPERIMENTAL

SiO$_2$ (Degussa 200m^2/g, amorphous), TiO$_2$ (Degussa, 50 m^2/g, mainly anatase),Al$_2$O$_3$ (Degussa, 100m^2/g, mainly γ form), Fe$_2$O$_3$ (Merck, < 5m^2/g) were all nonporous. Ortho-xylene (Fluka, purum > 99% and o-tolualdehyde (Aldrich, 99%) were analysed by GC before use. Helium (Carboxique HP, He+Ne > 99.995%) and oxygen (l'Air Liquide N48, > 99.998%) were used as received. For the experiments in gaseous phase, we employed a differential flow-photoreactor with a fixed-bed of powder oxide deposited on a membrane, as previously described (14). Experiments in liquid phase were carried out with a static photoreactor containing magnetically stirred suspensions of the oxides (15).

A Philips HPK 125W mercury lamp was employed. The radiant flux in the reactors was 5mW cm-2 after filtering (Corning 7-60) to limit the transmitted wavelengths to the 300-400 nm range. Shorter wavelengths are absorbed by o-xylene (λ max = 254 nm) and are rare in tropospheric conditions. All experiments were carried out at room temperature. A water-circulating cuvette was used to remove IR wavelengths.

Procedure

The differential photoreactor was fed with mixtures of O$_2$ (ca. 140 Torr), o-xylene (5 Torr) and He as a carrier gas. Typical gas-flow was 20 ml/min. In each experiment the amount of oxide employed was sufficient to completely absorb the incident photons. Organic products were monitored by GC using a flame ionisation detector equipped with a 2.5 m x 2.2 mm column of 15% carbowax 1540 on chromosorb WAW, 80-100 mesh. CO$_2$ and H$_2$O were analyzed with a catharometer detector using a 3m x 6.35 mm column of Porapak Q, 50-80 mesh.

The static photoreactor contained 80 mg of TiO$_2$ dispersed in either 25 ml of a 2 x 10^{-3} M aqueous solution of o-xylene or 10 ml of pure o-xylene. An experiment was also carried out with a suspension of Al$_2$O$_3$(83mg) in 2 x 10^{-3}M o-xylene (25ml). The suspensions were in contact with air. Products were monitored by GC using a 15 m x 0.522 mm fused-silica megabore column impregnated with DB-5+.

3. RESULTS AND DISCUSSION

3.1. Gas-phase photooxidation over SiO$_2$, Al$_2$O$_3$, Fe$_2$O$_3$ or TiO$_2$

No organic product nor CO$_2$ was detected after illuminating for 24 h when SiO$_2$ or Fe$_2$O$_3$ were used as solid beds. Similarly, it has been reported that SiO$_2$ does not affect the photodegradation rate of a variety of hydrocarbons (13,16,17) and a negligible heterogeneous degradation has been detected for aromatics in the presence of Fe$_2$O$_3$ (16). Therefore our results confirm that these major components of atmospheric aerosols do not play an important role in the degradation of o-xylene. However, a definitive conclusion cannot be drawn, since various samples of each oxide have not been tried ; in particular, the photocatalytic properties of Fe$_2$O$_3$ depend very much on the pretreatment.

A slow photooxidation, leading to o-tolualdehyde, was detected over alumina. Under our experimental conditions, a steady-state conversion of ca. 0.02% was reached after illuminating for 27 h. No reaction occurred in the absence of UV-illumination. Photoconductivity of alumina, photoadsorption of oxygen on this oxide (18), as well as a weak photocatalytic activity in the oxidation (19) and the dehydrogenation (15, 20) of alcohols have previously been observed.

Over TiO$_2$, o-tolualdehyde was again the only product found. A maximum conversion of 1.1% was obtained within 40 min (fig. 1). Afterwards the conversion decreased steeply and at the end of the experiment (24 h) TiO$_2$

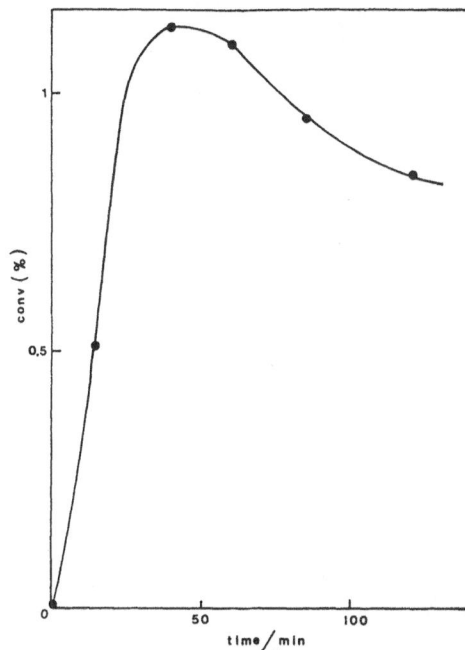

Figure 1. Conversion of o-xylene to o-tolualdehyde over TiO₂ in the gas-phase flow photoreactor versus illumination time.

was of a yellowish color. These features point to a poisoning of TiO_2 in the gas-phase reactor. However, no other product was discovered by capillary column GC analysis of the methanol extract of TiO_2.

The relative stability of o-tolualdehyde was shown by the absence of any product when the vapor of this aldehyde was used as a reactant instead of o-xylene over UV-illuminated TiO_2. However, o-toluic acid and other products of further oxidation of o-tolualdehyde were identified upon UV-illumination of TiO_2 dispersed in aerated liquid o-xylene in the static photoreactor, as will be reported elsewhere. Thus a slow photodegradation of o-tolualdehyde under atmospheric conditions cannot be ruled out.

The high activity of TiO_2 as a photocatalyst, owing to its semiconductor properties, is well known. Many compounds are photocatalytically oxidised over TiO_2 (21). It should be reminded that TiO_2 is present in the atmosphere in its pure state (22), even if it is in minor concentrations relative to those of other metal oxides. Therefore its impact on the fate of atmospheric organic pollutants cannot be neglected.

3.2. Aqueous-phase photo-oxidation in the presence of TiO_2 or Al_2O_3

In order to have an idea of the effect of water droplets on the degradation of o-xylene, experiments have been performed in a static photoreactor containing TiO_2 in 2×10^{-3}M aqueous solutions of o-xylene (pH 4.8). About 97% of the o-xylene introduced was adsorbed, which is not surprising in view of the hydrophobicity of aromatic compounds (the mixture of o-xylene and water was in fact an emulsion). The products

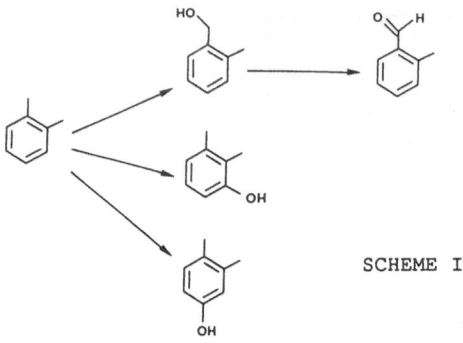

SCHEME I

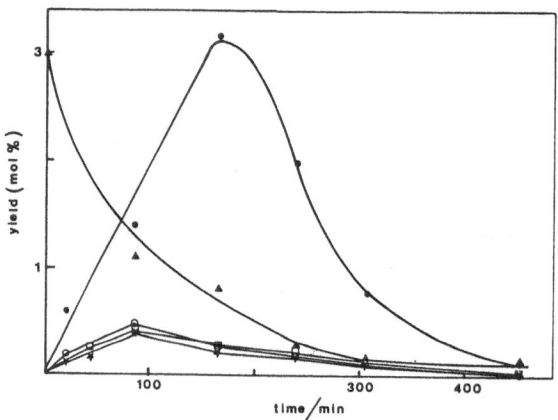

<u>Figure 2</u>. Kinetic profiles of aromatic compounds involved in
the photodegradation of o-xylene in aqueous TiO_2
suspension (see Scheme I):
▲ o-xylene ;
● o-tolualdehyde ;
○ 2-methylbenzylalcohol;
★ 2,3-dimethylphenol ;
□ 3,4-dimethylphenol.

obtained are shown in scheme I and their kinetics of appearance and
disappearance are presented in fig.2. All the aromatic products
progressively disappeared, which suggests the mineralization of o-xylene
in the presence of liquid water and TiO_2. The disappearance of o-xylene
does not show a first order kinetic, since after a certain reaction time
the rate slows down more than expected (fig.2), which could reflect
competition between the reactant and the intermediates. The same product
(plus methylbenzoic acid) were also detected by Hustert (23), under
similar conditions.The formation of phenols tends to show that OH·
radicals are involved. Water acts as a reactant in atmospheric oxidations
as already evidenced by the chalking of TiO_2-based pigments or polymer
charges (24).To sum up, TiO_2 is very active in the removal of aromatic
pollutants in the presence of water.

Another experiment which was carried out in aqueous medium with Al_2O_3 instead of TiO_2, showed that about 95% of o-xylene was adsorbed, but no product was found over an illumination period of 75 hours. This is not surprising, since there is no band-to-band transition of electrons in Al_2O_3 at the wavelengths employed, so that $OH\cdot$ radicals cannot be formed. These results also demonstrate that the presence of liquid water affects the mechanisms of photooxidation.

4.GENERAL CONCLUSION

Adsorbed o-xylene submitted to UV-illumination and oxygen can be either stable or transformed to o-tolualdehyde or completely degraded, depending not only on the adsorbent but also on the presence or absence of liquid water. This is obviously of great importance for the environment. These results contribute to demonstrate the necessity of systematic studies on the role of particulate oxides in the fate of chemicals in the atmosphere.

ACKNOWLEDGEMENTS

We thank Mr J. Disdier for his efficient technical asistance. A grant of the "Ministerio de Educacion y Ciencia" (Spain) to J.C. is acknowledged. This study was supported by the "Ministère de la Recherche et de la Technologie et le Secrétariat d'Etat à l'Environnement" (France) in the framework of the EUREKA-EUROTRAC-HALIPP program.

REFERENCES
(1) JAESCHKE, W., Ed. (1986). "The chemistry of multiphase atmospheric systems", Springer-Verlag, Berlin, 346.
(2) SEINFELD, J.H. (1989). Science, 43, 745.
(3) LAUNERYS, R. (1982). "Toxicologie industrielle et intoxications professionnelles". Ed. Masson, 2nd edition, Paris.
(4) PERRY, R.A., ATKINSON, R., PITTS Jr., J.N. (1977). J. Phys. Chem. 81, 296.
(5) ATKINSON, R. DARNALL, R., PITTS Jr., J.N. (1979). J. Phys. Chem. 83, 1943.
(6) TUAZON, E.C., Mac LEOD. H.,ATKINSON, R. CARTER W.P.L., (1986). Environ. Sci. Tech. 20, 383.
(7) TAKAGI, H., WASHIDA, N., AKIMOTO, H., NAGASAWA, K., USUI, Y., OKUDA, M., (1980). J. Phys. Chem. 84, 478.
(8) BANDOW, H., WASHIDA, N.,(1985). Bull Chem. Soc. Jpn. 58, 2541.
(9) SHEPSON, P.B. EDNEY, E.O., CORSE E.W. (1984). J. Phys. Chem. 88, 412.
(10) GERMAIN, J.E., (1969). "Catalytic conversion of hydrocarbons", Academic Press, New York, p. 256.
(11) BOAG, I.F. BACON, D.W., DOWNIE, J. (1975). J. Catal. 38, 375.
(12) GASIOR, M. GASIOR, I., GRZYBOWSKA, B. (1984). Appl. Catal. 10,87.
(13) BEHNKE, W., NOLTING, F., ZETZSCH, C. (1987). J. Aerosol. Sci. 18, 65.
(14) FORMENTI, M., JUILLET, F., TEICHNER, S.J. (1976). Bull. Soc. Chim. France, 1031.
(15) PICHAT, P., HERRMANN, J.M., DISDIER, J., COURBON, H., MOZZANEGA, M.N.(1981). Nouv. J. Chim. 5, 627.
(16) BEHNKE, W., NOLTING, F., ZETZSCH, C.(1986). Proceedings of the VIth inter. cong. of pesticide chemistry, Ottawa.
(17) GÜSTEN, H. (1986) in "The Chemistry of multiphase atmospheric systems", Jaeschke,W., Ed, Springer-Verlag, Berlin, 567.
(18) BALARD, H. MANSOUR, A.,PAPIRER, E., PICHAT, P. (1985). J. Chim. Phys. 82, 1051.
(19) MANSOUR, A., BALARD, H., PAPIRER, E. (1987). J. Chim. Phys. 84, 569.

(20) LAFRANCE, C.P., KALIAGUINE, S., ROBERGE, P.C., PICHAT, P. (1984) in "Catalysis on the energy scene", Elsevier, 309.

(21) See for example : PICHAT, P., FOX, M.A.(1988) in "Photoinduced electron transfer", Fox, M.A., Chanon, M., Eds., Elsevier, part D, pp. 241-302.

(22) FISHER, G.L., CHANG, D.P.Y., BRUMMER, M. (1976). Science 192, 553.

(23) HUSTERT, K., private communication.

(24) VÖLZ, H.G., KÄMPF, G. FITZKY, H.G., KLÄREN, A.(1981) : ACS Symposium Series n° 151, 163.

ETUDE EN ENCEINTE DE SIMULATION DE LA PHOTODEGRADATION DU NAPHTALENE EN PRESENCE DE PARTICULES.

P. FOSTER V.JACOB M.LAFFOND R.PERRAUD
Groupe de Recherche sur l'Environnement et la Chimie Appliquée
1, Rue F. Raoult. 38000 GRENOBLE

RESUME

L'étude générale entreprise vise à créer un modèle de réactivité physicochimique atmosphérique des HAP. Cette tentative de modélisation est effectuée en chambre réactionnelle. Le présent travail met en évidence l'importance d'intégrer dans les modèles de réactivité atmosphérique la notion de phase particulaire. En effet, nous avons mis en évidence que la présence d'un aérosol solide (ici muscovite, TiO_2 ou cendres volantes) induit, par rapport au milieu homogène correspondant, des modifications majeures d'ordre aussi bien cinétique que mécanistique. Nous avons notamment montré qu'en présence de radicaux OH., seul sensibilisateur du milieu réactionnel étudié, les cinétiques de réaction étaient ralenties en présence de différentes particules. Nous avons également mis en évidence que ces aérosols induisaient des phénomènes catalytiques conduisant à la formation de produits "nouveaux" non décelés en milieu homogène.

1. INTRODUCTION

Le devenir des hydorcarbures aromatiques polycycliques (HAP) reste une des préoccupations majeures des spécialistes de l'environnement. Leur nocivité, tant sur la santé que sur l'environnement, tout comme celle de certains de leurs produits de dégradation, leur confère un interêt particulier [1]; certains d'entre eux sont d'ailleurs désignés comme polluants prioritaires par l'EPA. De plus, les HAP réputés pour leur très grande réactivité [2,3] ont suscité un regain d'interêt du fait de l'amélioration des techniques expérimentales qui permettent aujourd'hui de les échantillonner et de les analyser de manière très précise [4,5,6,7]. De par ailleurs, le développement de méthodes non spécifiques de prélèvement et d'analyse de produits organiques à l'état de trace, ont permis d'accéder à la détection et à l'identification de composés sur un éventail aussi étendu que possible de produits.

Il est dès lors permis et intéressant de définir au mieux le processus selon lequel les HAP évoluent, et d'identifier les produits secondaires auxquels ils donnent naissance.

Mais la vitesse de dégradation de ces polluants primaires, comme la nature des sous-produits formés sont grandement tributaires du milieu dans lequel les HAP évoluent. En effet, les espèces initiatrices de leur transformation, mais également les caractéristiques physicochimiques du milieu réactionnel, telles que les particules, la température, l'irradiation, l'eau etc..., jouent un rôle primordial sur les phénomènes de dégradation.

En vue de modéliser la réactivité de quelques composés aromatiques à nombre de cycles carbonés réduit, tels que le naphtalène, l'anthracène, le phénanthrène, l'indane, l'indène, etc, nous avons entrepris un travail en enceinte de simulation, qui consiste à étudier la réactivité physicochimique d'un composé choisi comme modèle, à savoir le naphtalène.

Dans ce but les expériences suivantes ont été programmées:
- fiabilité de la technique expérimentale,
- dynamique du naphtalène dans l'enceinte de simulation,
- effet de l'irradiation sur la durée de vie du naphtalène,
- réactivité du naphatlène vis à vis de O_3, Cl_2, OH., NO_2, SO_2 pris séparément puis en mélange, sous irradiation.

Ces expériences effectuées dans un premier temps en phase homogène ont été réitérées en présence de particules de natures diverses. Ces dernières ont été choisies afin de recouvrir l'éventail possible de particules atmopshériques. Nous avons ainsi pris en compte la muscovite, le dioxyde de titane et les cendres volantes.

Le protocole expérimental employé (prélèvement et analyse) a été retenu d'une part pour sa fiabilité, et d'autre part pour sa simplicité et sa rapidité d'exécution. Notons toutefois que cette technique, qui se veut non exhaustive quant à la détection qualitative et quantitative des composés organiques, connait néanmoins quelques limites. En effet, l'analyse quantitative de composés très légers (contenant moins de trois carbones) ainsi que la détection de composés très lourds (HAP à plus de quatre noyaux) restent inaccessibles. Néanmoins, elle nous a permis de prendre en compte une multitude de produits provenant soit de la dégradation des composés étudiés, soit accompagnant initialement les produits que nous nous proposons d'étudier.

Au cours de ce travail, nous avons mis en évidence l'influence des divers paramètres pris en compte. Nous avons également établi des constatations importantes, en vue de la construction du modèle de réactivité des hydrocarbures polycycliques légers.

2. PARTIE EXPERIMENTALE

Nous disposons d'une enceinte de simulation, réacteur parallélépipédique de 2 m³, fermé et parfaitement agité, qui permet de créer l'atmosphère désirée [8]. Elle est équipée d'ouvertures permettant l'introduction de gaz, de liquides et de solides dans des concentrations voulues. Elle est munie d'un système d'irradiation qui comprend une lampe à vapeur de mercure d'une puissance de 700 Watts, ainsi qu'une batterie de douze néons d'une puissance totale de 400 Watts. L'irradiation que l'on reçoit au niveau du sol sous nos latitudes est ainsi approximativement recrée.

Un dispositif de prélèvement des gaz et des particules solides est prévu. Il est constitué d'une rampe de six prises d'échantillons. Sur chacune d'elle il est possible d'adapter : un filtre et une colonne de prélèvement.

D'autre part, l'accès à l'intérieur du réacteur est prévu par ouverture de sa base, ceci afin de recueillir les particules sédimentées et de permettre le nettoyage des parois.

L'échantillonnage des particules en suspension, en cours d'expérience est possible, mais il nécessite de prélever des volumes importants pour posséder suffisamment de matière pour l'analyse, ce qui induit des phénomènes perturbateurs du milieu réactionnel, de dilution notamment. Nous nous limitons donc volontairement, dans nos études à l'analyse de la fraction sédimentée, que l'on recueille en fin de simulation. D'ailleurs, ce choix se justifie par le fait que des études antérieures ont mis en évidence que la nature et la masse de produits analysés tant sur les particules en suspension que sur celles ayant sédimenté étaient pratiquement identiques.

L'échantillonnage de la phase gazeuse consiste en des prélèvements de deux litres du milieu réactionnel, collectés sur 100mg de tenax TA. La prise d'échantillon s'effectue de manière échelonnée sur une durée totale de simulation de six heures.

les échantillons gazeux et particulaires sont ensuite analysés en couplage Thermodésorption piègeage à froid- chromatographie gazeuse-spectrométrie de masse ou ionisation de flamme. Les rendements de thermodésorption sont estimés à 99%.

Dans les paragraphes qui suivent, nous allons donner les expériences effectuées et discuterons les résultats obtenus.

3. FIABILITE DE LA TECHNIQUE EXPERIMENTALE

Après étalonnage du chromatographe par injection de solutions étalons de naphtalène, nous avons testé la fiabilité du protocole expérimental.

L'introduction, dans l'enceinte réactionnelle de quantités croissantes connues de naphtalène nous a permis de vérifier :
- d'une part, la reproductibilité du système prélèvement-analyse
- d'autre part, la concordance avec la droite théorique (masse introduite = masse mesurée) que l'on a tracée.

Il s'avère que les concentrations mesurées sont identiques à celle introduites, et que la reproductibilité du protocole est de 95 %.

4. RESULTATS
4.1 DYNAMIQUE DU NAPHTALENE DANS L'ENCEINTE

Sous irradiation nulle, en présence ou en absence de particules (muscovite, TiO_2, cendres volantes), nous avons défini la dynamique du naphtalène dans l'enceinte de simulation. Par prélèvement d'aliquotes du milieu réactionnel, à des intervalles de temps choisis, nous avons établi que la loi cinétique de disparition du naphtalène est de la forme $d(N)/dt = k$, avec dans les différents cas de figure une constante identique: $k = 5,88 \ 10^7$ molécules $cm^{-3} s^{-1}$.

4.2 DUREE DE VIE DU NAPHTALENE SOUS IRRADIATION

En adoptant la procédure précédente, mais en irradiant le milieu, nous avons mis en évidence la non photodégradabilité du naphtalène.

4.3 REACTIVITE VIS A VIS D'ESPECES ACTIVES

Dans la mesure où nous avons cherché à recréer en simulation finale, des conditions proches de la réalité, nous avons choisi de travailler sur les espèces actives suivantes : NO_2, SO_2, Cl_2, O_3, OH.

Lors des premières simulations, la réactivité du naphtalène a été étudiée vis à vis des espèces prises isolément, puis vis à vis du mélange regroupant chacune d'entre elles. Ces expériences, effectuées dans un premier temps en phase homogène, ont été reproduites pour chaque type de particules étudié.

Elles nous ont permis :
- de déterminer la nature des sous-produits formés,
- de déterminer la vitesse de décomposition du naphtalène vis à vis de chaque espèce active,
- de calculer les constantes cinétiques de réaction, notamment par les techniques de cinétiques comparées,
- de déterminer les vitesses d'apparition et de disparition des produits principaux de réaction,
- de mettre en évidence le rôle des particules tant sur la nature des produits formés que sur les cinétiques de réaction.

Les résultats obtenus sont les suivants :
Les modifications inhérentes à la présence de particules diffèrent d'un sensibilisateur à l'autre. Elles portent toutefois tant sur les cinétiques de réaction que sur la mécanistique réactionnelle.

Nous avons tout d'abord pu constater que le naphtalène n'était dégradé ni par SO_2 ni par NO_2 durant six heures d'interaction, dans nos conditions expérimentales, que le milieu soit homogène ou hétérogène. En contre partie, en présence d'O_3, ou de OH., ou de Cl_2, le naphtalène disparait rapidement. La présence des diverses particules induit des modifications importantes, notamment de catalyse ou d'inhibition. Ces modifications sont d'autant plus marquées que le milieu est complexe et que les cinétiques sont rapides. Ces constatations sont illustrées par le milieu chloré où la constante cinétique de réaction (déterminée en émettant une hypothèse de dégénérescence de l'ordre) égale à $8,32\ 10^{-4}\ s^{-1}$ en absence de particules est environ deux fois inférieure à celles obtenues pour les différents milieux homogènes; dans le milieu où sont présents les mélanges d'oxydants, les modifications d'ordre mécanistiques sont les plus importantes. Nous ne traiterons ici qu'un seul exemple pour mettre en exergue ces phénomènes.

4.4 INTERACTIONS AVEC OH.

Les radicaux OH sont formés à partir de la photolyse du nitrite de méthyle, à une longueur d'onde supérieure à 290nm. L'enchaînement réactionnel est le suivant :

$$CH_3ONO + hu\ \text{-----}>\ CH_3O + NO$$
$$CH_3O + O_2\ \text{-----}>\ HCHO + HO_2$$
$$HO_2 + NO\ \text{-----}>\ OH. + NO_2$$

Afin de minimiser la formation de O_3 et de NO_3 pendant l'irradiation, NO est introduit dans le milieu réactionnel. Les concentrations initiales des diverses espèces sont les suivantes :

CH_3ONO : 10 ppm
NO : 5 ppm
Naphtalène : 300 ppb

4.5 ASPECT CINETIQUE

La constante de réaction du naphtalène avec les radicaux OH. est déterminée par la méthode des cinétiques comparées dont le principe est rappelé ici. Un composé référence, dont la constante cinétique de réaction avec les radicaux OH. est connue, est introduit dans l'enceinte réactionnelle à des concentrations voisines de celles du naphtalène.

En considérant que ces deux composés organiques réagissent exclusivement avec les radicaux OH., on peut écrire les réactions suivantes :

$$\text{Naphtalène} + OH. \xrightarrow{k1} P_1 \quad (1)$$
$$\text{Référence} + OH. \xrightarrow{k2} P_2 \quad (2)$$

En supposant que les phénomènes de dilution inhérents aux prélèvements sont négligeables, et que ni la référence ni le naphtalène ne sont photolysés, alors :

$$dLn(\text{Naphtalène})/dt = -k_1(OH.)$$

$$dLn(\text{Référence})/dt = -k_2(OH.)$$

Dans ces expressions k_1 et k_2 sont les constantes cinétiques des réactions 1 et 2. Ce système d'équations conduit à l'expression générale I:

$$d[(\text{Naphtalène})_{to}/(\text{Naphtalène})_t] = (k_1/k_2) \; \text{Ln}[(\text{Référence})_{to}/(\text{Référence})_t]$$

$(\text{Naphtalène})_{to}$ et $(\text{Référence})_{to}$ sont les concentrations du naphtalène et de la référence au temps to, $(\text{Naphtalène})_t$ et $(\text{Référence})_t$ étant leurs concentrations au temps t. Nous avons choisi de prendre comme référence le n-nonane dont la constante cinétique de réaction vaut $1,07 \; 10^{-8} \; cm^3 \; s^{-1}$ molécules $^{-1}$.

Le tracé de l'expression (I) doit être une droite de pente k_1/k_2 passant par l'origine;

5. RESULTATS
5.1 RESULTATS EN MILIEU HOMOGENE

Les données expérimentales recueillies sont en bon accord avec l'équation (I) (fig 1). La constante de disparition du naphtalène vaut $2,55 \; 10^{-11} \; cm^3 \; s^{-1}$ molécules $^{-1}$.
Ce résultat est comparé à ceux référencés dans la littérature et reportés dans le tableau suivant:

AUTEUR	REFERENCE ORGANIQUE	k s^{-1} molécule^{-1} cm^3
ATKINSON	n-NONANE	$2,42 \; 10^{-11}$
BIERMANN [9]	PROPENE	$2,35 \; 10^{-11}$
LORENZ [10]		$2,00 \; 10^{-11}$
GRECA	n-NONANE	$2,55 \; 10^{-11}$

5.2 RESULTATS EN MILIEU HETEROGENE

Nous avons dans un premier temps vérifié si la présence de particules modifiait la cinétique de réaction du nonane vis à vis des radicaux OH. pour déterminer s'il était possible d'effectuer des cinétiques comparées en utilisant la constante établie par ATKINSON en milieu homogène. Cette vérification a été effectuée en émettant l'hypothèse de dégénérescence de l'ordre par rapport aux radicaux OH..

Cette hypothèse conduit aux expressions générales suivantes :

$$dn/dt = -k_2(\text{nonane})(\text{OH.}) \quad (3)$$

En considérant la concentration en OH. constante tout au long de la simulation, l'équation 3 devient :

$$\text{Ln}[(\text{nonane})_t/(\text{nonane})_{to}] = -k_2(t-to)$$

Cette fonction a été tracée pour le milieu homogène ainsi que pour les trois milieux hétérogènes étudiés (fig 2). On a pu constater que les données expérimentales étaient en bon accord avec cette équation, et que la constante k'_2 était identique dans les quatre cas de figure envisagés.

Les particules ne modifient donc pas la cinétique de réaction du nonane avec les radicaux OH. La constante établie par ATKINSON a donc pu être utilisée dans la suite de ce travail.

En milieu hétérogène, les résultats restent toujours en excellent accord avec l'équation (I) (fig 3a, 3b,3c). Les constantes cinétiques calculées sont reportées dans le tableau suivant :

MILIEU	k cm^3 molécule^{-1} s^{-1}
HOMOGENE	$2,55\ 10^{-11}$
MUSCOVITE	$1,89\ 10^{-11}$
TiO$_2$	$2,11\ 10^{-11}$
CENDRES	$2,05\ 10^{-11}$

Nous pouvons constater que les particules tendent à ralentir les cinétiques de réactions du naphtalène avec les radicaux OH.

En considérant la concentration en OH. dans l'atmosphère égale à 1.10^8 molécules cm^{-3}, ces constantes cinétiques conduisent à des durées de vie du naphtalène dans l'atmosphère de :

MILIEU	DUREE DE VIE DANS L'ATMOSPHERE
HOMOGENE	9 HEURES
MUSCOVITE	7 HEURES
TiO$_2$	7,5 HEURES
CENDRES	7 HEURES

5.3 PRODUITS D'INTERACTION

Les tableaux suivants donnent la nature des produits obtenus lorsque le naphtalène est soumis seul à l'action des radicaux OH. Sont reportés les composés décelés dans la phase gazeuse de chaque milieu réactionnel ainsi que ceux adsorbés sur les différents types de particules.

- Composés identifiés dans les phases gazeuses :

HOMOGENE	MUSCOVITE	TiO$_2$	CENDRES
NAPHTALENE	NAPHTALENE	NAPHTALENE	NAPHTALENE
N.D	BENZENE	BENZENE	BENZENE
N.D	PHENYL ACETYLENE	N.D	N.D
BENZALDEHYDE	BENZALDEHYDE	BENZALDEHYDE	BENZALDEHYDE
TOLUALDEHYDE	TOLUALDEHYDE	TOLUALDEHYDE	TOLUALDEHYDE
NAPHTOQUINONE	NAPHTOQUINONE	N.D	NAPHTOQUINONE
ACETOPHENONE	N.D	ACETOPHENONE	ACETOPHENONE
N.D	N.D	BENZOPHENONE	N.D
PHENOL	PHENOL	PHENOL	PHENOL
ANHYDRIDE	ANHYDRIDE	ANHYDRIDE	ANHYDRIDE
PHTALIQUE	PHTALIQUE	PHTALIQUE	PHTALIQUE
N.D	N.D	N.D	ACIDE BENZOIQUE

- Composés adsorbés sur les différentes particules :

MUSCOVITE	TiO$_2$	CENDRES
NAPHTALENE	NAPHTALENE	NAPHTALENE
BENZENE	BENZENE	BENZENE
BENZALDEHYDE	BENZALDEHYDE	BENZALDEHYDE
N.D	N.D	TOLUALDEHYDE
FURFURAL	FURFURAL	FURFURAL
NAPHTOQUINONE	NAPHTOQUINONE	NAPHTOQUINONE
N.D	ACETOPHENONE	ACETOPHENONE
N.D	BENZOPHENONE	N.D
PHENOL	N.D	PHENOL
ANHYDRIDE	ANHYDRIDE	ANHYDRIDE
PHTALIQUE	PHTALIQUE	PHTALIQUE
ANHYDRIDE	ANHYDRIDE	ANHYDRIDE
ACETIQUE	ACETIQUE	ACETIQUE
PHTALIDE	PHTALIDE	PHTALIDE
PHTALATE	PHTALATE	PHTALATE
N.D	ACIDE BENZOIQUE	ACIDE BENZOIQUE

Deux types de produits sont obtenus au cours de ces simulations :
- les uns conservent la structure polyaromatique du naphtalène.
- les autres sont obtenus après rupture d'un des cycles du naphtalène.

Si parmi ces composés, la majorité d'entre eux se retrouvent dans les quatre simulations, certains sont spécifiques des milieux hétérogènes. Dans le premier cas de figure, la formation des produits est facilement justifiable par la réactivité classique des milieux homogènes. Dans le second cas, les mécanismes sont complexes et font appel à la réactivité en phase hétérogène qui est encore à l'heure actuelle très mal connue.

6. CONCLUSION

Nous avons constaté au cours de ce travail, que la présence de particules pouvait induire des modifications majeures dans l'évolution d'un milieu réactionnel. Ces modifications qui portent tant sur les cinétiques que sur les mécanismes de réaction, dépendent à la fois de la nature des particules et du type de sensibilisateur mis en jeu.

Nous avons constaté que les phénomènes de catalyse inhérents aux particules s'exprimaient par la présence de composés différents de ceux décelés en milieu homogène, mais également par une répartition différente des produits entre la phase gazeuse et la phase adsorbée. De plus, le rapport de certains produits majoritaires pris deux à deux et calculé pour la phase gazeuse et la phase particulaire, diffère d'un type de particules à l'autre.

Finalement, nous avons constaté à partir du bilan moléculaire que les particules agissent sur le taux de transformation du naphtalène.

TAUX DE TRANSFORMATION DU NAPHTALENE APRES SIX HEURES

MILIEU	Cl_2	OH.	Ox
HOMOGENE	99,9%	94,2%	99,6%
MUSCOVITE	99,9%	84,7%	99,5%
TiO_2	99,9%	93,8%	99,6%
CENDRES	99,2%	92,9%	99,6%

DEFICIT DU BILAN MOLECULAIRE APRES SIX HEURES

MILIEU	Cl_2	OH.	Ox
HOMOGENE	81,00%	69,80%	75,89%
MUSCOVITE	88,70%	19,46%	41,73%
TiO_2	92,53%	39,64%	93,97%
CENDRES	96,53%	44,50%	79,68%

REFERENCES

1- J.N. PITTS Jr., D.M. LOKENSGARD, P.S. RIPLEY,
   K.A. VAN CAUWENGERGHE, L. VAN VAECK, S.D. SHAFFER, A.J. THILL
   and W.L. BELSER Jr.
   Science 20: 1347-1349, 1980
2- T.L. GIBSON, P.E. KORSOG and G.T. WOLFF
   Atmospheric Environ. 20(8): 1575-1578, 1986
3- P. MASCLET, G. MOUVIER and K. NIKOLAOU
   Atmospheric Environ. 20(3): 439-446, 1986
4- B.B. WHEALS, C.G. VAUGHAN and M.J. WHITEHOUSE
   Chromatography 106: 109-118, 1975
5- J. MUELLER and E. ROHBOCK
   Talanta 27(8): 673-675, 1980
6- V. CANTUTI, G.P. CARTONI, A. LIBERTI and A.G. TORRI
   J. of Chrom. 17: 60-65, 1965
7- M. WITTENBERG, J. JAROSZ and L. PATUREL
   Anal. Chim. Acta 160: 185-196, 1984
8- P. DEHAULT
   Thèse de docteur ingégneur, 1983
9- H.W. BIERMANN, H. MAC LEOD, R. ATKINSON, A.M. WINER and
   J.N. PITTS Jr.
   Environ. Sci. Technol. 19(3): 244-248, 1985
10- R. ATKINSON and S.M. ASCHMANN
   Int. J. Chem Kinetics 18: 569-573, 1986

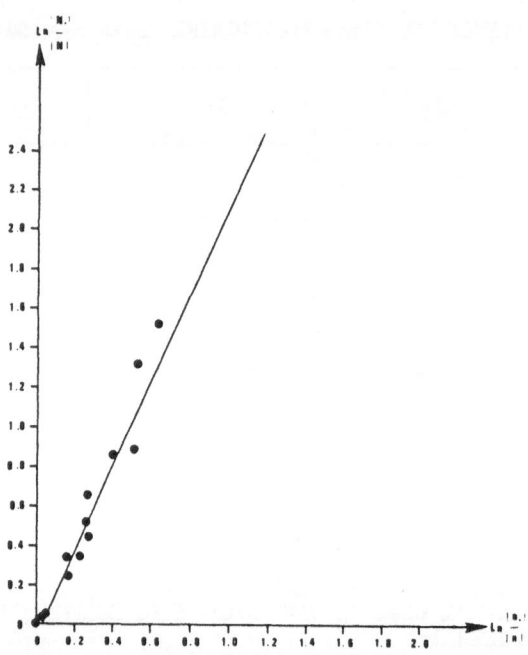

Figure 1: CINETIQUE COMPAREE EN MILIEU HOMOGENE

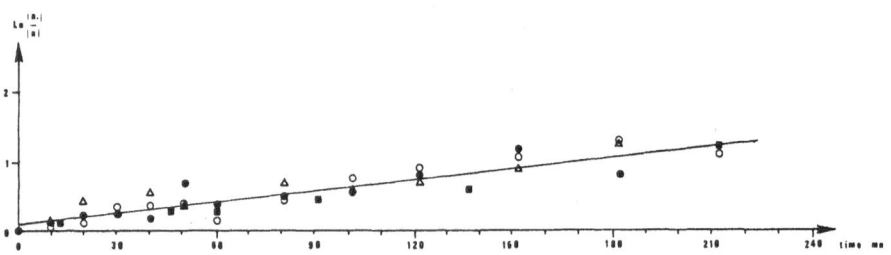

FIGURE 2: CINETIQUE DE DISPARITION DU n-NONANE EN MILIEU OH. (HYPOTHESE DI
DEGENERESCENCE DE L'ORDRE)

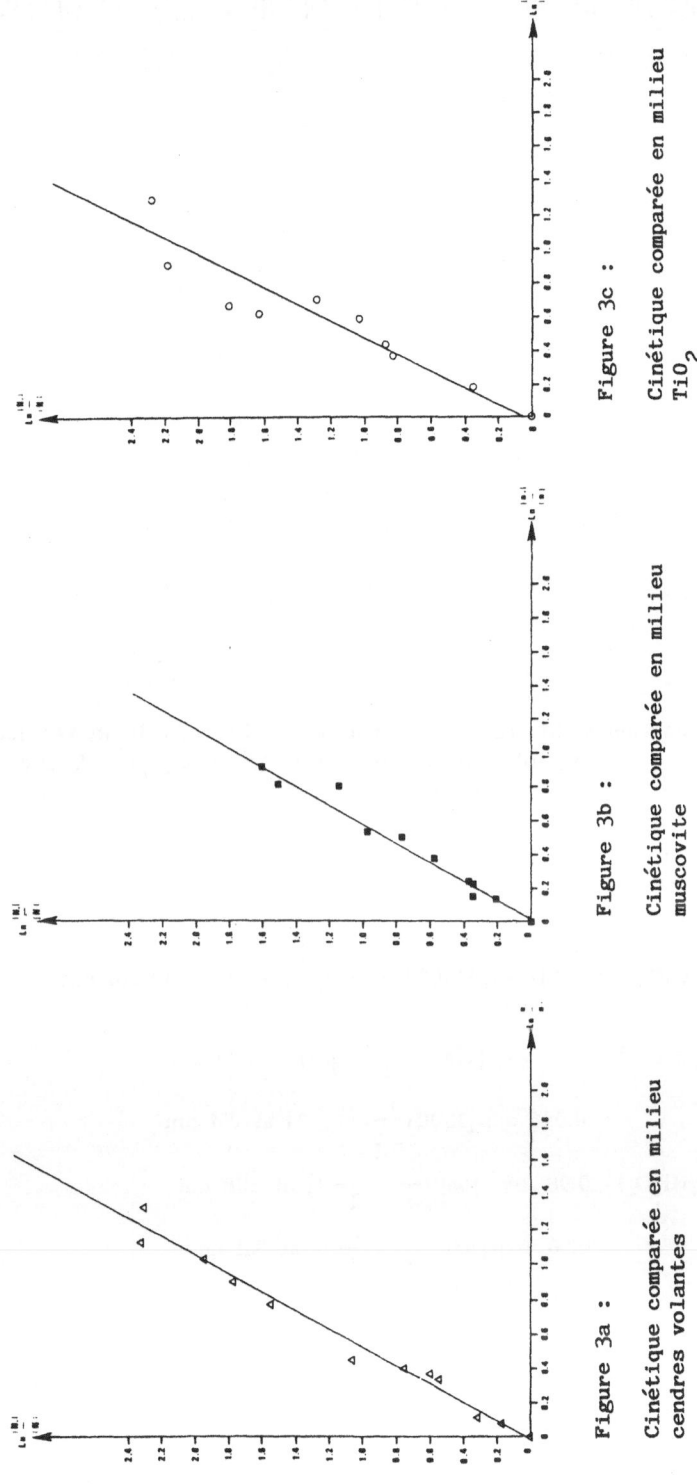

Figure 3a :

Cinétique comparée en milieu
cendres volantes

Figure 3b :

Cinétique comparée en milieu
muscovite

Figure 3c :

Cinétique comparée en milieu
TiO$_2$

FIGURE 3: CINETIQUE COMPAREE EN MILIEUX HETEROGENES

ABSOLUTE OH QUANTUM YIELDS IN THE LASER PHOTOLYSIS OF AQUEOUS SOLUTIONS OF NITRATE, NITRITE AND H$_2$O$_2$ AT 308 AND 351 NM.

R. ZELLNER

Institut für Physikalische Chemie und Elektrochemie,
Universität Hannover, 3000 Hannover

M. EXNER and H. HERRMANN
Institut für Physikalische Chemie,
Universität Göttingen, 3400 Göttingen

SUMMARY:

Absolute quantum yields for the formation of OH radicals in the laser photolysis of aqueous solutions of NO_3^-, NO_2^- and H_2O_2 at 308 and 351 nm and as a function of pH and temperature have been measured. A scavenging technique involving the reaction between OH and SCN^- ions and the time resolved detection by visible absorption of the $(SCN)_2^-$ radical ion was used to determine the absolute OH yields. The following results were obtained:

$$\Phi_{OH}(NO_3^-) = 0.017 \ \exp\left[1800 \ (\frac{1}{298} - \frac{1}{T})\right] \text{ at 308 and 351 nm}$$

$$\Phi_{OH}(NO_2^-) = 0.07 \ \exp\left[1560 \ (\frac{1}{298} - \frac{1}{T})\right] \text{ at 308 nm}$$

$$= 0.046 \ \exp\left[1800 \ (\frac{1}{298} - \frac{1}{T})\right] \text{ at 351 nm}$$

$$\Phi_{OH}(H_2O_2) = 0.98 \ \exp\left[660 \ (\frac{1}{298} - \frac{1}{T})\right] \text{ at 308 nm}$$

$$= 0.96 \ \exp\left[580 \ (\frac{1}{298} - \frac{1}{T})\right] \text{ at 351 nm}$$

Submitted to Vth European Symposium on Physico-Chemical Behaviour of Atmospheric Pollutants, Varese, September 1989

Together with the absorption coefficients and an assumed actinic flux within atmospheric droplets of 85% of the clear air value the partial photolytic lifetimes (τ_{OH}) of these molecules at 298 K are estimated as 24.8 d, 12.6 h and 35.9 h for NO_3^-, NO_2^- and H_2O_2, respectively. These lifetimes will increase by a factor of two (NO_3^-, NO_2^-) and by 15% (H_2O_2) at T = 278 K.

1. INTRODUCTION

Atmospheric trace gases soluble in aqueous droplet and with photochemical activity in the actinic wavelength region such as nitrate, nitrite and hydrogen peroxide, are potential sources of free radicals in the day-time troposphere.The radical species of primary interest in the tropospheric aqueous phase chemistry is the hydroxyl radical (OH).

It is widely accepted today [1] that OH radicals play an important role in oxidation processes in atmospheric aqueous systems (i.e. aerosols, clouds, fog and rain). Whereas oxidants such as ozone and H_2O_2 react with sufficient rates only with reagents such as S(IV)-compounds, there is a great variety of reagents including S(IV) and all soluble organic trace gases which can readily be oxidized in reactions with OH radicals. The product from these processes (i.e. S(VI), organic acids) will be removed from the troposphere by rain-out.

Present understanding of the sources of OH radicals in atmospheric aqueous droplets suggests that OH may either be scavenged from the ambient gas phase [2] or be formed via homogenous aqueous photo-dissociation processes [3]. The relative importance of these processes may be expected to be coupled in situation of high photo-chemical activity characterized by high gas phase OH concentrations and hence high gas phase formation rates of HNO_3, HNO_2 and H_2O_2. Under these circumstances both OH and the precursor molecules (HNO_3, HNO_2, H_2O_2) may be scavenged by the aqueous droplet. If on the other hand the liquid phase OH percursors are generated by heterogenous processes (i.e. hydrolysis of N_2O_5) we may expect the liquid phase photodissociation process to be the dominant OH source in the droplet.

Modelling of the OH source strength in aqueous droplets requires a knowledge of the quantum yields of the photodissociation processes of the main OH precursors within the actinic region of the solar spectrum. In the present work the absolute OH quantum yields for the processes

(1) $NO_3^- + H^+ + h\nu \longrightarrow OH + NO_2$

(2) $NO_2^- + H^+ + h\nu \longrightarrow OH + NO$

(3) $H_2O_2 + h\nu \longrightarrow 2\ OH$

have been determined at both 308 and 351 nm and as a function of pH and temperature.

2. EXPERIMENTAL

The experimental set-up used in this study is shown in figure 1. The photolysis light source is an excimer laser (Lambda Physik EMG 102) operating with XeCl at 308 nm (pulse energies typically 150 mJ) and with XeF at 351 nm (pulse energies typically 100 mJ). The laser beam is slightly focussed by a lens combination and allowed to illuminate the reaction cell through a beamsplitter and an aperture in front of the cell. The reaction cell is a cylindrical glass tube with an inner diameter of 18 mm and a length of 400 mm. It can be thermostated together with a reservoir volume by a heater/cooler-combination in the temperature range from 278 to 353 K. A small pump installed behind the cell allows the system to operate as a slow flow system. The analysing light source is a high pressure compact arc mercury xenon lamp (Hanovia), the light of which is focussed through the reaction cell and via the beamsplitter and a second lens onto the entrance slit of a 3 m monochromator (McPherson 218). To avoid photolysis of the solution by the analysing light beam a pair of UV blocking filters is placed between the lamp and the reaction cell.

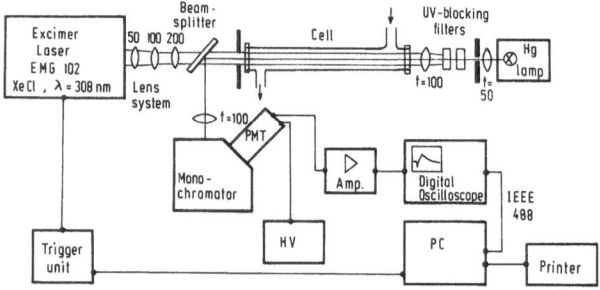

Fig. 1: Schematic representation of experimental set-up

The monochromator is equipped with an EMI 9789 QB photomultiplier tube. The PMT output is amplified and monitored with a 20 MHz digital storage oscilloscope (Gould 4035). The data of every single experiment are transferred via an IEEE bus to a pesonal computer. The computer is also connected via an RS 232 bus with a trigger device which allows the computer to trigger the laser after having received the oscilloscope data of the previous experiment. To improve the signal-to-noise-ratio typically 100 experiments were performed and averaged by the computer. The laser pulse energy was measured with a Gentec energy-meter before every set of experiments.

For the determination of the OH yields a method was chosen which makes use of the fast reaction of thiocyanate ion with OH radicals [4]. The simplified mechanism of this scavenging technique may be described as:

(a) $SCN^- + OH \longrightarrow SCN + OH^-$
(b) $SCN + SCN^- \longrightarrow (SCN)_2^-$
(c) $(SCN)_2 \cdot {}^- \longrightarrow$ products.

Its essential feature is the possibility of detection of $(SCN)_2 \cdot {}^-$ by time resolved absorption at 475 nm ($\mathcal{E}$ = 7650 l/mol·cm [5]. Fig 2. shows a typical absorption trace obtained following the photolysis of nitrite at 308 nm.

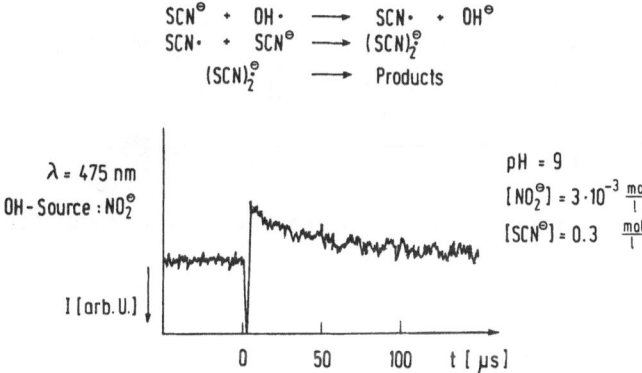

Fig. 2: Time resolved absorption profile of $(SCN)_2^-$ obtained following the photolysis of NO_3^- at 308 nm

In order to obtain initial OH concentration generated by the laser pulse the absorption profiles of $(SCN)_2 \cdot {}^-$ are extrapolated back to t = 0 and the stoichiometric relation $\Delta[SCN_2 \cdot {}^-] = \Delta[OH]$ is applied. Together with the number density of photons absorbed in the cell (ΔN) as determined from measurements of the laser pulse energy and the optical density of the solution the absolute OH quantum yields, $\Phi_{OH} = \Delta[OH]/\Delta N$, can then be calculated. Details of the experimental method and its analysis can be found elsewhere [6].

3. RESULTS AND DISCUSSION

All experiments were performed in solutions containing 3×10^{-3} mol/l of the precursor and 0.3 mol/l of the scavenger. The pH-value was adjusted using diluted $HClO_4$ or KOH-solutions. The results obtained for the photolysis at 308 nm are presented in Arrhenius form in Fig. 3.

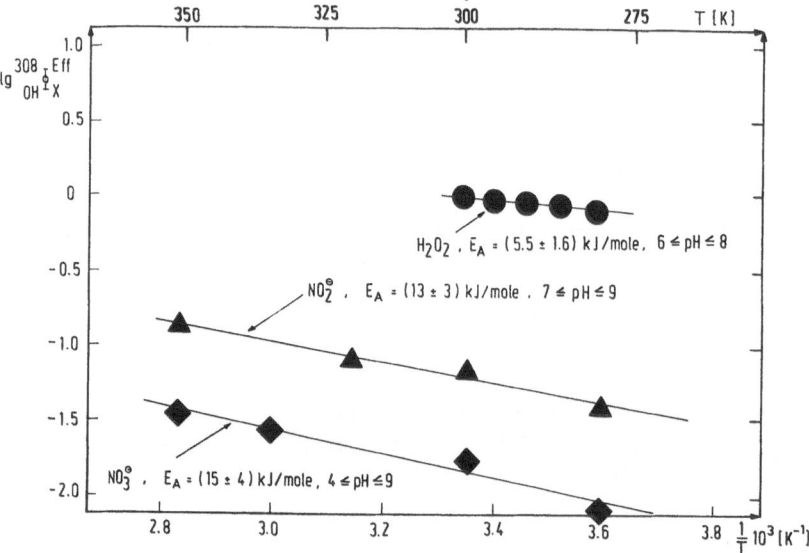

Fig. 3: Arrhenius representation of absolute OH-quantum yields in the photolysis of NO_3^-, NO_2^- and H_2O_2 at 308 nm.

(1) NO_3^--photolysis

The OH quantum yield in the photolysis of NO_3^- at 298 K and for 308 nm has been determined to be

$$\Phi_{OH}(298K) = 0.017 \pm 0.03$$

essentially independent on pH in the range $4 \leq pH \leq 9$.

There have been previous determinations [7,8] of this quantity, both using steady-state photolysis techniques with OH scavengers and final product analysis. Whereas Zepp et al. [7] obtained $\Phi_{OH} = 0.015$ consistent with our work, Warneck and Wurzinger [8] obtained a lower value of the OH quantum yield, $\Phi_{OH} = 0.0092$. The discrepancy with the latter data is presently unexplained.

The effect of temperature variation on the OH-quantum yield has been found to be relatively strong. From the data presented in Fig. 3 we obtain

$$\Phi_{OH}(T) = 0.017 \exp\left[1800(\frac{1}{298} - \frac{1}{T})\right]$$

corresponding to an apparent activation energy of 15±4 kJ/mol. As a consequence Φ_{OH} increase by almost a factor of 5 over the temperature region 278-353K.

(ii) NO_2^- -photolysis

The OH quantum yield in the photolysis of nitrite at room temperature has been determined to be

$$\Phi_{OH}(298K) = 0.07 \text{ and } \Phi_{OH} = 0.046$$

at 308 and 351 nm, respectively. Unlike NO_3^- the photolysis of NO_2^- therefore shows a strong decline of OH quantum yield with increasing wavelength. These results are entirely consistent with the recent findings of Zafiriou and Bonneau [4].
With regards to the temperature dependence of Φ_{OH} we obtain

$$\Phi_{OH}(T) = 0.07 \exp\left[1560\left(\frac{1}{298} - \frac{1}{T}\right)\right]$$

and

$$\Phi_{OH}(T) = 0.046 \exp\left[1800\left(\frac{1}{298} - \frac{1}{T}\right)\right]$$

at 308 and 351 nm, respectively. Therefore, the T-dependence of Φ_{OH} is essentially independent on wavelength and very similar to that found for NO_3^-.

(iii) H_2O_2-photolysis

The OH-quantum yield in the photolysis of H_2O_2 at 308 nm has been found to be

$$\Phi_{OH}(298K) = 0.98$$

with a temperature dependence of

$$\Phi_{OH}(T) = 0.98 \exp\left[660\left(\frac{1}{298} - \frac{1}{T}\right)\right]$$

The corresponding values at 351 nm are:

$$\Phi_{OH}(298K) = 0.96$$

and

$$\Phi_{OH}(T) = 0.96 \exp\left[580\left(\frac{1}{298} - \frac{1}{T}\right)\right].$$

As a consequence we conclude that the photolysis of aqueous H_2O_2 solutions at wavelengths > 300 nm produces OH radicals with nearly the maximum possible quantum yield and without negligible cage recombination. This result is not different from that of previous studies [9-11]. Although in each of these investigations quantum yields for the decay of H_2O_2 of unity - corresponding to $\Phi_{OH} = 2$ - were observed, it was concluded that due to apparent chain reactions the quantum yields should only be around 0.5. This is substantiated from the present work.

(iv) Atmospheric implications

The results obtained in their study together with the integrated absorptions of these species and the solar flux in the actinic wavelength region may be used to calculate photolysis frequencies (J) and photochemical lifetimes ($\tau=J^{-1}$). We obtain

	J/s^{-1}	τ
NO_3^-:	4.7×10^{-7}	24.8 d
NO_2^-:	2.2×10^{-5}	12.6 h
H_2O_2:	3.9×10^{-6}	71.8 h

As a consequence of the strong increase of the NO_2^- absorption near 350 nm, nitrite has by far the shortest lifetime of these species. These results apply to 298 K. Due to the temperature dependence of the quantum yields these lifetimes will increase by a factor of two (NO_3^-, NO_2^-) and by 15% (H_2O_2) at 278 K. The consequences of these photodissociation processes with regards to the source strength of OH formation in aqueous atmospheric droplets will be discussed elsewhere [6].

REFERENCES

(1) FINLAYSON-PITTS, B. and PITTS, J.N., (1986) "Atmospheric Chemistry", Wiley, New York

(2) CHAMEIDES, W. and DAVIS, D.D., (1982), J. Geophys. Res. 87, 4863

(3) GRAEDEL, T.E. and WESCHLER, C.J., (1981), Rev. Geophys. Space Phys, 19, 505

(4) ZAFIRIOU, O.C. and BONNEAU, R., (1987), Photochem. Photobiol., 45, 723

(5) BEHAR, D., BEVAN, P.L.T. and SCHOLES, G., (1972), J. Phys. Chem. 76, 1537

(6) ZELLNER, R., EXNER, M., and HERRMANN, H., (1989), J. Atmosph. Chemie, submitted for publication

(7) ZEPP, R.G., HOIGNE', J. and BADER, H., (1987), Environm. Sci. Technol. 21, 443

(8) WARNECK, P. and WURZINGER, C., (1988), J. Phys. Chem. 92, 6278

(9) HUNT, J.P. and TAUBE, H., (1952), J. Am. Chem. Soc. 74, 5999

(10) BAXENDALE, J.H. and WILSON, J.A., (1956), Trans. Far. Soc. 53, 344

(11) VOLMAN, D.H. and CHEN. J.C., (1959) J. Am. Chem. Soc. 81, 4141

SESSION II/C

ATMOSPHERIC CHEMICAL

AND

PHOTOCHEMICAL PROCESSES

JOINT EUROTRAC – COST 611

PROJECT LACTOZ

REACTIONS OF OH RADICALS WITH ALKYL NITRATES

O.J.NIELSEN
Chemistry Department, Risø National Laboratory, DK-4000 Roskilde, Denmark

M.DONLON and H.W.SIDEBOTTOM
Department of Chemistry, University College Dublin, Dublin, Ireland

J.J.TREACY
Department of Chemistry, Dublin Institute of Technology, Dublin, Ireland

SUMMARY

Rate constants for the reaction of OH radicals and Cl atoms with a series of n-alkyl nitrates have been determined at $295 \pm 3K$ and a total pressure of approximately 1 atm. The OH rate data were obtained using both the absolute rate technique of pulse radiolysis combined with kinetic spectroscopy and a conventional relative rate method. Chlorine atom rate constants were measured using only the relative rate method. The results of the present study indicate that at atmospheric pressure rate data for the reaction of OH radicals with alkyl nitrates from both direct and relative rate experiments are in good agreement. The data show that the nitrate group substantially decreases the rate constant for H atom abstraction from groups bonded to the $-ONO_2$ groups and also decrease those for groups in the β position. Similar results were found for the reaction of Cl atoms with these compounds. Evidence is also presented for an addition reaction of OH radicals with the nitrates.

1. INTRODUCTION

Alkyl peroxy radicals are important intermediates involved in the oxidation of volatile organic compounds in the troposphere [1-3]. The major sink process for alkyl peroxy radicals under polluted atmospheric conditions is reaction with nitric oxide. Although methyl peroxy radicals react almost exclusively to form NO_2 and the methoxy radical [4] reactions of larger peroxy radicals with NO have been shown to form stable alkyl nitrates in significant yields [5-7]. These compounds provide temporary reservoirs for nitrogen oxides and are expected to be involved in long range transport of NO_x. Reaction of these organonitrogen compounds with hydroxyl radicals has been suggested as an important factor in their atmospheric residence times, however, to date little mechanistic information concerning the formation and release of NO_x from these reservoirs is available. Further, there appears to be a major inconsistency between the direct measurements and relative rate techniques in the reported rate constants for the reaction of OH radicals with methyl nitrate. It would appear that at the higher pressures pertaining in relative rate studies [8] the measured rate constant is about an order of magnitude higher than that found using a low pressure discharge flow / resonance fluorescence technique [9]. Relative rate data for the reaction of OH radicals with a series of higher linear and branched chain alkyl nitrates suggests that the $-ONO_2$ group significantly decreases the reactivity relative to the corresponding unsubstituted alkanes [7,10].

The purpose of this work was to determine relative and absolute rate data for OH radical reactions with alkyl nitrates at atmospheric pressure. In order to provide more mechanistic information for these reactions rate constants for the reaction of Cl atoms with the alkyl nitrates were determined using the relative rate method.

2. EXPERIMENTAL

The apparatus for the pulsed radiolysis experiments has been described in detail elsewhere [11]. A single 30ns pulse of 2MeV electrons from a Febetron 705B field emission accelerator was used to irradiate mixtures of 12 torr H_2O and Ar at a total pressure of 1 atmosphere. Hydroxyl radicals are formed by collision with excited argon atoms produced via pulse radiolysis.

$$Ar^* + H_2O \qquad \rightarrow \qquad OH + H + Ar \qquad (1)$$

The kinetics of OH radicals were studied by monitoring the transient absorbance at 309 nm using a modified version of Beer's Law, $A = (\epsilon c l)^n$. This version of Beer's Law is required when the spectral band path is wide compared to the line width of the spectral features. The value of n was determined from the function $\log A = n \log \epsilon c l$ by vaying the optical path length and carrying out a linear regression best fit analysis. For a spectral band pass of 0.08 nm, a value of n = 0.70 was determined. The observed transient absorbance is a direct measure of the OH radical concentration at any time during the decay. The irradiation stainless steel cell has a volume of 1 litre and was mounted directly onto the accelerator. The analysing light source was a pulsed 150W high pressure xenon arc lamp coupled to a Hilger and Watts grating spectrograph. A Hamainatsu photo-multiplier and a Biomation 8100 waveform digitizer were used to detect and record the light intensity at the desired wavelength. Transfer and storage of raw data on a mainframe computer was controlled by a PDP11 minicomputer. The alkyl nitrates were premixed with Ar in a 50 litre FEP Teflon bag. Different partial pressures of the mixtures were added to 12 torr of water in the cell and made up with argon to a total pressure of 1 atm. All pressure measurements were made with an MKS Baratron 170 capacitance manometer.

The relative rate experiments were carried out at 730-750 torr in a 50 litre FEP Teflon cylindrical reaction chamber. Pressure measurements were made using an MKS Baratron 220A capacitance manometer. Measured amounts of the reactants were flushed from Pyrex bulbs by a stream of ultra-zero air (BOC) or ultra-pure nitrogen (BOC) into the reaction chamber, which was then filled to the full volume. All quantitative analyses were carried out by gas chromatography. Hydroxyl radicals were generated by photolysis of methyl nitrite in air at wavelengths longer than 290 nm:

$$CH_3ONO + h\nu \qquad \rightarrow \qquad CH_3O + NO \qquad (2)$$

$$CH_3O + O_2 \qquad \rightarrow \qquad CH_2O + HO_2 \qquad (3)$$

$$HO_2 + NO \qquad \rightarrow \qquad OH + NO_2 \qquad (4)$$

Irradiation was provided by 20 fluorescent lamps (10 blacklamps, Philips TL 20/08 and 10 sunlamps, Philips TL 20/09) giving a photolytic half-life for CH_3ONO of approximately 30 minutes. Chlorine atoms were generated from the photolysis of molecular chlorine.

Argon (AGA, 99.9%), chlorine (Matheson, 99.9%) and nitric oxide (BOC, 99.9%) were used as received. In all cases the H_2O was triply distilled. Methyl nitrite was synthesized from sodium nitrite and methanol. Strong acid nitration of the corresponding alcohol was used to prepare the alkyl nitrates. The nitrates and methyl nitrite were prepared immediately prior to their use. They were purified by trap-to-trap distillation with a middle fraction being retained and stored in the dark. The purity of each compound was checked by gas chromatography and infrared spectroscopy

3. RESULTS

The absolute reaction rates of alkyl nitrates with OH radicals were determined in a large excess of substrate. In this case simple first-order kinetics will be obeyed:

$$OH + S \qquad \rightarrow \qquad products \qquad (5)$$

$$\ln([OH]_o/[OH]_t) \quad = \quad k_5[S]t \quad = \quad k't$$

For the experimental conditions employed, all decays were exponential over at least three half-lives and the bimolecular rate constants k_5 were obtained from the slopes of k' versus [S] plots. The rate constants derived from the linear least squares fits of the straight lines, (error limits are $\pm 2\sigma$, are given in Table I.

In the relative rate studies OH radicals or Cl atoms are generated in the presence of the nitrate compound and a reference hydrocarbon.

$$OH + S \qquad\qquad \rightarrow \qquad products \qquad\qquad\qquad (5)$$

$$Cl + S \qquad\qquad \rightarrow \qquad products \qquad\qquad\qquad (6)$$

$$OH + RH \qquad\qquad \rightarrow \qquad products \qquad\qquad\qquad (7)$$

Providing the nitrate and the reference hydrocarbon are consumed only by reaction with OH radicals or Cl atoms k_5 and k_6 can be determined according to

$$\ln([S]_0/[S]_t) = k_5/k_7 \ln([RH]_0/[RH]_t) \qquad\qquad (8)$$

$$\ln([S]_0/[S]_t) = k_6/k_7 \ln([RH]_0/[RH]_t) \qquad\qquad (9)$$

Concentration-time data from the relative rate OH and Cl experiments plotted in the form of equations (8) and (9) gave straight line plots passing through the origin as predicted. The derived OH and Cl rate constants are given in Table I. The error limits are $\pm 2\sigma$ of the fit and the rate constant values are based on $k_7(OH + C(CH_3)_4) = 8.52 \times 10^{-13}$ cm^3 molecule^{-1} s^{-1} [12] and $k_7(Cl + C_2H_6) = 6.38 \times 10^{-11}$ cm^3 molecule^{-1} s^{-1} [13].

4. DISCUSSION

Rate constant data from pulse radiolysis and relative rate experiments for the reaction of OH radicals with n-alkyl nitrates at 1 atm total pressure are in good agreement. The rate constant for reaction with methyl nitrate is close to that previously reported for the OH+C_2H_6 reaction [12]. This suggests that replacement of a methyl by a nitrate group in ethane has very little effect on the reactivity with OH radicals. This result is somewhat unexpected, since if this reaction involved mainly H atom abstraction in the presence of the strongly electron withdrawing -ONO_2 group would be expected to considerably decrease the rate constant for abstraction by the electrophilic OH radical. The rate constant for the reaction of Cl atoms with CH_3ONO_2 determined in this work shows a decrease of about two orders of magnitude compared to the value for the Cl+C_2H_6 reaction [13]. The reaction of electrophilic Cl atoms with substituted organics is believed to involve a hydrogen atom abstraction process, hence the lack of any reduction in reactivity for the corresponding OH radical reaction indicates some mechanistic difference between the two systems. It is of interest to compare the results for the reactions of OH radicals and Cl atoms with CH_3ONO_2 to those obtained for the reactions with CH_3CN. Poulet et al. [14] have shown that the reactions of both OH and Cl with CH_3CN involve only H atom abstraction. In this case the electron withdrawing CN group greatly reduces both rate constants compare to those for the reaction with C_2H_6 despite a reduction in C-H bond dissociation energy of about 5 kcal mol^{-1} [15].

The reaction between OH radicals and methyl nitrate may involve both hydrogen abstraction and OH addition followed by rapid decomposition of the adduct.

$$OH + CH_3ONO_2 \qquad \rightarrow \qquad CH_2ONO_2 + H_2O \qquad\qquad (10)$$

$$OH + CH_3ONO_2 \rightarrow \quad CH_3O(OH)NO_2 \rightarrow \quad CH_3O + HNO_3 \quad (11)$$

From the above kinetic evidence it is proposed that the reaction of OH radicals with CH_3ONO_2 proceeds mainly via the addition channel at atmospheric pressure. Some support for this argument comes from the measurement of $k(OH + CH_3ONO_2)$ reported by Gaffney et al. [9] at pressures in the region of 2-3 torr. In this pressure region the rate constant is about an order of magnitude lower than observed at 1 atm in the present work. It is suggested that under low pressure conditions the $OH + CH_3ONO_2$ reaction involves an abstraction process whereas at higher pressures the addition complex is stable and the major channel is addition.

Further evidence for the addition channel can be seen from a comparison of the rate data for OH radicals and Cl atoms with the higher members of the alkyl nitrate series. Since both OH and Cl are electrophilic species a linear relationship between the logarithm of their rate constants with the alkyl nitrates would be expected if both sets of reactions involved the same reaction mechanism. Fig. 1 shows a free energy plot of the relevant rate constant data. The plot shows a distinct curvature indicating that the two systems are not entirely analogous. If it is assumed that addition is the major channel for the $OH + CH_3ONO_2$ reaction at atmospheric pressure, the rate constant for the abstraction process for each member of the higher n-alkyl nitrates may be estimated. A value of k_{add} for the $OH + CH_3ONO_2$ reaction may be derived from the available experimental data. Taking $k_{abs} = 0.34 \times 10^{-13}$ cm^3 $molecule^{-1}$ s^{-1} from the low pressure study of Gaffney et al. [9] a value of $k_{add} = 2.9 \times 10^{-13}$ cm^3 $molecule^{-1}$ s^{-1} is calculated for CH_3ONO_2 ($k_{obs} = k_{add} + k_{abs}$). It is unlikely that the length of the alkyl chain will influence the rate constant for the addition process. Hence the rate constant for abstraction for the n-alkyl nitrates can be derived from $k_{abs} = (k_{obs} - 2.9 \times 10^{-13})$ cm^3 $molecule^{-1}$ s^{-1}. The resulting linear free energy plot using the rate constant data for H atom abstraction by OH radicals from alkyl nitrates calculated in this way is shown in Fig. 1. There now appears to be an essentially linear correlation between the reactivities of OH radicals and Cl atoms towards the alkyl nitrates as expected if both processes involve direct H atom abstraction.

It is apparent from the rate constant data for hydrogen atom abstraction from n-alkyl nitrates by OH radicals that the $-ONO_2$ group has a strong deactivating effect for abstraction of H atoms positioned α and β to the $-ONO_2$ group. Bond dissociation energies are not available for these C-H bonds but it is unlikely that the $-ONO_2$ group will change the C-H bond energies greatly from those found in unsubstituted alkanes. Thus the reduction in reactivity at the βC atom and at least in part of the αC atom must result from inductive effects in the transition state involving polar repulsion between the electrophilic OH radical and the abstracted H atom.

The atmospheric lifetimes of the n-alkyl nitrates due to loss by reaction with OH radicals can be calculated from the rate constant data determined in this study. Assuming a tropospheric concentration of 1×10^6 molecules cm^{-3} for the OH radical, the relatively low reactivities of the nitrates give lifetimes in the range ca. 30 days for CH_3ONO_2 to ca 3 days for $n-C_5H_{11}ONO_2$. An estimate of the photodissociative lifetime for CH_3O-NO_2 of approximately 5 days has been reported in the lower troposphere based on absorption cross section data [16]. This estimate is based on a quantum yield for photo-decomposition of unity and is hence a lower limit. Thus these nitrates may act as temporary sinks for nitrogen in the troposphere and be important in long-range transport of odd nitrogen.

REFERENCES
1.   DEMERJIAN, K.L., KERR, J.A. and CALVERT, J.G., Adv.Environ, Sci. Technol., 4, 1 (1974)
2.   FINLAYSON-PITTS and B.J., PITTS, J.N. JR., Adv.Environ, Sci.Technol., 7,75 (1977)
3.   CARTER, W.P.L., LLOYD, A.C., SPRUNG, J.L. and PITTS, J.N. JR., Int.J.Chem.Kinet., 11, 45 (1979)

4. RAVISHANKARA, A.R., EISELE, F.L., KREUTTER, N.M. and WINE, P.H., J.Chm.Phys., 74, 2267 (1981)
5. ATKINSON, R., ASCHMANN, S.M., CARTER, W.P.L., WINER, A.M. and PITTS, J.N. JR., J.Phys.Chem., 86, 4563 (1982)
6. ATKINSON, R., CARTER, W.P.L. and WINER, A.M., J.Phys.Chem., 87, 2012 (1983)
7. ATKINSON, R., ASCHMANN, S.M., CARTER, W.P.L., WINER, A.M. and PITTS, J.N. JR., Int.J.Chem.Kinet., 16, 1085 (1984)
8. KERR, J.A. and STOCKER, D.W., J.Atmos.Chem., 4, 253 (1986)
9. GAFFNEY, J.S., FAJER, R., SENUM, G.I. and LEE, J.H., Int.J.Chem.Kinet., 18, 399 (1986)
10. ATKINSON, R., ASCHMANN, S.M., CARTER, W.P.L. and WINER, A.M. Int.J.Chem.Kinet., 14, 919 (1982)
11. HANSEN, K.B., WILBRANDT, R. and PAGSBERG, P., Rev.Sci.Instr., 50, 1532 (1979)
12. ATKINSON, R., Chem.Rev., 86, 69 (1986)
13. ATKINSON, R. and ASCHMANN, S.M., Int.J.Chem.Kinet., 17, 33 (1985)
14. POULET, G., LAVERDET, G., JOURDAIN, J.L. and LE BRAS, G., J.Phys.Chem., 88, 6259 (1984)
15. McMILLEN, D.F. and GOLDEN, D.M., Ann.Rev.Phys.Chem., 33, 493 (1982)
16. TAYLOR, W.D., ALLSTON, T.D., MOSCATO, M.J., FAZEKAS, G.B., KOZLOWSKI, R and TAKACS, G.A., Int.J.Chem.Kinet., 12, 231 (1980)

TABLE I: Rate constants for the reaction of OH radicals and Cl atoms with n-alkyl nitrates at room temperature and 1 atm[a].

Compound	$10^{13}\,k_{Cl}$(rel)	$10^{13}\,k_{OH}$(rel)	$10^{13}\,k_{OH}$(abs)
CH_3ONO_2	2.62 ± 0.02	3.3 ± 0.7 3.7 ± 0.9[b]	3.2 ± 0.5 0.34 ± 0.04[c]
$C_2H_5ONO_2$	42.7 ± 1.6	4.5 ± 0.2 4.8 ± 2.0[b]	5.3 ± 0.6
$n-C_3H_7ONO_2$	247 ± 15	7.6 ± 0.8 7.0 ± 2.2[b]	8.2 ± 0.8
$n-C_4H_9ONO_2$	924 ± 20	15.8 ± 0.7 13.9 ± 1.1[d]	17.4 ± 1.9
$n-C_5H_{11}ONO_2$	1571 ± 33	28.9 ± 0.9	33.2 ± 3.0

(a) In units of cm^3 molecule^{-1} s^{-1}
(b) Taken from ref (8)
(c) Taken from ref (9) at a total pressure of 2-3 torr
(d) Taken from ref (10)

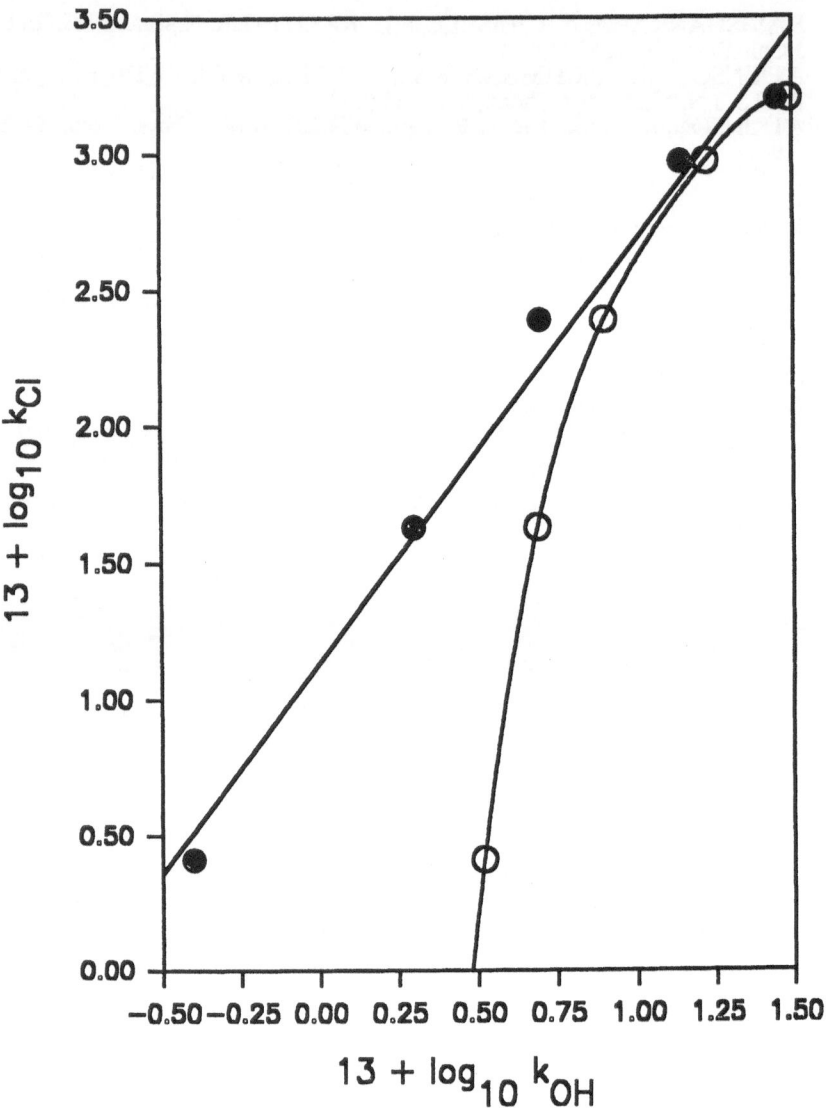

Fig.1   Linear free energy plot of the rate constant data for the reaction of OH and Cl with n-alkyl nitrates. Observed values of k_{OH} (O) and values of k_{OH} (●) corrected for the addition process

TROPOSPHERIC OXIDATION OF TOLUENE : REACTIONS OF SOME
INTERMEDIATE RADICALS.

A. Goumri, J-P. Sawerysyn, J-F. Pauwels et P. Devolder.
Laboratoire de Cinétique et Chimie de la Combustion de Lille - URA 876
Université des Sciences et Techniques de Lille Flandres Artois
59655 Villeneuve d'Ascq Cédex (France)

Summary
    The first step of tropospheric toluene oxidation by OH radicals
products either an adduct radical or a benzyl radical. We report an
estimation of the adduct + NO_2 rate constant ($4 \ 10^{-11}$ cm^3 $molec^{-1}$ s^{-1})
and direct measurements of the rate constants with O_2, NO and NO_2 of a
benzyl type radical (cm^3 $molecule^{-1}$ s^{-1} : O_2 : $8.3 \ 10^{-13}$, NO :
$10.3 \ 10^{-12}$, NO_2 : $4.9 \ 10^{-11}$))

1 - Introduction
        Aromatic hydrocarbons, which are produced by car engines, are
significant pollutants of urban atmospheres ; they are suspected to play
a prominent role in the mechanism of the photochemical smog (1) (2). The
tropospheric oxydation of aromatics is initiated by an initial attack of
OH radical which is the result of two competing processes : either ring
addition (with formation of a radical adduct) or abstraction . The
scheme I presents these two competing first steps for a typical
atmospheric molecule : toluene.
        For toluene, the rates of initial attack and the further various
possible reactions of the primary radicals have been reviewed by
Atkinson (1). Though the initial step of OH addition has been the
subject of various investigations, the subsequent reactions with O_2 or
atmospheric trace gas such as NO or NO_2 are much less known : Zellner et
al (3) have measured the rates of some OH-benzene adduct reactions. A
few authors (4) (5) (6) reported the rates of benzyl radical reactions
with O_2 or NO. Also, with the flash photolysis technique, Zetzsch et al
(7) (8) have directly measured the rates k_{-1} of the adduct unimolecular
decay for a few aromatic hydrocarbons ; furthermore, they recently
determined the rate of the reaction with NO of the important OH-toluene
adduct (9).
        Using the discharge flow technique, we have measured the rate of
the reaction with NO_2 of the OH-toluene adduct. We also report
measurements of the rate constants with O_2, NO and NO_2 of a benzyl type
radical : the p-fluorobenzyl radical.

2 - Toluene adduct reaction with NO_2
2-1 - Method : simulation of a simplified mechanism.

Zetzsch et al. (7)(8)(10) have presented a detailed and complete mechanism for the kinetics of OH-aromatics reactions ; from their exact analytical solution (7), they were able to derive accurate values of k_{1a} and k_{-1} in a range of temperatures for various hydrocarbons (7) (8). Following their framework, we consider the simplified following 3 parameter mechanism :

$$OH + toluene\ (+M) \qquad adduct\ (+M) \qquad k_1,\ k_{-1}$$
$$adduct + NO_2\ ----> products \qquad k_2$$

We have neglected wall recombinations of OH and adduct radicals and the minor contribution of abstraction. As for the general mechanism, it is possible to compute an exact analytical solution for the OH decay, provided the input parameters k_1, k_{-1} and k_2 are given. Figure 1 gathers the results of such computed OH decays for a range of values of the pseudo first order decay rate $K_2 = k_2[NO_2]$ and fixed values of k_1 and k_{-1} corresponding to 353 K. The value of k_1 is from our previous determination by discharge flow Resonance Fluorescence (12) whereas the value of k_{-1} is from the recent flash photolysis measurements of Witte and Zetzsch (10). As evidenced by fig.1, the OH decay curve is strongly dependant upon K_2. To derive the adduct bimolecular rate constant $k_2 = K_2/[NO_2]$, we have thus performed a series of experiments with known variable NO_2 concentrations and compared experimental with simulated decay curves.

2-2 - Results.

As for k_1 measurements, we have used the discharge flow technique with detection of OH by Resonance Fluorescence and photon counting (12). The NO_2 concentration is varied in the range 10^{-11} - 10^{14} molecule cm^{-3} with $[OH] \sim 2\ 10^{12}$ molecule cm^{-3}. We can thus infer the following value:
$$k_2 = (4 \pm 2)10^{-11}\ cm^3\ molecule^{-1}s^{-1}$$

3 - Parafluorobenzyl radical reactions with NO_2, NO, O_2

As a model for the benzyl radical, we have chosen the parafluorobenzyl (pFB) radical which exhibits very similar electronic properties (absorption and fluorescence slightly red-shifted) to benzyl (13). The experiment is performed at ~ 1 torr with the discharge flow set-up already described (12).

3-1 - Generation of pFB radical.

Fluorine atoms are produced at ~ 1 torr in the upstream part of the flow tube by a microwave discharge in a CF_4/He mixture. A double central injector allows successive introduction of the radical precursor p-fluorotoluene (p-FT) and the reactant X(X = O_2, NO or NO_2) :

$$\begin{array}{ll} & \text{fast} \\ F + p\text{-}FT\ ------> & p\text{-}FB + H_2O \qquad k_3 \\ ------> & other\ products \qquad k_4 \\ p\text{-}FB + X\ ------> & products \qquad k_5 \end{array}$$

3-2 - Detection of the radical.

The spectroscopic characteristics of the p-FB benzyl have been measured recently by Thrush et al (13). We probe the p-FB radical concentration by Laser Induced Fluorescence : a Yag-dye laser (Quantel) is Raman shifted through a hydrogen cell to the 0-0 absorption band of the radical (λ_{o-o} = 464.5 nm) ; the pulsed fluorescence is then sampled with a boxcar averager and displayed on a chart recorder.

3-3 - Rate constants results.

In fig. 2 are presented a few pseudo first -order decay plots of the p-FB radical concentration corresponding to a few added NO_2 concentrations. Fig. 3 shows the derivation of the bimolecular rate constant :

$$k_5 \text{ (reaction with } NO_2 \text{)} = 4.9 \ 10^{-11} \ cm^3 \ molecule^{-1} \ s^{-1}$$

To our knowledge, this is the first determination of a benzyl + NO_2 rate constant. Parallel measurements devoted to reactions of p-FB with O_2 and NO are gathered in table I, together with results for the benzyl radical from other investigations. Our values are in agreemement with the results of time resolved techniques (4)(5) but well above the recent determination by mass spectrometry (6).

3-4 - Branching ratio estimation.

Our present procedure also provides an estimation of another important parameter : the branching ratio between abstraction and addition $(R = k_{1b}/k_1)$. By adding either an excess flow of H_2O or an excess flow of Cl_2, the fluorine atoms are quantitatively converted to either OH radicals or Cl atoms via the following fast reactions :

$$F + H_2O \ ----\rangle \ OH + HF \qquad\qquad k_6$$
$$F + Cl_2 \ ----\rangle \ Cl + FCl \qquad\qquad k_7$$

Subsequent reactions with p-FT of OH or Cl provides p-FB radical concentrations respectively proportional to k_{1b} or $k_1 = k_{1a} + k_{1b}$ (the chlorine reaction with toluene is known to be very fast (6). The branching ratio is thus derived from the ratio p-FB radical concentration in presence of H_2O/p-FB radical concentration in presence of Cl_2. This procedure has been performed for both p-fluorotoluene and o-xylene with the following results

$$R(p\text{-}FT) = 0.3 \qquad\qquad R(o\text{-}xylene) = 0.11 \qquad\qquad (1)$$

Compared to indirect measurements ((1) : R   0.13), our direct measurement is in good agreement for o-xylene but well above for p-FT.

REFERENCES

(1)   R. Atkinson, (1986). Chem. Rev.Vol. 86. p. 69-201.
(2)   Leone, J.A., Seinfeld, J.H., (1984). Int. J. Chem. Kinet. Vol. 16,
       p. 159.
(3)   Zellner, R., Fritz, B., Preidel, M., (1985). Chem. Phys. Lett.
       Vol. 86, p. 412
(4)   Nelson, H.H., McDonald, J.R., (1982). J. Phys. Chem. Vol. 86, p.
       1242
(5)   Ebata, T., Obi, K., Tanaka, I., (1981). Chem. Phys. Lett. Vol. 77,
       p. 480.
(6)   Bartels, M., Edelbutter, J., Hoyerman, K., (1988) Twenty-second
       Symposium on Combustion Seatle, p. 43.
(7)   Witte, F., Urbanik, E., Zetzsch, C., (1986). J. Phys. Chem Vol.
       50, p. 3251.
(8)   Witte, F., (1987) Dissertation, Bochum.
(9)   Witte, F., Zetzsch and Devolder, P., (1989). 18.ᵗʰ Informal
       Conference on Photochemistry. January 9-13, Santa Monica (USA).
(10) Witte, F., Zetzsch, C., (20-22 sept. 1988). Joint Cost. Eurotrac
       meeting Norwich. po. 115.
(11) Bourmada, N., Devolder, P., Sochet, L-R., (1988) Chem. Phys. Lett.
       Vol. 149, p. 339.
(12) Bourmada, N., Carlier, M., Pauwels, J-F., Devolder, P. (1988). J.
       Chim. Phys. Vol. 85, p. 881.

(13) Charlton, T.R., Thrush, B.A., (1986) Chem. Phys. Lett. Vol. 125, p. 547.

(14) Hoffbauer, M. and Hudgen, J.W., (1985). J. Phys. Chem. Vol. 89, p. 5152.

	NO_2	NO	O_2	Réf.
	–	9.5	0.99	(5)
k	–	–	0.105	(4)
$(10^{-12}\ cm^3\ molec^{-1}\ s^{-1})$:	–	–	< 0.055	(6)
	49	10.5	0.83	this work

Table I : Rate constants of reactions of benzyl and p-fluorobenzyl with NO_2, NO and O_2 at   297 K.

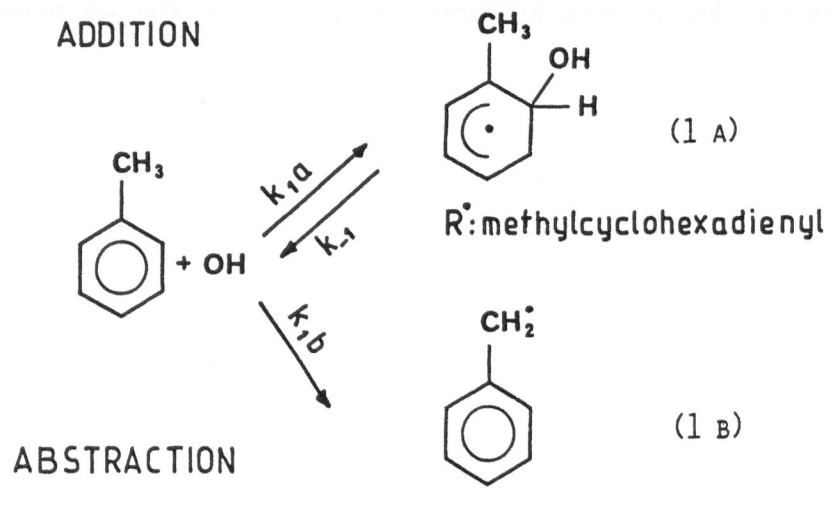

ADDITION

ABSTRACTION

(1 A)

R˙: methylcyclohexadienyl

(1 B)

R˙: benzyl

SCHEME I

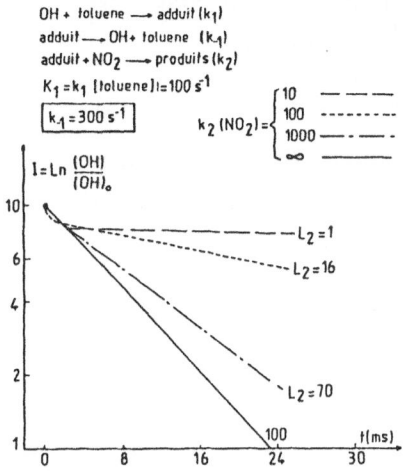

Figure 1

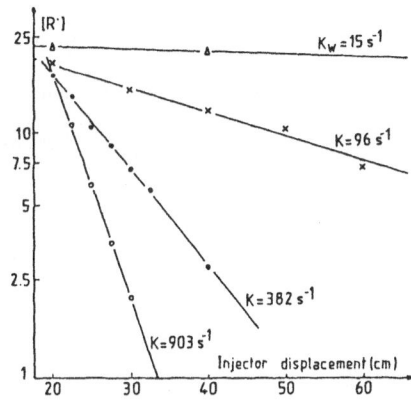

Figure 2

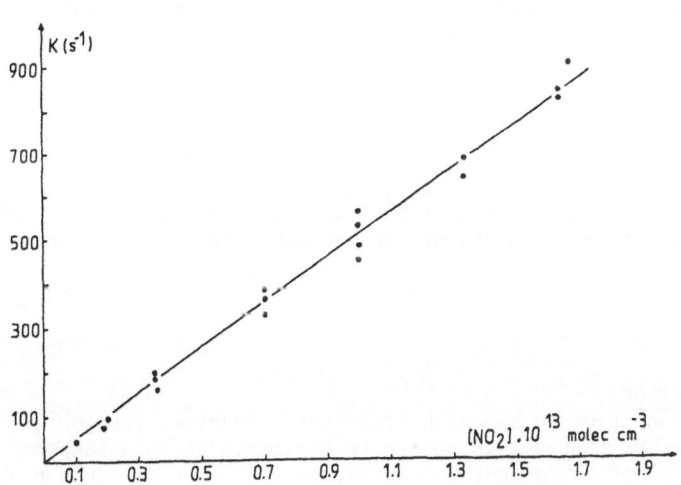

Figure 3

ADDUCT FORMATION OF OH WITH BENZENE AND TOLUENE
AND REACTION OF THE ADDUCTS WITH NO and NO$_2$

Cornelius Zetzsch, Rainald Koch, Manfred Siese and Franz Witte

Fraunhofer-Institut für Toxikologie und Aerosolforschung,
Nikolai-Fuchs-Straße 1,
D 3000 Hannover 61, FRG

and

Pascal Devolder

CNRS/Université des Sciences et Techniques de Lille, France

Summary

The behaviour of the adducts of OH with the aromatics benzene and toluene has been studied at 130 mbar in Ar in the presence of NO and NO$_2$ at four temperatures in the range from 300 to 350K. Biexponential decays of OH are observed in the absence of NO$_x$ and in the presence of NO. After eliminating NO$_2$ impurities from the NO sample, the reactivity of the adducts against NO was observed to vanish, and upper limits of $<10^{-14}$ and $<3\times10^{-14}$ cm^{3}s^{-1} are obtained for the reactions of benzene-OH and toluene-OH with NO from a quantitative treatment of the biexponential decays of OH. Triexponential decays of OH are observed in the presence of NO$_2$ and aromatics that can likewise be evaluated quantitatively. Values of (2.5 ± 0.6) and $(3.7\pm0.6)\times10^{-11}$ cm^{3}s^{-1} are obtained for the reactions of benzene-OH and toluene-OH with NO$_2$, independent of temperature. From the behaviour of OH in the presence of toluene, the rate constant for the addition channel is obtained to be k = $(1.9\pm0.5)\times10^{-13}$ exp $((+1040\pm70)K/T)$ cm^{3}s^{-1}, and the rate constant for the reverse reaction (the unimolecular decomposition of the adduct toluene-OH): k = $(2.5 \pm 0.7)\times10^{12}$exp $((-8040\pm120)K/T)$ s^{-1}, leading to a bond energy of OH to toluene in the adduct of (78 ± 1) kJ/mol. A rate constant of k = 2×10^{-12}exp $((-395\pm200)K/T)$ cm^{3}s^{-1} is obtained for the abstraction channel in the same temperature range.

1. INTRODUCTION

Addition of OH is the main atmospheric reaction path of aromatics (1), the abstraction channel can mostly be neglected at room temperature (1-2). The formation of the adduct is reversible (1-5), hence further reactions of the adduct (presumably with O$_2$, NO, NO$_2$ or O$_3$) are required to remove the aromatic pollutant from the atmosphere. The adduct of OH with benzene (the hydroxy-cyclohexadienyl radical, HCHD) has been detected by uv absorption (6), and preliminary information about the reactivity of HCHD against O$_2$, NO and NO$_2$ has been derived (7). In the present study, the kinetic behaviour of OH is investigated in an excess of the aromatics benzene and toluene at four temperatures each in the presence of various concentrations of NO and NO$_2$. The observations are described by the reaction scheme below.

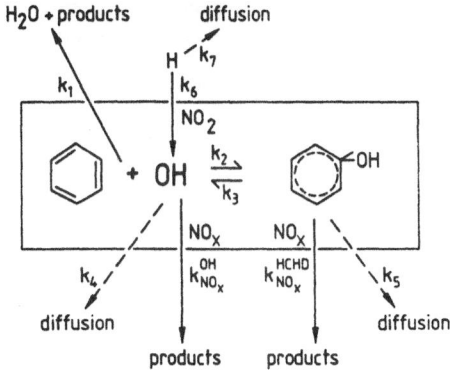

In the absence of NO_x, biexponential decays of OH are observed throughout (3-5), since the kinetic behaviour is dominated by the equilibrium in the frame. Similarly, biexponential decays are observed, if an excess of NO over OH is added to such systems in the absence of O_2. The abstraction (k_1) can be neglected for most aromatics at room temperature, and the initial decay of OH is mainly due to the formation of the adduct (k_2). The final decay (of the quasi-equilibrium concentration of OH) describes the decay of the adduct itself (due to unimolecular decay (k_3) and subsequent diffusion loss of OH (k_4) and diffusion loss of the adduct (k_5)). The photolysis of H_2O necessarily forms equal amounts of OH and atomic H. In the presence of O_2, the thus inherent production of HO_2 would complicate the kinetic behaviour in the presence of NO by far by the reaction $HO_2 + NO \longrightarrow NO_2 + H$, regenerating further OH. Hence it is required that O_2 is either well-known or preferably absent, i.e. that the inert gas is extremely pure and that air leaks into the resonance fluorescence cell are diminished. A similar complication occurs from the rapid reaction $H + NO_2 \longrightarrow NO + OH$ (k_6), leading to triexponential decays of OH in the presence of aromatics and NO_2.

2. EXPERIMENTAL

OH radicals are produced by pulsed vuv photolysis of H_2O and are monitored by resonance fluorescence of OH (3-5, 8). Mostly a flash energy of 2 J is employed (window material MgF_2 in experiments with benzene and SiO_2 with toluene) at partial pressures of H_2O of 0.13 mbar in the presence of Ar at 130 mbar, yielding initial OH densities of $2-5 \times 10^{10} cm^{-3}$. The signal is averaged 100 times at a dwell time of 2 ms in 256 channels. The inert gas, Ar, is purified by Oxisorb (Messer Griesheim), and the fluorescence cell is immersed into a flow of N_2, although its leak rate is less than 3×10^{-5} mbar l/s. The NO (0.2% premixed in Ar) is purified from NO_2 impurities by $Fe_2SO_4 \cdot 3 H_2O$. The flow of NO_2 is controlled using thermostated permeation tubes (Vici Metronics Dynacal, 350 and (1600 $\pm$ 30) ng/min at $30^{\circ}C$) in an Ar flow, that is further diluted appropriately with Ar.

3. EVALUATION OF KINETIC DATA

The system of homogeneous differential equations corresponding to the elementary reactions k_1-k_5 can be written in condensed form (3), and both reactions of NO_x, with OH ($k_{NO_x}^{OH}$) and with the adduct ($k_{NO_x}^{HCHD}$), can simply be incorporated as additional terms to the diffusion loss rates k_4 and k_5:

$$d[OH]/dt = -a[OH] + b[HCHD] \qquad \text{(Ia)}$$
$$d[HCHD]/dt = c[OH] - d[HCHD] \qquad \text{(Ib)}.$$

with the solution:

$$[OH(t)] = [OH]_0 \left\{ I_{01}\exp(-t/\tau_1) + I_{02}\exp(-t/\tau_2) \right\}, \qquad \text{(II)}$$

since the coefficients: $a = k_1 + k_2[\text{aromatic}] + k_4 + k_{NOx}^{OH}[NO_x]$, $b = k_3$, $c = k_2[\text{aromatic}]$ and $d = k_3 + k_5 + k_{NOx}^{HCHD}[NO_x]$ are constants. For the amplitude ratio of the two exponentials one obtains: $I_{01}/I_{02} = (\tau_1^{-1} - d)^2/bc$ and: $\tau_{1,2}^{-1} = (a+d)/2 \pm (0.25(a+d)^2 - ad + bc)^{1/2}$ for the decay rates.

The solution above holds for the absence and presence of NO. In the presence of NO_2, the reaction with H atoms (k_6) produces further OH. This adds an inhomogeneous contribution to the differential equation describing the temporal behaviour of OH. Since $[H]_0 = [OH]_0$, the production of OH by reaction 6 can be approximated by the additional term in equation (Ia): $+ a' \exp(-t/\tau_3) = k_6[OH]_0[NO_2]\exp(-t/\tau_3)$, yielding a third exponential term in the solution (II) with the constants I_{03} and $\tau_3^{-1} = k_6[NO_2] + k_7$. The general solution of this inhomogeneous system (9) is then:

$$[OH] = \frac{[OH]_0}{[bc+(\tau_1^{-1}-d)^2]} \left\{ (\tau_1^{-1}-d)^2 \left(1 - \frac{[NO_2]k_6}{\tau_1^{-1} - \tau_3^{-1}}\right) e^{-t/\tau_1} + bc\left(1 - \frac{[NO_2]k_6}{\tau_2^{-1} - \tau_3^{-1}}\right) e^{-t/\tau_2} \right.$$

$$\left. + \left(\frac{(\tau_1^{-1}-d)^2}{(\tau_1^{-1}-\tau_3^{-1})} + \frac{bc}{(\tau_2^{-1}-\tau_3^{-1})}\right) [NO_2]k_6 \, e^{-t/\tau_3} \right\} \qquad \text{(III)}$$

4. RESULTS

4.1 REACTIONS OF OH WITH NO_x

The reactions of OH with either NO or with NO_2 contribute to the observed decay rates of OH in the presence of aromatics. Therefore, the rate constant k_{NO}^{OH} was determined in the presence of Ar at 130 mbar. A very slight, negative temperature dependence was detected between 305 and 350K, and values of $1.05 \times 10^{-12} \text{cm}^3\text{s}^{-1}$ (at 320K) and $0.88 \times 10^{-12} \text{cm}^3\text{s}^{-1}$ (at 343K) were determined, yielding a weighted average of $0.95 \times 10^{-12} \text{cm}^3\text{s}^{-1}$, in good agreement with literature data (10). This value is used in the further evaluations of the rate constants for the reactions of the adducts benzene-OH and toluene-OH with NO. A value of $k_{NO2}^{OH} = (2.4 \pm 0.2) \times 10^{-12} \text{cm}^3\text{s}^{-1}$ was obtained for the reaction of OH with NO_2 in 133 mbar Ar at 319K, in reasonable agreement with literature values (10). This value should be considered as preliminary because of the low partial pressures of NO_2 accessible by the permeation tubes.

4.2 REACTIONS OF OH WITH THE AROMATICS IN THE ABSENCE OF NO_x

The behaviour of OH in the presence of benzene has been investigated in detail at various temperatures before, and the data of the present study derived at 305, 320, 333 and 349K agree well with the Arrhenius parameters of the previous work on benzene (5):

$$k_2 \pm 2\sigma = 2.3 \times 10^{-12}\exp((-190 \pm 60)\text{K/T}) \text{ cm}^3\text{s}^{-1} \qquad \text{(from 239-352K)}$$
and: $$k_3 \pm 2\sigma = 3 \times 10^{12}\exp((-8200 \pm 700)\text{K/T}) \text{ s}^{-1} \qquad \text{(from 299-352K)}.$$

For toluene, deviations from biexponential behaviour are observed, if OH is produced at 2 J flash energy with MgF_2 window. With a quartz window and at a low flash energy of 1.2 J such deviations could be avoided.

The results of the present study are determined from 33 biexponential decay profiles of OH at various concentrations of toluene at the temperatures 299, 310, 323 and 345K. In a first step, the biexponential decays are evaluated for the parameters τ_1, τ_2 and for the amplitude ratio of the exponentials, I_{01}/I_{02}. These parameters are used in a second step as the data set for a non-linear least squares fit (11) with the rate constants k_1 – k_5 as new parameters, where the rate constants k_1-k_3 are expressed in Arrhenius form and the temperature dependencies of the diffusion rates k_4 and k_5 are described by T^n.

In favorable cases the rate constants k_1 –k_5 can be completely separated because of their different contributions to the observed quantities τ_1, τ_2 and I_{01}/I_{02}. If the data set is incomplete or slightly inconsistent, the minor contributions from the abstraction and the diffusion channels have to be estimated and held fixed in the fit procedure (e.g., a temperature dependence of of T^2 for the diffusion channels).

By this method, the abstraction channel can be determined for the first time in the same temperature range as the other channels for toluene to be:

$$k_1 = 2 \times 10^{-12} \exp((-395 \pm 200)\text{K}/T)\ \text{cm}^3\text{s}^{-1}.$$

For the addition of OH to toluene, a rate constant of:

$$k_2 = (1.9 \pm 0.5) \times 10^{-13} \exp((+1040 \pm 70)\text{K}/T)\ \text{cm}^3\text{s}^{-1} \qquad \text{is obtained.}$$

The rate constant for the unimolecular decomposition of the adduct toluene-OH is observed to be:

$$k_3 = (2.5 \pm 0.7) \times 10^{12} \exp((-8040 \pm 120)\text{K}/T)\ \text{s}^{-1},$$

leading to a bond energy of (78 ± 1)kJ/mol. The diffusion loss rate of OH is simultaneously determined to be k_4 = 5-6 s^{-1} and the diffusion loss rate of the adduct, MHCHD, k_5= 3-4.5 s^{-1} with a reasonable temperature dependence of about T^2, as expected for the diffusion coefficient.

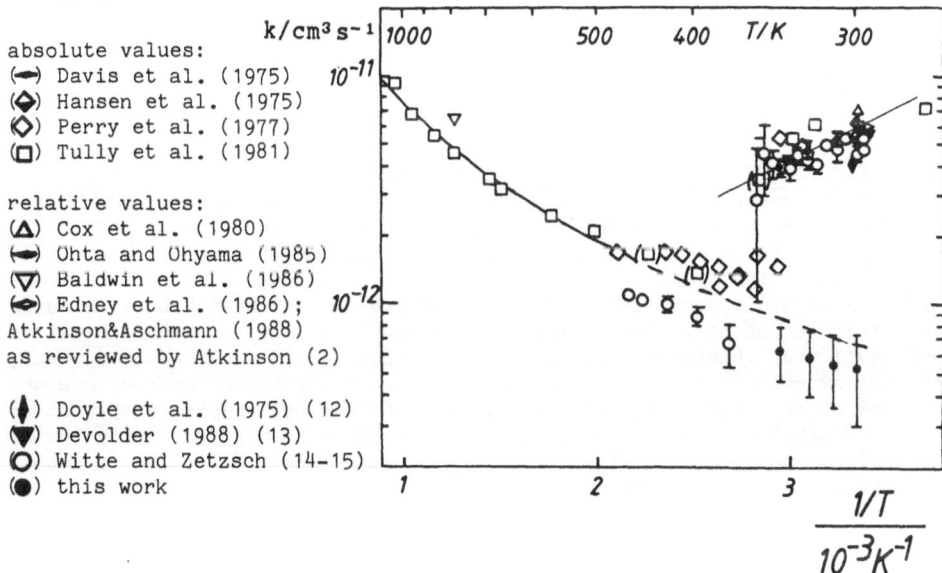

absolute values:
($\rightarrow$) Davis et al. (1975)
($\blacklozenge$) Hansen et al. (1975)
($\lozenge$) Perry et al. (1977)
($\square$) Tully et al. (1981)

relative values:
($\triangle$) Cox et al. (1980)
($\rightarrow$) Ohta and Ohyama (1985)
($\triangledown$) Baldwin et al. (1986)
($\leftrightarrow$) Edney et al. (1986);
Atkinson&Aschmann (1988)
as reviewed by Atkinson (2)

($\blacklozenge$) Doyle et al. (1975) (12)
($\blacktriangledown$) Devolder (1988) (13)
($\bigcirc$) Witte and Zetzsch (14-15)
($\bullet$) this work

Fig. 1. Arrhenius plot of the rate constants observed in the system OH + toluene in Ar at 130 mbar in comparison with other literature data and data from a recent review near the high pressure limit.

4.3 REACTION OF BENZENE-OH AND TOLUENE-OH WITH NO

The purity of the NO sample used for these measurements is of major importance, since the analogous reactions with NO_2 have been observed to be rather fast. Furthermore, maxima in the decay curves were observed with the unpurified NO sample. Computer simulations of the influence of NO_2, O_2 and NO by numerical integrations using the software package LARKIN (16) showed that the reaction of H atoms with NO_2 can hardly be distinguished from the influence of O_2, delivering OH by the reaction HO_2 + NO. Hence, a thourough leak testing of the equipment and a purification of the inert gas was performed. In the next step, a purification of the NO by cooling the flow of the diluted NO in Ar at 116K with isopentane slush at 2 bar total pressure lowered the determined values to less than 20%. By using a scrubber with $FeSO_4$, the reactivity of the adducts against NO was finally observed to vanish totally, as shown in fig. 2 for benzene.

Fig. 2:
Dependence of the final decay rate, τ_2^{-1}, of OH in the presence of benzene ($1.4 \times 10^{14} cm^{-3}$) on the concentration of highly purified NO.

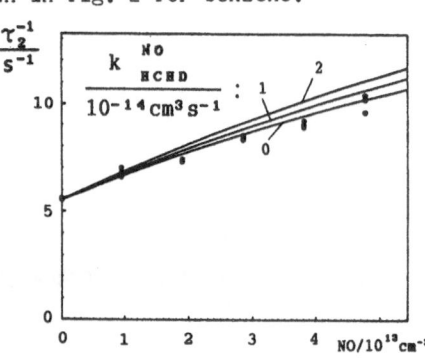

Figure 2 shows a comparison of the measured values with model calculations with the simple system of diffential equations (IIa,b) (note the slight curvature) with various assumptions for k_{NO}^{HCHD}. All three series of measurements indicate that the reaction of the adduct benzene-OH with NO can be neglected. An upper limit of $< 10^{-14} cm^3 s^{-1}$ is estimated from the behaviour in fig. 2, and similar data are obtained at 343K. The corresponding data for toluene-OH scatter slightly more, and an upper limit of $< 3 \times 10^{-14} cm^3 s^{-1}$ is obtained for the analogous reaction of MHCHD with NO.

4.4 REACTION OF BENZENE-OH AND TOLUENE-OH WITH NO_2

In the presence of NO_2 and aromatics, triexponential decays of OH are observed. Decays of OH in the presence of benzene in the absence and in the presence of two different concentrations of NO_2 are shown in fig. 3. As expected from the first experiments with NO_2-containing NO, the decays are complicated by a production of OH from the reaction of H with NO_2, leading to a third exponential term. These triexponential decays can be understood and evaluated quantitatively by using equation (III). On the other hand, the decay rate of the final exponential shows a dependence on the concentration of NO_2 similar to fig. 2. Values of 2.75+0.2 at305K, 2.45+0.06 at 320K, 2.5+0.2 at 333K and 2.5+0.2 at 349K are obtained from such data, summarized as a rate constant of $(2.5 \pm 0.6) \times 10^{-11} cm^3 s^{-1}$ for the reaction of benzene-OH with NO_2, independent of temperature.

Similar data for toluene-OH at 300, 311, 320, 338 and 353K resulted in values of 3.6+0.4, 3.6+0.2, 3.6+0.2, 4.0+0.6 and 4+2 ($\times 10^{-11} cm^3 s^{-1}$), respectively. Again the temperature dependence is less than the experimental scatter, and a value of $(3.7 \pm 0.6) \times 10^{-11} cm^3 s^{-1}$ is obtained by a comparison of model and experiment for the rate constant of the reaction of toluene-OH with NO_2. An example for a model calculation using equation

(III) is shown in fig. 4, where the concentration of NO_2 was varied by a factor of 300. The model calculation and the experimental results agree in that the rate constant can be extracted from the slope of plots of τ_2^{-1} versus NO_2, although the exponentials interfere each other in certain concentration ranges of NO_2.

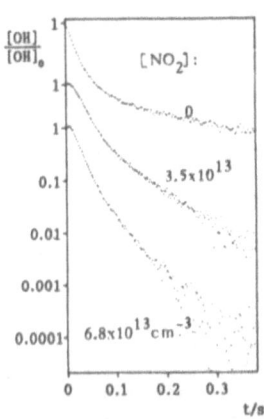

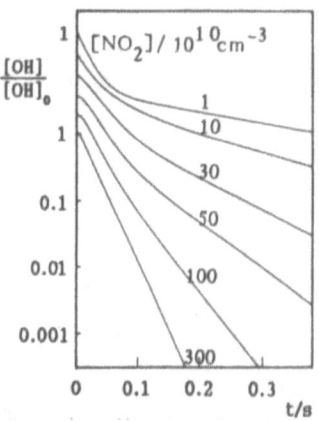

Fig. 3: Decays of OH in the presence of benzene ($3.2 \times 10^{13} cm^{-3}$) in the absence and presence of NO_2 at 305K.

Fig. 4: Model calculation for the triexponential decays of OH at 299K in the presence of toluene ($6.9 \times 10^{12} cm^{-3}$) and NO_2.

5. DISCUSSION

The behaviour of OH in the presence of toluene is in good agreement with previous literature data as can be seen in fig. 1. The improved technique of evaluating all decays at the different temperatures simultaneously enables us to determine the abstraction channel in the same temperature region as the addition channel. The results of the present study confirm our recent results in the temperature range from 380 to 480K (14, 15) which are lower by some 30-50% compared to previous data in the same temperature range (1).

It should be noted that the results of the present study on the reactions of the adducts with NO_x rely on the assumption that besides the unimolecular decay of the adducts and the reaction of H + NO_2 no further sources exist for a regeneration of OH in the system. Our present upper limits of $<10^{-14}$ and $<3 \times 10^{-14} cm^3 s^{-1}$ for the rate constants of the reactions of NO with the adducts benzene-OH and toluene-OH in Ar at 130 mbar disagree with previous measurements of the reaction of HCHD with NO, which resulted in values of $(1.0+0.5) \times 10^{-12} cm^3 s^{-1}$ at 298K in N_2 at 20 mbar and $(0.7+0.3) \times 10^{-12} cm^3 s^{-1}$ at 298K in O_2 at 20 mbar (6). This discrepancy may either be caused by an NO_2 impurity in the previous measurements or it may be interpreted in a way that the reaction of HCHD with O_2 is fast and that the measured reactivity could be assigned to the peroxi-radical, $HCHD-O_2$. Another determination from our own laboratory yielded already a lower value of $(5.5+0.5) \times 10^{-13} cm^3 s^{-1}$ at 318K in Ar at 130 mbar (14). Similarly, values of $(5.6+0.6) \times 10^{-13} cm^3 s^{-1}$ at 323K and $(6.0+0.8) \times 10^{-13} cm^3 s^{-1}$ at 333K have been obtained for MHCHD (15). By the already mentioned cooling of the flow of NO to 118K the values for the "reaction of NO with MHCHD" could be lowered to less than 20% of the previous ones. This lead us to speculate that an NO_2 impurity is the main reason for the discrepancy and to use a chemical method for the purification of NO by $FeSO_4$, resulting in the upper limits given above. Although NO might be consumed from the diluted gas flow by a chemical reaction of NO with the $FeSO_4$ to yield nitroso salts, this possibility is excluded by the reasonable agreement of our values for the reaction OH + NO + M with literature values (10).

The values of (2.5 ± 0.6) and $(3.7\pm0.6)\times10^{-11}$ cm^3s^{-1}, obtained for the reactions of benzene-OH and toluene-OH with NO_2 at 130 mbar in Ar, are in reasonable agreement with a literature value of $(0.85\pm0.2)\times10^{-11}$ cm^3s^{-1} for HCHD at 298K in N_2 at 20 mbar (7). Although one might speculate about a slight pressure effect of this reaction -if addition is at all the main path (7)- the stabilization of such polyatomic radicals should be fairly complete at a few mbar. On the other hand, absolute concentrations of highly diluted flows of NO_2 may be uncertain because of the corrosive behaviour of NO_2. Very recently, Devolder et al. determined a rate constant of $\sim 5\times10^{-11}$ cm^3s^{-1} for the rate constant of toluene-OH with NO_2 using the flow tube technique (17) in good agreement with our present result.

The role of these rate constants is discussed in the following section on the tropospheric fate of the adducts aromatic-OH. In the troposphere, the consecutive reaction of the adduct aromatic-OH with NO_2 has to compete with the strongly temperature dependent unimolecular decay. Since the reaction of O_2 with the adducts may be slow (7,18), the atmospheric fate and the lifetimes of aromatics are expected to be dependent on the pollution levels. The reactions of the adducts benzene-OH and toluene-OH with NO_2 will be more important than the reactions with NO, since the respective rate constants for the reactions with NO_2 are by more than three orders of magnitude higher. Furthermore, the concentrations of NO_2 are much higher than those of NO under most conditions of the free troposphere (with the exception of the direct vicinity of strong emitters like car exhausts or plumes of conventional power plants). The reaction of NO_2 with the adducts may produce nitroaromatics (18) and may compete with the reaction of O_2 with the adduct if the rate constants of the reactions with O_2 exceed values of 2×10^{-20} cm^3s^{-1} in a clean atmosphere ($NO_x \approx 100$ ppt) or values of 10^{-18} cm^3s^{-1} in a polluted atmosphere ($NO_x \approx 30$ ppb). Although such a reactivity against O_2 is not improbable, the reaction with NO_2 is considered exclusively in the summarizing estimate of tropospheric lifetimes (tab. I):

Tab. I: Degradation Pathways and Lifetimes of Toluene at $[OH] = 5\times10^5cm^{-3}$

Conditions	Dominating path	Corresponding rate expression	Lifetime
300K, 100ppt NO_2	MHCHD + NO_2 + abstraction	$k_{NO2}^{MHCHD}[NO_2] k_2[OH]/k_3$ + k1 [OH]	40 days
300K, 30ppb NO_2	MHCHD + NO_2	$k_{NO2}^{MHCHD}[NO_2] k_2[OH]/(k_3+k_{NO2}^{MHCHD}[NO_2])$	4 days
273K, 30ppb NO_2	MHCHD + NO_2	$k_2[OH]$	3 days

ACKNOWLEDGEMENT This work was supported by the Bundesminister für Forschung und Technolgie (grant 07EU705). The authors thank Dr. R. Atkinson, Riverside, and Dr. P. Devolder, Lille, for communicating unpublished results.

REFERENCES
1. PERRY, R.A., ATKINSON, R. and PITTS, Jr., J.N. (1977). Kinetics and mechanisms of the gas phase reaction of OH radicals with aromatic hydrocarbons over the temperature range 296-473K. J. Phys. Chem. 81, 296-304.
2. ATKINSON, R. (1989). Kinetics and mechanisms of the gas-phase reactions of the hydroxyl radical with organic compounds. J. Phys. Chem. Ref. Data (in press).

3. WAHNER, A. and ZETZSCH, C. (1983). Rate constants for the addition of OH to aromatics (benzene, p-chloroaniline, and o-, m-, and p-dichlorobenzene) and the unimolecular decay of the adduct. Kinetics into a quasi-equilibrium. 1. J. Phys. Chem. 87, 4945-4951.
4. WITTE, F. and ZETZSCH, C. (1984). The temperature dependence for the forward-backward reactions of the addition of OH to benzene, aniline and nitrobenzene. Proc. 3rd Europ. Sympos. on Physico-Chemical Behaviour of Atmospheric Pollutants (eds. B. Versino and G. Angeletti), pp. 168-176, Reidel, Dordrecht.
5. WITTE, F., URBANIK, E. and ZETZSCH, C. (1985). Temperature dependence for the addition of OH to benzene and to some monosubstituted aromatics (aniline, bromobenzene, and nitrobenzene) and the unimolecular decay of the adducts. Kinetics into a quasi-equilibrium. 2. J. Phys. Chem. 90, 3251-3259.
6. ZELLNER, R., FRITZ, B. and PREIDEL, M. (1985). A cw UV laser absorption study of the reactions of the hydroxyl-cyclohexadienyl radical with NO_2 and NO. Chem. Phys. Letters 121, 412-416.
7. FRITZ, B., HANDWERK, V., PREIDEL, M. and ZELLNER, R. (1985). Direct detection of hydroxy-cyclohexadienyl in the gas phase by cw-UV laser absorption. Ber. Bunsenges. Phys. Chem. 89, 343-344.
8. STUHL, F. and NIKI, H. (1972). Pulsed vacuum-uv photochemical study of reactions of OH with H_2, D_2, and CO using a resonance-fluorescent detection method. J. Chem. Phys. 57, 3671-3679.
9. STEPANOW, W.W. (1967). Lehrbuch der Differentialgleichungen. VEB Deutscher Verlag der Wissenschaften, Berlin.
10. PARASKEVOPOULOS, G. and SINGLETON, D.L. (1988). Reactions of OH radicals with inorganic compounds in the gas phase. Rev. Chem. Intermediates 10, 139-218.
11. PRESS, W.H., FLANNERY, B.P., TEUKOLSKY, S.A. and VETTERLING, W.T. (1986) Numerical recipes, the art of scientific computing. Cambridge Univ. Press, Cambridge.
12. DOYLE, G.J., LLOYD, A.C., DARNALL, K.R., WINER, A.M. and PITTS, Jr., J.N. (1975). Gas phase kinetic study of relative rates of reaction of selected aromatic compounds with hydroxyl radicals in an environmental chamber. Env. Sci. Technol. 9, 237-241.
13. DEVOLDER, P., BOURMADA, N., SAWERYSYN, J.P. (1988). Rate constants of the reactions of OH with benzene and toluene in the fall-off range. In: Air Pollution Research Report 17. Mechanisms of gas phase and liquid phase chemical transformations in tropospheric chemistry (ed. R.A. Cox). pp. 117-118, EUR 12035, CEC, Brussels.
14. WITTE, F. and ZETZSCH, C. (1988). Adduct formation and its unimolecular decay in the reaction of OH with toluene. In: Air Pollution Research Report 17. Mechanisms of gas phase and liquid phase chemical transformations in tropospheric chemistry (ed. R.A. Cox). pp. 115-116, EUR 12035, CEC, Brussels.
15. WITTE, F., and ZETZSCH, C. (1988). Adduct formation of OH radicals and consecutive reactions with O_x and NO_x. Annual Report of the Steering Committee of LACTOZ, a joint EUROTRAC/COST 611 project, pp. 62-66, EUREKA/CEC, Brussels.
16. DEUFLHARD, P. and NOWAK, U. (1986). Efficient numerical simulation and identification of large chemical reaction systems. Ber. Bunsenges. Phys. Chem. 90, 940-946.
17. DEVOLDER, P. (private communication).
18. ATKINSON, R., ASCHMANN, S.M., AREY, J. and CARTER, W.P.L. Formation of ring-retaining products from the OH radical-initiated reactions of benzene and toluene (submitted to Int. J. Chem. Kinet.).

TEMPERATURE DEPENDENCE OF REACTIONS OF THE NITRATE
RADICAL WITH ALKANES

Jane A. Bagley, Peter Biggs, Carlos Canosa-Mas,
Mark R. Little, A. Douglas Parr, Stuart J. Smith,
Steven J. Waygood, and Richard P. Wayne.

Physical Chemistry Laboratory, Oxford University,
South Parks Road, Oxford, OX1 3QZ, U.K.

SUMMARY

Rate constants have been determined as a function of
temperature for the reaction of NO_3 with n-butane,
i-butane (2-methylpropane), i-pentane (2-methylbutane)
(298-523K) and ethane (453-553K). Arrhenius expressions
for the reaction of NO_3 at primary, secondary and
tertiary hydrogen atom sites have been derived and used
to predict the rate of reaction between NO_3 and
i-pentane. The observed decays of NO_3 could be reproduced
by kinetic models using a reasonable mechanism.

1. INTRODUCTION

Reactions of the nitrate radical are of importance in the
night-time troposphere. Reaction with the radical is a
significant loss process for most volatile organic compounds,
and it is the major oxidation route for many unsaturated
organic species, including natural hydrocarbons. The radical
is a key intermediate (1) in the formation of nitric acid,
either via hydrolysis of N_2O_5 (resulting from the combination
of NO_2 with NO_3) or via abstraction of an H atom by NO_3.

There have been previous determinations (2,3) of the rate
of reaction of alkanes with NO_3 at room temperature, but this
is the first time that the temperature dependence of the
reactions has been studied. Since many urban areas contain
relatively high concentrations (4) of a wide variety of
alkanes, it is desirable that structure-reactivity
relationships be found that will assist understanding of the
urban night-time troposphere. Here we report the rate
constants for the reactions between NO_3 and ethane, n-butane,
i-butane (2-methylpropane) and i-pentane (2-methylbutane).

2. EXPERIMENTAL

The discharge-flow optical-absorption apparatus (5,6)
used in this work has been described previously. All
experiments were carried out at a pressure of ca. 2mmHg.

The minimum $[NO_3]$ detectable for a 1:1 signal-to-noise
ratio and a half second integration time was ca.
10^{12} molecules cm^{-3}. NO_3 was produced in the side-arm by the
reaction of F atoms with anhydrous HNO_3.

3. RESULTS

Concentrations of NO_3 were determined as a function of injector position. Plots of $\ln[NO_3]$ vs. contact time might be expected to be straight if there were no change in stoicheiometry with contact time. The plots for room temperature kinetic runs showed some curvature. For this reason, the observed rate constant, k_{obs}, was taken to be the slope of the plot between the first and second injector ports. At higher temperatures, where the first injector port was not always at the set temperature because of its proximity to the room temperature region of the detector, ports 2 and 3 were used instead.

The rate constant, k_1, for the initial abstraction step

$$RH \quad + \quad NO_3 \quad \longrightarrow \quad HNO_3 \quad + \quad R \qquad (1)$$

was obtained from the observed rate constant, k_{obs}, by assuming that $k_{obs}/k_1 = 2$. We have obtained evidence from other experiments that the reaction

$$R \quad + \quad NO_3 \quad \longrightarrow \quad products \qquad (2)$$

is fast, and thus on the timescale of our experiments at least two NO_3 molecules are lost for each abstraction step (reaction 1). The rate constants derived are shown in table I. The previous determinations of these rate constants by Atkinson et al (2,3), which were obtained using a different technique, are shown in brackets. The agreement is good, giving support for the hypothesis that $k_{obs}/k_1 = 2$. The possibility of further secondary reactions is considered later.

The Arrhenius plots for the reactions of ethane, i-butane and n-butane with NO_3 are shown in fig. 1. The Arrhenius parameters for the reaction with ethane can be taken as indicative of the rate of abstraction by NO_3 from primary H-atom sites. We expect, on the basis of these results, that the contribution of primary abstraction to the rate of reaction of NO_3 with the butanes at lower temperatures (less than 473K) will be negligible and hence that reaction at these temperatures is entirely due to abstraction by NO_3 of secondary and tertiary H-atoms in n-butane and i-butane respectively. The parameters derived from the Arrhenius plots are shown in table II. Curvature of the Arrhenius plots for the butanes at higher temperatures (>473K) cannot be explained by primary abstraction. The curvature may be due to a change in the stoicheiometric factor, k_{obs}/k_1. However, between 298K and 473K, the linearity of the Arrhenius plots suggests that k_{obs}/k_1 remains constant over this temperature range.

The Arrhenius parameters for primary, secondary and tertiary H atom abstraction derived from the plots were used to make a prediction for the rate of reaction of NO_3 with an alkane containing all three types of H atom. The alkane chosen was i-pentane.

Table I: Rate constants for the reaction of NO_3 with alkanes as a function of temperature. Comparisons with previous determinations are in brackets.

T/K:	298	348	373	423	473	523
$i-C_4H_{10}$	1.1±0.2 (0.97±0.25)[a]	4.5±1.6	8.0±0.8	23±4	54±12	130±24

T/K:	298	333	373	423	473	523
$n-C_4H_{10}$	0.45±0.06 (0.65±0.20)[a]	1.4±0.2	4.6±1.2	11.2±1.2	32±3	90±4

T/K:	298	323	373	423	473	523
$i-C_5H_{12}$	1.6±0.2	3.9±1.4	10±2	25±4	61±14	129±37

T/K:	453	473	523	553
C_2H_6	3.0±0.5	4.7±0.8	14±3	16±3

[a] Ref. (2)
All units cm^3 molecule^{-1} s^{-1}×10^{-16}.
All errors are 95% confidence limits

The Arrhenius plot and the prediction, (which illustrates the contribution from tertiary, secondary and primary abstraction) are shown in fig. 2.

4. SECONDARY REACTIONS

Curvature in the first-order plots is indicative of change in the stoicheiometric factor, k_{obs}/k_1, with contacttime. In an attempt to understand further the secondary reactions occurring, some experiments were performed with NO_3 and i-butane in the presence of excess O_2. It was found that k_{obs} was increased by a factor of approximatly 2.5 compared to the same reaction without O_2 present. A mechanism was postulated that contained 6 reactions including the reaction

Table II. Arrhenius parameters for the reaction of NO_3 with ethane, n-butane and i-butane.

Alkane	H-type	E/kJ mol^{-1}	A / cm^3 molecule^{-1} s^{-1}	A per H atom / cm^3 molecule^{-1} s^{-1}
C_2H_6	1°	36.8±2.8	(5.7±4.0)×10^{-12}	(9.5±6.8)×10^{-13}
$n-C_4H_{10}$	2°	26.7±0.7	(2.5±0.6)×10^{-12}	(4.1±0.1)×10^{-13}
$i-C_4H_{10}$	3°	24.6±0.7	(2.3±0.6)×10^{-12}	(2.6±0.1)×10^{-12}

All errors quoted are 1σ on the fit.

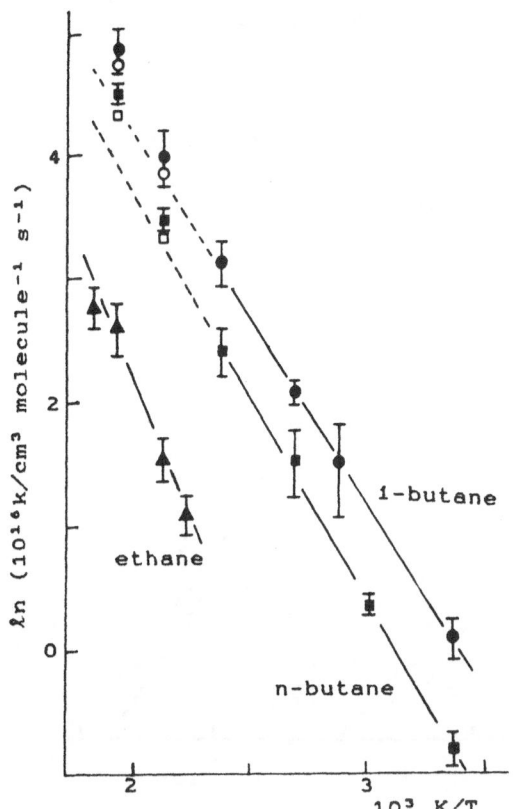

Fig. 1 Arrhenius plots for the reactions of NO_3 with ethane, n-butane and i-butane. Expected rate after subtraction of primary abstraction shown in unfilled symbols.

---

of NO_3 with RO and RO_2 (the self reactions of these species are unimportant under our experimental conditions). Numerical integration of the rate equations derived from the assumed mechanism could be made to give good visual fits if the branching ratios for certain of the reactions were chosen appropriately. For the experiments in the presence of O_2, the reaction

$$RO_2 \quad + \quad NO_3 \quad \longrightarrow \quad RO \quad + \quad O_2 \quad + \quad NO_2 \quad (3a)$$
$$\longrightarrow \quad RONO_2 \quad + \quad O_2 \quad \quad (3b)$$

gave the best fit with $k_{3a}/k_{3b}=1$ if the rate constants for the reactions

$$RO \quad + \quad NO_3 \quad \longrightarrow \quad ROONO_2 \quad \quad (4a)$$
$$\longrightarrow \quad RO_2 \quad + \quad NO_2 \quad (4b)$$

are such that $k_{4b} \gg k_{4a}$. If $k_{4a} \gg k_{4b}$, the $ROONO_2$ will decompose with $k = 2$ s^{-1} and the ratio k_{3a}/k_{3b} has to be increased to

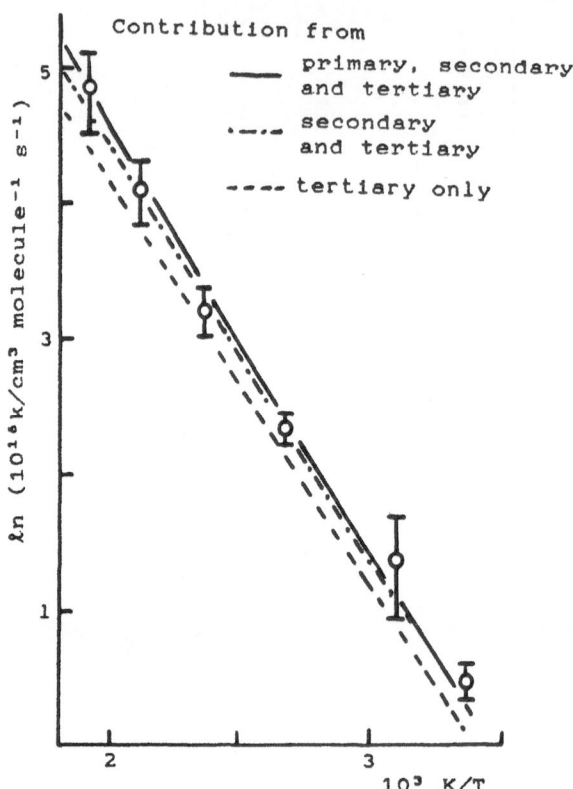

Fig. 2 Arrhenius plot for the reaction between NO_3 and i-pentane.

approximatly 9 to give good agreement with the experimental points.

Using the ratio $k_{3a}/k_{3b} = 1$ it was possible to simulate the experimental points for the the reaction of NO_3 with i-butane with no O_2 present and obtain an estimate of the branching ratio for the reaction

$$R \quad + \quad NO_3 \quad \longrightarrow \quad RONO_2 \qquad\qquad (2a)$$
$$\longrightarrow \quad RO \quad + \quad NO_2 \qquad (2b)$$

of $k_{2a}/k_{2b} = 0.05$.

5. DISCUSSION

The atmospheric half-lives with respect to reaction with NO_3 during the night of the alkanes studied at room temperature in this work are 4300, 1750 and 1200 hours for n-butane, i-butane and i-pentane ($[NO_3] = 10^9$ molecule cm^{-3} at night), compared to half-lives with respect to reaction with the OH radical of 77, 82 and 50 hours (7) ([OH] = 10^6 molecule cm^{-3} during the day). The rate of scavenging of these alkanes by NO_3 is less than 5% of that by the OH radical (although the values of atmospheric [NO_3] and [OH] are subject to review) so that the reactions studied in this work are not the major loss processes for the alkanes in the atmosphere.

The two most abundant hydrocarbons found in a survey (4) of air quality in 39 cities of the U.S.A. were i-pentane and n-butane. The median concentration of i-pentane was over 9 ppb. Total median alkane concencentrations (not including methane or ethane) were over 50ppb. Assuming an average rate constant for reaction with NO_3 of 1.2×10^{-16} cm^3molecule^{-1}s^{-1} (which is reasonable given our analysis of reactivity at different H atom sites), the rate of production of HNO_3 0.04ppb hr^{-1} due to alkane + NO_3 reactions in polluted atmospheres. The contribution to the production rate by reaction with i-pentane will typically be 0.005ppb hr^{-1}, making this molecule the most important atmospheric alkane in terms of HNO_3 production. These rates can be compared to a typical production rate of approximatly 0.25 ppb hr^{-1} (1) from the reaction of NO_3 with aldehydes.

REFERENCES

1. FINLAYSON-PITTS, B.J. and PITTS, J.N., Jr. (1986),
   Atmospheric chemistry, (Wiley, New York)
2. ATKINSON, R., ASCHMANN, S.M. and PITTS, J.N., Jr. (1988).
   Kinetics of the gas-phase reactions of the NO_3 radical
   with a series of organic compounds at 296±2K.
   J. Phys. Chem., 92, 3454.
3. WALLINGTON, T.J., ATKINSON, R., WINER, A.M.
   and PITTS, J.N., Jr. (1986). Absolute rate constants for
   the gas-phase reactions of the NO_3 radical with CH_3SCH_3,
   NO_2, CO, and a series of alkanes at 298±2K.
   J. Phys. Chem., 90, 4640.
4. SEINFELD, J.H. (1989) Urban air pollution:
   state of the science. Science, 10th Feb., p.745.
5. CANOSA-MAS, C.E., SMITH, S.J., TOBY, S. and WAYNE, R.P.
   (1988) Reactivity of the nitrate radical towards alkynes
   and some other molecules.
   J. Chem. Soc., Faraday Trans. 2, 84, 247.
6. CANOSA-MAS, C.E., SMITH, S.J., TOBY, S. and WAYNE, R.P.
   (1988) Temperature dependences of the reactions of the
   nitrate radical with some alkynes and with ethylene.
   J. Chem. Soc., Faraday Trans. 2, 84, 263.
7. ATKINSON, R. (1986) Kinetics and mechanisms of the
   gas-phase reactions of the hydroxyl radical with organic
   compounds under atmospheric conditions.
   Chem. Rev., 86, 69.

I. WÄNGBERG, S. LANGROVA and E. LJUNGSTRÖM
Department of Inorganic Chemistry
University of Göteborg and Chalmers University of Technology
S-412 96 Göteborg, Sweden

SUMMARY A flow reactor for studying heterogeneous
reactions of moderate rate has been built. The
decomposition of PAN was used to test the device. In the
presence of NO, the decomposition rate compares well with
literature values. An increase in surface to volume (S/V)
ratio does not influence the rate. When no NO is added,
an increased S/V ratio gives an increased rate of
decomposition.

1. INTRODUCTION

There are several reasons for atmospheric chemists to be
interested in surface reactions. Such processes may directly
influence the chemistry of the atmosphere. However, since the
surface is provided by the atmospheric aerosol and to some
extent by the ground itself, the surface to volume ratio (S/V)
is often very low. Thus, to be important, such a surface
reaction must proceed with high rate. In the real atmosphere,
these reactions are complicated by the complex and varying
character of the aerosol surface and the likely presence of a
liquid phase.

Another reason for interest in heterogeneous processes is
their possible influence on laboratory investigations. Rate
constants which are applied to atmospheric problems are most
often determined in laboratory equipment with an appreciable
S/V ratio. The possibility of heterogeneous side reactions is
always present and reactions which are negligible in the
atmosphere may be significant in the laboratory.

Peroxyacetyl nitrate (PAN) is an important, semi-stable
product which is formed when volatile hydrocarbons undergo
photochemical oxidation in the presence of nitrogen oxides.
PAN was first noted for its phytotoxicity and its strong eye
irritant power. When PAN was studied more closely it became
obvious that it may also serve as a reservoir for nitrogen
oxides. The thermal decomposition of PAN (I) is strongly
temperature dependent. PAN formed through photochemical
reactions at moderate temperature may be transported
considerable distances before an increase in temperature
causes significant decomposition and possible re-release of
NO_2.

$$CH_3C(O)OONO_2 \longrightarrow CH_3C(O)OO + NO_2 \qquad (I)$$

The rate constant of (I) and its temperature dependence is now quite well established through several investigations e.g. (1-4). Atkinson and Lloyd (5) recommend the following expression for calculating the rate constant of (I).

$$k_I = 1.95 \cdot 10^{16} \ e^{-13543/T} \ s^{-1}$$

The rate information for reaction (I) was usually obtained in experiments where an excess of NO was used to inhibit reaction (-I) by draining the peroxyacetyl radical via (II).

$$CH_3C(O)OO \ + \ NO_2 \ \longrightarrow \ CH_3C(O)OONO_2 \qquad (-I)$$

$$CH_3C(O)OO \ + \ NO \ \longrightarrow \ CH_3C(O)O \ + \ NO_2 \qquad (II)$$

In several papers dealing with PAN, claims are made that the decomposition is either sensitive or not sensitive to heterogeneous side reactions.(3,4,6)

The present investigation was undertaken to obtain additional information on the heterogeneous reactions of PAN and to characterize a flow reactor for studying heterogeneous reactions of moderate rate.

2. EXPERIMENTAL SETUP AND PROCEDURE

The experimental set-up is shown in Figure 1. The central part is a flow reactor of 3 l volume. To control the temperature of the reactor, liquid from a thermostat, capable of maintaining temperatures between -30 and +70° C is circulated through the reactor jacket. To obtain time resolution, a set of thin gas sampling tubes are positioned at various distances from the reactor inlet. The reactor may be used empty or filled with various packings to add catalytic surface and change the S/V ratio of the device. In the present investigation two S/V ratios were used. The empty reactor has a S/V of 92 m^{-1} and the reactor filled with ca 9 mm diameter soda glass beads has a S/V of 903 m^{-1}. The expected range of observable rates for a first- or pseudo-first order process is between $1 \cdot 10^{-5}$ and $5 \cdot 10^{-2}$ s^{-1} although results obtained at the extremes have high associated errors. For the present investigation, a Mattson Polaris FTIR spectrometer equipped with a small, flow-through gas cell with White optics, giving an optical path of 2.5 m, was used as detector. This detection system allowed the determination of the concentrations of PAN($7 \cdot 10^{13}$ molec.cm^{-3}, 2.8 ppm), NO_2($9 \cdot 10^{13}$ molec.cm^{-3}, 4 ppm), NO($8 \cdot 10^{14}$ molec.cm^{-3}, 33.3 ppm), CH_3NO_2($1 \cdot 10^{14}$ molec.cm^{-3}, 4.5 ppm), CH_3ONO_2($1 \cdot 10^{14}$ molec.cm^{-3}, 4.5 ppm), HCHO($1 \cdot 10^{14}$ molec.cm^{-3}, 4.5 ppm), and HONO($6 \cdot 10^{13}$ molec.cm^{-3}, 2.4 ppm) in the gas mixture leaving the reactor. The number in parentheses gives the approximate detection limit).

PAN was synthesized according to the procedure of Nielsen et al. (7) as modified by Senum et al. (4). PAN-containing gas mixtures were prepared by bubbling nitrogen through PAN containing dodecane solutions kept at 0° C.

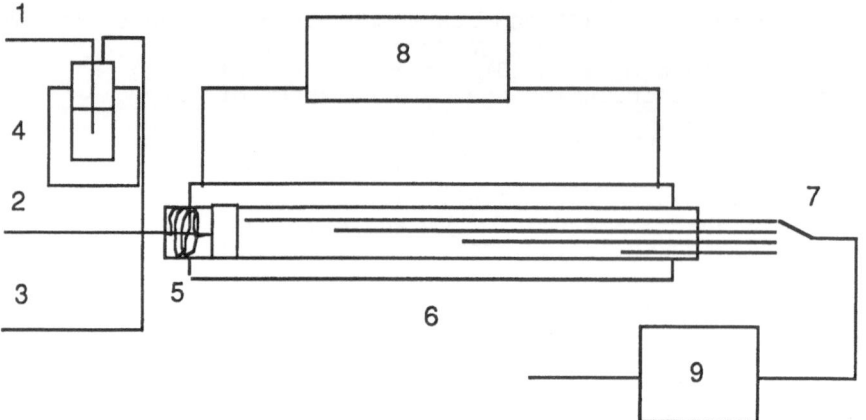

Figure 1. 1,2/ N_2 line, 3/ NO-N_2 or NO_2-N_2 line, 4/
thermostatted PAN bubbler, 5/ gas pre-heater and gas
distributor, 6/ flow reactor with jacket, 7/ sampling tubes,
8/ thermostat, 9/ FTIR spectrometer with White-optics flow
cell

The volume of the PAN containing liquid and the flow rate were
such that the concentration decreased less than 1% h^{-1}. The
concentrations used in the experiments were between $6.2 \cdot 10^{14}$
and $1.5 \cdot 10^{15}$ molecules cm^{-3} (ca 25-60 ppm). Nitrogen with less
than 5 ppm oxygen was used as matrix gas. The experiments were
made at atmospheric pressure.
    An experiment consisted of thermostatting the reactor at
the desired temperature and allowing the gas mixture to pass
the entire reactor. The exit concentrations of gases were
followed until they ceased to change. Then the next upstream
sampling tube was connected until a steady concentration
reading was obtained and so on.

3. RESULTS AND DISCUSSION

3A   Empty reactor, NO added
    A set of experiments with an empty reactor (S/V=92 m^{-1})
and addition of NO was made. The concentration of NO was such
that an additional increase in concentration did not increase
the PAN decomposition rate. Under these circumstances,
reaction (I) becomes rate determining and a first order PAN
decomposition is observed. The rate constants were obtained by
conventional ln(c)/time plots. To have a convenient rate of
decomposition, the experiments were made between 20 and 50° C.
Figure 2 is an Arrhenius plot, showing some of the earlier
results for reaction (I) together with the present data.
    The agreement between our results and the previously
determined ones is such that it is concluded that our hardware
is functioning properly. The individual rate constants are
given in Table 1. The temperature dependence between 20 and 50°
C is described by log A=15.0 s^{-1} and E_a=107.0 kJ.

Table 1 Observed first order rate constants for PAN decomposition in the presence and absence of NO.

T (K)	S/V (m^{-1})	NO	k (s^{-1})
293	92	X	$1.4(\pm 0.3) \times 10-4$
303	92	X	$7.0(\pm 1.0) \times 10-4$
313	92	X	$2.7(\pm 0.3) \times 10-3$
323	92	X	$7.7(\pm 0.9) \times 10-3$
303	903	X	$8.3(\pm 1.1) \times 10-4$
313	903	X	$2.5(\pm 0.3) \times 10-3$
323	903	X	$1.4(\pm 0.3) \times 10-2$
333	903	X	$3.5(\pm 0.7) \times 10-2$
323	92	–	$5.6(\pm 1.2) \times 10-5$
333	92	–	$4.1(\pm 0.5) \times 10-4$
313	903	–	$1.8(\pm 0.2) \times 10-4$
323	903	–	$7.6(\pm 1.1) \times 10-4$
328	903	–	$8.6(\pm 2.2) \times 10-4$
333	903	–	$1.5(\pm 0.2) \times 10-3$

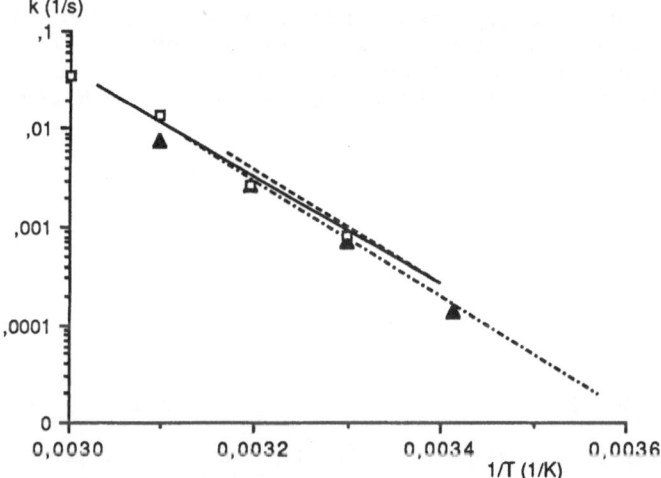

Figure 2. Temperature dependence of reaction (I). Solid line Cox & Roffey (1), dotted broken line Hendry & Kenley (2), broken line Schurath & Wipprecht (3), solid triangles present work S/V=92 m^{-1}, open squares present work S/V=903 m^{-1}.

The products formed in the 3A experiments were CO_2 and NO_2. CO_2 is formed in an amount somewhat less than PAN lost (molar basis). This is in disagreement with previous investigations e.g. (4) where, in addition to the CO_2, methyl nitrate was identified as a major product. As minor products reference (4) reports nitromethane, methane, formaldehyde and nitrous acid. The NO_2 formed was just over two times the amount of PAN lost. This is consistent with one molecule being formed through

reaction (I) and one formed through reaction (II). No other
product was found with the IR method. Some of the carbon
missing from the mass balance in our experiments could be
present in minor products below the detection limit. It is
clear however that methyl nitrate is not a major product in
our experiments.

3B Filled reactor, NO added

     To investigate whether reaction (I) was influenced by a
change in S/V ratio, some experiments were made with NO and
the filled reactor (S/V=903 m^{-1}). From Table 1 it is seen that
no significant change in rate was observed, except possibly at
323 K. The same product pattern was seen in the 3B and 3A
experiments. It is concluded that reaction (I), the
decomposition of PAN into NO_2 and the peroxyacetyl radical is a
gas phase reaction.

3C Empty and filled reactor, no NO added

     In the absence of NO, PAN should equilibrate with NO_2 and
the peroxyacetyl radical. The decomposition should stop unless
other routes for removal of the peroxyacetyl radical or direct
decomposition of PAN into stable products is possible. In
practice, the decomposition continues but the rate without NO
addition is considerably lower than when NO is added. Reaction
(I) is then never far from equilibrium and the [PAN]/[NO_2]
ratio controls the peroxyacetyl radical concentration.
     Possible reactions controlling the rate of PAN removal
are:

     PAN   --> CH_3ONO_2 + CO_2 (III)

     PAN   -->   ??   (IV)

     $CH_3C(O)O_2$ -->   ??   (V)

     2 $CH_3C(O)O_2$ -->   2 $CH_3C(O)O$ + O_2 (VI)

            NO_2
     PAN   -->   ??   (VII)

     Reaction (III) takes place in the gas phase, possibly via
a cyclic PAN intermediate with a rate [PAN]·$k_{(III)}$ (4).
Reaction (IV) is a heterogeneous, direct decomposition of PAN
with rate [PAN]·S/V·$k_{(IV)}$. Reaction (VI) is the gas phase self
reaction of peroxyacetyl radicals with rate
[$CH_3C(O)O_2$]^2·$k_{(VI)}$. Reaction (V) is a heterogeneous loss of of
peroxyacetyl radicals with rate [$CH_3C(O)O_2$]·S/V·$k_{(V)}$. The nature
of reaction (VII) is not known in detail but it is needed to
explain our experimental results. It proceeds with a rate
k_{VII}·[PAN]·[NO_2].
     Analysis of data from PAN decomposition experiments has
been made earlier, along similar lines, by Schurath and
Wipprecht (3). Their experiments with large NO_2 additions
showed that the decomposition rate of PAN was linear to

1/[NO$_2$]. Thus, it was concluded that heterogeneous
decomposition of peroxyacetyl radicals (reaction V) was the
main reaction route. A small contribution from a direct PAN
decomposition (reaction III or IV) was also observed. Senum et
al.(4) managed to analyse their decomposition data using a
simple first order approach. Our experiments without NO may
also be analyzed in this way, as exemplified in Figure 3 and
the results are given in Table 1.

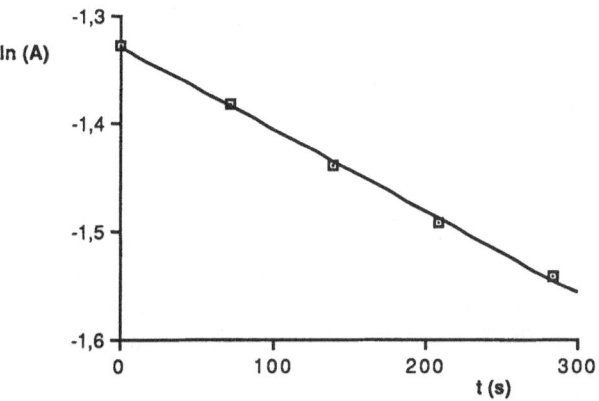

Figure 3. Experimental data from PAN decomposition at 50° C
and S/V=903 m^{-1}.

In experiments with various amounts of NO$_2$ added to the
PAN-containing gas we find a complex rate dependence on NO$_2$.
Figure 4 shows the decomposition rate, normalized by the PAN
concentration as a function of NO$_2$ concentration at 333 K, with
and without the extra reactor surface present.

The rate decreases with increasing NO$_2$ concentration in
the low NO$_2$ range. This is expected if the main reacting
species is the peroxyacetyl radical. The rate is also
sensitive to the S/V ratio of the reactor. This indicates that
reaction (V) is the principal path for the PAN decomposition
at low NO$_2$ concentration in the reactor. It is also in good
agreement with the results of Schurath and Wipprecht (3).
After passing a minimum around 100 ppm NO$_2$ the rate increases
slowly with increasing NO$_2$ concentration as depicted in
reaction (VII). The difference between the data from the empty
and filled reactor indicates that this also is a heterogeneous
process.

In the experiments with no NO$_2$ added we observe CO$_2$,
methyl nitrate and in the case with the filled reactor also
NO$_2$. The molar ratio (CO$_2$ formed)/(PAN lost) was considerably
less than one both with and without the extra reactor surface.
In the experiments made with the empty reactor, all nitrogen
lost with disappearing PAN was retrieved as methyl nitrate. In
the filled reactor experiments however, the nitrogen lost from

decomposing PAN was present in approximately equal parts as NO_2 and methyl nitrate.

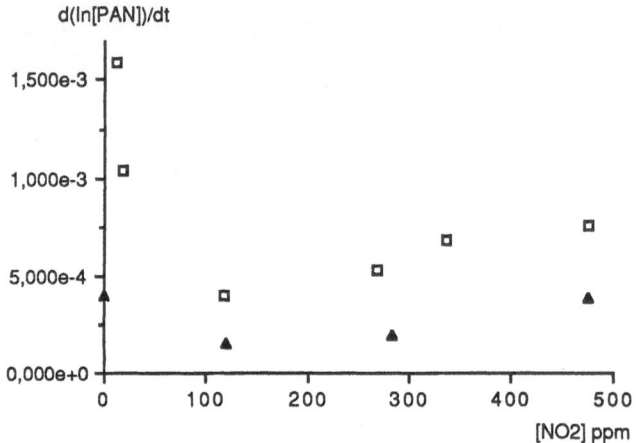

Figure 4   PAN decomposition rate at 333 K, normalized by the PAN concentration, as a function of NO_2 concentration. Solid triangles S/V=92 m^{-1}, open squares S/V=903 m^{-1}.

4. CONCLUSIONS
        From the work described above it is concluded that:
The decomposition of PAN into peroxyacetyl radicals and NO_2 is not effected by heterogeneous reactions.
In situations where the decomposition of PAN into peroxyacetyl radicals and NO_2 is not the rate determining step, both the S/V ratio and the NO_2 concentration affect the PAN disappearance rate and product distribution.
 At low NO_2 concentrations decomposition of peroxyacetyl radicals controls the overall PAN disappearance rate. At high NO_2 concentrations direct decomposition is the main channel.
It is unlikely that heterogeneous PAN decomposition has any effect on atmospheric chemistry.

REFERENCES
1. R.A. Cox and M.J. Roffey, Environ. Sci. Technol. 11 (1977) 900.
2. D.G. Hendry and R.A. Kenley, J. Am. Chem. Soc.,99 (1977) 3198.
3. U. Schurath and V. Wipprecht, "Proc.First European Symposium on Physico-chemical behaviour of Atmospheric Pollutants" B. Versino and H. Ott, Ed. EUR 6621, 1980, 157.
4. G.I. Senum, R. Fajer and J.S. Gaffney, J. Phys. Chem. 90 (1986) 152.
5. R. Atkinson and A.C. Lloyd, J. Phys. Chem. Ref. Data, 13 (1984) 315.
6. Y. Maeda, Y. Naka, T. Suetaka and M. Munemori, Chem. Express 2 (1987) 261.
7. T. Nielsen, A. Hansen and E. Thomsen, Atmos. Env. 16 (1982) 2447.

METHYLPEROXY SELF-REACTION :
PRODUCTS AND BRANCHING RATIO BETWEEN 223 AND 333 K

O.Horie, J.N.Crowley, G.K.Moortgat

Max-Planck Institut für Chemie,
(Division of Atmospheric Chemistry)
Saarstrasse 23,
Postfach 3060
D-6500 MAINZ.

SUMMARY.

Products from the Cl atom initiated oxidation of CH_4 were analysed by Matrix Isolation-Fourier Transform Infrared Spectroscopy (MI-FTIR) in order to determine the branching ratio k_{1a}/k_1 of the methylperoxy self-reaction.

$$2\ CH_3O_2 \quad \text{----------}> 2\ CH_3O + O_2, \qquad (1a)$$
$$\text{----------}> HCHO + CH_3OH + O_2 \qquad (1b)$$
$$\text{----------}> CH_3OOCH_3 + O_2 \qquad (1c)$$

The value $k_{1a}/k_1 = k_{1a}/(k_{1a} + k_{1b} + k_{1c})$ was found to vary linearly with temperature in the range 223 to 333 K, and is given by the relationship: $k_{1a}/k_1 = (2.3\pm0.1)\times10^{-3}\ T - (0.45\pm0.02)$ where T is the temperature in Kelvin.

1. INTRODUCTION.

The important role that peroxy radicals play in both atmospheric and combustion processes has stimulated a considerable research effort over the last 15 years. The results from research on the UV spectrum and kinetics of the self reaction of CH_3O_2 are summarized in a recent publication from this laboratory (1). It is now generally accepted that the products from the methylperoxy self-reaction arise as a result of intramolecular processes following formation of a transient intermediate $[CH_3O_4CH_3]$:(2,3)

$$CH_3O_2 + CH_3O_2 \ \text{---->}\ [CH_3O_4CH_3] \ \text{--->}\ 2\ CH_3O + O_2 \qquad (1a)$$
$$\text{--->}\ CH_3OH + HCHO + O_2 \qquad (1b)$$
$$\text{--->}\ CH_3OOCH_3 + O_2 \qquad (1c)$$

In the presence of excess oxygen, the methoxy radical (CH_3O) formed in reaction 1a is rapidly removed as follows:

$$CH_3O + O_2 \ \text{----->}\ HCHO + HO_2 \qquad (2)$$

The HO_2 thus produced may further react with CH_3O_2 to form molecular products:

$$HO_2 + CH_3O_2 \quad \text{-----> } CH_3OOH + O_2 \qquad\qquad (3a)$$
$$\text{-----> } HCHO + H_2O + O_2 \qquad\qquad (3b)$$

As is expected for a reaction that passes through an association complex, the overall reaction rate, $k_1 = k_{1a} + k_{1b} + k_{1c}$ shows a negative temperature dependence and the relative efficiencies of the three possible product channels also vary with temperature. To date, several studies of the room temperature branching ratio have failed to produce a consistent result. Of these studies only two have employed end-product analysis for determination of the branching ratio (4,5), the other having inferred a branching ratio from kinetic measurements (6). Two studies of the temperature dependence have recently appeared, (2,7) though in neither case did the temperature range correspond to that found in the stratosphere.

This paper describes a quantitative product analysis performed to accurately measure the relative importance of reactions 1a, 1b and 1c in the atmospherically relevant temperature range of -50 to +60°C (223 to 333 K).

2. EXPERIMENTAL

Products from the CH_3O_2 self reaction were monitored in a novel apparatus consisting of a temperature regulated quartz flow tube, a molecular beam generating module and a matrix isolation (MI)-FTIR sampling system (Figure 1). Methylperoxy radicals were produced by flowing gas mixtures containing Cl_2 (0.26-6.3 x 10^{15}), CH_4 (2-4 x 10^{18}) and O_2 (0.12-25 x 10^{18} molecules cm^{-3}) into the flow tube where they were subjected to ultraviolet photolysis (285 to 350 nm). The following reactions then occur:

$$
\begin{array}{llll}
Cl_2 + h\nu & \text{------>} & Cl + Cl & (4) \\
Cl + CH_4 & \text{------>} & CH_3 + HCl & (5) \\
CH_3 + O_2 + M & \text{------>} & CH_3O_2 + M & (6)
\end{array}
$$

The host matrix material employed was CO_2, therefore, in addition to the reactive gases, a flow of CO_2 was added which represented approximately 10% of the total flow. The gas mixture at atmospheric pressure, comprising both products and untransformed reactants, passed through the molecular beam generator before being directed onto a polished gold finger that could be held at temperatures between 5 and 300 K. For these experiments the cold finger was held at 50 K whilst the matrix was grown, and then cooled to 5 K for IR analysis. Depositing at 50 K ensured that no O_2 (which comprises 80% of the total flow) and only a small fraction of the CH_4 (comprising 8% of total flow) condensed. This had the effect of selectively concentrating the species of interest in the gas mixture, resulting in an estimated ten-fold increase in the concentration of the products trapped in the CO_2 matrix. Infra-red analysis of the trapped species was performed using a Bomem DA03.01 FTIR spectrometer. A typical spectrum is shown in Figure 2. The mixing ratio of the products in the gas phase was then determined by comparison with calibration spectra that were taken under identical conditions.

Experiments were carried out at 7 different temperatures, namely 223, 233, 253, 273, 293, 313 and 333 K. In all cases the total pressure was between 754 and 765 Torr.

The rate of chlorine photolysis, reaction 4, was determined by measuring the rate of change of absorption at 330 nm in the reaction cell when a static mixture of $Cl_2/CH_4/O_2$ was photolysed. The Cl_2 removal rate is an important parameter for the computer simulations that are described later.

Figure 1. Schematic Representation of Experimental Set-up.

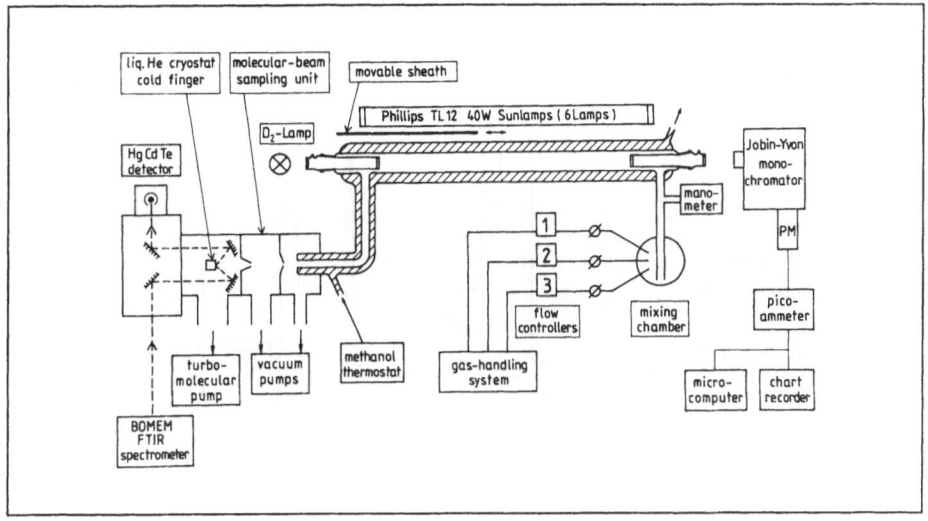

3. RESULTS.

Analysis of spectra obtained as described above revealed the presence of HCHO, CH_3OH, HCOOH and CO at their characteristic CO_2 matrix absorption frequencies of 1734.7, 1026.8, 1750.1 and 2139.2 cm^{-1}, respectively. The CO was present as an impurity in the CH_4.

Previous room temperature studies of the branching ratio (4,5) employed simple relationships between the HCHO and CH_3OH concentration to calculate the k_{1a}/k_{1b} ratio. This approach assumes that all HCHO product arises via reaction 1b and via reactions 1a and 2. Subsequent work has shown however that reaction 3b contributes ca 40% to overall reaction 3 at room temperature (8). In order to analyse product concentrations in terms of the branching ratio of reaction 1, reaction 3 and any others that produce or remove species such as HCHO or CH_3OH must be taken into account. To this end, the Facsimile simulation program (9) was employed to help extract branching ratio information from the experimentally obtained data. In these calculations, the input parameters k_{1a}, k_{1b} and k_{1c} were varied until the experimentally observed value of [HCHO]/[CH_3OH] was simulated for each experiment. The reaction scheme employed for this purpose included all of the important secondary reactions. Because there was no evidence for CH_3OOCH_3 in the spectra (the product of channel 1c) a contribution from this channel had to be estimated on the basis of previous work (4-7). Generally a value of less than 10% is chosen in the literature and calculations were therefore carried out in which the contribution from (1c) was equal to either zero or 10%. The results of these calculations are expressed graphically in Figure 3. Each point on this graph represents an averaged value of k_{1a}/k_1 at each temperature, and includes both the results for $k_{1c}= 0$ and $k_{1c}= 10\%$.

Figure 2. I.R. Spectrum of Products From Experiment at 223 K

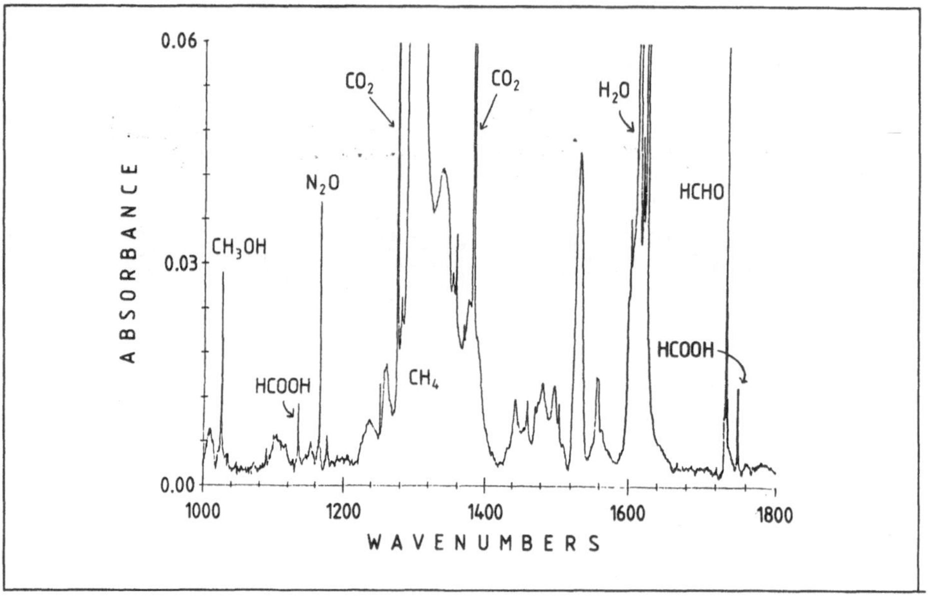

4. DISCUSSION.

Figure 3 reveals a linear temperature dependence for k_{1a}/k_1 in the temperatue range 223 to 333 K. The unweighted data from this plot were therefore fitted to an expression of the form Y = mX + C, giving :

$$k_{1a}/k_1 = (2.30\pm0.1) \times 10^{-3} T - (0.45\pm0.02)$$

From this relationship, a value of $k_{1a}/k_1 = 0.24\pm0.02$ is obtained for 298 K, which may be compared to previous room temperature determinations. Figure 3 shows all recent estimates of k_{1a}/k_1 in the temperature range relevant to this study. In each case all points were calculated from expressions given to describe the temperature dependence of the branching ratio rather than from raw data (except for the Niki et al.(5) and Kan et al.(4) results where only the room temperature branching ratio was measured). The results attributed to Niki and Kan in Figure 3 have been adjusted to take reaction 3b into account. It should be noted here that the dotted line representing the Lightfoot (2) work was derived from an expression that took into acount all previous measurements, and does not represent their own experimental data which was obtained in a different temperature range. Despite this, there appears to be excellent aggreement with this work. The adjusted Niki et al. result is also seen to be in good agreement.

The question why CH_3OOH, the third most important product after HCHO and CH_3OH, is not detected in these experiment is now addressed. The strongest IR gas phase absorption bands of CH_3OOH are at 2963.8 and 1330 cm^{-1}, with a further, weak absorption at 821.1 cm^{-1} having been used for its quantitative determination (5). An examination of Niki's spectrum (5) reveals that the relative IR cross-sections of HCHO and CH_3OOH at 1746 and 821 cm^{-1} respectively would rule out detection of CH_3OOH at this wavelength for the predicted concentrations in our system. Assuming that the gas phase and CO_2 matrix trapped IR absorptions of CH_3OOH are within a few cm^{-1} of each other, it is also very likely that the absorption bands at 2960 and 1330 cm^{-1} are obscured by the strong broad CH_4 absorptions in these regions.

Figure 3.   Comparison of Results in the Range 223 to 333 K.

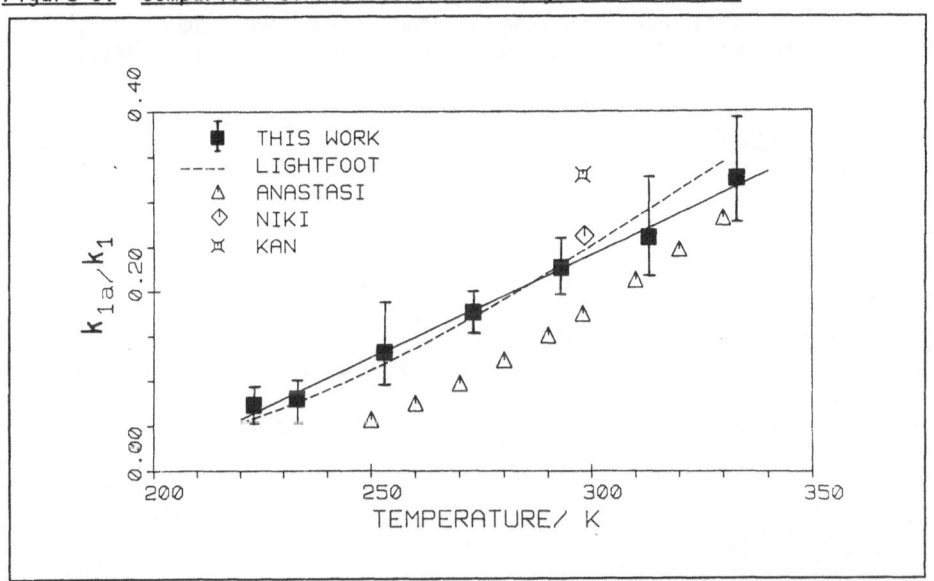

The analysis of the results has until now been carried out in terms of $HCHO/CH_3OH$ ratios at different temperatures. It is also an important test of the validity of the technique to compare measured and predicted absolute concentrations. For HCHO and CH_3OH there is reasonably good agreement, with no apparent systematic difference between predicted and observed concentrations. In contrast, the predicted HCOOH concentration is between 10 and 1000 times lower than measured, suggesting that there is an unexpected source in the system. One possibility considered is the heterogeneous decomposition of CH_3OOH to form H_2 and $HCOOH$ as found in the liquid phase (10).

A sensitivity analysis showed that the [HCHO]/[CH_3OH] ratio was sensitive to the input values for the k_1 branching ratio, the relative rates of reaction 3 and 1 at low temperature and the value selected for the k_{3a}/k_{3b} branching ratio. In the latter case a zero contribution from k_{3b} yielded a room temperature branching ratio for reaction 1 of approximately 0.3, in reasonable agreement with the older result of Niki et al.(5)

A combination of these results with those from reference 1, gives a final recommended value for the individual rate coefficients k_{1a}, k_{1b} and k_{1c} of 0.9, 2.4 and 0.3 x 10^{-13} cm^3 molecule^{-1} s^{-1} respectively.

5. REFERENCES.

1) F. Simon, W. Schneider and G.K. Moortgat, "UV spectrum of the methylperoxy radical, and the kinetics of its disproportionation reaction at 300 K". Submitted to Int. J. Chem. Kinetics, 1989

2) P.D. Lightfoot, R. Lesclaux and B. Veyret, "A flash photolysis study of the CH_3O_2 + CH_3O_2 reaction: Rate constant and branching ratios from 248 to 573 K". J. Phys. Chem., In Press.

3) P. Ase, W. Bock and A. Snelson, J. Phys. Chem., _90_, 2090 (1986).

4) C.S. Kan, J.G. Calvert and J.H. Shaw, J. Phys. Chem., _84_, 3411, (1980).

5) H. Niki, P.D. Maker, C.M. Savage and L.P. Breitenbach, J. Phys. Chem., _85_, 877, (1981).

6) D.A. Parkes, Proc. Int. Symp. Comb., _15_, 795, (1975).

7) C. Anastasi, P.J. Couzens, D.J. Waddington, M.J. Brown and D.B. Smith, "The self Reactions of Methylperoxy Radicals in the Gas Phase",Abstract available at 10th Int.Symp. on gas phase kinetics, Swansea, 1988.

8) M.E. Jenkin, R.A. Cox, G.D. Hayman and L.J. Whyte, J. Chem. Soc. Faraday Trans., 2, _84_, 913, (1988).

9) FACSIMILE Program, UK AERE Harwell, Computer Science and Systems Division, Didcot, Oxon, England.

10) A. Rieche and F. Hitz, Chem. Ber., _62_, 2458, (1929).

KINETIC STUDY OF THE EQUILIBRIA: CH_3O_2 + NO_2 <==> $CH_3O_2NO_2$ AND $CH_3C(O)O_2$ + NO_2 <==> $CH_3C(O)O_2NO_2$ IN THE GAS PHASE BY FLASH PHOTOLYSIS.

R. LESCLAUX, F. ZABEL, I. BRIDIER, F. CARALP, M.-T. RAYEZ, H. LOIRAT
and B. VEYRET
Laboratoire de Photophysique et Photochimie Moléculaire,
Université de Bordeaux I,
33405 Talence, France.

Summary

The equilibria between the CH_3O_2 and $CH_3C(O)O_2$ peroxy radicals and the corresponding peroxynitrates were studied in the gas phase by flash-photolysis UV-absorption. The knowledge of the stability of these molecules is very important in tropospheric chemistry because of their role as reservoirs for RO_2 and NO_x. The UV cross-section of $CH_3O_2NO_2$ was remeasured and the equilibrium constant was determined directly at around 353 K. The result was found to be in very good agreement with recommended values. The kinetic determination for the reaction forming PAN was done as a function of temperature and pressure and, combined with the results of a published study of the dissociation reaction, gave a value of the equilibrium constant.

1. INTRODUCTION

It is now well established that, in the troposphere, peroxynitrates (RO_2NO_2) can act as reservoirs of both NO_x and peroxy radicals. The importance of their roles depends on the values of the equilibrium constant, under conditions prevalent in the troposphere. We are presenting data concerning the two equilibria:

$$CH_3O_2 + NO_2 <==> CH_3O_2NO_2 \qquad\qquad [1, -1]$$
$$CH_3C(O)O_2 + NO_2 <==> CH_3C(O)O_2NO_2 \quad (PAN) \qquad [2, -2]$$

The flash-photolysis UV-absorption techniques were used for the rate constant measurements and, in the case of the first equilibrium, for a direct determination of the equilibrium constant value.

2. EXPERIMENTAL

All experiments were carried out using the flash photolysis apparatus described in detail elsewhere (1). Briefly, it consists of a 70 cm long thermostated Pyrex cell provided with a second outer evacuated jacket. The flash is generated by discharging two capacitors through external argon flash lamps. The analysing beam from a deuterium lamp passes twice through the cell and impinges onto a monochromator/photomultiplier unit. Individual experimental absorption curves are fed into a transient recorder and passed into a microcomputer for averaging and further data analysis.

The gas mixtures are flowed through the cell and regulated by Tylan flow controllers. Air was used as the carrier gas and the total pressure was 760 torr. The temperature of the cell was regulated with a thermostat at 298 K. Oxygen, and nitrogen (l'Air Liquide purity 99.5 % and 99.995 % respectively), synthetic air (same purities) and Cl_2 (l'Air Liquide 2% in N_2) were all used without purification. The Cl_2 concentration was measured repetively between experiments by optical absorption at 330 nm, using $\sigma(Cl_2) = 2.56 \times 10^{-19}$ $cm^2molecule^{-1}$.

Decay traces were analysed by non-linear least-squares (NL-LS), using numerical integration of the system of differential equations. RO_2 radicals were formed in the flash-photolysis of $Cl_2/(CH_4$ or $CH_3CHO)/$ Air mixtures:

$$Cl_2 + h\nu \dashrightarrow 2\ Cl$$
$$Cl + CH_4 \dashrightarrow HCl + CH_3 \qquad [3]$$
$$CH_3 + O_2 \dashrightarrow CH_3O_2 \qquad [4]$$
$$Cl + CH_3CHO \dashrightarrow HCl + CH_3CO \qquad [5]$$
$$CH_3CO + O_2 \dashrightarrow CH_3C(O)O_2 \qquad [6]$$

The concentration of NO_2 was monitored in situ after each experiment by its absorption at 439 nm where the cross-section is $\sigma = 6.33 \times 10^{-19}$ $cm^2molecule^{-1}$.

The NO_2 / RO_2 ratio was kept above 3 in order that the title reactions be the dominant pathways for removal of RO_2 radicals and below 10 because of the time constant of the acquisition system. However there are complications in the chemistry, due to the high concentration of radicals, that make a pseudo-first order analysis of the decay traces impossible. NO_2 photolysis can lead to ozone formation via the sequence:

$$NO_2 + h\nu \dashrightarrow O + NO \quad \text{and} \quad O + O_2 + M \dashrightarrow O_3 + M$$

the yield of O_3 was measured by flashing in situ NO_2 in air at atmospheric pressure. At the flash energy that was always used, $\Phi(O_3) = 0.15$ %.

3.RESULTS

$CH_3O_2NO_2$: The kinetics of reaction 1 was studied from 333 to 373 K and at atmospheric pressure. The experimental data were simulated using a full kinetic scheme. In these experiments it was necessary to measure accurately the cross-sections of $CH_3O_2NO_2$. This was done in the presence of an excess of NO_2 of a factor of 5 with respect to CH_3O_2 and the data are shown in Figure I along with previous published measurements. There are some discrepencies between the various spectra and the cross-section value at 250 nm was used in our kinetic simulations since there is a good agreement at this wavelength. The value of k_1 which was obtained was found to be in very good agreement with previous published values (2).

The equilibrium constant value K^*_{1c} was determined directly between 333 and 373 K, by measuring the ratios $[CH_3O_2NO_2]$ / $[CH_3O_2].[NO2]$ and k_1/k_{-1} independently . The agreement between our expression of K^*_{1c} (=1.12 $\times 10^{-28}$ exp(-11250/T)) and the recommanded value (2) is excellent.

$CH_3C(O)O_2NO_2$: The data of Reimar and Zabel (3) on the dissociation of PAN were combined with ours to yield the equilibrium constant K^*_2. The determination of k_{-2} had been done between 302 and 321 K and between 7 and 760 torr (3).

Our flash photolysis experiments were performed between 248 and 393 K and between 30 and 760 torr. A full kinetic simulation was carried out including reactions of the $CH_3C(O)O_2$, CH_3O_2 and HO_2 radicals with themselves and with NO_2. These results for the association reaction and those for the dissociation reaction were fitted with the Troe formula giving, within the common range of temperature, a value of $K^*_{2c} = 2.1 \times 10^{-8}$ cm^3 molecule^{-1} at room temperature. RRKM calculations were performed on reaction -2 and very good fits were obtained for both reactions 2 and -2 using the value of K^*_{2c} given above. The data points for both reactions and the Troe fall-off curves calculated to describe the RRKM results are given in Figure II. The corresponding expressions for the rate constants are, with Fc = 0.4:

association: $k_0 = 1.38 \times 10^{-28}$ $(T/300)^{-6.63}$ cm^{6}molecule^{-2}s^{-1}
$k_\infty = 11.2 \times 10^{-12}$ $(T/300)^{-1.03}$ cm^{3}molecule^{-1}s^{-1}

dissociation: $k_0 = 0.10 \times 10^{-2}$ exp$(-23330/RT)$ cm^{6}molecule^{-2}s^{-1}
$k_\infty = 3.65 \times 10^{16}$ exp$(-27080/RT)$ cm^{3}molecule^{-1}s^{-1}

4.DISCUSSION

Thermochemistry : For both peroxynitrates, the equilibrium constant determination allowed the calculation of ΔH°_{298} . The values of ΔS°_{298} were calculated with the MINDO/3 semi-empirical method and using the third law (4) method, the values of ΔH°_{298} were obtained (see Figure IV):

$CH_3O_2NO_2$: $\Delta H^\circ_{298} = -22.5 \pm 0.5$ kcal/mol
$CH_3C(O)O_2NO_2$: $\Delta H^\circ_{298} = -28.4 \pm 0.5$ kcal/mol

These values can be used to calculate the contribution C (RO_2) to the $O-(O)(NO_2)$ group of the additivity method by Benson (5):

$C(HO_2) = 3.7$ kcal/mol
$C(CH_3O_2NO_2) = 2.9$ kcal/mol
$C(CH_3C(O)O_2NO_2) = 4.1$ kcal/mol

giving an average of 3.6 ± 1.0 kcal/mol for the group value.

Atmosphere : There is a need for an accurate knowledge of the equilibrium constants for the reactions forming peroxynitrates. We report here two values that are important in tropospheric chemistry since PAN is very stable and can thus be transported over long distances and since methylperoxynitrate is abundant in the troposphere and is stable in the lower stratosphere. All necessary kinetic parameters are now available for the evaluation of the fate of $CH_3O_2NO_2$ in the atmosphere. In the case of PAN, our accurate determination of k_2 and K^*_2 will facilitate a similar evaluation. There is still a need for an accurate measurement of the rate constant for the reaction between the acetylperoxy radical and NO which is in competition with reaction 2.

5.CONCLUSION

Our direct study of K^*_1 is confirming previous evaluations and can be used to predict the stability of $CH_3O_2NO_2$. Our determination of K^*_2 was done by combining our direct kinetic study of reaction 2 with the recent work of ref. 3. The thermochemistry of both reactions is now well caracterised.

REFERENCES

(1) P.D. Lightfoot, B. Veyret and R. Lesclaux, (1989) to be published.

(2) NASA NASA Panel for Data Evaluation. "Chemical Kinetics and Photochemical Data for use in Statospheric modeling", Evaluation Nr.8, JPL Publication 87-41 (1987).
(3) A.Reimar and F.Zabel, proceedings of the IXth International Symposium on Gas Kinetics, Bordeaux, 1986.
(4) I.R.Slage and D.Gutman, J. Am. Chem. Soc., **1985**, *107*, 5342.
(5) S.W. Benson, Thermochemical kinetics, 2nd Ed; Wiley, New York, 1976.
(6) O.Morel, R.Simonaitis and J.Heicklen, Chem. Phys. Letters **1980**, *73*, 38.
(7) R.A.Cox and G.S.Tyndal, Chem. Phys. Letters **1979**, *65*, 357.
(8) S.P.Sander and R.T.Watson, J. Phys. Chem.,**1980**, *84*,1664-1674.

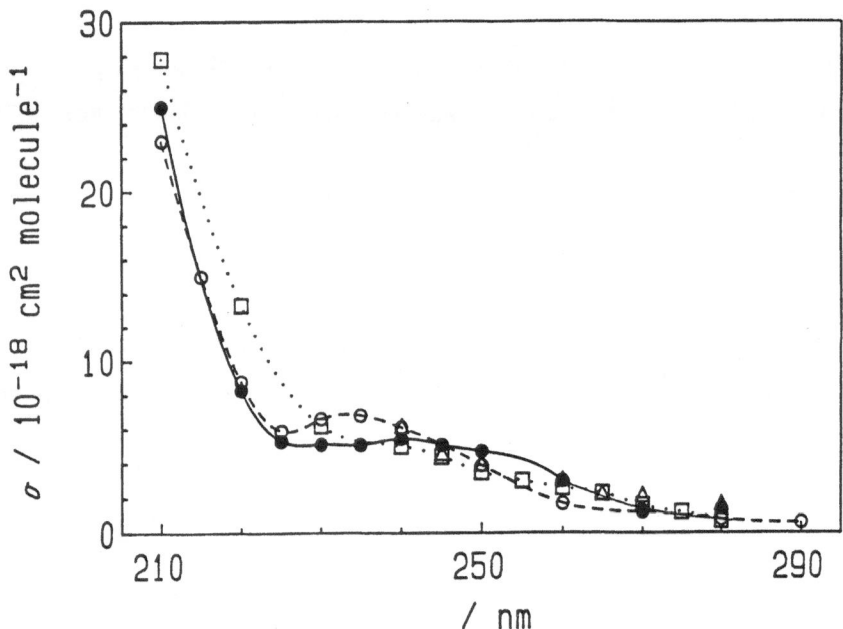

I. Cross-sections of $CH_3O_2NO_2$: filled-circle, our values; square (6); unfilled-circle (7); triangle (8).

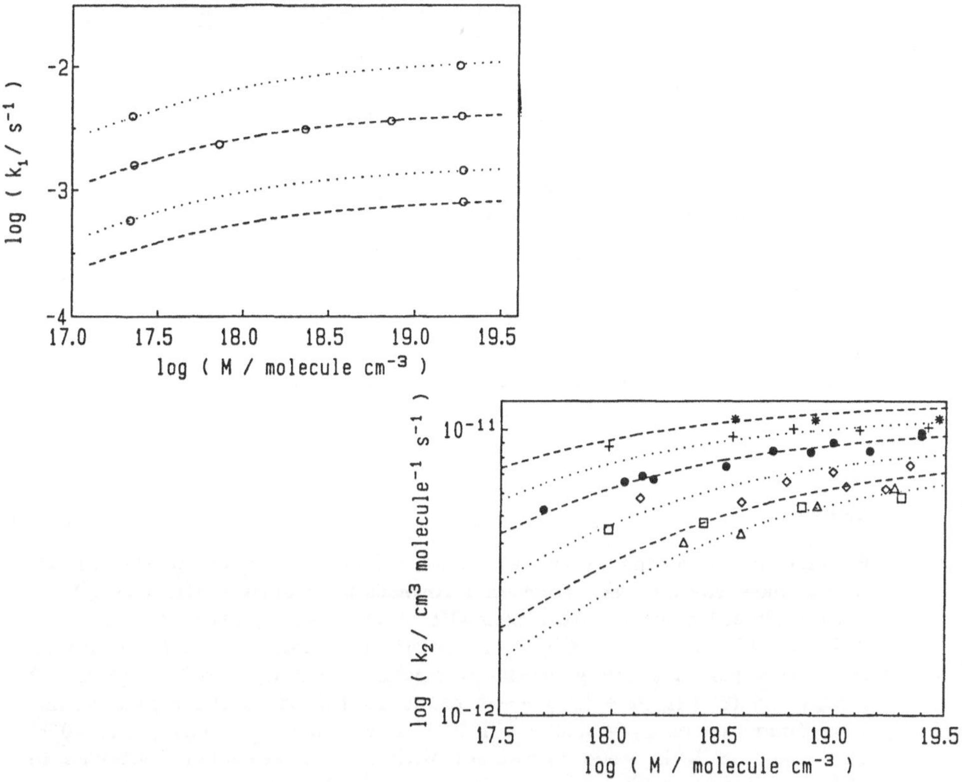

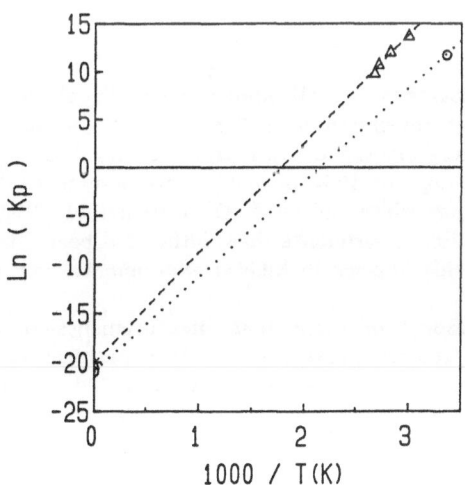

II. Top: Fall-off curves for reaction -2. The data points are from ref. 3 and the curves are RRKM calculations fitted with the Troe formula. Bottom: same figure for reaction 2. Data points are from this work.

III. Arrhenius plot of ln Kp vs. 1/T. Third law determination of   H°298 for CH3O2NO2 (dashed line) and CH3C(O)O2NO2 (dotted line).

LASER PHOTOFRAGMENT EMISSION: A NOVEL
TECHNIQUE FOR THE STUDY OF GAS PHASE
REACTIONS OF THE CH_3O_2 RADICAL

D. HARTMANN, J. KARTHÄUSER
Institut für Physikalische Chemie, Universität Göttingen,
D - 3400 Göttingen

and

R. ZELLNER
Institut für Physikalische Chemie und Elektrochemie, Universität Hannover,
D - 3000 Hannover

Summary

Product investigations of the 248 nm excimer photolysis of the CH_3O_2 radical have revealed the fractional formation of excited OH (A^2 Σ^+, v" = O) radicals. Emission from this OH state can be used as a selective and sensitive monitor of CH_3O_2 in kinetic experiments. We have applied this technique to a study of the reactions (1) CH_3O_2 + NO $\longrightarrow$ CH_3O + NO_2 and (2) CH_3O_2 + HO_2 $\longrightarrow$ products, for which the rate constant were found to be k_1 = (7.8 $\pm$ 1.7) 10^{-12} cm^3/s and k_2 = (5.2 $\pm$ 1.5) 10^{-12} cm^3/s at 298 K in good agreement with results previously obtained in different laboratories. An interference of the CH_3O_2 photofragment emission with HO_2 has not been noted.

1. INTRODUCTION

The CH_3O_2 radical has a well known near UV absorption with an absorption coefficient at maximum near 235 nm of σ = (4.8 $\pm$ 0.4) 10^{-18} cm^2 [1-3]. The corresponding electronic transition has not been fully characterized, but based on the analogy to HO_2 probably corresponds to $\tilde{X}^2A"$ $\longrightarrow$ $\tilde{A}^2A"$ which is both spin and orbital allowed. Observation of CH_3O_2 based on this transition in absorption experiments has hitherto been the most frequent detection method of this species in kinetic experiments. For a recent application see e.g. [4].

Contrary to absorption techniques measurements of the emission of electronically excited species provide generally a pathway to increased detection sensitivies. Resonance fluorescence and laser induced fluorescence present notable technical examples. In both cases the emitted radiation results directly from the same species. Because of the apparent predissociation of CH_3O_2 ($\tilde{A}$ ^{2}A") this technique fails for the CH_3O_2 radical. However, the energy available in CH_3O_2 upon absorption of a 248 nm photon is sufficient to create electronically excited fragments. The following processes are energetically possible:

$$\Delta\ H_R/KJmol^{-1}$$

$$CH_3O_2 + h\nu\ (248\ nm) \longrightarrow CH_3O\ (\tilde{X}\ ^2E) + O\ (^3P) \quad -234.0$$
$$\longrightarrow CH_3O\ (\tilde{X}\ ^2E) + O\ (^1D) \quad -44.3$$
$$\longrightarrow CH_3\ (\tilde{X}\ ^2A_2") + O_2\ (X^2\textstyle\sum_{g}^{-}) \quad -353.0$$
$$\longrightarrow CH_3\ (\tilde{X}\ ^2A_2") + O_2\ (a^1\triangle_{g}) \quad -258.7$$
$$\longrightarrow CH_3\ (\tilde{X}\ ^2A_2") + O_2\ (b\ ^1\textstyle\sum_{g}^{-}) \quad -196.1$$
$$\longrightarrow CH_2O\ (\tilde{X}\ ^1A_1) + OH\ (X^2\Pi) \quad -569.5$$
$$\longrightarrow CH_2O\ (\tilde{X}\ ^1A_1) + OH\ (A^2\textstyle\sum{}^{+}) \quad -183.1$$
$$\longrightarrow CH_2O\ (\tilde{A}\ ^1A") + OH\ (X^2\Pi) \quad -239.5$$

As a consequence we may expect O_2, OH and CH_2O excitation; formation of electronically excited CH_3O ($\tilde{A}^2A_1$) and CH_3 ($\tilde{A}\ ^2A"$) are endothermic by 143 and 553 kJ/mol, respectively, and can therefore not be obtained with 248 nm radiation.

Investigations of the photodissociation products of CH_3O_2 have to our knowledge not been performed previously. In the present paper we report the first study of this kind which has led to the detection of $OH(A^2\sum{}^{+})$ photofragment emission. Although the quantum yield of this channel is presently unknown, the resulting emission is sufficiently strong to enable kinetic experiments based on a detection of CH_3O_2 by means of this technique. Applications to studies of the reactions (1) $CH_3O_2 + NO \longrightarrow CH_3O + NO_2$ and (2) $CH_3O_2 + HO_2 \longrightarrow$ products are reported.

2. EXPERIMENTAL

In the present experiments we have used photodissociation techniques both for the generation and detection of CH_3O_2. Methylperoxi radicals are generated by the 193 nm excimer laser photolysis of azomethane in the presence of oxygen. With [Azomethane] $\sim$ 1.5 x 10^{13} cm^{-3} and a laser photon density in the order of 10^{16} cm^{-2} the typical initial CH_3-concentration amounts to $\sim 10^{12}$ cm^{-3}, which in the presence of a large excess of O_2 ($\sim$ 18 mbar) is readily ($\tau \sim$ 15 μs)converted to CH_3O_2. The detection of CH_3O_2 is achieved by a second (248 nm) excimer laser, triggered with a variable time delay ($\Delta\ t_1$) after the first photolysis laser. A schematic representation of the trigger sequence used in this technique is shown in Fig. 1. Both laser beams are directed through the reaction cell coaxially but counterpropagating, similar as in previous laser photolysis/LIF experiments performed in our laboratory [5,6]. The resulting emission from the center of the cell is collected via quartz optics and focussed onto a photomultiplier (EMI 9789 QB). For the purpose of identification of the emitted radiation a small monochromator (McPherson 218, 0.3 m) is placed in front of the photomultiplier; otherwise a simple cut-off filter (WG 280) is used to suppress laser stray light. The photomultiplier signal is detected by means of a box-car integrator with the gate width set at 0.5 μs.The signal obtained after about 500 individual laser pulses is displayed in analogue mode on a chart recorder. Details of this technique will be described elsewhere [7]. For the experiments on reaction (2) HO_2 radicals are generated simultaneously with CH_3O_2 using a cophotolysis of CCl_4 and azomethane at 193 nm. The conversion of the chlorine atome resulting from CCl_4 to HO_2 is achieved by adding small amounts of CH_3OH in the presence of excess O_2, viz:

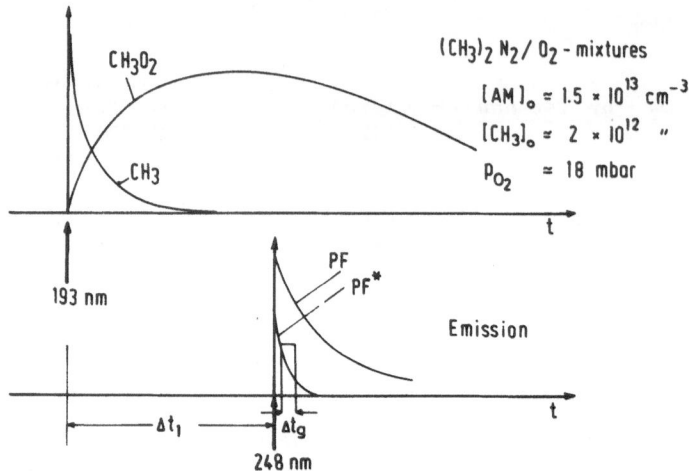

Fig. 1: Schematic representation of excimer laser trigger sequence in kinetic experiments of CH_3O_2 reactions using photofragment emission.

$Cl + CH_3OH \longrightarrow HCl + CH_2OH; \quad CH_2OH + O_2 \longrightarrow CH_2O + HO_2$. In some experiments $COCl_2$ instead of CCl_4 has been used as a primary Cl-atom source.

3. RESULTS AND DISCUSSION

3.1. Photodissociation of CH_3O_2 at 248 nm

The products of photodissociation of CH_3O_2 at 248 nm have been investigated using both emission and LIF studies. In the wavelength region accessible to us only the emission of OH (A ^{2} Σ^+, v" = O) could be detected. The resolution of the (0,0) band emission revealed high rotational excitation with an apparent temperature of T_{Rot} = 1100 $\pm$ 300 K [8]. A number of independent tests have shown that this emission must be ascribed to CH_3O_2 + hv. Although the dynamical details of this process remain presently somewhat speculative we suggest that OH (A $^2\Sigma^+$) is the result from the decomposition of CH_2OOH * which is formed by rapid isomerization of CH_3O_2 ($\tilde{A}$ ^2A"). A correlation diagram shows that this process is spin but not orbital allowed [7].

Among the ground state products resulting from CH_3O_2 + hv both OH $(X^2\Pi)$ and CH_3O (X ^2E) could be identified using LIF. Independent calibrations of the fluroescence intensities revealed quantum yields for the formation of these fragments of φ (OH) = 0.06 $\pm$ 0.03 and φ (CH_3O) = 0.2 $\pm$ 0.1. From simple mass balance considerations and with the assumption that the formation of OH($A^2\Sigma^+$) represents a minor channel of CH_3O_2 dissociation we conclude that CH_3 + O_2 (X,a,b) are the dominant photofragments of CH_3O_2 at 248 nm. This is in contrast to the photodissociation of HO_2, where OH + O $(^3P,^1D)$ dominate and the channel to H + O_2 is prevented by a high barrier [9,10].

3.2. Kinetics of the reactions of CH_3O_2 with NO and HO_2

The OH $(A^2\Sigma^+)$ photofragment emission of CH_3O_2 has been used to study the kinetics of the reactions of CH_3O_2 with NO and HO_2, both under pseudo-first order conditions using an excess concentration of the second reagent (NO, HO_2).

(1) $CH_3O_2 + NO \longrightarrow CH_3O + NO_2$

The study of this reaction was performed as a test of our method. The decay of the OH emission intensity in the presence of NO is found to decay exponentially (see Fig. 2) and to be proportional to the NO concentration. From this dependence a second order rate coefficient of
$$k_1 = (7.8 \pm 1.7)\ 10^{-12}\ cm^3/s$$
at 298 K ist obtained. This is in very good agreement with the presently accepted literature value of $k_1 = 7.6 \times 10^{-12}\ cm^3/s$ [11] and may be taken as

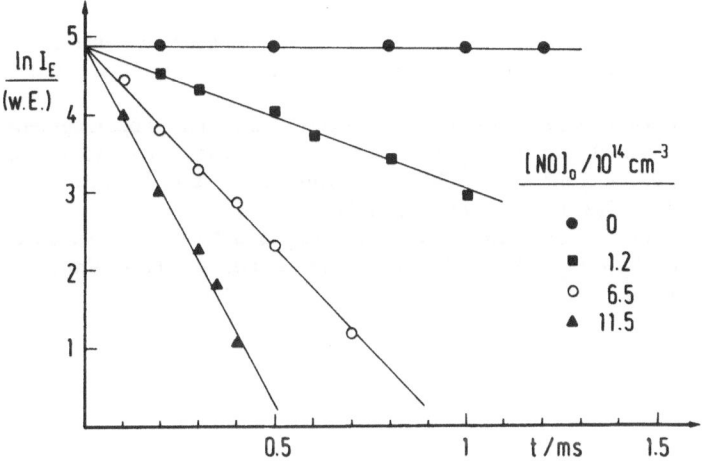

Fig. 2: Temporal behaviour of the OH emission intensity for different NO concentrations in studies of the reaction between CH_3O_2 and NO.

evidence for the suitability of the present detection technique of CH_3O_2.

(2) $CH_3O_2 + HO_2 \longrightarrow$ products

A much more challenging task compared to reaction (1) appeared the application of the present technique to the reaction between CH_3O_2 and HO_2 radicals. In the present study both reagents were generated by the 193 nm cophotolysis of azomethane / CCl_4 ($COCl_2$) / CH_3OH /O_2 mixtures. The concentration of HO_2 was in excess over CH_3O_2 and was varied between (1.0 - 5.5) 10^{13} cm^{-3} by changing the concentration of CCl_4 ($COCl_2$). It was calculated from the number of incident laser photons (N) and the concentration of the photolytic precursor and its absorption coefficient according to the methodes described previously [5.6].

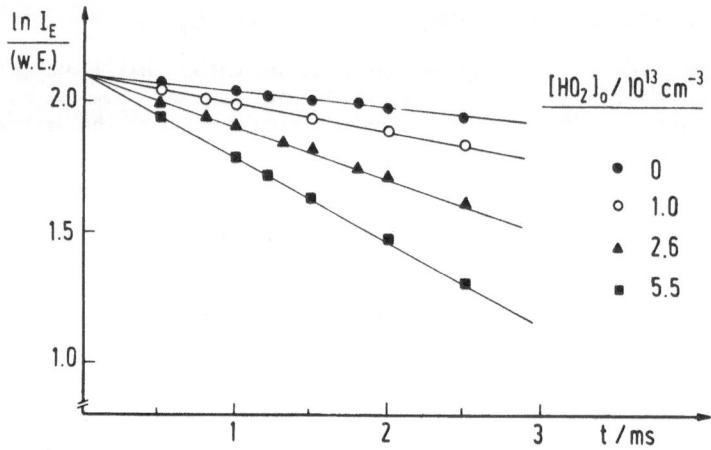

Fig. 3: Temporal behaviour of the OH emission intensity for different HO_2 concentrations in studies of the reaction between CH_3O_2 and HO_2.

For the conditions chosen in our experiments the OH photofragment emission intensity (see Fig. 3) is found to decay exponentially and to be proportional to the HO_2 concentration. The corresponding first order rate constants as a function of $[HO_2]$ are presented Fig. 4.

In order to prevent undue high losses of HO_2 due to the self-reaction HO_2 + $HO_2 \longrightarrow H_2O_2 + O_2$ (k = 1.7 x 10^{-12} cm³/s [11] we have limited the HO_2 concentration

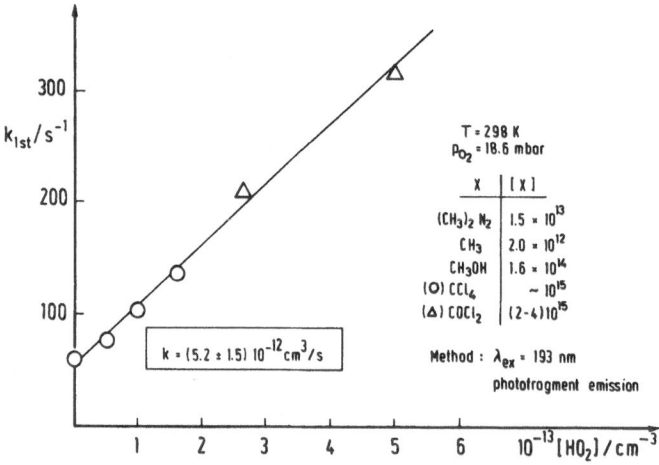

Fig. 4: Dependence of first order rate coefficients for the decay of CH_3O_2 as a function of HO_2

to $\leq 5.5 \times 10^{13}$ cm^{-3}. For the highest HO$_2$ concentration used its loss over half of the total reaction time (t = 1,25 µs) amounts to 18%. The first oder rate coefficient have been corrected accordingly.

From the slope of fig. 4 we obtain for the rate coefficient for reaction (2) at 298 K

$$k_2 = (5.2 \pm 1.5) \ 10^{-12} \ \text{cm}^3/\text{s}.$$

A comparison of this result with those from other studies is provided in table I. As can be seen our result is in very good agreement with most of the recent studies and with the value recommended by the NASA/JPL-evaluation ($k_2 = 6.0 \times 10^{-12}$ cm^3/s). This is particularly gratifying since our result is obtained by a completely different technique which does not rely on the absorption coefficients of the reagents.

Table I: Comparison   of room temperature rate coefficients for CH$_3$O$_2$ + HO$_2$
———> products

$10^{12}k/\text{cm}^3 \ \text{s}^{-1}$	Reference
6.5	Cox and Tyndall [12]
4.8	Moortgat et al. [13]
2.9	Dagaut et al. [14]
5.4	Jenkin et al. [1]
6.2	Lightfoot et al. [4]
5.2	this work

4. CONCLUSIONS

The photodissociation of CH$_3$O$_2$ at 248 nm has been shown to be associated with OH (A $^2\Sigma^+$) photofragment emission. This emission can be utilized as a detection method for CH$_3$O$_2$ in kinetic experiments. First applications to the reactions of CH$_3$O$_2$ with NO and HO$_2$ have confirmed the suitability of this technique. Extensions of this technique appear possible in the following directions:
   - studies of reactions of larger RO$_2$ radicals
   - detection of CH$_3$O$_2$ in the atmospheric environment.

Acknowledgement: Support of this work by "Deutsche Forschungsgemeinschaft" (SFB 93) and "Verband der Chemischen Industrie" is gratefully acknowledged.

REFERENCES

(1) JENKIN,M.A., COX, R.A., HAYMAN, G. and WHYTE, L.J. (1988), J. Chem. Soc. Faraday Trans. II. 84, 913
(2) SIMON, F., SCHNEIDER, W. and MOORTGAT, G.; (1989) Int. J. Chem. Kin. in press
(3) MOORTGAT, G., VEYRET , B. and LESCLAUX, R., (1989), J. Phys. Chem. in press
(4) LIGHTFOOT, P.D., LESCLAUX, R. and VEYRET. B, (1989), J. Phys. Chem., in press

(5) EWIG, F., RHÄSA, D. and ZELLNER, R. (1987). Ber. Bunsenges. Phys. Chem. 91, 708
(6) ZELLNER, R., HARTMANN, D., KARTHÄUSER,J., RHÄSA,D. and WEI-BRING, G. (1988) J. Chem. Soc. Faraday, Trans. II., 84, 549
(7) HARTMANN, D., KARTHÄUSER, J. and ZELLNER, R. (1989) to be published.
(8) HARTMANN, D., (1989); Ph. D. Dissertation, University of Göttingen
(9) LEE, L.C. (1982), J. Chem Phys. 76, 4909
(10) VASGUEZ, G.J., PEYERIMHOFF, S.D. and BUENKER, R.J. (1985), Chem. Phys. 99, 235
(11) DEMORE, W.B., MOLINA, M.J., GOLDEN, D.M., HAMPSON, R.F., KURY-LO, M.J., HOWARD, C.J., RAVISHANKARA, A.R. and SANDER, S.P. (1987) NASA/JPL-publication 87-41
(12) COX, R.A.. and TYNDALL, G. (1980)., J. Chem. Soc. Faraday Trans II.,76, 153
(13) MOORTGAT, G.K., BURROWS, J.P., SCHNEIDER, W., TYNDALL, G.S. and COX, R.A. (1986); IVth European Symp. on Physical and Chemical Behaviour of Atmosheric Pollutants, Stresa
(14) DAGAUT, P., WALLINGTON, T.J. and KURYLO, M.J. (1988), J. Phys. Chem. 92., 3833

ARRHENIUS PARAMETERS FOR THE GAS-PHASE REACTION OF OZONE WITH A SERIES OF 1-ALKENES OVER THE TEMPERATURE RANGE 240-324K

M.DONLON, D.J.O'FARRELL and J.J.TREACY
Department of Chemistry, Dublin Institute of Technology, Dublin, Ireland

H.W.SIDEBOTTOM
Department of Chemistry, University College Dublin, Dublin, Ireland

SUMMARY

Rate constants for the gas-phase reaction of ozone with ethene, propene, 1-butene, 1-pentene and 1-hexene have been determined at $298 \pm 1K$ and 1 atm total pressure using two different experimental procedures. The rate constants obtained were (in units of 10^{-18} cm^3 molecule^{-1} s^{-1}): ethene, 1.37 ± 0.04; propene 9.71 ± 0.43; 1-butene, 9.57 ± 0.52; 1-pentene, 9.20 ± 0.53; 1-hexene, 10.2 ± 0.4.
The following Arrhenius parameters were obtained:

$$k(\text{ethene}) = 5.06 \times 10^{-15} \exp[(-4861 \pm 179)/RT] \ cm^3 \text{ molecule}^{-1} s^{-1}$$
$$k(\text{propene}) = 4.88 \times 10^{-15} \exp[(-3691 \pm 138)/RT] \ cm^3 \text{ molecule}^{-1} s^{-1}$$
$$k(\text{1-butene}) = 3.7 \times 10^{-15} \exp[(-3579 \pm 308)/RT] \ cm^3 \text{ molecule}^{-1} s^{-1}$$
$$k(\text{1-pentene}) = 1.76 \times 10^{-15} \exp[(-3137 \pm 160)/RT] \ cm^3 \text{ molecule}^{-1} s^{-1}$$
$$k(\text{1-hexene}) = 1.25 \times 10^{-15} \exp[(-2856 \pm 113)/RT] \ cm^3 \text{ molecule}^{-1} s^{-1}$$

These data are compared with previously reported literature values, while this represents the first reported Arrhenius parameters for the gas-phase reaction of ozone with 1-pentene and 1-hexene.

1. INTRODUCTION

Ozone plays a major role in the chemistry of the Earth's atmosphere, its importance in the stratosphere and troposphere being particularly well documented [1,2]. Among the various classes of compounds present in the troposphere, unsaturated hydrocarbons are unique in exhibiting significant reactivity towards ozone as well as towards the hydroxyl radical. Numerous potentially important roles of the ozone-olefin reaction have been recognised since these reactions can provide mutual sinks for both ozone and olefins and concomitantly serve as sources of partially oxided compounds e.g. CO, aldehydes, ketones and organic acids.

The kinetics and mechanisms of the gas-phase reactions of ozone with organic compounds have recently been comprehensively reviewed by Atkinson and Carter [3]. Most studies on the reactions of ozone with unsaturated hydrocarbons have been limited to room temperature, the primary purpose being to elucidate the role of these reactions in atmospheric pollution. The authors point to the limited number of temperature dependence studies which, for the most part, are limited to acyclic alkenes and that even for these compounds many of the Arrhenius parameters are uncertain.

In this study, as part of a larger investigation into the effects of the number and orientation of various substituents on the Arrhenius pre-exponential factors, which reflect steric and orientation effects, and the Arrhenius activation energies, which reflect energetic factors, rate constants for the gas-phase reaction of ozone with selected unsaturated hydrocarbons over the atmospherically important temperature range 240-324K are reported.

2. EXPERIMENTAL

The experimental technique is based upon observing the increased rate of ozone decay in the presence of known excess concentrations of reactant hydrocarbon. Two distinct methods of mixing the reactants were employed.

Method 1. Temperature dependence studies were carried out in a 60 dm^3 FEP Teflon reaction chamber housed in a commercial deep-freeze cabinet. Sub-ambient temperatures were achieved using a modified temperature control unit, while higher temperatures were obtained using hot air blowers. A uniform temperature was maintained by positioning fans on either side of the bag, with the reaction temperature being monitored by a thermocouple placed in the centre of the bag. Reactants entered the chamber through a ¼" o.d. Teflon tube, placed along the centre of the bag, which was plugged at one end and perforated along its length. This allowed for rapid mixing of reactants (< 1 min) as determined by gas-chromatographic analysis of a test hydrocarbon. Ozone was prepared by passing zero grade air (Air Products) through an ozone generator (Monitor Labs) directly into the bag. Accurate concentrations of hydrocarbon were added by placing a known pressure into a calibrated volume and sweeping the reactant into the bag with diluent gas. The total volume was determined by timing the flow through a calibrated rotameter. Before the start of each run the bag was purged and then filled to a known volume with ozonized air giving a final ozone concentration < 1ppm. The reactant hydrocarbon was added and the contents allowed to mix for apporximately 2 min before sampling. Temperature equilibration was established well before the addition of the hydrocarbon. In most runs the alkene was added just before the ozone decay measurements were started but the order of addition of reactants had no detectable effect upon the ozone decay rates. Background ozone decays in the absence of hydrocarbon were determined periodically and shown to be negligible (< 10^{-4} s^{-1}) compared to the observed ozone decay rates in the presence of hydrocarbon. Ozone concentration was continuously monitored during the reaction by a Monitor Labs Model 8810 U.V. photometric ozone analyser. The signal was fed to a potentiometric chart recorder and to a BBC microcomputer. Ethene was quantitatively monitored by gas chromatography (Perkin Elmer F-11) with flame ionization detection using a 4'x1/8" o.d. stainless steel column packed with 80/100 mesh Porapak Q and operated at 50°C. All other hydrocarbon reactants were analysed on a 12'x1/8" stainless steel column packed with 20% DC 200 on 80/100 mesh chromosorb W, operated at 60°C. The hydrocarbons were monitored prior to and after reaction.

Method 2. Room temperature rate constants were also obtained in a 200 dm^3 FEP Teflon reaction chamber and used for comparison purposes. The procedure for these experiments was as prescribed by Pitts and co-workers [4], except that ozone decays were followed by U.V. absorption instead of chemiluminescence. The bag was divided into two sub-chambers with ozone being injected into one section and the hydrocarbon into the other, using ultrahigh purity air as the diluent gas. The contents of the two sub-chambers were mixed by removing the divider and pummelling the bag for about 30 seconds. After mixing the initial ozone and hydrocarbon concentrations were in the ranges 0.05-0.5 ppm and 1-10 ppm respectively.

Ethene, propene and 1-butene (Messer Grieshiem) had a stated purity >99.95% and were used as received. 1-Pentene and 1-hexene (Aldrich Chemical Company) has a stated purity of 99%+ and were trap to trap distilled before use.

3. RESULTS

Second-order rate constants were obtained by monitoring the increased rates of ozone decay in the presence of known excess concentrations of the hydrocarbons. In the presence of a hydrocarbon, the processes for removing ozone are:

$$O_3 \quad + \quad \text{wall} \quad \rightarrow \quad \text{loss of } O_3 \qquad (1)$$
$$O_3 \quad + \quad \text{hydrocarbon} \quad \rightarrow \quad \text{products} \qquad (2)$$

and hence

$$\frac{-d[O_3]}{dt} = (k_1 + k_2[hydrocarbon])[O_3] \qquad I$$

where k_1 and k_2 are the rate constants for reactions (1) and (2) respectively. For [hydrocarbon]:[O_3] $\geq$ 10 the hydrocarbon concentration remains approximately constant throughout the reaction and eqn I may be rearranged to yield:

$$\frac{-d\ln[O_3]}{dt} = k_2[hydrocarbon]) + k_1 \qquad II$$

Since the background ozone decay, k_1, was shown to be negligible compared to the decay rates observed in the presence of hydrocarbon, the second order rate constant k_2 can be readily obtained from the dependence of the ozone decay rate. $-d\ln[O_3]/dt$, on the initial hydrocarbon concentration. Excellent first order plots were obtained from the measured rates of ozone decay in all the experiments. This is in keeping with the previous work of Niki and co-workers [5,6,7] where it has established that the optimum experimental conditions for determining rate constants involve carrying out the reactions in the presence of > 50 ppm O_2, at high [hydrocarbon]:[O_3] ratios and monitoring the ozone decay. Plots of the resulting first-order rate constants against hydrocarbon concentration were linear as expected and the rate constants, k_2, shown in Table I were obtained from least squares analyses of the slopes. Fig. 1 shows the data for ethene, propene, 1-butene, 1-pentene and 1-hexene plotted in Arrhenius form over the temperature range 240-324K. The Arrhenius parameters are given in Table I together with the room temperature rate constants determined in this work and the available literature values.

4. DISCUSSION

Room temperature rate constants for the gas-phase reaction of ozone with ethene, propene, 1-butene and 1-hexene are shown in Table I, together with the more recent literature values [5,6,7,8,9,10,11,12,13]. The earlier reported rate constants of Becker et al [14], Cadle and Schadt [15] and Buffalini and Altshuler [16] have not been included since their reliability is open to question, as discussed by Atkinson and Carter [3]. Rate data obtained in this work using the divided bag technique (Method 2) described by Pitts and co-workers [4] are in excellent agreement with those determined at 298K using the perforated tube mixing method (Method 1) developed in this laboratory. This agreement lends confidence to the results which have been obtained in the temperature dependent studies using Method 1.

The results given in Table I indicate that ethene is approximately five times less reactive towards ozone than propene at 298K. Further increases in the chain length in the 1-alkene series does not lead to increased reactivity within experimental error. Arrhenius parameters derived from the rate data obtained in this work for the reaction of ozone with this series of 1-alkenes are shown in Table I. Ethene was included among the alkenes studied since it has been the subject of a number of previous investigations the results of which are in reasonable agreement [9,10,11]. The A-factor and activation energy of 5.06×10^{-15} cm^3 molecule^{-1} s^{-1} and 4.9 kcal mol^{-1} respectivley are in excellent agreement with the values reported by both De More [10] and Herron [11], even though the temperature range of this work does not overlap with the former. The activation energy of 5.8 kcal mol^{-1} reported by Kerr et al [9] would seem to be a little high. This value was obtained over a narrow temperature range and at only two temperatures and must be regarded as approximate. The increased reactivity of propene over ethene is due to the decrease in the activation energy. The lowering of the activation energy more than compensates for a slight decrease in the pre-exponential factor. The data indicate that in going from propene to 1-hexene there is a continuing small systemmatic decline in the value of E. However, this decrease appears to be offset by a corresponding decrease in the A-factor resulting in similar room temperature rate constants for this series of compounds. Whether these trends in the Arrhenius parameters are real or due to some experimental artifact is difficult to ascertain from the available data. The A-factors for the 1-alkenes are all in the region of 5×10^{-15} cm^3 molecule^{-1} s^{-1}, which is in keeping with transition state theory calculations based on a five membered ring transition state formed by 1,3 dipolar cycloaddition of ozone [10].

REFERENCES

1.   WAYNE, R.P., Chemistry of Atmospheres, Oxford University Press, Oxford (1985)
2.   BECKER, K.H. and COX, R.A., Report AP/54/84 of the Commission of European Communities (1986)
3.   ATKINSON, R. and CARTER, W.P.L., Chem.Rev., 84, 437 (1984)
4.   PITTS, J.N. JR., WINER, A.M., FITZ, D.R., ASCHMANN, S.M. and ATKINSON, R., "Experimental Protocol for Determining Ozone Reaction Rate Constants", EPA 600/53-81-024 (May 1981)
5.   STEDMAN, D.H., WU, C.H. and NIKI, H., J.Phys.Chem., 77, 2511 (1973)
6.   JAPAR, S.M., WU, C.H. and NIKI, H., J.Phys.Chem., 78, 2318 (1974)
7.   JAPAR, S.M., WU, C.H. and NIKI, H., J.Phys.Chem., 80, 2057 (1976)
8.   COX, R.A. and PENKETT, S.A., J.Chem.Soc. Faraday Trans. 1, 68, 1735 (1972)
9.   ADENIJI, S.A., KERR, J.A. and WILLIAMS, M.R., Int.J.Chem.Kinet, 13, 209 (1981)
10.   DE MORE, W.B., Int.J.Chem.Kinet, 1, 209 (1969)
11.   HERRON, J.T. and HUIE, R.E., J.Phys.Chem., 78, 2085 (1974)
12.   ATKINSON, R., ASCHMANN, S.M., WINER, A.M. and PITTS, J.N. JR., Int.J.Chem.Kinet, 13, 1133 (1981)
13.   SU, F., CALVERT, J.G. and SHAW, J.H., J.Phys.Chem., 84, 239 (1980)
14.   BECKER, K.H., SCHURATH, U. and SEITZ, H., Int.J.Chem.Kinet, 6, 725 (1974)
15.   CADLE, R.D. and SCHADT, C.J., J.Am.Chem.Soc., 74, 60002 (1952)
16.   BUFALINI, J.J. and ATTSCHULER, A.P., Can.J.Chem., 43, 2243 (1965)

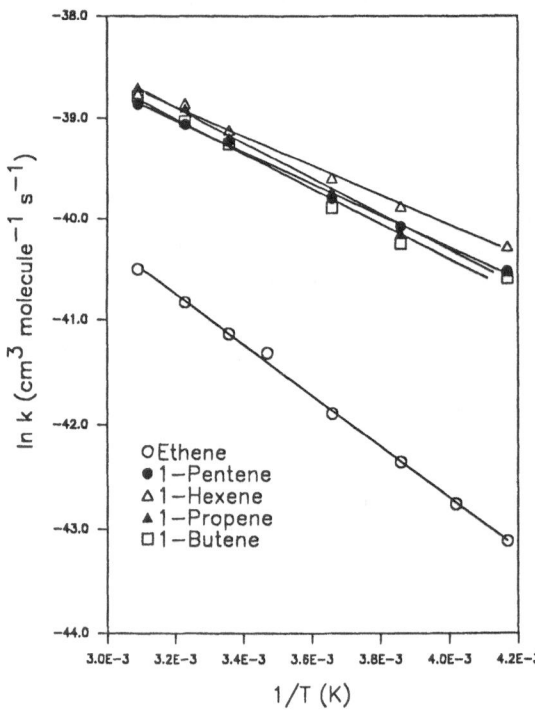

Fig 1.   Arrhenius plots of the rate constants for reaction of ozone with a series of 1-alkenes over the temperature range 240-324K

Table I: Rate constants and Arrhenius parameters for the reaction of O_3 with acyclic monoalkenes

alkene	A, cm^3 molecule^{-1} s^{-1}	E, cal mol^{-1}	k_{298}, cm^3 molecule^{-1} s^{-1}	temp. range K	technique	ref
ethene	3.3×10^{-15}	4700 ± 200	1.18×10^{-18}*	178-233	S-UV	10
	9.0×10^{-15}	5081 ± 332	1.69×10^{-18}*	235-362	SF-MS	11
		~ 5800	1.6×10^{-18}	260-294	S-CL	9
			$(1.9 \pm 0.1) \times 10^{-18}$		S-CL	7
			$(1.43 \pm 0.19) \times 10^{-18}$		S-CL	12
			$(1.80 \pm 0.11) \times 10^{-18}$		S-FTIR	13
			$(1.55 \pm 0.15) \times 10^{-18}$		S-CL	5
	1.20×10^{-14}	5232 ± 117	1.75×10^{-18}*		E	3
	5.06×10^{-15}	4861 ± 179	1.37×10^{-18}*	240-324	S-UV	M1
			$(1.59 \pm 0.11) \times 10^{-18}$		S-UV	M2
propene	6.14×10^{-15}	3769 ± 217	1.06×10^{-17}*	250-362	SF-MS	11
			1.26×10^{-17}		S-CA/CL	8
			$(1.25 \pm 0.10) \times 10^{-17}$		S-CL	5
			$(1.32 \pm 0.03) \times 10^{-17}$		S-CL	7
			$(1.04 \pm 0.14) \times 10^{-17}$		S-CL	12
	1.32×10^{-14}	4182 ± 648	1.13×10^{-17}*		E	3
			$(0.97 \pm 0.04) \times 10^{-17}$		S-UV	M2
	4.88×10^{-15}	3691 ± 138	0.96×10^{-17}*	240-324	S-UV	M1
1-butene	2.93×10^{-15}	3350 ± 40	1.03×10^{-17}*	225-363	SF-MS	11
			$(1.23 \pm 0.04) \times 10^{-17}$		S-CL	6
	3.46×10^{-15}	3403 ± 325	1.10×10^{-17}*		E	3
			$(0.96 \pm 0.05) \times 10^{-17}$		S-UV	M2
	3.73×10^{-15}	3579 ± 308	0.88×10^{-17}*	240-324	S-UV	M1
1-pentene			$(1.07 \pm 0.04) \times 10^{-17}$		S-CL	6
			1.2×10^{-17}		E	3
	1.76×10^{-15}	3137 ± 160	0.92×10^{-17}*	240-324	S-UV	M1
			$(0.95 \pm 0.03) \times 10^{-17}$		S-UV	M2
1-hexene			1.36×10^{-17}		S-CA/CL	8
			$(1.10 \pm 0.15) \times 10^{-17}$		S-CL	5
			$(1.11 \pm 0.03) \times 10^{-17}$		S-CL	6
			$(1.21 \pm 0.28) \times 10^{-17}$		S-CL	12
			1.17×10^{-17}		E	3
	1.43×10^{-15}	2938 ± 139	1.02×10^{-17}*	240-324	S-UV	M1
			$(1.10 \pm 0.07) \times 10^{-17}$		S-UV	M2

* Calculated from the Arrhenius expression
S-CA, static - chemical analysis
S-UV, static - ultraviolet absorption
S-CL, static - chemiluminescence
SF-MS, stopped-flow - mass spectrometry
S-FTIR, static - Fourier transform-infrared
E, data evaluation

ref. M1, This work - method 1
ref. M2, This work - method 2

PHOTOOXIDATION STUDY OF METHYLETHYLKETONE AND METHYLVINYLKETONE

W.Raber, K.Reinholdt, G.K.Moortgat.

Max-Planck Institut für Chemie
(Division of Atmospheric Chemistry)
Saarstrasse 23, D-6500 MAINZ

Summary

The photooxidation of methylethylketone (MEK) and methylvinylketone (MVK) has been studied by FTIR-spectroscopy at 298 K both in air in the pressure range from 50 to 760 Torr and in N_2/O_2 mixtures with varying oxygen partial pressure. For MEK, observed products were CO, CO_2, CH_3CHO, HCHO, CH_3OH, CH_3OOH, CH_3COOH and HCOOH. In MVK photolysis, CO, CO_2, HCHO, HCOOH, CH_3COOH and C_3H_6 were detected. A mechanistic interpretation of the results is presented. Estimated overall quantum yields at 760 Torr were 0.3 and < 0.1 for MEK and MVK, respectively.

1.0 Introduction

Photodissociative processes play an important role in atmospheric chemistry as they govern the production of free radicals which are responsible for the chemical transformation of several atmospheric trace gases. The photolysis of carbonyl compounds is an important example of such a process.

Carbonyl compounds in the atmosphere occur primarily as intermediate products from the photooxidation of both naturally occurring and anthropogenic hydrocarbons. A very important natural hydrocarbon is isoprene, a compound emitted by plants (1). Laboratory experiments have shown formaldehyde, methacrolein, methylvinylketone, and methylglyoxal to be products resulting from the oxidation of isoprene (2-4). In contrast, the photolysis of aromatic hydrocarbons such as toluene and xylene, which have almost strictly anthropogenic origins, results in products such as glyoxal, methylglyoxal, and biacetyl. The photooxidation of higher alkanes results in the production of both acetone and methylethylketone (5).

All of the above mentioned carbonyl compounds may be further transformed, partially through reaction with OH and NO_3 radicals and/or O_3, and partially through solar photolytic decomposition. Photolysis processes in particular provide a pathway that can result in the formation of tropospheric ozone. In order to develop more accurate models of the atmosphere it is thus important to characterize the stoichiometries and kinetic pathways following photolysis of various carbonyl substances. The photolytic parameters of a few of the smaller carbonyl compounds such as formaldehyde and acetaldehyde have been obtained in the past (6-9). Photochemical data for the other atmospheric aldehydes and ketones however, are still relatively sparse; a few primary processes have been mentioned by Calvert and Pitts (10).

The aim of the present study, was to investigate the photooxidation mechanism of MEK and MVK under atmospheric conditions.

2.0 Experimental

The experiments to determine the product distribution of the photolyis of various carbonyl compounds were carried out in a long-path multiple-reflection cell coupled to a Bomem DA3 Series FTIR spectrometer as described previously (8). The cell consisted of a 1.4 m long quartz tube with an inner diameter of 200 mm and a volume of 44.2 l. The number of passes made by the IR-beam (36) was optimized to give a maximum signal-to-noise ratio, this corresponding to a path length of 43.2 m.

The cell was surrounded by 7 radially mounted Philips TL12 Sunlamps. These lamps emit in the wavelength region 280 to 350 nm with a broad maximum near 310 nm. MEK absorbs between 230 and 320 nm with a maximum at 275 nm ($\sigma=6\times10^{-20}$ cm^2 $molecule^{-1}$). The MVK absorption spectrum extends from 260 to 400 nm, $\sigma_{max}=7\times10^{-20}$ cm^2 $molecule^{-1}$ at 330 nm (see Figure 1).

Figure 1: <u>UV absorption spectra of MEK and MVK compared to TL12 sunlamp emission spectrum (arbitrary units)</u>

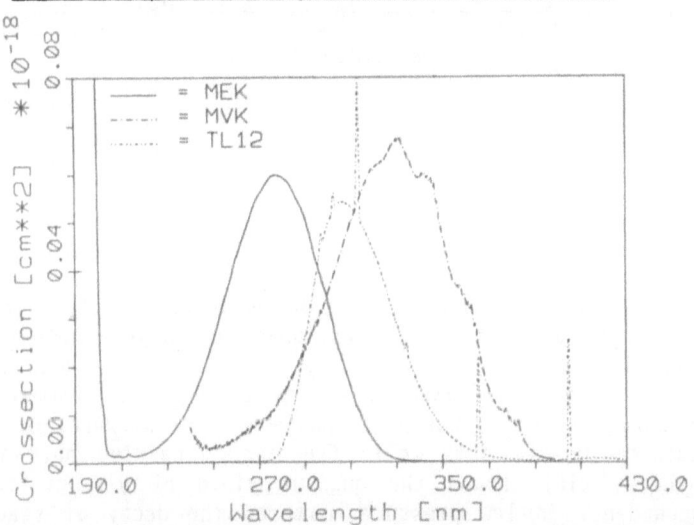

Gases were admitted through a specially designed gas inlet system providing full instantaneous mixing. A Cu:Ge detector was used for the acquisition of infrared spectra in the 450 to 3940 cm^{-1} range. Spectra were measured at 1 cm^{-1} resolution.

In order to perform both qualitative and quantitative analysis of the reaction mixture, calibration spectra of MEK, MVK and expected products were measured in the same apparatus. MEK (Merck, analytical grade) and MVK (Aldrich,99% quoted purity) were purified by double distillation and degassing immediately prior to dosing into the cell.

Oxygen (Linde, 5.0), nitrogen (Linde, 5.0) and synthetic air (Linde 5.0) were used without further purification.

The concentrations of the reactants were varied between 0.6 and 9.0 x 10^{15} molecules cm^{-3} in mostly 760 Torr air. In a few experiments the total pressure (50 to 760 Torr) or the oxygen partial pressure was varied (between 4 and 760 Torr).

The total irradiation time amounted to 50 minutes, spectra were taken at 5 minutes intervals.

Figure 2: FTIR spectrum obtained after 35 min photolysis of MEK

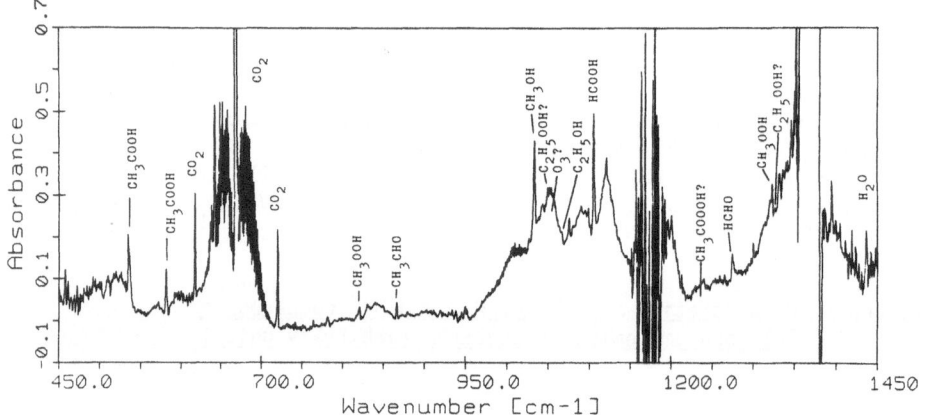

3.0 Results

3.1 MEK

A typical FTIR-spectrum resulting from photolysis of 7.2 x 10^{15} molecules cm^{-3} of MEK in air after 35 minutes is displayed in Figure 2.

Identified products are CO_2, CO (not shown), H_2O, HCHO, CH_3CHO, CH_3OH, CH_3OOH, HCOOH and CH_3COOH. Other bands in the spectrum (designated by question marks) are tentatively assigned to C_2H_5OH, C_2H_5OOH, O_3 and CH_3COOOH.Quantitative analysis of all products was performed by either measuring peak heights of suitable absorption bands by comparison with calibration curves or by computer stripping (BOMEM software) of reference spectra measured in the same cell. Experiments carried out at various total pressures (air) showed the quantum yield of dissociation to be pressure dependent. In low pressure mixtures the decay of reactant and formation of products is significantly greater (by a factor of two) than in measurements at atmospheric pressure, this due to reduced quenching of the excited ketone at lower pressures.

The temporal behaviour of MEK and some of the products is shown in Figures 3a (50 Torr air) and 3b (760 Torr air). Those products not shown i.e. HCOOH and CH_3OOH are seen to increase steadily in concentration during the course of the experiment. The concentration of these compounds after 40 minutes photolysis were 4.8x10^{14} and 3.3x10^{14} molecules cm^{-3} (50 Torr) and 1.9x10^{14} and 1.4x10^{14} molecules cm^{-3} (760 Torr), respectively.

Figure 3: Concentration versus time profile for MEK photolysis

3a): 226 mT MEK in 50 Torr air 3b): 227 mT MEK in 760 Torr air

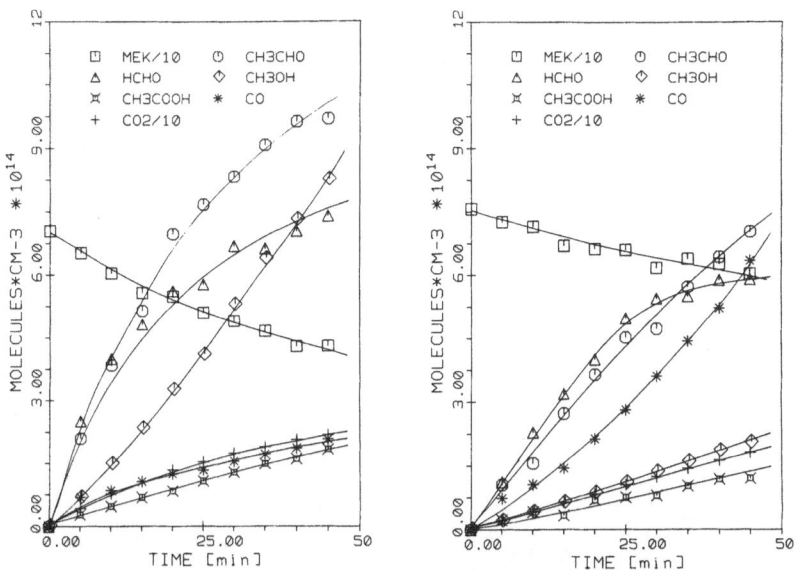

In experiments where the oxygen partial pressure was low, removal of CH_3O occurs predominantly via reaction with RCHO, resulting in enhanced methanol production (RCHO = CH_3O, CH_2O, CH_2OH, CH_3CHO).

$$CH_3O + O_2 \quad ----> \quad HCHO + HO_2$$
$$CH_3O + RCHO \quad ----> \quad CH_3OH + RCO$$

3.2 MVK

Figure 4: FTIR spectrum obtained after 40 min photolysis of MVK

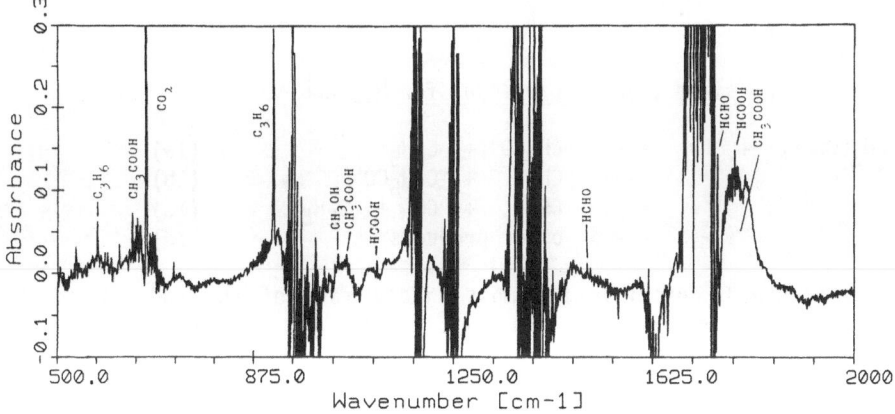

Figure 4 shows an IR spectrum obtained after 40 minutes photolysis of MVK, initial concentration 7.9×10^{15} molecules cm^{-3}. Detectable products were CO, CO_2, HCHO, HCOOH, CH_3COOH and C_3H_6.

Like MEK, the rate of photodissociation of MVK was found to be pressure dependent. At lower pressures, C_3H_6 becomes the main product.

Quantitative analysis led to the concentration-time profiles displayed in Figures 5a and 5b (50 and 760 Torr air); not shown are CH_3COOH and HCOOH, with concentrations (after 40 minutes irradiation) of 2.7×10^{13} and 1.5×10^{13} molecules cm^{-3} (50 Torr) and 1.3×10^{13} and 5.7×10^{12} molecules cm^{-3} (760 Torr), respectively.

Figure 5: Concentration-time profile for MVK photolysis
5a): 254 mT MVK in 50 Torr air 5b): 239 mT MVK in 760 Torr air

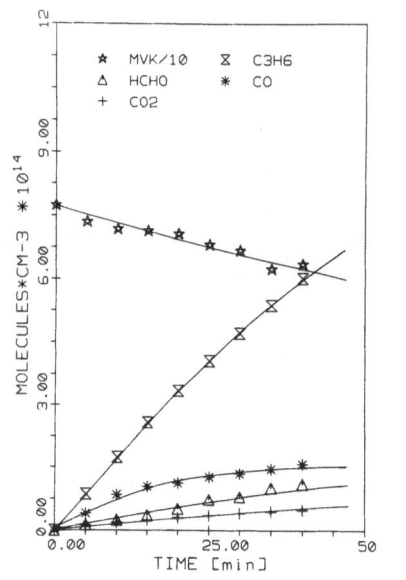 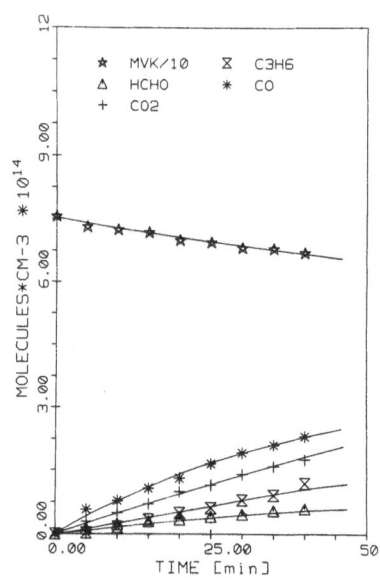

4.0 Discussion

The possible primary processes for MEK are :

$$CH_3COC_2H_5 + hv \quad ----> \quad CH_3CO + C_2H_5 \qquad (1a)$$
$$----> \quad CH_3 + C_2H_5CO \qquad (1b)$$
$$----> \quad CH_3 + CO + C_2H_5 \qquad (1c)$$
$$----> \quad \text{other products} \qquad (1d)$$

Figure 6 describes the most likely fate of the radical species produced in MEK photolysis in the presence of oxygen. Stable IR detectable compounds are shown in boxes.

The mechanistic interpretation of the results described above has been carried out with the help of the computer simulation program FACSIMILE.

The photolytic removal rates of both MEK and MVK were determined by calculating their overlap integrals with the known emission spectrum from the TL12 lamps, and comparing this with the overlap integral for Cl_2/TL12. This yielded a relative rate of photolysis for MEK and MVK with Cl_2 whose photolysis rate was measured directly by photolysing static mixtures of Cl_2, CH_4 and O_2 in the same apparatus.

Figure 6: Proposed mechanism in MEK photolysis

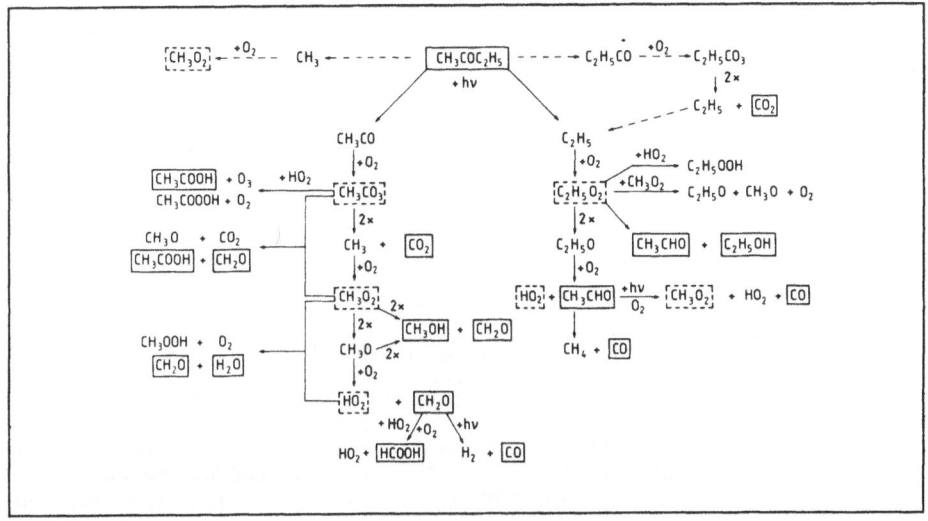

According to computer simulations which included all known secondary chemistry, reaction 1a is the major photodissociative pathway for the removal of MEK. Simulations of the data enabled calculation of the overall quantum yield of photodecomposition in the photoactive wavelength region in this experiment. For MEK, the values obtained were $\phi \approx 0.3$ at 760 and ≈ 0.6 at 50 Torr. The simulated results showed good agreement with the experimental data for MEK, CH_3CHO, HCHO, CH_3OOH, CH_3COOH and HCOOH; CO, CO_2 and methanol are predicted to be present in lower concentrations than observed.

For **MVK** the possible primary photolysis processes are:

$$CH_3COCH=CH_2 \ + \ h\nu \quad ----> \quad CH_3CO \ + \ CH=CH_2 \qquad (2a)$$
$$----> \quad CH_3 \ + \ CH_2=CHCO \qquad (2b)$$
$$----> \quad CH3 \ + \ CO \ + \ CH=CH_2 \qquad (2c)$$
$$----> \quad \text{other products} \qquad (2d)$$

Figure 7 describes a mechanistic interpretation of the appearance of the observed products. An intramolecular rearrangement of the electronically excited MVK eventually leading to propene formation is proposed. The significantly enhanced production of propene at 50 Torr can be explained by assuming that different excited states lead either to C_3H_6 production or direct photodissociation.

Figure 7: Proposed mechanism in MVK photolysis

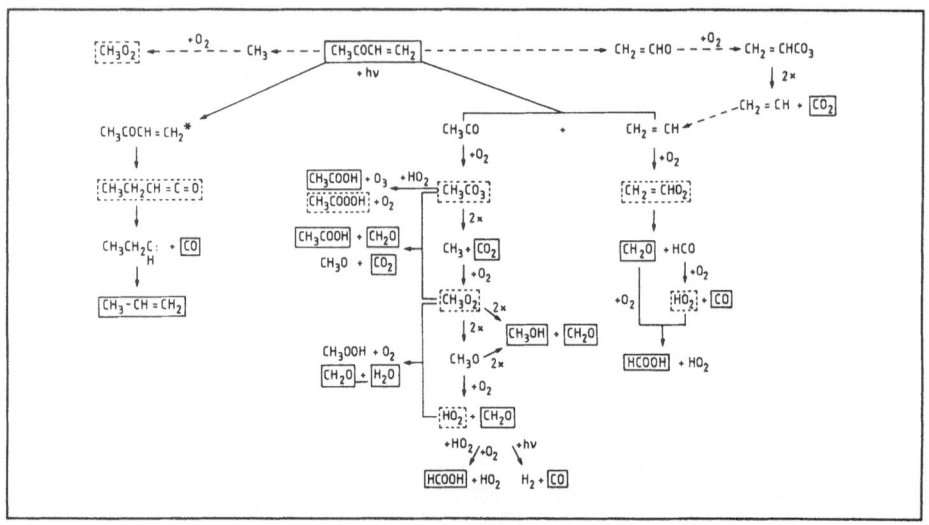

With use of computer simulations, a value for the overall photodissociation quantum yield of less than $\phi < 0.1$ was estimated.

The low conversion of MVK in these experiments may preclude detection of other products present in lower concentrations.

5.0 References

1)   R.A. Rasmussen, J. Air Poll. Control Assoc., 22, 537, (1972).
2)   R.A. Cox, R.G. Derwent and M.R. Williams, Environ. Sci. Technol., 14, 57, (1980)
3)   A.C. Lloyd, R. Atkinson, F.W. Lurmann and N. Nitta, Atm. Env., 17, 1931, (1983).
4)   J.P. Killus, G.Z. Whitten, Env. Sci. Technol., 18, 142, (1984).
5)   J.G. Calvert and S. Madronich, J. Geophys. Res., 92, 2211, (1987).
6)   R. Atkinson and A.C. Lloyd, J. Phys. Chem. Ref. Data. 13,315,(1985).
7)   G.K. Moortgat, W. Seiler and P. Warneck, J. Chem. Phys. 78 , 1185, (1983).
8)   G.K. Moortgat, R.A. Cox, G. Schuster, J.P. Burrows and G.S. Tyndall, J. Chem. Soc., Faraday Trans. 2, 85, 809, (1989).
9)   H. Meyrahn, J. Pauly, W. Schneider and P. Warneck, J. Atmos. Chem., 4, 277, (1986).
10) J.G. Calvert and J.N. Pitts Jr., "Photochemistry", J. Wiley and sons inc., (1966).

ROOM TEMPERATURE RATE COEFFICIENT FOR THE REACTION BETWEEN CH_3O_2 AND NO_3

J.N. Crowley[*], J.P. Burrows, G.K. Moortgat,
G. Poulet[#] and G. LeBras[#]

Max-Planck-Institut fur Chemie,
(Division of Atmospheric Chemistry)
Saarstrasse 23,
Postfach 3060
D-6500 Mainz

\# Centre de Recherches sur la Chimie de la Combustion et des Hautes
Températures, CNRS-45071 Orléans-Cedex, FRANCE.

SUMMARY

The molecular modulation spectroscopic technique was employed to study the kinetics of NO_3 radicals produced in the 253.7 nm photolysis of flowing gas mixtures of $HNO_3/CH_4/O_2$ at room temperature. By computer fitting of the NO_3 temporal behaviour, a rate coefficient of $(2.30 \pm 0.64) \times 10^{-12}$ cm^3 molecule^{-1} s^{-1} was obtained for the reaction between CH_3O_2 and NO_3.

1.0 Introduction

The nitrate radical absorbs strongly in the visible region and is photolysed by light at wavelengths less than 630 nm [1,2]. Consequently, its concentration in the atmosphere increases at night where it is formed by the reaction between NO_2 and O_3:

$$NO_2 + O_3 \ ---> NO_3 + O_2$$

The ozonolysis of organic unsaturated compounds [3] and the reactions of NO_3 with hydrocarbons are believed to produce organic peroxy radicals, RO_2 [4-6]. The peroxy radicals formed by such reactions at night in the troposphere are likely to react with NO_3.

Although significant amounts of the simplest and smallest organic peroxy radical, CH_3O_2, are unlikely to be produced from the slow oxidation of CH_4 by NO_3, it may be generated in the NO_3 induced oxidation of non methane hydrocarbons at night in the troposphere. However, the reactivity of CH_3O_2 towards NO_3 may be considered to be representative of several larger peroxy radicals present in the nighttime troposphere.

In this work, the reaction between NO_3 and CH_3O_2 was studied in the modulated photolysis of mixtures of HNO_3, O_2 and CH_4.

2.0 Experimental

The apparatus used in this investigation has been described elsewhere
(7). Flows of HNO_3, CH_4 and O_2, maintained by calibrated flow meters, were
mixed prior to entering the reaction vessel. This vessel consisted of
three concentric quartz tubes, providing a reaction zone (inner volume)
and two jackets. Thermostatted water was flowed through the inner jacket
to maintain the cell at 298 K, and the outer jacket was evacuated. The
reaction volume was 150 cm in length and 5 cm internal diameter.

The cell was surrounded by six photolysis lamps (Hg, 253.7 nm) mounted
inside a reflecting aluminium cylinder. The ratio of lights-on to lights-
off time for the lamps was variable between 0.1 and 9.

Quartz windows inset into the ends of the cell provided a single pass
optical path length of 132 cm. Two spectroscopic light sources were
employed: collimated beams from either a tungsten lamp or a D_2 lamp,
passed through the cell and were focussed onto the entrance slit of a
Jobin-Yvon H25 double monochromator. A slit width giving a 1.5 nm spectral
resolution was used. For the measurement of NO_3 absorption at 623 nm, two
600 nm cut-off filters were used to reduce stray light.

A steady flow of HNO_3 through the cell was maintained by passing O_2
over a reservoir of HNO_3 held in a temperature regulated bath. Variation
of the bath temperature and therefore the HNO_3 vapour pressure enabled the
partial pressure of HNO_3 in the cell to be changed. The concentration of
HNO_3 in the cell was determined from its absorption at 220 nm; CH_4 and O_2
concentrations were calculated from their flow rates and the total
pressure in the cell, which was measured by a MKS Baratron capacitance
manometer. HNO_3 was synthesised under vacuum by the reaction between H_2SO_4
and $NaNO_3$ and purified by trap to trap distillation. CH_4 (Linde 3.5) and
O_2 (Linde 5.0) were used without further purification.

3.0 Results.
3.1 Experiments in the absence of CH_4

Photolysis of mixtures of HNO_3 in oxygen leads to NO_3 production via
reactions 1 and 2:

$$HNO_3 + hv \longrightarrow OH + NO_2 \qquad\qquad (1)$$
$$OH + HNO_3 \longrightarrow NO_3 + H_2O \qquad\qquad (2)$$

Reaction with NO_2, generated in reaction 1 and present as impurity in the
HNO_3, represents the major loss of NO_3 in the system:

$$NO_3 + NO_2 + M \longrightarrow N_2O_5 + M \qquad\qquad (3)$$

The time dependent behaviour of NO_3 in the photolysis of HNO_3/O_2 mixtures
at a total pressure of 20±2 Torr was monitored by optical density changes
at 623 nm. An example of an NO_3 absorption trace obtained in this manner
is displayed in Figure 1. In the initial part of the curve, absorption

increases as NO_3 is produced via reactions 1 and 2. After circa 600 ms the NO_3 signal reaches a steady concentration determined by reactions 1,2, and 3. When the lights are switched off the NO_3 decays via reaction 3. Analysis of these curves enabled the absorption cross-section for NO_3 at 623 nm and the NO_2 concentration to be calculated. The determination of $\sigma(NO_3)$ at 623 nm, the concentration of NO_2 present as impurity in the HNO_3 and the photolytic removal rate of HNO_3 are described below:

3.1.1 HNO_3 Photolysis Rate.

Use of an NOCl actinometer enabled the HNO_3 photolysis rate to be calculated. Dissociation of one NOCl molecule by 253.7 nm radiation leads to the removal of two molecules of NOCl via reactions i and ii.

$$NOCl + h\nu \ ----> \ NO + Cl \qquad\qquad\qquad i.$$
$$Cl + NOCl \ ----> \ Cl_2 + NO \qquad\qquad\qquad ii.$$

The rate of removal of NOCl in the system was calculated by monitoring its absorption at 210 nm with the photolysis lights on. Knowledge of the relative absorption cross sections of HNO_3 (8) and NOCl (9) at 254 nm enabled a value for the photolysis rate, $k_1 = 2.8 \times 10^{-4}$ s^{-1}, to be determined.

3.1.2 NO_3 Cross-section at 623 nm.

In order to accurately measure NO_3 concentrations, it is necessary to know $\sigma(NO_3)$ at the monitoring wavelength. In this case, where the spectral resolution is low (1.5 nm) an effective $\sigma(NO_3)$ must be calculated. The rate of increase of absorption at 623 nm when the lights are on is a function of NO_3 concentration, effective $\sigma(NO_3)$ and path length, i.e.

$$d \ (OD)/dt = \sigma \times 1 \times d \ [NO_3]/dt \qquad\qquad (A)$$

The rate of NO_3 production depends on the removal rate of HNO_3:

$$d[NO_3]/dt = -k_1[HNO_3] \qquad\qquad (B)$$

Combining equations (A) and (B) enabled an effective value of $\sigma(NO_3) = 1.0 \times 10^{-17}$ cm^2 molecule^{-1} to be calculated from measurements of the initial rate of change of absorption $(d(OD)/dt)_0$. This is slightly lower than the previously determined value $(1.19 \times 10^{-17}$ cm^2 molecule$^{-1})$ obtained in this laboratory (10), which was measured at 1 nm resolution.

3.1.3 NO_2 Concentration.

The NO_2 impurity in the HNO_3 was generally present at too low a concentration to be measured precisely by optical techniques in this system. In order to determine the NO_2 concentration in the absence of CH_4, mixtures of HNO_3/O_2 were photolysed. The major NO_3 loss process in these mixtures is reaction 3. In such experiments the NO_2 concentration controls the NO_3 steady state concentration, and the rate of decay during the dark phase. Simulations of the NO_3 traces thus obtained were made using the FACSIMILE chemical simulation program (11). In this manner, an initial NO_2 concentration was obtained and thus the HNO_3/NO_2 ratio for a given batch of HNO_3. In order to simulate a slight decay of NO_3 during the steady state period in experiments where methane was either present or absent, it was necessary to introduce a first order loss of 0.2 s^{-1}. Experiments in which a 630 nm cut-off filter prevented photolysis of NO_3 in the 623 nm absorption band reproduced this effect, proving that the analysis light did not contribute to the decay. The products of this decay, i.e. either $NO + O_2$ or $NO_2 + O$ did not effect the simulations. A small first order loss of NO_3 has been observed previously in studies of this species (12).

Figure 1

NO_3 absorption profile (623 nm) and Facsimile Simulations. (A, Experimental curve, no methane. B, Experimental curve, methane present. C,D,E simulations, with k_6 equal to $1x10^{-13}$, $1.51x10^{-12}$ and $1x10^{-11}$ cm^3 molecule^{-1} s^{-1} respectively.

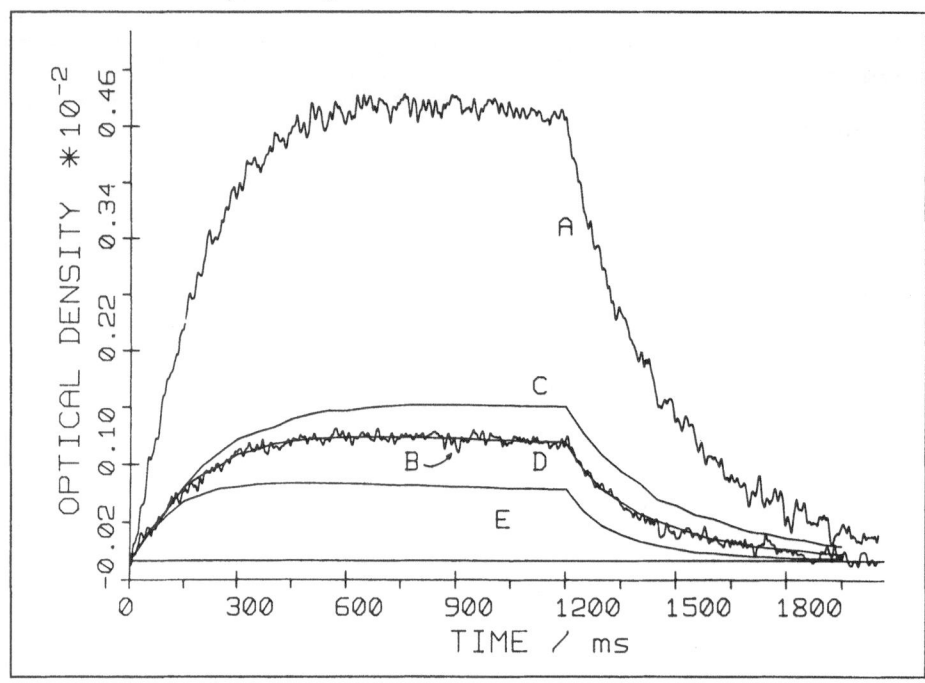

3.2 $CH_3O_2 + NO_3$ rate coefficient.

When CH_4 is added to the HNO_3/O_2 gas mixture, both the time required for NO_3 to reach steady state and the steady state concentration itself were significantly reduced as observed in Figure 1. These changes are explained by two effects: firstly the OH radicals are partitioned between reactions with HNO_3 and CH_4 resulting in a lower NO_3 production rate, and secondly, the production of CH_3O_2 via reactions 4 and 5 provides a new removal process, namely reaction 6:

$$OH + CH_4 \ ------> \ CH_3 + H_2O \qquad\qquad (4)$$
$$CH_3 + O_2 + M \ ------> \ CH_3O_2 + M \qquad\qquad (5)$$
$$NO_3 + CH_3O_2 \ ------> \ Products. \qquad\qquad (6)$$

For the NO_3 time dependence to be sensitive to reaction between NO_3 and CH_3O_2, it is important that competitive reactions of the two species are kept to a minimum. For both NO_3 and CH_3O_2, the most important competitive reactions involve NO_2, producing N_2O_5 and $CH_3O_2NO_2$ respectively, emphasing the importance of keeping the NO_2 impurity in HNO_3 as low as possible. NO_3 absorption traces thus obtained were then analysed using FACSIMILE to extract a rate coefficient for the $CH_3O_2 + NO_3$ reaction.

Table 1 lists the conditions and results of 9 experiments which were used to determine k_1. Combining the results from these experiments gives an average value for k_1 of $(2.30 \pm 0.64) \times 10^{-12}$ cm^3 molecule^{-1} s^{-1}.

Table 1.

expt.	[concentration/molecules cm^{-3}]			Pressure	k_1 #
	[HNO_3]	[NO_2]	[CH_4]		
	($\times 10^{16}$)	($\times 10^{12}$)	($\times 10^{17}$)	Torr	($\times 10^{-12}$)
A	2.03	1.13	2.53	19.5	3.00
B	1.70	1.20	2.49	19.1	2.56
C	1.30	1.14	2.14	17.8	1.60
D	2.63	1.75	2.14	17.8	1.51
E	2.29	2.70	2.19	18.2	3.30
F	2.09	2.14	2.51	17.3	2.50
G	1.10	0.99	1.57	12.8	2.14
H	1.80	0.91	2.23	20.2	2.73
I	2.20	0.90	2.58	21.0	1.43

\#   Units of k_1 are: cm^3 molecule^{-1} s^{-1}.

4.0 Discussion.

In this work, the first direct measurement of the rate coefficient between NO_3 and an organic peroxy radical has been presented. The rate coefficient obtained, $(2.30 \pm 0.64) \times 10^{-12}$ cm^3 molecule^{-1} s^{-1}, is seen to be in reasonable agreement with the result of Hjorth et al. who have recently estimated a value of 4.0×10^{-12} cm^3 molecule^{-1} s^{-1} for the reaction between NO_3 and the peroxy radical produced from the NO_3 induced oxidation of methyl ethylene [13]. It is also similar in magnitude to the rate coefficient for the reaction between NO_3 and HO_2 which has been measured to be 4.5×10^{-12} cm^3 molecule^{-1} s^{-1} (14).

There are several possible product channels for the reaction between NO_3 and CH_3O_2. The exothermic channels are listed below.

$$\Delta H \text{ / kJ mole}^{-1}$$

$$
\begin{array}{lll}
NO_3 + CH_3O_2 \longrightarrow CH_3O + NO_2 + O_2 & -39 & (6a) \\
\longrightarrow HO_2 + HCHO + NO_2 & -150 & (6b) \\
\longrightarrow HONO + HCHO + O_2 & -275 & (6c) \\
\longrightarrow HNO_3 + HCHO + O & -82 & (6d)
\end{array}
$$

Although the least exothermic, reaction 6a involves the least molecular rearrangement and by comparison to the dominant channel in the reaction between NO_3 and HO_2 (14) (where the products are $HO + NO_2 + O_2$) is expected to be the most important. The computer simulations showed that the final reaction rate obtained was essentiallly insensitive to the products of the reaction.

References

1) R.A. Graham and H.S. Johnston, J. Phys. Chem., 82, 254, (1978).
2) F. Magnotta and H.S. Johnston, Geophys. Res. Lett., 7, 769, (1980).
3) O. Horie and G.K. Moortgat, Chem. Phys. Lett., 156, 39, (1989).
4) H. Bandow, M. Okuda and H. Akimoto, J. Phys. Chem. 84, 3604, (1980).
5) C.A. Cantrell, J.A. Davidson, W.P.L. Carter and J.G. Calvert, J. Geophys. Res., 91, 5347, (1986).
6) J. Hjorth, G. Restelli, F. Cappellani and H. Skov, work presented at the EEC COST 61a Workshop September 1988, University of East Anglia, Norwich, U. K. (1988).
7) R.J. Singer, J.N. Crowley, J.P. Burrows, W. Schneider and G.K. Moortgat, J.Photochem., in press, (1989).
8) W.B. Demore, M.J. Molina, S.P. Sander, D.M. Golden, R.F. Hampson, M.J. Kurylo and C.J. Howard, A.R. Ravishankara, Chemical Kinetics and Photochemical Data for use in Stratospheric Modelling, Evalution No 8, JPL publications 87-41, (1987).
9) G.S. Tyndall, K.M. Stedman, W. Schneider, J.P. Burrows and G.K. Moortgat, J. Photochem., 36, 133, (1987).
10) J.P. Burrows, G.S. Tyndall and G.K. Moortgat, J. Phys. Chem. 89, 4848, (1985).
11) FACSIMILE program, U.K. AERE Harwell, Computer Science and Systems Division, Didcot, Oxon, England.
12) C.E. Canosa-Mas, M. Fowles, P.J. Houghton and R.P. Wayne, Faraday Trans. 2, 83, 1465, (1987).
13) J. Hjorth, G. Restelli and F. Cappellani, In press.
14) A. Mellouki, G. Le Bras and G. Poulet, J. Phys. Chem., 92, 2229, (1988).

THE REACTION OF OH WITH C_2H_2 AND THE SUBSEQUENT REACTION
OF HYDROXYVINYL WITH O_2

Z.Y. Zhang* and J. Peeters
Department of Chemistry, Katholieke Universiteit Leuven, Belgium

Summary

The kinetics of OH/C_2H_2 and $OH/C_2H_2/O_2$ systems have been studied at $T = 288$ K and $p = 2$ torr (He bath gas) using the discharge-flow technique combined with Molecular Beam Sampling Mass Spectrometry.
The $OH + C_2H_2$ reaction yields both the stabilized hydroxyvinyl adduct and the fragmentation products $CH_2CO + H$:

$$OH + C_2H_2 \rightarrow CH_2CO + H \qquad \text{<r.1a>}$$
$$\rightarrow HCCHOH \qquad \text{<r.1b>}$$

The total rate constant k_1 was determined from OH decays at excess C_2H_2 : $k_1 = (2.4 \pm 0.15)10^{-13}$ $cm^3molecule^{-1}s^{-1}$. Upon addition of NO_2, k_{1b} was derived to be $(1.4 \pm 0.4)10^{-13}$ $cm^3molecule^{-1}s^{-1}$.
In the presence of O_2, the subsequent reaction of the hydroxyvinyl radical leads mainly to regenerated OH together with glyoxal; two less important routes produce formic acid and formaldehyde, respectively :

$$HCCHOH + O_2 \rightarrow CHOCHO + H \qquad \text{<r.2a>}$$
$$\rightarrow HCOOH + CHO \qquad \text{<r.2b>}$$
$$\rightarrow HCHO + CO + OH \qquad \text{<r.2c>}$$

The branching fractions were found to be $k_{2a}/k_{2b}/k_{2c} \simeq 0.75 : 0.13 : 0.12$.
Aspects of the reaction mechanisms were clarified by using OD in place of OH.

1. INTRODUCTION

The reaction of the hydroxyl radical with acetylene is of importance from the point of view of both combustion and atmospheric chemistry.
At flame temperatures the major reaction channel is thought to be direct H-abstraction (1-3) :

$$C_2H_2 + OH \rightarrow C_2H + H_2O \qquad \text{<r.1c>}$$

Under atmospheric conditions on the other hand, the reaction has been proposed to proceed via an addition followed by either redissociation or stabilization (4-6) :

$$C_2H_2 + OH \rightleftharpoons HCCHOH^|$$
$$HCCHOH^\dagger + M \rightarrow HCCHOH + M$$

Measurements of the pressure- and temperature dependence of the rate coefficient support the view that the overall process

$$C_2H_2 + OH \rightarrow HCCHOH \qquad \text{<r.1b>}$$

is the dominant atmospheric channel (4-8)
At very low pressures however, Kanofsky et al. detected CH_2CO as the major reaction product (9) :

$$C_2H_2 + OH \rightarrow CH_2CO + H \qquad \text{<r.1a>}$$

---

* presently at NIST, Gaithersburg, Maryland, USA.

Reaction <1a> was also invoked by Vandooren and Van Tiggelen in their study of acetylene flames ($T_f \leq 1400$ K) (10). Hack et al. confirmed both channels <1a> and <1b> by direct product analysis in a T = 300 K discharge-flow investigation (11).

Possibly, ketene arises as a decomposition product of the initial HCCHOH[†] adduct, after rearrangment by 1,3H-migration.

The prime objective of the present study is to obtain measurements of the separate rate constants k_{1a} and k_{1b} at room temperature and at low pressure. Our second aim is to clarify the mechanism of CH_2CO formation.

Thirdly, we will address the subsequent reaction of the stabilized HCCHOH adduct with O_2, a process of direct interest to atmospheric chemistry. Schmidt et al. (7) observed significant slowing down of OH decays in $C_2H_2/OH/O_2$ systems; it is ascribed to OH regeneration, as evidenced by the formation of OD when substituting C_2D_2 for C_2H_2. Led by the appearance of vinoxy radicals (CH_2CHO) and of glyoxal, they suggest the reactions

$$HCCHOH^{\dagger} \xrightarrow{1,3H} CH_2CHO$$
$$CH_2CHO + O_2 \rightarrow CHOCHO + OH$$
$$HCCHOH + O_2 \rightarrow CHOCHO + OH \qquad\qquad <r.2a>$$

In an analysis of $C_2H_2/OH/O_2$ systems under atmospheric conditions, Hatakeyama et al. (12) conclude that reaction

$$HCCHOH + O_2 \rightarrow HCOOH + CHO \qquad\qquad <r.2b>$$

occurs beside reaction <2a>, with branching fractions $k_{2a}/k_{2b} = (0.7\pm0.3):(0.4\pm0.1)$. However, contrary to Schmidt et al. (7), Hatakeyama et al. observe the hydrogen atom in the regenerated hydroxyl of reaction <2a> to come from the hydroxyl reactant.

2. EXPERIMENTAL

Experiments were carried out with two discharge-flow reactor systems, both coupled to Molecular Beam sampling Mass Spectrometers; the two instruments have been described earlier (13,14).

The OH + C_2H_2 kinetics were investigated using apparatus A (see e.g. Ref. 13), with OH generated by the H + $NO_2 \rightarrow$ OH + NO reaction (H being produced in a microwave discharge through H_2/He gas mixtures). Wall-loss of OH was minimized by a teflon sleeve inserted in the quartz reactor tube. Excess acetylene was added through an axially movable central injector tube. The reaction time is determined by the distance from the OH-C_2H_2 mixing point to the sampling probe and by the flow velocity of the gas mixture. The C_2H_2-concentration, accurately determined from measured flow rates, ranged from 2 to 7 x 10^{14} molecule cm^{-3}; [OH] at the mixing point was about 0.6 to 3 x 10^{12} molecule cm^{-3}. The flow velocity was about 2000 cm s^{-1}. The high $[C_2H_2]/[OH]_0$ ratio (100 to 1000), combined with scavenging of the primary product HCCHOH by excess NO introduced at concentrations around 5 10^{14} molecule cm^{-3} via the central inlet tube, was in most instances sufficient to suppress OH-removal by the secondary OH + HCCHOH reaction. Still, some OH-decay experiments were contaminated by secondary reactions, as evidenced by a slight increase of the OH-decay rate at larger reaction times; in such cases a parabolic fit ln i(OH) = ln i(OH)$_0$ - k_{obs} t - qt^2 allowed k_{obs} to be extracted with sufficient precision. All experiments where the term in t^2 exceeded 15 % of the first order term were discarded. As a standard procedure, the OH-termination on the reactor walls in the kinetic region was duly accounted for by substracting the OH-decay rate without C_2H_2 from the decay rate in the presence of C_2H_2. Obviously, such a procedure yields a "decay constant" $k_{obs} = k_1[C_2H_2] + (k_w^w - k_w^0)$ where k_w^w is the OH-wall loss constant with the reactor surface covered by C_2H_2 and k_w^0 the wall loss constant ($\simeq$ 20 s^{-1}) in the absence of C_2H_2.

For the determination of the total rate constant k_1, the final NO_2-concentration in the kinetic region was carefully kept below 5 10^{10} molecule cm^{-3} so as to avoid OH-regeneration by H + $NO_2 \rightarrow$ OH + NO (with H-atoms arising via reaction <1a>).

In a second series of OH-decay experiments however, excess $[NO_2] \simeq 2\ 10^{13}$ molecule cm^{-3} was deliberately added so as to take advantage of the above OH-regeneration reaction to completely suppress reaction <1a> (yielding $H + CH_2CO$) as a *net* OH-sink. Indeed, at $[NO_2] = 2\ 10^{13}$, the H-conversion rate amounts to $\sim 2500\ s^{-1}$, far outrunning the OH-decay rate by reaction <1a>, which was well below 200 s^{-1}. In these circumstances, *net* OH-decay can only be attributed to reaction <1b>.

Some of the data on the subsequent reaction of HCCHOH with O_2 were collected with instrument B (see e.g. ref. 14). Fluorine atoms are generated from F_2/He mixtures in an Al_2O_3 side-arm equipped with a microwave cavity; subsequently, the F-atoms entering the quartz reactor tube are reacted with a large excess of H_2O fed in through an outer central axially movable tube, yielding OH by the process $F + H_2O \rightarrow OH + HF$. Downstream, the OH radicals are mixed with C_2H_2 and O_2, both entering the reactor through an inner central injector tube, which likewise is axially movable.

In some experiments, OH was replaced by OD as reactant; it was created either by $D + NO_2$ or by $F + D_2O$.

In all experiments, the bath gas was helium. Total pressure was always 2.0 torr; the temperature was 288 K.

In order to avoid interferences by fragment ions, the energy E_{el} of the ionizing electrons in the mass spectrometer ion source was always chosen about 3 eV above the ionization potential of the species under investigation.

3. RESULTS AND DISCUSSION

3.1 Products of the C_2H_2 + OH reaction

The nature of the primary reaction products at p = 2 torr (He) and T = 288 K was established by reacting $\sim 10^{12}$ molecule cm^{-3} of OH, generated by $H + NO_2$, with $\sim 4\ 10^{15}$ molecule cm^{-3} C_2H_2. Thus fast and complete conversion of OH radicals is ensured while at the same time secondary reactions with OH will be negligible. In these conditions, at $E_{el} = 17$ eV, two product peaks were observed : CH_2CO^+ and $C_2H_3O^+$. The CH_2CO^+ signal exhibits a slight rise from t = 2 ms to t = 12 ms, whereas the $C_2H_3O^+$ signal decreases significantly from t = 2 to 12 ms. The difference in the profile shapes shows that CH_2CO^+ is not a fragment ion of C_2H_3O. The species CH_3, CO, C_2H and H_2O were not detected. Hence, it can be concluded that in the given experimental conditions channels <1a> and <1b> are the dominant reaction routes.

Confirmation was provided by product analysis of OD/C_2H_2 systems where strong $C_2H_2DO^+$ and $CHDCO^+$ peaks were detected together with a weaker CH_2CO^+ peak (~ 5 times smaller than $CHDCO^+$). Our observation that ketene arising in OD/C_2H_2 systems is mainly CHDCO strongly supports the mechanism

$$OD + HC \equiv CH \rightarrow H\dot{C} - C \begin{subarray}{c}OD \\ H\end{subarray} \xrightarrow{1,3D} \begin{subarray}{c}D \\ H\end{subarray} C = C \begin{subarray}{c}O \\ H\end{subarray} \longrightarrow \begin{subarray}{c}D \\ H\end{subarray} C = C = O + H$$

3.2 Rate constants of the C_2H_2 + OH reaction

As outlined in section 2, the overall rate constant k_1 was determined from OH-decays at large excess C_2H_2. Figure 1 shows a plot of the primary reaction OH-decay constants versus $[C_2H_2]$ at p = 2.0 torr (He) and T = 288 K. The data yield : $k_1 = (2.4 \pm 0.14)10^{-13}\ cm^3 molecule^{-1}s^{-1}$.

Figure 2 shows the primary reaction OH-decay constants versus $[C_2H_2]$ in the presence of added $NO_2 \simeq 2\ 10^{13}$ molecule cm^{-3}, where only reaction <1b> leads to a net loss of OH. Only five measurements were made; the reason for the lower precision of each single measurement and the larger scatter of the data - as compared to the OH-decay constants in the absence of added NO_2 - is not clear. The weighted Fig.2 data give $k_{1b} = (1.4 \pm 0.4)\ 10^{-13}\ cm^3\ molecule^{-1}\ s^{-1}$. Combination of the k_1 and k_{1b} values gives : $k_{1a} = (1.0 \pm 0.4)10^{-13}\ cm^3 molecule^{-1}s^{-1}$.

Our total rate constant agrees fairly well with the value of Schmidt et al. : $k_1 =$

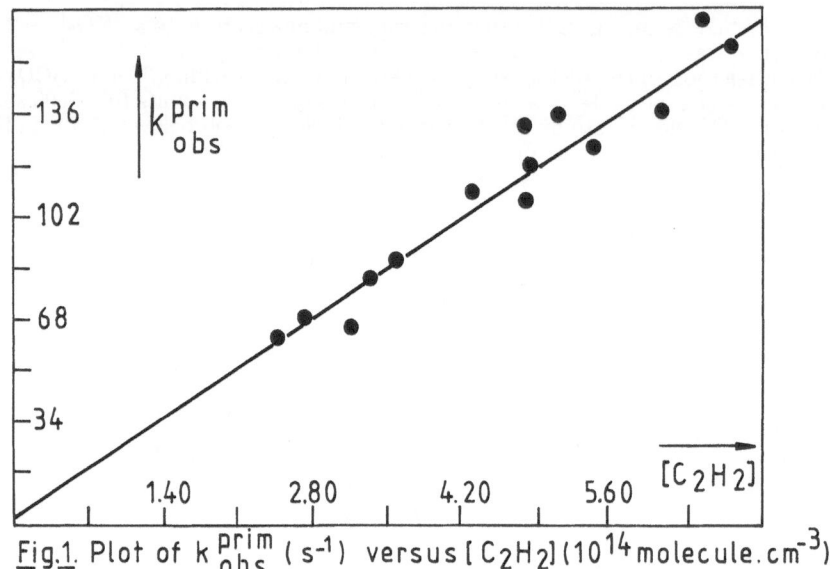

Fig.1. Plot of k^{prim}_{obs} (s^{-1}) versus [C$_2$H$_2$] (10^{14} molecule.cm^{-3})
The slope of the straight line obtained by regression
represents $k_1.P_{tot}$ = 2.0 torr (He); T = 288K

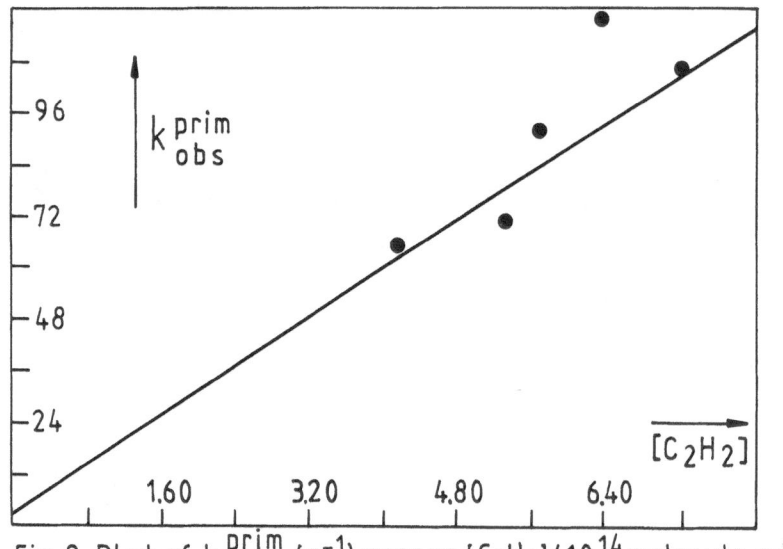

Fig.2. Plot of k^{prim}_{obs} (s^{-1}) versus [C$_2$H$_2$] (10^{14} molecule.cm^{-3})
in the presence of added [NO$_2$] ≃ 2.10^{13} molecule cm^{-3}.
The slope of the straight line obtained by regression
represents $k_{1b}.P_{tot}$ = 2.0 torr; T= 288 K.

$(1.7 \pm 0.5)10^{-13}$ (7); measurements of k_{1b} at low pressure have not been reported, but our k_{1a} can be compared with the "bimolecular rate constant" derived by extrapolation of k_1 to zero pressure : "k_{bi}" $\simeq 0.5 \; 10^{-13}$ (7), a factor 2 lower than our k_{1a} result. It should be pointed out however that k_{1a}, the rate constant of the $CH_2CO + H$ formation channel, is not necessarily independent of pressure if the reaction proceeds through a $HCCHOH^\dagger$ intermediate.

3.3 Mechanism of the HCCHOH + O$_2$ reaction

When $\sim 4 \; 10^{15}$ molecule cm^{-3} O_2 is added to the OH/C_2H_2 systems described in section 3.1, the OH decay is markedly retarded. The concentration of CH_2CO shows a more pronounced rise between t = 2 and 12 ms and the final concentration is nearly 3 times higher than in the absence of O_2. Moreover other reaction products appear : HCHO, CHOCHO and HCOOH with a signal ratio 16:4:1 at t = 12 ms for E_{el} = IP + 3 eV in each case.

With OD as reactant, in the presence of O_2 slowing-down of OD decays as well as formation of OH are both observed. Also the concentrations of stable products detected in OD/C_2H_2 systems are raised. In addition, deuterated formaldehyde, glyoxal and formic acid are found along with the respective undeuterated products; undeuterated formaldehyde and glyoxal are produced in much larger quantities than their deuterated varieties while deuterated formic acid is more abundant than its isotopic counterpart.

The findings are explained by the reactions

$$HCCHOH + O_2 \rightarrow CHOCHO + OH \qquad <r.2a>$$
$$\rightarrow HCOOH + CHO \qquad <r.2b>$$
$$\rightarrow HCHO + CO + OH \qquad <r.2c>$$

with routes <2a> and <2c> resulting in OH regeneration which will also cause an increased formation of CH_2CO and of H by reaction <1a>.

The HCCHOH + O$_2$ reaction starts with

The subsequent reaction of I most likely proceeds via 1,5 H migration followed by dissociation of OH :

The transition state for the 1,5 H shift involves little bending. The barrier to OH dissociation is probably low, since the step is exothermic for $\sim$ 28 Kcal/mol. With OD as primary reactant one expects nondeuterated glyoxal, in agreement with our observations, which confirm the viewpoint of Hatakeyama et al. (12).
The subsequent reaction of I' may be inferred from the reaction of the vinyl radical with O_2, which yields exclusively HCHO + CHO (15); most probably it involves a four-centre intermediate :

With OD one expects deuterated formic acid, as observed. Structure I is unlikely to follow the example of I' because of steric hindrance of the OH which is in cis with the

O_2-group; moreover, the 1,5 H-migration path open to structure I is energetically more favourable.

Several mechanisms can be suggested for the formation of HCHO; one possibility is decomposition of vibrationally excited glyoxal formed in the overal process <1a> : $CHOCHO^\dagger \rightarrow HCHO + CO$.

It should be pointed out that another important source of HCHO in $OH/C_2H_2/O_2$ systems is the sequence

$$H + C_2H_2 \rightarrow C_2H_3 \qquad\qquad <r.3>$$
$$C_2H_3 + O_2 \rightarrow HCHO + CHO \qquad\qquad <r.4>$$

with H formed in reaction <1a>. Also, there is a second mechanism for OH regeneration :

$$CHO + O_2 \rightarrow CO + HO_2 \qquad\qquad <r.5>$$
$$HO_2 + H \rightarrow OH + OH \qquad\qquad <r.6>$$

Regarding HCCO formation, the relative importance of reactions <2c> and <4> can be derived from the shape of the [HCHO] vs. time profile : the first process is kinetically controlled by the $OH + C_2H_2$ reaction, the second by $H + C_2H_2$; the [OH] and [H] profile shapes differ sharply.

Precise profiles of HCHO and also of CH_2CO, CHOCHO and HCOOH were measured with $[OH]_0 = 0.4$ to 1.2×10^{13} and both $[C_2H_2]$ and $[O_2] = 3.5$ to 6×10^{15} molecule cm^{-3} (OH generated by $F + H_2O$). The branching ratio k_{2a}/k_{2b} was determined from the experimental CHOCHO and HCOOH yields using relative sensitivies based on group additivity; corrections for removal by OH were made iteratively by computer modeling of the full reaction scheme, simultaneously optimizing the parameter k_{2c}/k_2 for a best fit with the experimental HCHO profiles. Our own k_{1a} and k_{1b} values were used; k_3 was taken equal to $8 \cdot 10^{-15}$ (16); other rate constants were taken from the recent litterature, including the non-critical k_2 value $5.6 \cdot 10^{-12}$ (7).

Our result obtained in this way : $k_{2a}/k_{2b}/k_{2c} \simeq 0.75:0.13:0.12$ agrees with the Hatakeyama value (12) for reaction <2a>, but differs for the other channels.

The research was sponsored by the Fund for Collective Fundamental Research, Belgium. Z.Y.Z. gratefully acknowledges the financial support of the Belgian and P.R.C. governments and of the K.U. Leuven Research Council. J.P. is a Research Associate of the National Fund for Scientific Research, Belgium.

REFERENCES

(1) Warnatz, J. in "Combustion Chemistry", p. 290, Springer Verlag, 1984.
(2) Jachimowsky, J.C. Combust. Flame 29, 55 (1977).
(3) Levy, J.M. Combust. Flame 46, 7 (1982)
(4) Perry, R.A. et al. J. Chem. Phys. 67, 5577 (1978)
(5) Michael, J.V. et al. J. Chem. Phys. 73, 6108 (1980)
(6) Perry, R.A. and Williamson, D. Chem. Phys. Lett. 93, 331 (1982)
(7) Schmidt, V. et al. Ber. Bunsenges. Phys. Chem. 89, 321 (1985)
(8) Wahner A. and Zetzsch, C. Ber. Bunsenges. Phys. Chem. 89, 323 (1985)
(9) Kanofsky, J.R. et al. J. Phys.Chem. 78, 311 (1974)
(10) Vandooren, J. and Van Tiggelen, P. 16th Int. Symp. Comb., 1143 (1967)
(11) Hack, W. et al. Oxid. Commun. 5, 101 (1983)
(12) Hatakeyama, S. et al. J. Phys. Chem. 90, 173 (1986)
(13) Fonderie, V. and Peeters, J. in "Physicochemical Behaviour of Atmospheric Pollutants", II, p. 165, Reidel, 1982
(14) Fonderie, V. et al. in "Physicochemical Behaviour of Atmospheric Pollutants", III, p. 274, Reidel, 1984
(15) Gutman, D. and Nelson H.H. J. Phys. Chem. 87, 3902 (1983)
(16) Boullart, W. and Peeters, J. unpublished results.

LABORATORY STUDIES OF $HOCH_2CH_2O_2$ RADICALS PRODUCED BY THE PHOTOLYSIS OF 2-IODOETHANOL IN THE PRESENCE OF O_2

M.E. JENKIN and R.A. COX
Chemical Physics Group, B551 Harwell Laboratory
Didcot, Oxon OX11 ORA, UK

Summary

The molecular modulation technique coupled with uv absorption spectroscopy has been used to investigate the uv spectrum and kinetics of reactions of the $HOCH_2CH_2O_2$ radical (formed in the atmospheric oxidation of ethene) at 298K. The radicals were generated by the 254 nm photolysis of $HOCH_2CH_2I$ in the presence of O_2. At 760 Torr, $HOCH_2CH_2O_2$ was found to obey second order kinetics owing to a series of reactions initiated by its self reaction:

$$HOCH_2CH_2O_2 + HOCH_2CH_2O_2 \rightarrow \text{products} \qquad (3)$$

Values of k_{3obs}/σ were measured for $205 \leqslant \lambda/nm \leqslant 240$, with the value at 230 nm being $(6.81 \pm 0.4) \times 10^5$ cms^{-1}. This corresponds to σ (230 nm) = $(2.35 \pm 0.26) \times 10^{-18}$ cm^2 $molecule^{-1}$ and $k_{3obs} = (1.60 \pm 0.17) \times 10^{-12}$ cm^3 $molecule^{-1}$ s^{-1}. At lower pressures, HO_2 was also generated in the system in significant quantities enabling measurement of the rate constant for reaction (4):

$$HO_2 + HOCH_2CH_2O_2 \rightarrow \text{products} \qquad (4)$$

At 10 Torr, k_4 was found to have a value of $(4.8 \pm 0.5) \times 10^{-12}$ cm^3 $molecule^{-1}$ s^{-1}.

1. INTRODUCTION

It is well established that the oxidation of organic compounds in the atmosphere occurs by free radical driven mechanisms in which organic peroxy radicals (RO_2) participate (1,2). In the presence of NO_x, O_3 is formed as a by product of these reaction schemes, since RO_2 radicals react with NO to produce NO_2, which is photolysed rapidly to yield O_3. In the continental boundary layer, concentrations of NO_x are generally sufficiently high that RO_2 radicals react exclusively with NO leading to O_3 formation. In the remote continental boundary layer, the marine boundary layer and the background troposphere, however, concentrations of NO_x are much lower, and the reactions of RO_2 radicals with HO_2 can compete with the reactions with NO, thereby inhibiting the production of O_3 to a certain extent.

The oxidation of CH_4 is known to be a major contributor to the photochemical production of O_3 in the background troposphere (2).

Non-methane hydrocarbons, although present in much lower concentrations, may also be significant since the rate coefficients for their reactions with OH, which usually initiate the oxidation, are much larger (3). C_2H_4 may be particularly significant, since it has been observed at concentrations up to 0.5 ppb in the background troposphere (4). It is clear, therefore, that the chemistry of the derived peroxy radical, $HOCH_2CH_2O_2$, in particular the rates of reaction with NO and HO_2, is of interest from the point of view of O_3 production in the background troposphere.

In this paper we present the results of experiments performed to investigate the 254 nm photolysis of 2-iodoethanol as a source of $HOCH_2CH_2O_2$ radicals:

$$HOCH_2CH_2I + hv \ (\lambda = 254 \ nm) \ \rightarrow \ HOCH_2CH_2 + I \qquad (1)$$

$$HOCH_2CH_2 + O_2 + M \ \rightarrow \ HOCH_2CH_2O_2 + M \qquad (2a)$$

The photolysis of $HOCH_2CH_2I$ in the presence of O_2 was found to produce a radical, with a UV spectrum characteristic of a peroxy radical, which we assign to $HOCH_2CH_2O_2$. Values for its self reaction rate constant (reaction (3)) and its reaction with HO_2 (reaction (4)) were determined:

$$HOCH_2CH_2O_2 + HOCH_2CH_2O_2 \rightarrow products \qquad (3)$$

$$HOCH_2CH_2O_2 + HO_2 \rightarrow products \qquad (4)$$

2. EXPERIMENTAL

The experiments were performed in a 1.2 l cylindrical quartz reaction vessel, 120 cm in length. The vessel was surrounded by a jacket through which ethanol was circulated, allowing temperature regulation of the vessel. All experiments were performed at 298K. Kinetics measurements were made using the molecular modulation technique described previously (5) to observe the time resolved behaviour of transient species generated during the intermittent photolysis of gas mixtures made up of ca.1-5 x 10^{15} molecule cm^{-3} of $HOCH_2CH_2I$ in 10 Torr of O_2. N_2 was added in order to vary the pressure in the range 10-760 Torr. Photolysis of the iodide was achieved using up to 5 low pressure mercury lamps (Philips TUV 40W) emitting mainly at 253.7 nm.

Reaction mixtures were monitored using a collimated beam from a deuterium lamp which was focussed onto the slit of a 0.75 m monochromator (Spex 1700), followed by detection on a photomultiplier (EMI, 9783B). Modulated absorption signals due to transient species (typically ca. 10^{-3}) were accumulated and averaged in the manner described previously (5). Additional spectral measurements were made by dispersion of the monitoring beam on a 0.5 m spectrograph (B and M Spectronik) and detection on a 1024 channel photodiode array (Reticon), giving a spectral coverage of ca. 80 nm. The detector was coupled to a multichannel analyser (Tracor Northern TN1710) allowing storage and manipulation of the data. This arrangement has been described in detail elsewhere (6).

Kinetics measurements were made using flowing gas mixtures to minimise the accumulation of reaction products. The residence time in the vessel was ca. 15s. The flows of the constituent gases (N_2 and O_2) were regulated using mass flow controllers (MKS, Type 261). A proportion of the gas flow was diverted through a jacketted bubbler containing $HOCH_2CH_2I$. The temperature of the bubbler could be adjusted in order to vary the concentration of the iodide in the gas flow. The concentration of the iodide in the reaction vessel was determined by conventional absorption spectroscopy at 260 nm. $\sigma(260$ nm) for $HOCH_2CH_2I$ (1.22 (± 0.24) x 10^{-18} cm^2 molecule^{-1}) was measured independently by admitting known pressures of the iodide (up to 0.5 Torr) into a 10 cm path length absorption cell and measuring the extinction. Accurate pressure measurement was achieved using an MKS Baratron (170M-6B). The UV spectrum was also characterised in the wavelength range 210-290 nm using diode array spectroscopy as described above. The resultant absorption cross sections are presented in Table 1.

N_2 (Air products, High Purity), O_2 (BOC, Breathing Grade) and $HOCH_2CH_2I$ (Aldrich, 99%) were used as received.

3. RESULTS
The photolysis of $HOCH_2CH_2I$ in the presence of O_2 and N_2 at 298K, and at 760, 100 and 10 Torr total pressure

a) 760 Torr experiments: The variation of modulated absorption signal with wavelength was investigated over the range 205 to 310 nm for experiments in which $HOCH_2CH_2I$ was photolysed in the presence of O_2 and N_2 initially at a total pressure of 760 Torr. The results are displayed in Fig. 1, and show that significant absorption due to transient species was observed over the entire wavelength range. At wavelengths below ca. 240 nm, the waveforms obtained obeyed second order kinetics, and it seems reasonable that absorption at these wavelengths was due to the $HOCH_2CH_2O_2$ radical formed by the reaction sequence (1) followed by (2a), and removed by a series of reactions initiated by the self reaction of $HOCH_2CH_2O_2$:

$$HOCH_2CH_2O_2 + HOCH_2CH_2O_2 \rightarrow HOCH_2CH_2O + HOCH_2CH_2O + O_2 \qquad (3a)$$

$$\rightarrow HOCH_2CH_2OH + HOCH_2CHO + O_2 \qquad (3b)$$

Detailed analysis of the waveforms using the methods described previously (7) provided a value of $k_{3obs}/\sigma(230$ nm) = (6.81 $\pm$ 0.4) x 10^5 cms^{-1} at 298K. The first order decay of $HOCH_2CH_2I$ during photolysis at 760 Torr was measured in a separate experiment performed on a static gas mixture (k = 1.08 x 10^{-2} s^{-1}, 5 lamps). If the observed decay is attributed to photolysis alone (reaction (1)), then radical production rates calculated using this value lead to k_{3obs} = (1.60 $\pm$ 0.17) x 10^{-12} cm^3 molecule^{-1}s^{-1} and $\sigma(230$ nm) = (2.35 $\pm$ 0.26) x 10^{-18} cm^2 molecule^{-1}. At wavelengths above 240 nm, the observed modulated waveforms had a contribution from a kinetically distinct transient species. The behaviour of this species was very similar to that obsered during the photolysis of CH_3I in the presence of O_2 in this laboratory (8), where the spectrum obtained was made up of contributions due to CH_3O_2 and another transient believed to be CH_3OI. By analogy, the additional transient in the present work may be due to $HOCH_2CH_2OI$, since its spectrum appears to be very similar. If the spectrum tentatively assigned to CH_3OI previously (8), is subtracted from the transient spectrum obtained during $HOCH_2CH_2I$ photolysis (760 Torr), then the resultant spectrum is attributed to the $HOCH_2CH_2O_2$ radical with the corresponding cross section values presented in Table 1. These values were calculated relative to that measured at 230 nm.

A further series of experiments was carried out using diode array spectroscopy to measure the yield of HCHO from $HOCH_2CH_2I$ photolysis at 760

Torr. Measurements of HCHO in its UV bands (250-360 nm) indicated that its yield relative to $HOCH_2CH_2I$ removed was ca. 0.3. The production of HCHO in the system can be accounted for by decomposition of the $HOCH_2CH_2O$ radical formed from reaction (3a), as previously described by Niki et al. (9):

$$HOCH_2CH_2O + M \rightarrow HCHO + CH_2OH + M \qquad (5)$$

$$CH_2OH + O_2 \rightarrow HCHO + HO_2 \qquad (6)$$

Since reaction channel (3b) does not result in the production of HCHO, the measured HCHO yield can be used to obtain the branching ratio for reaction (3), which was calculated to be $k_{3a}/(k_{3a} + k_{3b}) = (0.18 \pm 0.02)$ at 298K and 760 Torr. HO_2 produced simultaneously with HCHO reacts further with $HOCH_2CH_2O_2$ (reaction (4)). Consequently, the observed second order constant (k_{3obs}) referred to above is greater than the true elementary constant k_3. Since 18% of the $HOCH_2CH_2O_2$ self reaction produces HO_2, $k_{3obs} = 1.18\ k_3$. Therefore k_3 has a value of $(1.36 \pm 0.21) \times 10^{-12}\ cm^3$ $molecule^{-1}s^{-1}$.

b) <u>100 Torr and 10 Torr experiments</u>: Modulated absorption signals obtained over the wavelength range 205 to 320 nm at 100 Torr and 10 Torr total pressure are displayed in Fig.1 along with those obtained at 760 Torr. It is apparent that the shape of the transient spectrum changed dramatically as the pressure was lowered. The signals at the high end of the wavelength range were enhanced and, at the lower wavelengths, the maximum of the spectrum shifted from 240 nm at 760 Torr to ca. 225 nm at 10 Torr. Inspection of the time resolved absorption waveforms also indicated that more than two kinetically distinct transient species were absorbing. In particular, at wavelengths below 240 nm, the waveforms could no longer be attributed to a single absorbing species. It seems likely that $HOCH_2CH_2O_2$ is still produced at low pressure, but a further species absorbing at lower wavelengths must also be produced. Fig 2 shows the result of subtracting the 760 Torr transient spectrum (ie. $HOCH_2CH_2O_2$) from that obtained at 10 Torr, for the wavelength range 205 to 250 nm. The resultant spectrum matches that of the HO_2 radical extremely well, as shown in the diagram, so it appears that significant quantities of HO_2 are being produced at low pressure, in addition to $HOCH_2CH_2O_2$.

There are several ways that HO_2 may be produced in this system. We have already seen that it is produced in relatively small amounts as a secondary radical, even at 760 Torr, following the self reaction of $HOCH_2CH_2O_2$. It was clear, however, from the time resolved absorption waveforms, that more direct sources of HO_2 also need to be invoked. One possibility is that the reaction of $HOCH_2CH_2$ with O_2 has a second channel which becomes important at low pressures:

$$HOCH_2CH_2 + O_2 \rightarrow HO_2 + HOCHCH_2 \qquad (2b)$$

This is entirely analagous to that observed for C_2H_5 at low pressures, since 6% of the reaction of C_2H_5 with O_2 has been observed to produce HO_2 and C_2H_4 at 10 Torr (10).

Another potential source of HO_2 arises if $HOCH_2CH_2$ formed from the photolysis of $HOCH_2CH_2I$ is sufficiently energy rich for decomposition to occur before stabilisation and subsequent reaction with O_2:

$$HOCH_2CH_2^{\neq} \rightarrow OH + CH_2CH_2 \qquad (7)$$

OH radicals would attack $HOCH_2CH_2I$ in this system, with two potentially important pathways:

$$OH + HOCH_2CH_2I \rightarrow HOCH_2CH_2 + HOI \qquad (8a)$$

$$\rightarrow HOCHCH_2I + H_2O \qquad (8b)$$

There is little data on the reactions of OH with alkyl iodides, but available information suggests that both pathways will occur to a certain extent (11, 12). Channel (8a) produces stabilised $HOCH_2CH_2$ which would lead to the production of $HOCH_2CH_2O_2$, whereas channel (8b) produces $HOCHCH_2I$ which would almost certainly react with O_2 to produce HO_2:

$$HOCHCH_2I + O_2 \rightarrow HO_2 + ICH_2CHO \qquad (9)$$

The production of HO_2 by either this route, or by reaction (2b) would both be expected to become more important as the pressure is lowered, but these two sources of HO_2 are indistinguishable by molecular modulation studies alone since in either case $HOCH_2CH_2$ reacts partially to produce HO_2 and partially to produce $HOCH_2CH_2O_2$. Detection of the products CH_2CH_2, $HOCHCH_2$ or HOI would provide further information on these aspects of the decomposition of $HOCH_2CH_2I$ at low pressure.

The time resolved absorption waveforms obtained at 220 nm, 230 nm, 240 nm and 250 nm at 10 Torr were simulated using the Harwell FACSIMILE program (13). This allowed optimisation of certain parameters within an assumed chemical scheme by non-linear least square fitting. The optimised parameters were as follows:

(i)   The proportion of $HOCH_2CH_2$ leading to HO_2 production either by direct reaction with O_2 (reaction (2b)) or following decomposition to produce OH (reaction sequence (21), (8b), (9)).

(ii) The branching ratio for the $HOCH_2CH_2O_2$ self reaction $k_{3a}/(k_{3a} + k_{3b})$ which influences the secondary production of HO_2.

(iii) The rate constant for the reaction of HO_2 with $HOCH_2CH_2O_2$ (reaction (4)).

The absorption cross sections used for $HOCH_2CH_2O_2$ were those displayed in Table 1. Values for HO_2 were measured in an additional series of modulated photolysis experiments in which Cl_2 was photolysed in the presence of H_2 and O_2 at 760 Torr. The measured cross sections are also listed in Table 1.

The modulated waveforms obtained experimentally at the four wavelengths are shown in Fig.3 along with the fitted waveforms. At 220 nm, the absorption was predominantly due to HO_2, whereas at 250 nm, $HOCH_2CH_2O_2$ was the major absorber. The size and shape of the waveforms were well described by the model for the complete wavelength range, and the following best fit values were obtained for the varied parameters (10 Torr, 298K):

$$k_4 = (4.8 \pm 0.5) \times 10^{-12} \ cm^3 \ molecule^{-1}s^{-1}$$
$$k_{3a}/(k_{3a} + k_{3b}) = (0.8 \pm 0.2)$$
proportion of $HOCH_2CH_2$ reacting to produce HO_2 radicals = (0.41 ± 0.03)

4. DISCUSSION

The photodecomposition of $HOCH_2CH_2I$ in the presence of O_2 is expected to produce $HOCH_2CH_2O_2$ radicals by reactions (1) and (2a). At 760 Torr,

the observed transient spectrum and is attributed to $HOCH_2CH_2O_2$ and an iodine containing transient absorbing at higher wavelengths. The absorption spectrum of $HOCH_2CH_2O_2$ given in Table 1 is typical of an organic peroxy radical, although the absorption maximum is shifted to higher wavelengths by several nm compared with simple alkyl peroxy radicals. The values of the absorption cross sections are also significantly different from those expected by comparison with other peroxy radicals, being smaller by almost a factor of two (eg. $\sigma(230 \text{ nm})$ = (2.35 ± 0.26) x 10^{-18} cm^2 molecule^{-1} for $HOCH_2CH_2O_2$; $\sigma(230 \text{ nm})$ = (4.40 ± 0.49) x 10^{-18} cm^2 molecule^{-1} for CH_3O_2 (8)). The quoted error limits are believed to be representative of the accuracy of the determination. However there is an uncertainty arising from the assumption of $HOCH_2CH_2O_2$ production at 760 Torr being equal to the rate of loss of 2-iodoethanol. If, for example there was significant occurrence of direct HO_2 production (reaction 2b) or secondary attack on the precursor by OH, then the above value of σ would be less than the true value.

The self reaction of $HOCH_2CH_2O_2$ appears to be significantly faster than for alkyl peroxy radicals. The value of k_3 determined here, (1.36 ± 0.21) x 10^{-12} cm^3 molecule^{-1} s^{-1}, is about a factor of 15 faster than the analogous reaction of $C_2H_5O_2$, 8.6 x 10^{-14} cm^3 molecule^{-1} s^{-1}(14), but is similar in magnitude to that observed for the $C\ell CH_2CH_2O_2$ radical, 3.57 x 10^{-12} cm^3 molecule^{-1} s^{-1}(15), suggesting that the OH and Cl substituents have an activating effect. In contrast, however, the rate constant determined for the reaction of HO_2 with $HOCH_2CH_2O_2$, k_4 = (4.8 ± 0.5) x 10^{-12} cm^3 molecule^{-1} s^{-1}, from modelling the 10 Torr experiments, is comparable with that for the analagous reaction of $C_2H_5O_2$, 5.8 x 10^{-12}cm^3 molecule^{-1} s^{-1}(14), and is therefore not enhanced by the presence of the OH group. It should be emphasised that the error limits on these rate coefficients, k_3 and k_4, do not take into account errors arising from the uncertainty in the value of $\sigma(HOCH_2CH_2O_2)$ (see above) which would lead to underestimation of the values of k_3 and k_4.

5 OZONE PRODUCTION FROM THE ATMOSPHERIC OXIDATION OF ORGANIC COMPOUNDS

The background troposphere and the marine and remote continental boundary layers are characterised by relativley low concentrations of NO_x, as indicated in the introduction, and the reactions of organic peroxy radicals with HO_2 are able to compete with the reactions with NO, thereby influencing the potential of a particular organic compound for the production of O_3.

Kinetic data for the reactions of organic peroxy radicals with both HO_2 and NO are limited to the radicals CH_3O_2, $C_2H_5O_2$, $HOCH_2O_2$, CH_3CO_3 and $HOCH_2CH_2O_2$. The rate coefficients for the reactions of these radicals with HO_2 and NO are listed in Table 2. Also presented are the relative rate constants k_{NO}/k_{HO_2} which are found to vary from 0.46 for $HOCH_2O_2$ to 3.6 for $HOCH_2CH_2O_2$ on the basis of currently available rate constants. This suggests that, under 'low NO_x' conditions, the potential of parent organic compounds for O_3 production can be inhibited to different extents by the reaction of the derived peroxy radical with HO_2. It is perhaps more appropriate, therefore, that the ranking or organic compounds for the production of O_3 in low NO_x scenarios should incorporate a factor such as k_{NO}/k_{HO_2} for the derived peroxy radical in addition to the rate coefficient for reaction of OH radicals with the parent organic compound.

ACKNOWLEDGEMENT

This work was carried out as part of the EUROTRAC project LACTOZ, and was funded partially by the UK Department of the Environment, and partially by the Commission of the European Communities.

REFERENCES

1) Atkinson and Lloyd (1984). J. Phys. Chem. Ref. Data, 13, 315-444.

2) WMO (1985). Report No. 16, "Atmospheric ozone - Assessment of our understanding of the processes controlling its distribution and change". Vol. 1, Chap. 4.

3) Atkinson, R., (1986). Chem. Rev. 86, 69-201.

4) Rudolph, J., KFA Julich, private communication.

5) Jenkin, M.E., and Cox, R.A. (1985). J.Phys. Chem., 89, 192-199.

6) Cox, R.A. and Hayman, G.D. (1988). Nature, 332, 796-800.

7) Jenkin, M.E., Cox, R.A., Hayman, G.D. and Whyte, L.J. (1988). J.Chem. Soc., Faraday Trans. 2, 84, 913-930.

8) Jenkin, M.E. and Cox, R.A., J. Phys. Chem., submitted.

9) Niki, H., Maker, P.D., Savage, C.M. and Breitenhach, L.P. (1981) Chem. Phys. Lett. 80 499-503.

10) Plumb, I.C. and Ryan, K.R. (1981). Int. J. Chem. KJinet., 13, 1011-1028.

11) Brown, A.C., Canosa-Mas, C.E. and Wayne, R.P. (1989). Atmospheric Environment, in press.

12) Garraway, J. and Donovan, R.J. (1979). J.C.S. Chem. Commun. 1108.

13) Curtis, A.R. and Sweetenham, W.P. (1987). "FACSIMILE/CHECKMAT user manuals". AERE Report R-12805 (HMSO London).

14) Atkinson, R., Baulch, D.L., Cox, R.A., Hampson, R.F., Kerr, J.A. and Troe, J. (1989). Int. J. Chem. Kinet., 21, 115-150.

15) Burrows, J.P., Moortgat, G.K., Tyndall, G.S., Cox, R.A., Jenkin, M.E., Hayman, G.D. and Veyret, B. (1989). J.Phys. Chem., 93, 2375-2382.

16) Moortgat, G.K., Veyret, B. and Lesclaux, (1989). J.Phys. Chem., 93, 2362-2368.

17) Veyret, B., Rayez, J-C. and Lesclaux, R. (1982). J.Phys. Chem., 86, 3424-3430.

18) Schmidt, V.H., Yuhui Su., Becker, K.H. and Brockman, K. (1986). Proceedings of the 4th European Symposium on the Physico-chemical behaviour of atmospheric pollutants. Stresa, 23-25th September.

TABLE 1

Measured Absorption Cross Sections for $HOCH_2CH_2I$, $HOCH_2CH_2O_2$

and HO_2 (in units 10^{-18} cm^2 molecule^{-1})

λ (nm)	$HOCH_2CH_2I$	$HOCH_2CH_2O_2$	HO_2
200	-	-	4.24*
205	-	0.79	4.32
210	0.090	1.12	4.18
215	0.093	1.47	3.86
220	0.118	1.79	3.40
225	0.170	2.10	2.89
230	0.279	2.35	2.31
235	0.480	2.55	1.71
240	0.783	2.64	1.18
245	1.118	2.59	-
250	1.350	2.45	0.45
255	1.387	-	-
260	1.220	1.98	0.13
265	0.935	-	
270	0.644	1.22	
275	0.399	-	
280	0.235	0.58	
285	0.133	-	
290	0.077	0.21	

* cross sections measured in system where HO_2 removed by its self reaction $HO_2+HO_2 \rightarrow H_2O_2+O_2$. Measured rate constants for this reaction at 298K/760 Torr, were $(2.44 \pm 0.20) \times 10^{-12}$ cm^3 molecule^{-1} s^{-1} (Bath gas O_2) and $(2.85 \pm 0.30) \times 10^{-12}$ cm^3 molecule^{-1} s^{-1} (Bath gas 80% N_2 20% O_2)

TABLE 2

Rate constants for reactions of organic peroxy radicals with NO
and with HO$_2$ at 298K

RO$_2$	$10^{12}k_{HO_2}$	$10^{12}k_{NO}$	k_{NO}/k_{HO_2}
CH$_3$O$_2$	4.9[a]	7.6[a]	1.5
C$_2$H$_5$O$_2$	5.8[a]	8.9[a]	1.5
HOCH$_2$O$_2$	12.0[b]	5.6[e]	0.46
CH$_3$CO$_3$	13.0[c]	14.0[a]	1.1
HOCH$_2$CH$_2$O$_2$	4.8[d]	17.0[f]	3.6

[a] Atkinson et al (14)

[b] Burrows et al (15)

[c] Moortgat et al (16)

[d] This work

[e] Veyret et al (17)

[f] Schmidt et al (18)

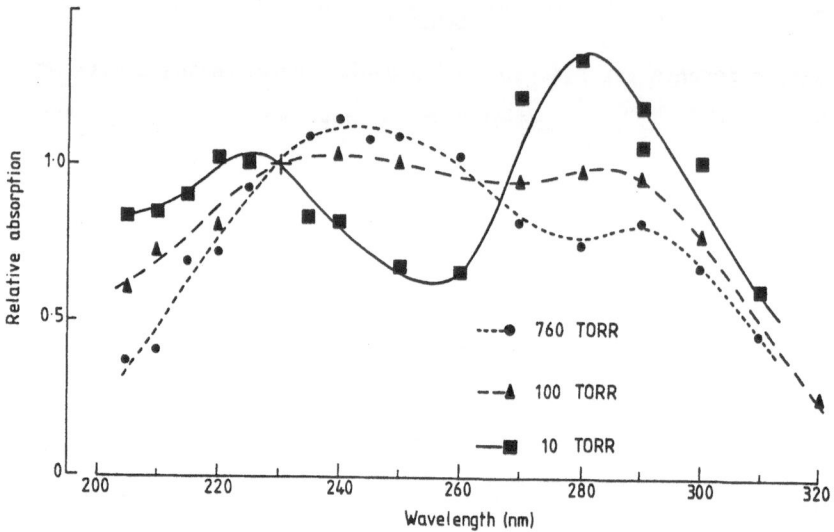

FIG. 1. TRANSIENT ABSORPTION SPECTRA MEASURED AT 760 TORR (--•--), 100 TORR (--▲--) AND 10 TORR (--■--) DURING THE PHOTOLYSIS OF $HOCH_2CH_2I$ IN THE PRESENCE OF O_2 AND N_2. MEASUREMENTS ARE RELATIVE TO SIGNAL OBTAINED AT 230nm (+)

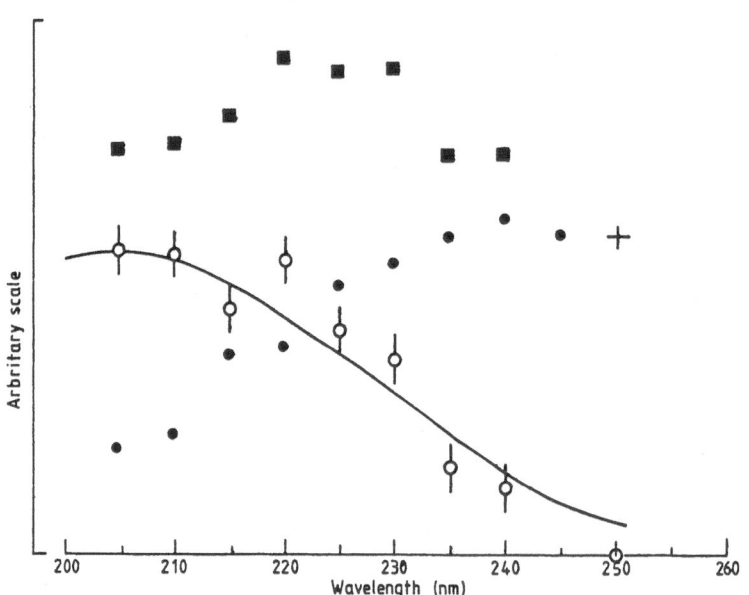

FIG. 2. $HOCH_2CH_2I$ / O_2/N_2 SYSTEM : COMPARISON OF HO_2 SPECTRUM (——) WITH DIFFERENCE SPECTRUM (O) OBTAINED BY SUBTRACTION OF 760 TORR DATA (•) FROM 10 TORR DATA (■) AFTER NORMALISATION AT 250nm (+)

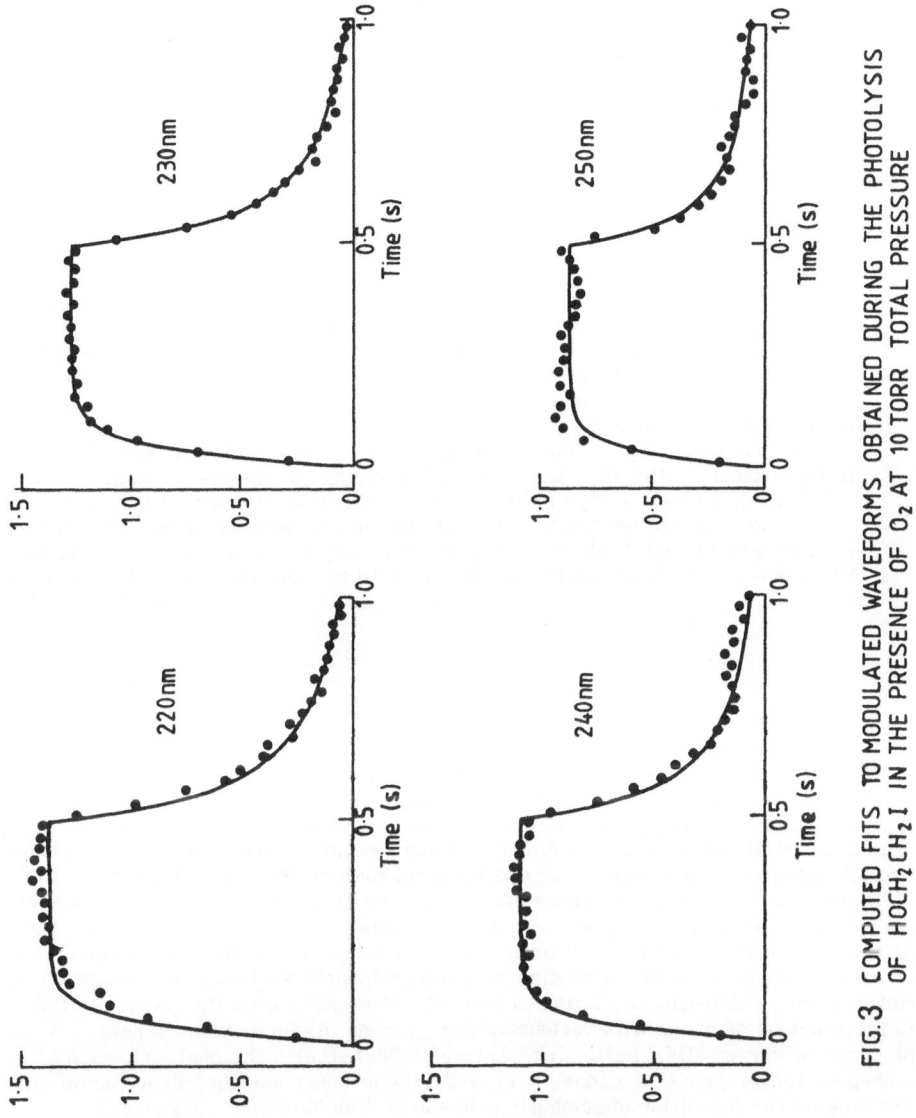

FIG.3 COMPUTED FITS TO MODULATED WAVEFORMS OBTAINED DURING THE PHOTOLYSIS
OF HOCH$_2$CH$_2$I IN THE PRESENCE OF O$_2$ AT 10 TORR TOTAL PRESSURE

REACTION OF MONOTERPENE HYDROCARBONS WITH O_3, SO_2 AND NO_2 - FORMATION OF ACIDIC COMPOUNDS

D. Kotzias, M. Duane, B. Nicollin and H. Schlitt
COMMISSION OF THE EUROPEAN COMMUNITIES,
Joint Research Centre, Ispra (VA) - Italy.

Summary

Monoterpenes are the principal volatiles emitted in large quantities by conifers. They appear to play a significant role in growth regulation, insect attraction, desease resistance and human allergies. Because of their chemical nature, mostly unsaturated hydrocarbons, monoterpenes react very fast with atmospheric constituents e.g. O_3, OH-radicals (rate constants of the reactions of most terpenes with O_3 range from ~ 10^3 to ~ 10^6 l / mol. sec, with OH-radicals from ~ 10^{10} l / mol. sec to 14×10^{10} l / mol. sec) and may form various oxidation products partly as aerosols.
In the framework of our investigations on the role of monoterpene hydrocarbons in the dry deposition of atmospheric pollutants (e.g.O_3, SO_2, NO_x), we have been studying the gas-phase reactions of selected monoterpenes with ozone in the presence of SO_2 (β -pinene) as well as in the presence of SO_2/NO_2 (α-pinene, β-pinene and limonene). Emphasis was given in identifying the main reaction products and in quantifuing the formed sulphuric acid aerosol.

1. INTRODUCTION

Terpenes and isoprene are emitted from conifers and deciduous trees in rather significant quanitites. Total worldwide emissions of terpenes are about 9×10^8 tons per year (Zimmermann, 1978). Because of their reactive behaviour, most of them are unsaturated hydrocarbons, they undergo many reactions with various atmospheric constituents, especially with ozone and hydroxyl radicals (Yokuchi, 1985; Gäb, 1985; Stieglitz and Jay, 1987). Westberg and Rasmussen (1972) as well as Gay and Arnts (1977) also clearly demonstrated that monoterpenes are very reactive and that their products from the reaction with ozone form aerosols. The ability of monoterpenes upon reaction with ozone to oxidize SO_2 into sulphuric acid was first reported by Hatakeyama, Kobayashi and Akimoto in 1984. They estimated the yields of H_2SO_4 in the reaction of ozone with various olefins - among others α- and β-pinene - in the presence of excess SO_2. Nolting and Zetzsch (1988) studied the photooxidation of α-pinene in the presence of ozone, NO_x and SO_2 in smog chamber experiments and reported on the formation of sulphuric acid and of "sulphate".

Several studies exist on the oxidation of SO_2 into sulphuric acid aerosol. Cox and Penkett (1972), Calvert et al. (1978) Martinez and Herron (1981,1983) reported on this reaction and extensively discussed possible reaction mechanisms. Additional studies were made in order to clarify the role of humidity on the H_2SO_4 aerosol formation in the SO_2 - NO - C_3H_6 - air system (Izumi et al., 1987).

To our knowledge, no report has been concerned with the effect of NO_2 on the H_2SO_4 - forming potential of Criegee intermediates originally produced by the terpene /O_3 reaction *in the dark*. The scope of this study, therefore, was to examine the role of NO_2 on the reaction of selected monoterpenes with ozone in the presence of SO_2. Assuming high terpene emissions close to tree surfaces and considerable amounts of O_3, SO_2 and NO_2 in ambient air, a possible environmental significance of the SO_2/NO_2 oxidizing process (Stangl, Kotzias, Geiss, 1988) could be expected.

2. EXPERIMENTAL SECTION

The experiments were carried out in Teflon bags (800 l - 1m^3) made out of 0,02 mm thick FEP - Teflon® (a fluorinated ethylene-propylene copolymer, manufactured by Du Pont) with a surface to volume ratio of ~0,05 cm^{-1}. The bags are divided into two halves of ~ 400 - 500 l (using a metal partition) filled with purified (zero) air and the reactants. This device allows to separate reactive components before starting the experiment (e.g. in the system terpene/O_3/SO_2/NO_2 one half was filled with zero air, O_3 and SO_2, since the reaction of O_3/SO_2 is known to be very slow (k = 1,5 × 10^{-8} ppm^{-1} min^{-1}), the other half was filled with zero air, the terpene and NO_2 since the thermal reaction between olefins and NO_2 is very low, k ≤ 1,1 × 10^{-20} cm^3/molecules · sec (Niki et al., 1984). The experiment was started by thorough mixing of the two parts. Recently, Jay and Steiglitz (1988) could identify several products upon reaction of NO_2 and selected terpenes (camphene, α- and β-pinene) in the dark. This is probably due to the high concentrations used for their experiments (~ 20 ppmV of terpenes and 50 - 100 ppm of NO_2). With respect to our initial concentration conditions, no considerable amounts of terpene-NO_2 products could be expected.

In order to determine the rates of disappearance of the components due to absorption on the bag wall or permeation, several test runs were conducted. We found that the SO_2 wall decay was about 1% h^{-1}. No great differences between dry and humidified purified air were found. The ozone concentration (s. also Stangl et al., 1987) as well as the NO_2 and terpene concentrations remain stable over a period of 4-6 h.

The desired concentration of the terpenes was prepared by a two step dilution procedure from ppm to ppb levels. The compounds (6-10 μl liquid) were injected during filling with zero air into a 150 l teflon bag. Then by using gas-tight syringes, aliquot parts were taken from this chamber and injected into the 400-500 l half-bag, filled with zero air. The concentration of β-pinene and other selected terpenes in the bag was controled by GC. For the analysis of the reaction products activated charcoal containing tubes have been used. After collecting (~ 500l air) the absorbed organic compounds were eluted with hexane or CH_2Cl_2 and further prepared for GC-MS analysis.

3. RESULTS AND DISCUSSION

Table 1 summarizes the experimental data on the yield of H_2SO_4 obtained in the reaction of β-pinene with ozone in the presence of SO_2. In order to study the humidity effect, the experiments were conducted under different r. h. conditions while the other conditions were held constant.

We found that, also for pinenes, the gas-phase oxidation of SO_2 was influenced by the presence of H_2O vapour. The total oxidation [SO_2]$_{ox}$ was found to decrease as relative humidity (r. h.) increases (s. run 1-2, 4-5, 7-8, table 1). At 6% r. h. about 14% of the initial concentration of SO_2 is oxidised within 4 hours; when r. h. is ~45% an overall SO_2 oxidation of only 7% was found (s. run 1 and 2, table 1). Assuming the Criegee-intermediates as the oxidizing species the retarded SO_2 oxidation found may be explained as resulting from their depletion due to the reaction with the H_2O molecules (eq. 1).

$$R\ \overset{.}{C}HOO\cdot + H_2O \rightarrow RCOOH + H_2O \quad (eq.\ 1)$$

Apart from the sulphuric acid aerosol, no other sulphur containing compounds could be detected with our detection techniques. The H_2SO_4 yield with resepct to SO_2 disappearance is varying between 60 ~ 95%. This variation is likely to be caused by particle formation and deposition process on the wall. Nopinone was identified as the main volatile product. Its concentration, however, does not exceed 2% of the converted ß-pinene.

Table 2 summarizes the results of the α-, ß-pinene and limonene reactions with ozone/NO_2/SO_2. Addition of NO_2 to the system terpene/O_3/SO_2 terpene-degradation rates. The ratio Δterpene/[terpene]$_0$ was for ß-pinene in most of the cases more than 0,90 after 4 hours, for α-pinene and limonene more than 0,95 after 1 hour reaction time. The corresponding terpene degradation rate in the system terpene/O_3/SO_2 was for e.g. ß-pinene in the range between 0,45 - 0,76. The ratio ΔH_2SO_4 / ΔO_3, however, is significantly lower (~70%) compared to similar reactions without NO_2 addition (e.g. run 1 and 3 table 1 and run 1 and 5 table 2). Our findings for the reaction of terpenes with ozone in the presence of NO_2 and SO_2 can be well explained assuming the following reaction sequences:

$$PIN + O_3 \quad \xrightarrow{K_2} \quad \text{Criegee intermediates} \quad (\text{eq. 2})$$

$$K_2 = \sim 10^{-17} \, cm^3/\text{molec. sec.}$$

$$NO_2 + O_3 \quad \xrightarrow{K_3} \quad NO_3 + O_2 \quad (\text{eq. 3})$$

$$K_3 = \sim 10^{-17} \, cm^3/\text{molec. sec.}$$

$$PIN + NO_3 \quad \xrightarrow{K_4} \quad \text{products} \quad (\text{eq. 4})$$

$$K_4 = \sim 10^{-12} \, cm^3/\text{molec. sec.}$$

$$INT + NO_2 \quad \xrightarrow{K_5} \quad NO_3 + \text{products} \quad (\text{eq. 5})$$

$$K_5 = \sim 10^{-13} \, cm^3/\text{molec. sec.}$$

PIN = terpene
INT = Criegee intermediates

Reaction 4 accelerates terpene degradation, while reaction 5 represents an additional sink for the Criegee intermediates as the responsible oxidizing species for SO_2. With the addition of NO_2 to the system ß-pinene/O_3/SO_2 the influence of relative humidity and the relationship between total SO_2 oxidation and the O_3/terpene ratio are less significant than without.

We could identify nopinone and pinonaldehyde as the main volatile reaction products upon reaction of ß-pinene and α-pinene with O_3/SO_2/NO_2 respectively by means of gas chromatography - mass spectrometry and by comparison with known spectra.

The reaction of limonene leads to an unidentified volatile product with molecular weight M^+ 134. The amount of these reaction products remains less than 2-3% of the total terpene degraded. It seems however, that both SO_2 and NO_2 addition causes a pronounced effect with respect to the nature of the reaction products.

4. CONCLUSIONS

The gas-phase reaction of selected monoterpenes with O_3, SO_2 and NO_2 was investigated. It was found that the SO_2 oxidation yield and the subsequent H_2SO_4 formation was significantly influenced by the presence of H_2O vapour in the system ß-pinene /O_3/SO_2. As r. h. increases the SO_2 oxidation yield decreases. Addition of NO_2 to the system terpene/O_3/SO_2 causes high terpene conversion rates. However, the SO_2 oxidation is lower than by similar reaction without NO_2 additions.

Our findings are in close agreement with the results of Hatakeyama et al., 1984 regarding the differences in SO_2 oxidation between exocyclic and endocyclic double-bond compounds. ß-Pinene reaction with ozone, SO_2 and NO_2 leads to higher H_2SO_4

yields, than the similar reaction of α-pinene. The reaction of the emitted monoterpene hydrocarbons with ozone/SO_2-NO_2 leads to an overall increasing of acidity - due to the formation of inorganic and at higher humidity organic acids - in forest areas.

5. REFERENCES

1.   Cox, R. A. and Penkett, S. A., (1972). *Aerosol formation from sulphur dioxide in the presence of ozone and olefinic hydrocarbons.* I Chem. Soc. Faraday Trans. I, 68, 1735-1753.

2.   Gäb, S., Hellpointner, E., Turner, W. V. and Korte, F., (1985) *Hydroxymethyl hydroperoxide and bis (hydroxymethyl) peroxide from gas-phase ozonolysis of naturally occuring alkenes.* Nature, Vol. 316, No. 6028, pp. 535-536.

3.   Gay, B. W.Jr. and Arnts, R. R., (1977). *The chemistry of naturally emitted hydrocarbons.* ETPA - 600/3 - 77 - 001b.

4.   Hatakeyama, S., Kobayaski, H. and Akimoto, H., (1984). *Gas-phase oxidation of SO_2 in the ozone-olefin reactions* J. Phys. Chem., 88, 4735 - 4739

5.   Izumi, K., Mizuochi, M., Murano, K., Fukuyama, T., (1987). *Humidity effects on photochemical aerosol formation in the SO_2 - NO - C_3H_6-Air system.* Atmospheric Environment, Vol. 21, No. 7, pp. 1541 - 1553.

6.   K. Jay and L. Stieglitz *Proceedings of: "3. Statuskolloquium des PEF".* Karlsruhe, FRG., 10 - 12.3.1987

7.   K. Jay and L. Stieglitz *Proceedings of: "4. Statuskolloquium des PEF".* Karlsruhe, FRG., 8 - 10.3.1988

8.   Martinez, R. I. and Herron, J. T., (1981). *Gas phase reaction of SO_2 with a Criegee-intermediate in the presence of water vapour.* J. Environ, Sci. Health, A 16 (6), 623 - 636.

9.   Martinez, R. I. and Herron, J. T., (1983). *Acid precipitation: The role of O_3 - alkene - SO_2 systems in the atmospheric conversion of SO_2 to H_2SO_4 aerosol.* J. Environ. Sci. Health, A 18(6), 739 - 745.

10. F. Nolting and C. Zetzsch *Proceeedings of: "4. Statuskolloquium des PEF".* Karlsruhe, FRG., 8 - 10.3.1988

14. Stangl, H. Kotzias, D. and Geiss, F., (1988). *How forest trees actively promote acid deposition.* Naturwissenschaften, 75, 42 43.

15. Westberg, H. H. and Rasmussen, R. A., (1972). *Atmospheric photochemical reactivity of monoterpene hydrocarbons.* Chemosphere 4, 163 - 168.

16. Yokouchi, V. and Ambe, Y. (1985) *Aerosols formed from the chemical reaction of monoterpenes with ozone.* Atmospheric Environment, Vol. 19, No. 8, pp. 1271 - 1276.

17. Zimmermann, P. R., Chatfield, R. B., Fishman, J., Crutzen, P. J. and Hanst, P. L., (1978). *Estimates on the production of CO and H_2 from the oxidation of hydrocarbon emissions from vegetation.* Geophy. Res. Lett., 5, 676 - 682.

Table 1: Reaction of ß-Pinene with O₃/SO₂

Run	$[O_3]_0$ ppb	$[\text{ß-pinene}]_0$ ppb	$[SO_2]_0$ ppb	r.h. %	$[SO_4^{2-}]_f$ µg/m³	$\dfrac{\Delta H_2SO_4}{\Delta O_3}$	$\dfrac{[SO_2]ox}{[SO_2]_0}$	$\dfrac{\Delta H_2SO_4}{\Delta SO_2}$
1	205	100	110	6	63,0	0,33	0,14	0,75
2	205	90	120	45	42,0	0,18	0,07	0,89
3	244	190	122	6	92,4	0,36		0,95
4 a	100	94	99	6	33,6	0,44	0,11	0,72
5 a	103	100	103	48	23,1	0,28	0,08	0,61
6	110	132	100	6	44,1	0,34		0,65
7	97	61	51	10	n. m.	0,26 $\frac{\Delta SO_2}{\Delta O_3}$	0,10	
8	98	58	52	63	n. m.	0,16 $\frac{\Delta SO_2}{\Delta O_3}$	0,06	
9	102	61	51	9	n. m.	0,30 $\frac{\Delta SO_2}{\Delta O_3}$		
10	200	109	100	15	63,0	0,19		
11	250	96	102	9	63,0	0,22		0,78
12	194	56	52	9	16,8	0,13		0,57

[ ]₀ = initial concentration   a = reaction time 2h
                                     for all other runs the reaction time was 4h

[ ]f = final concentration   n. m. = not measured, $\Delta H_2SO_4/\Delta O_3$ was calculated as $\Delta SO_2/\Delta O_3$

$[SO_2]ox$ = SO₂ oxidized

Table 2: Reaction of α-pinene, β-pinene and limonene with O_3, SO_2, and NO_2

	Run	$[O_3]_0$ ppb	$[terpene]_0$ ppb	$[SO_2]_0$ ppb	$[NO_2]_0$ ppb	r.h. %	$[SO_4^{2-}]_f$ µg/m³	$\dfrac{\Delta H_2SO_4}{\Delta O_3}$	$\dfrac{\Delta H_2SO_4}{\Delta SO_2}$
β-pinene	1	199	99	103	95	10	39,9	0,08	0,86
	2	205	98	107	105	51	31,5	0,07	0,60
	3 a	305	103	103	100	14	43,2	0,10	0,93
	4	295	105	101	102	57	34,8	0,08	0,83
	5	200	161	108	97	14	62,1	0,13	0,79
	6	200	160	106	100	53	60,0	0,13	0,81
α-pinene	7 b	200	102	103	100	15	10,5	0,03	0,5
	8 b	220	87	106	98	53	6,3	0,01	0,31
	9 b	299	110	107	108	14	13,4	0,03	0,80
limonene	10 b	193	194	200	215	15	37,8	0,09	0,66
	11 b	215	190	200	200	50	14,7	0,03	0,87
	12 b	200	96	103	115	15	8,4	0,03	0,96
	13 b	195	50	101	108	15	6,3	0,02	0,92

[ ]$_0$ =initial concentration

[ ]$_f$ = final concentration

a = reaction time was 2 h

b = reaction time 1h, for all other runs 4 h

PRODUCTS AND MECHANSISM OF THE GAS PHASE REACTION
BETWEEN THE NITRATE RADICAL AND ARENES

F. Cariati° and B. Rindone*

 Università di Milano
°Dipartimento di Chimica Inorganica e Metallorganica
*Dipartimento di Chimica Organica e Industriale
 20133 Milano - Italy

J. Hjorth and G. Restelli

Commission of the European Communities
Joint Research Centre - Environment Institute
21020 Ispra (Va) - Italy

Summary

The reaction between the nitrate radical NO_3 and the aromatic hydrocarbons toluene and xylenes has been studied in the gas phase in a 480 L reaction chamber. The reactions were followed by in situ infrared spectroscopy and the products investigated using chromatographic techniques by sampling the gas mixture. Benzaldehyde, benzylnitrate and their methylsubstitutes were observed as main products. It appears that in the gas phase hydrogen abstraction is the first step of the reaction in accordance with what suggested on the basis of kinetic studies. The mechanisms following the initial formation in air of the peroxyradicals and leading to the observed products are likely to follow schemes known from the study of the reaction between NO_3 and alkenes.

1. INTRODUCTION
     The nitrate radical NO_3, generated by the reaction between NO_2 and O_3 is now recognized to be, during the night, a common constituent of the lower troposphere where it can play a significant role as a sink for organic species present in air (1). Its reactivity towards organic compounds and the reaction products have been the subject of extended investigations for some years now.
     The nitrate radical is known to react with aromatic hydrocarbons with rate constants in the range from $< 2 \times 10^{-17}$ cm^3 $molec^{-1}$ s^{-1} (benzene) to

5.6×10^{-16} $cm^3 molec^{-1} s^{-1}$ (1,2,3 trimethylbenzene) (2). Information on reaction products and mechanisms is essentially limited to the hypothesis, based on kinetic studies, that the reaction proceeds by H-abstraction from substituent methyl groups (3-5).

In this study the nitrate radical was obtained in equilibrium with N_2O_5 and NO_2 in a reaction chamber and reacted with vapours of benzene, toluene, para- and ortho-xylene in zero air. The reaction was followed by in situ FT-IR spectroscopy. Bands attributable to benzaldehyde were observed but other information on the formation of nitrates were obscured by the absorption of HNO_3 formed in the cell from the reaction of N_2O_5 with H_2O. The reaction products were trapped on suitable absorbing columns and subsequently analyzed by chromatographic techniques.

Benzene was too slowly reacting to produce any measurable product. The methyl substituted benzenes were found to form products in the gas phase originating from the H-atom abstraction from the alkyl group. In air the benzylradical should lead to the formation of the observed products, aldehydes, nitrates and, in minor amounts, alcohols, according to schemes known from the study of the reaction of NO_3 with alkenes.

Reactions of arenes with NO_3 in the liquid phase were performed using cerium(IV) ammonium nitrate in acetonitrile. The main goal was to use these reactions for setting up the analytical techniques used subsequently for the study of the gas phase reaction. The same reaction products were observed in the liquid phase reaction.

2. EXPERIMENTAL

The reactions were performed in purified air, in the dark, in a 480 L reaction chamber, Telfon-coated and equipped with a multiple reflection White type mirror system (total beam pathlength 81 m) connected to a Bruker IFS 113 V interferometer.

N_2O_5 was synthesized in the chamber by adding O_3 to excess NO_2; vapours of the aromatic hydrocarbons in purified air were subsequently added and the reaction was allowed to proceed until N_2O_5 was totally consumed. Reaction times of the order of 5-6 hours were used, mainly set by the heterogeneous transformation of N_2O_5 into HNO_3. Typical initial concentrations were 20 ppmv for N_2O_5, a few ppm for NO_2 and from few tens of ppmv for xylenes up to several hundreds of ppmv for benzene.

Samples to be used for product analysis were obtained by pumping known volumes ($\sim$ 40 L) of the reaction mixture through glass columns packed with either 750 mg Amberlite XAD-2 or 150 mg of cocoanut charcoal. The desorption was performed by extraction with 0.5 ml dichloromethane under mechanical agitation for about one hour.

The products were eventually identified by analysis of the dichloromethane solution without further preconcentration, applying two analytical techniques:

– 401 –

1) GC-MSD analysis, using a HP 5890 gaschromatograph with a 30 m capillary column (SPB5, 0.25 mm) interfaced with a quadrupolar detector (HP 5970) operating in Electron Impact mode at 70 eV;

2) GC-FTIR analysis, using the same gaschromatograph with a different capillary column (JXR, 0.32 mm) interfaced with a FTIR detector (HP 5965A).

Injections were performed in the "on column" mode.

Two of the charcoal samples were analyzed by thermal desorption at 200°C for 20 seconds using a Dynatherm 850 thermal tube desorber, connected on line with the GC-MSD system.

The identification of the products was aided by computerized library search including the MS and the FTIR spectra, and finally confirmed by comparison with spectra of pure reference compounds (commercially available samples). Benzyl nitrates were synthetized by reaction of the corresponding chlorides with silver nitrate (6).

The reactions in solution were performed by mixing equimolecular amounts of cerium(IV) ammonium nitrate and of the hydrocarbon in acetonitrile. The typical initial concentrations were about 1 mM. The reaction mixture was left in a nitrogen atmosphere for two weeks. Acetonitrile was then evaporated and the residue wad dissolved in diethyl ether and filtered. After evaporation of the solvent, the residue was dissolved in acetonitrile and analyzed by HPLC, using a Variant 5000 instrument equipped with a Reverse Phase C 18 Column with an acetonitrile-water gradient elution. A HP Diode Array Detector was used. The identification of the products was accomplished using also GC-MSD analysis and by comparison with authentic specimens.

3. RESULTS AND DISCUSSION

In Table I reactants and products are numbered and for sake of simplicity these numbers will be used in the following discussion. Benzene was essentially unconverted under the experimental conditions applied.

In the gas phase reaction of toluene 2 (Fig.1 shows the GC-MS analysis of the material eluted with dichloromethane after activated carbon entrapment), the products observed were benzaldehyde 2a, benzylnitrate 2b, the isomeric mixture of nitrotoluene 2c and toluene dimers 2d. The structural assignments were confirmed by GC-FTIR and by GC and HPLC comparison with authentic specimens. Fig.2 shows the GC-FTIR spectrum of the nitrate 2b.

With para-xylene 3, elution of the XAD-2 resin showed that reaction products were 4-methylbenzaldehyde 3a, 4-methylbenzyl nitrate 3b and minor amounts of 4-methylbenzyl alcohol 3c and the nitroderivative. The GC-FTIR profile is shown in Fig.3. Elution of the activated carbon trap with dichloromethane gave similar results.

With ortho-xylene 4, the elution of the XAD-2 resin with dichlormethane revealed in the GC-MS analysis the presence of 2-

methylbenzaldehyde 4a, 2-methylbenzyl nitrate 4b and 2-methylbenzyl alcohol 4c and the isomeric 2-methylnitrotoluenes 4d.

Analysis of the IR spectra recorded in the gas phase in the reaction chamber did not provide clear information except for the presence of bands tentatively attributed to aldehydes. The spectral regions useful for nitrate bands identification were mostly obscured by the build-up of HNO_3 (Fig.4).

The study of the reaction of toluene 2 with CAN in acetonitrile led to the identification of benzaldehyde 2a, benzyl nitrate 2b, the isomeric mixture of nitrotoluenes 2c, and some benzoic acid 2e.

The reaction of para-xylene 3 with CAN in acetonitrile gave a very similar result. Again, 4-methylbenzaldehyde 3a, 4-methylbenzyl nitrate 3b, 4-methylbenzyl alcohol 3c were found together with the nitroderivative 3d.

These results suggest that in the gas phase reaction of toluene and xylenes with NO_3, hydrogen abstraction on the methyl group by the nitrate radical to give benzyl radicals and nitric acid is occurring as the first step. This is in agreement with the hypothesis supported by the kinetic studies. In air, insertion of oxygen gives rise to the peroxy radical 6 and its reduction product, the alkoxy radical 7 (from $2RO_2 \rightarrow 2RO+O_2$ or $RO_2+NO_3 \rightarrow RO+NO_2+O_2$). The peroxyradical 6 can generate by disproportion the aldehyde 9 and the alcohol 8; the oxy-radical can lead to formation of the aldehyde following H-abstraction by O_2 and in the presence of NO_2 of the benzylnitrate 10 by NO_2 addition (Fig.5,). All these mechanisms, after the initial formation of the peroxyradicals, follow schemes known from the study of the reaction between NO_3 and alkenes (2,7).

The products 1a, 2c, 2d, 3d and 4d observed in the GC analysis may not have been formed in the gas phase reaction; heterogeneous reactions in the column absorber cannot, in fact, be excluded.

A fairly extensive literature exists (see, e.g., Refs.8-10) on the study of the liquid phase reaction of the nitrate radical with aromatics, performed using CAN. The experimental conditions are however not exactly the same as those used in this study. In spite of the similarity of the pattern of the reaction products observed, an interpretation of the mechanism occurring in this second case is not straightforward. More work is in progress to better characterize the reaction in the gas phase.

ACKNOWLEDGEMENTS

We thank Mr. G. Ottobrini and our students S. Polesello and V. Paratici for technical support.

REFERENCES

(1) WINER, A.M., ATKINSON, R and PITTS, G.N. Jr. (1984). Science, 224, 156.

(2) FINLAYSON-PITTS, B.J. and PITTS, J.N. Jr. (1986). In: Atmospheric Chemistry, J. Wiley and Sons, London.

(3) ATKINSON, R., CARTER, W.P.L., PLUM, C.N., WINER, A.M. and PITTS, J.N. Jr. (1984). Int. J. Chem. Kinetics, 16, 887.

(4) ATKINSON, R., ASHMANN, S.M. and WINER, A.M. (1987). Env. Sci. Technol., 21, 1123.

(5) ATKINSON, R. and ASCHMANN, S.M. (1988). Int. J. Chem. Kinetics, 20, 513.

(6) DESSEIGNE, G. (1946). Bull. Soc. Chim. France, 98, 343.

(7) HJORTH, J., LOHSE, C., NIELSEN, C.J., SKOV, H. and RESTELLI, G., submitted for publication.

(8) DINCTURK, S. and RIDD, J.H. (1982). J. Chem. Soc. Perkin Trans. II, 961, 965.

(9) BACIOCCHI, E., DEL GIACCO, T., ROL, C. and SEBASTIANI, G.V. (1985). Tetrahedron Letters, 541.

(10) BACIOCCHI, E., DEL GIACCO, T., MURGIA, S.M. and SEBASTIANI, C.V. (1987). J. Chem. Soc. Chem. Commun., 1246.

TABLE I: Reactants and reaction products.

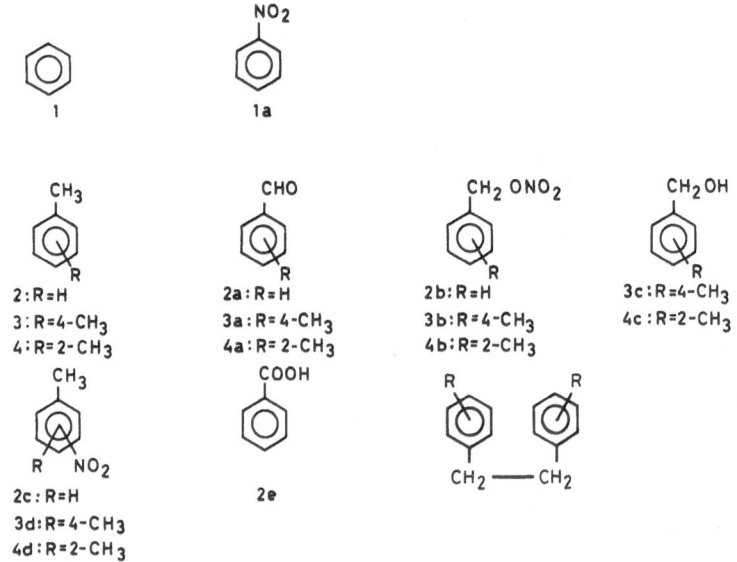

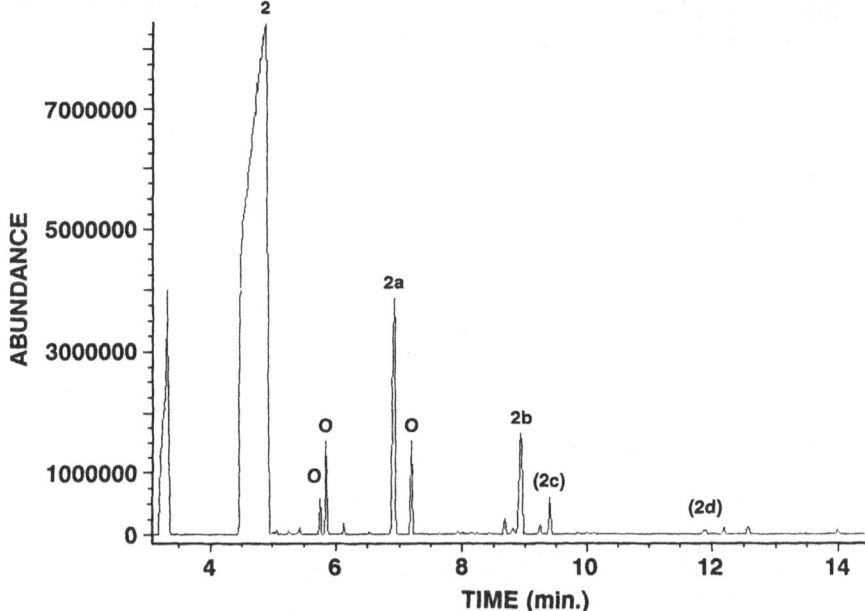

Fig.1: Products from the gas phase reaction toluene + NO_3; elution after activated carbon entrapment. O indicates identified contaminants.

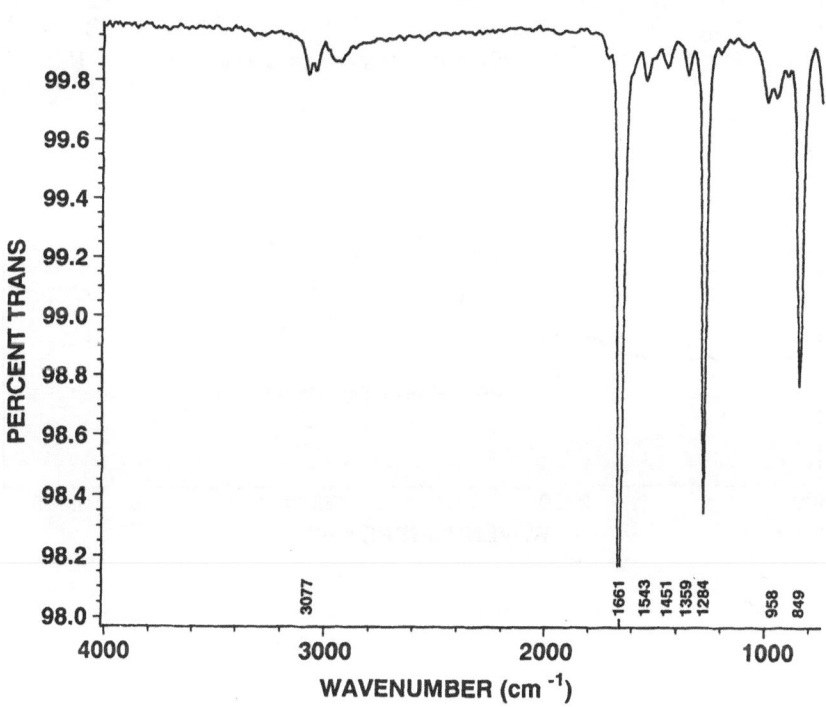

Fig.2: GC-FTIR products analysis: spectrum of benzylnitrate 2b.

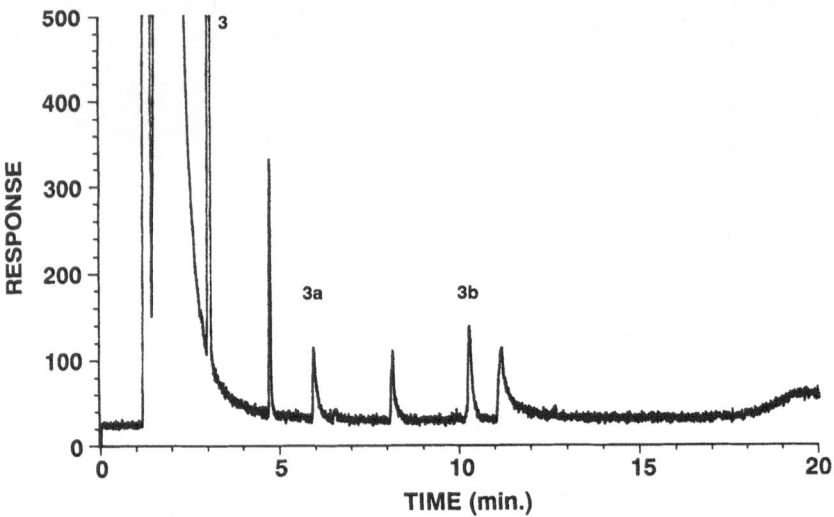

Fig.3: GC-MS analysis of the products of the gas phase reaction of ortho-xylene + NO$_3$; elution after entrapment in XAD-2 resin.

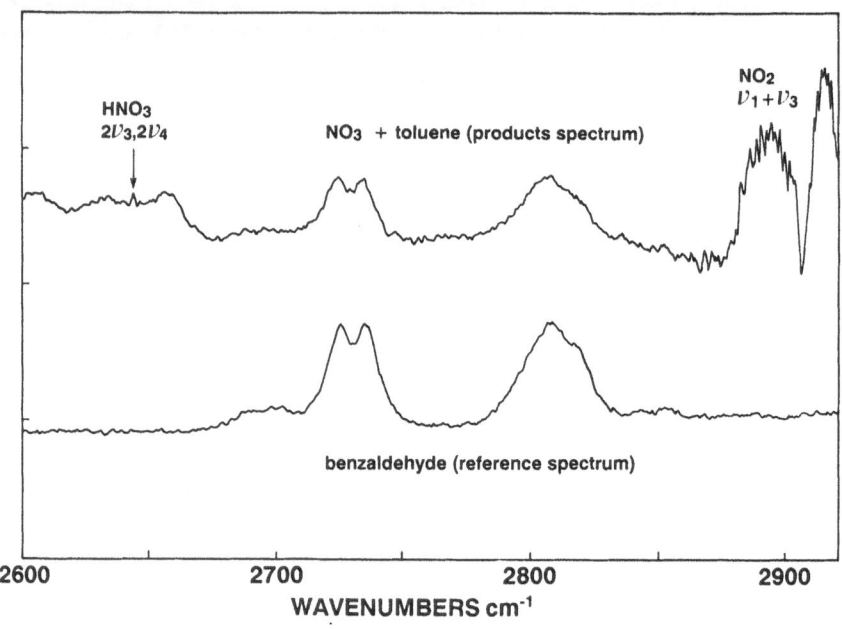

Fig.4: IR spectrum of the products of the reaction NO$_3$ + toluene. Bands attributed to benzaldehyde are compared to an authentic spectrum.

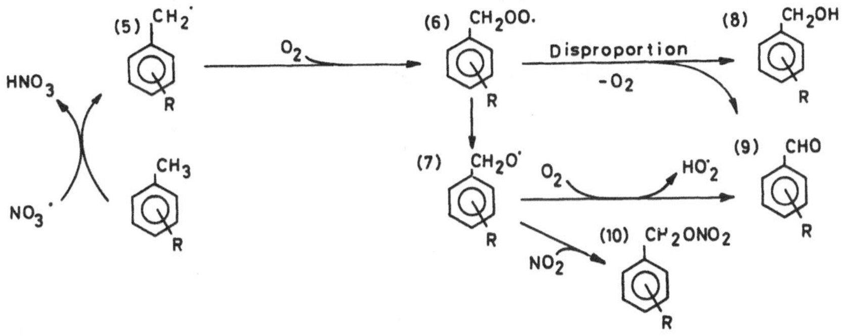

Fig.5: Possible reaction pathway according to identified products.

THE SELF—REACTION OF THE NO_3 RADICAL

P. Biggs, C.E. Canosa–Mas, M. Joseph, P.S. Monks and R.P. Wayne

Physical Chemistry Laboratory, University of Oxford,
South Parks Road, Oxford, OX1 3QZ, UK

SUMMARY

   This paper describes two separate techniques for the measurement of
slow reactions. The use of these techniques is exemplified by the derivation
of the rate of the self–reaction of the NO_3 radical.
   The first technique is a laser flash photolysis one, where the inherently
low rate of the decomposition of the radical is used to allow the build–up of
a larger concentration of NO_3 than is produced in a single flash. The rate of
the disappearance of the radical on cessation of photolysis is used to
determine the rate of the self–reaction. The second technique is a gas–phase
stopped–flow system. Here, a conventional flow system is used to generate a
steady state concentration of the radicals. At some time, a portion of the
flow is isolated, and the concentration of the radicals in this portion is
monitored with time.
   The rate constant at room temperature for the reaction

$$NO_3 + NO_3 \longrightarrow products$$

was found to be $(3.3\pm0.5) \times 10^{-16}$ cm^3 molecule^{-1} s^{-1} as measured by the
flash photolysis method and $(2.5\pm0.6) \times 10^{-16}$ cm^3 molecule^{-1} s^{-1} when
measured by the stopped–flow technique.

1. INTRODUCTION

   The nitrate radical is now well established as an important oxidant in the
troposphere at night. Reaction with the radical is a significant loss process for
most volatile organic compounds, and it is a major oxidation route for many
unsaturated organic species, including natural hydrocarbons. The radical is a key
intermediate in the formation of nitric acid, either by abstraction of an H atom by
NO_3 or indirectly by hydrolysis of N_2O_5.
   The slowness of the reaction of NO_3 with the smaller alkanes (e.g. methane
and ethane) and other compounds such as halocarbons makes the measurement of
rate constants for the interactions difficult with conventional flash or flow
techniques. It is the purpose of the experiments described here to demonstrate that
the modified techniques proposed for the study of these slow reactions are viable
and to characterize some slow reactions of the nitrate radical, such as its wall loss
and its self reaction.

2. LASER FLASH PHOTOLYSIS

Laser flash photolysis was used to generate NO_3 radicals by the reaction of F atoms with HNO_3, the F atoms being generated by the the photolysis of molecular fluorine by the $\lambda=308$ nm radiation from a XeCl excimer laser.

$$F_2 + h\nu \ (\lambda=308 \text{ nm}) \longrightarrow F + F$$

$$F + HNO_3 \longrightarrow NO_3 + HF$$

The NO_3 concentration was monitored by the optical absorption of the laser radiation from an argon ion / dye laser combination, operated at $\lambda=623$ nm, in a 20 pass White cell of total path length 4.2 m.

The cell was filled with a mixture of He, F_2 and HNO_3 and photolysed at such a repetition rate that NO_3 built up to a steady state concentration. Photolysis rates of up to 30 Hz were used, generating concentrations of NO_3 in the range $10^{14} - 10^{15}$ molecules cm^{-3}. On cessation of photolysis, the NO_3 was seen to decay over a period of up to 50 seconds.

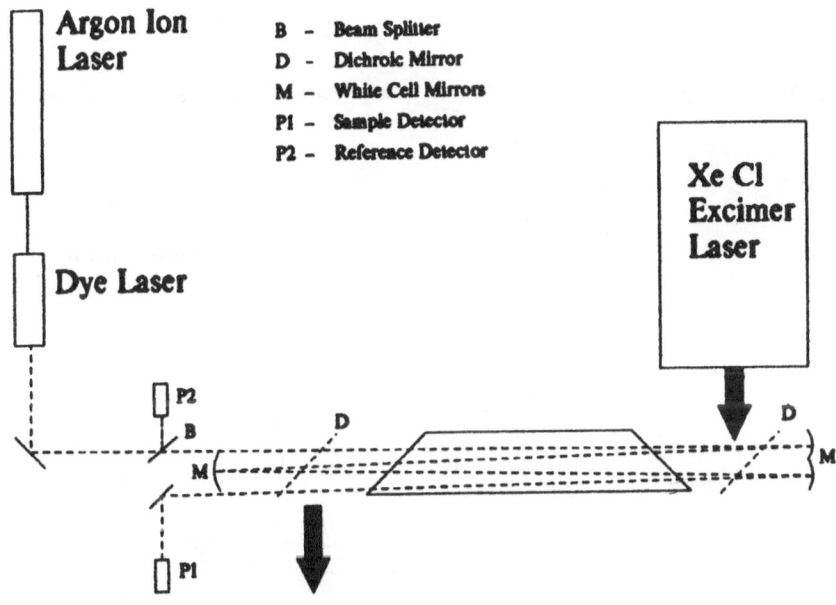

Figure 1.   The laser flash–photolysis system

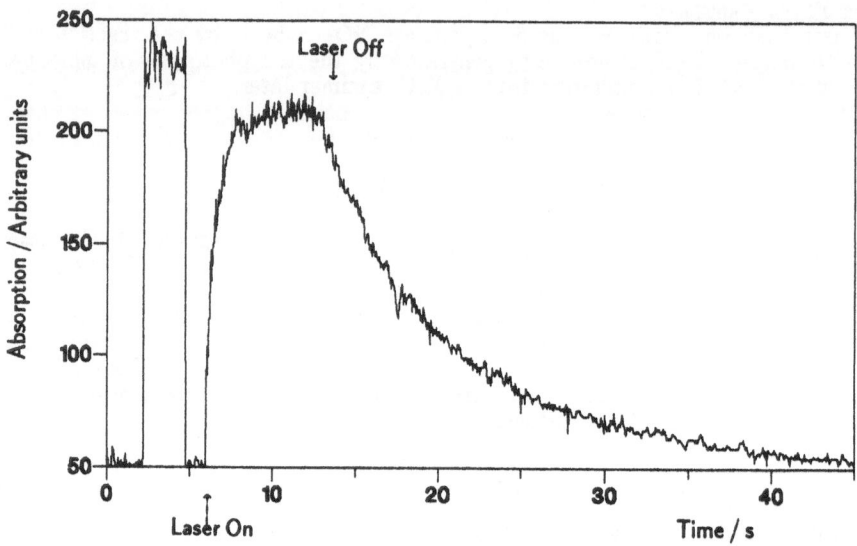

Figure 2. A typical trace obtained on photolysing $F_2 + HNO_3$ mixtures.

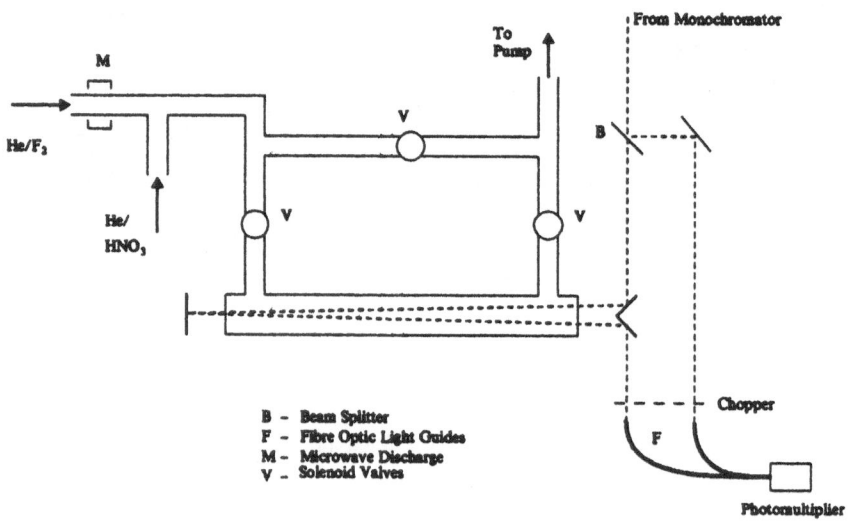

B – Beam Splitter
F – Fibre Optic Light Guides
M – Microwave Discharge
V – Solenoid Valves

Figure 3. The stopped–flow system.

3. The Stopped—Flow System

Stopped flow experiments were performed in the system shown in figure 3. NO_3 was again made by the $F + HNO_3$ reaction, the F atoms this time being generated by passing molecular fluorine through a microwave discharge.

The NO_3 was detected by optical absorption at $\lambda=662$ nm in a two–pass absorption cell using a dual–beam spectrometer arrangement with a total path length of 0.6 m. The wavelength of the interrogating beam was selected by a monochromator.

A flow of gas was set up through the absorption cell and a steady state of NO_3 was observed as in a normal flow system. Once the concentration of NO_3 had stabilized, the cell was isolated by solenoid valves. The NO_3 concentration was once again seen to decay over a period of 10's of seconds

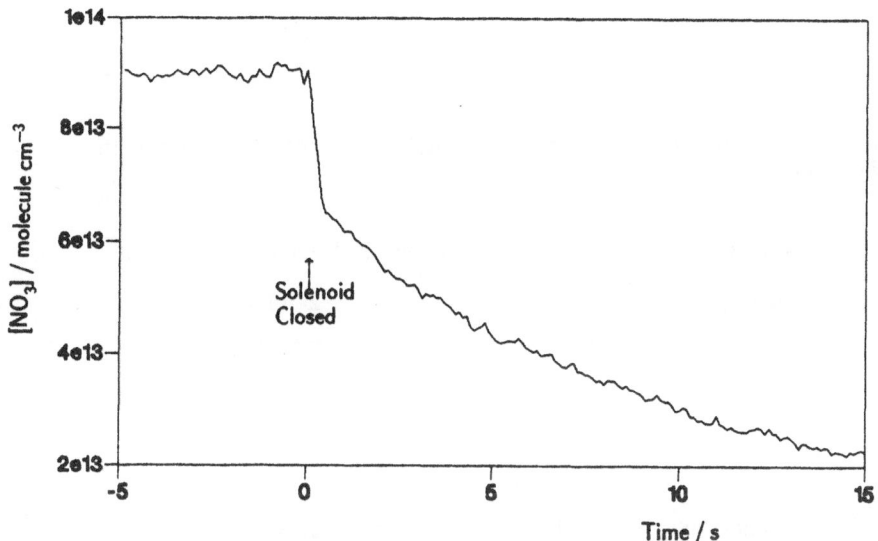

Figure 4. A typical decay trace from the stopped–flow system

4. Results

Figure 5 shows two decay curves from the flash photolysis experiment for different starting values of $[NO_3]$. The curves were simulated by numerical integration of the kinetic equations derived from the reaction scheme below.

$$NO_3 \xrightarrow{\text{wall}} NO_2 + \tfrac{1}{2}O_2 \qquad (1)$$

$$NO_3 + NO_3 \longrightarrow 2NO_2 + O_2 \qquad (2)$$

$$NO_3 + NO_2 \rightleftharpoons N_2O_5 \qquad (3)$$

The values of k_1 and k_2 obtained from a number of experiments are shown in table 1.

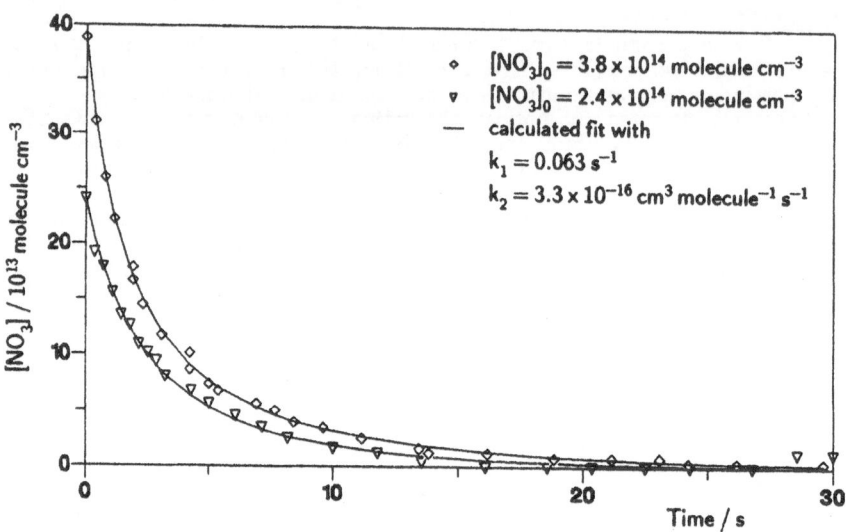

Figure 5. Decay profiles from the flash–photolysis experiments showing the modelled fits.

Table 1. Summary of results from the flash photolysis system

$[NO_3]_0/cm^{-3}$	k_1/s^{-1}	$k_2/cm^3\ s^{-1}$
2.4×10^{14}	0.072	3.4×10^{-16}
3.3×10^{14}	0.075	3.3×10^{-16}
3.9×10^{14}	0.058	3.2×10^{-16}
4.4×10^{14}	0.039	4.2×10^{-16}
4.5×10^{14}	0.072	2.6×10^{-16}
Average	0.063 ± 0.013	$(3.3 \pm 0.5) \times 10^{-16}$

A fit to the data shown above derived from the stopped–flow experiments is shown in figure 6. The sharp drop at the beginning is thought to result from NO_2 present at the start of the decay. Indeed, this initial rapid decay, although not shown on the diagram, can be succesfully modelled.

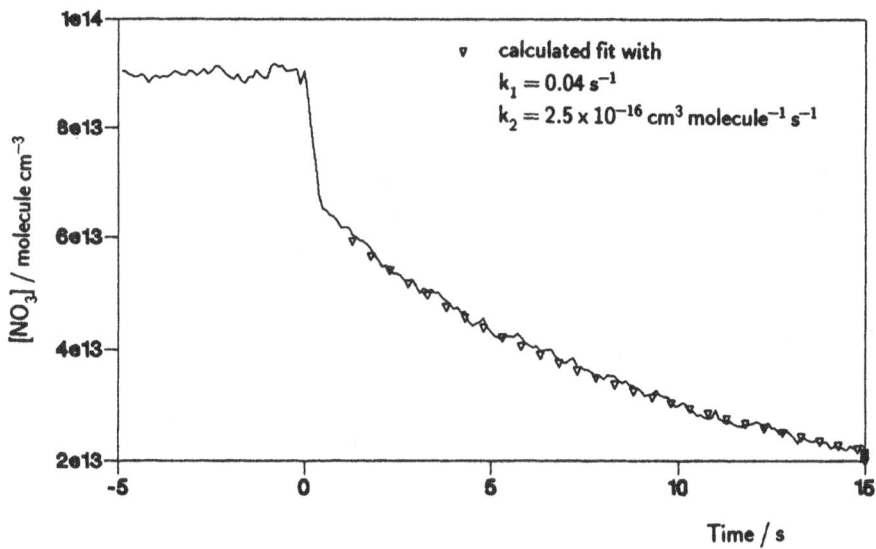

Figure 6. Diagram showing the fit to the stopped–flow data.

5. SUMMARY

Despite the fact that these experiments are in their very early stages of development, it has been shown that valid, consistent results can be obtained from them. The results presented above should be compared with the rate constant obtained by Graham and Johnston in 1978 (J. Phys. Chem. 254, $\underline{82}$ (1978)), who obtained a value for k_2 at 298 K of 2.3×10^{-16} cm^3 molecule^{-1} s^{-1}.

Work has already started on measuring the rates of other slow reactions in these systems, and initial results are very promising. A <u>preliminary</u> value for an upper limit on the room–temperature rate constant of the reaction of NO$_3$ with ethane was measured to be 1.1×10^{-17} cm^3 molecule^{-1} s^{-1}.

SESSION III/A

FIELD MEASUREMENT

AND

THEIR INTERPRETATION

SUMMARY OF SESSION III
FIELD MEASUREMENT AND THEIR INTERPRETATION

S. BEILKE, Umweltbundesamt, Offenbach,
Federal Republic of Germany

Working group 3 of COST 611 is concerned with field measurements and their interpretation of trace gases and aerosols in
the troposphere and stratosphere on different scales in space
and time ranging from local to global.
Over the last century the emphasis of air pollution problems
has moved from local and urban scales to regional and global
ones.Today the main global air pollution problems are a.) the
possible climatic change due to an increase of greenhouse gases
and b.) the depletion of ozone in the stratosphere due to an
increase of CFCs' and other long-lived trace gases.
On the regional scale of Europe the main air pollution problems
are a.) the photochemical oxidant generation in the surface-near
troposphere and b.) acid deposition.
Within Working group 3 all of these air pollution problems are
addressed but special emphasis is placed on gases which are
involved in acid deposition and ozone formation mainly in Europe.

To place the present position of Working group 3 in context and
to emphasize the future direction of scientific research,a brief
historical look at these two environmental problems in Europe is
useful.

In contrast to acid deposition which has been known as a local
urban air pollution problem for more than 130 years(ref.1),first
photochemical pollution episodes in Europe were reported in the
early 1970s(ref.2).Today both acid deposition and photochemical
oxidants are encountered in large areas of Europe depending on
emission characteristics and meteorological conditions under
which the precursors may react.
For acid deposition the main pollutants involved are SO_2,
followed by NO_x,NH_3 and VOC.The main precursors of ozone are
NO_x and VOC.

In general,the anthropogenic emissions of these main precursors
have considerably increased in Europe as a whole since the onset
of industrialisation in the middle of the 19 th century.There
are,however,more or less important differences in the long-term
trends as well as regarding the geographical areas and gases
concerned.
As far as SO_2 is concerned indications are strong that the
emissions increased by a factor of two between 1950 and 1970 in
most of Europe and the same general trend holds for NO_x(ref.3).
In most countries of the CEC SO_2 emissions have generally
levelled off in the 1970s and are on a declining trend since
then mainly as a result of the environmental control on emissions
(ref.3).In the period 1980-1987 emissions in western Europe decreased by 35 - 40 % (ref.4) but in Europe as a whole the SO_2 -
decrease is much smaller(ref.5).

In contrast to SO_2,the NO_x emissions have continued to increase in western Europe at least until 1980.

Emissions of NH3 are likely to have increased by ca. 50 % in
Europe as a whole between 1950 - 1980 (ref.3).The information
about ammonia is,however,rather scare and only minor changes
are expected in the period 1980-1987(ref.4).

Emissions of VOC are rather uncertain and are believed to have
increased by more than a factor of two in Europe between 1950
and 1970(ref.3).A quantification of the VOC emission change
from 1980 is difficult to make.

An example for long-term trends is shown in figure 1 estimating
in some detail the emissions of SO2 and NOx for the territory
of the Federal Republic of Germany(ref.6,7).

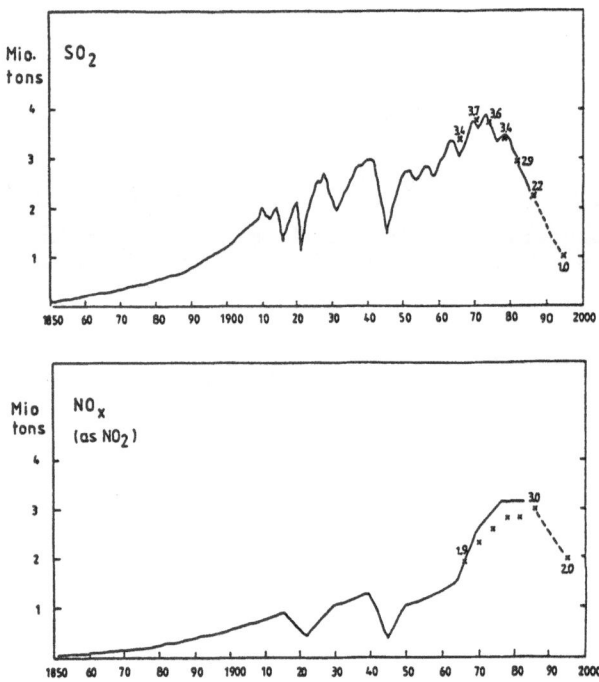

FIGURE 1 : Long-term emission curves for SO_2 and NO_x for the
territory of the Federal Republic of Germany.

solid line from 1850-1982.(ref.6)
crosses:emission figures for 1966,1970,1974,1978,
1982,1986 and 1995 (ref.7)

Indications are strong that the SO2 and sulfate concentrations
in the air are also on a declining trend in some areas of western
Europe over the past few years(ref.5,8).
An interesting result in this context is shown in figure 2

indicating a pronounced decrease of SO_2 concentrations in the FRG in 1988 compared to the corresponding concentrations of the previous years.The decrease measured by the 13 background monitoring stations of the Federal Environmental Agency ranges from 28% in the east up to 70% in the western part.

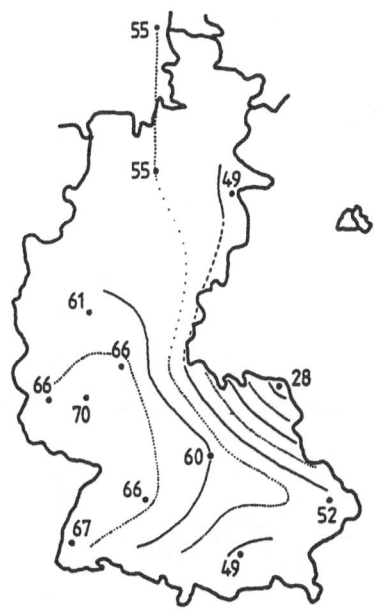

FIGURE 2 : Decrease of yearly mean SO_2 concentrations in 1988
at 13 rural stations in the FRG compared to
1980-1987(ref.8) in percent.

The decrease can be interpreted in terms of a combined effect of an emission reduction in the FRG and its western neighbours and two relatively mild winters and a cool summer in between with air flow mainly from western directions(ref.8).

Another interesting example is shown in figure 3 summarizing the results of aircraft measurements carried out over the FRG between 1-3 February 1989 during a synoptic weather situation characterized by a stable anticyclone centred over Central Europe. A strong temperature inversion between 400-800 m was observed connected with very low wind speed within the mixing layer. The concentrations and amounts of SO_2 within the mixing layer were generally very low and much smaller than the corresponding values for NO_x(ref.9).Some decades ago such a meteorological situation was characterized by high SO_2 and much lower NO_x concentrations.As transboundary transport has hardly influenced those concentrations the profiles should reflect the emission characteristics in the FRG as well as the different physico-chemical behaviour of both gases under the meteorological conditions encountered.

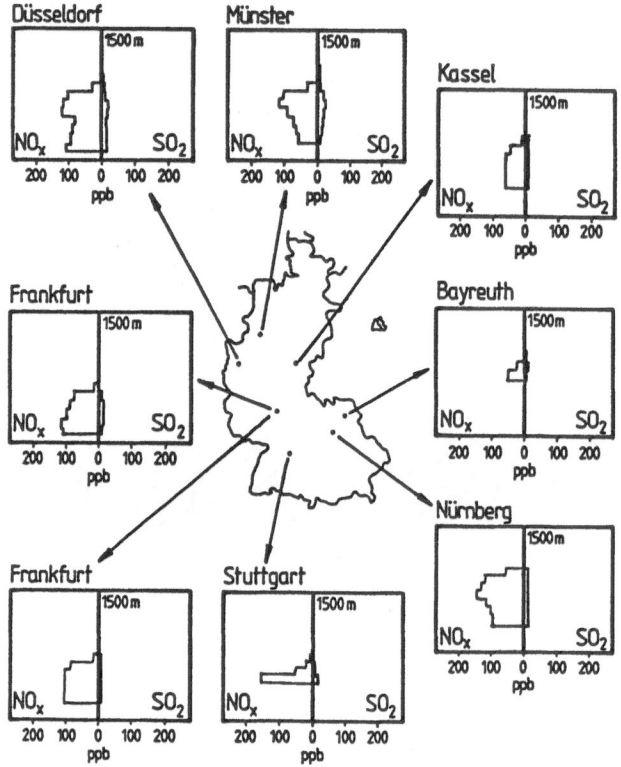

FIGURE 3 : Vertical concentration profiles of SO₂ and NOₓ within the mixing layer over different places in the FRG measured by aircraft on 1-3 February 1989 during a high pressure system centred over central Europe. The wind velocity was very low and a strong temperature inversion occurred between 400 and 800 m above the ground.
Profiles over Düsseldorf(1 February 1989),Münster 1 Feb.),Kassel(2 Feb.),Bayreuth(2 Feb.),Nürnberg(2 Feb. Stuttgart(2 Feb.),Frankfurt(2 and 3 Feb.) ref.9.

In spite of the remarkable reduction of SO₂ emissions in most countries of western Europe the problem of acid deposition will still remain an environmental issue of concern in large areas of Europe at least over the next decade.

As far as acid deposition research is concerned,the most important research needs have moved from SO₂ to the other gases involved in acid deposition(NOₓ,NH3,VOC).

The main reason is that our knowledge about the physico-chemical behaviour of SO_2 in the atmosphere is much better than our knowledge about the behaviour of NO_x, NH_3 and VOC. The important work carried out under the CEC environmental projects COST 61a* and COST 61a bis** has contributed a lot to fill some of the fundamental knowledge gaps in the field of SO_2.

Another environmental issue of concern in Europe is the problem of photochemical oxidants in the lower troposphere. There is growing evidence of an increase of background surface ozone over Europe over the last decades(ref.2). Superimposed on this long-term increase, episodes with elevated ozone concentrations have been observed over large areas for some decades. In central and northern Europe such episodes occur under the influence of large anticyclonic high pressure systems when there is a well defined boundary layer with a persistent subsidence inversion where emissions accumulate and react through photochemical processes.

Models for episodes with high ozone concentrations in the tropospheric boundary layer are reasonably well defined but only a few field measurements have been carried out to validate such models. Therefore from the point of view of Working group 3 the most important research need is to study the budget of photooxidants and related compounds during episodes of high ozone concentrations through well defined measuring campaigns using instrumented aircrafts and monitors at ground.

In contrast to acid deposition which is on a slightly declining trend in various European regions, the surface background ozone appears to be increasing or to remain at the high level observed today. Although predictions of further ozone trends are more difficult to make than trends for acid deposition, indications are strong that the problem of photooxidants becomes more important relative to the acid deposition problem in Europe over the next decades.

The shift of these environmental problems is to some extent reflected by the new research inventory of Working group 3. Most of the 63 projects assigned up to now to Working group 3 are concerned either directly or indirectly with the problem of photooxidants in the lower troposphere followed by acid deposition research projects.

Furthermore, most of the 12 priority projects on tropospheric research work in the R & D programme on environmental protection (STEP = Science and Technology for Environmental Protection, 1989-1992) are concerned with problems of photochemical oxidants and acid deposition.
The following priority research projects which are mainly concerned with field measurements and modelling activities come under Working group 3(not in the order of importance):

---

* COST 61 a : Physico-chemical behaviour of SO_2 in the atmosphere(1972-1976)

** COST 61 a bis: Physico-chemical behaviour of atmospheric pollutants

a.) European joint project on the budget of photooxidants and related species over the North Sea region

b.) Source-receptor relationship of NO$_x$ in Europe

c.) Natural emissions of VOC in Europe

d.) Regional cycles of air pollutants in the Mediterranean area

e.) Historical records of atmospheric trace substances.

The objective of the North Sea project(a.) is to quantify the budget of ozone and its precursors over the southern North Sea region as a budget area for episodes with high oxidant concentrations by means of field measurements(aircraft and ground based measurements) and models.

Regarding the NO$_x$ project(b.) the most important scientific question concerns the physico-chemical behaviour of NO$_x$ emitted by low level sources taking into account that in the west European OECD countries between 30-75 % of total NO$_x$ emissions are released by mobile sources.Of particular interest is to determine how much of the low level NO$_x$ emissions in Europe are available for atmospheric boundary layer chemistry,how much enters the free troposphere and how much is removed locally.

As far as the natural VOC emissions in Europe are concerned(c.) it is believed that they contribute significally to the total emissions on this continent but little hard information is available to substantiate this.Of particular interest is to measure natural VOC emissions by different but typical vegetation types in several parts of Europe during episodes of high ozone concentrations.

In the years to come,more emphasis will be given to air pollution problems in the southern Mediterranean countries taking into account the differences in meteorology as compared to North and Central Europe which requires to some extent new approaches to characterize pollutant cycles in the southern regions.
The main objective of the Mediterranean project(d.) is to quantify the diurnal cycles of air pollutants(O3,NOx,SO2,ect.) over the Iberian peninsula on an episodic basis during summer when the thermal low forms.Other budget areas for investigating air pollution cycles under similiar meteorological conditions are Italy and Greece.Under such meteorological conditions pollutants are injected into the mid-troposphere during the day where they become available for long-range transport during the night.In this way a substantial fraction of ozone and other pollutants originating from the Mediterranean region may end up over central Europe.

The overall aim of the project on historical records of atmospheric trace substances(e.) is to study the long-term changes in the composition of the atmosphere in the Northern Hemisphere and especially over Europe from pre-industrial times until today on the basis of ice core analyses and other information.

It is hoped that some of the most urgent scientific questions in the field of tropospheric chemistry can be answered through the activities of the new Working group 3.

References:

(1)   Smith,R.A.(1872)
        Air and rain-the beginnings of a chemical climatology.
        London,Longmans,Green and Co.,1872 (6oo pages).

(2)   Hov,Ø.;Becker,K.H.;Builtjes,P.;Cox,R.A. and Kley,D.(1986)
        Evaluation of the photooxidant-precursor relationship in
        Europe.CEC,Brussels(Air Pollution Research Report 1),
        project COST 611.

(3)   Hov,Ø.;Allegrini,J.;Beilke,S.;Cox,R.A.;Eliassen,A.;
        Elshout,A.J.;Gravenhorst,G.;Penkett,S.A.and Stern,R.(1987)
        Evaluation of atmospheric processes leading to acid
        deposition in Europe.CEC project COST 611,Air Pollution
        Research Report 10.

(4)   Schneider,T. and Bresser,A.H.M.(1988)
        Evaluation of phase I of the Dutch priority programme on
        acidification.Paper presented at the 3.meeting of acidi-
        fication research co-ordinators(Marc III),Bracebridge,
        Canada,September 1988.

(5)   Mylona,S.N.(1989)
        Detection of sulphur emission reductions in Europe during
        the period 1979 - 1986,EMEP MSC-W Report 1/89.

(6)   Häberle,M. and Herrmann,K.(1984)
        Entwicklung von Emissionen und Immissionen wichtiger
        Luftschadstoffe,WBL Zeitschrift für Umwelttechnik,
        August 1984.

(7)   Umweltbundesamt(1988)
        Luftreinhaltung 88.Tendenzen-Probleme-Lösungen.
        Materialien zum 4.Immissionsschutzbericht der Bundesre-
        gierung an den Deutschen Bundestag nach § 61 Bundes-
        immissionsschutzgesetz.Erich Schmidt Verlag Berlin,
        716 pages.

(8)   Umweltbundesamt(1989)
        a.) Rückgang der Schwefeldioxid-Belastung in der Bundes-
            republik Deutschland,Presse-Information 12/89
        b.) Fricke,W.(personal communication)

(9)   Umweltbundesamt(1989)
        Mehr NO$_x$ als SO2 bei einer windschwachen und austausch-
        armen Wetterlage.Monatsberichte aus dem Meßnetz 2/89.

A TWO YEARS STUDY OF OXIDANTS (PAN AND OZONE) IN A FORESTED AREA

P. PERROS, A. PROYOU and G. TOUPANCE
L.P.C.E. Université Paris-Val de Marne - F 94010 CRETEIL - FRANCE

SUMMARY
Measurements of PAN and ozone in the ambient air were made for two years at Donon in a forested area. Background levels and monthly mean values are discussed. Relation between wind directions and concentrations are also considered. The study of a photochemical episode shows long range transports and local effects.

INTRODUCTION
Forest decline in Europe has been ascribed to a number of factors among which atmospheric pollutants and particularly photooxidants play an important part. Ozone and PAN not only are phytotoxic but also lead to irritation of mucous membranes. Ozone is a precursor of reactive radicals as OH which play a dominant role in oxidation processes in the atmosphere. All these reasons lead to a considerable interest in tropospheric oxidant climatology.
We present here ozone and PAN measurements made at Donon during 1987 and 1988. Donon is a forested medium-altitude area (750 m asl) in the Vosges mountains touched by the forest decline, located at 48°30'23" N, 7°9'2" E. We have also used the ozone data of a neighbouring station (Aubure) of higher altitude (1100 m asl).

EXPERIMENTAL
Ozone was measured by a commercial UV analyser (Environnement SA Mod. 1003 AH). PAN measurements were made using a gas chromatograph equipped with an ECD. The analytical processes, calibration methods and instrument automatization have been described in detail elsewhere (1, 2, 3).

RESULTS AND DISCUSSION
1 - GENERAL DESCRIPTION OF THE OZONE AND PAN BEHAVIOUR.
The statistical behaviour of ozone and PAN is specified on the figure 1, as monthly variations of the mean concentration (horizontal dashes) and the percentiles 98, 90, 50, 10 and 2. The maxima of ozone mean concentrations are observed between April and September, while very low values (5ppb) could be observed during winter. For PAN, the highest mean concentrations are obtained in April and July and are associated with high values of percentiles 98 and 90. This observation possibly suggest that monthly means of PAN are influenced, in a large extent, by the occurrence of photooxidant pollution episodes.

2 - OXIDANT LEVEL IN THE FREE TROPOSPHERE AND LOCAL BACKGROUND LEVEL.
2.1 Background level. In order to evaluate the influence of photochemical activity we have been induce to determine a reference local concentration. This concentration, that we call here "background concentration", is defined as the ozone or PAN concentration observed on the site when the atmospheric transport processes allow a permanent supply to the site by air non directly influenced by the ground. As this situation is particularly encountered when the wind velocity is high, we have calculated, for each site, the monthly mean concentrations of ozone and PAN which are associated with different wind speed classes. A general tendency to an asymptotic value, different from month to month, is observed when the wind speed is increasing, let us say above 4 m/s. This suggests that the lower atmosphere, in the studied area, can be considered as

relatively well mixed for wind velocities higher than 4 m/s. This threshold is compatible with the criteria of a well mixed atmosphere given by Pasquill (4). In such conditions, the asymptotic value obtained for wind velocities higher than 4 m/s represents what we have defined as the "background concentration".

Following this methodology, we have determined the background levels of PAN and ozone at Donon as monthly means for the years 1987 and 1988 (figure 2). It could be noticed that the ozone background is generally in the order of 20 to 30 ppb. The maximum (> 35 ppb) observed during spring has been found as well in 1987 as in 1988. The PAN background level (Fig. 5) is maximum from spring to early summer (0,45 ppb) and minimum during fall and winter (0,13 ppb).

2.2 Free troposphere level. On a natural site, located far away from any anthropogenic source, surface ozone is mainly influenced by the tropospheric ozone reservoir (5). So, the free troposphere can be considered as an ozone reservoir regarding to the mixing layer and it is fitting to evaluate its concentration which could be obtained through direct measurement in the free atmosphere using altitude stations as Zugspitze (6) or vertical sounding (lidars, Brewer-Mast...). Having nothing comparable in the Vosges area we have tried to determine the concentration of ozone in the free atmosphere above the Vosges mountains by using ground level measurements.

If the local deposition and trapping processes are negligible, the background concentration values previously determined for wind speeds > 4m/s will approach the free troposphere concentration. For this purpose it is interesting to compare our result on the background concentrations with the results reported by REITER (6) for measurements directly obtained in the free atmosphere at an altitude of 2964 m asl at Zugspitze (FRG) (figure 2). At Zugspitze, the maximum of the ozone concentration is practically reached in April and persists with few variations until August. Our background concentration at Aubure in April is roughly similar to the Reiter determination but is clearly depressed in late spring and summer. This effect is more intense at Col du Donon. We had to come to the conclusion that, though corresponding to relatively well mixed atmosphere situations, the background concentration calculated at Donon is much lower (35 %) than the free atmosphere concentration determined as well at Zugspitze as at Aubure. We conclude that, concerning ozone level, the Col of Donon is poorly influenced by free atmosphere even for high wind velocity.

Regarding PAN there is, as far as we know, no monthly concentration profiles available in the free atmosphere to be compared with our own profiles.

3 - ANNUAL VARIATION OF MONTHLY MEAN.

The annual monthly mean profile for ozone at Col du Donon is presented on figure 3 by reference to the profiles observed for Hohenpeissenberg and Garmisch (6, 7). A bimodal behaviour is exhibited at Donon with two maxima, one during spring and another at the end of summer. The principal minimum appears in winter. These results are close to those reported for similar altitude (Garmisch) except for the late summer maximum which is not observed at Garmisch (Fig. 3). The ozone profile at 1000 m asl (Hohenpeissenberg) is given as comparison of profile obtained at a higher altitude.

The difference between background concentration and monthly mean is shown on figure 4 for the Col du Donon : during summer, an ozone overproduction, by respect to background concentrations, is obviously present while an ozone consumption is evidenced during winter. In addition, it must be pointed out that the summer ozone overproduction leads to a level exceeding the spring background maximum. This suggests that local photochemical production is important during summer at Donon.

Concerning PAN (Fig. 5) we notice, during spring and late summer, that the mean concentration is very significantly higher than the background concentration . This remark supports the idea that the spring maximum of PAN is more determined by processes occurring in the lower parts of the atmosphere than by transport from the upper troposphere.

The origin of the ozone maximum observed in natural areas is actually the subject of some controversies (7). The conventional view is that stratospheric intrusions are most effective during late winter and early spring (8). However, several authors suggest a tropospheric origin (1, 9, 10) with the argument that the spring PAN maximum is also observed in natural areas. As PAN is only formed by photochemical reactions in the troposphere, they speculate, as a consequence, that the ozone maximum is also mainly due to photochemical reactions in the troposphere. In addition, it has been recently proposed that PAN, ozone and their precursors may accumulate in air masses at high latitudes during winter (10). In this theory, these compounds could react from early spring, as soon as there is enough UV-light available. This hypothesis seems to be at variance with the fact presented above showing that the spring PAN maximum has its origin in the lower layers of the atmosphere.

4 - OXIDANT VARIATIONS WITH METEOROLOGICAL FACTORS.
Wind directions data locally recorded at Col du Donon have been used in order to find an eventual relation between this variable and the photooxidants. We have considered only the hourly ozone values recorded from 10.00 h to 17.00 h (T.U.), and we have checked the wind directions associated with each hourly value. In order to eliminate the non reliable directions we have used only the values associated with a wind velocity higher than 1 m/s.

The wind rose for Col du Donon is presented on fig 6 with the mean ozone concentration recorded for each wind sector. The wind rose is completely governed by the orientation NW-SE of the saddle. A detailed examination of the local topography indicates that this wind rose just discriminates roughly the opposition of the "west" and "east" origins of the air masses. It is interesting to point out that the variation of the ozone mean concentrations with wind directions is very small.

For the "crisis" situations, defined as situations where the ozone concentration exceeds the percentile 95, a very significant increase of the East sector is observed on the wind roses (Fig.7).

We have also studied the wind rose which corresponds to "crisis" situations for PAN (PAN > 1,4 ppb) (Fig 8). East sector is again particularly in evidence (70 % of crisis episodes). Contrary to ozone, this strong dissymmetry is also observed on the pollution rose for PAN (Fig. 9).

It is important to remind that easterly winds are generally associated, in the east of France, to anticyclonic situations which are favourable to clear and sunny sky, high temperature and low wind velocity. All these factors are favourable to photooxidant production. It is therefore unsuitable to associate prematurely sources located eastward of our site to the fact that high ozone levels are preferably recorded by easterly winds (11).

5 - STUDY OF A PHOTOCHEMICAL EPISODE.
During the period studied here, numerous episodes have shown high photooxidant levels. We have chosen to study, as an example, the episode extending from 16th to 19th of August.

The high values recorded (Fig. 10), as well for ozone (>80 ppb) as for PAN (1.5 ppb), have lead us to search the origin of the air masses reaching the site during this period. We have also used the data of a neighbouring site (Aubure) located at a higher altitude (1100 m asl).

Taking into account the complexity of the orographic system, we have used both the 850 hPa-backtrajectories ending at Col du Donon, provided by the French Météorologie Nationale, in order to identify possible long range transport, but also the local wind direction in order to take into account the local influences.

It is possible to split up the episode in four distinct parts.

5.1 At the beginning of the episode (15-16 th of August) the air mass reaching the site originates in SW sector, poor in pollutant sources, and ends at Donon by a valley situated in the NW direction. The ozone concentrations are close to background levels (40-50 ppb). It is fitting to mention that this level is reached even during the night at Aubure. The important atmospheric turbulence on this site (wind speed > 4 m/s) leads to a continuous restock of air from the free troposphere. In Col du Donon, the smaller turbulence (wind speed < 1 m/s) favours the ground deposition.

5.2 **On and after the 17th of August** the meteorological situation moves with a rotation of the air mass trajectories accompanied by a shift of local wind to SE. The pollutant levels recorded results from the superposition of two phenomena :

- Long range transport : the air mass reaching Donon during the morning has flown over the Paris area during the previous night. The relatively small emission of ozone precursors leads to a moderate increase of ozone and PAN. During the afternoon the air mass flies over the whole industrialised areas of the Seine valley and of NE of France.

- Local influences and particularly Strasbourg : the air mass, at the end of its trajectory, penetrates into the Alsace plain by its northern end and reaches Donon by the valley situated at its SE. The important ozone decrease observed at night is probably due to the ozone titration by NO emitted in the Strasbourg urban area during the evening, as well to ground deposition due to night stability at Donon. At sunrise, a sudden photochemical production of ozone and PAN is observed, up to 1.2 ppb PAN and 78 ppb O_3, which witness to an important injection of photooxidant precursors along the trajectory.

The nocturnal situation on the 18 th is governed on one hand by the flight of the air mass over south of England, Benelux and then Strasbourg at the end of the trajectory and ,on the other hand, by the low atmospheric turbulence at Aubure (wind < 1m/s) and the higher one at Donon (wind : 3 m/s). During all the day the area is influenced by possible long range transports from London area and Benelux and by short range influences (Strasbourg). It could be pointed out that this situation corresponds to the highest concentrations of PAN (1.5 ppb) which have been recorded during the episode.

5.3 **During the afternoon of 18 th** the local wind direction shifts to N and then to W in the evening. The immediate consequence is that Donon is no longer directly influenced by the Alsace plain. The fact that ozone reaches its maxima values (82 ppb) during this period, indicates that long range transport of ozone and precursors could have occurred. Simultaneously and apparently paradoxically, the PAN concentrations begin to decrease. This might evidence that PAN cannot be transported over long distances during summer. In fact, for the temperatures observed during this episode (25 °C), the PAN life time, considering only its thermal decomposition, is only 30 mn (12).

5.4 **The end of the episode the 19 th** of August still corresponds, during the morning, to high levels : though other factors could be put forward, we could point out that the air mass has flown over Paris area and the Seine valley within the previous 36 hours. During the afternoon and the following days the air mass comes from SW sector, poor in oxidant precursors emission sources : the PAN and ozone concentrations drop down to their background level .

CONCLUSIONS

We have shown that Donon is a site poorly influenced by the free troposphere and that local influences coming from the Alsace valley are frequently observed. The differences between monthly mean concentrations and background concentrations show an ozone overproduction during summer, an ozone consumption during winter and a regional production of PAN during spring. The wind direction study indicates that the highest levels of ozone and PAN are mainly observed by easterly winds. The study of a photooxidant pollution episode shows that PAN can be used as a good indicator for short range transport during summer. In a saddle, the pollutant concentrations are largely influenced by air arising from the valleys and by the detailed orientation of the network of valleys and crests. In such a location, 850 hPa-backtrajectories are insufficient to explain the oxidant concentration variation and local wind directions must be also considered in order to take into account short range transport.

ACKNOWLEDGEMENTS
We would like to thank the French Ministry of the Environment (SRETIE) for financial support, ASPA for local assistance and for ozone and meteorological data supply. One of us (A. P.) wishes to thank Air Quality Agency for grant.

BIBLIOGRAPHY

1 - PERROS P., TSALKANI N., TOUPANCE G., 1988, "PAN measurements in a forested area", *Environ. Sci. Technol.*, **9**, p 351-359
2 - TSALKANI N., PERROS P., TOUPANCE G., 1988, "Continuous measurements of PAN in a meriterranean site (Athens, Greece)", *Environ. Sci. Technol.*, **9**, p 143-152
3 - TSALKANI N., PERROS P., TOUPANCE G., 1987, "High PAN concentrations during non-summer periods : a study of 2 episodes in Creteil (PARIS), France", J. Atmos. Chem., **5**, 291-299.
4 - PASQUILL F., SMITH F., 1983, "Atmospheric Diffusion", J. Wiley & sons.
5 - SINGH H.B., VIEZEE W., JOHNSON W.B., LUDWIG F.L., 1980, "The Impact of Stratospheric Ozone on Tropospheric Air Quality", *J. Air Pollution Control Ass.* **30**, p. 1009-1017.
6 - REITER R.,SLADKOVIC R., KANTER H.J., 1987, "Concentration of Trace Gases in the Lower Troposphere Simultaneously Recorded at Neighboring Mountain Stations ; Part II : Ozone", *Meteorol. Atmos. Phys.*, **37**, 27-47.
7 - LOGAN J.A., 1985, "Tropospheric Ozone: Seasonal Behaviour, Trends and Anthropogenic Influence", *J. Geophys. Res.* **90**, p. 10463-10482.
8 - DANIELSEN E.F., 1968, "Stratosphere-troposphere exchange based on radioactivity, ozone and potential vorticity", J. Atmos. Sci. **25**, 502.
9 - PENKETT S.A., 1989, "Trends in atmospheric trace gases", *Atmospheric Ozone Research and its Policy Implications, p. 159, T. Schneider et al. Ed.; Elseviers Pub.*
10 - GUICHERIT R., 1989, "Concentrations and patterns of ozone in western Europe", *Atmospheric Ozone Research and its Policy Implications, p. 167, T. Schneider et al. Ed.; Elseviers Pub.*
11 - TOUPANCE G., PERROS P., SOEDOMO M., TSALKANI N., 1988, "Continuous field measurements of oxidants (PAN and Ozone) in the Paris basin." in *"Field Measurements and their interpretation"*, CEC Air Pollution Research Report 14, S. BEILKE, J. MORELLI and G. ANGELETTI Edit, , 173-184.
12 - SCHURATH U., KORTMANN U., GLAVAS S., 1984, "Properties, formation and detection of peroxyacetyl nitrate", *3 rd Europ. Symp. Physico-Chemical Behaviour of Atmospheric Pollutants*, B. Versino & G. Angeletti, Eds., D. Reidel, p. 27.

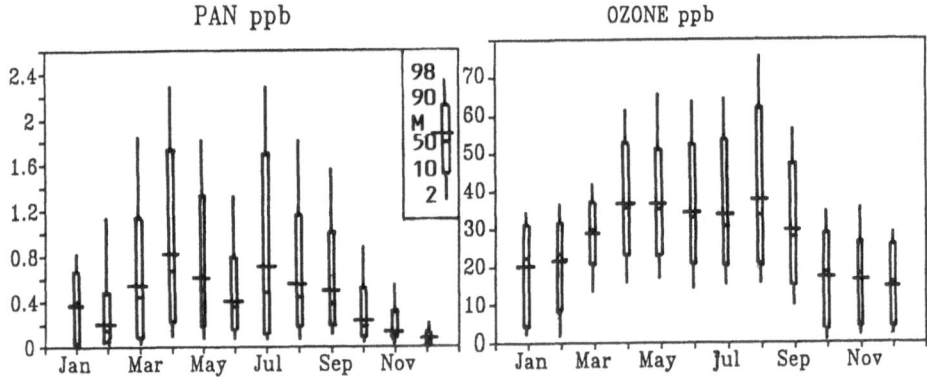

Figure 1 : Monthly variation of percentiles (98, 90, 50, 10, 2) and means (M) of Ozone (**a**) and PAN (**b**).

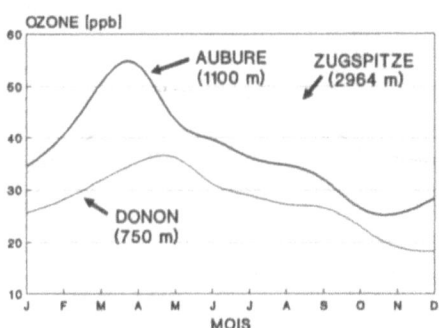

Figure 2 : Annual variation of background level.

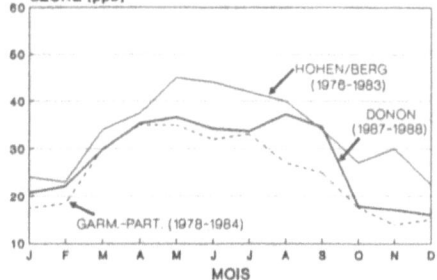

Figure 3 : Annual variation of monthly mean.

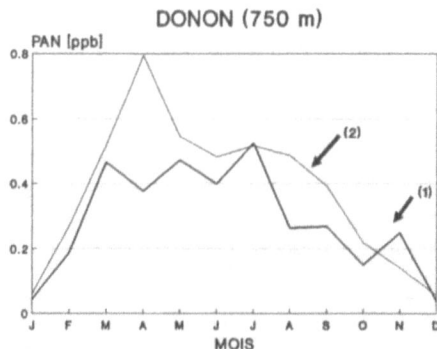

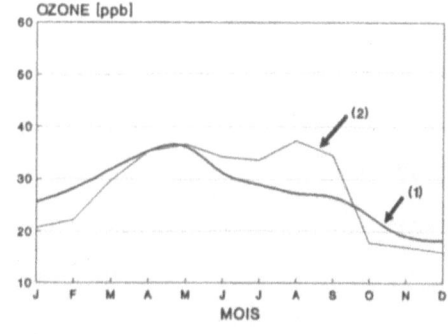

Annual variation of background level (**1**) and monthly mean (**2**)
of Ozone (Figure 4) and PAN (Figure 5)

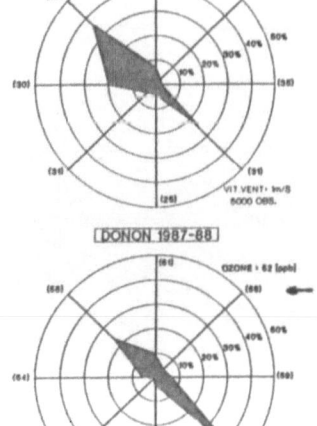

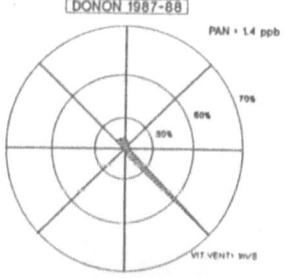

Figure 6 : Wind rose at Donon.

Figure 7 : Wind rose at Donon
of Ozone crisis episods.

() Ozone concentrations
associated with each wind sector.

Figure 8 : Wind rose at Donon
of PAN crisis episods.

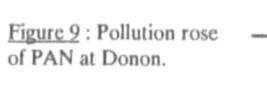

Figure 9 : Pollution rose
of PAN at Donon.

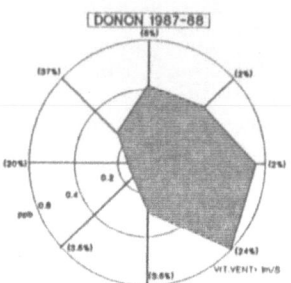

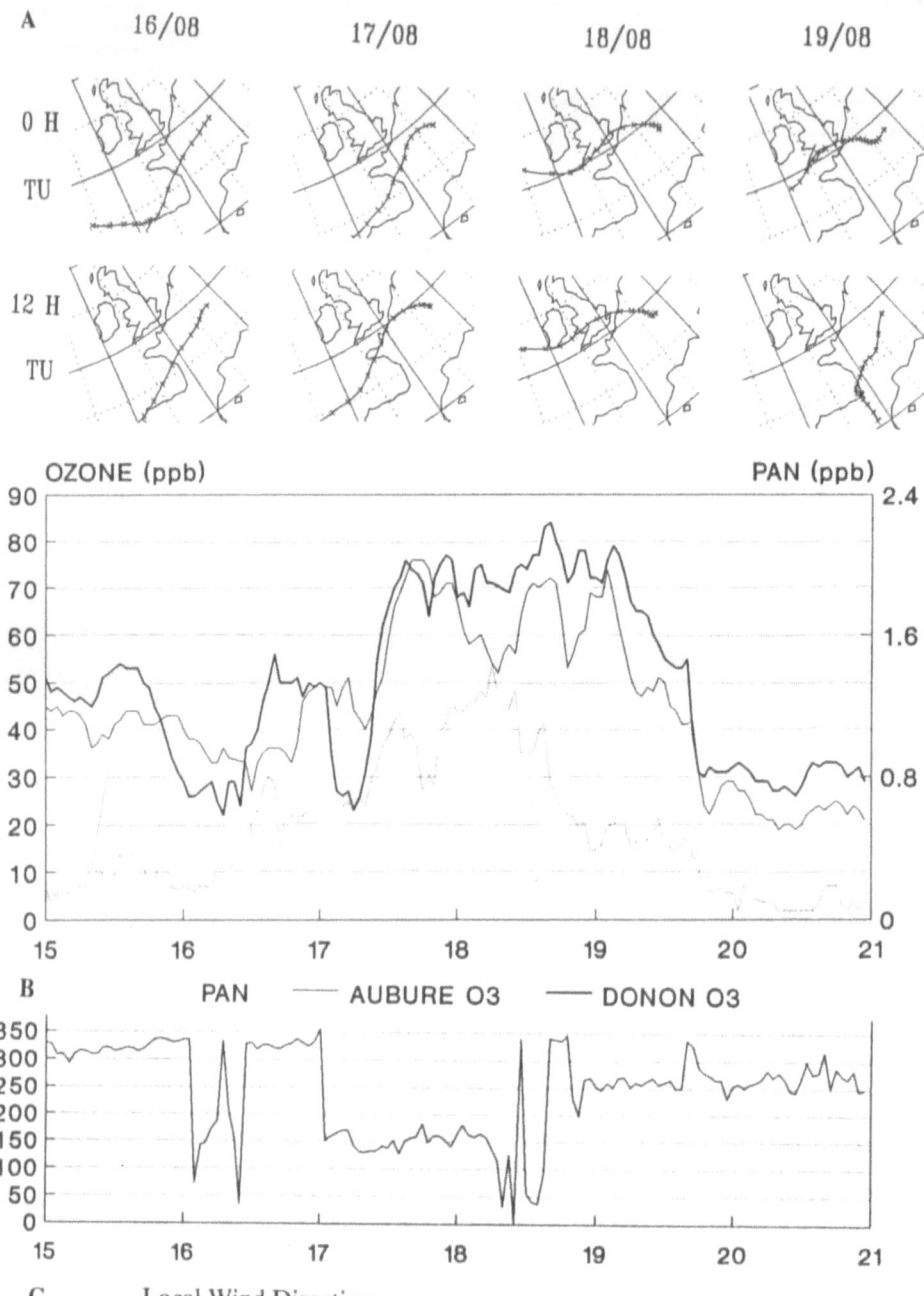

Figure 10 : **Photochemical episode** (15 - 21 August 1988):
    a - 850 hPa-bactrajectories.
    b - Oxidants profiles.
    c - Wind direction profile.

STUDY OF RAIN AND FOGS COLLECTED IN THE VOSGES REGION (1986-1988)

Ph. DEREXEL[*], P.MASNIERE[**]

*   Université Louis Pasteur, 2 rue Blaise Pascal 67008 STRASBOURG

** ELECTRICITE DE FRANCE, Direction des Etudes et Recherches, 6 Quai
    Watier 78400 CHATOU

SUMMARY

An automatic opening sequential rainfall detector has been installed at the
Donon site to study the physicochemical characteristics of rainfall in a
forest area of the Vosges mountains affected by "acid rain". Sequential
sampling line (valves, proximity detectors, sample, distribution arm and
physicochemical measurements are controlled by an automatic device
connected to a printer. The sampling capacity is around 200 fractions.
From July 1986 to October 87, 11 rainfall sequences (corresponding to 20
events) have been sampled per 0.1 mm interval. Rainfall has been classified
in relation to the origin of precipitations and the trajectory of the cloud
formation in question. It is thus possible to follow the change in
physicochemical characteristics (pH, conductivity, ionic contents) of
precipitations as a function of atmospheric circulations. Depending on the
latter, the change in characteristics can be more or less acute during
rainfall ; in any event, mean pH values are the lowest for Eastern and North-
Eastern circulations as well as for orographic type precipitations. An
assessement of wash-out have been undertaken, on the basis of change in the
ionic content of the sampled fractions.
Measurements of the chemical composition of fog water have been made,
concerning a series of fog events sampled with a system based on the
principle of inertial impaction.

1. INTRODUCTION

De nombreuses études ont été réalisées ces dernières années dans le cadre
du programme français DEFORPA concernant les dommages forestiers attribués à la
pollution atmosphérique. Le massif des VOSGES est considéré comme l'une des
montagnes françaises les plus affectées (1). C'est ainsi qu'une étude fine des
pluies et des brouillards a été effectuée près de la station de mesure du col du
Donon (727 m) à l'aide de collecteurs mis au point par EDF (DER) et le
Laboratoire de Thermodynamique de l'Université de STRASBOURG ; des mesures de
brouillards ont également été réalisées sur d'autres sites de la région. Les
résultats partiels qui sont présentés concernant la période 1986-1988.

Les pluies sont échantillonnées à l'aide d'un pluviomètre séquentiel (pas
de 0,1 mm de hauteur de pluie) à ouverture automatique, piloté par un automate
programmable ; sa conception tient compte d'une maintenance épisodique. Les
gouttelettes de brouillard sont collectées à l'aide d'appareils fonctionnant
sur un mode d'impaction passive.

Une fraction seulement des précipitations affectant le site du Donon, ont été recueillies durant les périodes de fonctionnement (1/8/86 - 22/11/86 et 1/5/87 - 15/11/87) présentées ici. En effet, 11 séquences pluvieuses (20 événements) ont été échantillonnés, alors que 28 séquences (comprenant éventuellement des précipitations orographiques) ont été relevées par un poste pluviométrique de la Météorologie Nationale proche du Donon. En outre, 15 séquences de précipitations neigeuses n'ont pas été prises en compte.

Les mesures effectuées sur les fractions séquentielles de pluie ont permis à la fois de suivre l'évolution des caractéristiques physicochimiques en fonction des flux synoptiques et d'évaluer le phénomène de lessivage des polluants par la pluie ("wash-out") et, indirectement, celui de l'incorporation des polluants dans les gouttelettes de nuages ("rain-out").

La caractérisation et la classification d'une série d'épisodes de brouillards ont permis de lier mesures chimiques en phase aqueuse et paramètres météorologiques.

2. TECHNIQUES D'ECHANTILLONNAGE DES PLUIES ET DES BROUILLARDS

2.1. Pluviomètre séquentiel à ouverture automatique

Cet appareil, décrit par ailleurs (1) collecte les pluies par pas de 0,1 L/m2 de précipitation (correspondant à un volume de 40 mL pour chaque fraction).

Un automate programmable assure la détection des pluies, l'ouverture et l'obstruction de la surface de captation ; il gère la commande des divers actionneurs (déplacement du couvercle, électrovannes, déplacement du bras distributeur d'échantillon au-dessus des tubes destinés à les conserver dans une unité de congélation, pompe à eau de rinçage) et la mesure de la conductivité et du pH sur les prélèvements successifs. Compte tenu des séquences de rinçage, plus de 200 prélèvements séquentiels de pluies peuvent être recueillis et conservés plusieurs semaines dans l'unité de congélation afin de limiter les possibilités des transformations chimiques et de développements d'algues. Ces prélèvements séquentiels sont analysés ultérieurement en laboratoire (chromatographie ionique, absorption atomique, mesures de pH et conductivité). L'approvisionnement prévu en air comprimé et le nombre de tubes d'échantillonnage, permettent d'envisager une autonomie de fonctionnement supérieure (en général) à 3 semaines. Les batteries de secours permettent d'autre part un fonctionnement de l'appareil (y compris la fin du cycle de prélèvement), durant 15 minutes en cas de coupure d'alimentation de réseau électrique.

2.2. Collecteur de brouillard

Cet appareil est constitué d'une nappe conique de fils de nylon espacés de quelques dixièmes de millimètres. Les gouttelettes percolent à la surface des 1200 mètres de fils constituant la nappe et sont recueillies par gravité dans un cône de réception. L'impaction réalisée ainsi est passive, mais il est possible d'équiper l'appareil d'une aspiration forcée afin d'augmenter les quantités recueillies durant un certain laps de temps. L'ensemble du système n'est découvert qu'au cours des séquences de brouillard étudiées. Les prélèvements sont effectués généralement durant la période diurne, lorsque la visibilité est inférieur à 1 kilomètre et l'humidité proche de 100 % ; l'arrêt est conditionné par la dispersion du brouillard.

3. EVOLUTION DES CARACTERISTIQUES PHYSICOCHIMIQUES DES PLUIES
   EN FONCTION DES CIRCULATIONS ATMOSPHERIQUES
   (Juillet 1986 - Octobre 1987)

Les valeurs extrêmes concernant les teneurs en divers ions, le pH, la conductivité des échantillons séquentiels relevés lors des 20 événements pluvieux pris en compte sont rassemblées dans le tableau 1 ; les valeurs moyennes du pH (calculées à partir des moyennes de concentrations en ions H^+) et de la conductivité sont également indiquées pour chaque événement.

Chaque averse est associée à un flux synoptique qui est calculé pour le niveau 850 le hPa.

Bien que les flux de Nord-Est et Est ne représentent qu'un cinquième du volume global des précipitations, ils sont en moyenne responsables de 50 % du dépôt des sulfates.

Par ailleurs, un seul événement de précipitation à caractère orographique (provoquée par une ascendance forcée liée au relief) a pu être échantillonné ; les hauteurs précipitées par ce type d'averse sont généralement faibles (inférieures à 1,0 mm) les teneurs en sulfates et nitrates sont élevées et le pH moyen est faible (3,50).

4. EVOLUTION DU LESSIVAGE DE L'ATMOSPHERE SOUS L'EFFET DES
   PRECIPITATIONS

Le lessivage de l'atmosphère peut être mis en évidence pour des précipitations qui surviennent après une période sèche plus ou moins prolongée (2, 3, 4, 5, 6). Les premières fractions collectées sont caractérisées par des charges en éléments dissous relativement importantes. En moyenne, nous avons trouvé qu'environ 80 % des éléments présents dans l'atmosphère sont entraînés au cours des 30 premières minutes de l'averse, passé ce délai le pluviomètre recueille de la pluie qui n'est plus enrichie par lessivage. Ce phénomène est illustré sur les figures 1 et 2.

Afin de calculer la part due au lessivage de l'atmosphère, on procède par l'intégration de la surface comprise sous la courbe des ions considérés ; dans la figure 2, le trait pointillé correspond à ce calcul pour les sulfates.

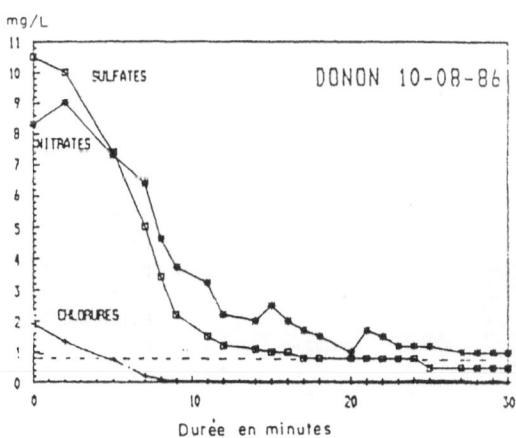

On a représenté dans la figure 3 les quantités des divers ions majeurs déposées au sol, dues uniquement au lessivage de l'atmosphère, en fonction des périodes sèches précédant l'événement pluvieux (de 0 à 10 jours). Le dépôt total, qui correspond à la somme des apports au sol pour les différents ions est relativement faible pour des périodes inférieures à 2 jours en ce qui concerne le site du Donon.

Figure 1 : Evolution de la composition des
précipitations en fonction du temps.
Evénement pluvieux du 10 août 1986
après 5 jours de période sèche.

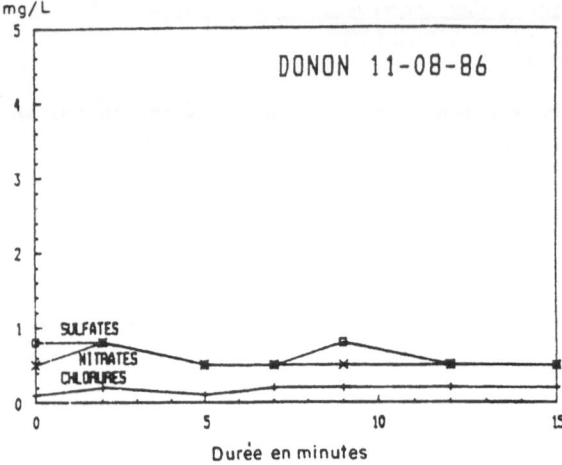

mg/L

DONON 11-08-86

SULFATES
NITRATES
CHLORURES

Durée en minutes

Figure 2 : Evolution de la composition des
précipitations en fonction du temps.
Evénement pluvieux du 11 août 1986 après
19 heures de période sèche.

Selon la durée de la période sans précipitation, le comportement des ions $SO_4^=$, NO_3^-, NH_4^+ s'écarte notablement de celui des ions Cl^-, Na^+, K^+. Le premier groupe d'éléments est caractérisé par des accroissements importants du dépôt en fonction de la durée de la période sèche ; les accroissements moyens sont les suivants (μMole/m2/ jour) : NO_3^- : 9,1 ; $SO_4^=$: 10,7 ; NH_4^+ : 9,7. Le deuxième groupe d'éléments présente des accroissements plus faibles, respectivement : 1.3, 0,9 et 4,7 μMole/m2/jour pour K^+, Na^+, et Cl^-. Le dernier élément, Cl- représente toutefois un cas intermédiaire.

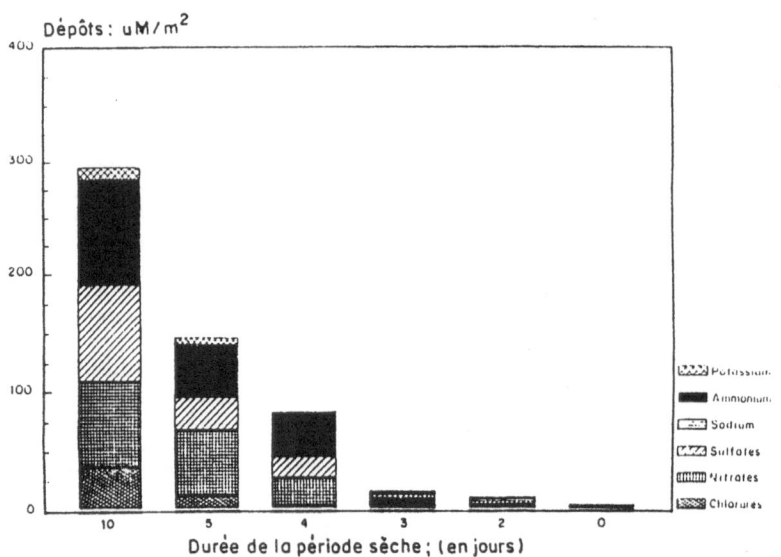

Dépôts : uM/m^2

Durée de la période sèche ; (en jours)

Potassium
Ammonium
Sodium
Sulfates
Nitrates
Chlorures

Figure 3 : Evolution des dépôts issus du lessivage pour les
différents ions en fonction de la durée de la période
sèche (en micromoles/m2).

Il n'est pas surprenant que les quantités de NH_4^+, NO_3^- et SO_4^{--} augmentent en fonction de la durée des périodes sèches. En effet, ces périodes sont liées à la présence d'un anticyclone qui, en recouvrant la France, empêche la circulation des courants perturbés d'Ouest.

5. EVOLUTION DES CARACTERISTIQUES PHYSICOCHIMIQUES DES GOUTTELETTES DE NUAGE EN FONCTION DES FLUX SYNOPTIQUES

Des mesures aéroportées de la phase aqueuse des nuages constitueraient l'approche la plus adéquate du "rain-out". Faute de tels moyens, on peut cependant, à partir des teneurs ioniques calculées au sol pour les échantillons séquentiels de pluie en tenant compte du "wash-out", apprécier l'importance du "rain-out" au niveau du Donon et lier son évolution aux trajectographies des masses d'air.

La composition chimique des gouttelettes de nuage s'obtient en retranchant à la composition des précipitations la part due au lessivage. Elle présente une variabilité assez importante, en fonction du parcours emprunté par les systèmes nuageux.

Les résultats obtenus sont regroupés dans la figure 4

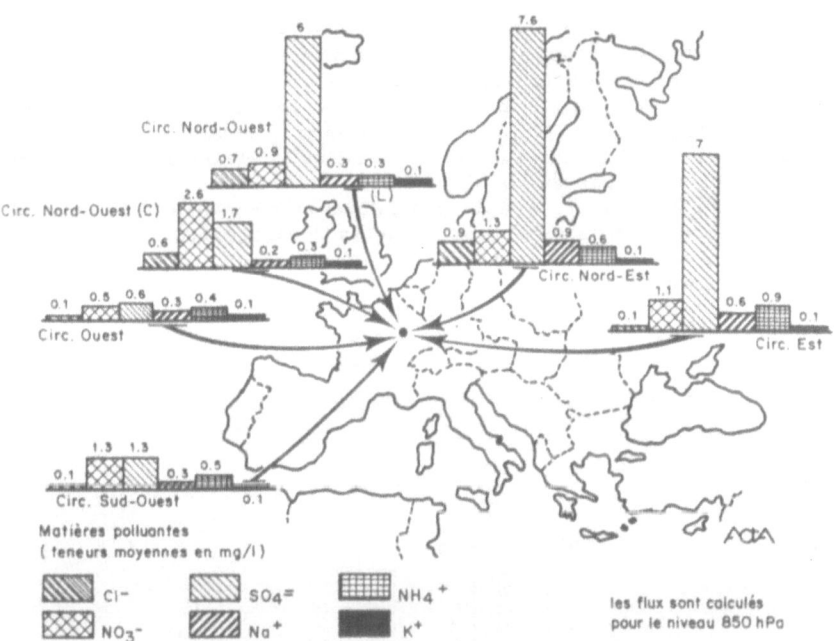

Figure 4 : Evolution des caractéristiques physico-chimiques des gouttelettes de nuage en fonction des flux synoptiques.

Les sulfates représentent la presque totalité des éléments dissous pour les flux d'E, NE et NW à longue trajectoire : 71,4 % à 72,3 % des éléments dosés sont des sulfates, la dispersion des valeurs est faible. Ces trois trajectographies mettent en évidence l'origine anthropogénique de ces dépôts. L'opposition est marquée avec les trajectoires océaniques qui sont moins chargées en sulfates ; la part de ceux-ci est en effet comprise entre 30,0 % et 34,2 %.

La contribution des nitrates est diversifiée suivant les flux. Un premier groupe présente des valeurs proches de 10 %, ce qui correspond à une constante pour les zones continentales. Un deuxième groupe comprend les circulations de NW à courte trajectoire et celles de SW dont les parts respectives s'élèvent à 47,3 % et 34,2 %. On peut expliquer ces valeurs plus élevées en raison de l'origine terrigène de certains composés (ces 2 trajectoires couvrent le territoire français).

Dans le cas des circulations d'Ouest, la part du sodium est assez nettement marquée. Celle-ci représente 15 % des éléments dosés présents en solution, et correspond majoritairement à des apports d'origine marine.

6. CARACTERISTIQUES PHYSICOCHIMIQUES DE QUELQUES BROUILLARDS

Les brouillards sont mis en cause dans certaines régions présentant un dépérissement forestier notable, en raison de leur caractère particulièrement agressif lié à une acidité qui peut être élevée (7). Les résultats présentés concernent les analyses réalisées sur la phase aqueuse collectée sur un site du massif vosgien, les Trois-Fours (1230 m).

Dans les Vosges, le brouillard dans les secteurs d'altitude est peu fréquent (10 à 20 jours par an aux Trois-Fours), ce qui fait son originalité par rapport au brouillard de rayonnement qui dans la Plaine d'Alsace se manifeste en moyenne 70 jours par an.

Entre 1986 et 1988, 34 épisodes de brouillards ont été échantillonnés, qui peuvent être classés essentiellement en 3 catégories :
- n° 1 : abaissement de stratus (20,6 % des cas),
- n° 2 : advection (2,9 %) ;
- n° 3 : base de nuage au niveau de la station de mesure (76,5 %).

Chacun des trois types de brouillard se distingue par des pH et des teneurs moyennes en nitrates, sulfates et ammonium différentes (fig. 5).

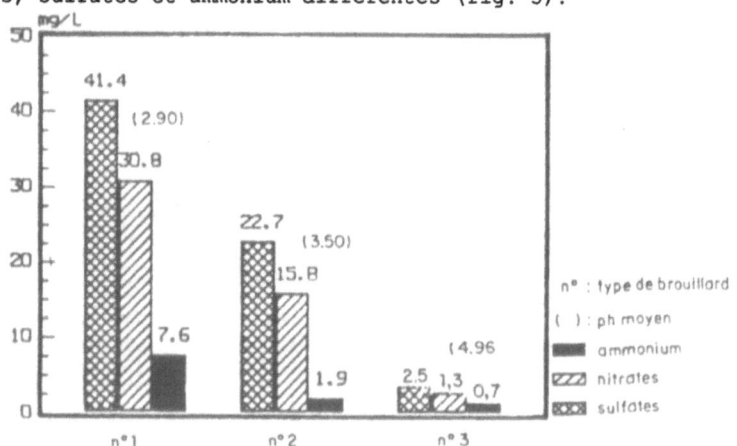

Figure 5 : Teneurs moyennes en divers ions et pH des 3 types de brouillards échantillonnés.

23 % des épisodes de brouillards collectés au cours de cette étude présentent un pH inférieur à 3,5 et des concentrations élevées en nitrates et sulfates ; ils correspondent aux cas n° 1 et 2.

A partir des teneurs relevées en ions H^+, NO_3^+ et $SO_4^=$ dans la phase aqueuse, on peut estimer à 93 % la part de nitrates et sulfates sous forme d'acide nitrique et sulfurique pour le cas n° 1 ; elle n'est que 44 % pour le cas n° 2. Les différences de teneur en eau (0,35 g/m3 air en moyenne pour le cas n° 1 et 0,8 g/m3 air pour le cas n° 2) expliquent en partie cet écart.

Ces 2 cas sont liés à des masses d'air en provenance du secteur Est.

La majorité des épisodes de brouillard correspond à la base de systèmes nuageux associés à des perturbations en provenance de l'Océan Atlantique ; les teneurs en eau sont plus élevées (1,1 g/m3 air). Dans ce cas, seuls 10 % des nitrates et sulfates sont sous forme d'acide nitrique et sulfurique.

7. CONCLUSION

L'utilisation d'un collecteur séquentiel (pas de 0,1 mm de hauteur de pluie), permet une étude très fine des précipitations et permet en particulier de différencier qui est dû au lessivage local de l'atmosphère et ce qui est dû à la composition des nuages.

Pour l'ensemble des fractions, le pH est compris entre 3,4 et 5,4. Le pH, de même que les diverses teneurs en ions mesurés, sont fonction des trajectoires des masses d'air et de la charge en aérosols dans l'atmosphère au niveau du site susceptible d'être partiellement lessivée lors de l'averse.

La masse totale déposée pour les différents anions et cations étudiés augmente en fonction de la durée de la période sèche précédant la précipitation (au-delà de 2-3 jours). Cet accroissement se différencie suivant l'origine naturelle ou anthropogénique des ions : il est plus élevé pour les sulfates, nitrates et l'ammonium (origine surtout anthropogénique) que pour les chlorures (origine mixte), le sodium et le potassium (origine naturelle).

A partir des études des trajectographies, il apparaît que les acidités les plus marquées sont généralement liées à des circulations d'Est et Nord-Est (pH généralement inférieur à 3,8). La contribution des sulfates à la charge ionique est alors nettement prépondérante.

Les analyses de la phase aqueuse des 34 brouillards échantillonnés impliquent une nette distinction entre brouillards d'advection et d'abaissement de stratus d'une part et de base de nuage au niveau de la station d'autre part. Les premiers cas, liés à des masses d'air issues de flux de section Est, présentent des pH inférieurs à 3,5 dûs à de fortes teneurs en acides sulfuriques et nitriques.

REFERENCES

[1] J. DUBOIS, P. MASNIERE
     Description et fonctionnement d'une station automatique de prélèvements séquentiels et d'étude de pluies. Pollution Atmosphérique, avril-juin 1986.
[2] H.W. GEORGII, E. WEBER
     Investigation on tropospheric wash-out. Technical final - report ; contract Air Force 61 (052) - 249 : p. 55 1964.

[3] G.L. PELETT, R. BUSTIN, R.C. HARRIS
Sequential sampling and variability of acid precipitation in Hampton, Virginia - Wat. air soil pollut. N° 21 ; 1984.

[4] S. KUMAR,
A Eulerian model for scavenging of polluants by rain drops - Atmospheric Environment. 19 ; 1985.

[5] M.D. SEYMOUR, T. STOUT,
Observation on the chemical composition of rain using short sampling times during a single event - Atmospheric Environment. 17 ; 1983.

[6] W.H. CHAN, D.H.S. CHUNG
Regional scale precipitation scavenging of SO_2, SO_4, NO_3 and HNO_3 - Atmospheric Environment, 20, 1986.

[7] B. HILERAN
Acid fig - Environ. Sci. Technol. vol 17, n° 3. 1983.

Type de circulation atmosphérique	Date	pH	cond.	Cl⁻	NO₃⁻	SO₄=	Na⁺	NH₄⁺	K⁺
Nord-Ouest	24.08.86	(3,90) 3.49-4.45	(16,2) 8.0-18.0	0.1-8.2	0.6-3.7	1.0-5.3	0.1-1.6	0.2-1.1	0.1-0.1
	29.08.86	(3,80) 3.50-4.40	(13,6) 8.0-18.0	0.2-0.3	1.7-2.3	1.2-2.0	0.1-0.3	0.7-0.9	0.1-0.1
	03.09.86	(4,20) 3.60-4.49	(12,4) 8.0-30.0	0.1-0.6	1.0-5.6	0.6-6.7	0.1-0.5	0.3-2.2	0.1-0.2
	21.06.87	(4,10) 3.59-4.45	(17,5) 7.2-37.5	0.2-1.2	3.8-10.7	2.9-12.3	0.1-0.4	2.7-5.5	0.1-0.1
	23.06.87	(4,00) 3.45-4.10	(16,4) 10.0-35.0	0.1-0.8	0.2-4.8	0.3-3.6	0.1-0.4	0.1-2.9	0.1-0.2
	24.06.87	(3,99) 3.40-4.50	(18,2) 6.0-45.0	0.5-2.2	3.0-4.8	0.7-2.6	0.1-0.4	0.1-0.5	0.1-01
	01.08.87	(4,03) 3.87-4.15	(15,5) 7.5-21.0	0.2-2.1	0.2-2.1	0.9-7.1	0.1-0.5	0.4-1.5	0.1-0.2
	02.08.87	(3,98) 3.65-4.32	(18,3) 7.5-20.5	0.1-1.2	0.9-3.7	1.1-4.5	0.1-0.8	0.2-0.8	0.1-0.1
Ouest	24.08.86	(4,10) 3.56-4.68	(3.1) 2.1-15.0	0.1-1.5	0.1-1.4	0.1-1.4	0.1-1.4	0.1-1.8	0.1-0.2
	07.10.87	(4,31) 3.96-4.62	(2,8) 2.0-10.5	0.1-0.8	0.1-1.1	0.1-0.9	0.1-1.0	0.1-0.7	0.1-0.2
	20.11.87	(4,34) 4.06-4.57	(17,2) 10.0-40.0	0.1-7.3	0.1-0.8	0.8-1.7	0.5-1.4	0.1-1.2	0.1-1.2
Sud-Ouest	10.08.86	(4,15) 3.70-5.30	(14,5) 5.0-45.0	0.1-1.9	0.1-10.5	1.0-9.0	0.1-0.6	0.3-2.0	0.1-1.0
	11.08.86	(4,95) 4.40-5.45	(8.6) 5.6-18.0	0.1-2.0	0.5-0.8	0.5-0.8	0.5-2.3	0.1-0.2	0.1-0.2
	28.08.86	(4,02) 3.71-4.51	(13.1) 12.0-14.5	0.5-2.0	1.4-4.6	1.4-4.8	0.5-0.6	0.7-0.8	0.1-0.1
	19.10.86	(3,83) 3.43-4.65	(11,2) 3.0-120.0	0.1-3.6	0.1-11.6	0.3-19.6	0.1-0.5	0.2-1.9	0.1-1.0
	22.11.86	(3,88) 3.79-4.00	(65.1) 40.5-85.0	0.1-0.1	0.8-1.8	1.0-1.9	0.3-1.1	0.6-3.3	0.1-0.3
Sud	03.05.87	(3.92) 3.89-3.98	(65.0) 62.0-68.0	0.7-1.6	0.8-1.2	6.0-7.0	0.3-0.8	0.3-0.9	0.1-0.1
Est	05.05.87	(3,85) 3.70-3.91	(64.0) 50.0-74.0	0.1-0.3	0.5-3.0	6.0-14.1	0.3-1.2	0.5-2.0	0.1-0.3
Nord - Est	08.05.87	(3.78) 3.75-3.80	(75.2) 70.0-85.0	0.7-3.3	1.2-1.6	5.5-13.6	5.5-13.6	0.5-2.1	0.1-0.1
Précipitations à caractère orographique	02.09.86	(3,50) 3,48-3,5	(48,9) 37,5-55,0	06-0.9	7,6-9,9	7,0-9,6	0,4-0.9	1,8-2,3	0,1-0,2

conductivité exprimée en µS/cm   teneurs exprimées en mg/l

( ) : valeur moyenne

Tableau I - Caractéristiques physico-chimiques des précipitations échantillonnées séquentiellement en 1986-1987 au col du DONON

COXY's IN THE ATMOSPHERE

Günter Helas, Gerhard Schuster and Stephen Wilson*
Max-Planck-Institut für Chemie, Otto-Hahn-Institut
D - 6500 Mainz, Federal Republic of Germany

* present address: Cape Grim Baseline Air Pollution Station, Smithton, Tasmania, Australia

Abstract

A brief overview on the origin of the halogenated carbonyls $COCl_2$, $COClF$ and COF_2 in the atmosphere is given. A description of the measurement technique, matrix-isolation-spectroscopy, which was used for the determination of these compounds, and results obtained from several flights over northern Europe will be discussed. While $COCl_2$ appears to be present in both troposphere and stratosphere, the two other compounds were detected only in the stratosphere.

Key words : Carbonyl halides, $COCl_2$, $COClF$, COF_2, measurement, atmosphere

1. Introduction

The chlorofluorocarbons (CFC) are known to be very stable in the atmosphere [WMO,1985]. Their major decomposition pathway is photolysis above the ozone layer. The formed radicals are subsequently oxidized by atmospheric oxygen producing halogenated carbonyl compounds, COXYs. Another production pathway for COXYs is provided by not fully halogenated CFCs. These are susceptable to OH attack and release an hydrogen atom during this reaction. Carbonyl fluoride (COF_2) is produced primarily from F-12 (CCl_2F_2) [Jayanty et al.,1975] and F-22 ($CHClF_2$), whereas carbonyl chlorofluoride ($COClF$) is formed by oxidation of F-11 (CCl_3F) [Jayanty et al., 1975]. Our experiments suggest that F-21 ($CHCl_2F$) can act as precursor also (see experimental section). Phosgene ($COCl_2$) can be formed from various molecules, eg. $CHCl_3$, CH_3CCl_3, C_2Cl_4, C_2HCl_3, and CCl_4 [Singh,1976; Crutzen et al.,1978]. Ultimately, the carbonyl halides should decay either by photolysis or hydrolysis.

Phosgene was first measured near the surface by Singh [1976] and Singh et al. [1977]. We have recently reported on measurements of phosgene in the upper troposphere and lower stratosphere [Wilson et al.,1988]. In the same samples we have also identified $COClF$ [Wilson et al.,1989]. COF_2 was detected in spectra from the ATMOS experiment [Rinsland et al.,1986; Raper et al., 1987]. This stimulated us to check our matrix samples for the presence of COF_2. In this contribution we summarize the determinations of carbonyl halides which we retrieved from near tropopause samples.

2. Experimental

The measurement technique used for determining the carbonyl halides was Matrix-Isolation-Spectroscopy. This technique is described in detail elsewhere [Griffith and Schuster,1987]. The method has been used for determining a variety of atmospheric trace components [Helas et al.,1989; Schuster et al.,*this volume*; Wilson et al.,1988; Wilson et al.,1989]. Thus only a brief outline is given here. The air to be probed is cryogenically sampled at the temperature of liquid argon (87 K). The collected material consists mainly of CO_2 and H_2O, with the trace components embedded within the solid. The sample is transferred into a Fourier-transform infrared spectrometer, where the glassy matrix is measured. Commonly used analysis is performed on the spectra, where peak height and/or area of the species of interest are compared with the amount of CO_2 present.

The technique utilizes two characteristic advantages. First, the trace components to be measured are preconcentrated by a factor of roughly 3000, because they are sampled in atmospheric CO_2. Second, the IR-active rotational-vibrational bands of the polyatomic molecules degenerate to pure vibrational bands due to the embedding in the solid. This feature provides very high and extremely small absorption bands making them exceptionally characteristic. The measurement procedure is calibrated against laboratory standards using the same method as for outdoor samples.

Laboratory standards of COF_2 and COClF had to be prepared for identification and quantification in the matrix samples. Though $COCl_2$ is commercially available, this compound was also made in order to confirm the preparation methods of all the three compounds. Chloroform ($CHCl_3$), F-21 ($CHCl_2F$), and F-22 ($CHClF_2$) were individually mixed with CO_2 and Cl_2 in synthetic air. Irridiation with black light produced chlorine radicals which extracted the hydrogen atom from the H-containing halomethanes. The intermediates were subsequently oxidized to the respective carbonyl halides $COCl_2$, COClF, and COF_2. The build-up of these compounds was monitored and they were identified by their known gas phase absorption frequencies [Nielson et al.,1952]. The build-up of the halocarbonyls and the decay of the parent compounds were inversely proportional to each other. The irridiated mixtures were aliquoted, individually diluted and frozen out. The absorption bands used for identification in the matrices are presented in Table I. The detection limits are approximately 4 pptv for $COCl_2$ and COClF, and 2 pptv for COF_2. The overall error in sampling and analysis is estimated to 20%.

Air samples were collected aboard an aircraft (Learjet 25D), which was modified with an extra inlet for contaminant free sampling. The samples were taken from above the cloud layer (w 4 km) up to 14 km between Germany and northern Scandinavia. A Dasibi-1008 monitor was used to determine ozone mixing ratios simultaneously to the grab sampling.

3. Results and discussion

The determined mixing ratios of the carbonyl halides are combined in Figure 1. They range from 10 to 29 pptv for $COCl_2$, 4 to 21 pptv for COClF, and 2 to 14 pptv for COF_2. The lower values coincide with our lower limits of detection of the respective compounds. The mixing ratios are plotted relative to the tropopause altitude. The position of the tropopause was determined by interpolating radiosonde information to the sampling location and this was supported by simultaneously measured ozone mixing ratios. In general, only phosgene was measured in the troposphere, whereas the other two compounds were only observed in the stratosphere.

The results of a flight LJ8 (21-Feb-87) are shown in Figure 2. During this flight the greatest elevation above the tropopause was reached. The mixing ratios are plotted versus the simultaneously measured ozone. This method establishes a relative altitude scale, which takes individual struc-

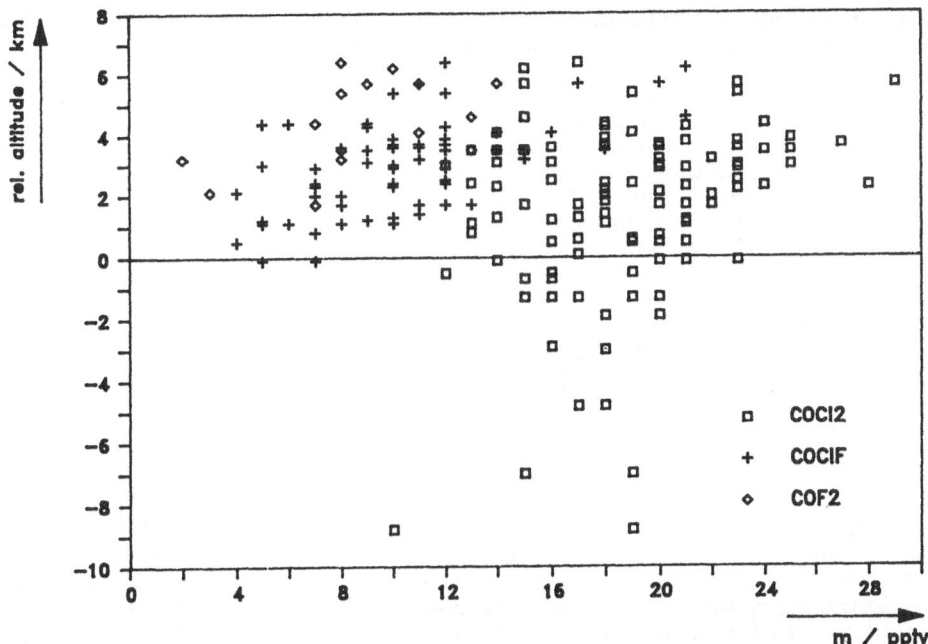

Fig. 1: Mixing ratios of COCl₂, COClF and COF₂ versus altitude relative to the tropopause as measured above northern Europe.

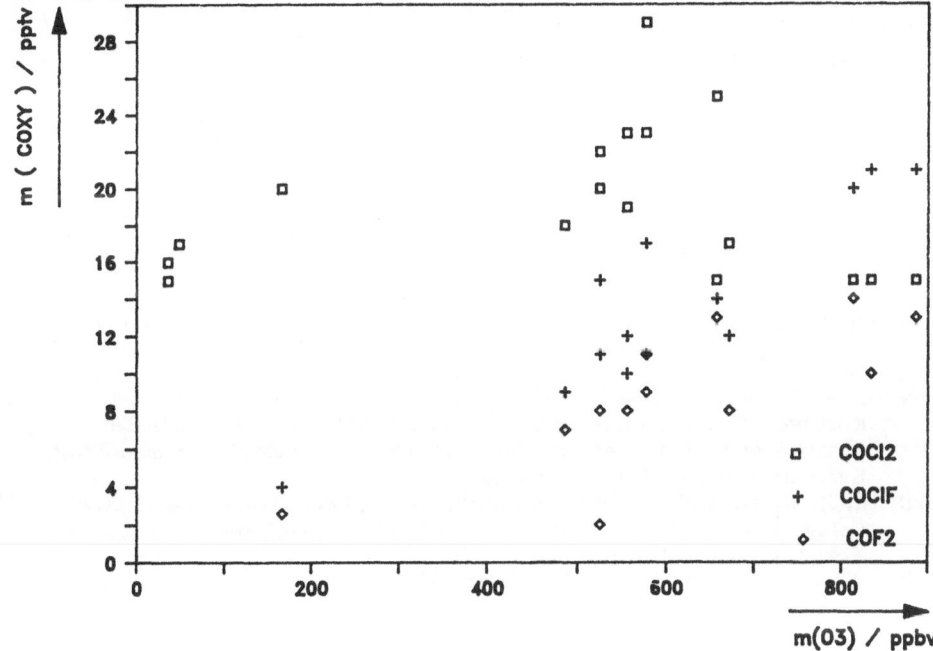

Fig. 2: Mixing ratios of COCl₂, COClF and COF₂ versus ozone mixing ratios of an individual flight.

tures in the stratosphere into account. Both figures indicate an increase of mixing ratios of all the three compounds with altitude in the stratosphere.

At present it is only possible to relate our carbonyl halide determinations to the data of COF_2 given by Rinsland et el. [1986]. However, their determinations apply to regions higher up in the stratosphere. Nevertheless an extrapolation of their curve to our altitudes would yield mixing ratios of less than 10 pptv. This is not in contrast to our data.

Model calculations of concentration profiles of carbonyl halides have been made by Crutzen et al. [1978]. Though their calculations apply for the year 1976, the general trend should still be valid. One would expect a parallel shift to larger mixing ratios on the basis of larger source strengths. They calculate an increase of mixing ratios of these compounds directly above the tropopause. However, as they do not take tropospheric sources of phosgene into account, they underestimate it's mixing ratios in this region [Wilson et al., 1988,1989]. Another, more important difference is apparent in the lower mixing ratios of COF_2 compared to those of COClF. As shown above, the parent compounds of COF_2 and COClF are mainly F-12 and F-11. Their ratio is w 1.7, both on the basis of emission strength [Crutzen et al.,1978 as well as on the basis of mixing ratios [unpublished results;WMO,1985], whereas our ratio of COF_2 to COClF is $w0.6$. This discrepancy is not yet understood. It could be due to either an overestimation of COF_2, or an underestimation of COClF. As the calibration procedures are straightforward, this feature must be real. The photolytic stability of COF_2 should be greater than that of COClF [DeMore et al,1987]. The relative hydrolysis rates of these two carbonyl halides are unknown at present. If this decay process were the only cause for COF_2 being less abundant than COClF, then the hydrolysis rates must be distinctly different.

Acknowledgements

We thank G. Moortgat for advice and F. Simon for experimental help with the generation of the carbonyl halides. We also wish to thank Flight Captains Grtner and Lechner' for cooperation during the flights. This work was in part supported by the Bundesministerium fr Forschung und Technologie through the grant ATP-88.

Literature

Crutzen, P. J., I. S. A. Isaksen, and J. R. McAfee, 1978. The impact of the chlorocarbon industry on the ozone layer. *J. Geophys. Res. 83*, 345 - 363.

DeMore, W. B., M. J. Molina, S. P. Sander, D. M. Golden, R. F. Hampson, M. J. Kurylyo,, C. J. Howard, and A. R. Ravishankara, 1987. Chemical kinetics and photochemical data for use in stratospheric modelling, Evaluation No. 8, JPL publication 87-41.

Griffith, D. W. T., and G. Schuster, 1987. Atmospheric trace gas analysis using matrix isolation-Fourier transform infrared spectroscopy. *J. Atmos. Chem. 5,*59 - 81.

Helas, G., G. Schuster, and S. R. Wilson, 1989. Measurements of ozone and other trace components near the tropopause over northern Europe. Quadrennial Ozone Symposium 1988, in: *Proceedings of the International Ozone Symposium 1988*, ed. R. D. Bojkov and P. Fabian, A. Deepak Publ., Hampton, 1989.

Jayanty, R. K. M., R. Simonaites, and J. Heicklen, 1975. The photolysis of chlorofluoromethanes in the presence of O_2 or O_3 at 213.9 nm and their reaction with O (^{1}D). *J. Photochem.* 4, 381 – 398.

Nielson, A. H., T. G. Burke, P. J. H. Woltz, and E. A. Jones, 1952. The infrared and Raman spectra of F_2CO, FClCO, and Cl_2CO. *J. Chem. Phys.* 20, 596 – 604.

Raper, O. F., C. B. Farmer, R. Zander, and J. H. Park, 1987. Infrared spectroscopic measurements of halogenated sink and reservoir gases in the stratosphere with the ATMOS instrument. *J. Geophys. Res.* 92, 9851 – 9858.

Rinsland, C. P., R. Zander, L. R. Brown, C. B. Farmer, J. H. Park, R. H. Norton, J. M. Russel III, and O. F. Raper, 1986. Detection of carbonyl fluoride in the stratosphere. *Geophys. Res. Lett.* 13, 769 – 772.

Schuster, G., S. Wilson, and G. Helas, 1989. Formaldehyde measurements in the free troposphere. *This volume.*

Singh, H. B., 1976. Phosgene in ambient air. *Nature 264*, 428 – 429.

Singh, H. B., L. Salas, H. Shigeishi, and A. Crawford, 1977. Urban–nonurban relationships of halocarbons, SF_6, N_2O, and other atmospheric trace constituents. *Atmos. Environ.* 11, 819 – 828.

Wilson, S. R., P. J. Crutzen, G. Schuster, D. W. T. Griffith, and G. Helas, 1988. Phosgene measurements in the upper troposphere and lower stratosphere. *Nature 334*, 689 – 691.

Wilson, S. R., G. Schuster, and G. Helas, 1989. Measurements of COFCl and $COCl_2$ near the tropopause. Quadrennial Ozone Symposium 1988, in: *Proceedings of the International Ozone Symposium 1988*, ed. R. D. Bojkov and P. Fabian, A. Deepak Publ., Hampton, 1989.

World Meteorological Organization, 1985. Atmospheric Ozone 1985, report no 16, Geneva, Switzerland.

Table I: Absorption bands of COXYs in CO_2 matrices

Compound	Wavenumber in cm^{-1}	Detection limit
$COCl_2$	851.0	4 pptv
	852.1	
	858.2	
	859.3	
	860.3	
	1808.9	
	1816.5	
COClF	1074.8	4 pptv
	1088.0	
	1858.2	
	2165.3	
COF_2	972.1	2 pptv
	1245.6	

THE DEPOSITION OF ACIDIFYING COMPONENTS ON WET SURFACES
- occurrence and chemical composition of dew -

F.G. RÖMER, B.H. TE WINKEL and L.H.J.M. JANSSEN

N.V. KEMA, Joint Laboratories and Other Services
of the Dutch Electricity Supply Companies
Environmental Research Department
P.O. Box 9035, 6800 ET ARNHEM, The Netherlands

Summary

The contribution of dew to acid deposition has been investigated by
means of measurements as well as modelling. Frequency and duration
of the occurrence of dew have been calculated by a dew model. The
model is based on an analysis of the energy balance of a surface and
only requires the input of a limited number of synoptic data. Model
results proved to be in fairly good agreement with measurements of
dew using an automatic sequential dew sampler. The number of dewy
nights and hours was calculated for the year 1987 and amounted to
about 225 nights and about 1600 hours. Thus dew is a frequently
occurring phenomenon at night.
Dew has been sampled in different –urban and rural– locations and
the chemical composition of these samples was compared with the
composition of rain.
Contrary to rain, a strong influence of local air pollutant
concentrations on dew composition was observed. The mass fluxes of
the main components and some trace elements in dew are small in
comparison with rain. Furthermore, also in contrast with rain the
unstable and reactive anions sulphite, nitrite and organic
substances are practically always present in dew in high
concentrations. Dissolved sulphur dioxide appeared to be mainly
present as hydroxymethane sulphonic acid. Contrary to the deposition
of species from the gas phase, it was established that aerosol
fluxes deposited on the upper side of a surface are higher than on
the lower side. The deposition velocity of sulphur dioxide has been
calculated from the measured concentrations of sulphur components in
air and dew and shows a strong correspondence with the amount of dew.

1. INTRODUCTION

In the Netherlands the removal of air pollutants by dry deposition
is quantitatively more important than by wet deposition (1). Dry
deposition depends strongly on atmospheric conditions and on the
characteristics of the receptor surface (2). By day the deposition
velocities will be enhanced due to atmospheric turbulence which reduce
the surface resistance against the uptake of the various pollutants. As a
consequence lower deposition rates may be expected by night. However, the

meteorological conditions at night may lead to the formation of dew and thus change the receptor surface resistances, especially for substances which are soluble in water. Data from literature suggested a frequent occurrence of dew at our latitude (3). After dew formation evaporation mostly occurs early in the morning. This will temporarily lead to high concentrations of the various dissolved components, which may cause damage to the natural (vegetation) or artificial receptor surfaces. Therefore, our main interest was to quantify the physical as well as the chemical aspects of dew. Therefore, it was decided to make an inventory of dew occurrence in the Netherlands. To this end a dew model, based on the energy balance of a surface, was developed which utilises a limited number of synoptic meteorological data.

Preliminary measurements in 1985 on the chemical composition of dew have demonstrated the presence of a variety of stable and unstable components. To gain more insight into dew chemistry, dew composition was studied systematically in the period up to 1988.

This paper provides a brief description of the dew model and model results. Besides, the chemical composition of dew is discussed in some detail.

2. MATERIALS AND METHODS

Sampling and chemical analysis (2)

Dew samples were collected from horizontal glass plates (0.25 m^2) cleaned in advance. Samples were collected above a grass field (height 1 m) at an urban site (Arnhem) from 16h00-08h00. The samples were applied for the determination of:
- their chemical composition
- the deposition fluxes on one or both side(s) of a (tip-up) plate.

Two identical automated sequential collectors (glass surfaces 0.30 m^2), equipped with sprinklers and wiping mechanisms for surface cleaning, were developed and applied to investigate the variation in dew concentrations during the night and the effect of local air pollutant concentrations on the chemical composition of dew, i.e. at a rural (Renkum) and an urban site (Arnhem). The distance between the samplers was 12 km. The sampling periods were 2 and 3 hours respectively.

Aerosols were sampled with an automatic dichotomous sampler (Sierra 245) provided with filters (Sierra, C-240-T2).

All dew and aerosol samples were stored in a refrigerator prior to chemical analyses.

Dew was analysed for main components, intermediates and some organic substances. Aerosols were analysed for ammonium, chloride, nitrate and sulphate. Chemical analysis was performed within a week after sample collection. The number of components to be analysed depended to a large extent on the sample volumes needed for chemical analysis.

Ionchromatography with conductivity detection was used for the determination of chloride, nitrite, sulphite, nitrate and sulphate. Continuous flow analysis was applied for the determination of all other non-metals. The detection methods used were spectrophotometry (hydronium, ammonium, hydroxymethane sulphonic acid (HMSA), methanal), potentiometry (fluoride), chemiluminescence (hydrogen peroxide). Metal ions and trace elements were determined by atomic absorption spectrometry (sodium, potassium, As, Cd, Cr, Fe, Cu, Mn and V) and inductively coupled plasma spectrometry (magnesium, calcium).

SO_2 was measured using a pulsed fluorescent (Thermo Electron type 43A) monitor.

Dew model

Our interest was primarily to determine the occurrence (frequency and duration) of dew over a year. The model is based on the surface energy balance of a surface which is given by:

$$H + \lambda E + G + M = Q$$

with λ the latent heat of vaporisation and H and λE the fluxes of sensible and latent heat respectively (defined positive upwards), M the change in heat capacity of the dew plate, G the heat flux from the underlying material to the dew plate and Q the net radiation (all units $W.m^{-2}$). The latter two energy densities are defined positive downwards.

The net radiation Q is the net radiative energy loss that cools the surface in relation to the air and the material below.

Deposition of dew occurs as the conditions $\lambda E < 0$, i.e. $H < Q$ and $G \sim 0$ and $M \sim 0$ are satisfied.

In our model it is assumed that dew deposition occurs, i.e. $\lambda E < 0$, if:

$$u^* < (1.5/dq) . A$$

with u^* the friction velocity, dq the specific humidity deficit and A a factor which depends on cloud cover.

For the calculation of dew occurrence only easily obtainable synoptic data are needed. This means the calculation of dq from air temperature and relative humidity, the determination of cloud cover for the calculation of u^* and, finally, the establishment of dew formation if the condition $u(meas) < 10u^*$ is met, with $u(meas)$ as the wind velocity.

The details of the model and of the comparison of measurements and model calculations is published elsewhere (4).

Meteorology

Model calculations needed the input of synoptic data of nearby meteorological stations. The data of interest are hourly values of cloud cover, relative humidity, air temperature, wind velocity and deposition. A data base was placed at our disposal by the kind permission of the Agricultural University of Wageningen.

3. RESULTS

The model

Results of model calculations of dew occurrence in De Bilt are given in Table I.

Table I. Monthly data concerning the number of dewy nights, dewy hours, hours of rain and hours without wet deposition (rain or dew) in 1987. The last column refers to the total number of hours during which the radiation balance was negative.

month	dewy nights	dewy hours	hours of rain	dry	total
Jan	18	151	148	140	439
Feb	24	193	73	20	286
Mar	17	122	68	29	219
					(continued)

(continued)

month	dewy nights	dewy hours	hours of rain	dry	total
Apr	15	101	37	75	213
May	14	86	60	58	204
Jun	22	93	50	26	169
Jul	19	89	71	49	209
Aug	19	125	62	57	244
Sep	20	154	108	44	306
Oct	18	170	116	95	381
Nov	19	173	189	73	435
Dec	19	170	180	85	435
total	224	1627	1162	751	3540

The model calculations for the frequency and duration of dew showed a fairly good agreement with the measurements of the automatic sequential dew samplers and are discussed in detail (4). An example is given in Figure 1.

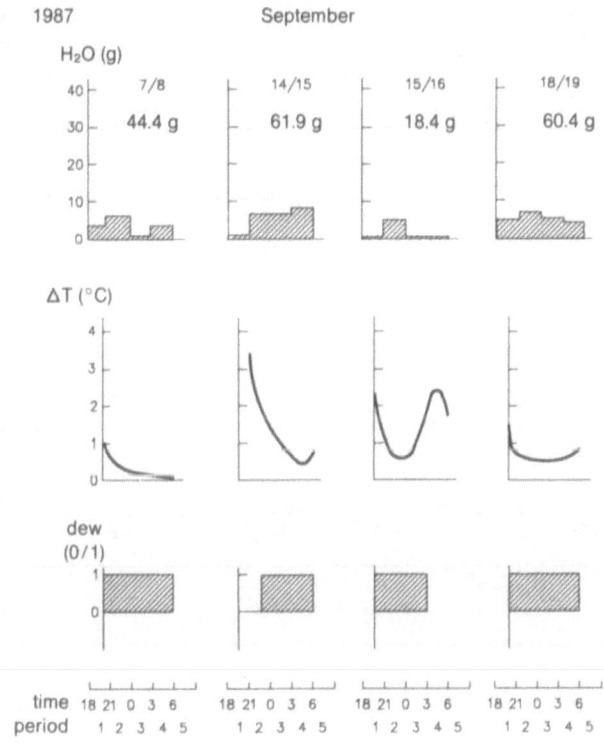

Fig. 1. Dew deposition. Results of measurements (g.0.3 m^{-2}) and model calculations for dew deposition in September 1987.
(ΔT = temp. difference between air and dew point).

Chemical composition

In the period from 1985 to 1988 dew was sampled mainly in autumn. Amounts of dew sufficient for chemical analysis occurred in about 50% of the nights (n = 111). With the automatic sequential samplers a mean dew rate of 0.012 mm/h and a maximum of 0.032 mm/h was found.

An overview of the concentration levels of the main components, unstable components and trace elements in dew as collected with the automatic sequential samplers is presented in Table II.

Table II. Chemical composition of dew collected at an urban site; ranges, weighted mean values, x (μmol/L) and number of samples, n.

main components				unstable components				trace elements		
	range	x	n		range	x	n		x	n
pH	4.9– 7.0	6.1	112	NO_2^-	6– 99	28	122	Cd	0.01	9
NH_4^+	50 –550	211	84	$SO_3^=$	1–160	27	134	Cu	0.28	29
Na^+	27 – 87	73	61	CH_2O	1– 21	9	38	Cr	0.03	7
K^+	10 – 58	24	60	HMSA	10–140	52	14	Fe	0.39	20
Ca^{++}	42 –175	58	61	H_2O_2	0– 3.5	0.7	31	Mn	0.52	24
Mg^{++}	6 – 23	15	61					V	0.23	24
Cl^-	13 –536	118	126					As	<0.05	4
NO_3^-	10 –319	82	132							
$SO_4^=$	11 –283	75	135							

Table II shows that significant amounts of unstable components are found in dew. Especially the high concentrations of nitrite, sulphite and hydroxymethane sulphonic acid are remarkable. These substances are not normally present in rain. To distinguish between the chemical compositions of dew and rain, detailed measurements were carried out in Arnhem. In Table III the total deposition fluxes of some main components measured in dew and in rainwater in October 1986 are presented.

Table III. Total deposition (mm) and deposition fluxes ($mmol/m^2$) for a number of main components in rain and in dew.

	deposition	NH_4^+	Cl^-	NO_3^-	$SO_4^=$	Na^+	K^+
dew	1.4	0.45	0.09	0.15	0.29	0.06	0.02
rain	100	6.3	13.9	2.6	3.6	11	0.3

The figures in Table III show the differences in deposition fluxes between dew and rainwater. The deposition flux of rainwater is much higher and varies from one to three orders of magnitude over the various components.

To investigate the effect of local air pollutant concentrations on the chemical composition of dew, some experiments were carried out in

Arnhem (urban area) and Renkum (rural area). It was found that the deposition fluxes of sulphite, nitrite and methanal at the urban site were higher than at the rural site. For ammonium, pH and chloride the opposite was observed. These differences were tested for significance with a sign test (90% confidence interval).

The effect of the position of a receptor surface on the deposition of substances from the gas phase and aerosol phase was determined. The data were obtained with the tip-up sampler in 1985 and 1986. It is found that for practically all components that deposit from the aerosol phase the concentrations on the upper side of the glass plate exceed those on the lower side. This effect is visualised for sulphur compounds in Figure 2. The differences in concentrations between upper and lower side were tested (sign test, 95% confidence interval). For sodium, potassium, magnesium, calcium, sulphate, ammonium, nitrate and chloride significant differences were found. For substances depositing from the gas phase (e.g. methanal, ammonia, sulphur dioxide) no systematic differences between upper and lower side of the receptor were found.

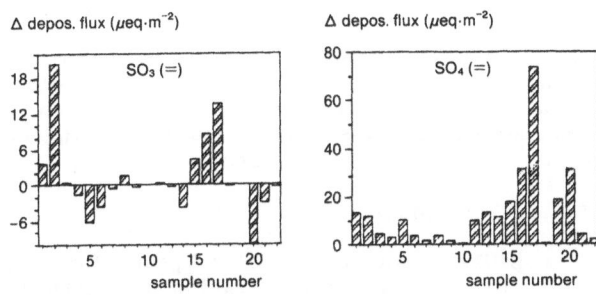

Fig. 2. Deposition differences between upper and lower side of a glass plate (μeq/m^2) for sulphur components depositing from the gas ($SO_2 \rightarrow SO_3^=$) and the aerosol phase ($SO_4^=$).

4. DISCUSSION AND CONCLUSIONS

Results of the model calculations show that dew is a very frequently occurring phenomenon in the Netherlands. In our dew model heat capacity of the dew plate and heat flux from the ground are assumed to be negligible. The model is therefore only applicable to thin surfaces which are badly connected (low conductance for heat transport) to a larger body, as is the case with e.g. grass or leaves of vegetation. Taking both hours of dew and of rain in 1987 into account it appears that a surface is wet for about 75% of the nocturnal hours. This result is important in quantifying the deposition, in particular of easily soluble air pollutants which may cause injury to vegetation.

Considering the chemical composition of dew at the measuring sites, it appears that the mean concentrations of most of the main components are in the same order of magnitude as the concentration values usually found in rainwater. However, systematic differences are detected for potassium and calcium, which are present in higher concentrations in dew, and for hydronium which was only present in low quantities (pH $\geq$ 6). In contrast to concentration values, dry deposition fluxes of substances to wet surfaces amounted only to a small percentage of wet deposition fluxes

(Table III).

Unlike main components, the unstable components such as sulphite and nitrite anions occur in relatively high concentrations in dew (Table II). These substances are normally not present in rainwater or only in low concentrations. Furthermore it appeared that the concentration of free sulphite is only a small fraction of the concentration of sulphur IV. Due to the omnipresence of methanal it appeared that about 80% of sulphur IV occurs as the adduct HMSA.

As dew occurs so frequently, the dew phenomenon may be important in the effects of acid deposition. After dew formation evaporation mostly occurs early in the morning. Finally, this will lead to temporarily high concentrations on the receptor surface of both stable and reactive substances. Effects are expected to be dependent on speciation rather than on concentration. In view of this assumption some substances are suspect, especially the oxidisable nitrite and sulphite, the oxidant hydrogen peroxide (Table II) and hydronium.

The results of some experiments strongly suggest that the chemical composition of dew is determined to a large extent by local air pollutants (measurements at an urban and a rural site).

On wet receptor surfaces the sulphur dioxide deposition velocities are increased (2). However, under atmospheric conditions favourable for dew formation deposition velocities for sulphur dioxide are low. Calculations were made to establish the deposition velocity from concentrations of sulphur species (S(IV) and S(VI)) in dew and air. The deposition velocity (v_d) was calculated from:

$$v_d = (F\ S(IV) + F\ S(VI))/c(SO_2)$$

with (F S(IV) + F S(VI)) being the total sulphur flux over the sampling period and $c(SO_2)$ the mean gas concentration over the sampling period (receptor surface height = 1 m; gas concentration measurement 1.50 m). The sulphate flux was corrected for the aerosol contribution. The deposition velocity of 0.1 cm/s for aerosol sulphate was assumed to be constant and was taken from literature (6). The results were plotted and the plots are presented in Figure 3.

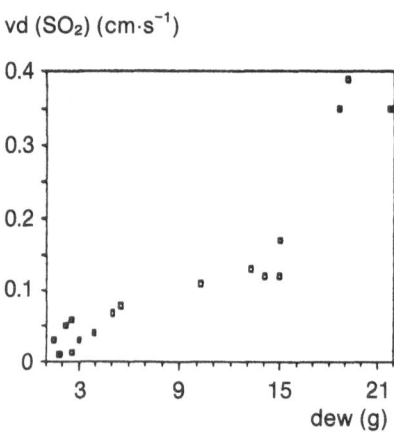

Fig. 3 Relation between the deposition velocity vd (cm/s) of sulphur dioxide and the amount of dew (g) (two and three hour mean values).

Figure 3 shows clearly that the deposition velocity of sulphur dioxide depends on the amount of dew and amounted to a maximum vd of 0.4 cm/s. The deposition velocities found in our experiments are in the same order as given in literature (6). The linear relation between wetness and deposition velocity is quite remarkable and has not been described in literature so far. Moreover, it is expected that the deposition velocity will become constant when maximum wetness is attained.

ACKNOWLEDGEMENT
This study has been supported by the Dutch government and was carried out in the framework of The Dutch Priority Programme on Acidification.

REFERENCES
(1) Schneider, T., and Bresser, A.H.M., 1988. Dutch Priority Programme on Acidification, Evaluatierapport Verzuring (in Dutch), report no. 00-06, RIVM, Bilthoven, The Netherlands.
(2) Voldner, E.C., et al., Atmos. Environ., 20 (1986), pp. 2101-2123.
(3) Hofmann, G., 1955. Die Thermodynamik der Taubildung (in German). Berichte des Deutschen Wetterdienstes, nr. 18.
(4) Janssen, L.H.J.M., and Römer, F.G., 1989. Paper submitted for publication.
(5) Römer, F.G., and Te Winkel, B.H., 1988. KEMA-report 50583-MOL 87-3170 (in Dutch).
(6) Mulawa, P.A., et al., Atmos. Environ., 20 (1983), pp. 1389-1396.

A COMPARATIVE STUDY OF TWO MICROPHYSICAL SCHEMES FOR MESOCALE MODELING OF POLLUTANT TRANSPORT AND DEPOSITION

N. CHAUMERLIAC, E. RICHARD and R. ROSSET
LAMP/OPGC: 12, avenue des Landais, 63000 Clermont-Ferrand, France

Summary

Acid deposition results from very tight and complex interactions between microphysical, dynamical and chemical processes. For instance, coalescence mechanisms determine the partition of chemical soluble species and then the partition of acidity between cloud and rain. Gas dissolution and mass transfer processes are depending on drop sizes so that a detailed microphysical scheme is needed. The currently used microphysical scheme of Kessler (1969) and parametrisations derived from Berry and Reinhardt's work (1973) are compared. They are based upon different spectral distributions for raindrops (Marshall-Palmer and log-normal distributions respectively). They are using different formulations for the autoconversion, evaporation processes as well as for the fall velocity of raindrops. A comparative study of these two schemes is performed under various meteorological conditions: a mountain wave simulation and a frontal situation. Differences in the spatial and time evolution of microphysical fields are investigated in order to evaluate their possible impact upon wet deposition mechanisms. Then, the two schemes are compared on simple chemical scenarios: gas dissolution in cloud and rain, gas scavenging by raindrops and wet deposition, vertical cloud transport of an insoluble tracer.

1. INTRODUCTION

Results obtained from two microphysical schemes capable of simulating cloud and precipitation processes in a mesoscale model (Kessler, 1969; Berry and Reinhardt, 1974a; Berry and Reinhardt, 1974b) have been compared for various orographic situations in a β-mesoscale model (Richard and Chaumerliac, 1989). In view of the major differences found in raindrop size distributions as well as in the rates associated with various microphysical processes, two types of problems have been addressed. First, the impact of differences in raindrop spectra upon wet deposition mechanisms is investigated over a mountain wave scenario. Then, to extend the comparative study restricted to orographic precipitation, the two schemes are tested in case of frontal rain. Vertical transport of carbon monoxide from the planetary boundary layer to the free troposphere is studied in connection with the crucial role of evaporation and water loading effects which affect the dynamics of moist frontogenesis.

2. OROGRAPHIC SITUATION

In order to compare the two schemes, the model has been run over a mountain wave simulation. Results from these comparisons are given in Richard and Chaumerliac (1989). We just recall on Figure 1 the vertical cross-sections of cloud water mixing ratios, rainwater mixing ratios and precipitation rates computed after six hours of simulation for Kessler and Berry and Reinhardt.

Those results deal with the mountain wave simulation and have been obtained with the same conditions of rain production for the two schemes (similar cloud water mixing ratios). The main point is that precipitation intensities are comparable between Kessler and Berry and Reinhardt but the precipitation extends further downwind in case of Berry and Reinhardt. In contrast, the corresponding rainwater mixing ratios are quite different. This comes from the fact that rainwater mass is carried by much larger raindrops in Kessler and that these larger raindrops precipitate with larger fall velocities than in the case of Berry and Reinhardt.

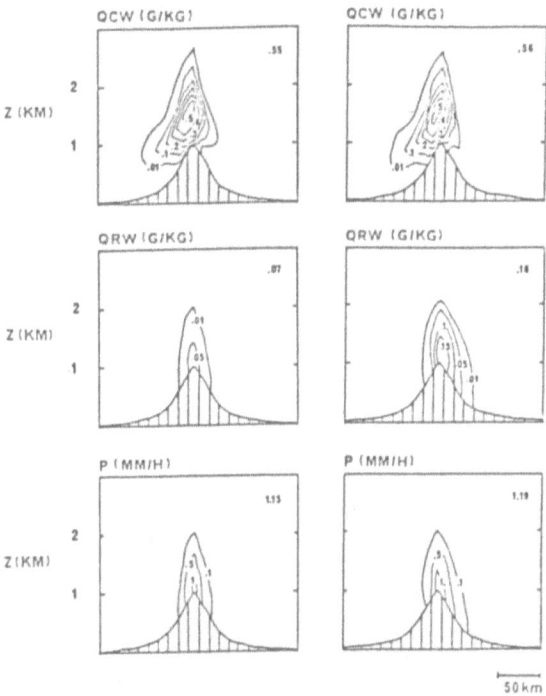

Fig. 1. Vertical cross-sections of cloud water mixing ratio (q_{cw}), rainwater mixing ratio (q_{rw}) and preciptation rate (P).

On this mountain wave simulation, a series of experiments have been performed and consisted in transporting a soluble, non reactive gas. Three runs have been performed: one with highly soluble gas ($H = 10^5$ M/atm), the second one with the same gas but with excluding direct scavenging by raindrops and the last one with less soluble gas ($H = 10^3$ M/atm). We have found large differences between the two schemes for the three runs, showing that a spectral information for raindrops is required to adequately describe direct scavenging of gases by raindrops. To summarize the main results, vertical cross-sections of deposition rates are presented in Figure 2 for the two parameterizations and the three tests quoted above.

Without direct scavenging of gases by raindrops, we get similar deposition rates but with greater extension on the downwind side. When direct scavenging is considered, deposition rates are about three times greater in the case of Berry and Reinhardt than in Kessler. Smaller raindrops are represented in the scheme of Berry and Reinhardt: they are more sensitive to advection and wind drift than for Kessler and have longer residence time in the atmosphere which allows for more time for scavenging gases. For less soluble gases, the same differences in deposition rates are noticed between the two parameterizations. But this time, the drift effect is greater and the maximum of deposition rate is displaced downwind from the mountain top for the case of Berry and Reinhardt.

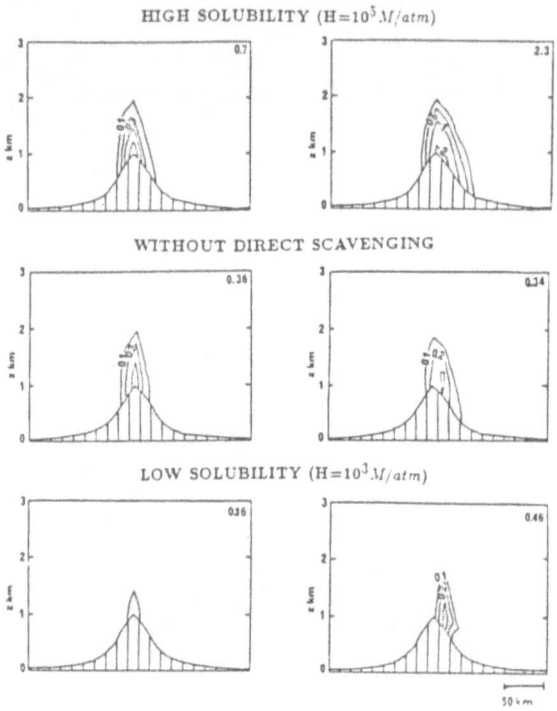

HIGH SOLUBILITY (H=10³ M/atm)

WITHOUT DIRECT SCAVENGING

LOW SOLUBILITY (H=10³ M/atm)

50 km

Fig. 2. Vertical cross-sections of deposition rates.

3. FRONTAL SITUATION

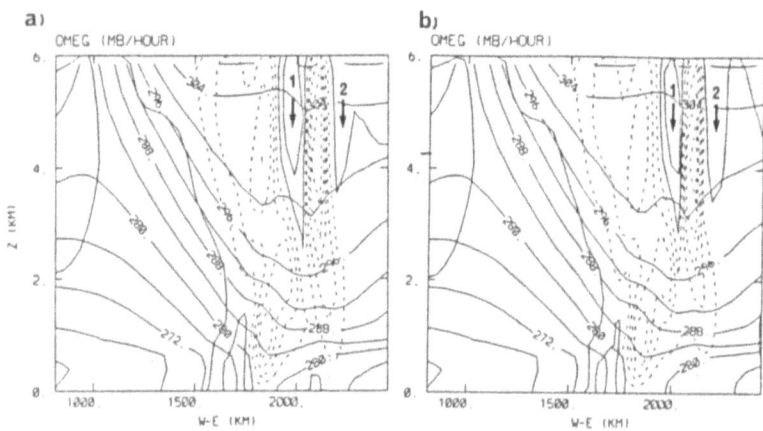

Fig. 3. Vertical cross-sections at 81h of vertical velocity (contour interval 10 mb/h) superimposed with potential temperature (K) in case of a) Kessler and b) Berry and Reinhardt.

Numerical simulations of frontogenesis in a moist atmosphere have been performed after Hsie et al. (1984). A two-dimensionnal version of the model (Nickerson et al., 1986) has been run for 24h including planetary boundary layer physics (Blackadar, 1978) and diabatic effects for an Eady wave evolution. The mesoscale model is initialized with the solutions from the analytical model of Keyser and Anthes (1982). Results for dynamical and thermodynamical fields are shown in Figure 3 after 21h simulation time in case of Kessler and Berry and Reinhardt. They compare well with the results from Hsie and al. (1984) with a banded structure of the vertical velocity in the warm sector. An ascending vertical jet arises in the atmospheric boundary layer ahead of the front: it is very distinct in the first km of the atmosphere on the potential temperature field. This jet is due to horizontal convergence forced by surface frictions. The two others ascending cells are related to moist convection with a subsidence zone in between due to evaporation. In the upper part of these cells, there are some differences between the two microphysical schemes. The updrafts and downdrafts in this region have contrasted intensities in the two cases due primarily to the fact that Kessler raindrops are larger and evaporate less efficiently than in the case of Berry and Reinhardt. Then water loading effects are acting quite differently in the two schemes, since the partition between precipitating and precipitated water is different as seen in Figure 1. There are two opposite effects of liquid water on the development of convection: one is the evaporation intensifies the downdrafts and subsequently the development of convection. The other is the water loading effect, which reduces the intensity of the updraft. In Figure 4 are presented the evolution over 24h of carbon monoxide vertical profiles. Initially, CO is produced in the boundary layer, it is sparingly soluble in water and can be chosen as a primary tracer of air motion through clouds (Dickerson et al., 1987). Carbon monoxide is transported from the boundary layer to the free troposphere along the frontal surface. In Figure 4a and 4b, the two evolutions of CO profiles are selected in the two downdraft regions indicated with arrows in Figure 3. The first downdraft is influenced essentially by water loading effects, more important in the case of Berry and Reinhardt and CO is transported at higher levels than in Kessler case. In Figure 4b, the second downdraft, driven by evaporation at the eastern border of the rainbands has the effect of transporting CO downwards in the case of Berry and Reinhardt (if we follow the 160ppb isocontour) while the opposite effect is obtained in Kessler case.

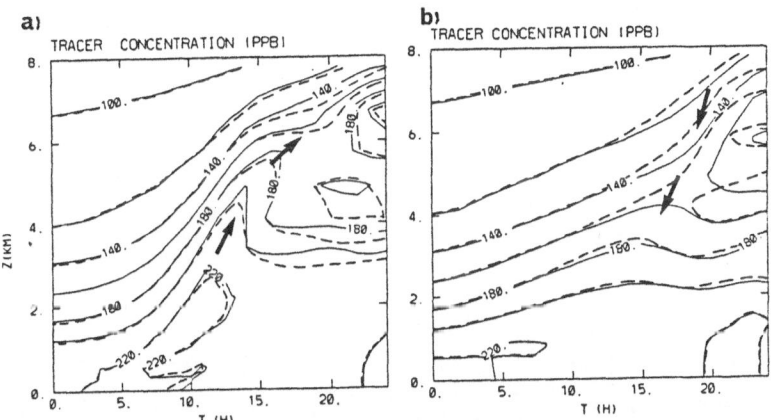

Fig. 4. Evolution of vertical profiles of CO concentration (ppb) over 24h for Kessler case (dashed lines) and Berry and Reinhardt case (solid lines) in a) downdraft region 1 and b) downdraft region 2 indicated in Figure 3.

4. CONCLUSION

The influence of diabatic effects and microphysics on pollutant distribution has been shown through two type of studies. One is an orographic situation for which gas dissolution and wet deposition mechanisms are found to be drop size dependent processes, emphasizing the need for detailed spectral information in mesoscale modeling. The second study on frontal rain has shown that pollutant transport by clouds is sensitive to diabatic effects so that it is necessary to have a good representation of water loading effects which are key processes to couple dynamics and microphysics of moist fronto-genesis. Further investigations are needed to couple these studies since transport of gases to the upper troposphere has important implications regarding ozone production and acid deposition.

REFERENCES

(1) Berry, E.X., and R.L. Reinhardt, 1974a: An analysis of cloud drop growth by collection: Part II. Single initial distributions. J.A.S., 31, 1825-1831.
(2) Berry, E.X., and R.L. Reinhardt, 1974b: An analysis of cloud drop growth by collection: Part III. Accretion and self-collection. J.A.S., 31, 2118-2126.
(3) Blackadar A.K., 1978: High-resolution models of planetary boundary layer. Advances in Environmental Science and Engineering, Vol. 1, J.R. Pfafflin and E.N. Ziegler, Eds., Gordon and Breach, 50-85.
(4) Dickerson R.R., G.J. Huffman, W.T Luke, L.J. Nunnermacker, K.E. Pickering, A.C.D. Leslie, C.G. Lindsey, W.G.N. Slinn, T.J. Kelly, P.H. Daum, A.C. Delany, J.P. Greenberg, P.R. Zimmerman, J.F. Boatman, J.D. Ray, and D.H. Stedman, 1987: Thunderstorms: An important mechanism in the trans-port of air pollutants. Science, Vol. 235, 460-465.
(5) Hsie E.U., R.A. Anthes, and D. Keyser, 1984: Numerical simulation of frontogenesis in a moist atmo-sphere. J.A.S., 41, 2581-2594.
(6) Kessler E., 1969: On the distribution and continuity of water substance in atmospheric circulations. Meteor. Monog., 10, N° 32, 84pp.
(7) Nickerson E.C., E. Richard, R. Rosset, and D.R. Smith, 1986: The numerical simulation of clouds, rain and airflow over the Vosges and Black Forest Mountain: a meso-β model with parameterized microphysics, M.W.R., 114, 398-414.
(8) Richard E., and N. Chaumerliac, 1989 : Effects of different parameterizations on the simulation of mesoscale orographic precipitation, accepted for JAM.

DAILY PROFIL STUDY OF THE ATMOSPHERIC DMS CONCENTRATION NEAR TO AN INTENSE COASTAL SOURCE AT THE "POINTE DE PENMARC'H"

S. PASHALIDIS , P. CARLIER, G.MOUVIER

Laboratoire de physico-chimie de l'atmosphère
Université Paris VII , 2 place Jussieu , 75251 Paris
cedex 05

SUMMARY

During a campaign for the study of the chemistry occuring in the oceanic air masses reaching the Western Europe at the "Pointe de Penmarc'h" at July 1988, we have mesured the atmospheric DMS concentrations in proximity of an important coastal algae field. The measurements were carried out at two levels (5m and 47m) over the ocean surface with a sampling step of 2 hours.
For this study we have used the sampling and analysis developed by our laboratory and already applied in the field studies on the Polarstern in September 1988. The range of the measured concentrations varies between 180 ng(DMS)m^{-3} and 1500 ng(DMS)m^{-3} at the low level (5m) and between 35 ng(DMS)m^{-3} and 135 ng(DMS)m^{-3} at the high level. Our measurements indicate a very strong vertical gradient, which depends on the speed of local wind due to the intensity of the local DMS source.
Finaly the dependence of the emission flux in comparison with certain parameters (as low tide, high tide,sunshine) has been proved clearly.

1.INTRODUCTION

The european program OCEANO-NOX (1) is devoted to the study of the chemical oxidation mechanisms of trace compounds in the lower troposphere under the particular conditions of oceanic environments without disturbances. As they constitute a essential data basis for the air quality of western Europe we focused our study on the example of the masses of atlantic air reaching the european coasts.
The problem that we will discuss here is the dimethylsulfide one which, as we already reported in our paper relative to the Polarstern crossing has not only a major impact on the acidity of oceanic atmospheres but may be susceptible through one of its oxidation products, the methanesulfonic acid, to have a significant impact on the global radiative budget (3).

2.SAMPLING SITE

To carry out all the field experiments in the
framwork of the project "OCEANO NOX" we selected the site
of the "Pointe de Penmarc'h" at the end of the Bigouden
country in Britany (see map, figure 1).

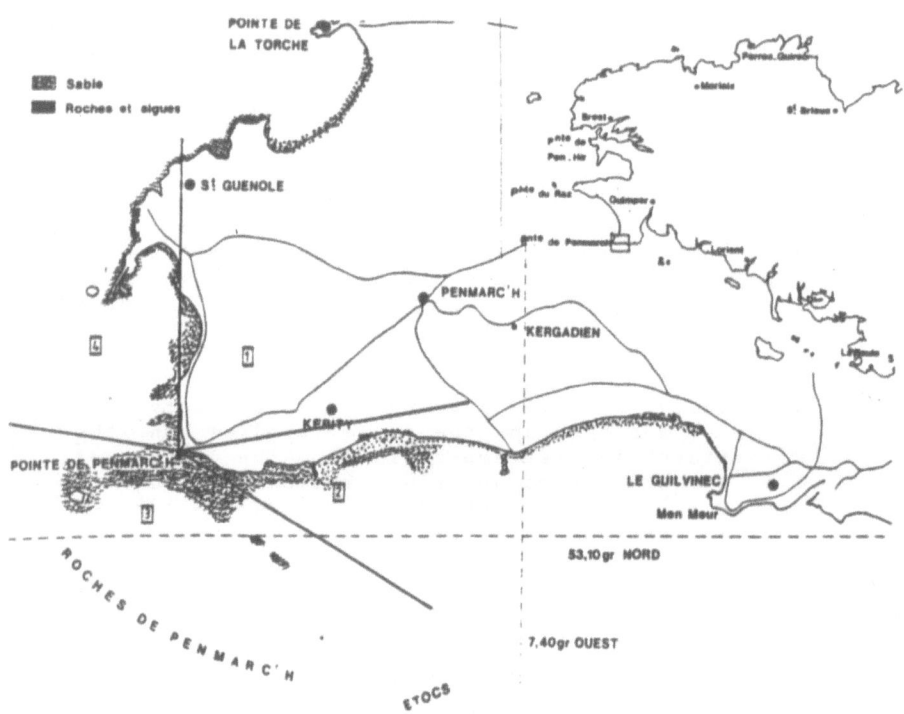

fig.1   Sampling site

This site presents the following advantages :
- it is largely open to the ocean in the direction of
prevailing winds ;
- it is a rocky flat coast where the flow of the air
coming from the sea is not locally disturbed by relief
effects ;
- it is situated in an area far from big agglomerations
and important industrial zones.
- an extended coastal algae field form there a intense
local source of organic compounds of biogeochemical
origin where the time of residence in the atmosphere we
are looking for determining ;
- finally, the old lighthouse of Penmarc'h today not
working and set graciously in our disposition from the
"administration des phares et balises" form an
observation tower, ideal and free of charge.
On the other hand in the low rooms of this old lighthouse

we could install all the physicochemical measuring instruments necessary to the totality of the experiences
    The sampling of the sulphur compounds were carried out :
- on the one hand at the ground level of the signal station of "Marine Nationale" (Semaphore de Saint-Pierre, Penmarc'h) about 5m above the level of reference of maps of the IGN ;
- on the other hand at the top of the old lighthouse about 47m above the same reference level.

3. EXPERIMENTAL
    The sampling and analysis methods are identical to those used during the Polarstern crossing and described in the related paper (2).
During the same campaign, DMS measurements were carried out simultaneously by NGUYEN et al. and we could confirm a good agreement of the results.

4. RESULTS

    We present here the results obtained in the field campaign carried out in july 1988.
    During the periode of measurements from 10 to 15/7/88 we sampled with sea winds generally a little strong including between 6 and 12 m/s. No one phenomenon of breeze was observed during this periode.
The weather was cloudy without a rain recorded. The consequence of this is that during this periode measured concentration will essentially depend on the flux emission by reason of the good ventilation and the bad vertical dispersion.
    The whole of the results are presented in the figure 2 (signal station and old lighthouse). The concentrations are graduated from 180 to 1500 ng (DMS)/m3 at the signal station and 35 to 135 ng (DMS)/m3 at the top of the old lighthouse.
The magnitude of these concentrations and the importance of the vertical gradient confirm the predominance of the local source (algae field) upon a poor ocean background, as it was shown from a previous campaign (4). Anyway, the increased strength of winds and the bad irradiation explain that we had not the strong concentrations that we had during september, 1983.

5. DISCUSSION

a. Influence of solar cycles and tides

    Comparing the periods of day and night or of high tide and low tide, the other parameter remaining invariable, it appears clearly that the emissions are greatest (in maximum) at the beginning of the afternoon and in low tide.

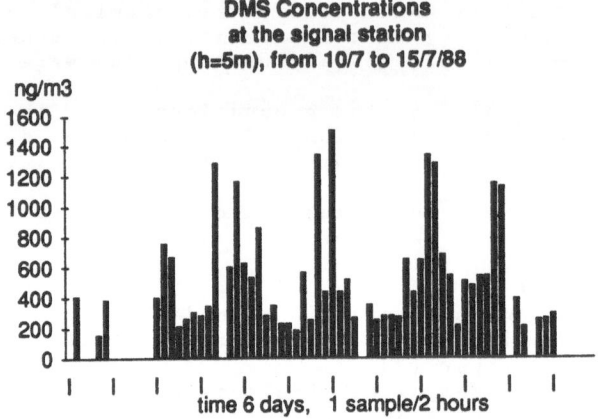

**DMS Concentrations
at the signal station
(h=5m), from 10/7 to 15/7/88**

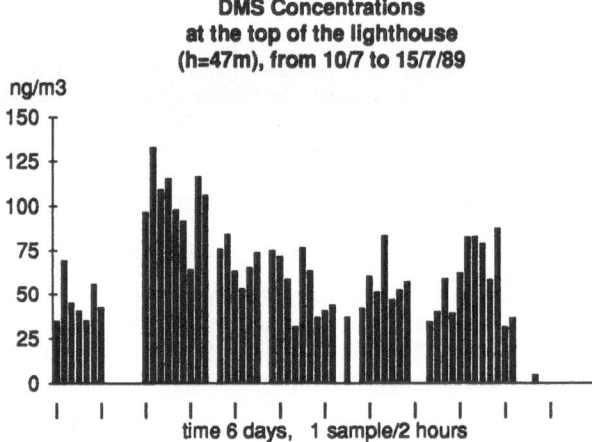

**DMS Concentrations
at the top of the lighthouse
(h=47m), from 10/7 to 15/7/89**

fig. 2   DMS concentrations

b. Vertical gradient

The observed vertical gradient between the ground
level and the top of the old lighthouse is very directly
depending on the speed of the wind (figure 3). It is
resulting that this situation give to us more
informations for the vertical dispersion than for the
chemical degradation.

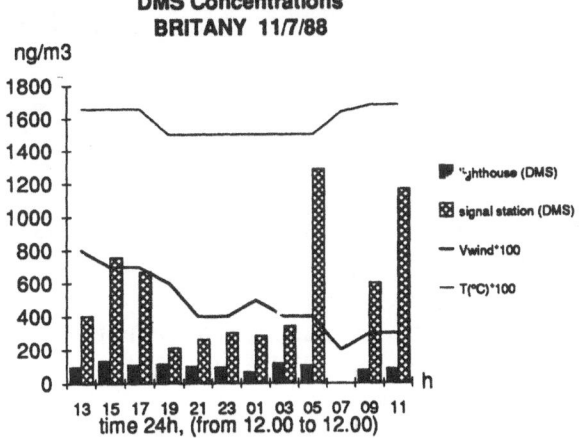

fig.3   Relation between wind speed and concentrations

6. CONCLUSION

The method of micrometrology of organo-sulphur compounds which we developped and tested in Penmarc'h in July 1988 is turned out to be correct for sampling in such type of rather biologicaly rich site and permit to treat a high number of samples (max 50/day) in regard of the difficulties inherent in such microanalysis technics which is depending on a very precise protocol that is reserved to laboratory specialized in the ultraces analytical chemistry. It has been possible to us to acquire daily profiles (with a sampling step of 2h in two points). To know much more precisely the DMS distribution around a local source (Pointe de Penmarc'h) it will possible in the next campaigns to dispose a third point, for example, situated some km in land coast.

7. ACKNOWLEDGEMENTS

The researches were carried out in the context of the european project "OCEANO NOX" which has received the support of the Commission of European Communities. The authors would like to thank, on the one hand, the National Navy and the Service of the ligth houses for the authorization which were given to them for working and on the other hand, the personnel of these establishements who were received them sympathetically and helped them for the implantation of the experience.

REFERENCES

1.P. CARLIER
Synthetic   presentation   of the european project   OCEANO-
NOX.
COST 611 WP III workshop,   3-4 may 1988, Villefranche sur
mer, -in " field measurements and their interpretation" ,
S BEILKE ,J MORELLI, G ANGELETTI ed. C.E.E. Air Pollution
Research No 14, 1988, p 55

2.S.PASHALIDIS, P.CARLIER, G.MOUVIER
DMS distribution study over the Atlantic ocean during the
Polarstern crossing from Bremerhaven to Rio Grande do Sul
 (Ib in dem)

3.R.J. CHARLSON,J.E. LOVELOCK,M.O. ANDREAE ,S. WAREN
 Oceanic phytoplankton,   atmospheric sulphur,cloud albedo
and climate
Nature, 326, 655,1987

4.C. LUCE
THESE, Université Paris 7,1984

PRINCIPAL COMPONENT ANALYSIS OF WET DEPOSITION IN FINLAND

Tuomas Laurila
Finnish Meteorological Institute, Air Quality Department
Sahaajankatu 22E, SF-00810, Helsinki, Finland

Summary

Correlation and factor analysis was applied to daily wet deposition observations which were obtained during 1988 at three Finnish EMEP background stations. At the two southernmost stations, where the deposition levels are higher, three factors explained about 90 % of the total variance. In Virolahti the correlations were high between the first factor and anthropogenic precursors and alkaline particulates (SO_4^{2-}, NO_3^-, NH_4^+, Ca^{2+}, Mg^{2+}, K^+). The second factor reflected marine ions (Cl^-, Na^+, Mg^{2+}, K^+). At the island station Utö the factor related to marine ions was more important. The third factor was related to free acidity. The correlation between the third factor and H^+ and NO_3^- was positive and between Ca^{2+} it was negative. In central Finland (station Ähtäri) the three first factor components accounted for 85 % of the total variance. The correlations were high between the first factor and ions which have anthropogenic precursors (H^+, NO_3^-, SO_4^{2-}, and NH_4^+), the second factor and terrigeneous particulates (Ca^{2+}, K^+, and Mg^{2+}), and the third factor and sea or road salts (Cl^-, Na^+).

1. INTRODUCTION

As a consequence of human activities, the acidity of precipitation in Finland is higher than would be expected at these northern latitudes. Yearly pH values of precipitation-weighted average concentrations range between 4.3 - 4.7 at the three Finnish EMEP stations. Anthropogenic emissions of sulphur and nitrogen oxides are dominant precursors of sulphate and nitrate, which are the main acid ions in precipitation. However, the correlation coefficients between hydrogen ion concentrations and sulphate and nitrate concentrations are low at Finnish EMEP stations. The correlation between sulphate and hydrogen ion is 0.68 at Ähtäri and insignificant at Utö and Virolahti. Correlations between nitrate and hydrogen ion are slightly higher, 0.46, 0.30, and 0.75, at Utö, Virolahti and Ähtäri, respectively. Consequently, the acidity of precipitation does not depend only on sulphate and nitrate, but is governed by all ions, both acidic and alkaline, in a complex way. Thus, there is a need to find out which ions contribute most to the acidity and which ions are responsible for neutralizing the sample. This presentation will focus on this question. Multivariate statistical methods will be applied to daily observations carried out in 1988 at Finnish EMEP stations.

Factor analysis is widely used when aerosol composition and origin are studied. Only a few studies have been made to characterize event samples of precipitation. Factor analysis of event samples from Illinois has been done by Gatz (1), from Minnesota by Krupa et al. (2), and from Spanish Basque country by Ezcurra et al. (3). Gatz found four ionic groups having a common source: crustal dust (Ca^{2+}, Mg^{2+}, K^+), a 'gaseous precursors' factor (NH_4^{2-}, NO_3^-, SO_4^{2-}), sea and road salt (Na^+, Cl^-), and a strong acids factor (H_2SO_4, HNO_3). Krupa et al. had six stations: at one station H^+ loaded with SO_4^{2-} and NO_3^- at one station H^+ loaded with SO_4^{2-} and at four stations H^+ did not load with any of the anions. Ezcurra et al. (1988) found that there was three components explaining the variability of ionic concentrations of precipitation in Spanish Basque country. All ions contributed to the first

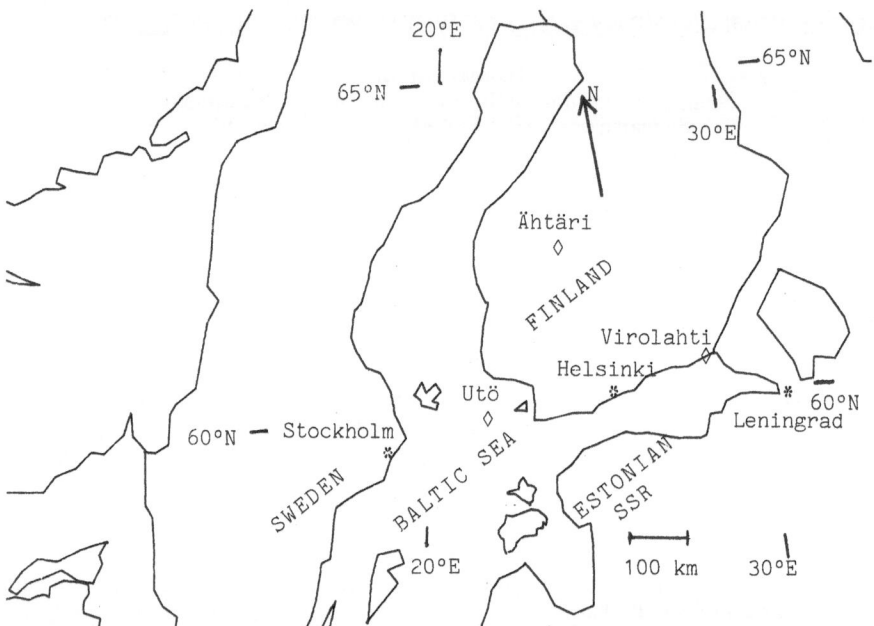

Fig. 1. Locations of the measuring sites (Utö, Virolahti, and Ähtäri) and the major cities.

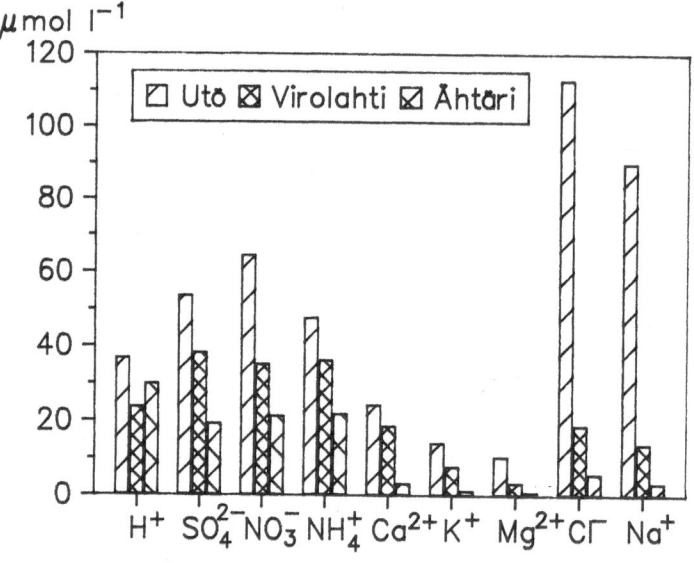

Fig. 2. Volume-weighted averages of the concentrations in Utö, Virolahti, and Ähtäri in 1988.

component (SO_4^{2-} and NH_4^+ had the highest and H^+ the lowest correlation coefficients). The second factor was sea salt (Cl^-, Na^+, Mg^{2+}). The correlation between this factor and H^+ and NO_3^- was negative. Third component showed that H^+ correlated positively with NO_3^- but was neutralized by Ca^{2+}.

2. THE DATA AND STATISTICAL METHODS

Monitoring of acid rain on a daily basis has been conducted since the seventies at these three stations. Utö is a treeless island in the Baltic Sea located about 80 km southwest of mainland Finland. Virolahti is near the south coast only 150 km from Leningrad and Ähtäri is in central Finland where the terrain is mainly forested (Fig. 1.). Precipitation was collected in open polyethylene buckets which were changed and washed with deionized water every morning at 6.00 UTC. This study covers only 1988 because then all major ions have been analyzed. pH was measured potentiometrically. SO_4^{2-}, NO_3^-, and Cl^- were determined by ion chromatography. NH_4^+ was analyzed by the indophenol colorimetric method and Ca^{2+}, K^+, Mg^{2+}, and Na^+ by atomic absorption spectroscopy. Interrelations of ions were studied using correlation matrices of log-transformed concentrations and subsequent factor analysis and varimax rotation. Factor analysis were done with commercially available SPSS statistical package.

3. RESULTS

The statistical summary of precipitation-weighted concentrations in Fig. 2 shows that the most important ions are SO_4^{2-}, NH_4^+, NO_3^-, and H^+. In equivalents the precipitation weighted average SO_4^{2-}/NO_3^- and SO_4^{2-}/NH_4^+ ratios are about 2 and NO_3^-/NH_4^+ ratio is about unity. At the marine station Utö Cl^- and Na^+ concentrations are very high. On the average 12 % of sulphate is from sea salts in Utö. This figure is below 2 % in Virolahti and Ähtäri. In Virolahti, Ca^{2+} is also significant. Average annual precipitation amounts are 510, 672, and 680 mm in Utö, Virolahti, and Ähtäri, respectively. Inorganic acids in precipitation are rather neutralized in Finland. The equivalent ratio of $H^+/(SO_4^{2-} + NO_3^-)$ is 0.21 in Utö and Virolahti and 0.5 in Ähtäri. In the eastern US these acids are much less neutralized, for example the same ratio was 0.79 in the data of Gatz (1). This ratio was 0.18 in Spanish Basque country (3).

Correlation matrices of log-transformed concentrations (Table 1) show that (1) NO_3^- is more frequently observed in acid samples than SO_4^{2-}. (2) NH_4^+ is often observed in acid samples. (3) SO_4^{2-} is neutralized by NH_4^+ and Ca^{2+}. In Ähtäri and Utö NH_4^+ is the dominant neutralizing cation while it is Ca^{2+} in Virolahti. (4) In Ähtäri, where neutralization is lower, SO_4^{2-} contributes more to acidity of precipitation. (5) In Utö, Na^+, Cl^-, and Mg^{2+} are solely of marine origin but in Virolahti and Ähtäri correlations indicate also terrigeneous and anthropogenic sources.

Principal component analysis and subsequent varimax rotation was applied to the data shown in the correlation matrices. Three component was retained because then the physical interpretation was straightforward. The third component had an eigenvalue of 1.097, 1.163, and 0.848 in Utö, Virolahti, and Ähtäri, respectively. The first three components accounted for 89, 90, and 85 % of the total variance at these three stations, respectively. The communalities were the poorest for K^+ (0.59-0.69) and H^+ (0.63-0.78), and the best for Cl^- (0.99-1.00) and Ca^{2+} (0.84-0.95).

In the marine station Utö, the first factor correlated with marine salt ions (Cl^-, Na^+, Mg^{2+}, and K^+) (Fig.3). The second factor has typical anthropogenic ions (SO_4^{2-}, NH_4^+, NO_3^-, and Ca^{2+}) and a slight correlation with H^+. The third factor is an acidity factor which correlated positively with NO_3^- and negatively with Ca^{2+}.

Table 1. Matrix correlation coefficients of log-transformed concentrations at a 99 % significance level.

a) Utö (87 cases)

	H+	SO$_4$$^{2-}$	NO$_3$-	NH$_4$+	Ca^{2+}	K+	Mg^{2+}	Cl-	Na+
H+	1,000		,460	,323					
SO$_4$$^{2-}$		1,000	,682	,860	,640	,469	,342		
NO$_3$-	,460	,682	1,000	,764	,555	,449			
NH$_4$+	,323	,860	,764	1,000	,493	,450			
Ca^{2+}		,640	,555	,493	1,000	,686	,576	,441	,449
K+		,469	,449	,450	,686	1,000	,586	,679	,699
Mg^{2+}		,342			,576	,586	1,000	,909	,902
Cl-					,441	,679	,909	1,000	,992
Na+					,449	,699	,902	,992	1,000

b) Virolahti (126 cases)

	H+	SO$_4$$^{2-}$	NO$_3$-	NH$_4$+	Ca^{2+}	K+	Mg^{2+}	Cl-	Na+
H+	1,000		,302	,237					
SO$_4$$^{2-}$		1,000	,841	,799	,867	,650	,739	,485	,264
NO$_3$-	,302	,841	1,000	,786	,789	,636	,684	,538	,352
NH$_4$+	,237	,799	,786	1,000	,652	,422	,498	,203	
Ca^{2+}		,867	,789	,652	1,000	,760	,840	,551	,293
K+		,650	,636	,422	,760	1,000	,789	,709	,583
Mg^{2+}		,739	,684	,498	,840	,789	1,000	,816	,614
Cl-		,485	,538	,203	,551	,709	,816	1,000	,688
Na+		,264	,352		,293	,583	,614	,688	1,000

c) Ähtäri (153 cases)

	H+	SO$_4$$^{2-}$	NO$_3$-	NH$_4$+	Ca^{2+}	K+	Mg^{2+}	Cl-	Na+
H+	1,000	,685	,750	,486	,222	,222			
SO$_4$$^{2-}$	,685	1,000	,731	,827	,651	,604	,408	,231	
NO$_3$-	,750	,731	1,000	,718	,527	,361	,294	,292	
NH$_4$+	,486	,827	,718	1,000	,548	,440	,381	,292	
Ca^{2+}	,222	,651	,527	,548	1,000	,715	,713	,312	,340
K+	,222	,604	,361	,440	,715	1,000	,600	,413	,483
Mg^{2+}		,408	,294	,381	,713	,600	1,000	,676	,660
Cl-		,231	,292	,292	,312	,413	,676	1,000	,821
Na+					,340	,483	,660	,821	1,000

In Virolahti, the first factor correlated with typical anthropogenic products and also with particulate cations (Ca^{2+}, Mg^{2+}, and K+). The second factor reflected marine or road salt ions. The third factor explained the H+ variance by correlating positively with NO$_3$- and negatively by Ca^{2+} and Mg^{2+}.

In Ähtäri, the major anthropogenic group (factor 1) included also H+. The second factor was characterized by terrigeneous particulates (Ca^{2+}, Mg^{2+}, and K+) while the third factor had high correlations with sea or road salt (Cl-, Na+).

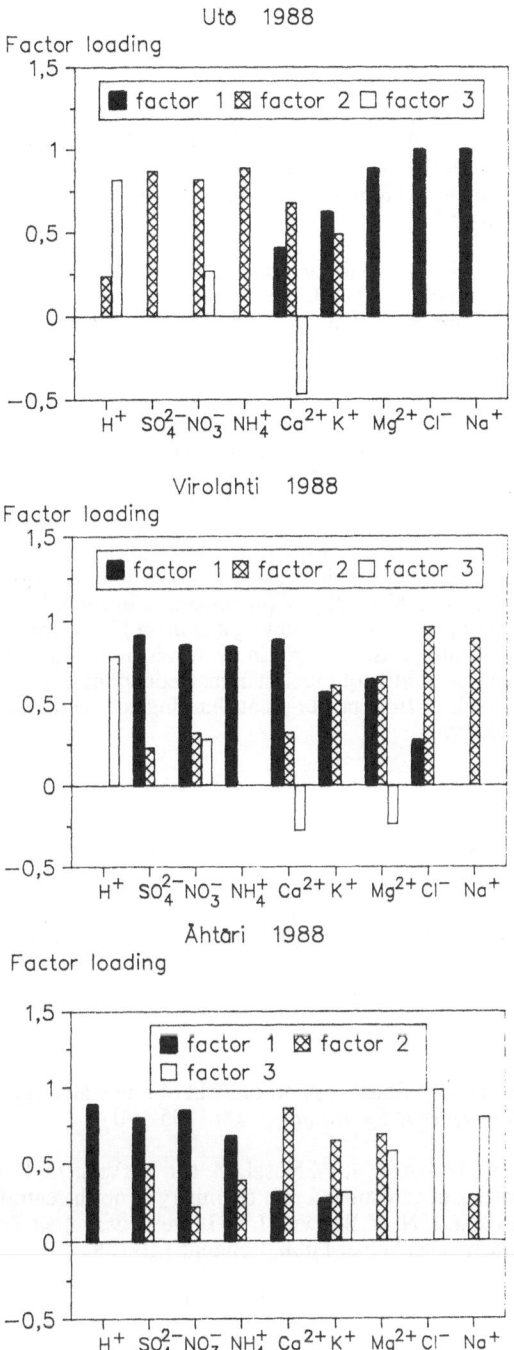

Fig. 3. Varimax rotated factor loadings for concentrations in Utö, Virolahti, and Ähtäri.

4.CONCLUSIONS

The chemical composition of daily wet deposition observations made at three Finnish EMEP stations during one year were studied by correlation and factor analysis. Four kinds of factors were found: (1) Sea or road salt (Cl^-, Na^+, Mg^{2+}, K^+), (2) Terrigeneous (Ca^{2+}, K^+, Mg^{2+}), (3) Main anthropogenic (SO_4^{2-}, NH_4^+, NO_3^-). These anthropogenic ions were observed in Ähtäri with H^+ but in Virolahti and Utö they were connected with Ca^{2+}. (4) Independent H^+ factor which was found in Utö and Virolahti was correlated positively with NO_3^- but negatively with Ca^{2+}.

These factors and their interpretation are very similar to those observed by Ezcurra et al.(3) and to those of Gatz (1). In the eastern US the acid (H^+) factor was correlated with SO_4^{2-} and not NO_3^-, but the neutralization is much lower there.

The marine component was most important at the island station, where it accounted for 50 % of the total variance. In the coastal area, sea or road salt explained 18 % and in the inland station 9 % of the total variance. In coastal and inland areas K^+ and Mg^{2+} are mainly of terrigeneous origin while the sea is the most important source at the island station.

Sulphate is very often neutralized, especially in the most southern parts of Finland, by soil particulates (Ca^{2+}, K^+, Mg^{2+}) or NH_4^+. In this area acidity has not a simple linear relation to sulphate deposition. This stresses the emission reductions of NO_x because nitrate is more clearly related to the acidity of precipitation. In central parts of Finland, where the overall neutralization of precipitation is lower, it is more likely that lower SO_2 emissions will lead to lower H^+ concentrations. Changes in the levels of neutralizing cations (especially NH_4^+ and Ca^{2+}) will change acidity of precipitation. Reductions of particulate emissions, for example from unpaved roads or from power plants burning coal or oil shale, will probably increase precipitation acidity.

ACKNOWLEDGEMENTS

I am grateful to Dr. Sylvain Joffre and Dr. John Ogren for the critical reading of the manuscript and their valuable comments. This work was funded by the Acidification Research Project (HAPRO) of the Ministry of Agriculture and Forestry and the Ministry of the Environment.

REFERENCES

(1)   Gatz, D. F. (1984). Source apportionment of rain water impurities in central Illinois. *Atmospheric Environment,* **18,** 1895-1904.

(2)   Krupa, S. V., Lodge Jr., J. P., Nosal, M. and McVehil G. E. (1987). Characteristics of aerosol and rain chemistry in north central USA. In: R. Perry, R. M. Harrison, J. N. B. Bell and J. N. Lester (eds.): *Acid Rain: Scientific and technical advances.* Selper Publ., London, 121-128.

(3)   Ezcurra, A., Casado, H., Lacaux, J. P. and Garcia, C. (1988). Relationships between meteorological situations and acid rain in Spanish Basque country. *Atmospheric Environment,* **22,** 2779-2786.

NITROUS ACID AT MAINZ: OBSERVATION AND IMPLICATION FOR ITS FORMATION MECHANISM.

G. Lammel*, D. Perner and P. Warneck

Max Planck Institut für Chemie, Saarstrasse 23, 6500 Mainz
* KfK Karlsruhe, Laboratorium für Aerosolphysik und Filtertechnik,
Postfach 3640, 7500 Karlsruhe

Summary

The formation of HNO_2 was found to be first order in NO_2 and its average formation rate was 0.3 % h^{-1} in Mainz. This data can be explained by heterogeneous reactions of NO_2 with the ground surface using a rate constant of 3.7×10^{-8} ppm^{-1} min^{-1} m for pyrex or stainless steel surfaces and a surface/volume ratio of 0.2 m^{-1}. For the estimate a mixing layer of 50 m was assumed together with multiplying the ground surface by 10 to account for structures like buildings. HNO_2 formation rates as found at various places show variations from 0-2.7 % h^{-1}. A reaction at the ground requires the largest rate constants ever observed in the laboratory.
As high values usually are observed during strong haze, the possibility of an aerosol involvement in the heterogeneous reaction must be considered. In Mainz no correlation between the HNO_2 formation rate and the elemental composition of the aerosol was found. However, certain conditions the aerosol surface could be a source for HNO_2. In this case a large "inner" aerosol surface and/or a higher reactivity of "activated" surfaces are necessary.

Whereas emission control strategies in Germany have succeeded in an abatement of atmospheric SO_2 levels, the situation for NO_x ($NO + NO_2$) remains unsatisfactory. The increased consumption in urban areas of fossil fuels, especially by motor vehicles, causes the NO_2 mixing ratios to remain constant or even to increase slightly. Therefore, the fate of NO_x and the effects of the immediate reaction products, such as HNO_2, still require considerable attention.

Already during the first spectroscopic detection of HNO_2 [Perner and Platt, 1979] in ambient air its heterogeneous formation was suspected. Subsequently many investigations have confirmed this presumption. The interaction of NO_x with surfaces leads ultimately to nitrogen oxyacids, especially gaseous nitrous acid, from field [Platt, 1986] and laboratory studies. The basic route of formation was found to be

$$NO_2 + NO_2 + H_2O = HNO_{2g} + HNO_{3ad}$$

Other potential sources for HNO_2 are combustion processes. Yet fossil fuel burning in large furnaces has shown not to yield HNO_2 so far. The observation of HNO_2 was reported recently during biomass burning in the savanna by Rondon and Sanhueza, 1989. Internal combustion engines are the only known appreciable source. The exhaust of Otto engines, when operated in the normal mode and without catalyst, contained under high load conditions <0.01% of HNO_2 with respect to total NO_x emitted. In Diesel engine exhaust HNO_2 made up 1 to 3% of total NO_x. In West Germany, the emission from traffic with a typical mix of combustion engines contains no more than

0.2% of HNO₂/NOₓ [Perner et al., 1987]. As a consequence the often observed HNO₂/NOₓ in excess of 0.2% is definitely due to secondary formation.

The hydrolysis of pernitric acid [Lammel et al., 1989] and similarly that of peroxyacetylnitrate [Grosjean et al., 1984] leads to HNO₂ at high pH. Such events may occur in photochemical active air masses.

It appears that direct emissions are too small and homogeneous gas reactions are too slow to explain the observed HNO₂ formation rates [Kessler, 1984]. Therefore special attention was given to the possibility of aerosols being involved in the formation process [Lammel, 1988].

Experimental Results

During 1986 and 1987 measurements by means of long path differential optical absorption spectroscopy (DOAS) have been performed in Mainz to investigate the process of HNO₂ formation. The HNO₂ mixing ratios were highly variable and sometimes HNO₂ was below the detection limit especially during intensive air exchange. In Figure 1 typical nighttime profiles for a number of trace gases is shown.

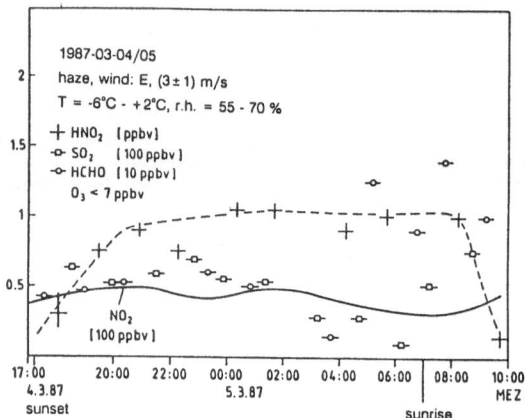

Figure 1: Nighttime profile of atmospheric trace gases

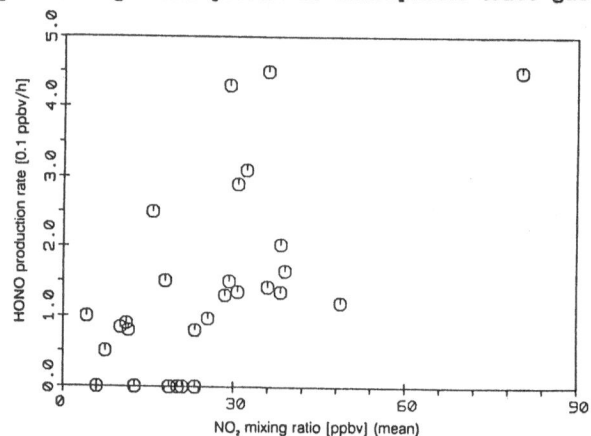

Figure 2: Dependence of nitrous acid formation rate on NO₂-mixing ratio

The HNO₂ formation rate shows a first order correlation with NO₂ (Figure 2). However, the correlation of the maximum HNO₂ mixing ratio with NO₂ is much better (Figure 3). This confirms a previous report by Kessler (1984).
The average conversion rate for NO₂ to HNO₂ was 0.3 % h⁻¹· This value compares with the previous value of 0.6 in Jülich [Kessler and Platt, 1984].

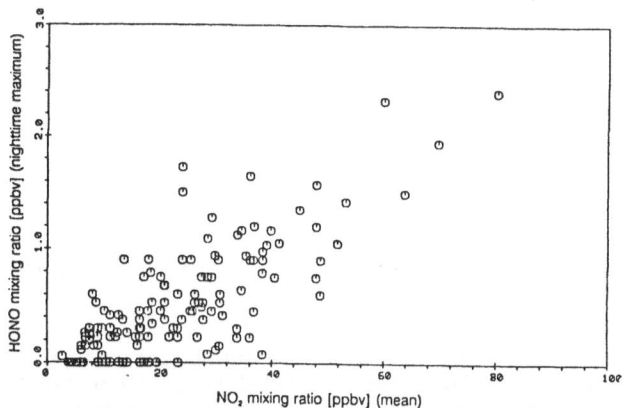

Figure 3: Dependence of maximum nitrous acid mixing
ratio on NO₂-mixing ratio

The correlation with H₂O is not so obvious Figure 4. This also agrees with the observations by Kessler (1984).

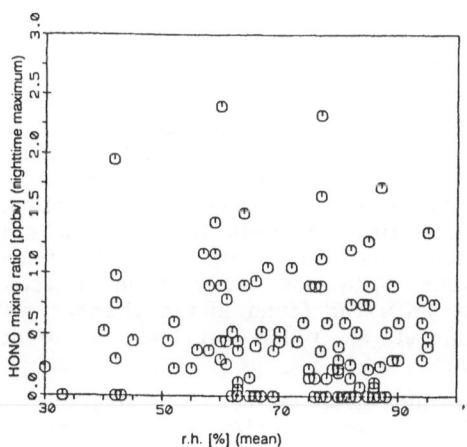

Figure 4: Dependence of maximum nitrous acid mixing ratio
on relative humidity

Some of the higher conversion rates are given in Table 1 and are compared with those from other areas.

Table 1: Rates of HNO_2 formation from growth of $[HNO_2]$ in field observations after sunset.

No.	Place	Date	Temp.	H_2O	mixing height[*]	conversion rate[**]
			K	ppm	m	% h^{-1}
1	Mainz	12.12.86	277	6500	50	0.66 (a)
2	Mainz	4.3.87	272	3600	50	0.69 (a)
3	Mainz	31.3.87	278	6500	50	1.7 (a)
4	Los Angeles	7.8.80	295	14000	50	0.54 (b)
5	Montelibretti	22.9.88	295	23000	50	1.3 (c)
6	Israel	17.5.88	295	18000	100	2.7 (d)

[*] estimated
[**] $100 \times [HNO_2]/(2[HNO_2] + [NO_2])$
(a) Lammel, 1988; (b) Harris et al., 1982; (c) Perner et al., 1989; (d) Perner et al., unpublished results

Aerosol particles were collected and the anion composition of its soluble fraction analysed. In addition, the elemental composition was determined by x-ray fluorescence for 20 samples. No correlation of aerosol composition with maximum nighttime HNO_2 levels was observed. High rates of HNO_2 formation sometimes occurred when the content of Cr and Ti was low.
Whenever nitrite was detected on the aerosol, its concentration exceeded the value given by the gas-liquid solution equilibrium. Consequently, the aerosol must then have been a source for HNO_2. However, its source strength could not be evaluated.

Formation Reactions.

The rate of NO_2 disappearance is found to be first order in NO_2 and in H_2O in the majority of laboratory reports over the full range of atmospheric mixing ratios [Sakamaki et al., 1983; Pitts et al., 1984; Svensson et al., 1987; Jenkin et al., 1988]. The initial rate of gas phase HNO_2 formation is at most 50 % of the NO_2 disappearance rate. A persistent production of NO, although small compared with NO_2 disappearance, is observed [England and Corcoran, 1974; Sakamaki et al., 1983; Pitts et al., 1984; Jenkin et al., 1988]. NO_2 is produced as well in those systems. Though its formation is not so obvious, with prolonged reaction times it leads to smaller NO_2 disappearance rates and in the end to a maximum in HNO_2 mixing raio [Jenkin et al.].
The associated gaseous HNO_3 has never been detected [Jenkin et al.]. Moreover, the missing HNO_3 was found at the chamber walls by De Santis et al., 1987 and Jenkin et al. which provides additional evidence for a heterogeneous process.

In denuders the stoichiometric factor for conversion of NO_2 into nitrite sometimes is one. This is common with gualacol-coated denuders used for ambient NO_2 detection [Buttini et al., 1987]. However it also occurs on sodium carbonate coatings at ambient NO_2 levels [De Santis et al.]. If the common cause were some organic reducing agent taking part in the heterogeneous reaction such a process would very likely occur on ambient surfaces as well.

The apparent second order rate constants in ppm^{-1} min^{-1} as given by most

$$d[NO_2]/dt = 2\ k_1\ [NO_2][H_2O]; \qquad d[HNO_2]/dt = k_2\ [NO_2][H_2O]$$

authors from their laboratory results for NO_2 disappearance or HNO_2 formation have to be modified according to the surface to volume ratio of their containment, S/V m^{-1}. The dimension of k_{2het} is then ppm^{-1} min^{-1} m. Table 2 shows laboratory rate constants as given in the literature.

Table 2

Temp K	S/V m^{-1}	wall	k_1 ppm^{-1} min^{-1}	k_2 ppm^{-1} min^{-1}	k_{2het} $ppm^{-1}min^{-1}$ m	
303	3.7	Tefl.	2.5×10^{-8}	1.0×10^{-8}	2.7×10^{-9}	Sakamaki et al.,1983
300	3.4	Tefl.	8.0×10^{-9}	–	2.3×10^{-9}	Pitts et al., 1984
295	2	Poly-ethene	–	–	1.4×10^{-8}	Kessler, 1984
280	14	Tefl.	8.8×10^{-8}	1.3×10^{-7}	$9.3 \times 10^{-9*}$	Svensson et al.,1987
295	14	Tefl.	4.1×10^{-8}	4.2×10^{-8}	$3.0 \times 10^{-9*}$	"
323	14	Tefl.	3.3×10^{-8}	1.2×10^{-8}	$8.6 \times 10^{-10*}$	"
295	14	Glass rins.	5.3×10^{-6}	2.8×10^{-6}	2.0×10^{-7}	"
295	14	Stainl. Steel	4.6×10^{-7}	5.4×10^{-7}	3.9×10^{-8}	"
296	13	Pyrex	5.3×10^{-7}	4.8×10^{-7}	3.7×10^{-8}	Jenkin et al., 1988
303	6.9	Tefl.	2.8×10^{-8}	1.4×10^{-9}	1.4×10^{-9}	Akimoto et. al. 1987
303	6.9	Tefl.	–	4.3×10^{-9}	4.3×10^{-9}	"

* passivated surface

The large variation of k_{2het} is mostly due to variation in the surface activity. Remarkable is also the negative temperature dependence of k_{2het} which seems to be somewhat at variance with field observations where the highest mixing ratios are observed in summer.

Destruction

HNO_2 reaches a plateau value (Fig. 1) which implies a limited lifetime for HNO_2. Jenkin et al., found 30 min lifetime in their cell which amounts to $k_{het} = 2.3 \times 10^{-3}$ min^{-1} m. The mechanism should be:

$$HNO_{2g} + HNO_{3ad} = 2\ NO_2 + H_2O$$

because of complete mass balance. The small production of NO is probably due to heterogeneous reaction

$$HNO_2 + HNO_2 = NO + NO_2 + H_2O$$

Discussion

Observation of HNO_2 during heavy rain in Köln [Kessler, 1984] indicated a formation on dry surfaces of buildings. In such a case when the ground is involved, the apparent rate of formation for HNO_2 depends on the height of the mixing layer. In Table 1 the mixing height for most field measurements is estimated to be 50 m which seems reasonable for a nighttime stratified airmass. In addition the lightbeam ran in most experiments 15 to 40 m above the surface. The apparent surface to volume ratio then reaches 0.02 m^{-1}. In

Israel where the lightbeam averaged 100 m above the ground 0.01 m^{-1} is assumed. Structures at the ground increase the actual surface area compared with that of a flat surface by an assumed factor of 10. The resulting S/V ratios are given in Table 3.

Table 3: Rates of HNO_2 formation (% h^{-1}): Comparison of measurements data (Table 1) with calculations based on heterogeneous laboratory reaction rate constants (Table 2) and the ground surface.

No.	S/V m^{-1}	measured	calculated % h^{-1} a	calculated % h^{-1} b	calculated % h^{-1} c
1	0.2	0.66	0.024	0.29	1.6
2	0.2	0.69	0.01	0.16	0.8
3	0.2	1.7	0.02	0.29	1.6
4	0.2	0.54	0.05	0.58	3.2
5	0.2	1.3	0.08	1.0	5.6
6	0.1	2.7	0.04	0.44	2.4

a 3.9x10^{-9} ppm^{-1} min^{-1} m, Svensson et al., 1987, passivated surface
b 3.7x10^{-8} ppm^{-1} min^{-1} m, Jenkin et al., 1987
c 2.0x10^{-7} ppm^{-1} min^{-1} m, Svensson et al., 1987

The calculated rates for a passivated surface (column a) are too small. The rate constant for stainless steel or Pyrex (column b) describes only some of the observations. Yet the highest value, for rinsed glass, reported so far is able to describe also the largest formation rates found in Mainz and Israel, experiments number 3 and 6, respectively.

What might be the role of the aerosol then? For comparison, a clean air aerosol has a S/V of 10^{-4} m^{-1} and even in very polluted air with a visibility of about 1 km S/V is only 10^{-2} m^{-1}, which seems to exclude a major aerosol effect.
However, Kessler, 1984 has found a correlation between HNO_2 formation rate and visibility. When he changed the height of the light beam above ground he observed the same HNO_2 mixing ratios which also contradicts a flux of HNO_2 from the ground. Again the highest HNO_2 production rate observed was observed under conditions of low visibility and thus high aerosol load (Table 1, No. 6).
In this study the light path was 13 km and for the assumed S/V= 0.001 m^{-1} a heterogeneous rate constant of k_{het} = 2.5x10^{-5} ppm^{-1} min^{-1} m would be required to describe the high conversion rate. This is two orders of magnitude larger than the fastest reported value (Table 2). Though, in principle, larger rate constants – Takagi et al., 1986 reported 5.2x10^{-6} ppm^{-1} min^{-1} m for the formation of methylnitrite from NO_2 and CH_3OH – cannot be ruled out they seem first of all improbable for the HNO_2 formation.
Yet, shape and composition of the aerosol particles is extremely variable. The inner surface may sometimes be much larger than given by the outside geometry and a sponge–like structure can have a surface area by orders of magnitude larger. Secondly the surface may be highly activated especially for charge–transfer type reactions by adsorbed ions. Those conditions occasionally may combine leading to those surprisingly high conversion rates as observed in the field. This scheme explains also the small HNO_2 mixing ratios observed at high relative humidities when condensation blocks the large inner surface.

Mechanism

The mechanism most in line with all the observations especially also including isotope experiments [Sakamaki et al., 1983; Svensson et al., 1987] is an adsorbed NO₂ molecule of nitrosylnitrate structure reacting with gaseous water.

Conclusion

Though under most circumstances the surface of the ground may be responsible for HNO₂ formation, some results from field studies definitively require a participation of the aerosol. Under those circumstances only highly structured particles, exceeding the outside surface area, combined with a higher surface reactivity than found in the laboratory, can provide the necessary conversion rates.

Until now, mostly the morning pulse of OH from the photolysis of HNO₂ accumulated at nighttime was considered in the OH budget. Yet the observed high conversion rates of NO₂ to HNO₂ imply an appreciable and continuing OH-radical source during the day. Under suitable conditions very large OH concentrations close to the producing surfaces are expected. These conversion rates may be even underestimated if a photolytic enhancement occurs similar to that recently observed by Akimoto et al., (1987).

Literature

Akimoto, H., H. Tagaki, and F. Sakamaki, Photoenhancement of the nitrous
acid formation in the surface reaction of nitrogen dioxide and water vapor: Extra radical source in smog chamber experiments, Int. J. Chem. Kinetics, 19, 539-551, 1987

Buttini, P., V. Di Palo, and M. Possanzini, Coupling of denuder and ion
chromatogrphic techniques for NO₂ determination in air, Sci. Total Environ., 61, 59-72, 1987

De Santis, F., A. Febo, and C. Perrino, Nitrite and nitrate formation on a
sodium carbonate layer in the presence of nitrogen dioxide, Ann. Chimica, 763-768,1987

England, C., and W.H. Corcoran, Kinetics and mechanisms of the gas-phase
reaction of water vapor and nitrogen dioxide, Ind. Eng. Chem. Fundam., 13, 373-384, 1974

Grosjean, D., K. Fung, J. Collins, J. Harrison, and E. Breitung, Portable
generator for on-site calibration of peroxyacetyl nitrate analysers, Anal. Chem., 56, 569-573, 1984

Harris, G.W., W.P.L. Carter, A.M. Winer, J.N. Pitts, U. Platt, and D. Perner,
Observations of nitrous acid in the Los Angeles atmosphere and implications for the predictions of ozone-precursor relationships, Environ. Sci. Technol, 16, 414-419, 1982

Jenkin, M.E., R.A. Cox, and D.J. Williams, Laboratory studies of the kinetics
of formation of nitrous acid from the thermal reaction of nitrogen dioxide and water vapor, Atmosph. Environm., 22,487-498, 1988

Kessler, C., Dissertation, Gasförmige Salpetrige Säure in der belasteten
  Atmosphäre, Köln, 1984

Kessler, C., und U. Platt, Nitrous acid in polluted air masses – sources and
  formation pathways, CEC, Proc. 3rd Europ. Symp. Physico-Chem.
  Behaviour Atmos. Poll., Varese, Italy, 10-12 April, pp. 412-422, 1984

Lammel, G., Dissertation, Salpetrige Säure und Nitrit im Mehrphasensystem
  der belasteten Atmosphäre, Mainz, 1988

Lammel, G., D. Perner, and P. Warneck, On the decomposition of pernitric
  acid in aqueous solution, submitted J. Phys. Chem., 1989

Perner, D., C. Kessler, and U. Platt, HNO₂, NO₂ and NO measurements in
  automobile engine exhaust by optical absorption. Proc. Int. Symposium,
  Monitoring of gaseous pollutants by tunable diode lasers, Freiburg,
  13-14 Nov. 1986, CEC, Eds. R. Grisar, H. Preier, G. Schmidtke, G.
  Restelli, Reidel Publ. Co. 1987

Perner, D., U. Parchatka, H.-J. Karbach, and I.C. Eslick, Determination of
  atmospheric trace gases by long path absorption measurements, Report
  on Field Intercomparison exercise on nitric acid and nitrate
  measurement, Area della Ricerca di Roma, CNR, Montelibretti, CEC
  Project Cost 611, 1989

Pitts, J.N., S. Sanhueza, R. Atkinson, W.P.L. Carter, A.M. Winer, G.W. Harris,
  and C.N. Plum, An investigation of the dark formation of nitrous acid in
  environmental chambers, Int. J. Chem. Kinetics., 16, 919-939, 1984

Platt, U., The origin of nitrous and nitric acid in the atmosphere, NATO ASI
  Series, Vol. G6, Chem. of multiphase atmospheric systems, Ed. W.
  Jaeschke, Springer Verlag, 1986

Rondon, A., and E. Sanhueza, High HNO₂ atmospheric concentrations during
  vegetation burning in the tropical savannah, Tellus, 41B, 474-477, 1989

Sakamaki, S,. S. Hatakeyama, and H. Akimoto, Formation of nitrous acid and
  nitric oxide in the heterogeneous dark reaction of nitrogen dioxide and
  water vapor in a smog chamber, Int. J. Chem. Kinetics., 15, 1013-1029,
  1983

Svensson, R., E. Ljungström, and O. Lindquist, Kinetics of the reaction

INTERREGIONAL MODELING OF AIR POLLUTION IN PORTUGAL: PRELIMINARY RESULTS

C.BORREGO, M.COUTINHO and J.RUA
Departamento de Ambiente e Ordenamento
Universidade de Aveiro, 3800 AVEIRO, Portugal

Summary

This paper presents the results of the preliminary stage of an interregional modeling plan for atmospheric pollutants over the continental territory of Portugal. This project involves a coordinated study of several models and will provide an important tool to evaluate the effects of the introduction of new industrial sources and future emission control strategies.
These preliminary calculations were extremely important since the main obstacle to air pollution modeling was surpassed: the build up of the meteorological and emission data bases needed for the model input.
A comparison of the simulation results with the scarce air quality data is tried in this paper. Daily average concentration isolines and the study of temporal evolution of concentration for some particular points might give some information on the validity of the results.

1. INTRODUCTION

The first air pollution modeling work at an interregional level over Portugal was tried through the application of the MesoPort model (1). The MesoPort model is a Gaussian variable trajectory puff superposition model developed through the adaptation of the Mesopac meteorological processor and the Mesopuff II dispersion model (2) to the portuguese case. This model is suitable for modeling the transport, diffusion and removal of air pollutants from multiple point and area sources at transport distances beyond the range of conventional straight-line Gaussian plume models. Its meteorological pre--processor creates fields of wind components, mixing height and stability class from twice-daily radiosonde data. MesoPort accommodates up to 5 pollutants: SO_2, $SO_4^=$, NO_x, HNO_3 and NO_3^-

These results represent the first phase of a wider modeling project (3) and will be used as a reference for the implementation of a more complex model which includes photochemical simulations: the ArPort model. The ArPort model will be adapted to the portuguese territory and data bases, and structured over the Airshed Model of Systems Applications, Inc.(4). This model, as developed by SAI, is an episodic eulerian long range transport model of medium complexity, specially designed for the simulation of non-linear chemical processes such as those occuring during photochemical episodes.

2. DESCRIPTION OF THE STUDY
An important feature that has to be considered before the implementation of an interregional model is the large amount and detail of the input data required. Collection and preparation of these input data might become extremely critical in countries with develloping environmental policies and incipient monitoring networks.

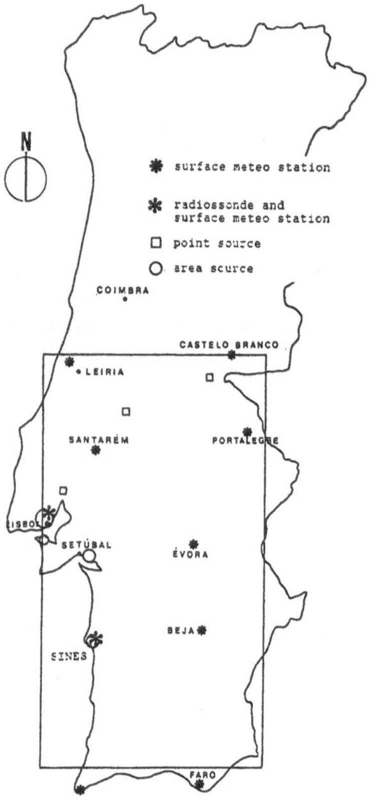

The modeling area was choosen taking into account the geographical distribution of the main industrial sources, the topographical features of the portuguese territory and the spatial availability of radiosonde data. The southern region of Portugal, with an area of about 350x225 km^2, was choosen due to its rather flat terrain, being the location place of important industrial sources. Three different episodes of consecutive days in February, October and November 1985 were studied, simulating the behaviour of the SO$_2$ and NO$_x$. A radiosonde measurement campaign was accomplished during August 1988 to cover the lack of information on Summer weather conditions.

Emission and air quality data was provided by the Directorate General of Environmental Quality (DGQA). Information on the 1985 yearly average emissions of SO$_2$, NO$_x$ and VOC, calculated through emissions factors, were available. Air quality data consisted of ground-level SO$_2$ and NO$_x$ concentration measurements collected by DGQA at the Sines region.

As a result of this work several data-bases were build up:
- 10 meteorological surface stations.
- 2 radiossonde stations.
- emissions of 4 thermal-power plants, 5 chemical plants, 3 paper-mill plants, 1 oil refinery, 1 steel plant and the traffic of Lisbon.
- land cover information.
The collection of this data was important, as the first attempt for a coordination between the national institutions

Fig.1 Modeling area

* surface meteo station
* radiossonde and surface meteo station
□ point source
○ area source

COIMBRA
CASTELO BRANCO
LEIRIA
SANTARÉM
PORTALEGRE
LISBOA
SETÚBAL
ÉVORA
BEJA
SINES
FARO

responsible for the meteorological and air quality planning.

3. **SIMULATION RESULTS**
Output obtained by the MesoPort model consists of hourly average concentrations for 12.5x12.5 km² grid points over the study area. In order to minimize the amount of data to analyse, daily average concentration isolines were drawn (5). This paper presents the results for two consecutive days episodes in October and November 1985.

October - the meteorological situation over the modeling region during this period was characterized by fair weather, with light wind from Northeast during daytime and North during the night. Ground-level hourly average SO_2 concentrations show peaks between 50 and 60 μg.m⁻³. During the night, simulation of atmospheric stability drops down the concentrations values. Figure 2 shows SO_2 daily average simulation results for the October episode.

November - similar to October weather situation with slightly stronger North-Northwest winds. Weather maps indicate daytime temperatures around 13-15°C and around 5-7°C during the night. Pollutant plumes are clearly evident about 70 to 90 km south and southeast of the emission areas, with atmospheric impact at a regional level. Ground-level hourly average SO_2 concentrations show maxima around 30-40 μg.m⁻³. Daily average SO_2 concentrations isolines are drawn in figure 4.

4. **EVALUATION ATTEMPT**
Portuguese air quality network is quite incipient, almost restricted to measuring stations located in the vicinities or in the interior of industrial complexes. The air quality data used in this study was obtained by the Sines network. These semi-automatic measuring stations are located close to important point sources and strongly influenced by their emissions. In fact, calculated concentrations fail to reproduce measured values, reflecting the difficulty of comparing point measurements with calculated box averages (6).

Comparison of temporal evolution for the calculated and measured concentrations was also tried for both case studies. Apparently, the meteorological pre-processor overestimates the radiation effects on the atmospheric stability, leading to extremely low concentrations during the night. This effect is particularly important during the October episode, as shown in figure 3. Air quality data considered in this comparison is the maximum concentration of 3 air quality measuring station located around Sines (7).

As a result of spatial averaging, modeling calculations underestimate SO_2 ground-level concentrations in the Sines region. Figure 5 shows observed and calculated concentration evolution for a single station (Monte Chãos). Moreover, daily average concentrations show small agreement with the measured values for the complete set of days studied.

Air quality data from the Barreiro-Seixal network (in the south shore of the Tejo river, in front of Lisbon) will be available, for further evaluation studies. Anyway, this set of measuring stations will probably show the same problems than the one used in this paper.

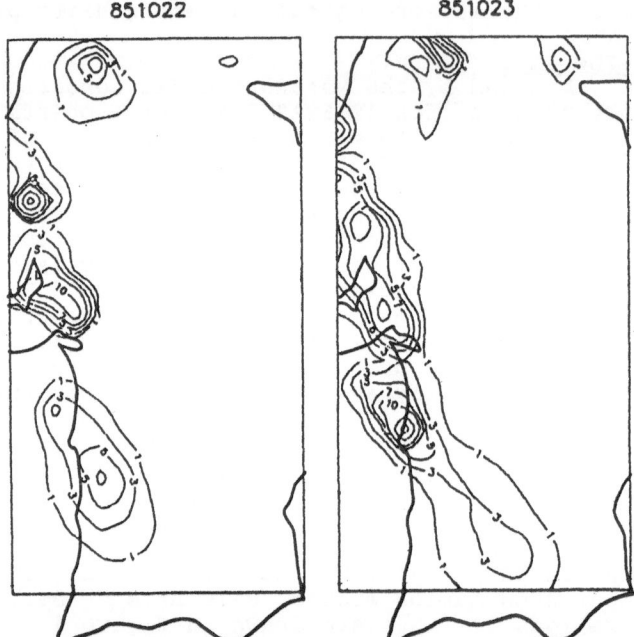

Fig.2 SO₂ ground-level daily average concentrations in µg/m³
for 22 and 23 October 1985. Isolines are drawn over the
portuguese coast and boarder lines.

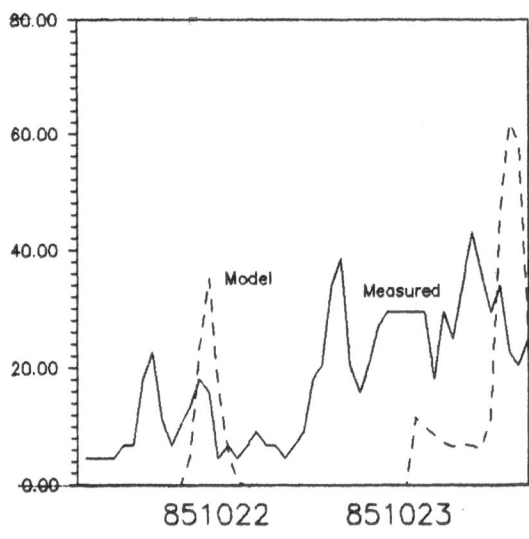

Fig.3 Computed and observed SO₂ concentrations (µg/m³) at the
Sines station, for October episode.

851120   851121

Fig.4 SO₂ ground-level daily average concentrations in µg/m³
for 20 and 21 November 1985. Isolines are drawn over the
portuguese coast and boarder lines.

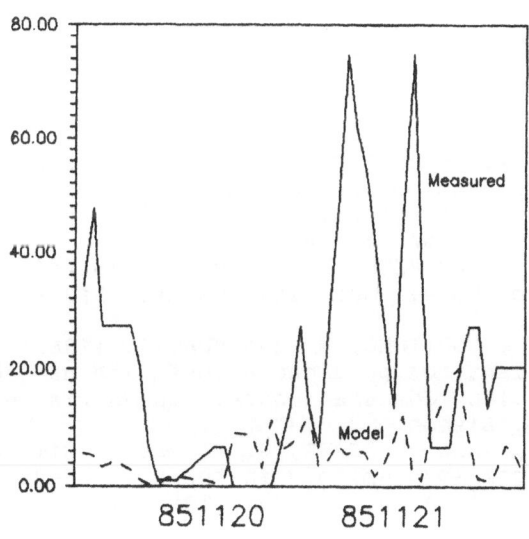

Fig.5 Computed and observed SO₂ concentrations (µg/m³) at the
Sines station, for November episode.

5. CONCLUSIONS

This paper includes the SO_2 dispersion modeling results over the Southern region of Portugal, using the MesoPort model. An evaluation attempt, comparing these results with the air quality data available is made. Due to the scarcity of air quality data no final conclusions should be drawn from the results of this first evaluation.

Source plume fusion appears to be an important phenomenum for the understanding of certain simulation results. Care should be drawn on the concentration maxima analysis, because some of the peaks might appear over zero-wind grid points as a result of the wind interpolation scheme. This problem will be considered in future applications of the model.

Problems faced during the evaluation of the modeling results, stress the need of creating an efficient air quality network in Portugal. Air quality stations should also be located in rural areas, where concentrations are independent from local effects. The lack of this environmental data can be a strong limitation to the application of atmospheric pollutants dispersion models.

ACKNOWLEDGEMENTS

The authors wish to express their appreciation to Dr. Renato de Carvalho of the National Institute of Meteorology and Geophysics. This study was sponsored by the Directorate General of Environmental Quality of the Portuguese Ministry of Planning and Territory Administration.

REFERENCES

(1)   BORREGO, C., RUA, J. and COUTINHO, M. (1988). Avaliação preliminar da poluição atmosférica em Portugal a nível inter-regional - Modelo MesoPort, AMB-QA(4)/88, Dep. Ambiente e Ordenamento, Univ. Aveiro.
(2)   SCIRE, J., WORMANN, F., BASS, A. and HANNA, S. (1984). User's guide to the Mesopuff II model and related processor programs, EPA 600/8-84-013, U.S. Environmental Protection Agency.
(3)   COUTINHO, M., BORREGO, C., RUA, J. and COSTA, M.J. (1989). Application and implementation of an atmospheric pollution interregional model to Portugal, 8th World Clean Air Congress, The Hague.
(4)   REYNOLDS, S. and REID, L. (1979). An introduction to the SAI-Airshed Model and its usage, EF78-53R, Systems Applications Inc..
(5)   BORREGO, C., COUTINHO, M. and RUA, J. (1989). Modelização puff-gaussiana da poluição atmosférica na região Sul de Portugal, 1as Jornadas sobre Indústria e Ambiente, LNETI/DGQA, Lisbon.
(6)   STERN, R., and SCHERER, B. (1982). Simulation of a photochemical smog episode in the Rhine-Rhur area with a three dimensional grid model, 13th ITM on Air pollution modelling and its applications, NATO/CCMS.
(7)   BARROSO, J. (1988), A qualidade do ar na região de Sines: a poluição pelo dióxido de enxofre - 1983-86, 1a Conf. Nacional sobre a qualidade do Ambiente, Univ.Aveiro.

AN ADVANCED MODEL FOR THE DESCRIPTION OF CONVERSION PROCESSES OF
NITROGEN OXIDES IN PLUMES OF LARGE POINT SOURCES

P. BANGE[1], L.H.J.M. JANSSEN[1] and F.T.M. NIEUWSTADT[2]

1 Environmental Research Department, N.V. KEMA, P.O. Box 9035,
6800 ET Arnhem, The Netherlands

2 Laboratory of Aero- and Hydrodynamics, Rotterdamseweg 145,
2628 AL Delft, The Netherlands

Summary

A description is given of a model which is used to predict the
oxidation rate of nitrogen oxides in plumes of large point sources
such as power plants. Results of this model can be used both to
calculate concentrations of NO and NO_2 in the vicinity of the
source and for sub-grid parametrisation of chemical reactions in
meso- or large-scale models. To model the fast non-linear oxidation
reaction of NO, we focus on the dispersion and mixing of the plume
and the chemical reactions occurring simultaneously. The dispersion
for the first 30 – 60 seconds after emission from the source is
described by a jet phase module. For distances up to about 25 km,
dispersion parameters of an instantaneous plume are used which were
derived from SF_6 tracer gas experiments. The effect of
concentration fluctuations on the oxidation rate was also taken into
account. The model is validated by observed airborne NO_x
conversion data. Furthermore a sensitivity analysis was carried out.
The latter showed that the highest uncertainties are caused by
errors in the size of the instantaneous plume. This may explain also
differences between measurements and model calculations.

1. INTRODUCTION
    Nitrogen oxides and their conversion products cause air pollution
problems on several scales. Standards of NO_2 can be exceeded locally in
the vicinity of NO sources. On a larger scale nitrogen oxides contribute
to acid deposition and photochemical air pollution.
    The rate at which NO is converted into the toxic NO_2 in exhaust
plumes is limited by the mixing rate of the plume with its surrounding
air. This surrounding air contains ozone, which is the oxidizing agent in
the chemical reactions. Mixing has to be studied on various scales: not
only the broadening of the plume is of importance, but also its internal
mixing. This effect of small-scale processes influencing large-scale
processes such as photochemical oxidant formation in which nitrogen
oxides participate on a larger – i.e. regional – scale. Builtjes and
Stern [1] showed that O_3 concentrations are underestimated near source
areas when a large scale Eulerian grid model is used to calculate
photo-oxidant formation during photochemical episodes. In these models no
individual plumes are modelled because of the large size of the cells in
which calculations are carried out. Pollutants are assumed to be

uniformly distributed in one cell.

In this paper it will be shown that if the instantaneous (i.e. not time averaged) dispersion parameters of the exhaust plume are known in sufficient detail, a satisfactory estimate of the reaction of NO and ozone to NO_2 may be given. Results of our model may also be used to predict conversion rates inside cells of large grid models to reduce the discrepancies between measurements and model calculations as pointed out by Builtjes and Stern [1].

In general reactive plume modelling starts with the atmospheric diffusion equation for reacting species. For the concentration c_i of species i one can write:

$$\frac{\partial c_i}{\partial t} + \frac{\partial}{\partial x_j} u_j c_i = M\frac{\partial^2 c_i}{\partial x_j \partial x_j} + R_i(c_1,\ldots c_n,T) + S_i(x,t) \tag{1}$$

in which u_i is the velocity in the i-th direction, t is time, M the molecular diffusion coefficient, R_i the amount of species i formed by reactions of all the other constituents, T the temperature and S_i the source term for the species i [2].

A common way to tackle this equation is to apply Reynolds' decomposition. The concentrations c_i and the velocities u_i are split into a mean and a fluctuating part:

$$c_i = \bar{c_i} + c_i' \qquad \text{and} \qquad u_i = \bar{u_i} + u_i' \tag{2}$$

These are substituted into the diffusion equation. Thus we get:

$$\frac{\partial c_i}{\partial t} + \frac{\partial}{\partial x_j} \bar{u_j}\bar{c_i} + \frac{\partial}{\partial x_j} \overline{u_j'c_i'} = M\frac{\partial^2 \bar{c_i}}{\partial x_j \partial x_j} + R_i(\bar{c_1}+c_1',\ldots\bar{c_n}+c_n',T) + S_i(x,t) \tag{3}$$

Defining relations for the newly introduced fluctuation terms (finding a solution for the closure problem) determines part of the difference between various models. These terms may be estimated, ignored, measured under field or laboratory conditions or derived from computer simulations of turbulent flows. If second-order cross terms (with $c_i'x_j'$) are taken into account, we achieve second-order closure.

The following reactions are modelled in all NO oxidation models:

$NO + O_3 \rightarrow NO_2 + O_2$	(at reaction rate k_1)	(4)

and its reverse reaction

$NO_2 + O_2 + u.v.light \rightarrow NO + O_3$	(at reaction rate k_2)	(5)

sometimes extended by:

$2NO + O_2 \rightarrow 2NO_2$	(at reaction rate k_3)	(6)

Our model is based on the models of Varey et al. [4], Richard et al. [5] and Janssen et al. [6]. It is a second order closure model: closure is achieved by the use of dispersion parameters of an instantaneous Gaussian plume which are based on tracer gas dispersion experiments and an estimation of the effect of concentration fluctuations on the basis of wind tunnel data as proposed by Builtjes & Talmon [7]. No chemical equilibrium is assumed, and reactions (4), (5) and (6) are taken into account. Richard et al. [5] used the dispersion parameters of a time-averaged plume and Varey et al. [4] assumed constant diffusion. Neither model accounted for concentration fluctuations or for the initial

– 484 –

phase of plume dispersion.

2. THE MODEL

During the initial phase of plume development we assume that up to a certain time τ mixing is fast and complete due to strong internal turbulence. We assume that together with the effect of higher temperatures, this will cause all ozone that mixes into the plume to react with the NO emitted. See Figure 1.

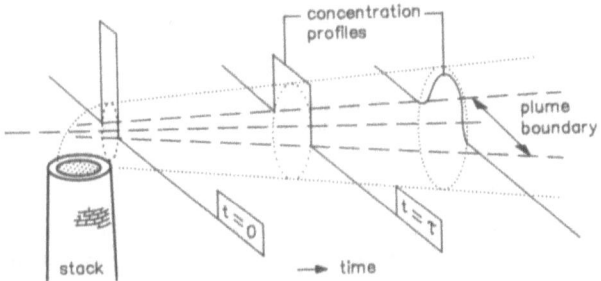

Figure 1. The modelling of the initial phase of plume dispersion. During the jet time τ mixing is assumed to be complete. A flat concentration profile is assumed for a plume width of the 2σ value of the instantaneous plume (see Figure 3).

For time values greater than τ the coupled equations for NO, NO_2 and O_3 are solved using an implicit finite difference method. Radial symmetry of the plume is assumed, so that these equations can be written in cylindrical coordinates. For NO_2 we have:

$$\frac{\partial[NO_2]}{\partial t} = D\{\frac{\partial^2[NO_2]}{\partial r^2} + \frac{1}{r}\frac{\partial[NO_2]}{\partial r}\} + k_1[NO][O_3] + k_3[O_2][NO]^2 - k_2[NO_2] \tag{7}$$

The equations for NO and O_3 are equivalent. The turbulent diffusion coefficient D is assumed to be equal for all components and is only dependent on the distance x from the source. D is chosen in such a way (see [5]) that when a point source is assumed and no jet phase is modelled, a Gaussian plume will result for an inert component. With the introduction of a finite source and a jet phase the NO_x profile will differ from a Gaussian profile at short distances from the source. The instantaneous dispersion parameters which have been used to calculate D will be discussed in the next paragraph.

The effect of concentration fluctuations for time values greater than τ is represented by the second-order cross term arising from Reynolds' decomposition of equation (7); it is modelled according to [7]. See Figure 2.

3. MEASUREMENTS
3.1 Dispersion parameters

Figure 3 shows a log-log plot of data of instantaneous plume dimensions derived from KEMA tracer gas dispersion experiments. SF_6 was released at a height of 100 m, and concentration profile recordings were made with a measuring van at different distances from the source. The plot also shows the best fit to the data and a line drawn according to Pasquill's [8] plume dispersion parameters. The dashed lines are

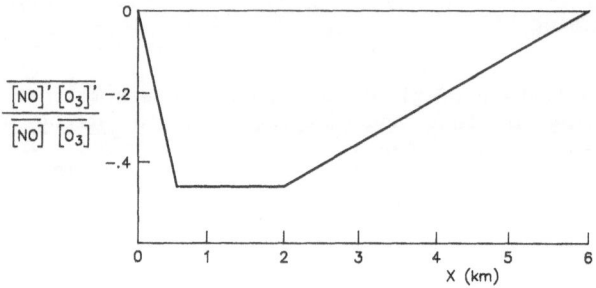

Figure 2. Concentration fluctuations derived from wind tunnel experiments according to [7]. These concentration fluctuations led to an apparent reduction of reaction rate k_1 during the first 6 kilometers of plume dispersion.

approximately the 95% uncertainty limits. At 10 km from the source the instantaneous plume width is about 50% of the (hourly mean) Pasquill width.

Figure 3. Measurements of plume dispersion by using SF_6 tracer gas. A best fit through the measurements and the curve describing the dispersion according to [8] have also been plotted.

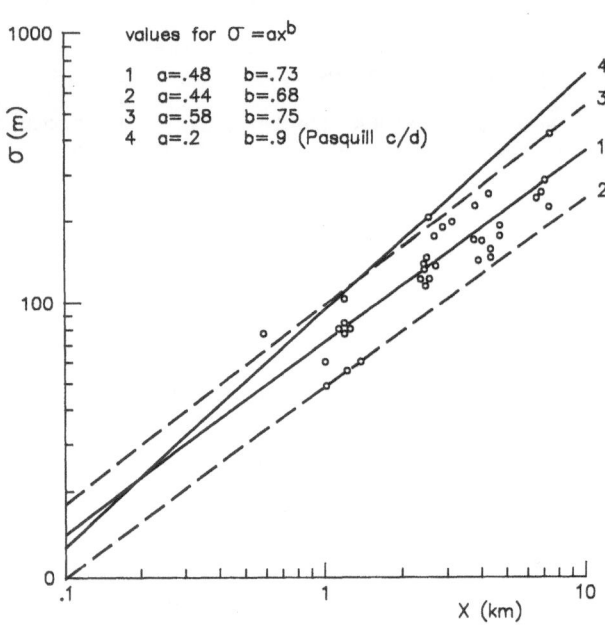

values for $\sigma = ax^b$

1	a=.48	b=.73
2	a=.44	b=.68
3	a=.58	b=.75
4	a=.2	b=.9 (Pasquill c/d)

3.2 NO/NO$_2$ measuring flights in power plant plumes

From 60 flights carried out to measure daytime NO_x conversion in power plant plumes 10 were selected that were made under the following conditions: the season was autumn or spring, wind speed at plume height 5-15 m/s, Pasquill stability class c/d and ozone background concentration 40-60 ppb. Continuous recordings were made of NO, NO_2 and O_3. The plumes were crossed 5 to 10 times at the same distance from the stack. Profiles were integrated and a mean value of the NO_2/NO_x ratio was calculated for various crossing distances. These measured conversion

rates are shown together with results of model calculations in Figures 4
to 7.

4 RESULTS
The input parameters for the model calculations are given in
table I. Lower and upper boundaries are given, which were used in a
sensitivity analysis of the model.

Oxidation rates measured and calculated are plotted as a function of
the distance from the stack. In all figures the oxidation rates
calculated using mean input values are plotted as a continuous line.
Curves with other input data are represented by dashed lines. Figures 4
to 7 show results of the sensitivity analysis.

It appeared that if results of measuring flights were compared with
results of model calculations oxidation rates are generally
underestimated at distances smaller than 5 km and overestimated at larger
distances.

Table I. Input parameters of the model and their values as used
in the sensitivity analysis. Some of the effects are shown in
Figures 4 to 7. Parameters are sorted in descending order of
effects.

Input parameter	input values			effect	
	mean	low	high	(unit)	
Dispersion a	.48	.44	.58	(m)	Figure 4
parameters / b	.66	.61	.68	(-)	
NO_x emission	1000	400	2000	(kg/h)	
Wind speed	10	5	15	(m/s)	Figure 5
Ambient ozone	40	30	50	(ppb)	
Missing of centre of the plume	0	1	2	(sigma)	Figure 7
Temperature (effects on k_1 and k_2)	20	10	30	(°C)	
k_3	.25	.15	.35	(min^{-1})	
Jet phase time	0	30	60	(sec)	
Conc. fluctuations	0	(see Figure 2)			Figure 6

5 CONCLUSIONS
Results of calculations with a computer model on the basis of
instantaneous plume dispersion parameters that are derived from tracer
gas experiments, were compared with results of measured airborne NO_x
conversion rates. A sensitivity analysis was carried out which showed the
following results:
- the largest uncertainties in conversion rates are due to little
knowledge of the instantaneous plume, followed in descending order of
importance by: -changes in the power plant exhaust -changes in wind
velocity - higher measured oxidation rates because the plume is not
crossed through its centre - variations in the ozone background. Other
effects like those caused by the variation in ozone background
concentration, higher temperatures and mixing rate during the first stage
of plume development (the jet phase) and the effect of concentration
fluctuations are less important.

Figure 4. Effects of uncertainties in the plume dimensions on the conversion rate of NO to NO₂, as compared with measurements. Taking dispersion parameters according to Pasquill leads to conversion rates slightly higher than the upper curve.

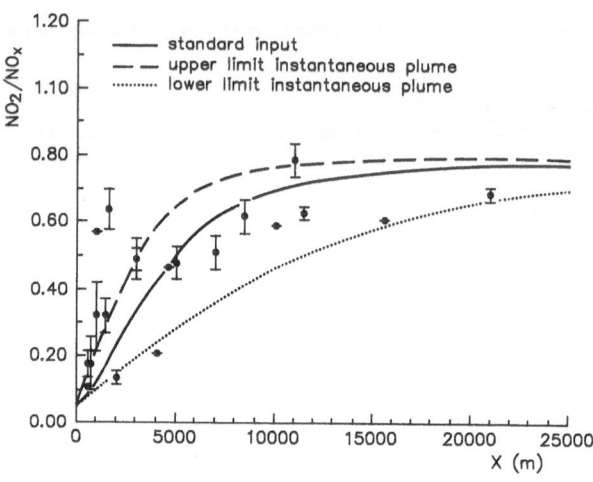

Figure 5. The effect of changing wind velocities. Effects of varying the NO$_x$ exhaust were almost similar. Effects of varying the ozone background, changing the temperature and sunshine (k$_3$) are somewhat smaller.

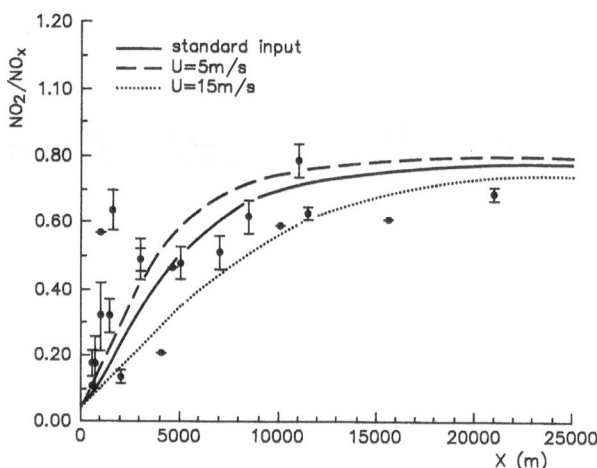

Figure 6. The effect of concentration fluctuations. Effects of a 30- or 60-second jet phase on the oxidation rate are even smaller.

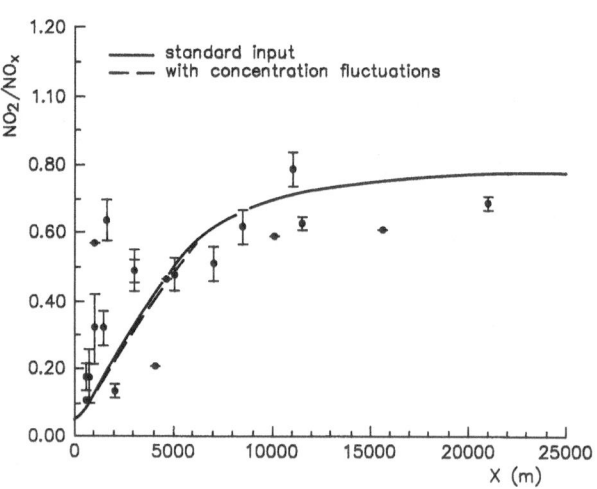

Figure 7. Crossing the plume with a measuring aircraft at some distance above or under the centre of the plume also leads to differences in measured oxidation rates. Near the edge of the plume conversion rates are higher because of higher ozone concentrations.

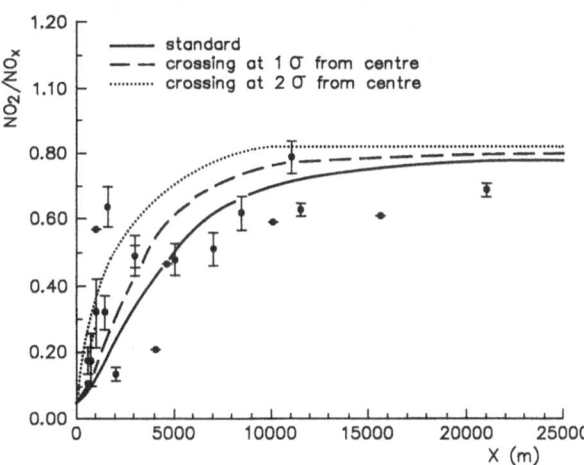

When model calculations are corrected for the fact that the oxidation measurements are generally made at lines that do not cross through the centre of the plume, near-source conversion rates are predicted well, but conversion rates at distances greater than 5 km are overestimated. It is most likely that this is caused by errors in the dispersion parameters. Therefore, a better understanding and description of the instantaneous plume has the highest priority.

REFERENCES

[1] BUILTJES, P.J.H. and STERN, R.M. (1988). Photochemical processes: emissions and atmospheric transport. Paper presented at the European Convention: Combustion-Pollution Reduction: New techniques in Europe. May 16, 17, Hamburg, FRG.
[2] SEINFIELD, J.H. (1986). Atmospheric chemistry and physics of air pollution. John Wiley and Sons (Chichester).
[3] JANSSEN, L.H.J.M. (1986). Mixing of ambient air in a plume and its effects on the oxidation of NO. Atm. Env., vol 20, 2347-2357.
[4] VAREY, R.H.H., SUTTON, S. and MARSH, A.R.W. (1984). A numerical model for the production of nitrogen dioxyde in power plant plumes. Env. pol. ser. B., vol 7, 107-127.
[5] RICHARD, H.G., SCHNEIDER, F. and JANICKA, J. (1985). Ausbreitung und Reaktion von Stickoxiden in Abgasfahnen von Punktquellen. Staub. Reinhaltung der Luft, Band 45, nr 2, 74-79.
[6] JANSSEN, L.H.J.M., VAN HAREN, F., VAN DUUREN, H. and VAN WAKEREN, J.H.A. (1989). Conversion processes of nitrogen oxides in daytime and at night. Measurements and modelling in the plumes of power plants. Submitted to Atmospheric Environment.
[7] BUILTJES, P.J.H. and TALMON, A.M. (1987). Macro- and microscale mixing in chemical reactive plumes. Boundary Layer Meteo. 41, 417-426.
[8] PASQUILL, F. (1974). Atmospheric diffusion. John Wiley & Sons (Chichester).

CONTRIBUTION D'UN PANACHE DE CENTRALE THERMIQUE AUX DEPOTS ACIDES. CALCULS A L'AIDE DU CODE PARADE DE DEPOTS HUMIDES ET DE COMPOSITIONS CHIMIQUES DE PLUIES

Eric Joos

Electricité de France, Direction des Etudes et Recherches, 6 quai Watier, 78400 Chatou

Résumé

Le code PARADE a été développé par EDF afin de quantifier la contribution aux dépôts acides des panaches de centrales thermiques alimentées au charbon. Parmi les configurations soumises au modèle, celles faisant intervenir des nuages précipitants sont particulièrement importantes car elles sont à l'origine de dépôts locaux de sulfates et nitrates plus élevés que ceux issus de configurations sans pluie. Différents scénarios faisant intervenir deux types de nuages précipitants, stratus et cumulus, ont été simulés afin de comparer les valeurs des dépôts acides provenant du panache, et leur évolution en fonction de la disponibilité d'H_2O_2 et du pH dans le nuage ainsi que de sa nature convective et des caractéristiques des précipitations. La composition chimique des pluies au sol est également étudiée et des comparaisons avec des valeurs de dépôts issues de la littérature sont effectuées.

1. INTRODUCTION

Le dioxyde de soufre et les oxydes d'azote émis par les centrales thermiques alimentées au charbon, au cours de leur transport et de leur dispersion dans l'atmosphère, retombent au sol. Mais, en altitude, ils sont soumis à un certain nombre de processus physicochimiques aboutissant à leur oxydation partielle en espèces acides (aérosols de sulfate et de nitrate, acide nitrique) qui se déposent également.

Afin de quantifier la contribution de ses panaches aux dépôts acides globaux, Electricité de France, en collaboration avec des spécialistes externes, a développé le code de calcul PARADE (PAnache Réactif en Atmosphère avec DEpôts) pour simuler à l'échelle de la journée les configurations les plus fréquentes en France et appliquer ensuite le modèle à un cas concret représentatif, intégrant ces configurations sur une période de temps allant de la semaine au mois.

Parmi les situations simulées, celles caractérisées par l'interaction du panache avec des nuages précipitants sont essentielles à prendre en compte car elles peuvent aboutir localement à des dépôts acides nettement plus importants que les dépôts secs se produisant sans la présence de nuages.

2. TRAITEMENT DES INTERACTIONS PHYSICOCHIMIQUES AVEC DES NUAGES PRECIPITANTS ET DES DEPOTS HUMIDES

Le modèle PARADE est basé sur le Reactive Plume Model (1) appartenant à la catégorie des modèles gaussiens divisés. Ce code lagrangien traite initialement la dispersion horizontale du panache et la chimie en phase gazeuse dans une rangée de cellules symétriques, 2x5 au maximum, transporté par le vent moyen et dans lesquelles la concentration d'une espèce est supposée homogène avec l'altitude. Le schéma cinétique chimique utilisé est le Carbon-Bond Mechanism, CBM IV.

Dans une première version de PARADE, le traitement de la chimie dans le panache a été étendu à la formation des aérosols en atmosphères non saturées, avec calcul de leur composition chimique et determination de leur distribution en taille (2).

La prise en compte des interactions physicochimiques du panache avec des nuages précipitants a nécessité de nombreux ajouts au code incluant le traitement de la résolution et du transport vertical, de la microphysique des nuages, de la chimie en phase aqueuse et des dépôts humides.

La résolution verticale a été incorporée au moyen de couches superposées, 10 au maximum, depuis la surface du sol jusqu'au sommet de la section du panache. D'une couche à l'autre, les concentrations des espèces chimiques et les valeurs des variables météorologiques (température, humidité relative, pression, contenu en eau liquide...) ainsi que l'épaisseur de la couche peuvent être différentes. Cependant, cette dernière demeure fixe durant une simulation.

Le transport vertical à travers ces couches consiste en la diffusion verticale, le transport convectif et les précipitations. La diffusion verticale est traitée à l'aide de coefficients de diffusion dépendant de la stabilité du cas traité. La convection est prise en compte à l'aide d'une vitesse ascensionnelle qui peut varier avec la hauteur, provoquant alors un entraînement ou un "détraînement" de l'air ambiant dans la section du panache. Le taux de précipitation est calculé par le module de microphysique et varie également avec la hauteur.

La simulation de la microphysique des nuages consiste à calculer la distribution de l'eau entre les quatre phases suivantes : vapeur, eau liquide nuageuse, eau liquide de pluie, glace (ou neige). Le traitement de la microphysique de l'eau liquide est issu de la formulation de Kessler (3).

Le traitement de la chimie en phase aqueuse est basé sur le schéma réactionnel de Seigneur et Saxena (4) qui comporte 14 équilibres de phases gaz/liquide, 16 équilibres de dissociation ionique ainsi qu'une trentaine de réactions chimiques irréversibles concernant 40 espèces qui décrit la formation des sulfates et des nitrates et inclut la chimie des radicaux. Pour certaines espèces, lorsque cela est nécessaire, une limitation du transfert massique dans l'une ou l'autre des deux phases a été introduite.

Les dépôts humides de sulfates et de nitrates, mais aussi de H^+ et de SO_2 - incluant les espèces de dissociation ionique $HSO3^-$ et $SO3^{--}$ - sont calculés à partir des résultats des modules de microphysique des nuages et de chimie en phase aqueuse. La quantité d'eau de pluie est déterminée à partir de la simulation des processus microphysiques ainsi que la vitesse de chute des gouttelettes. Leur composition chimique est calculée dans le sous-programme de chimie en phase aqueuse avec toutefois l'hypothèse que celle-ci est la même, dans la cellule d'une couche considérée, que celle des gouttelettes du nuage éventuellement présent.

Deux sortes de valeurs sont calculées par PARADE pour les dépôts humides : un taux de dépôt local sous la coupe du panache, à un instant et donc à une distance donnée de la centrale; un taux de dépôt moyenné entre le temps/endroit initial et le temps/endroit considéré. Dans la présente communication, nous ne nous intéresserons qu'aux dépôts locaux.

3. SIMULATIONS D'INTERACTIONS AVEC DES NUAGES PRECIPITANTS

Les simulations effectuées sont basées sur une même configuration du panache correspondant à un cas concret représentatif, observé expérimentalement en octobre 85 lors d'une campagne d'étude de l'évolution physicochimique du panache de la centrale du Havre. Notons qu'en raison de l'importance en France de l'énergie d'origine nucléaire, et d'une plus faible consommation durant les mois de printemps et d'été, les centrales alimentées au charbon fonctionnent surtout en automne-hiver.

La coupe initiale, située à 2 km de la centrale, comporte cinq couches, le panache étant localisé dans la seconde à partir du sol, entre 300 et 700 m. Chaque couche contient quatre cellules. L'évolution de la largeur mesurée du panache est fournie au modèle, 1 km à la distance initiale, 10 et 23 km à respectivement 20 et 60 km de la centrale et 37 km pour la dernière coupe effectuée à 90 km de la centrale.

La charge de la centrale correspondant à ce cas est de 1100 MW et la vitesse du vent, relativement stable durant l'étude expérimentale, de même que sa direction, peut être considérée comme constante au cours de la simulation et égale à 5,5 m/s.

Les concentrations en ppbV des principaux composés gazeux, dans la coupe initiale du panache (pan) et dans l'air ambiant (amb), pour la cas de base, sont les suivantes :

$[SO_2]_{pan} = 200$; $[SO_2]_{amb} = 0,1-0,5$; $[NO]_{pan} = 150$; $[NO]_{amb} = 2$

$[NO_2]_{pan} = 40$; $[NO_2]_{amb} = 5$; $[O_3]_{pan} = 20$; $[O_3]_{amb} = 35$

$[NH_3]_{amb} = 3$; $[H_2O_2]_{amb} = 0,2$; $[NMHC] = 63$ ppbC

Pour les sulfates et nitrates, les concentrations ambiantes sont respectivement égales à 4 et 2 $\mu g/m^3$.

Nous avons supposé que les profils de concentration dans l'air ambiant sont homogènes avec l'altitude. La température au sol est de 10 °C et un coefficient atténuateur des constantes photolytiques a été introduit dans PARADE de manière à tenir compte de la présence des nuages.

Afin de déterminer pour chacune des simulations la contribution nette du panache aux dépôts, nous en avons effectué une seconde, dans les mêmes conditions, mais en remplaçant dans la coupe initiale du panache les concentrations de SO_2 et des NOx par les concentrations ambiantes. Les résultats de cette simulation sont ensuite soustraits de ceux de la première.

L'heure de début de simulation est 13h00 TU.

Deux types de nuages précipitants, stratus et cumulus, ont été étudiés dans différents scénarios.

3.1 Scénarios avec stratus

Le stratus est situé dans les quatre couches supérieures d'épaisseurs suivantes : 400, 600, 800, 900 m. Ses caractéristiques sont fournies dans le tableau 1. Nous avons supposé que son interaction avec le panache commence dès le début de la simulation pour durer au moins deux heures correspondant à une distance de 40 km.

La possibilité d'apparition dans la phase aqueuse de sulfates et, dans une moindre mesure de nitrates, dépend de la concentration des espèces oxydantes dans l'air ambiant mais aussi du pH. Ainsi la constante de Henry de SO_2, compte tenu de sa dissociation en HSO_3^-, varie notablement en fonction du pH. De même, si la constante cinétique de la réaction d'oxydation de S(IV) par H_2O_2 dépend faiblement du pH, il n'est est pas de même pour celles des réactions impliquant O_3 ou O_2, cette dernière étant catalysée par les ions métalliques Fe^{3+} et Mn^{2+}. La concentration de H_2O_2 en phase gazeuse disponible pour l'oxydation du SO_2 du panache dans le nuage dépend de la concentration du SO_2 ambiant, celui-ci consommant après dissolution tout ou partie du H_2O_2.

La figure 1 permet de comparer l'évolution avec le temps des dépôts humides de sulfates provenant du panache pour 3 atmosphères différentes. Le cas de base C1 correspond à un air ambiant présentant des concentrations de 0.1 ppb en SO_2, de 0,2 ppb en H_2O_2 et un pH moyen, dans le nuage hors panache, de 4,6. Le cas C2 diffère de C1 par une concentration ambiante de SO_2 légèrement plus élevée et égale à 0,5 ppb. Le cas C3 est identique au cas C1 avec toutefois un pH moyen hors panache égal à 6 correspondant à une concentration d'ammoniac plus élevée de 1 à 2 ppb.

La comparaison des cas C1 et C2 montre que la contribution du panache à la formation et au dépôt de sulfates est non seulement fonction de la concentration ambiante de H_2O_2 mais également de celle du SO_2 susceptible de réagir avec le peroxyde d'hydrogène dans les gouttelettes du nuage. Dans le cas C2, compte tenu de l'excès de SO_2 ambiant, il n'y a pas de H_2O_2 résiduel. La formation des sulfates provient des autres réactions aboutissant à un dépôt relativement stable d'environ 1.10^{-4} kg/ha/min. Pour C1, la concentration de peroxyde d'hydrogène dans les gouttelettes de nuage hors panache, qui est de 13 umole/l, a pour conséquence une formation et un dépôt de sulfates plus importants, notamment dans les trente premières minutes, restant par la suite environ deux fois supérieur à celui du cas C2. Cet excès de H_2O_2 explique également les dépôts de H^+ plus élevés (figure 2) ainsi qu'une contribution du panache à l'acidité des précipitations légèrement plus importante (tableau 2). On constate également que, pour les deux cas, la contribution relative du panache aux dépôts de sulfates est plus faible que celle aux dépôts de H^+. Cependant, si

	Stratus	Cumulus
Hauteur de la base du nuage (m)	300	700
Hauteur du sommet du nuage (m)	2700	1700
Vitesse ascensionnelle (m/s)	0,2	0,1-2
Contenu en eau liquide (g/m^3)	0,55	2,0
Gradient de température (°C/100m)	-0,5	-0,8
Taux de précipitation (mm/h)	2,6	4,9

Tableau 1. Caractéristiques du stratus et du cumulus

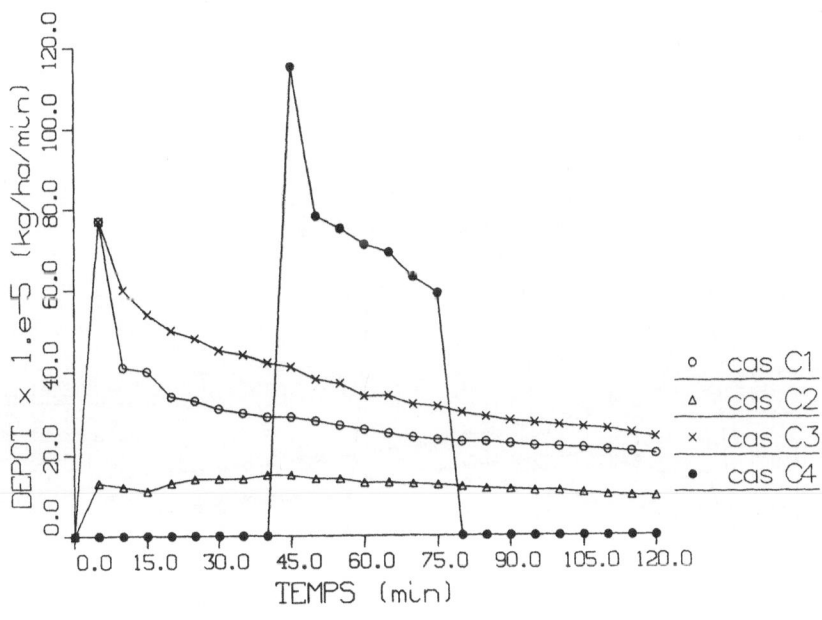

Figure 1 . Evolution avec le temps des dépôts de sulfates

	Cas C1	Cas C2	Cas C3	Cas C4
pH des précipitations sans panache*	4,77	4,77	6,35	5,42
pH des précipitations avec panache	4,43	4,54	4,82	4,58
SO_4^{--} (10^{-6} mole/l) sans panache*	46	51	46	6
SO_4^{--} (10^{-6} mole/l) avec panache	54	54	57	15
Contribution du panache* au dépôt de SO_4^{--} (%)	14	6	18	59
Contribution du panache* au dépôt de H^+ (%)	60	42	97	85

* à 30 minutes pour les cas C1, C2, C3 et à 60 minutes pour le cas C4

Tableau 2. Concentration dans les précipitations et contribution relative du panache pour SO_4^{--} et H^+

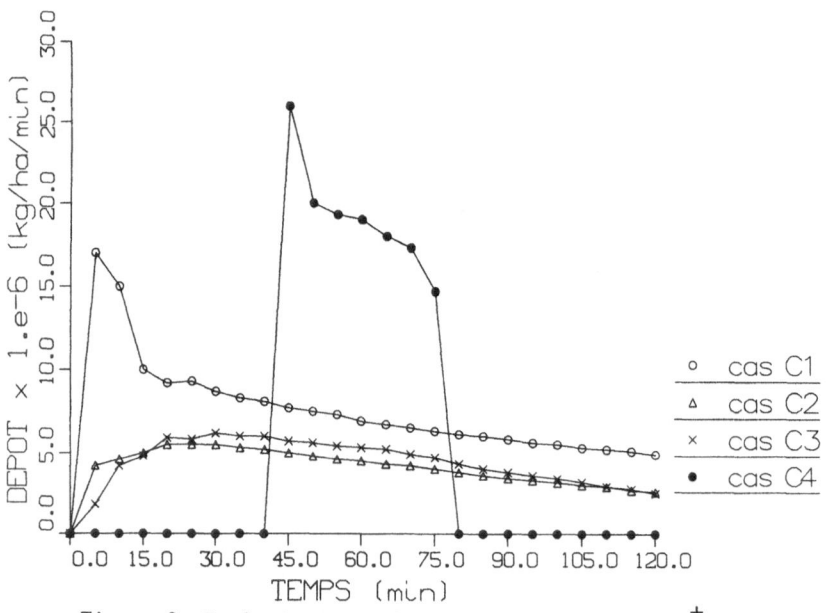

Figure 2. Evolution avec le temps des dépôts de H^+

cette dernière varie peu sur les deux heures de simulation, elle double dans le cas des sulfates.

Pour le cas C3, qui constitue un cas limite, le pH élevé dans le nuage hors panache peut être dû, localement, à une concentration d'ammoniac plus importante ou à la présence de pollution alcaline (Ca^+, Mg^{2+}, K^+). La contribution du panache à l'acidité des précipitations est alors plus forte avec cependant un pH plus élevé que pour C1 et C2. La quasi totalité des dépôts de H^+ proviennent du panache. Une autre conséquence du pH ambiant élevé est une contribution accrue du panache aux dépôts de sulfates résultant d'une augmentation de la constante cinétique de la réaction d'oxydation de S(IV) par O_3.

En ce qui concerne les nitrates, qui proviennent quasi entièrement de la dissolution de l'acide nitrique, l'activité photolytique réduite a pour conséquence une contribution relative du panache aux dépôt inférieure à 2%, avec un maximum de 2.10^{-5} kg/ha/min, et des dépôts globaux décroissant, entre 30 et 120 minutes, de 1.10^{-3} à 3.10^{-4} kg/ha/min. Après 30 minutes, la concentration dans les précipitations en $NO3^-$ est de 37 µmole/l.

3.2 Scénario avec cumulus

Le cumulus impliqué dans le cas C4 est situé dans les trois couches supérieures d'épaisseurs suivantes : 300, 300, 400 m. Ses caractéristiques sont fournies dans le tableau 1. La composition chimique de l'atmosphère de ce cas est celle du cas C1.

Si la durée de l'interaction du cumulus avec le panache - 30 minutes, trois quarts d'heure après le début de la simulation - est nettement plus faible que celle du stratus, ses conséquences sur les dépôts de sulfates et d'H^+ sont plus importantes comme l'indique la figure 2. L'entraînement de l'air ambiant résultant de la variation de la vitesse ascensionnelle, ainsi que le contenu en eau liquide et le taux de précipitation plus élevés, aboutissent à une contribution du panache aux dépôts de sulfates et d'H^+ supérieure à celle observée dans le cas du stratus au même moment de la simulation. Lorsque le panache interagit avec le cumulus, le pH décroit de presque une unité par rapport à la simulation sans panache. La valeur maximale, qui est atteinte au début de l'interaction, est de 4,2.

Notons que les dépôts de S(IV) provenant du panache représente, pour les cas C1 et C4, 10 à 20% du soufre total, alors que dans le cas C3, où le pH plus élevé favorise une dissolution du SO_2, cette proportion peut être supérieure à 30%, notamment dans les trente premières minutes.

La contribution du panache aux dépôts de nitrates est inférieure à 2% avec une valeur maximale de 8.10^{-5}. Après une heure de simulation, la concentration en nitrates dans les précipitations est de 6 µmoles/l.

CONCLUSION

A l'aide de trois cas d'interaction du panache avec un stratus précipitant nous avons mis en évidence les conséquences, sur les dépôts locaux de sulfates et de H^+, ainsi que sur la composition chimique des précipitations, de la disponibilité en phase aqueuse de H_2O_2, et d'une augmentation locale du pH du nuage. Nous avons d'autre part montré, sur un scénario mettant en jeu un cumulus précipitant, que la contribution du panache à ces dépôts était alors plus importante, bien que nettement plus brève, avec une acidité des précipitations renforcée par la présence du panache.

En ce qui concerne les dépôts provenant du panache si l'on retient pour les sulfates une valeur de 3.10^{-4} kg/ha/min, nous obtenons une valeur annuelle d'environ 5 gS/m^2/an. Ce chiffre est proche des valeurs moyennes annuelles de dépôts humides provenant du réseau EMEP en 1984, environ 1gS/m^2/an (5). Pour les nitrates nous pouvons fixer une valeur de 1.10^{-5} kg/ha/min soit 0.1 gN/m^2/an proche de la valeur moyenne annuelle provenant du même réseau, environ 0,4 gN/m^2/an. Pour les dépôts de H^+, nous retiendrons une valeur de 5.10^{-6} kg/ha/min ce qui donne une valeur annuelle de 2.6 kg/ha/an beaucoup plus élevée que les valeurs moyennes annuelles des régions polluées des Etats-Unis soit 50g/ha/an (6). Toutes ces valeurs extrapolées à l'année n'ont évidemment pour but qu'une comparaison préliminaire. Nous allons à présent appliquer PARADE à un cas concret représentatif et calculer les dépôts acides provenant du panache, sur une période allant de la semaine à plusieurs mois, en prenant en compte des variables météorologiques tels que les changements de direction et de vitesse du vent, la probabilité de présence de nuages et leur type, la durée d'éventuelles précipitations ainsi que leurs caractéristiques. Seule une telle étude permettra d'obtenir des valeurs permettant une interprétation par comparaison avec des valeurs expérimentales de dépôts globaux.

REFERENCES

(1) STEWART, D.A. and LIU, M.K. (1981). Development and application of a reactive plume model. Atmospheric Environment, 15, 11, 2377.

(2) JOOS, E., MENDONCA, A. and SEIGNEUR, C. (1987). Evaluation of a reactive plume model with power plant plume data - Application to the sensitivity analysis of sulfate and nitrate formation. Atmospheric Environment, 21, 6, 1331.

(3) KESSLER, E. (1969). On the distribution and continuity of water substances in atmospheric circulations. Meteorol. Monogr. No. 32, 84 pp.

(4) SEIGNEUR, C., SAXENA, P. (1988). A theoretical investigation of sulfate formation in clouds. Atmospheric Environment, 22, 1, 1988.

(5) SCHAUG., J., PACYNA, J., HARSTAD, A., KROGNES, T., SKJELMOEN, J.E. (1987). EMEP/CCC-Report 1/87, NILU, LILLESTROM, Norvège.

(6) BERNABO, J.C., SMYTHE, K.D. (1988). Material Degradation and Acidic Deposition. Rapport EPRI EA-5424.

ESTIMATION OF WET AND DRY DEPOSITION
OF Pb AND Cd OVER THE LIGURIAN SEA

C. MIGON [1,2], J. MORELLI [3], L. ELEGANT [2], P. COURAU [1]
and E. NICOLAS [1]

(1) Laboratoire de Physique et Chimie Marines, La Darse, BP 8,
   06230 VILLEFRANCHE-SUR-MER, FRANCE

(2) Laboratoire de Thermodynamique Expérimentale, Université de Nice,
   Parc Valrose, 06034 NICE CEDEX, FRANCE

(3) Institut de Biogéochimie Marine, Ecole Normale Supérieure,
   1, rue Maurice Arnoux, 92120 MONTROUGE, FRANCE

ABSTRACT

Measurements of atmospheric deposition were carried out at Cap Ferrat on the Southeastern coast of France. Pb and Cd were sampled in the atmospheric aerosol and in rainwater, during 1986 and 1987. Aerosols were analysed by graphite furnace atomic absorption spectrophotometry ; dissolved matter, which represents almost all of heavy metals in rainwater, was analysed by differential pulse anodic stripping voltammetry on hanging mercury drop electrode.

On the basis of annual observations, the following results were highlighted :

- The atmospheric concentrations of Pb and Cd are in the ranges of 5-100 and 0,05-1,5 ng.m^{-3} respectively.

- The concentrations of Pb and Cd in rainwater are in the ranges of 1-30 and 0,05-1,5 µg.l^{-1} respectively.

-The annual dry deposition fluxes of Pb and Cd, calculated from measured concentrations and dry deposition rates found in literature, are 3.2 and 3.5.10^{-3} kg.km^{-2}.yr^{-1} respectively.

-Assuming that the Cap Ferrat is representative of the Ligurian Sea, these results are extrapolated to this surface area. So that, an estimation of total atmospheric deposition is given (Pb : 6.3 ; Cd : 173.6.10^{-3} kg.km^{-2}.yr^{-1}) and the relative importance of wet and dry inputs is discussed.

1. Introduction

It is now well known that the atmosphere plays a major role in the transport of trace elements from anthropogenic and natural land-based sources to the sea (NRIAGU and PACYNA, 1988 ; NRIAGU 1989), particularly in the case of the Mediterranean Sea (GUERZONI *et al.*, 1988 ; MIGON, 1988). For most of toxic metals, the natural fluxes are small compared with anthropogenic emissions and human activities have strong impacts on the biogeochemical cycles of these elements.

A very few data is available now about the heavy metal atmospheric deposition on the Mediterranean Sea. Pollutant emissions are increasing in the Mediterranean environment and more studies are necessary for assessing the extent of regional and global effects of toxic metals.

This work focuses on the budgets of atmospheric lead and cadmium over the Ligurian Sea (see Figure 1), in order to quantify the present-day anthropogenic discharges on the Mediterranean Sea. If these metals do not play any biological role in the growth of aquatic organisms, (ROMEO, 1985) their toxicity has been well studied already (NORDBERG *et al*, 1985).

The strong variability of elemental concentrations in the Western Mediterranean atmosphere implies that a continuous sampling methodology is necessary for any reliable budget calculation. Then, measurements of atmospheric deposition were carried out at Cap Ferrat, including aerosol and rainwater collection, during 1986 and 1987.

2. Experimental

Atmospheric samples were collected at Cap Ferrat (43° 41' 10" N, 7° 19' 30" E), on the Southeastern coast of France. Such a location is generally exposed to marine winds, originating most of the time from Southeast. Moreover, the study of rainfall at Cap Ferrat, from 1970 to 1980, shows a value which is 10% lower than at the airport of Nice, situated on the shore line. This suggests the marine character of the site. The sampling site was established at 130 m above the sea level, very closely to the Cap Ferrat signal-station, which provided complete meteorological data.

Aerosol samples were collected at the top of a meteorological mast. Filters were cellulose acetate membranes Sartorius SM 11106 (porosity 0.45µm, diameter 47 mm). Sampling duration was 4 to 8 hours, with a nominal air flow of 1m3.h^{-1}. 32 samples were collected without any particle size distinction. This could seem a very few data, but the aerosol study was carried out in agreement with different and typical meteorological situations, according to their frequency. The chemical composition of aerosols being strongly influenced by the origin of the incoming air mass to the samling site (MORELLI *et al*, 1983 ; DULAC *et al*, 1987), some air mass trajectories were studied with the aim to evidence such an influence. These 32 samples were analysed in order to determine mean aerosol concentration, from which the dry material inputs were calculated, using dry deposition rates found in the literature. Aerosol analysis was carried out by graphite furnace atomic absorption spectrophotometry. Aliquots of the membrane (3 mm in diameter) were punched out of the filters and introduced into the carbon-rod atomizer. The homogeneous distribution of particulate matter on filters permits such an experimental approach. However, this method has been already used for seawater analysis (COPIN-MONTEGUT *et al*, 1986). Carbon cups were used for Pb while carbon tubes (2 to 3 times more sensitive) were used for Cd. The presentation of the detection limits would not be useful here, as they are well below the blanks levels. Reproducibility was always below 10 % and blank levels from filters were low if compared with the aerosol concentration range, i.e. 0 to 7 % for Pb and 1 to 16 % for Cd. The spectrophotometer was a Varian Techtron AA 1275 equipped with a CRA 90 atomizer and Ultra Carbon pyrolytic crucibles. Atomization was carried out under an argon atmosphere (Liquefaction de l'air).

Wet deposition was continuously collected with a rain collector used and described elsewhere (NURNBERG *et al*, 1984 ; MIGON, 1988) which is open only when it rains, and rainwater is automatically filtrated. Thus, only dissolved matter was collected. Anyway, for Pb and Cd, particulate matter represents only 0 to 10 % of total wet deposition, according to the pH, as shown by several filter analysis. This is in agreement with the results published by NGUYEN *et al.* (1979). Rainwater was analysed by differential pulse anodic stripping voltammetry on hanging mercury drop electrode. Measurements were performed with an EG & G Princeton Applied Research 264 A polarographic analyser in conjonction with a 303 A static mercury drop electrode. A medium mercury drop (1.6 mm2) was used.

All the analysis were carried out under laminar airflow benches fitted with high efficiency particulate filters. Sampling, storage and analysis methodology are discussed in detail in previous papers (MIGON, 1988 ; MIGON and CACCIA, 1989).

3. Results

After two years of observations, Pb and Cd concentration ranges can be given for both aerosol and rainwater (See Table 1). Bulk data can be found in previous work (MIGON, 1988). The variability of the atmospheric metal concentrations is high on a daily time scale in the Northwestern Mediterranean (CHESTER *et al.*, 1984 ; DULAC *et al.*, 1987). Nevertheless, a seasonal pattern can be observed (MIGON, 1988 ; BERGAMETTI *et al.*, 1989) and strong variations are rather characteristic of elements associated with mineral particles, as Al or Si, especially when Saharan events occur (BERGAMETTI *et al.*, 1988). Pb and Cd do not exhibit sporadic high concentrations. Then, 32 aerosol samples and a two years-continuous rainfall sampling should allow to estimate both dry and wet yearly deposition.

Following the assumption that the concentration of atmospheric Pb and Cd is the sum of two contributions (i.e. anthropogenic and natural emissions) whose distribution is log-normal, a mean aerosol concentration was calculated (MIGON and CACCIA, 1989. See Table 1). These concentrations are to be multiplied by the dry deposition rates found in literature (DULAC, 1986) to give dry fluxes. Then, the yearly dry flux is obtained by multiplying the dry fluxes by the mean number of dry days in a year, calculated from 1986 and 1987 (sea Table 1). Let us notice that some uncertainties are associated with such a calculation since about two days are necessary after a rain event to reload the atmosphere in such a Mediterranean location (BERGAMETTI, 1987).

All the rainy events were collected at Cap Ferrat during 1986 and 1987. For each event, a wet flux was calculated by multiplying the metal concentration by the rainfall amount. Then, the sum of all these fluxes was divided by two and a mean yearly wet flux was obtained, which is also presented in Table 1. In fact, rainfall is less intense in the open sea than on the coasts, but the number of rainy days is fairly the same. A lower rainfall is partially compensated by higher concentrations and one can admit that the wet fluxes are very similar on the shore line and in the open sea. The yearly total flux is obviously the sum of the dry and wet contributions. The comparison of dry and wet fluxes shows very different results for the two metals studied here (see Table 2). In the case of lead, the dry contribution is slightly higher than that of rainfall, while the dry way represents only 2 % of the total deposition of Cd. In a Western Mediterranean environment, the dry deposition rate is higher for lead (DULAC, 1986 ; see Table 1). This should be explainable by the existence of heavier Pb particles associated with short-scale transport, including natural lead and lead from gasoline combustion. Indeed, the coasts surrounding the Ligurian Sea are rather poorly industrialized but much inhabited. In the case of Cd, the natural contribution should be very low since more than 60 % of the atmospheric natural emissions are due to volcanoes (NRIAGU, 1989) and MARTIN *et al.*, (1984) have shown that the Western Mediterranean basin is practically not affected by volcanic plumes. Then, the atmospheric cadmium over the Ligurian Sea should be essentially originating from human activities and associated with long-range transport, i.e. associated with thin particles. Such a result is in agreement with a previous work (MIGON and CACCIA, 1989) where it was found that anthropogenic Cd concentrations are 100 times higher in the Cap Ferrat environment than natural Cd levels. The total atmospheric flux at Cap Ferrat is to be compared with that of other areas. In Table 2 also can be found values of Pb and Cd fluxes for the total Mediterranean Sea (GUERZONI *et al*, 1988), the North Sea (BUAT-MENARD, 1986), and, as an example of remote location, Enewetak Atoll, in the Tropical North Pacific (ARIMOTO *et al*, 1985). This last case shows that the Ligurian Sea, as well as the whole Mediterranean Sea, is obviously submitted to numerous land-based pollutant sources. Indeed, the Cd and Pb levels are respectively 17 and 90 times higher than those of Enewetak Atoll. The maps published by PACYNA (1984) show that many strong anthropogenic emissions are close to the Ligurian Sea. Some meteorological situations, particularly in winter, with Northwestern winds, can make strong anthropogenic inputs to reach this marine area. For example, on 12.03.86, the highest Pb flux (480 µg.m^{-2}) associated with a rain event, during this study, occured. The air mass trajectories, arriving at the 925 and 700 hPa barometric levels (See Figure 2) are in agreement with the metal concentrations found and show that several European industrialized areas can have an important influence on the Ligurian shore line. Nevertheless, the atmospheric flux of heavy metals is lower than in the North Sea. For the whole Mediterranean Sea, the Cd atmospheric value is very similar to that of the Ligurian Sea but logically lower, since the Eastern Mediterranean basin should be less contaminated by human activities. In the case of lead, the mean atmospheric flux is higher for the total Mediterranean Sea, perhaps because of the influence of cities like Cairo, Istanbul or Athens, where the automobile traffic may be very determinant for Pb emissions. However, a great uncertainty is due to the dry deposition rate of lead particles, which is very difficult to estimate, taking into account the interaction of lead aerosol with fog or dew in the coastal atmosphere and the subsequent effects on particle size and deposition velocity (BERGAMETTI, 1987 ; MIGON, 1988. For example, GUERZONI *et al* (1988) propose for the Mediterranean Sea an evaluation of Pb dry inputs (30 % of the total Pb inputs) which is lower than the value given in Table 2 (51.5 %).

Assuming that the surface area of the Ligurian Sea is 5.3.10^4 km^2 (See Figure 1) the total atmospheric deposition should be 328.5 tons per year for lead and 9.2 tons per year for cadmium. Let us recall that particule matter in rainwater was not considered but, taking into account this parameter, the atmospheric deposition should become at maximum (i.e. if particulate matter was always about 10 % of total wet inputs, and this is not the case) : 340 T.yr^{-1} for Pb and 10 T.yr^{-1} for Cd.

4. Conclusion

The aim of this work was to estimate the atmospheric deposition for lead and cadmium over the Ligurian Sea. Most of uncertainties probably come from dry deposition, as for lead, and it seems necessary to carry out new measurements of dry deposition rates in order to lower such errors. However, this study should be a first evaluation of atmospheric deposition budget of toxic metals on the Ligurian Sea. Other measurements in other sites should allow to accurate such an assessment.

Moreover, the use of unleaded petrol should be soon generalized in Europe and the results from Cap Ferrat are to be compared with further similar studies, where atmospheric Pb concentrations should decrease.

References

ARIMOTO R., DUCE R.A., RAY B.J. and UNNI C.K., 1985. Atmospheric trace elements at Enewetak Atoll : 2 : Transport to the ocean by wet and dry deposition, J. Geophys. Res., 90, 2391-2408.

BERGAMETTI G., 1987. Apports de matière par voie atmosphérique à la Méditerranée Occidentale : aspects géochimiques et météorologiques. PhD Thesis, Université Paris VII, 302 pp.

BERGAMETTI G., BUAT-MENARD P. and MARTIN D., 1988. Trace metal in the Mediterranean atmosphere, In : Air Pollution Report n° 14 on Field measurements and their interpretation (Proceedings of the European Workshop organized in the framework of the COST 611 Concerted Action "Physico-chemical behaviour of atmospheric pollutants", in Villefranche-sur-Mer, France, on 3-4 May 1988), Ed. S. BEILKE, J. MORELLI and G. ANGELETTI, Commission of the European Communities, 88-95.

BERGAMETTI G., DUTOT A.L., BUAT-MENARD P., LOSNO R. and REMOUDAKI E., 1989. Seasonal variability of the elemental composition of atmospheric aerosol particles over the Northwestern Mediterranean, Tellus, 41B, 353-361.

BUAT-MENARD P., 1986. Air to sea transfer of anthropogenic trace metals, In : "The role of air-sea exchange in geochemical cycling", Ed. P. BUAT-MENARD, Reidel Publishing Company, 477-496.

CHESTER R., SHARPLES E.J., SANDERS G.S. and SAYDAM A.C., 1984. Saharan dust incursion over the Tyrrhenian Sea, Atmos. Environ., 18, 920-935.

COPIN-MONTEGUT G., COURAU P. and NICOLAS E., 1986. Distribution and transfer of trace elements in the Western Mediterranean, Mar. Chem., 18, 189-195.

DULAC F., 1986. Dynamique du transport et des retombées d'aérosols métalliques en Méditerranée Occidentale, PhD Thesis, Université de Paris VII, 241 pp.

DULAC F., BUAT-MENARD P., ARNOLD M. and EZAT U., 1987. Atmospheric input of trace metals to the Western Mediterranean Sea : 1. Factor controlling the variability of atmospheric concentrations, J. Geophys. Res., 92, 8437-8453.

GUERZONI S., LENAZ R. and QUARANTOTTO G., 1988. Field measurements at sea : atmospheric trace metals "end-members" in the Mediterranean, In : Air Pollution Report n° 14 on Field measurements and their interpretation, (Proceedings of the European Workshop organized in the framework of the COST 611 Concerted Action "Physico-chemical behaviour of atmospheric pollutants" in Villefranche-sur-Mer, France, on 3-4 May 1988), Eds. S. BEILKE, J. MORELLI and G. ANGELETTI, Commission of the European Communities, 96-100.

MARTIN D., CHEYMOL D., IMBARD M. and STRAUSS B., 1984. Climatology of forward trajectories of Mount Etna plume over a 18-year period, Bull. Volcanol. 4714, 1115-1123.

MIGON C., 1988. Etude de l'apport atmosphérique en métaux-traces et sels nutritifs en milieu côtier méditerranéen ; implications biogéochimiques, PhD Thesis, Université de Nice, 217 pp.

MIGON C. and CACCIA J.L., 1989. Separation of anthropogenic and natural emissions of particulate heavy metals in the Western Mediterranean atmosphere, Atmos. Environ., to be published.

MORELLI J., MARCHAL T., GIRARD REYDET L., CARLIER P., PERROS P., LUCE C. and GIRARD R., 1983. Variations des concentrations en soufre particulaire dans un environnement atmosphérique côtier en relation avec le changement d'origine des masses d'air, J. Rech. Atmos., 17, 3 259-271.

NGUYEN K.D., VALENTA P. and NURNBERG H.W., 1979. The determination of toxic trace metals in rainwater and snow by differential pulse stripping voltammetry, Sci. Total Environ., 12, 151-167.

NORDBERG G.F., GOYER R.A. and CLARKSON T.W., 1985. Impact of effects of acid precipitation on toxicity of metals, Environ. Health Perspec., 63, 169-180.

NRIAGU J.O., 1989. A global assessment of natural sources of atmospheric trace metals, Nature, 338, 47-49.

NRIAGU J.P. and PACYNA J.M., 1988. Quantitative assessment of worldwide contamination of air, water and soils by trace metals, Nature, 333, 134-139.

NURNBERG H.W., VALENTA P., NGUYEN K.D., GODDE M. and URANO DE CARVALHO E., 1984. Studies on the deposition of acid and ecotoxic heavy metals with precipitates from the atmosphere, Fresenius Z. Anal. Chem., 317, 314-323.

PACYNA J.M., 1984. Estimation of the atmospheric emissions of trace elements from anthropogenic sources in Europe, Tellus, 36B, 163-178.

ROMEO M., 1985. Contribution à la connaissance des métaux-traces (Cd, Cu, Hg, Pb, Zn) dans l'écosystème marin au niveau du plancton ; approches analytiques et expérimentales, PhD Thesis, Université de Nice, 172 pp.

	concentration ranges		Dry deposition rates [1] (cm.S-1)	Dry fluxes (Kg.km-2.yr-1)	Wet fluxes (Kg.km-2.yr-1)
	Aerosol (ng.m-3) [mean value]	Rainwater (µg.l-1)			
Pb	5-100 [34.2]	1-30	0.41	3.2	3
Cd	0.05-1.5 [0.29]	0.05-1.5	0.053	3.5.10-3	0.17

(1) After DULAC (1986)

Table : 1 Concentration ranges and estimation of dry and wet fluxes for Pb and Cd.

	Dry contribution %	Wet contribution %	Cap Ferrat : Yearly total flux (Kg.km-2.yr-1)	Mediterranean Sea[1](Kg.km-2.yr-1)	North Sea[2] (Kg.km-2.yr-1)	Tropical North Pacific [3] (Kg.km-2.yr-1)
Pb	51.5	48.5	6.2	10	26	0.07
Cd	2	98	173.5.10-3	130.10-3	430.10-3	10.10-3

(1) GUERZONI *et al.*, 1988

(2) BUAT-MENARD, 1986

(3) ARIMOTO *et al.*, 1985

Table 2 : Respective importance of dry and wet deposition and comparison of total atmospheric deposition on the Ligurian coast with other sites

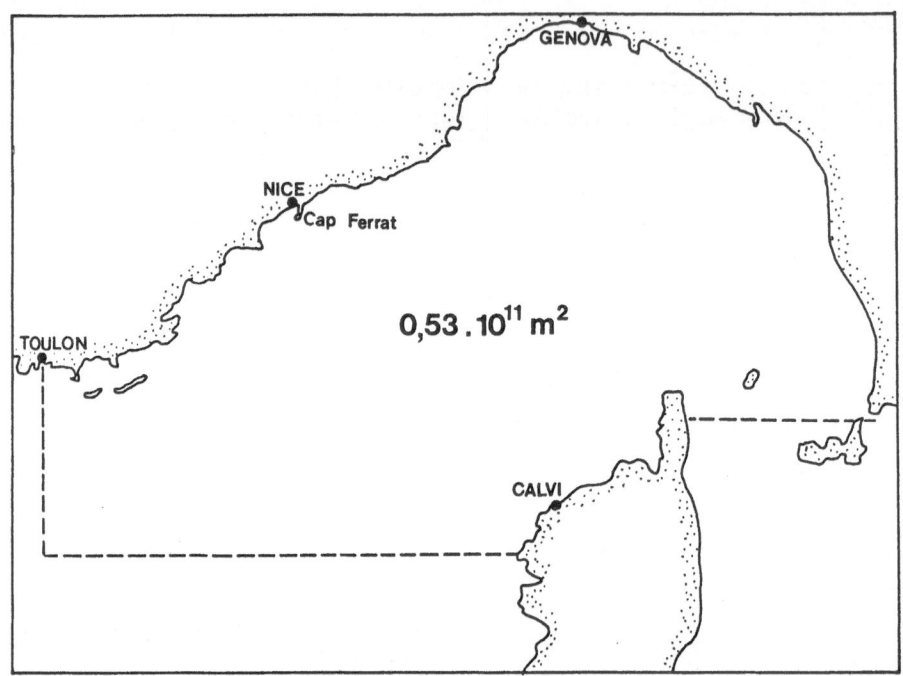

Figure 1 : Limits of the Ligurian Sea and sampling location

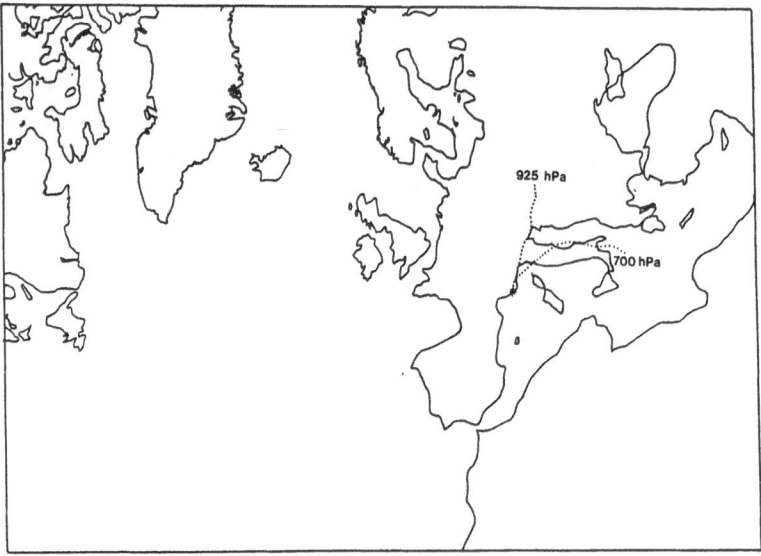

Figure 2 : Back trajectories (4 days) of air masses arriving in the sampling zone at two different barometric levels on the Cap, Ferrat (12 March 1986 at 12.00 UT)

LOW MOLECULAR WEIGHT ORGANIC COMPOUNDS IN THE PO VALLEY FOG WATER

M.C. FACCHINI, S. FUZZI, J. LIND and G. ORSI

Istituto FISBAT - C.N.R.
Via de' Castagnoli 1 - 40126 Bologna, Italy

Summary

Organic compounds which play a role in the chemistry of the atmospheric dispersed liquid phase (fog, cloud, precipitation) must be of sufficiently high polarity and low molecular weight to render them soluble: i.e. carbonylic and carboxylic compounds with less than three carbon numbers. Carbonyls can interact with S(IV) species in solution to form hydroxyalkanesulfonic acids, adducts which effectively increase the solubilities of both carbonyl and S(IV) species and so constitute an important aqueous phase reservoir. Organic acids can contribute significantly to the acidity of fog and cloud droplets in the absence of H_2O^+ provided by strong, inorganic acids. Identification and quantification of species within these two classes of compounds were performed on fog water samples collected at a field station in the Po Valley. Several carbonyl compounds were detected, though formaldehyde was by far the most abundant species present, with measured comcentrations up to 567 µmol/l. The percentage of formaldehyde bound as hydroxymethanesulfonic acid (HMSA) ranged from 35% to 99%. Acetaldehyde was present at concentrations of about one order of magnitude less than formaldehyde. Traces of acrolein were also present in these fog samples. The average concentrations of formic and acetic acid for all samples were 127 µmol/l and 104 µmol/l respectively, but traces of propionic acid were also present.

1. INTRODUCTION

Several classes of organic compounds exist in the atmosphere but only a few of them are present in significant concentration in the atmospheric liquid phase (cloud and fog droplets, precipitation).
The role of organic compounds in the atmospheric dispersed liquid phase is determined by their solubility. The major physico chemical properties affecting the solubility of chemical compounds in water are molecular weight and polarity. The solubility of a particular organic compounds is direcly proportional to the polarity and inversely proportional to the molecular weight. Carbonyl compounds and carboxylic acids with a carbon number less than three are for these reasons the main organic constituents of the atmospheric liquid water droplets (1).
Partitioning of chemical compounds between gas and liquid phase depends on Henry's law equilibium:

$$RCHO(g) \rightleftharpoons RCHO(aq)$$

$$RCOOH(g) \rightleftharpoons RCOOH(aq)$$

but also on the major chemical reactions occurring in the water medium (reactions with the solvent H_2O and/or other compounds). Carbonyl solubility is primarily a function of hydration constant

$$RCHO(aq) \overset{Kh}{\rightleftharpoons} RCH(OH)$$

In addition, the reaction of carbonyls with S(IV) in solution to form hydroxyalcansulfonates

$$RCHO(aq) + HSO_3^- \rightleftharpoons RCH(OH)SO_3^-$$

also enhances their solubility with respect to the predictions of Henry's law.

The carboxylic acid distribution between the phases is, on the other hand, strongly pH dependent due to their dissociation in water:

$$RCOOH(aq) \overset{Ka}{\rightleftharpoons} \cdot H^+ + RCOO^-$$

At pH < pKa the phase distribution equilibria are entirely shifted towards the gas phase, whereas at pH > pKa the acids are partitioned in the liquid phase in the dissociated form (2).

The importance of carbonyl compounds in the atmospheric liquid phase chemistry is mainly connected with the possibility of forming the above mentioned hydroxyalcansulfonates which inhibit the S(IV) to S(VI) liquid phase conversion and, at the same time, represent a source of acidity in solution (3). The oxidation reaction of formaldehyde with OH radicals also represents an important source of formic acid in the droplets (2,4).

Organic acids account for an important fraction of the acidity measured in precipitation samples collected in remote areas of the world (5,6).

In this paper we report on measurements of carbonyl compounds and carboxylic acids in radiation fog water samples collected in the Po Valley. The radiation fog system is an ideal environment for studying the distribution of chemical species between gas and liquid phase, due to the high stability which lets us reasonably assume equilibrium conditions.

2. EXPERIMENTAL

2.1 Fog water collection

Fog samples were collected at the field station of S. Pietro Capofiume in the eastern part of the Po Valley during the period January-March 1989. This period was charaterized by a particularly high fog occurence: 95 samples were collected during 31 fog episodes of different time duration. The fog collector used in this study has been described elsewere (7).

2.2 Analytical procedures

Formaldehyde, acetaldehyde and acrolein were determined by HPLC as dinitrophenylhydrazone-derivatives (8) and spectrophotometric detection (λ = 360 nm). Following the same procedure described in (9) we determined free and total carbonyl concentration after breaking the possible S(IV) adducts by NaOH addition. The S(IV)-carbonyl adduct concentration was then indirectly determined as the difference between total and free carbonyl concentration, on the assumption that hydroxyalcansulfonates are by far the most important adducts in fog droplets.

Formic, acetic and propionic acids were determined by ion exclusion chromatography and conductivity detection. The analyses were performed shortly after collection, at least within 24 hours after sampling, to avoid possible decomposition due to microbial activity.

3. RESULTS AND DISCUSSION

A wide range of pH and total ionic strength of the droplet solutions were found in the different samples and this allows us to evidence some important aspects of the chemistry of carbonyl compounds and carboxylic acids in fog. Table 1 reports concentration range and quartiles for these organic compounds detected in fog water; pH and total ionic strength of the fog water samples are also reported in the same fashion.

	pH	$\Sigma c_i z_i$	$HCHO_t$	CH_3CHO_t	$Acrol._t$	$HCOOH_t$	CH_3COOH_t	$CH_3CH_2COOH_t$
MIN	3.41	1000	16	DL	DL	12.9	13.1	DL
25 perc.	4.65	2300	99	DL	DL	46.9	41.7	3.9
50 perc.	5.57	6200	130	6	DL	81.5	76.9	7.9
75 perc.	6.49	10310	171	12	DL	178.2	153.7	14.8
MAX	7.10	31480	567	24	4	296.8	296.7	24.0

Table 1 - Statistic summary of the concentration of carbonyl compounds and carboxylic acids in the fog water samples (μmol/l). pH and total ionic strength ($\Sigma c_i z_i$) data are also reported. The index "t" (total) means for carbonyls the sum of free+S(IV)-bound compounds, while for organic acids it means dissociated+undissociated species. DL means "below detection limit".

3.1 Carbonyl compounds

Formaldehyde is the major carbonyl compound with a concentration 10-20 times greater than acetaldehyde. Acrolein is present only in traces. In the absence of concurrent gas phase measurement for these compounds we can nevertheless observe that the difference in the liquid phase concentration of the various carbonyls could completely be justified on the grounds of a very different hydration constant:

$$Kh_{(HCHO)} = 2.5 \cdot 10^3$$

$$Kh_{(C_2-C_3)} \simeq 1$$

As previously reported, we determined the fraction of carbonyl compounds bound in the form of hydroxyalcansulfonates. While free formaldehyde represents only a minor fraction of the total concentration (15% on average), acetaldehyde and acrolein do not appear to form stable adducts, as theoretically predicted by Betterton et al.(12). In addition, the ratio bound/total formaldehyde appears to be strongly pH dependent, as reported in Fig. 1. This ratio is close to 1 between pH 4 and 5 and decreases at a lower or higher pH than this interval. This result is in agreement with experimental and theoretical kinetic studies in literature which report that the maximum HMSA stability is in the same pH range (13,14,15). Possible decomposition processes could be responsible for the higly scattered data in the pH range above 5.

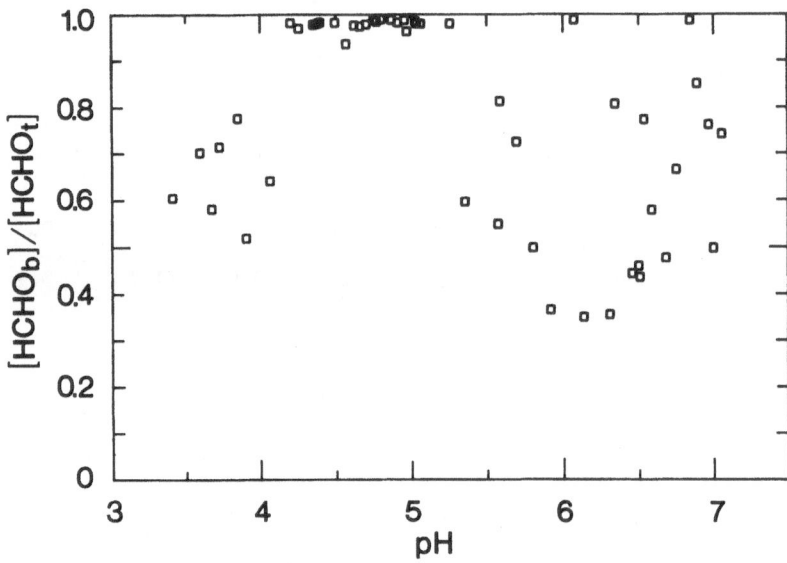

Fig. 1 - Trend of the ratio bound to total formaldehyde for the collected fogwater samples, as a function of pH.

3.2 Carboxylic acids

Formic and acetic acids represent the major carboxylic compounds in the Po Valley fog water samples with a minor contribution of propionic acid. Occasionally traces of other unidentified acids appear in the analysis. Formic and acetic acid concentrations do not exhibit any definite trend as a function of pH. On the other hand, the lack of gas phase measurements does not allow a complete treatment of this subject.

Organic acid concentrations are much greater than those reported in literature for precipitation samples (5,6,16). On the other hand organic acids contribute only 0.01 to 3.4% of the total ionic strength of the fog water samples.

In addition, formic and acetic acid are present in comparable concentration in our samples ($HCOOH/CH_3COOH$ = 1.35, variance = 0.66). This testifies to a substantial contribution of anthropogenic sources in the concentration of these species in the Po Valley fog water (16,17).

Acknowledgements - A. Correggiari and S. Miserocchi are acknowledged for their assistance during the field operations. This study was sponsored jointly by CEC (Contr. EV4V-0084-C) and ENEL-CRTN (Contr. 2RTII0326).

REFERENCES

(1) GRAEDEL T.E. and C.J. WESCHLER (1981). Chemistry within aqueous at-
     mospheric aerosols and raindrops. Rev. Geophys. Space Phys., 19,
     503-539.
(2) JACOB D.J. (1986). Chemistry of OH in remote clouds and its role in
     the production of formic acid and peroxymonosulfate. J. Geophys.
     Res., 91, 9807-9826.
(3) ADEWUY Y.G., S.Y. CHO, R.P. TSAY and G.R. CARMICHAEL (1984). Impor-
     tance of formaldehyde in cloud chemistry. Atmos. Environ., 18,
     2413-2420.
(4) CHAMEIDES W.L. and D.D. DAVIS (1983). Aqueous-phase source of formic
     acid in clouds. Nature, 304, 427-429.
(5) KEENE W.C., J.N. GALLOWAY and J.D. HOLDEN Jr. (1983). Measurement of
     weak organic acidity in precipitation from remote areas of the world.
     J. Geophys. Res., 88, 5122-5130.
(6) KEENE W.C. and J.N. GALLOWAY (1984). A note on acid rain in a Amazon
     rain forest. Tellus, 36B, 137-138.
(7) FUZZI S., G. CESARI, F. EVANGELISTI, M.C. FACCHINI and G. ORSI
     (1989). An automatic station for fog water collection. Submitted to
     Atmospheric Environment.
(8) FACCHINI M.C., G. CHIAVARI and S. FUZZI (1986). An improved HPLC
     method for carbonyl compound speciation in the atmospheric liquid
     phase. Chemosphere, 15, 667-674.
(9) FACCHINI M.C., J. LIND, G. ORSI and S. FUZZI (1989). The chemistry of
     carbonyl compounds in the Po Valley fog water. Sci. Tot. Environ., in
     press.
(10) SCHECKER H.G. and G. SCHULZ (1969). Untersuchungen zur Hydratations-
     kinetik von Formaldehyd in wä Briger Lösung. Z. Phys. Chem., 65, 221.
(11) BUSCHMANN H.J., H.H. FÜLDNER and W. KNOCHE (1980). The reversible
     hydration of carbonyl compounds in aqueous solution. Part I. The
     keto/gem-diol equilibrium. Ber. Bunsenges. Phys., Chem., 85, 41.
(12) BETTERTON E.A., Y. EREL and M.R. HOFFMANN (1988). Aldehyde-bisulfite
     adducts: prediction of some of their thermodynamic and kinetics pro-
     perties. Environ. Sci. Technol., 22, 92-99.
(13) DASGUPTA P.K., K. DE CESARE, J.C. ULLREY (1980). Determination of
     atmospheric sulfur dioxide without tetrachloromercurate and mechanism
     of the Shiff reaction. Anal. Chem., 52, 1912-1922.
(14) KOK G.L., S.N. GITLIN and A. LAZRUS (1986). Kinetics of formation and
     decomposition of hydroxymethansulfonate. J. Geophys. Res., 91, 2801-
     2804.
(15) DONG S. and P.K. DASGUPTA (1986). On the formaldehyde bisulfite hy-
     droxymethanesulfonate equilibrium. Atmos. Environ., 20, 1635-1637.
(16) TALBOT R.W., K.M. BEECHER, R.C. HARRISS and W.R. COFER III (1988).
     Atmospheric geochemistry of formic and acetic acids at a mid-latitude
     temperate site. J. Geophys. Res., 93, 1638-1652.
(17) WINIWARTER W., H. PUXBAUM, S. FUZZI, M.C. FACCHINI, G. ORSI, N.
     BELTZ, K. ENDERLE and W. JAESCHKE (1988). Organic acid gas and liquid
     phase measurements in the Po Valley fall-winter conditions in the
     presence of fog. Tellus, 40B, 348-357.

HIGH CONCENTRATIONS OF SULFUR DIOXIDE IN THE MIDDLE AND UPPER TROPOSPHERE - CASE STUDIES AND METEOROLOGICAL ANALYSIS

G.E.F. Ockelmann[*] and H.W. Georgii
Institut für Meteorologie und Geophysik
der Universität Frankfurt
Feldbergstraße 47
D 6000 Frankfurt/M 1
F. R. G.

Summary

Aircraft measurements of atmospheric SO_2 were performed during the STRATOZ experiments in June 1984 and in March 1985 on board of a Caravelle 116 aircraft in altitudes between 1 and 12 km. For the detection of SO_2 a sensitive chemiluminescence method was used which has a detection limit of 10 pptv. On several flights in the northern hemisphere high concentrations in the order of several hundred pptv of SO_2 were detected in the middle and upper troposphere. The results are presented in form of vertical distributions and discussed in the light of meteorological analyses which were performed for each of the flights. It is demonstrated that high temporal changes of SO_2 can be attributed to atmospheric transport from different source regions.

1. INTRODUCTION

Sulfur dioxide is an important pollutant in urban areas of the earth and its abundance in the atmosphere stands for a series of environmental problems like acid precipitation, the forest decline and the decrease of pH in freshwater lakes. The vertical distribution of SO_2 in the atmosphere over the continents is usually characterized by a sharp decrease frome some tenths of ppbv in the boundary layer to about 100 pptv in altitudes of 3000 m (1). The vertical gradient is the result of effective atmospheric sink mechanisms in the troposphere like homogeneous reactions with OH radicals (2, 3) and heterogeneous oxidation processes in clouds and precipitation (4). It was therefore suggested, that only small amounts of the mainly anthropogenic produced SO_2 could penetrate in the middle and upper troposphere or into the stratosphere (5).
In the following aircraft measurements of SO_2 between 1 and 12 km altitude are presented which show large variations within a time of a few days. The SO_2 results are discussed in the light of meteorological analyses.

[*]present adress: Institut Fresenius GMBH Im Maisel 14
D 6204 Taunusstein-Neuhof

2. EXPERIMENTAL

2.1 SULFUR DIOXIDE MEASUREMENT TECHNIQUE
For the SO_2 measurement a sensitive chemiluminescence method was used which consists of wet chemical filter sampling and consecutive chemiluminescence analysis (6). The filtermaterial (Microsorban 98S) is impregnated by a 0.1 m TCM-solution. Sampling is done at a flow rate of about 10 l/min. The typical sampling volume is 120 l S.T.P., corresponding to a sampling time of 12 minutes. For the analysis the filters are washed out with TCM and analysed by the chemiluminescence technique. The principle of the measurement is based on the oxidation of SO_2 by an acid $(8*10^-6m)$ potassium permanganate solution (pH=2.50) This reaction is accompanied by a chemiluminescence. The light yield of the chemiluminescence which is proportional to the SO_2 present in the sample is detected by a photomultiplier/photoncounter system. The detection limit of the method defined as three times the standard deviation of the filter's blanks is about 10 pptv.

2.2 THE CARAVELLE AIRCRAFT
The measurements of SO_2 were performed during the atmospheric research flights STRATOZ III/S on board of a Caravelle 116 aircraft owned by the Centre d'Essais en Vol in Bretigny (France). The Caravelle 116 is a twin-engine jet. In its civil version it has a capacity of 120 passengers. Maximum altitude is 38000 ft (11700 m).The whole equipment for sampling and analysis of SO_2 was installed on the aircraft. The analyses of filter samples was performed at the ground just after each flight.

3. SULFUR DIOXIDE MEASUREMENTS

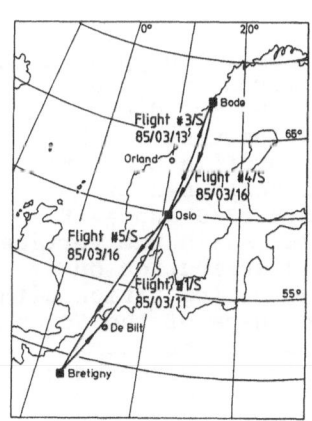

Fig. 1 Flight-route

In the following results of the experiment STRATOZ III/S are presented. The experiment started in Bretigny near Paris on March 11, 1985. The flight route which is shown in Fig. 1 lead to Oslo and on March 13 to Bodo in Norway (Latitude: 67°N). The return flights to Bretigny were made both on March 16. As the flight tracks from Bretigny to Bodo and back to Bretigny were nearly identical, the vertical SO_2 distributions measured on these flights can be compared with respect to temporal variability. Fig. 2a shows the SO_2 results of the flights #1/S and #5/S between Bretigny and Oslo. Horizontal bars indicate the error of the measurement, deduced from two analyses of the same filter sample. Obviously there is a large discrepancy between the SO_2 distributions measured on March 11 and

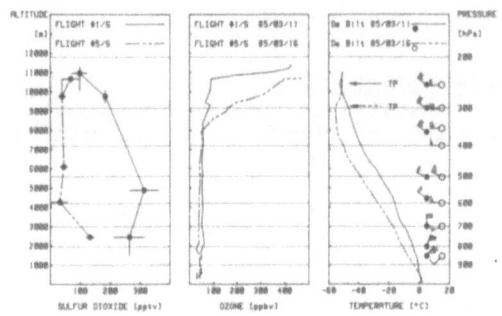

Fig. 2 Results of flights #1/S and #5/S

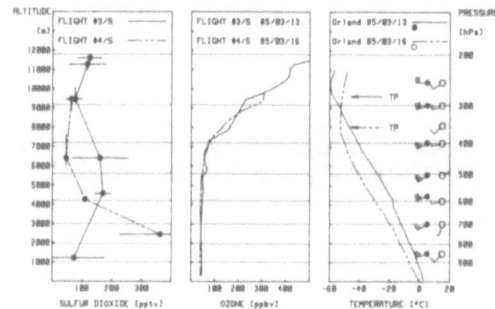

Fig. 3 Results of flights #3/S and #4/S

March 16. On March 11 high SO_2 mixing ratios of several
hundred pptv were found throughout the free troposphere. The
SO_2 maximum of 300 pptv was measured at an altitude of 5000 m
decreasing to 200 pptv in 10000 m and to about 100 pptv at the
tropopause level in 11000 m altitude. On March 16 the SO_2
mixing ratios were up to six times lower. The SO_2 decreased
rather rapidly with height in the first kilometers in the
troposphere and the general shape of this profile confirms
well with results of earlier measurements performed by
Meixner[1].SO_2 data measured on the flights #3/S and #4/S
between Oslo and Bodo are shown in Fig. 3a. Again the vertical
profiles of SO_2 are quite different. The distribution on
flight #3/S from March 13 shows increasing mixing ratios with
height which peak at 4500 m altitude. The shape of the SO_2
profile is similar to the result of flight #1/S between
Bretigny and Oslo, whereas the SO_2 mixing ratio measured on
flight #4/S strongly decreased with height. The mixing ratios
in altitudes of 6000 m and above were well below 100 pptv and
therefore comparable to the result of flight #5/S between Oslo
and Bretigny on the same day. Consequently it can be stated
from the SO_2 data measured on these flights that the temporal

variability of SO_2 was much higher than spatial variations over a horizontal distance of about 2000 km.

4. METEOROLOGICAL DATA AND TRAJECTORY CALCULATIONS

In Fig. 2c and 3c rawinsonde data from De Bilt (Fig. 2c) and Orland (Fig. 3c) are shown. The location of the stations is indicated in Fig. 1. The wind data measured at De Bilt during flights #1/S and #5/S on March 11 and March 16 were similar in the middle and upper troposphere. However the temperatures measured on March 16 were nearly 10 K lower than on March 11. The height of the tropopause decreased from 10400 to 9200 m during that time. This is in agreement with the heights of the tropopause deduced from the ozone measurements which were performed on the aircraft (see Fig. 2b). Similar temperature changes as already shown for De Bilt were reported from the rawinsonde measurements in Orland for the time between March 13 and March 16 .

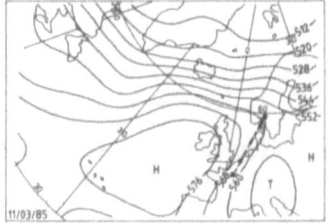

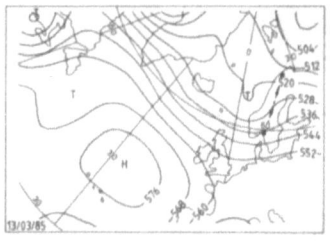

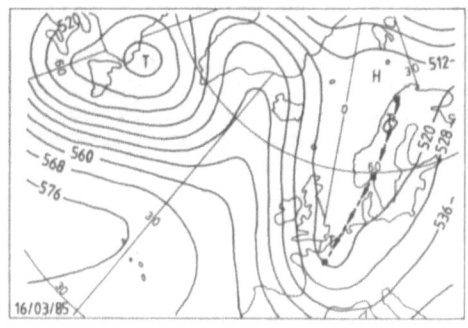

Fig. 4a, 4b and 4c Isobaric 500 hPa-maps for March 11, 13, 16

In Fig. 4a - 4c isobaric 500 hPa-maps are shown for the days the flights were performed. The maps from March 11 and March 13 show a high pressure system east of the Acores. North of the anticyclone the isolines are directed zonally, which causes a transport from west to east. On March 16 the situation is quite different. Due to a series of troughs and ridges the isolines show a wavelike structure which causes a meridional exchange of air. On the basis of these contour charts isobaric trajectory calculations were performed for each flight starting from the stations De Bilt and Orland. The time step for the trajectories calculated backward over 60 hours was six hours. Fig. 5a - 5c show the trajectories at 500, 300 and 200 hPa for De Bilt on March 11 which represent the situation for flight #1/S between Bretigny and Oslo. It is evident from the trajectories that there was a strong transport from the west. Air masses were transported from

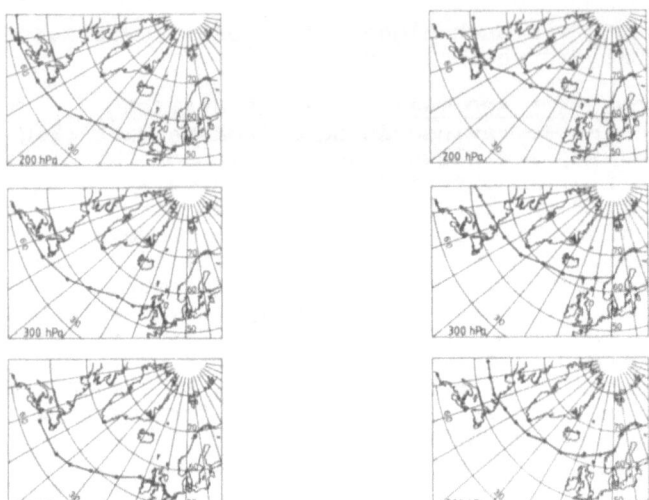

Fig. 5 (left) and 6 (right) Isobaric trajectories calculated
for De Bilt on March 11 (left) and Orland on March 13 (right)

North-America to northern Europe within 3 days. The same is
verified for Orland on March 13 (see Fig. 6a-6c). The origin
of the trajectories arriving at Orland however was 10° north
of the trajectories calculated for De Bilt.

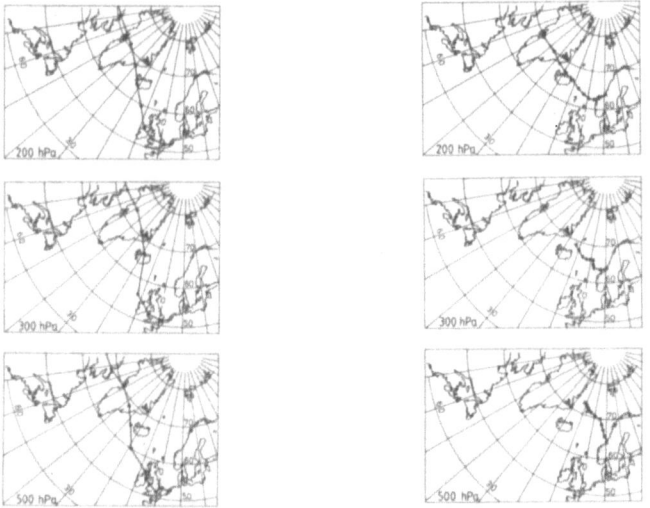

Fig. 7 (left) and Fig. 8 (right) Isobaric trajectories
calculated for De Bilt (left) and Orland (right) for March 16

The trajectory calculations for De Bilt on March 16 (see Fig.
7a-7c) demonstrate that the distances covered by these
trajectories were much smaller than for March 11 and they
indicate that airmasses out of arctic regions were advected.

This was already verified from the changes in temperature between March 11 and March 16 at De Bilt. The trajectories calculated for Orland on March 16 also demonstrate that airmasses which arrived at Orland on that day had their origin in high latitudes.
As a conclusion it can be stated that the free tropospheric SO_2 distribution in temperate latitudes of the northern hemisphere can be strongly changed by meteorological transport out of different source regions.

REFERENCES

(1) Meixner, F.X. (1984). The vertical sulfur dioxide distribution at the tropopause level. J. Atm. Chem., 2, 175-189.
(2) Cox, R.A., Sheppard, D. (1980). Reactions of OH radicals with gaseous sulphur compounds. Nature 284, 330-331.
(3) Barnes, I., Bastian, V., Becker, K.H. (1986). Products and kinetic of the OH initiated oxidation of SO_2, CH_3SH, DMS, DMDS and DMSO. Proc. 4th Europ. Symp. on Physico-chemical Behaviour of Atmospheric Pollutants. G. Angeletti, G. Restelli (ed). D. Reidel Publ. Comp., Dordrecht, 327-337.
(4) Beilke, S., Gravenhorst, G. (1978). Heterogenous SO_2 oxidation in the droplet phase. Atm. Environ. 12, 231-239.
(5) Crutzen, P.J. (1976). The possible importance of CSO for the sulfate layer of the stratosphere. Geophys. Res. Lett., 3, 73-76.
(6) Meixner, F.X., Jaeschke, W.A. (1981). The detection of low atmospheric SO_2 concentrations with a chemiluminescence technique. Int. J. Environ. Anal. Chem., 10, 51-67.

ION DEPOSITION DUE TO FOG WATER INTERCEPTION AT HIGH ELEVATIONS

G. Kroll, P. Winkler

Deutscher Wetterdienst Meteorologisches Observatorium Hamburg
Frahmredder 95, D-2000 Hamburg 65

Summary

The ion deposition via fog interception has been estimated using a one dimensional cloud droplet deposition model and measured concentrations of trace constituents. The model uses observed meteorological data as input parameters. The fog deposition becomes important for mountain regions which rise more than 600 m above the surrounding flat terrain. The ion deposition via fog becomes as important or more important than wet deposition above that heights. The regional variation of ion deposition via fog is much higher than that of precipitation due to large variations in the fog respectively cloud water composition. Although the present assessment is still relatively uncertain, we can conclude that a cut of extreme concentrations reduces the ion deposition over proportionally.

1. Introduction

The importance of the contribution of the fog water interception by trees for the hydrological cycle has been recognized early (Linke, 1916; Grunow, 1957). It can therefore be expected that appreciable amounts of trace substances are deposited via fog interception. Among the hypothesis, which have been discussed in connection with the forest damages, acid deposition has proved to be substantial. While acid deposition via rain can hardly explain the regional variation of the damage pattern, acid deposition via fog interception is locally very variable. The intercepted material is dripping to the ground only under the trees and can therefore damage the ground exactly at the same place where trees are rooting.

The deposition process by fog interception is highly complex because it is influenced by numerous parameters as the wind speed, liquid water content and drop size distribution, the geometry of the trees, the wind and turbulence profile in the stand, the capture efficiency of fog droplets and the chemical composition of the fog water, which in turn may depend on wind direction and the relative distribution of pollutant sources.

In our study we calculate the fog water deposition by interception by means of a one dimensional resistance model (Lovett, 1984, 1986). In order to obtain realistic results, we used observed meteorological input data from routine weather observations. Additionally, we collected fog- respectively cloud water at several mountain stations and analyzed the samples for chemical constituents. By combining the calculated water fluxes to the model forest with the measured chemical constituents, ion depositions were obtained. Some preliminary results have been published in Kroll and Winkler, 1988.

2. Chemical composition of fog water

Cloud water has been collected with an impactor sampler (Winkler, 1986). The calculated 50 % cut off radius is 2.3 μm which has been confirmed by experiments in a fog wind tunnel. At the six stations:

> GA: Großer Arber (Bavarian forest),
> HP: Hohenpeissenberg (pre-alpian region),
> SL: Schauinsland (black forest),
> KF: Kleiner Feldberg (Taunus, near Frankfurt),
> WK: Wasserkuppe (Rhön),
> KA: Kahler Asten (Sauerland, near Ruhr Area),

a total of 400 samples has been collected and analyzed for $SO_4^=$, NO_3^-, Cl^- (ion chromatography), pH, electrical conductivity, NH_4^+ (ion sensitive electrode), Pb, and Mn (AAS). The following table 1 shows the average concentrations. For reasons of comparison the average rain water composition of Brotjacklriegl in the bavarian forest is included. The liquid water content (LWC) was determined from the sampled air and water volumes. Note, that the presented concentrations have been normalized to a uniform LWC. To obtain absolute concentrations multiply the indicated values with 0.1/LWC.

From table 1 we see that the composition of fog- respecitvely cloud water shows an appreciable regional variation which is larger than that of rain water. This phenomenon may be confined to the region near the cloud base, because mountain stations do usually not rise much above the cloud base. The average ion concentration in fog water is higher than in rain with enrichment-factors varying between 2.5 ($SO_4^=$) and 5.2 (Cl^-). Note, that in fog water the NO_3^- concentration exceeds that of $SO_4^=$, opposite to rain water at most sites. The ion concentration of individual samples shows larger scatter as is demonstrated in fig. 1 for $SO_4^=$ at the station GA. It

Table 1: Average concentrations of trace constituents in fog water. For reasons of a better comparability all concentrations have been calculated for a standard liquid water content (LWC) of 0.1 g/m³. For reasons of comparison the average rain water composition at Brotjacklriegl (BJ), also situated in the Bavarian forest, is added.

STATION		HP	KA	WK	SL	GA	BJ
conductivity	(μS/cm)	85	206	70	73	164	32
pH-value		4.82	3.70	4.21	4.07	3.83	4.40
sulfate	(mg/l)	5.3	10.5	5.2	3.5	12.5	5.0
nitrate	(mg/l)	6.3	14.2	5.1	2.9	13.8	4.4
chloride	(mg/l)	1.1	2.1	1.7	0.9	2.6	0.5
ammonium	(mg/l)	3.1	5.8	2.5	2.4	5.6	1.4
lead	(μg/l)	77	86	57	116	115	
manganese	(μg/l)	5	13	7	7	12	
number		38	126	20	22	195	
LWC	(g/m³)	0.025	0.057	0.055	0.042	0.127	

should be emphasized that at this station supercooled fog water was also collected during winter. No yearly cycle can be observed. Extremes exceed the average concentration by more thana factor of 10. By means of trajectory analysis we found that extreme concentrations at GA were cau sed by the industries in DDR and CSSR but also by large cities like Munich or Nürnberg. It is also remarkable that the station KA near the Ruhr area shows the highest average concentrations.

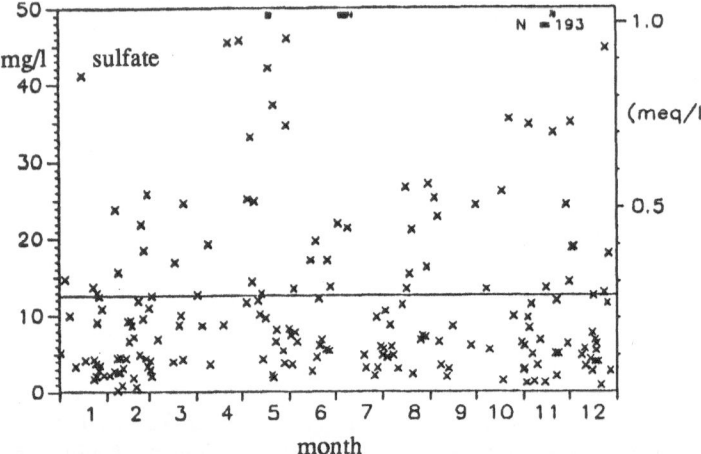

Fig. 1: $SO_4^=$ concentration of fog water samples collected at Grosser Arber, bavarian forest.

3. Calculation of the water flux

The deposited amount of fog water is calculated by means of the one dimensional resistance model of Lovett (1984). The model needs wind velocity and liquid water content (LWC) above the canopy as input parameters. The model forest is a balsam fir forest of 10.5 m height to which we have adhered for various reasons. The canopy is subdivided into layers of 1 m height. The transport from layer to layer is calculated by evaluating the wind and turbulence profiles, which are exponentially decreasing functions of the cumulative leaf area index. In each layer the water flux to seven vegetation components is calculated: to three classes of needle bearing twigs, three classes of bare twigs, and boles. The resistances to these components are depending on a relation between the capture efficiency and the Stokes number, which was determined experimentally (Thorne et al., 1982).

While Lovett (1984, 1986) used climatological average data to asses the cloud water deposition we are trying to become more realistic in that we use observed data from meteorological routine observations as input parameters. Since the LWC and the drop size distribution is not available we use the observed visibility to parameterize both parameters. By means of extensive sensitivity studies we proved that the parameterization of the size distribution is not critical (Kroll and Winkler, 1988, 1989). Different types of size distributions result in differences of the calculated water fluxes of only 10 %. The relation between visibility and LWC has been chosen in such a way that the parameterization is adjusted to observations of Garland, (1971), Chylek (1978), and Pinnick (1979).

With the modified model and observed meteorological data deposition rates of cloud water to the model forest have been calculated. Table 2 shows the calculated average deposition of water for the years 1982-1988.

Table 2: Average water deposition via fog interception

Station	GA	WK	Fe	KF	KA	HP
station height	1090	720	1060	650	640	480
fog deposition mm/a	1550	1140	990	520	270	70
rain deposition mm/a	1300	1140	1953	1730	1500	1203

The deposited amounts vary appreciable between the stations. This is caused not only by differences in the frequency of fog but also due to fog density and combinations of dense fogs with high wind velocities. At the stations GA and WK the fog deposition is as high or higher than the rain water amount. At the HP is makes up 6 % and the other stations between 18 an 51 %.

Nevertheless a correlation between the deposited fog water and the height of the station can be recognized. In fig. 2 we have not chosen the height above sea level but the height of a mountain above the average level of the flat terrain in that region (e.g. 450 - 500 m in Bavaria). Such a correction is necessary in order to obtain a meteorologically relevant height. We can see that the fog water deposition becomes an important factor if a mountain rises more than 600 m above the average area level.

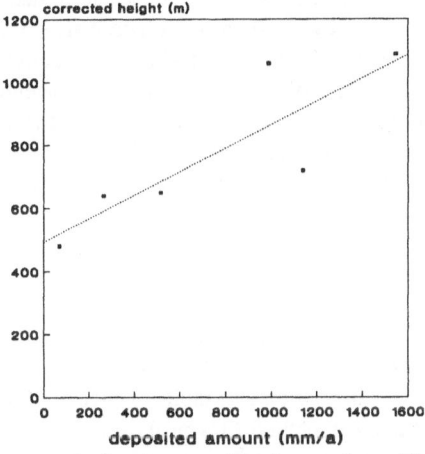

Fig.2: Calculated fog water deposition for different stations as function of the station height above the average level of the surrounding region.

In the course of the year, the highest deposition rates occur during the winter.

The figures presented in Table 2 are to be considered as a first assessment which are still affected with large uncertainties:

1) The weather observations are not available for every hour. We have assumed that the fog frequency during the missing hours is the same as during hours with observations. However, the missing hours occur during the night where the fog frequency is higher than during the day.

2) The visibility is reported in classes and the width of the classes is too broad for our purpose especially at low visibilities which are connected with high deposition rates. An representative average visibility has to be assumed in such a class. At stations where low visibilities occur very frequently (GA) the uncertainty in the calculated deposition rate due to the assumption of a representative value may become as large as 30 %.

3) No uniform relation between the capture efficiency for fog droplets and the Stokes number is reported in the literature. In fig. 3 we show three relations given by Thorne et al. (1982), Bache (1979), and Jonas (1984). Note that the range of Stokes numbers which is most relevant for the deposition of fog droplets, is between 0.2 <Stk<5. If we would use the relation of Jonas (1984) instead of the relation of Thorne et al. (1982) we come up with deposition rates which may be more than 40 % higher as the numbers given in table 2 (station GA).

4. Assessment of the ion deposition

A first assessment of the ion deposition can be obtained by a combination of the calculated fog water deposition with the measured average concentration belonging to the measured average liquid water content (table 1). If we use the meteorological data of the year 1985 we obtain for the two stations GA and KA the figures in colums 1 of table 3.

Table 3: Yearly ion depositions calculated with the Lovett model using various calculation modes. (For explanation see text). Yearly deposited amount in kg ha^{-1}yr^{-1}

		calculation procedure			
station	ion	1	2	3	4
GA	NH_4^+	64	37	38	25
	$SO_4^=$	143	89	92	58
	NO_3^-	159	101	105	61
	Cl^-	29	21	21	16
KA	NH_4^+	21	22	24	15
	$SO_4^=$	37	42	46	29
	NO_3^-	51	60	65	38
	Cl^-	8	10	11	6

The real deposition process is much more complex because the trace substance content in fog varies not only from event to event but also within the event and also with drop size. Since we do not have continuous measurements we apply the following approach in order to simulate the reality to a certain degree. For every day with fog we take one set with measured concentrations from our data pool of the respective station. From this set we calculate for each hour with fog a quasi-actual concentration by using the LWC which is derived from the actually determined visibility. By this procedure we can vary the LWC during the day corresponding to the observed visibility and consequently the concentrations of the ions also. At the next day with fog we take the next data set for the trace constituents and vary it again during the day corresponding to the LWC variations and proceed until every set from the pool has used once. If more fog days occur than data sets are available we use the pool another time starting with the first set again. By this procedure we use all measured values of trace constituents in the fog water corresponding to the observed frequency. By summing up the calculated hourly deposition values the yearly deposition is obtained. This method leads to the values in column 2 of table 3 which are lower by 30-40 % as compared with the values obtained by the first method. Obviously, extreme values affect the average more than the consideration of their frequency.

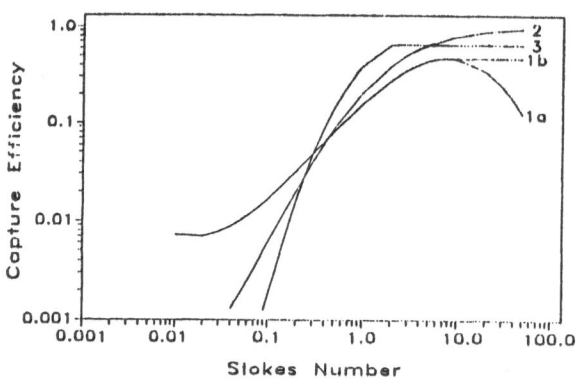

Fig. 3: Different relations between capture efficiency and Stokes parameter. 1: Thorne et al. (1984), 2:Bache (1979), 3: Jonas (1984)

We also assumed a radius dependent concentration where the small droplets are higher concentrated by a factor of 2 than the larger droplets. From fig.3 which depicts the data for SO_4^- and NO_3^- we see that this causes only a slight increase of the ion deposition which demonstrates that the deposition of small droplets is of minor importance.

Finally a scenario was assumed in which the highest concentrations (15% of all values) were cut and the remaining concentrations were reduced by 10%. The corresponding deposition values are shown in column 4 of table 3, demonstrating that the deposition does not react linearly. Avoiding peak concentrations would reduce the total deposition more than proportionally.

5. Conclusions

The ion deposition by fog interception is a very important process. Even at the station KA where the deposited fog water amount makes up only 20 % of the rain amount the ion concentration is so high that the deposition via rain and fog are of similar order of magnitude.

The fog deposition becomes important for mountains rising above the surrounding flat terrain by more than 600 m.

At present the assessment is still very uncertain because of large differencies of the capture efficiencies reported in the literature. If we would accept higher literature values the deposited amounts would rise by more than 40 %.

The fog- respective cloud water composition shows appreciable local and regional variations. The concentrations were usually higher than in rain water. Extreme concentrations exceed the average by more than a factor of 10.

For a realistic assessment of the ion deposition via fog interception the fluctuations of the meteorological parameters and the variations of the fog water composition have to be taken into consideration. Using climatological averages does not result in reasonable values.

A cut of peak concentrations reduces the ion deposition overproportionally.

Acknowledgment: This work was financially supported by the Bundesministerium für Forschung und Technologie under grant 07431018

6. Literature

Bache, D. H.: Particulate transport within plant canopies. I: A framework for analysis. Atm. Environ. **13** (1979) 1257-1262. II: Prediction of deposition velocities. Atm. Environ. **13** (1979) 1681-1687.

Chylek, P.: Extinktion and LWC in fogs and clouds. J. Atm. Sci. **35** (1978) 296-300.

Garland, J. A.: Some fog drop size distributions obtained by an impaction method. Quart. J. Roy. Met. Soc. **97**(1971) 483-494.

Grunow, J.: Probleme der Niederschlagserfassung und ihre Bedeutung für die Wirtschaft. Met. Rdsch. **9** (1956) 62-68.

Jonas, R.: Ablagerung und Bindung von Luftverunreinigungen an Vegetation und anderen atmosphärischen Grenzflächen. KFA Jülich GmbH (Abt. Sicherheit und Strahlenschutz) (1984) S. 1949.

Kroll, G., Winkler, P.: Estimation of wet deposition via fog.
In: K. Grefen, J. Löbel (eds.). Environmental Meteorology, (1988) 227-236.

Kroll, G., Winkler, P.: Influence of meteorological parameters on interception of cloud troplets in a coniferous forest. Cont.Atm. Phys. (1989), in press

Linke, F.: Niederschlagsmessungen unter Bäumen. Meteorol. Zeitschr. (1916) 141.

Lovett, G.M.: Rates and mechanisms of cloud water deposition to a subalpine balsam forest. Atm. Environ. **18** (1984) 361-371.

Lovett, G. M., Reiners, W.A.: Canopy structure and cloud water deposition in subalpine coniferous forests.Tellus **38B** (1986) 319-327.

Pinnick, G. R., Jennings S. G., Chylek, P., Auvermann, H. J.: Verification of a linear relation between IR extinction, absorption and LWC of fogs. J. Atm. Sci. **36** (1979) 1577-1586.

Thorne, P. G., Lovett, G. M., Reiners, W. A.: Experimental determination of droplet impaction of canopy components of Balsam fir. Journ. Appl. Met. **21** (1982) 1413-1416.

Winkler, P.: Observation on fog water composition in Hamburg. In: Atmospheric Pollutants, H.-W. Georgii (ed.). D.Reidel Publ. Comp., Dordrecht/Holland (1986) 143-151.

AIRCRAFT MEASUREMENTS OF HYDROGEN PEROXIDE OVER THE NORTHEASTERN UNITED STATES

W. Junkermann and F.Slemr
Fraunhofer-Institut für Atmosphärische Umweltforschung,
Garmisch-Partenkirchen, FRG

Summary

Spatial distribution of hydrogen peroxide and of some other atmospheric trace gases (NO, NO_2, NOy, O_3, SO_2) was measured from an aircraft over the northeastern U.S. during an extended field campaign in August and September 1988. Hydrogen peroxide was measured with an improved instrument based on the enzyme catalyzed fluorometric technique.
The observed spatial distribution of hydrogen peroxide and of other trace gases was strongly dependent on meteorological conditions. High mixing ratios of hydrogen peroxide were found in dry and slightly polluted air masses. Moist and more strongly polluted air masses were characterized by low mixing ratios of hydrogen peroxide despite the indication of high photochemical activity by high ozone mixing ratios.

Introduction

Hydrogen peroxide is considered to be one of the most important oxidizing reagents in the atmosphere and one of the major secondary pollutants produced by photochemical reactions. The reaction of H_2O_2 with SO_2 in clouds seems to be the most important oxidation pathway from SO_2 to sulfate. H_2O_2 is produced by recombination of hydrogen radicals and, consequently, its measurement can yield information about the budget of these radicals. For these reasons H_2O_2 is considered to be one of the most critical species in the diagnostic evaluation of Eulerian models. We report here preliminary results from aircraft measurements of the spatial distribution of H_2O_2 and some other gases made for the U.S. Environmental Protection Agency above the Northeastern U.S. in order to provide a data base for an acid deposition model evaluation.

Experimental

The H_2O_2 measurements were made with an instrument based on the fluorometric technique described by Lazrus et al. (1986). This technique employs the formation of a fluorescent dimer by the reaction of hydrogen peroxide with p-hydroxyphenylacetic acid catalyzed by horseradish peroxidase. The instrument used in this study has been designed specifically for operation on airplanes with pressurized cabins. Thermal mass flow controllers allow a pressure independent operation down to 500 hPa, corresponding to a ceiling height of approximately 5500 m.

The ambient air sample for all instruments was pumped in through a Pitot tube and a manifold at a rate of more than 50 l(STP)/min. From this manifold, an air sample was drawn into the instrument through a 1.5 m long 1/4" PFA tubing at a rate of 2 l(STP)/min. The loss of H_2O_2 within the inlet tubing was checked using a built in permeation source, and was found to be less than 3%. As the instrument is a computer controlled unit and has an internal set of reagents, it can be operated unattended for up to eight hours. It can be calibrated during the flight either using liquid standards or a calibration gas mixture from a built in permeation device.

The noise of the improved detection electronics was below 20 pptv H_2O_2. At such low H_2O_2 levels the detection limit is H_2O_2 produced by the reactions of ozone with water within the stripping coils. This ozone interference is smaller than 30 ppt for an ozone mixing ratio of 100 ppb (Lazrus et al., 1986). The instrument response, which has a time constant of 25 s, is delayed by 150 s due to the time required for the completion of the reaction and for the transport of the liquids from the sampling coil to the fluorescence detector.

The H_2O_2 measurements were only a small part of the program which included continuous monitoring of ozone by ethylene chemiluminescence and UV photometry, SO_2 by pulsed fluorescence, total sulfur by flame photometry, NO and NOy by ozone chemiluminescence, and NO_2 by luminol chemiluminescence. The program also took grab samples of air and filter samples of aerosols for analysis of hydrocarbons, halocarbons, and major ions.

Flight patterns

The flights were made from Columbus, Ohio, between August 15 and September 27, 1988. The flight altitudes varied between 150 and 3200 m above ground. Four different flight patterns have been flown to cover a wide range of meteorological conditions and different levels of air pollution. These were:

a) "Zipper/Curtain" pattern (Fig. 1a). This pattern was designed to provide moderate spatial coverage over a broad region covering both source and receptor areas. It was flown several times between Columbus, Ohio and upstate New York up to Whiteface Mountain, as indicated in Fig. 2.

b) "High Resolution Box" (HRB): This pattern, shown schematically in Fig. 1b, was flown three times in a given day, in order to provide data with high density coverage within a small area, for testing models of air chemistry leading to the formation of oxidants and other secondary pollutants. The position of the HRB is shown in Fig. 2.

c) "Long Narrow Corridor" and "Pre-/Post-Frontal" patterns were flown to obtain data from differently polluted air masses and to investigate washout of trace species and regeneration of secondary pollutants. The long corridor extended from upstate New York to Atlanta, Georgia.

Results

a) "Zipper/Curtain" flight on August 31, 1988:
Fig. 3 shows the distribution of H_2O_2 as a function of altitude above MSL and of the airplane position. During the first part of the flight only slightly polluted moist air (relative humidity 70-100%, scattered cumulus clouds) was found with a layer of dry air above 2700 m. H_2O_2 was fairly evenly distributed in the moist air and its mixing ratio was about 200 pptv. In the dry air above 2700 m, the H_2O_2 mixing ratio increased to 400-500 pptv. The accompanying measurements of NO_y, SO_2 and O_3 did not show any vertical structure during this part of the flight either in moist or in dry air.

The situation changed in the western part of the flight. The air masses below 2500 m were heavily polluted, as indicated by high mixing ratios of NO_y, SO_2 and O_3, and the H_2O_2 mixing ratios decreased to values below 100 pptv in these air masses despite the indication of high photochemical activity by high ozone mixing ratios. Between 2500 and 3200 m dry warm air masses dominated, which were less polluted and contained up to 1.1 ppbv H_2O_2. One area within the polluted air mass with a peak value of H_2O_2 of 1.2 ppbv could be explained by an intrusion of air from higher altitudes caused by a cumulus cloud.

b) "High Resolution Box" flights on September 1, 1988
The HRB-pattern was flown three times with a time lag of approximately 2 h between the flights. The data shown in Fig. 4 are averaged over all 16 vertical profiles of one box flight. The broad bands for the individual components indicate the scattering of the concentrations within the two hours flying time. The vertical structure of the air mass was characterized by a strong inversion layer at 2200 m which remained stable during the day.

During the first flight (Fig. 4a) the H_2O_2 mixing ratios reached a maximum value of 1.2 ppbv just above the inversion layer. They decreased continuously to about 0.4 ppbv above the ground and to about 0.7 ppbv at 3200 m. High pollutant concentrations were found only up to 1600 m. During the second flight (Fig. 4b) the peroxide in the uppermost level increased by up to 30 %. A slight decrease of H_2O_2 was apparent at altitudes of about 1000 m which was correlated with an increase in SO_2.

During the third flight (Fig. 4c) H_2O_2 was almost evenly distributed over all altitudes. Below the inversion layer this seems to be the result of strong vertical mixing which is also indicated by the reduction of the vertical ozone gradient observed earlier. Above the inversion layer in an area with high solar radiation fluxes no diurnal variation of H_2O_2 mixing ratios was apparent.

c) "Pre-/Post-Frontal" flights on Sept. 2, 6, and 8, 1988
On September 2 the pre-frontal flight was made along the track shown by the dotted line in Fig. 2. The same flight track was flown on September 6 and 8 after the passage of a

low pressure system. No significant change in H_2O_2 between flights before and after the frontal passage could be observed. The mixing ratios of NOy and SO_2 in the pre-frontal and post-frontal air masses were also not much different. However, the ozone mixing ratios increased after the passage of the front by 30% to approximately 75 ppbv. Two days later the H_2O_2 mixing ratio increased by a factor of two in the same region but no change in the average mixing ratios of the other gases was found.

Summary

Distribution of hydrogen peroxide was measured with high spatial and temporal resolution over the Northeastern U.S. from August 15 to September 27, 1988. High mixing ratios of H_2O_2 were found only in dry slightly polluted air masses. Low mixing ratios were observed in moist and more strongly polluted air masses even where high ozone mixing ratios indicated high photochemical activity. The variability of H_2O_2 mixing ratios seems to reflect more the transport processes than the short time photochemical reactions.

References

A.L. Lazrus, G.L. Kok, J.A. Lind, S.N. Gitlin, B.G. Heikes and R. Shetter; Automated fluorometric method for hydrogen peroxide in air, Anal. Chem. **58**, 594-597, 1986.

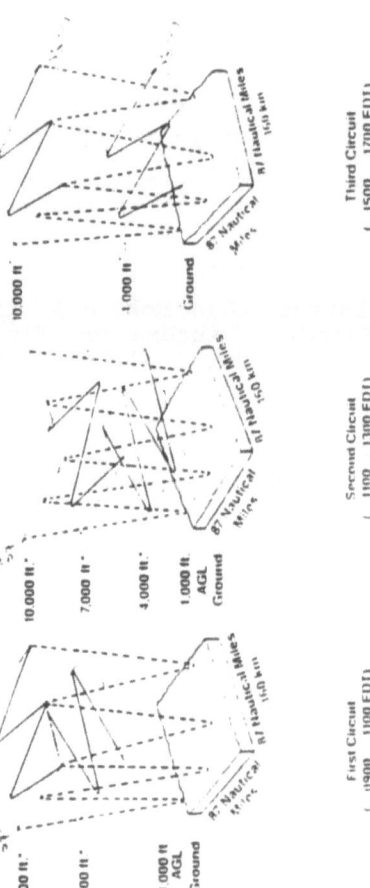

Figure 1 : Aircraft Sampling Patterns

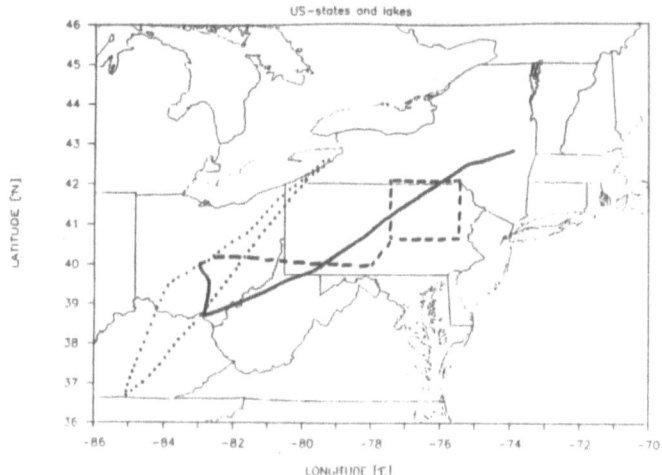

Fig.2: Flight tracks for different patterns
Solid line -> Zipper/Curtain, dashed line ->
High Resolution Box, dotted line -> Pre/Post
Frontal

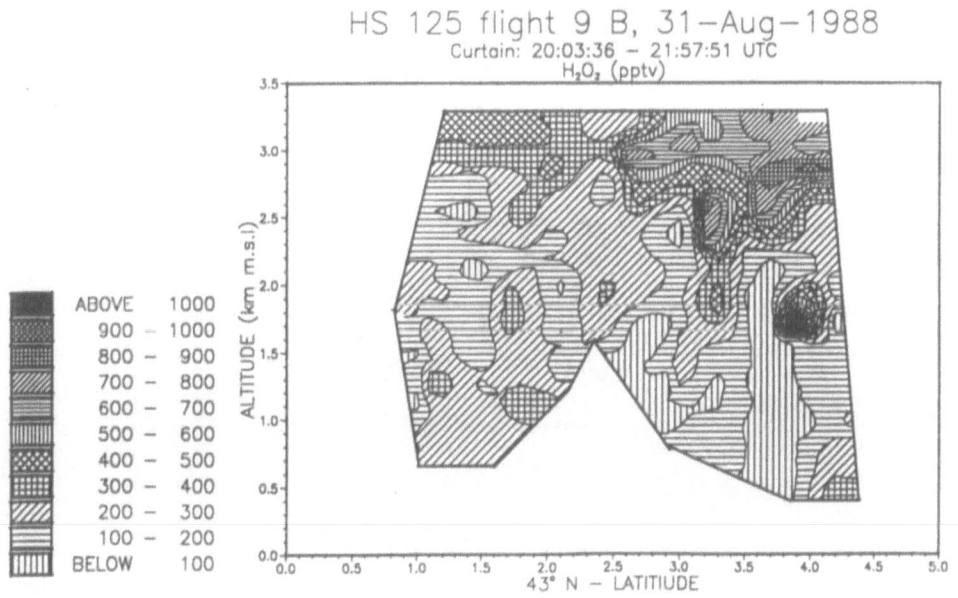

Fig.3: Vertical distribution of H_2O_2, Curtain, Aug.31, 1988

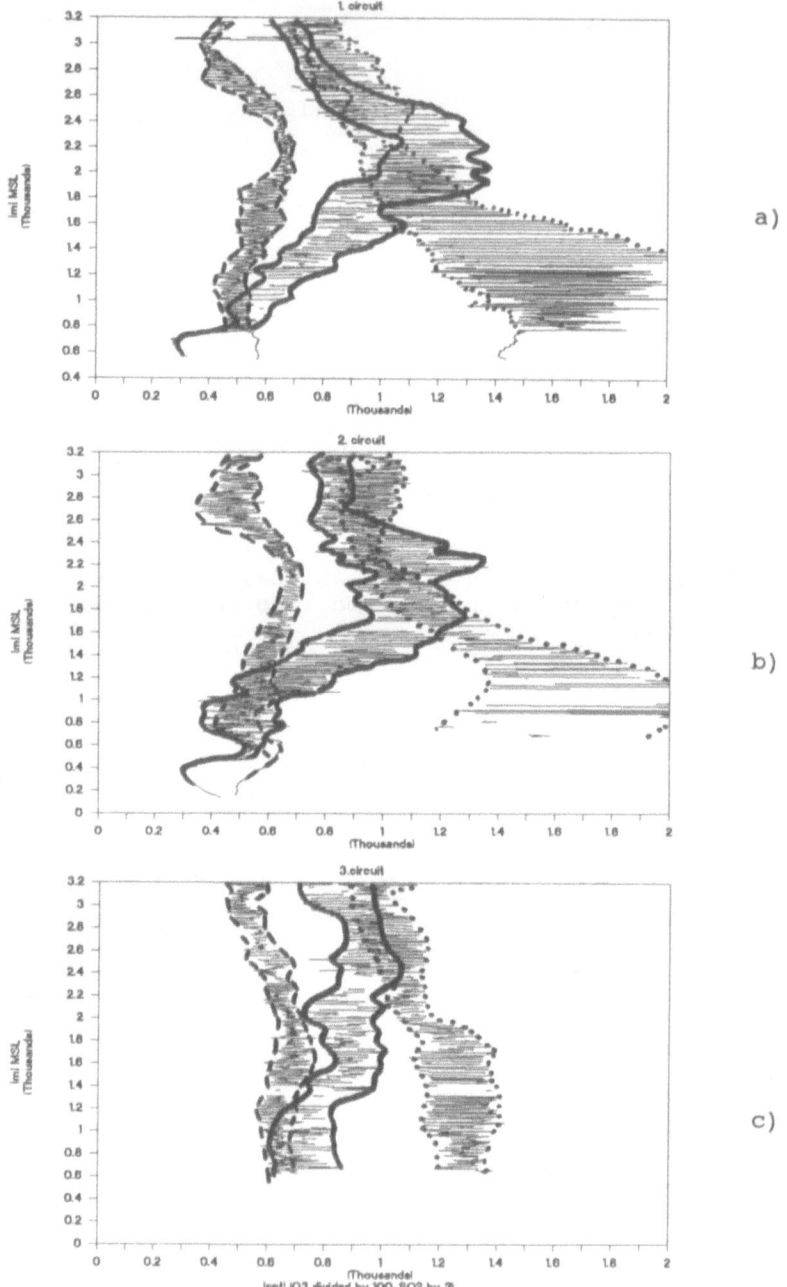

Fig.4: Concentration of H_2O_2, SO_2 and O_3 in the High Resolution Box, averaged over 16 individual profiles for each circuit. Solid line H_2O_2, dotted line SO_2, dashed line O_3. a) 8:40 -10:15 am, b) 10:15 -11:52 am, c) 1:43 -3:30 pm.

RESULTS OF AIRCRAFT MEASUREMENTS OF TRANSBOUNDARY MASS FLUXES OF AIR POLLUTANTS OVER THE FEDERAL REPUBLIC OF GERMANY

D. Paffrath and F.M. Rösler
German Aerospace Research Establishment (DLR)
Institute of Atmospheric Physics

SUMMARY

Airborne concentration measurements of selected pollutants were carried through during a period of two years from 1987 to 1988 over the Federal Republic of Germany (FRG) to determine transboundary mass fluxes of SO_2, NO, NO_x, O_3, and particles, and of some anion and cation components of the aerosol collected on filters. The concentrations of the trace gases were measured real-time on board the DLR Environmental Research Aircraft as well as meteorological parameters, while particle samples taken during the flights were analysed afterwards in the laboratory. In this paper examples of the results of concentration distributions along the boundaries of the Federal Republic of Germany and of transboundary mass fluxes of air pollutants are presented and discussed.

1. INTRODUCTION

The air pollution impact on a receptor at the ground is a function of the immission concentrations from local or near-by sources as well as from remote sources (long-range and transboundary transport).

The German Umweltbundesamt designed a project for the quantitative measurement of transboundary mass fluxes of air pollutants at the boundaries of the Federal Republic of Germany. The measurements were carried through by GEOSENS bv, the Netherlands, and by the German Aerospace Research Establishment (DLR).

Aircraft measurements of transboundary mass fluxes of SO_2, NO, and NO_2 at the border between Bavaria and the CSSR were performed by Paffrath (1985a) (1). The airborne instrumentation is described by Paffrath (1985b) (2). The results were compared with measurements of mass fluxes obtained by a mobile COSPEC remote sensing instrument on a van from the German Umweltbundesamt (Beilke, Berg, and Grosch, 1985) (3) on the one hand and of the Bavarian Landesamt für Umweltschutz (Rabl, Beilke, and Paffrath, 1987) (4) on the other hand. Aircraft measurements of pollutant mass fluxes along the whole border of the F.R. Germany were then carried through by GEOSENS (1988) (5) and results were published by Lelieveld, Jansen and van Dop (1989) (6). The German Aerospace Research Establishment (DLR) employed an instrumented airplane to measure mass fluxes of various pollutants along the borders of the F.R. of Germany (e.g. Paffrath, Peters, Rösler, and Baumbach, 1987) (7).

2. MEASUREMENT PROCEDURE

The airborne equipment (220 kg) was installed in a twin-engined propeller-driven Queen Air Be-65. The instrumentation allows the real-time measurement and recording of the meteorological parameters air temperature, air pressure, relative humidity, dew point temperature, and visibility (scattering coefficient), of the trace gases SO_2, NO_x, NO, O_3, of the concentrations of the particle mass (deduced from scattering coefficient) and of aitken nuclei. Furthermore, aerosol was collected on membrane filters, and the samples were analysed for anions CL^-, NO_3^- and SO_2^- and for cations Na^+, K^+, and NH_4^+ afterwards in the laboratory.

The SO_2 was measured by a flame photometer detector, detection limit = 1 ppb, NO, NO_x by a ozone-chemiluminescence analyzer, detection limit = 2 ppb, and the ozone by a ethen-chemiluminescence monitor, noise = 3 ppb. The scattering coefficient was determined by an integrating nephelometer, and the aitken particles by a condensation nucleus counter. The time constant of the instruments was in the range between 2 and 5 seconds, the flame photometer detector about 10 seconds and more, depending on magnitude of concentration variation.

The well-known method of mass flux calculation is based upon the simultaneous measurement of the concentration and the wind field within a vertical plane through which the flux is to be determined according to the equation:

$$\int_{\xi_2}^{\xi_1}\int_{z=0}^{z_1} c(z, \xi) \cdot v(z, \xi) \cdot \cos \varphi(z, \xi) \cdot d\xi \cdot dz$$

where c is the concentration, v the wind velocity, φ the angle between the normal unity vector of the plane element $dF = d\xi \cdot dz$ and the wind vector v, ξ the coordinate in flight direction, and z the coordinate perpendicular to the earth surface. While the concentrations were measured by the airborne equipment the wind informations were usually taken from the routine measurements of the German Weather Service (DWD) supplemented by pilot balloone soundings of the DLR mobile ground station (measuring van) which was on its way along the flight path.

The measuring flights were performed along quasi-horizontal flight paths near the borders at the lowest possible altitude in the atmospheric mixing layer. The bounder-parallel quasi-horizontal traverses were supplemented by vertical ascents and descents of the airplane. From the traverses the horizontal concentration distributions of the pollutants were obtained while the vertical profiles yielded informations on the vertical structure of the atmosphere and the vertical distribution of the air pollutans. Since the vertical structure of the atmosphere is a large-scale phenomenon and - during stationary conditions - varies only little from place to place, it seems to be appropriate to measure vertical profiles every 100 or 150 km, which was practicable accordings to the speed and the range of the aircraft. The vertical profile of all pollutants at each point of the flight path was determined by linear interpolation between the two neighbouring profiles really measured. In the first step this procedure yields a normalized vertical profile of the concentration. We may obtain good approximations of the real concentration profiles if in a second step the normalized profiles are multiplied by the real concentration values c(H) measured at flight altitude. This is true if the atmospheric boundary layer is well mixed and the concentration at flight level is not influenced by near-by sources the immissions of which may not have been dispersed sufficiently within the mixing layer. Such effects, of course, have to be taken into account.

On the basis of this procedure we obtained the mass fluxes for SO_2, NO, NO_x, O_3, and particles. The particle mass concentration was calculated from the scattering coefficient b_{scat} by the equation:

$$c_{part} = 3.8 \cdot 10^5 \cdot b_{scat.}$$

For ion components mass flux values could be estimated from ion concentration measurements with a very sensitive ion chromatography technique and a special procedure.

Since the ion components to be measured are constituents of the aerosol which is collected on the filters, it is, therefore, a plausible approximation that the ion concentration is proportional to the aerosol mass concentration. This is, of course, not true for horizontal variations of the ion concentration depending on the geometric distribution of the sources. If, however, the immissions are well mixed within the boundary layer, then we may assume that

$$c_{ion}(h) = k \cdot c_{part}(h) = k \cdot 3.8 \cdot 10^5 \cdot c_{part}(h)$$

The proportionality factor k can be determined from the ratio of the ion concentration $c_{ion}(H)$ at flight level H to the particle mass concentration $c_{part}(H)$ at the same point. Consequently the mass flux of any ion component F_{ion} may be calculated from the mass flux ΔF_{ion} per km:

$$\Delta F_{ion} = \sum_{i=1}^{n} \frac{c_{ion}(H)}{b_{scat}(H)} \cdot b_{scat}(h_i) \cdot v(h_i) \cdot \cos \varphi(h_i) \cdot \Delta s \cdot \Delta h_i$$

with $\Delta s = 1000$ m. It should be noted, however, that looking at the results, we should be aware of the fact that the values are only raw estimations.

3. RESULTS

The overall results of the investigation are documented by Paffrath, Peters and Rösler (1989) (8). Only a few examples can be presented in this paper. In fig. 1 the concentration distributions of SO_2 and NO_x along the easterly border of the F.R. of Germany are shown together with the calculated mass fluxes at various sections as well as the total mass fluxes. Fig. 2 shows the results at a very similar weather situation two days later at the westerly border. The highest mass fluxes of SO_2 were found at the border section between Weiden (Northeast Bavaria) and Kassel which means that the main sources for this heavy input of SO_2 into Germany are situated in the southern part of the German Democratic Republic (GDR) and the northern part of Czekoslovakia (CSSR). The mass flux in this section is 80 % of the total mass flux of 526 t/h. The NO_x mass fluxes (total 266 t/h) are nearly half of the SO_2 and the GDR/CSSR contribution to the total NO_x flux is only 50 %. Fig. 2 shows that the SO_2 concentration along the westerly border is more evenly distributed and that the total mass flux (601 t/h) is only slighlly higher than that at the easterly border. (Comparing the two figures note that the scale in fig. 2 is another one that in fig. 1). On the other hand the output flux of NO_x at the westerly border of Germany is about two times higher than the input flux at the easterly border. This means that SO_2 is transported across the FRG without any considerable change, or with other words, that the deposition of SO_2 over the area of the FRG is just compensated by the emissions from the sources in the FRG. Even the SO_2 contribution of the Ruhr industrial area is small compared with the whole mass flux. NO_x, however, is produced to a large extent in the FRG such that more NO_x is exported than imported. This is mainly due to the widespread and dense automobile traffic all over the country and to the fact that with all energy conversion processes fuel is burned at high temperatures.

The table shows the total values of mass fluxes along the flight routes together with the date of flight, the starting and ending point, the length Δs of the section, and with the mean wind direction during the flight (plus = import into the FRG, minus = export out of the FRG).

table: total values of mass fluxes along border sections (t/h)

date	section Δs	Δs km	wind	SO₂	NO	NOₓ	O₃	part.
09.03.87	Teisendorf-Bd.Hersfeld	599	ENE	427	12.3	91	759	1231
10.03.87	Bd.Hersfeld-Teisendorf	599	ENE	480	18.4	81	n.m.	675
12.03.87	O'hofen-Bd.Hersfeld	318	E	840	25.6	124	322	1116
12.03.87	Giessen-Hornisgrinde	260	E	-567	-14.7	-123	-253	-605
06.08.87	Teisendorf-Braunschwg.	747	W	-67	-32.9	-133	-602	-353
07.08.87	Braunschwg.-Teisendorf	747	SSW	-8	-3.2	-18	-272	-102
24.08.87	Teisendorf-Hof	348	SE/SSE	2	4.5	16	17	34
31.08.87	Teisendorf-Braunschwg.	747	NNW	-26	-19	-84	-238	+97
01.09.87	Teisendorf-Braunschwg.	747	S	64	7.5	0	-217	-701
11.09.87	Teisendorf-Braunschwg.	747	WSW	-85	-34.5	-173	-1488	-980
12.09.87	Braunschwg.-Teisendorf	747	SW	-134	-129	-511	-2439	-2049
05.04.88	Teisendorf-Braunschwg.	747	ENE	526	22.1	266	1198	2651
06.04.88	Münster-Freiburg	556	ENE	-1028	-34.5	-598	-1660	-3759
07.04.88	Freiburg-Bremerh.	768	ENE	-601	-43.5	-505	-1363	-2735
08.04.88	Fallingbostel-Haßfurt	330	NNW	-19	-1.4	-23	+67	+423
06.09.88	Norderney-Freiburg	800	NW	118	40.7	555	1451	722
08.09.88	Teisendorf-Bremerh.	947	NE	228	11.4	72	1549	814
08.09.88	Bremerh.-Freiburg	947	SE	-259	-31.2	-179	-1505	-1166
13.09.88	Karlsruhe-Bremerh.	648	WNW	206	62	215	1500	822
13.09.88	Bremerh.-Teisendorf	947	W	-81	-29.9	-136	-1350	-763

To compare mass flux values of different flights one has to take the different lengths of the sections into account. If we do so by forming the specific mass fluxes in t/h/km it can be seen that the values are similar for unique wind directions: the standard deviations are below 33 %, although the standard deviations considering all wind directions amount to 600 - 1000 %.

We want to acknowledge that the investigations were financially supported by the German Umweltbundesamt, Berlin and that Dr. S. Beilke gave valuable suggestions. We are grateful to Dipl.-Phys. W. Peters for his particular engagement in the processing of the data.

4. REFERENCES

(1) PAFFRATH, D., (1985a). Flugzeugmessungen der grenzüberschreitenden Luftverschmutzung im Raum Weiden-Hof/Bayern. DFVLR-FB 85-25.

(2) PAFFRATH, D., (1985b). DFVLR-Meßsystem zur Erfassung der räumlichen Verteilung von Umweltparametern in der Atmosphäre mit mobilen Meßträgern. DFVLR-FB 85-08.

(3) BEILKE, S., BERG, R., GROSCH, W., (1985). Preliminary Results on Measurements of Transboundary Fluxes of Air Pollutants across the Border between the FRG and its Eastern Neighbours. NATO/CCMS Pilot Study on Air Pollution Control Strategies and Impact Modelling. First Follow-Up Report, Experiences with the Application of Advanced Air Pollution Assessment Methods and Monitoring Techniques. No. 153.

(4) RABL, P., BEILKE, S., PAFFRATH, D., (1987). Messungen grenzüberschreitender Schadstofftransporte nach Nordost- Bayern. In: Lufthygienische Situation in Nordostbayern. Bayerisches Landesamt für Umweltschutz.

(5) GEOSENS, (1988). Messungen des Flusses von Luftverunreinigungen entlang ausgewählter Grenzabschnitte. Forschungs- und Entwicklungsvorhaben des Umweltbundesamtes, Nr. 104 04 221/01.

(6) LELIEVELD, J., JANSEN, F.W., VAN DOP, H., (1989). Assessment of Pollutant Fluxes across the Frontiers of the Federal Republic of Germany on the Basis of Aircraft Measurements. Atmosph. Environ. 23, No. 5, 939-951.

(7) PAFFRATH, D., PETERS, W., RÖSLER, F.M., BAUMBACH, G., (1987). Fallstudie über den Beitrag des Ferntransports von SO_2 zur lokalen Luftverschmutzung in der Bundesrepublik Deutschland. Staub - Reinhaltung der Luft 47, H. 7/8, ML 35-41.

(8) PAFFRATH, D., PETERS, W., RÖSLER, F.M., (1989). Messungen des Massenflusses von Luftverunreinigungen entlang ausgewählter Grenzabschnitte in der Bundesrepublik Deutschland. Abschlußbericht für das Umweltbundesamt (final report), to be published.

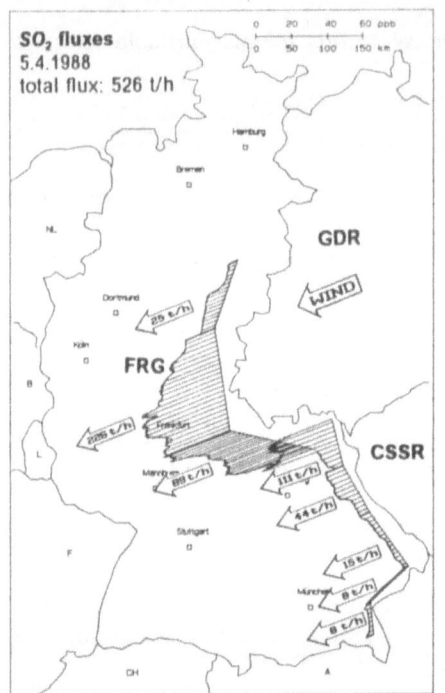

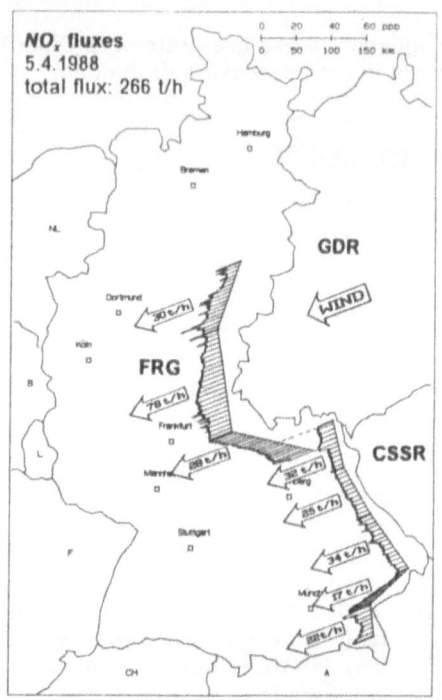

fig. 1

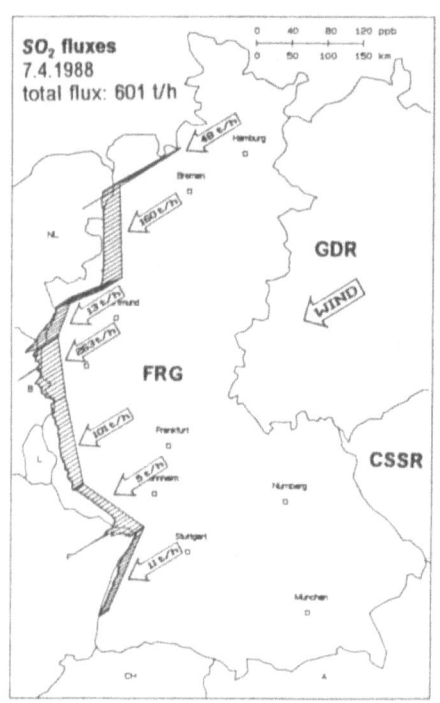

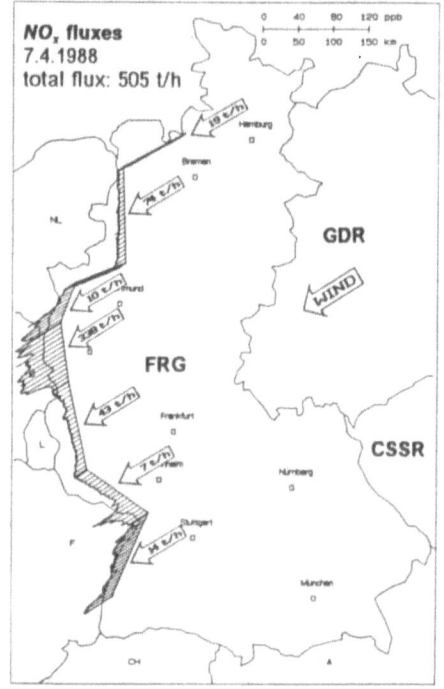

fig. 2

MEASUREMENTS OF GASEOUS HYDROGEN PEROXIDE IN

SOUTHERN GERMANY

L. Kins

Meteorologisches Institut der Universität München
Barbarastraße 16, D-8000 München 40,FRG

SUMMARY

Since 1987 different experiments are performed to study the behaviour of gaseous hydrogen peroxide (H_2O_2) in the atmospheric boundary layer. These experiments should give information about the formation and transportation processes of H_2O_2. The measurements comprise main gaseous species known to influence the H_2O_2-formation, such as ozone and nitrogen oxides, as well as detailed meteorological measurements like temperature, rel. humidity, windspeed and -direction, solar radiation and photolysis rates in different heights (0.2-50m) in the atmospheric boundary layer.

The daily cycle of the H_2O_2 mixing ratios with maxima during the early afternoon can be related to photochemical activity. Occasionally H_2O_2 mixing ratios during the night reach the values of the daily maxima. This observation can only be explained by transport processes.

1. INTRODUCTION

Hydrogen peroxide is an important component, both for the photochemistry of the troposphere and for the oxidation of sulphur dioxide in the aqueous phase.

The principle source of hydrogen peroxide is the combination of two hydroperoxy (HO_2) radicals

$$HO_2 + HO_2 \xrightarrow{H_2O,M} H_2O_2 + O_2$$

with the HO_2 radicals being generated through reactions of OH
radicals with hydrocarbons.

In presence of large amounts of nitrogen oxides (NO,
NO_2) the formation of hydrogen peroxide might be suppressed
by the reacions

$$HO_2 + NO \ \text{-->} \ OH + \ NO_2$$
and $$OH + NO_2 \ \text{-->} \ HNO_3$$

thus, both H_2O_2 and HNO_3 can be considered as sinks of atmospheric free radicals.

Measurements of the daily variations of peroxide and
nitric acid concentrations in air, together with the trace
gases (e.g. NO, NO_2) and the atmospheric processes (e.g.
solar radiation) known to influence the H_2O_2 and HNO_3 concentrations, may give some valuable insight into the free radical chemistry driving the production of photochemical oxidants.

2. EXPERIMENT

Measurements of H_2O_2, HNO_3, NO, NO_2, O_3, O_3-photolysis
rates and meteorology (Temp., r.H., wind, solar radiation)
are performed at the meteorological observatory at Garching,
north of the city of Munich, FRG.

During south to southwesterly winds air masses reaching
the location are transported across the city of Munich. North
and southeast of the experiment site only small towns with no
major industries are located within a distance of 100 km. One
kilometer to the west of the sampling site is a highway with
much traffic.

Gaseous H_2O_2 is collected by cryo sampling and analyzed
by chemiluminescence technique subsequently (Jacob et al.
1986). Figure 1 shows the sampling and analytical device for
the hydrogen peroxide measurements.

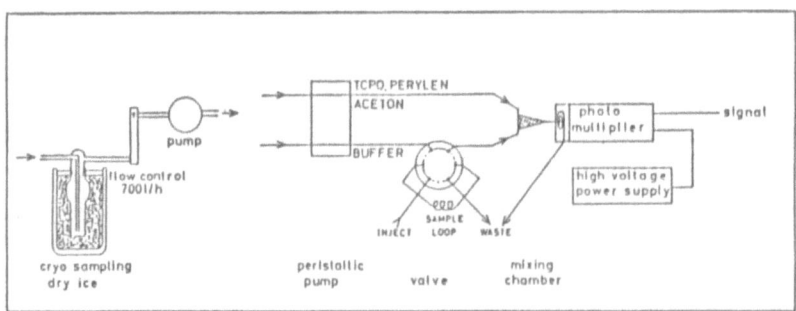

fig. 1: *H_2O_2 cryo sampling and analytical technique.*

Nitric acid is collected on filterpacks consisting of a teflon prefilter (0,2 μm pore size) for particle separation followed by two non-impregnated nylon filters for adsorbance of nitric acid (Meixner et al. 1985). The filters are analyzed by ion-chromatography (DIONEX 2020i), ozone and nitrogen oxides are continuously monitored by commercial instruments, O_3-photolysis rates are measured by photoelectric detectors developed by Junkermann et al. 1989.

The analytical techniques, their detection limits and the sampling times of the instruments used in our experiments are shown in table 1.

Tab. 1: Analytical techniques, their sampling rate and detection limits used during the experiments.

	technique/ instrument	sample (integration) interval	detection limit ppbv
H_2O_2	cryosampling/ chemiluminescence	30 - 60 min.	0,025 25°C/60% rh 0,005 0°C/60% rh
HNO_3	filterpack/ ion Chromatography	120 min.	0.025
O_3	UV-absorption Monitor Labs 8850	10 min.	2
NO,N_2	chemiluminescence Thermo Electron	10 min.	1

3. RESULTS

Two measurement campaigns were performed in 1988 (April and September).

Exemplary, some of the results obtained during these two periods are shown in the figures 2 - 6. Due to analytical problems the nitric acid measurements in April are scarce. Weather conditions were similar during both periods, with low windspeed, only few clouds, relatively high solar radiation. Rain occured only at the end of the measurement periods shown here.

The following features of the results are striking:

- H_2O_2 and HNO_3 mixing ratios show pronounced diurnal variation with maxima during noon to late afternoon;
  this reflects the photochemical production of both gases
- H_2O_2 mixing ratios in September are significantly higher
  than in April, HNO_3 mixing ratios are in the same concentration range with one exception during both periods. But
  NO_x is higher in September and ozone is lower in September
  than in April, also the radiation maxima are lower in September.

According to simple model predictions one expects under
these chemical and atmospheric conditions:
a) lower H_2O_2 concentrations than HNO_3 concentrations in
September
b) higher H_2O_2 concentrations than HNO_3 concentrations in
April
The measurements show exactly the contrary of these expecta-
tions.

Nocturnal increase of H_2O_2 mixing ratios occured several times,which can be attributed to advective effects,
either horizontal or vertical. In the example here it cor-
relates with increase in NO_x and a change of wind direction
from E to W prior to a rain event.

4. REFERENCES

Jacob P.,Tavares T.M.,Klockow D.,1986, Methodology for the
determination of gaseous hydrogen peroxide in ambient air,
Fres.Z .Anal. Chem. 325, 359-365

Junkermann W.,Platt U.,Volz-Thomas A., 1989, A photoelectric
detector for the measurement of photolyses frequencies of
ozone and other atmospheric molecules. J. Atmos. Chem. 8,
203-227

Meixner F.X., Müller K.P., Aheimer G., Höfken K.D., 1985,
Measurements of gaseous nitric acid and particulate nitrate,
in: F.A.A.M. Leeuw, N.D. van Egmond (eds.). Proceedings of
the COST 611 meeting "Pollutant cycles and transport:
modelling and field experiments", RIVM Bilthoven, The Netherlands.

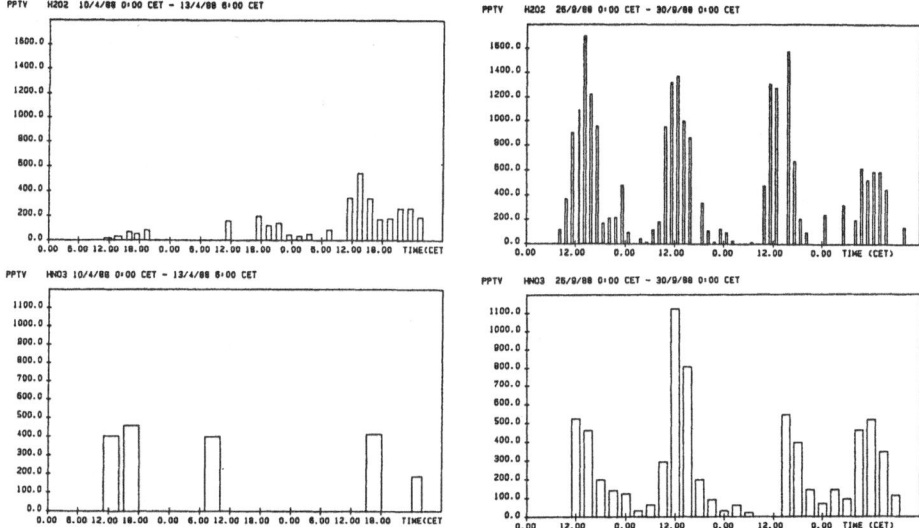

Fig.2: H_2O_2 and HNO_3 mixing
ratios measured between
10/4/88 and 13/4/88.

Fig.3: H_2O_2 and HNO_3 mixing
ratios measured between
26/9/88 and 30/9/88.

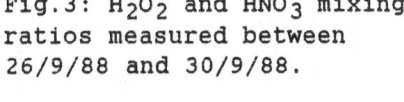

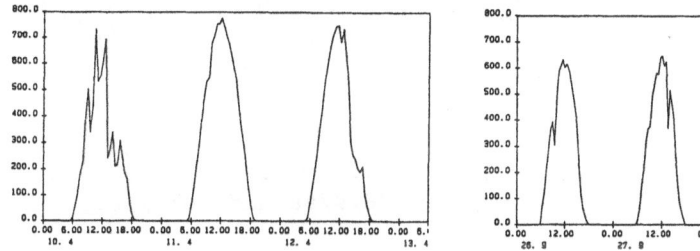

Fig.4: Solar radiation during the April (A) and the September
(B) measurements.

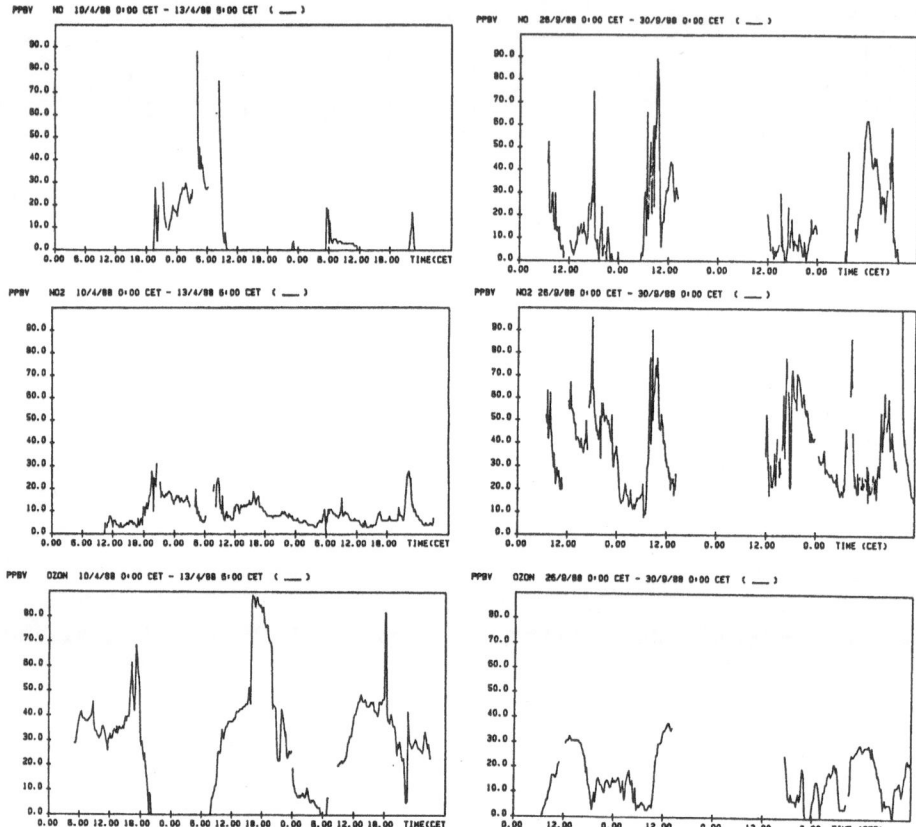

Fig.5: NO, NO$_2$, O$_3$ mixing ratios measured between 10/4/88 and 13/4/88.

Fig.6: NO, NO$_2$, O$_3$ mixing ratios measured between 26/9/88 and 30/9/88.

DATA INTERPRETATION FROM A TRANSPORTABLE LABORATORY IN A COMPLEX COASTAL SITE

J. Plaza, M.D. Andrés, M. Martín and M. Millán

CIEMAT-IPRYMA, Ed. 3b, Avda. Complutense 22, E-28040, Madrid

Summary

In July and August 1988 an experimental campaign was launched in the surroundings of the Castellón industrial area, located at the spanish East coast and considered as a complex coastal site. The campaign objective was to complement the mosaic on meso-scale flows over the whole Iberian peninsula, to determine the effects of these flows on the pollutant concentration fields.

A careful study of pollutant concentration cycles related to meteorological parameters enables the characterization of concentration patterns at the chosen location. Surface atmospheric dynamics, during the summer in this area, are controlled by local cycles of NW nocturnal drainage and E-SSE diurnal sea-breeze flows. These cycles are also affected by the formation of a thermal low over the center of the peninsula on summer days.

A transportable laboratory was used to register values averaged every ten minutes of O_3, NO, NO_2 and SO_2. This unit was placed in the zone of influence of the coastal industries during the sea-breeze regime. The temporal evolution of the pollutant concentrations was analyzed and shows the existence of repeated patterns of ground level concentrations (GLCs) under similar meteorological conditions. High ozone values have been detected to be associated with the sea-breeze onset.

On the basis of short-term backward trajectories, it has been possible to identify the source of the SO_2 and NO_2 peaks. The obtained concentration values can have a broad origin: rural, marine, traffic, oil fired power plant and refinery emissions, and urban plume. Besides, the meteorological transition periods can be characterized by changes on pollutant concentrations.

INTRODUCTION

The city of Castellón, 5 Km. inland from the Eastern Mediterranean Spanish coast, is placed in the northern part of the Mijares River flood plain (fig.1).

The Castellón industrial area is located at the shoreline, SE of the city. It includes an oil-fired power plant, which has two 540 Mw. units, each feeding a 150 m. stack. This plant operates intermittently, either for peak loading or backup purposes (only eight days over the whole campaign). It also includes an oil refinery, close to the power plant, in continuous operation, with a large number of emission points, none of them exceeding 50 m. height.

A site 8 Km. downwind ESE of the industrial area was chosen to place the transportable laboratory, wich provides data averaged every ten minutes of meteorological parameters and ground level concentrations (GLCs) of NO, NO_2, SO_2 and O_3.

DATA ANALYSIS

The wind rose obtained from autotransportable laboratory data (fig. 2) shows two main direction intervals: the first one includes E to SSE directions, approximately 50% of the time, and winds of variable intensity (1-7 m/s); the second interval includes W to NNW directions, 40% of the time, and weaker winds (1-3 m/s).

Previous work in this coastal site and analysis of former meteorological data allow us to identify these two intervals (1). The first one corresponds to directions involving sea-land breeze, and the second one to nocturnal drainage of the Mijares valley or land-sea breeze.

This type of local breeze-drainage circulation, daily cyclic, occurs without any variation during 95% of the experimental period.

The general pattern of pollutants is strongly influenced by both the daily meteorological cycle and the monitoring site. Fig. 4 represents the daily profiles from ten minute values of measured trace gases, averaged over the whole experimental campaign. Sulfur dioxide, involving slower transformations, presents noticiable concentrations during breeze and early drainage periods, probably due to a re-entry cycle. Fig. 3 presents GLCs of pollutants, averaged by direction sectors.

In general, a high averaged $[NO_2]/[NO]$ ratio might indicate clean air masses (not detectable NO), or recently oxidized by available ozone, whereas a lower ratio might indicate local injections of NO or an inhibition of mixing conditions that prevents NO oxidation.

Maximum $[NO_2]/[NO]$ ratio averages correspond to direction sectors of low frequency, ENE and SSW, and this fact could be explained considering that these values reflect short transition periods between predominant flows: diurnal sea breeze that introduces low concentrations of NO, therefore high NO_2/NO ratio, and nocturnal drainage that introduces higher concentrations of NO_x and low NO_2/NO ratio.

The sum of oxidant species ($[NO_2] + [O_3]$), averaged by direction sectors, could give some information about the potential foto-oxidant capacity of air masses that reach the monitoring site. In a HCs-free atmosphere, if the fotostationary equilibrium is established, ($[NO_2] + [O_3]$) would remain constant. Direct comparison of averaged values in this oxidant species rose could be made in a relative way. Several factors must be taken into account: HCs emissions that can shift this equilibrium, the mixing height, the night-time chemistry, the dry deposition rates of NO_2 and O_3, etc. However, the difference between averaged values of ($[NO_2] + [O_3]$) from marine origin and other directions is quite appreciable. This would implicate the existence of greater concentration of oxidants, mainly ozone at this time of the year, associated to sea-breeze directions (2). Fig. 5 shows a weekly data series (8-14 August), obtained from hourly averaged values. This series corresponds to the greatest ozone GLCs measured during the campaign. Fig. 6 and 7 summarize the main GLCs patterns and its associated meteorological parameters during the experimental time:
1. At night, the combined effect of drainage flow (NW to N) and surface

thermal inversions produces high NO concentrations, probably from A-7 highway traffic, and NO_2 concentrations, originated from NO conversion with near surface O_3 concentration.

2. Early in the morning (from 7 to 8 h. approx., 9 August), the solar radiation leads to the drainage-breeze transition: low wind speed, erratic wind direction and high temperature gradient. GLCs reflect a high O_3 concentration gradient, and a NO decrease. Although the NO_2 concentration does not vary, the NO_2/NO ratio also increase.

3. Eventually during the sea breeze, for example, from 11 to 12 h. the 9th August, a SO_2 and NO_x (mainly NO_2) increase is noticiable, associated to a O_3 decrease. The air mass back-trajectory (3) corresponding to this time period shows the monitoring site downwind the industrial area (fig. 8: outer trajectories were made using standard deviation of direction wind by a factor two). The origin of GLCs measured seems to be the oil refinery, since the power plant was not in operation. NO_x emissions reaching the laboratory are quite oxidized. Fig. 9 shows thirteen short-term back-trajectories of these NO_2 concentration increases, none of them exceeding 40 ppb.

4. Most part of the time during the sea-breeze regime, the ratio $NO/NO_2/O_3$ is constant. In these cases, wind direction never cross the ESE sector (fig. 6). The associated O_3 values could correspond to natural background at this time of the year or to the generation and transport of ozone from other areas (4), which adds to background concentration.

5. The SO_2 and NO_2 increases corresponding to operational days of the power plant are always associated to a sea-breeze rotation that cross the ESE sector. Fig. 10 shows eleven back-trajectories of chosen SO_2 peaks, all of them larger than 40 ppb. Some of these trajectories shift towards S direction, due to the difference between wind direction at the surface and at the emission height.

6. Fig. 11 shows back-trajectories of air masses that correspond to four ozone concentrations peaks, exceeding 90 ppb. These maximum O_3 values occur before midday, and are coincident with NO_2 increases. All these factors could suggest the relationship between maximum O_3 GLCs and precursor emissions (HCs and NO_x) from the oil refinery, that could increase O_3 concentration in approx. 15-20 ppb over its concentration in the sea-breeze and above layers. These two added phenomena, both the thermal convection and the HCs and NO_x emissions from the industrial area, could explain these early maximum values (5).

REFERENCES

(1) M. Millán et al., "Field measurements of plume dispersion in a complex coastal site", COST 611, 1988.

(2) B.S. Gimeno, V. Bermejo et al., "Efectos de la calidad del aire en rendimiento y calidad de cultivo de sandías", Internal Report Ciemat, abril 1989.

(3) A. Stern, H.C. Wohlers, R.W. Boubel, H.P. Lowry, "Fundamentals of Air Pollution", Academic Press, First Edition, 1973

(4) M.D. Andrés et al., "Análisis de los valores de concentración de contaminantes en un emplazamiento costero complejo: Castellón", Internal Report Ciemat, julio 1989.

(5) H. Gusten, G. Heinrich et al., "Photochemical formation and transport of ozone in Athens, Greece", Atmospheric Environment, 22, 7, 1335-1346, 1988.

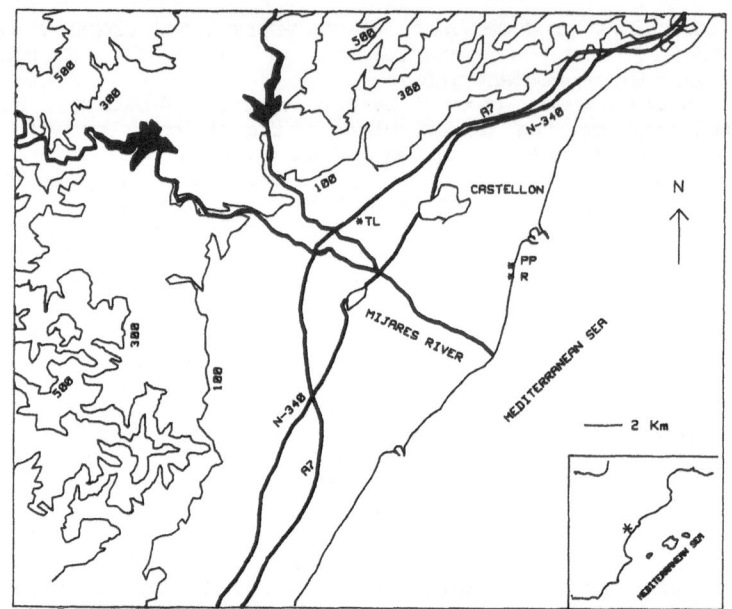

Fig. 1 General view of the area and location of Castellón.
TL: transportable laboratory. PP: power plant. R: refinery

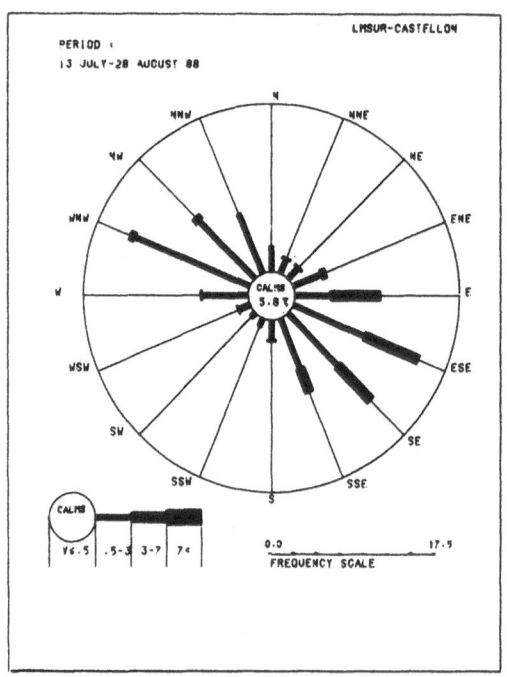

Fig. 2 Wind rose at the transportable laboratory site

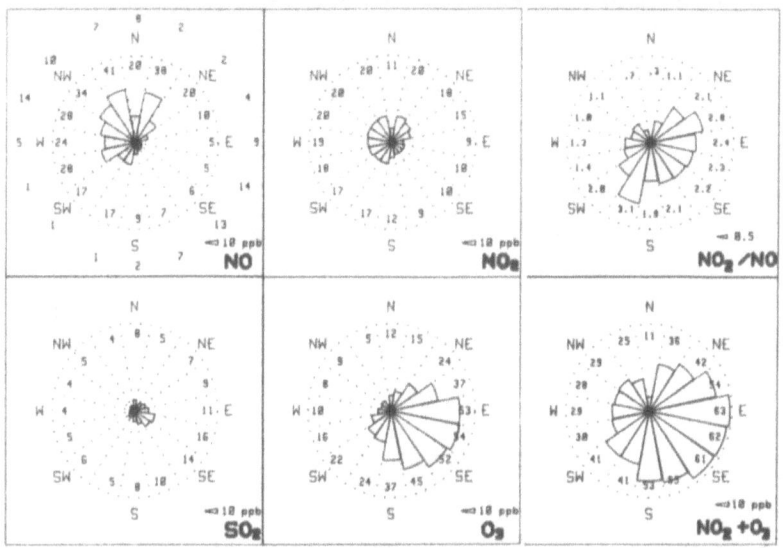

Fig. 3 Pollutant roses. Numbers inside the circles represent averaged values. Outside numbers are relative frequencies (%) of each direction sector

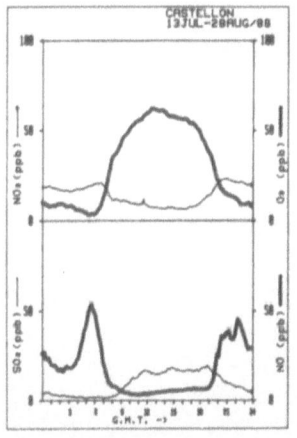

Fig. 4 Daily evolution average of pollutants

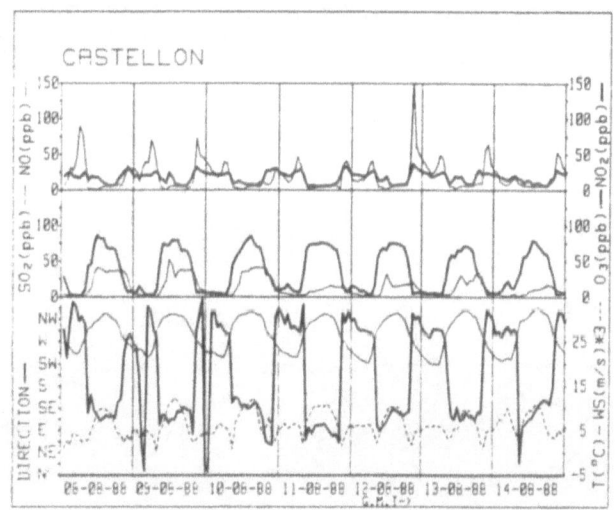

Fig. 5 Weekly series of hourly averaged values

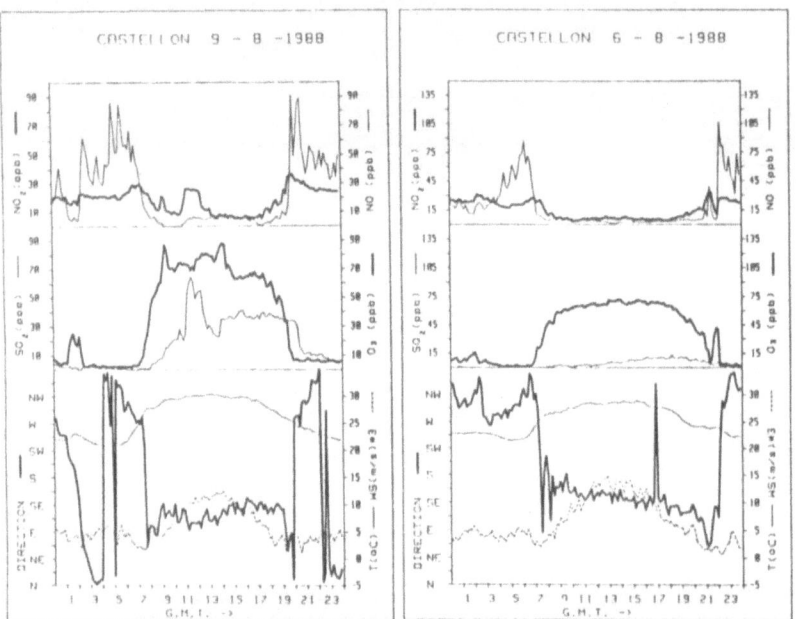

Fig. 6 and 7 Daily records from ten minute averages

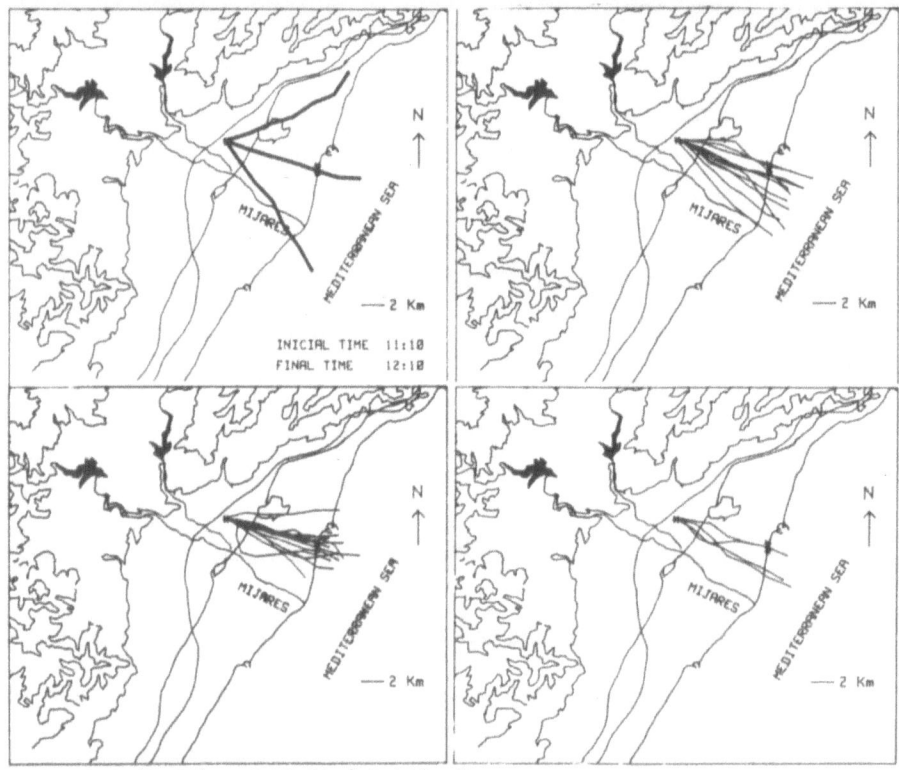

Fig. 8-9 (left) and 10-11 (right) Short-term back-trajectories

MEASUREMENTS OF NO AND NO$_y$ BETWEEN 0-14 KM ALTITUDE OVER EUROPE

D. BRÜNING and F. ROHRER

Institut für Atmosphärische Chemie
Kernforschungsanlage Jülich GmbH
Postfach 1913, D-5170 Jülich, F.R.G.

SUMMARY

Vertical profiles of NO and NO$_y$ between 0-14 km altitude were measured over Europe in July 1988 using a chemiluminescence detector. Measurements are presented from four aircraft flights made over the Atlantic coast of France and Ireland and over Central Europe. The vertical profiles of the mixing ratios exhibited a C shape with low values in the middle troposphere and high values near the boundary layer and the tropopause.

1. INTRODUCTION

Nitrogen oxide plays a controlling role in the chain of catalytic reactions which produce tropospheric ozone. This role is most easily illustrated by the oxidation of carbon monoxide. In the presence of sufficient NO it takes the path,

$$
\begin{array}{llll}
CO & + OH & \rightarrow H & + CO_2 \quad R1 \\
H & + O_2 + M & \rightarrow HO_2 & + M \quad\quad\quad R2 \\
HO_2 & + NO & \rightarrow OH & + NO_2 \quad\quad R3 \\
NO_2 & + hv & \rightarrow NO & + O \quad\quad\quad R4 \\
O & + O_2 + M & \rightarrow O_3 & + M \quad\quad\quad R5 \\
\hline
CO & + 2 O_2 & \rightarrow CO_2 & + O_3
\end{array}
$$

which results in a net production of O$_3$. In the absence of NO, the oxidation of CO takes a different path:

$$
\begin{array}{llll}
CO & + OH & \rightarrow H & + CO_2 \quad\quad R1 \\
H & + O_2 + M & \rightarrow HO_2 & + M \quad\quad\quad R2 \\
HO_2 & + O_3 & \rightarrow OH & + 2 O_2 \quad\quad R6 \\
\hline
CO & + O_3 & \rightarrow CO_2 & + O_2
\end{array}
$$

In this case, ozone is destroyed. At the "critical" volume mixing ratio of NO of 10 pptv, production (by R1 to R5) and destruction (by R1, R2, R6) are equal. Above this value, the oxidation of CO leads to a photochemical production of ozone.

To estimate the global tropospheric production of O$_3$, information on the global background concentration of NO is required. We have therefore recently started to measure the NO volume mixing ratio on a global scale (1,2). Here, we report vertical profiles of the NO and NO$_y$ volume mixing ratios made aboard a small jet-aircraft (Cessna Citation II), which was able to climb up to 14.5 km altitude.

A number of other components (non methane hydrocarbons, PAN, CO , CH₄) were measured during the campaign by another group of our laboratory .

2. EXPERIMENTAL

NO and NO$_y$ were measured by a chemiluminescence detector described earlier (3). We define here NO$_y$ as the sum of NO, NO$_2$, NO$_3$, 2 x N$_2$O$_5$, HNO$_2$, HNO$_3$, HNO$_4$, PAN and nitrate-aerosol (4).
The instrument flown is shown schematically in Fig. 1. It consisted of two detectors operated in parallel. NO was detected by the chemiluminescence reaction between NO in the sample air and ozone added from an ozone generator.

$$NO \quad + O_3 \qquad -> NO_2^* \quad + O_2 \qquad\qquad\qquad R7$$
$$NO_2^* \qquad\qquad -> NO_2 \quad + hv \qquad\qquad\qquad R8$$

NO$_y$ was measured after conversion to NO in a converter (4) consisting of a heated (400°C) gold tube (l = 0.3 m, d = 6 mm). A low amount of CO (0.1 %) was added to the sample air for a complete reduction of the various NO$_y$-components. The addition of CO was giving a "fake" NO$_y$-signal of about 50 ppt, which was determined each day at the time of calibration and was subtracted from the NO$_y$-signals.
The fluorescence was detected by a photomultiplier tube (EMI 9658 R, cooled with dry ice) through a red filter (RG 610). To increase the sensitivity, the reaction chamber consisted of gold-plated aluminium.

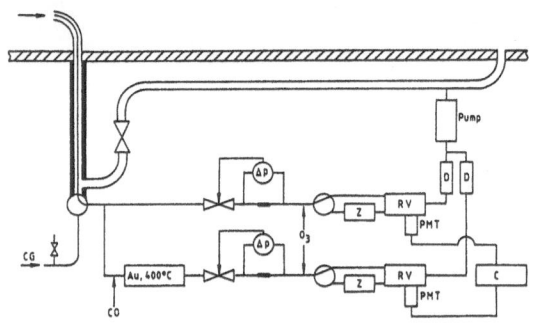

Fig. 1. Schematic drawing of the chemiluminescence detector system
RV reaction vessel   Z zeroing volume
D ozone destroyer   C computer
CG calibration gas

The dark current of the PMT and the interfering chemiluminescences were measured in a zero-mode by passing the sample air-ozone mixture through a Teflon-coated relaxation volume. The NO-signal was calculated from the difference between the measure-mode (without passage through the relaxation volume) and the average of the two neighboring zero modes. The signal from the PMT's was fed to photon counters, which were controlled by a computer. The counters were read out each 10 seconds with a mode switching time of 80 seconds.
The sensitivity of the two detectors was determined prior to each flight with a calibrated NO/N$_2$-mixture (AIRCO) and synthetic air. The calibration mixture was prepared by using two thermostated mass flow controllers (Advanced Semiconductor Materials), which had been calibrated by soap bubble flow meters. The NO-detector showed a sensitivity of 4 counts s^{-1} ppt^{-1} (NO), the NO$_y$-detector 2 counts s^{-1} pptv^{-1} (NO$_y$). The calibrations showed day to day variations of less then 5 %.
The detection limit was calculated from the frequency distribution of the differences of the measured and calculated zero mode signals. These distributions were gaussian and showed standard deviations of 8 ppt NO and 20 ppt NO$_y$ (for a 10 second integration time).
The reaction mixture was pumped through the reaction vessel at 10 mbar pressure and a volume flow rate of 100 l min^{-1}. The flow was controlled by measuring the pressure drop across a capillary in the inlet. That signal which was obtained with the help of differential pressure gauges, was used to drive a

Teflon needle valve which regulated the flow to a constant rate. The pressure drop was calibrated against the flow controllers of the gas calibration system prior to each flight. With this equipment, the flow rate could be held constant to better than 2 % even in a fast descent of the aircraft.

The inlet system consisted of a 1 m long Teflon tube with an inner diameter of 4 mm, which was cooled by flowing outside air over its outer surface. The air intake was located 0.20 m above the hull of the aircraft to prevent contamination.

Auxiliary information namely altitude and outside temperature was supplied by the aircraft navigation system. The temperature and pressure was used to calculate the potential temperature (without correction for water content).

3. SITE AND METEOROLOGICAL SITUATION

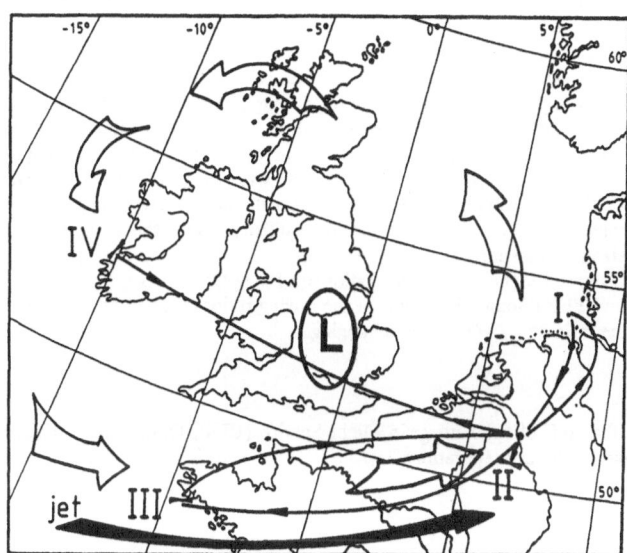

Fig.2 shows the locations of and the meteorological situation during the flights. The latter was dominated by a low pressure system, which remained fixed over Britain for a few days in advance and during the campaign . The tropopause was folded leading to tropopause hights around 8.5 km. At those conditions, the aircraft was able to fly right into the stratosphere. The locations of the flights and the weather conditions are summarized in the next paragraph.

Fig. 2. Meteorological situation . Open arrows indicate the wind direction, the filled arrow indicates the position of a 'jet stream'

I.   Transfer flight over Central and Northern Germany, tropopause height at 11.2 km, wind direction southwest with 20-40 knots, airmasses were carried over industrialized regions in West Germany, date 1.7.1988 (results see fig.3)

II.   Flight near Jülich at the border between Belgium and West Germany, tropopause hight 8.5 km, wind direction west to southwest with 20-40 knots increasing to 70-100 knots over the tropopause ('jet stream'), airmasses probably originated from polar zones and then transported over France, date 2.7.1988 (results see fig.4)

III. Flight near Brest at the French Atlantic coast, tropopause hight 9.5 km, airmasses probably originated from polar zones, wind direction west with 30-40 knots increasing to 70-100 knots over the tropopause ('jet stream'), date 3.7.1988 (results see fig.5)

IV.   Flight near Shannon at the Atlantic coast of Ireland, tropopause hight 8.5 km, wind direction northwest with 20-30 knots, airmasses were transported over Britain and Scotland, date 4.7.1988 (results see fig.6)

4. RESULTS AND DISCUSSION

Fig. 3 to 5 are showing the results of the four flights, namely the mean NO and NO_y values averaged over 5 minute intervals. During those intervals PAN- and NMHC-samples were taken and the aircraft did not change altitude. The standard deviations of the data points were clearly a measure for the natural variability of the concentrations. Normally, the deviation consisted of the sum of 5ppt NO (or 30 ppt NO_y) and of 5% of the corresponding signal. At a few locations, the variability was very much greater. Especially the data points at 6 km altitude in fig.6 (1050 ppt NO and 4000 ppt NO_y) had a standard deviation of 900 ppt NO and 1800 ppt NO_y. Looking through the raw-data, those values could be attached to a short event of 20 s duration. The raw data neighboring the event were very close to the values at 5 km and 6.5 km altitude. Similar conclusions could be drawn for the data points at 4 km altitude in fig.3. Nevertheless, for a better comparison with the PAN data, the presented means were calculated out of the whole set of available data points.

Generally, the vertical profiles of NO and NO_y presented in fig. 3 to 6 exhibited a C shape with low values in the middle troposphere and high values near the boundary layer and the tropopause. But there are a few points which should be discussed in more detail. In fig.4 and fig.5, the measured NO mixing ratios in the troposphere were very low (around 10-30 ppt). The samples were taken apparently in polar airmasses. The ratio NO_y/NO was very high (around 100) indicating a long 'photochemical history'. The NO mixing ratio at 1 km altitude was very low in fig.5 and very high in fig.4 respectively, due to the absence respective presence of a NO ground level source. The NO mixing ratio was sharply increasing by a factor of five at the tropopause. At the same time, the ratio NO_y/NO was going down to 10-20. The reason for this sharp change in NO_y/NO is probably due to the fact that the aircraft trajectory touched the 'jet stream' (fig.2) which carried airmasses of a very different history. Thus those figures are illustrating the influence of fast horizontal transport.

The profiles in fig.3 and fig.6 showed higher NO mixing ratios (60-100 ppt) in the troposphere. The calculated ratio NO_y/NO was around 20. Although the values of the mixing ratios changed by more than a factor of two at the tropopause, the ratio NO_y/NO was not changing.

ACKNOWLEDGEMENTS

The authors wish to thank the staff of the European Flight Service (EFS), Düsseldorf, for their kind assistance during the flights and the flight preparations.

REFERENCES

(1) Drummond,J.W., Ehhalt,D.H., and Volz,A.
Measurements of nitric oxide between 0 - 12 km altitude and 67°N - 60°S latitude obtained during STRATOZ III.
J. Geophys. Res. 93, 15831-15849, 1988.
(2) Rohrer,F., Ehhalt,D.H.
Measurements of NO and NO_y obtained during TROPOZ I
Proceedings of the Quadrennial Ozone Symposium, Workshop on Tropospheric Ozone
Göttingen 1988
(3) Drummond,J.W., Ehhalt,D.H., and Volz,A.
An optimized chemiluminescence detector for tropospheric NO measurements.
J. Atmos. Chem. 2, 287-306, 1985.
(4) Fahey,D.W., Eubank,C.S., Hübler,G., and Fehsenfeld,F.C.
Evaluation of a catalytic reduction technique for the measurement of total reactive odd-nitrogen NO_y in the atmosphere.
J. Atmos. Chem. 3, 435-468, 1985.

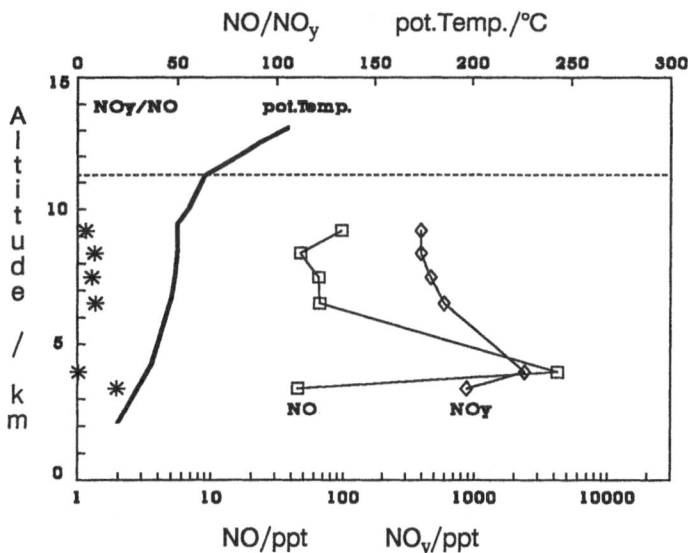

Fig. 3   Vertical profiles of NO, NO$_y$, potential temperature and the ratio NO$_y$/NO over West Germany, 1.7.1988 , 15:00 - 18:00 MEZ , ascent , (see position I in fig.2) dashed line indicates tropopause

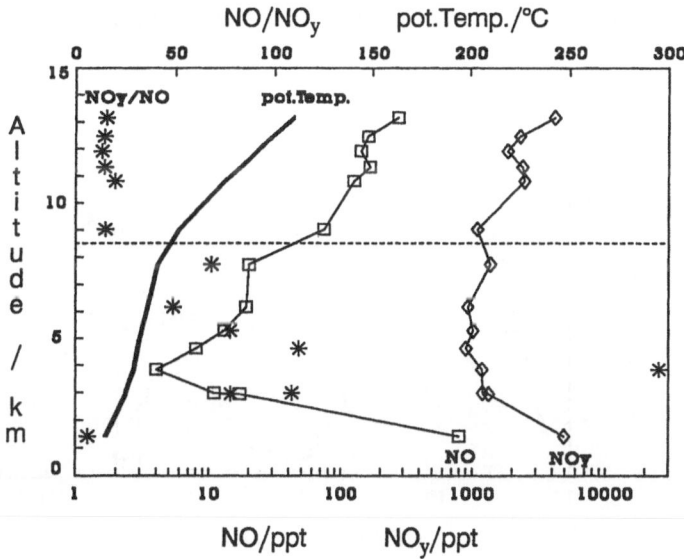

Fig. 4   Vertical profiles of NO, NO$_y$, potential temperature and the ratio NO$_y$/NO over Jülich (Central Europe), 2.7.1988 , 14:20 - 16:40 MEZ , ascent , (see position II in fig.2) dashed line indicates tropopause

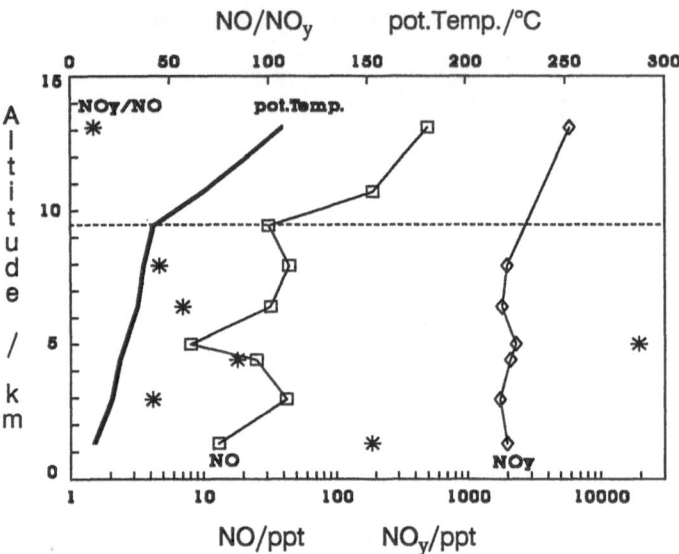

Fig. 5   Vertical profiles of NO, NO$_y$, potential temperature and the ratio NO$_y$/NO over the French Atlantic coast near Brest, 3.7.1988 , 17:00 - 18:40 MEZ , descent , (see position III in fig.2) , dashed line indicates tropopause

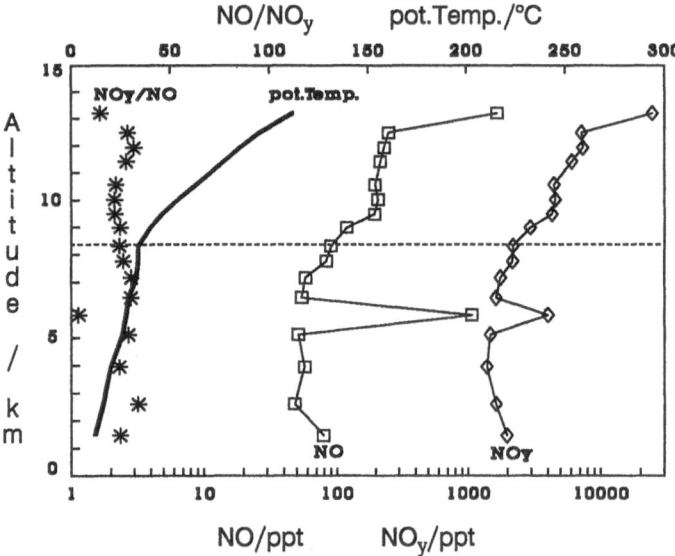

Fig. 6   Vertical profiles of NO, NO$_y$, potential temperature and the ratio NO$_y$/NO over the Irish Atlantic coast near Shannon, 4.7.1988 , 13:40 - 15:05 MEZ , descent, (see position IV in fig.2) , dashed line indicates tropopause

FORMALDEHYDE MEASUREMENTS IN THE FREE TROPOSPHERE

G. Schuster, S. Wilson[+], and G. Helas[*]
Max-Planck-Institut für Chemie, Otto-Hahn-Institut
D - 6500 Mainz, Federal Republic of Germany

[+] present address: Cape Grim Baseline Air Pollution Station, Smithton, Tasmania, Australia
[*] to whom correspondence should be addressed

Abstract

Determinations of formaldehyde have been performed in tropospheric air above northern Europe. Mixing ratios obtained range from 23 to 142 pptv with an average of 72 pptv and 31 pptv standard deviation of a total of 13 samples. Formaldehyde mixing ratios decrease during the dark time as expected.

Key words : Formaldehyde, HCHO, measurement, troposphere

1. Introduction

Formaldehyde is a relatively stable intermediate in the oxidation chain of hydrocarbons in the atmosphere. It is an important species to start radical chain reactions by photolysis. In the free troposphere one would expect methane to be the most important precursor. This molecule has, in time scales compared to the lifetime of formaldehyde, relatively constant mixing ratios. So it should be possible to assess the mixing ratio of formaldehyde by the formula (see for example Warneck [1988])

$$m(HCHO) = \frac{k_1[OH] \ m(CH_4)}{j(HCHO) + k_2 \cdot [OH]}$$

which expresses, under steady state conditions, the ratio of the source term as oxidation of methane and of the sink term photolysis and reaction of formaldehyde with OH. This leads, with appropriate numbers, to a mixing ratio of ca. 200 pptv. Mixing ratios of formaldehyde in this order of magnitude have been observed by several authors in the unpolluted troposphere: Platt et al. [1979], Platt and Perner [1980], Lowe et al. [1980], Fushimi and Miyake [1980], Zafiriou et al. [1980], Neitzert and Seiler [1981], Lowe and Schmidt [1983]. This formula implies the decrease of formaldehyde mixing ratios during the night due to the lowered source term. This has been confirmed in marine air by Lowe and Schmidt [1983], who have shown a diurnal variation with lower values during the night. It also implies a decrease of HCHO mixing ratios with height as the water vapor pressure, essential for forming OH, decreases. This also has been shown tentatively by Lowe et al. [1980]. They measured a drop in mixing ratios from the ppbv range, which is typical for continental air, to lower than 100 pptv in heights between 4 and 6 km. The mixing ratios mostly were determined by derivatizing the formaldehyde with phenylhydrazine, subsequent separation on HPLC columns and absorption detection. So we found it useful to compare the reported data with results obtained with a totally different technique.

2. Experimental

The method applied for determining formaldehyde mixing ratios was matrix-isolation-spectroscopy (MIS). The method has been described in detail elsewhere [Griffith and Schuster, 1987] and was succesfully used for parallel sampling and measurement of several trace components in the free troposphere and lower stratosphere [Wilson et al.,1988; Wilson et al.,1989, Helas et al.,1989]. Briefly, air samples were collected aboard a small aircraft (Learjet 25D), which was modified for contaminant-free sampling. Six flights were performed between Germany and Spitzbergen during which samples were taken between 4 and 14 km height with 13 taken in the troposphere of a total of 55 samples. Air samples were cryogenically trapped at the temperature of liquid argon (87 K). The material frozen out consisted primarily of CO_2 and water. When appropriately handled, the CO_2 formed a glassy matrix, with the condensable trace gases isolated within the solid. The IR-spectrum was then measured with a FTIR-spectrometer. Usual analysis was performed on the IR-spectra, where the peak heights and/or areas were measured relative to the amount of CO_2 present. The measurement procedure was calibrated against laboratory standards using the same sample handling as for the flight samples. The overall error of the handling and measurement procedure may be estimated to 20%.

The definition of the tropopause for the presented samples used here is set to a corresponding ozone mixing ratio of 50 ppbv. This artificial limit is due to the sample handling with the MIS as will be shown below. Interpolations of tropopause heights as given from radiosonde data, however, justify this procedure.

3. Results

The Fig. 1 shows the summary of all tropospheric data. The average is 73 pptv and a standard deviation of 33 pptv of 13 samples ranging from 23 to 142 pptv for a height range between 4 and 12 km. The plot shows no obvious trend with height. Though this data set is rather limited, it is interesting to separate the individual samples to their respective sampling conditions, as some where collected during daytime whereas others during the polar night. A higher average of 88 pptv was found, when the sun was above the horizon, and a lower average of 60 pptv, when the sun was below, which is displayed in Fig. 2.

4. Discussion

The mixing ratios of formaldehyde encountered in air masses between 50 and 75 °N in heights from 4 to 12 km are in the expected order of magnitude and cover the same range as is given by Lowe et al. [1980] and Lowe and Schmidt [1983], who sampled over the Eifel, Germany. Our values do not show a vertical trend, which may be due to the scarcity of the data and it should be noted, that totally different air masses have been probed in different seasons of the year.

The influence of light is visible in the data. As discussed above, the main source of formaldehyde is the reaction of OH radicals with methane. So

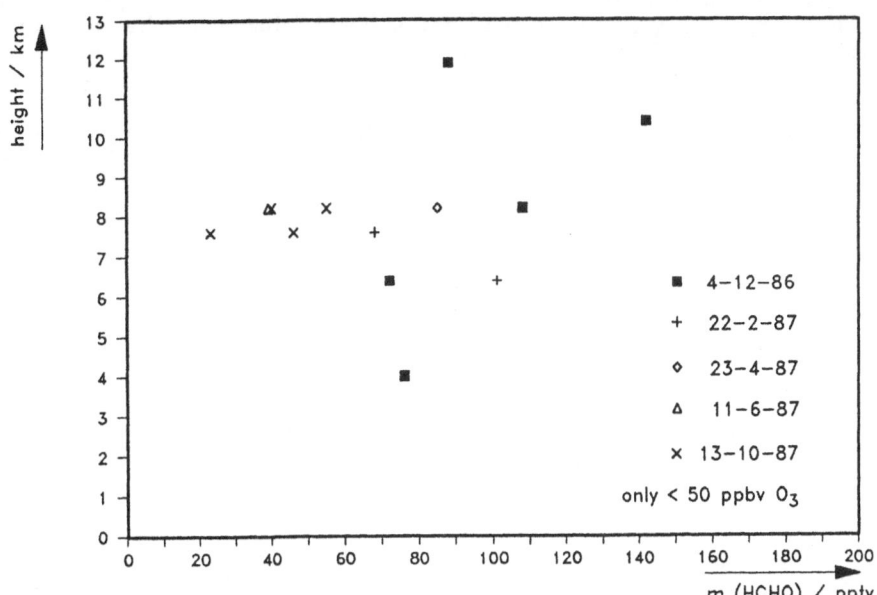

Fig. 1
Mixing ratios of formaldehyde sampled between 4 km and tropopause versus
altitude as determined by matrix-isolation-spectrocopy from samples which
were collected between Germany and Spitzbergen in different seasons of the
year

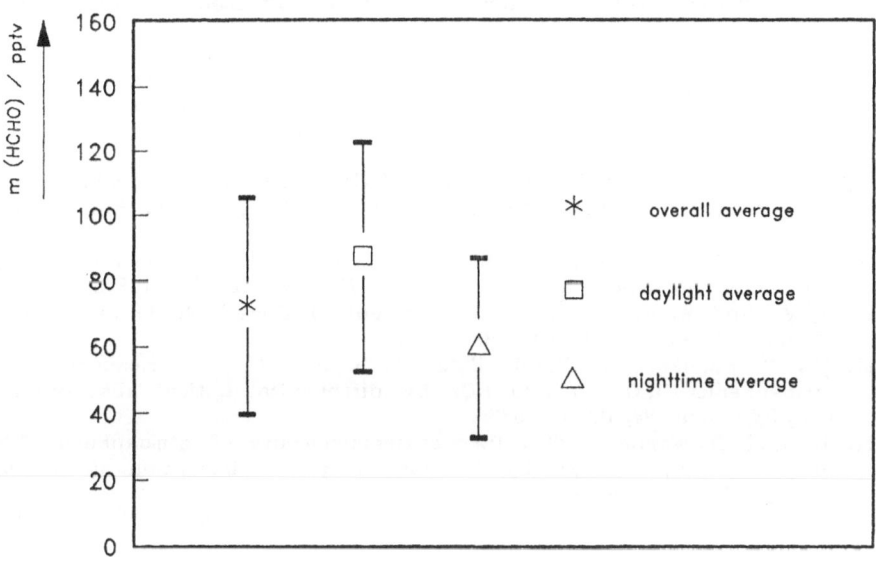

Fig. 2
Mixing ratios of formaldehyde as averages and standard deviation bars at
different conditions as stated.

the lowering of the OH concentration will diminish the amount of formaldehyde. Additionally, however, the sink terms simultaneously are minimized. Therefore, the mixing ratios of HCHO seem to be 'frozen in' at a lower level during night time, if there were no additional losses. The averages of the daylight data and of the nighttime data differ in the expected manner, though the change is not large. A close inspection reveals in the nighttime data rather high values for February and April. For these, the NO_x mixing ratios measured in parallel point to slightly polluted air. Basically, however, these results confirm for the troposphere the above shown assessment of HCHO mixing ratios and agree well with previous measurements.

Acknowledgement

We thank captains Gärtner and Lechner of the FVG, Bad Aibling for cooperative assistance during the flights. This work was in parts sponsored by the Bundesministerium für Forschung und Technologie via the grant ATP 88.

Literature

Fabian, P., R. Borchers, G. Flentje, W. A. Matthews, W. Seiler, H. Giehl, K. Bunse, F. Müller, U. Schmidt, A. Volz, A. Khedim, and F. J. Johnen, 1981. The vertical distribution of stable trace gases at midlatitudes, *J. Geophys. Res. 86*, 5179 - 5184.
Fushimi, K. and Y. Miyake, 1980. Contents of formaldehyde in the air above the surface of the ocean. *J. Geophys. Res. 85*, 7533 - 7536.
Griffith, D. W. T., and G. Schuster, 1987. Atmospheric trace gas analysis using matrix isolation-Fourier transform infrared spectroscopy. *J. Atmos. Chem. 5*, 59 - 81.
Helas, G., G. Schuster, and S. R. Wilson, 1989. Measurements of ozone and other trace components near the tropopause over northern Europe. Quadrennial Ozone Symposium 1988, in: *Proceedings of the International Ozone Symposium 1988*, ed. R. D. Bojkov and P. Fabian, A. Deepak Publ. Hampton, 1989.
Lowe, D. C., U. Schmidt, and D. H. Ehhalt, 1980. A new technique for measuring tropospheric formaldehyde (CH_2O). *Geophys. Res. Lett. 7*, 825 - 828.
Lowe, D. C., and U. Schmidt, 1983. Formaldehyde (HCHO) measurements in the nonurban atmosphere. *J. Geophys. Res. 88*, 10844 - 10858.
Neitzert, V., and W. Seiler, 1981. Measurement of formaldehyde in clean air. *Geophys. Res. Lett. 8*, 79 - 82.
Platt, U., D. Perner, and H. W. Pätz, 1979. Simultaneous measurement of atmospheric CH_2O, O_3 and NO_2 by differential optical absorption. *J. Geophys. Res. 84*, 6329 - 6335.
Platt, U. and D. Perner, 1980. Direct measurements of atmospheric CH_2O, HNO_2, O_3, NO_2, and SO_2 by differential optical absorption. *J. Geophys. Res. 85*, 7453 - 7458.
Schmidt, U., A. Khedim, D. Knapska, G. Kulessa, and F. J. Johnen, 1984. Stratospheric trace gas distributions observed in different seasons. *Adv. Space Res. 4*, 131 - 134.
Warneck. P., 1988. Chemistry of the natural atmosphere, Academic Press.

Wilson, S. R., P. J. Crutzen, G. Schuster, D. W. T. Griffith, and G. Helas, 1988. Phosgene measurements in the upper troposphere and lower stratosphere. *Nature 334*, 689 - 691.

Wilson, S. R., G. Schuster, and G. Helas, 1989. Measurements of COFCl and COCl₂ near the tropopause. Quadrennial Ozone Symposium 1988, in:*Proceedings of the International Ozone Symposium 1988*, ed. R. D. Bojkov and P. Fabian, A. Deepak Publ., Hampton, 1989.

Zafiriou, O., C., J. Alford, M. Herrera, E. T. Peltzer, and R. B. Gagosian, 1980. Formaldehyde in remote marine air and rain: flux measurements and estimates. *Geophys. Res. Lett. 7*, 341 - 344.

Formic and acetic acid in presence of cloud and dew water

W. R. Hartmann, M. O. Andreae, and G. Helas
Max-Planck-Institut für Chemie, Otto-Hahn-Institut
Division of Biogeochemistry
P. O. Box 3060, D-6500 Mainz, Federal Republic of Germany

Abstract

The partitioning of formic and acetic acid between liquid and gaseous phase are computed at various liquid water contents. It can be shown, that at low liquid water supply distinctly more formic than acetic acid will be dissolved. A selective transport by washout by clouds is concluded. Further an example is given on the ability of dew to act as a temporary sink and source for organic acids. The assessment shows that the diurnal variation of mixing ratios often found during surface measurements to a great extent will be due to dissolution in dew and subsequent evaporation on the next day.

Key words: gas-liquid partitioning, formic acid, acetic acid, transport

1. Introduction

Organic acids have been reported as constituents of rain water [Keene et al., 1983]. Of these, formic and acetic acid are most prominent, while other acidic organic compounds like propionic, oxalic, pyruvic and lactic acid and others are only found as minor constituents [Likens and Galloway, 1983; Norton et al., 1983; Guiang et al., 1984; Kawamura et al., 1985; Andreae et al. 1987, 1988, 1989]. Though in anthropogenically polluted regions the contribution of organic acids to the total anionic charge is only a few percent, it can account for a considerable amount of the acidity observed in remote regions [Keene et al., 1983; Andreae et al., 1988]. Most measurements of organic acids in precipitation show formic acid to be present at higher concentrations than acetic acid [Keene and Galloway, 1986; Andreae et al., 1987; Keene and Galloway, 1988; Elbert and Andreae, 1989]. Though this may be due to different source strengths for both compounds, it also suggests that physico-chemical processes govern the ratio of formic over acetic acid in hydrometeors. This partitioning between liquid and gas phase should be likewise for cloud and fog droplets and for dew.

In order to analyze the gas/liquid partitioning of the organic acids, we use a simple model which assumes "local" equilibrium governed by the Henry's law constants of formic and acetic acid and estimated liquid water amounts present in the atmosphere. The assumption of local equilibrium implies that transport phenomena are not taken into account.

2. Description of the calculation

The fraction n of an airborne species present in hydrometeors is given by:

$$n = \frac{v_s}{v_s + v_g} = \frac{1}{1 + HM_w/R_gTL}$$

Here v_s and v_g are the number of moles of the substance in the solution and gas phase, respectively. L, T, R_g, and M_w are the liquid water content in g m^{-3}, the absolute temperature in K, the gas constant in 8.3×10^{-5} m^3 bar deg^{-1} mol^{-1}, and the molecular weight of water [Warneck, 1988]. Here a pseudo Henry's law constant H^* is necessary to account for dissociation and pH:

$$H^* = H \left(1 + \frac{K_a}{[H^+]} \right)$$

The expressions for the temperature dependent Henry's law constant and the dissociation constant of formic and acetic acid have been taken from Winiwarter et al. [1988].

HCOOH: $H = 5600 \exp [5736 (1/T - 1/298)]$
             $K_a = 1.8 \times 10^{-4} \exp [151 (1/T - 1/298)]$
CH$_3$COOH: $H = 8800 \exp [6391 (1/T - 1/298)$
             $K_a = 1.7 \times 10^{-5} \exp [50 (1/T - 1/298)]$.

Dimensions are given in mol dm^{-3} atm^{-1} and mol dm^{-3}, respectively. In order to investigate the behavior of the system in response to environmental conditions, liquid water content, pH and temperature are varied.

Initial mixing ratios of formic and acetic acid are set to 1 ppbv. This selection provides a simple system and falls within the range of typically measured values. The pH-values adopted are 3.5, 4.5, and 5.5. A value of about 5.5 would result, if atmospheric CO_2 were the single solute, but lower values may be more realistic. Ideal solution behaviour is assumed and any influences of drop sizes are neglected. Reflecting cloud layer conditions, 5.5 °C and a pressure of 850 hPa are adopted, corresponding to an altitude of ca. 1500 m.

3. Results and discussion

The following three figures show the results of the calculations. Fig. 1 displayes the ratio of formic to acetic acid present in the atmosphere as a function of the liquid water content (LWC). It should be noted that the LWC shown here ranges from 0.001 g m^{-3} H$_2$O, corresponding to the water content of aerosol particles, through 0.1 g m^{-3}, typical for fog, to 2 g m^{-3} and more, characteristic for precipitating clouds [Winkler, 1988; Mason, 1971; Pruppacher and Klett, 1980]. The range of LWC is chosen to demonstrate the extreme range of possible conditions. Especially at the lower end of the LWC-range, the validity range of Henry's law will be exceeded. However, with the simple scheme presented here it is not possible to cover effects like salting-out and precipitation inside the droplets. Fig. 1 shows, that in moderate to clean air masses formic acid depletes more in the gas phase than acetic acid when hydrometeors are present. Fig. 2 shows the corresponding ratio of formic to acetic acid in the liquid phase. Besides the case of very acidic hydrometeors, formic acid is much more abundant in the droplets.

When applied to the atmosphere, the calculations predict a stronger flux of formic acid relative to acetic acid into hydrometeors. So a selective transport by washout by clouds is concluded. This has not been observed directly for clouds. It would necessitate a Langrangian experimental approach, in which the gas phase and liquid phase concentrations were measured as the air masses were moving. However, the formation of dew provides an equivalent liquid water reservoir, which is build up temporarily. In order to assess how much formic and acetic acid can be stored in dew, the amount of precipitable water is determined under adiabatic conditions when the temperature is lowered overnight. Then the decrease of gas phase mixing ratios resulting from the dissolution of organic acid in the dew is evaluated. Though other source and sink processes should not be neglected, qualitatively this approach would explain the frequently found diurnal behavior of gas phase organic acids in surface measurements with distinctly lower mixing ratios during night time [e.g. Andreae et al, 1988; Talbot et al., 1988]. Here, as an example, a diurnal cycle of measurements in the Bavarian forest is shown in Fig. 3. The mixing ratios of formic and acetic acid as well as the ratio of both is given for the course of the day. The gas phase samples were taken on a tower at 51 m elevation above ground. The surrounding area within several km range is covered with firs up to 30 m heigh. Though the influence of biospheric emissions on the mix-ing ratios found is obvious, the ratio of both acids decreases over night as expected. The amount of dew formed was noticeable. Unfortunately analysis of dew water was not performed. We can, however, estimate for the condi-tions during sampling 0.5 g m^{-3} H$_2$O as precipitable water. For a pH of slightly less than 6 the decrease of mixing ratios would be explained solely by dissolution. As the dew will reevaporate the following morning, the formic and acetic acid will be set free again. Thus "effective" source and sink strengths of these compounds can be much lower.

Acknowledgement

We thank G. Enders for assistance when using the measurement site in the Bavarian forest of the Institut für Bioklimatologie und Angewandte Meteorologie der Universität München.

Literature

Andreae, M. O., R. W. Talbot, and S. M. Li, 1987. Atmospheric measurements of pyruvic acid. *J. Geophys. Res. 92*, 6635 - 6641.

Andreae, M. O., R. W. Talbot, T. W. Andreae, and R. C. Harriss, 1988. Formic and acetic acid over the central Amazon region, Brazil. 1. Dry season. *J. Geophys. Res. 93*, 1616 - 1624.

Andreae, M. O. R. W. Talbot, H. Berresheim, and K. M. Beecher, 1989. Precipitation chemistry in central Amazonia, *submitted to J. Geophys. Res.*

Elbert, W., and M. O. Andreae, 1989. Deposition of organic anions at a semirural site in central Europe. In: *Mechanism and effect of pollutant-transfer into forests.* ed. H.-W. Georgii. Kluwer, Dordrecht. in press.

Guiang, III, S. F., S, V, Krupa, and G. Pratt, 1984. Measurements of S(IV) and organic anions in Minnesota rain. *Atmos. Environment 18*, 1677 - 1682.

Kawamura, K., L. L. Ng, and I. R. Kaplan, 1985. Determination of organic acids in the atmosphere, motor exhausts, and engine oils. *Environ. Sci. & Technol. 19*, 1082 - 1086.

Keene, W. C., and J. N. Galloway, 1986. Considerations regarding sources for formic and acetic acids in the troposphere. *J. Geophys. Res. 91*, 14466 - 14474.

Keene, W. C., and J. N. Galloway, 1988. The biogeochemical cycling of formic and acetic acids through the troposphere: an overview of current understanding. *Tellus 40B*, 322 - 334.

Keene, W. C., J. N. Galloway, and J. D. Holdren, Jr., 1983. Measurement of weak organic acidity in precipitation from remote areas of the world. *J. Geophys. Res. 88*. 5122 - 5130.

Likens, G. E., and J. N. Galloway, 1983. The composition and deposition of organic carbon in precipitation. *Tellus 35B*, 16 - 24.

Mason, B. J., 1971. *The physics of clouds*. 2nd ed., Oxford University Press (Clarendon), London.

Norton, R. B., J. M. Roberts, and B. J. Huebert, 1983. Tropospheric oxalate. *Geophys. Res. Lett. 10*, 517 - 520.

Pruppacher, H. R., and J. D. Klett, 1980. *Microphysics of clouds and precipitation*. Reidel, Dordrecht, Holland.

Talbot, R. W., K. M. Beecher, R. C. Harriss, and W. R. Cofer III, 1988. Atmospheric geochemistry of formic and acetic acids at a mid-latitude temperate site. *J. Geophys. Res. 93*, 1638 - 1652.

Winiwarter, W., H. Puxbaum, S. Fuzzi, M. C. Facchini, G. Orsi, N. Beltz, K. Enderle, and W. Jaeschke, 1988. Organic acid gas and liquid-phase measurements in Po valley - winter conditions in the presence of fog. *Tellus 40B*, 348 - 357.

Warneck, P., 1986. The equilibrium distribution of atmospheric gases between the two phases of liquid water clouds. in: NATO ASI Series, Vol. G6, *Chemistry of Multiphase Atmospheric Systems*, ed. W. Jaeschke, Springer, Berlin.

Winkler, P., 1988. The growth of atmospheric aerosol particles with relative humidity. *Physica Scripta 37*, 223 - 230.

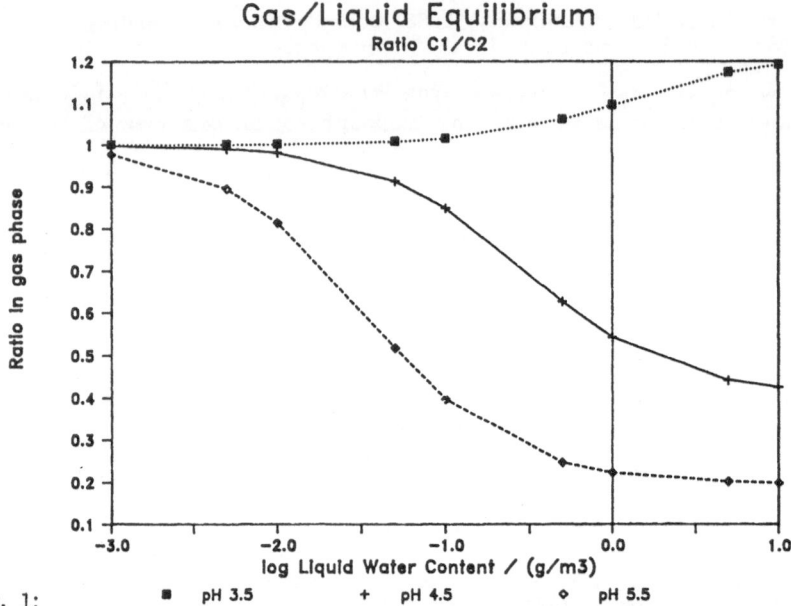

Fig. 1:
Ratios of the gas phase concentrations of formic and acetic acid in the presence of hydrometeors at varied liquid water content (LWC) and different pH. Conditions are given in the text.

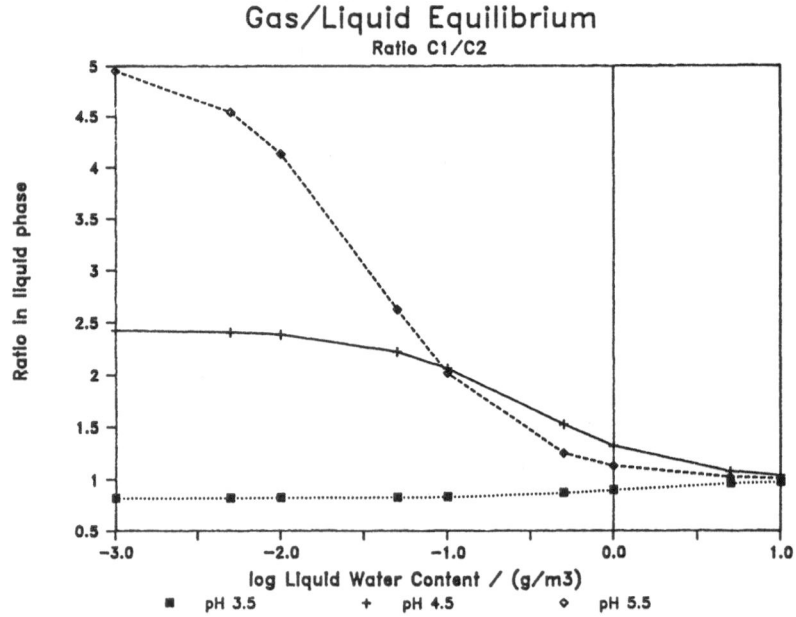

Fig. 2:
Ratios of the liquid phase concentrations of formic and acetic acid in the presence of hydrometeors at varied liquid water content (LWC) and different pH. Conditions are given in the text.

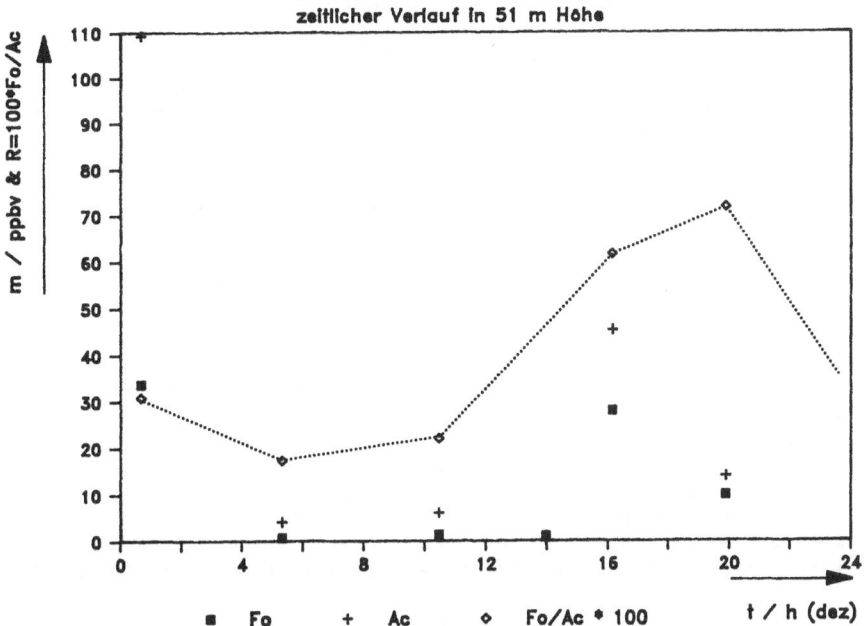

Fig. 3:
Gas phase mixing ratios of formic and acetic acid and their ratio as measured in the course of the day at the Bavarian forest in July, 1988. Sampling altitude was 51m.

ONE-YEAR MEASUREMENTS OF PAN IN THE PARIS BASIN: TEMPERATURE AND WIND DIRECTION EFFECT.

N. TSALKANI*, P. PERROS, A.L. DUTOT and G. TOUPANCE#
Laboratoire de Physicochimie de l'Environnement
Université Paris-Val de Marne - 94010 Créteil - France.

SUMMARY
PAN has been measured continuously for one year in Créteil, a south-eastern suburb of Paris. Correspondence analysis was used to analyse the important PAN data base obtained during this period, in relation to associated meteorological parameters and the location of precursors sources. This analysis revealed a complex relation (U-shaped curve) between PAN and temperature, with a minimum in the range of 3-5°C to 10-12°C and an increase in the lower and higher temperatures, arising from the competition between photochemical production and thermal decomposition. Long range transport of PAN is associated with easterly winds and occurs under anticyclonic conditions in all seasons, partic-ularly in winter. The highest PAN values are observed in summer. They have a local origin and are associated with NW to SW winds, revealing the net impact, in summer, of nearby sources such as the urban con-glomeration of Paris on PAN levels recorded in the surrounding areas.

1. INTRODUCTION
Photochemical air pollution has become nowadays a major environmental problem in most important urban and industrial centres all over the world. Peroxyacetyl nitrate (PAN) is one of the main components of photochemical smog and is generally considered as an excellent indicator of photochemical activity. However, the data available for ambient air PAN concentrations, compared to ozone, are still relatively sparse. The great majority of PAN measurements in ambient air have been performed in the U.S.A (1-3), Canada (4-6) and some countries of north-western Europe (7-9).

Prior to our study, no data were available in France on atmospheric PAN concentrations. Field measurements of PAN in other European cities have shown substantial PAN levels, that reached several ppb (7,10). These PAN values, however, are much lower compaired to those observed in California (1,2). This paper presents the first systematic observations of ambient PAN concentrations in the greater Paris area, performed during the period October 1985 - October 1986. The results are discussed in relation to asso-ciated meteorological parameters, with the use of correspondence analysis as statistical tool.

2. EXPERIMENTAL PROCEDURES
PAN in ambient air was determined with an automated gas chromatograph (DELSI INSTRUMENTS, model 120 FB) equipped with an electron capture dete-

---

*Present address: Ministry of the Environment, Planning and Public Works - Division of Air Pollution Control - 147 Patission str. - 11251 Athens - Greece.
#Author to whom correspondence should be addressed.

ctor. The detector had a 15 mCi Ni-63 source of β-particles and operated in the constant frequency mode. A pneumatically operated 6-port 2-position gas sampling valve with a 4 ml stainless steel sample loop was used to switch air samples onto the GC column on an hourly basis, by means of an electrical timer with a 60-min cycle. Ambient air was drawn through a 1.50 m Teflon tube by means of an air pump.

The PAN peak was separated on a 0.3 cm id x 60 cm glass column, packed with a mixture of 4.8% QF1 and 0.2% Diglycerol on 80-100 mesh Chromosorb G AW DMCS. The column and injector were maintained at 30°C in order to minimize thermal decomposition of PAN, while the detector operated at 60°C for self-cleaning purposes. A mixture of 90% Ar -10% CH_4 was used as carrier gas at a flow rate of 25 ml min^{-1} in the column and 40 ml min^{-1} in the detector. Under the above conditions, PAN eluted in ca. 4.5 min.

The chromatograph was calibrated by two independent methods, in an attempt to minimize calibration uncertainties. The first one was based on infrared spectroscopy (11), while the second made use of the quantitative formation of nitrite ions during alkaline hydrolysis of PAN and susequent colorimetric determination of the nitrite content. Pure gaseous PAN samples used for calibration were prepared by the method of Gaffney et al. (12). The two calibration methods showed a difference of 10%, which is considered as very satisfactory, taking into account the difficulties in the preparation of pure PAN analytical standards, as well as the inherent thermal instability of this compound.

3. RESULTS AND DISCUSSION

PAN concentrations were measured from October 1985 to October 1986 in Créteil, a south-eastern suburb of Paris located at about 10 km from the centre of the town. The urban density decreases rapidly in the NE, E and SE, while it is elevated in the south (Seine valley), west and north. The sampling point was situated in the University PARIS -VAL DE MARNE, about 8 m from the ground in a well ventilated zone.

The daily mean PAN concentrations for the period October 1985-October 1986 were analysed in relation to the main meteorological parameters, namely the temperature, wind direction and velocity and solar radiation. The meteorological data used were obtained from two stations of the National Meteorology network: Trappes (30km west of Créteil) for solar radiation and Melun (35 km east of Créteil) for the other three parameters.

Correspondence analysis (13,14) was used to determine the relations and correlations between PAN concentrations and meteorological conditions. The results of the analysis, applied to our set of data, revealed a parallel evolution of PAN, solar radiation and temperature, which indicates the existence of a positive correlation between PAN and these two meteorological parameters. However, an "anomaly" was observed concerning the class of low temperatures (-10°C to -3°C) which was associated with elevated PAN levels. On the contrary, a negative correlation was observed between PAN levels and wind velocity, while the wind direction showed, unexpectedly, no clear correlation with PAN concentrations.

A quantitative aspect of these results can be given by considering the relative contribution coefficient of each variable. In our case, the mean value of this coefficient for each variable is as follows:

PAN: 42%, solar radiation: 37%, temperature: 35%, wind velocity: 33% and wind direction: 8%.

This means that the meteorological parameters which influence the concentration of PAN in the air are, in decreasing order, the solar radiation, the temperature, the wind velocity and, in a much lesser extent, the wind direction.

These results can be explained in terms of mechanisms of generation and removal of PAN in the atmosphere. The preponderant role of solar radiation on PAN formation is to be expected, taking into account the series of photochemical reactions between primary pollutants (hydrocarbons and NOx) leading to PAN. Globally, solar radiation seems to be a limiting factor in the formation of PAN in the Paris basin. This is in agreement with the results found in Edmonton, Canada (6) but in contrast with those from the Athens basin (8,9), where evidence indicates that solar radiation is not a limiting factor.

The decrease of PAN levels during strong winds (negative correlation between PAN and wind velocity) is due to the dispersion of primary pollutants, which leads to a decrease in the concentration of secondary pollutants, and therefore to lower PAN concentrations.

The overall positive correlation between PAN and temperature, issued from correspondence analysis, deserves further study since from a first point of view it appears in contrast with the inherent thermal instability of PAN. To get a better insight of this phenomenon, we plotted in fig. 1 the daily mean PAN concentration versus the daily average temperatures for the whole period of the study. A positive correlation between PAN and temperature appears clearly in the range of elevated temperatures (greater than 10-12°C), and a negative correlation can be observed in the lower temperatures (below 0°C). In the range of intermediate temperatures (0°C to 10-12°C) no clear trend can be distinguished. It can be noticed, however, that two groups of points (A and B in fig. 1) deviate from the general trend: they correspond, in fact, to two particular episodes, recorded on 14-17 February 1989 (group A) and on 14-17 March 1986 (group B). One of them has been discussed in details elsewhere (15).

The relation between PAN and temperature is thus expressed by a U-shaped curve, which reflects the mean behaviour of PAN in the atmosphere with respect to temperature, with the exception of some episodes which are governed by different factors.

It should be noted, however, that the positive correlation in the range of elevated temperatures may just be a secondary correlation. What is in reality observed is rather a correlation between PAN and solar radiation. This explains the positive correlation between PAN and elevated temperatures, since the latter correspond to the summer period when solar intensity and duration are high. The minimum observed in the medium temperatures (0°C to 10°C) is due to the cloudy and rainy weather generally associated to such temperatures, which does not favour the formation of PAN. The lower temperatures, which correspond to the winter period, are generally associated to clear skies; photochemical processes are then active but limited, since solar intensity and duration are low. Thus, below 0°C the increase in the mean PAN concentration with decreasing temperatures is rather due to the enhanced thermal stability of PAN which increases towards the lower temperatures.

The bad correlation, issued from correspondence analysis, between wind direction and PAN levels led us to examine separately these two parameters. For this purpose, the hourly PAN data were classified according to 8 wind sectors (N, NE, E, SE, S, SW, W and NW) and the average PAN concentration was calculated for each sector. The resulting "pollution roses" are presented in fig. 2 for the winter and summer period, during day- and night-time. We have also noted, in parentheses, the number of hourly PAN measurements corresponding to each sector.

In winter, during the day as well as during the night, the highest PAN levels are recorded by easterly winds. Taking into account the absence of any major emission sources in the eastern vicinity of the Paris basin, this suggests a long-range transport of pollutants from continental Europe. The SE wind sector, associated to an average PAN concentration of 3.87 ppb, is not significant because of the limited number of situations corresponding to this sector. The easterly winds arise generally from the combination of an anticyclone, often situated in winter over central Europe or over northern Scandinavia, with the low pressure systems developing over western France. The sky is generally clear and solar radiation approaches its winter maximum. These easterly winds bring to the Paris basin air masses enriched in primary pollutants during their way over the continent. The photochemical transformation of the primary pollutants to secondary pollutants explains the important PAN levels observed in Créteil by easterly winds. Moreover, the enhanced thermal stability of PAN at the low winter temperatures leads to a lifetime of 1 to 2 days for PAN, which is consistent with the hypothesis of long-range transport.

The westerly directions (NW, W and SW) are often associated to a cloudy weather and to strong winds creating a turbulent atmosphere. The dispersion of the primary pollutants, combined with the low solar radiation, can explain the low PAN concentrations measured in winter when the wind comes from these directions.

In summer, a similar situation is observed at night, with very low PAN levels by westerly winds and higher PAN concentrations by easterly flow. It is noteworthy that the annual average of the night-time PAN values by easterly winds is close to 2 ppb, both in summer and in winter (fig. 2 a,c). This elevated value is not affected by emissions from the Paris agglomeration, since Créteil is situated at the eastern limits of the city; it indicates, therefore, that the background PAN level in the continental air masses may often be in this order of magnitute.

At daytime in summer a marked increase of the PAN concentrations corresponding to the wind sectors W, NW, N and NE is observed compared to the winter period. Taking into account the location of Creteil at the southeast of the Paris agglomeration, these sectors cover the air masses passing over the urban zone before arriving at Créteil. The urban emissions of Paris are therefore responsible for the formation of these substantial PAN levels (up to 2.5 ppb) in such occasions. Important PAN levels, reaching an average of 3.2 ppb, are also recorded during the day when the wind blows from the south. Emissions from the urban and industrial zone of the Seine valley, situated at the south of Paris, can explain the formation of these elevated PAN concentrations by southerly winds.

4. CONCLUSION

The thorough analysis of the important number of PAN field measurements which were performed during one year in the Paris basin, led to the following results:
- a positive correlation between PAN and solar radiation, which means that solar intensity is a limiting factor in the formation of PAN in the Paris basin.
- a negative correlation between PAN and wind velocity.
- a complex dependence of PAN on temperature, with a minimum in the range of 0 to 10-12°C and an increase in the lower and higher temperatures.
- a long range transport of PAN in western Europe by eastern flow during anticyclonic conditions, particularly in winter.
- a net impact, in summer, of nearby sources such as the urban conglomeration of Paris on PAN levels recorded in the surrounding areas (PAN levels as high as 20 to 30 ppb).

REFERENCES

1) Lonneman W.A., Bufalini J.J. and Seila R.L. (1976) PAN and oxidant measurement in ambient atmospheres. Envir. Sci. Technol. 10 (4), 374-380.

2) Grosjean D. (1983) Distribution of atmospheric nitrogenous pollutants at a Los Angeles Area smog receptor site. Envir. Sci. Technol. 17(1), 13-19.

3) Singh H.B. and Salas L.J. (1983) Peroxyacetyl nitrate in the free troposphere. Nature 302 (5906), 326.

4) Peake E., MacLean M.A. and Sandhu H.S. (1983) Surface ozone and peroxyacetyl nitrate (PAN) observations at rural locations in Alberta, Canada. J. Air Pollut. Control Ass. 33, 881-887.

5) Corkum R, Giesbrecht W.W., Bardsley T. and Cherniak E.A. (1986) Peroxyacetyl nitrate (PAN) in the atmosphere at Simcoe, Canada. Atmos. Envir. 20, 1241-1248.

6) Peake E., MacLean M.A., Lester P.F. and Sandhu H.S. (1988) Peroxyacetyl nitrate (PAN) in the atmosphere of Edmonton, Alberta, Canada. Atmos. Envir. 22, 973-981.

7) Brice K.A., Penkett S.A., Atkins D.H.F., Sandalls F.J., Bamber, D.J., Tuck A.F. and Vaughan G. (1984) Atmospheric measurements of peroxyacetyl nitrate (PAN) in rural, south-east England: Seasonal variations, winter photochemistry and long-range transport. Atmospheric Environment 12, 2691-2702

8) Tsalkani N., Perros P. and Toupance G. (1988) Continuous atmospheric measurements of peroxyacetyl nitrate (PAN) in a mediterranean site (Athens, Greece). Environ. Technol. Lett. 9, 143-152.

9) Tsani-Bazaca E., Glavas S. and Gusten H. (1988) Peroxyacetyl nitrate (PAN) concentrations in Athens, Greece, Atmos. Envir. 22, 2283-2286.

10) Nielsen T., Samuelsson U., Grennfelt P. and Thomsen E.L. (1981) Peroxyacetyl nitrate (PAN) in long-range transported polluted air, Nature 293 (5833), 553-555.

11) Tsalkani N. and Toupance G. (1989) Infrared absorptivities and integrated band intensities for gaseous peroxyacetyl nitrate (PAN), Atmos. Envir. (in press).

12) Gaffney J.S., Fajer R. and Senum G.I. (1984) An improved procedure for high purity gaseous peroxyacetyl nitrate production: Use of heavy lipid solvents. Atmos. Envir. 18, 215-218.

13) Benzecri J.P. and F. (1980) Pratique de l' Analyse des Donnees. Dunod Ed., Bordas.

14) Dutot A.L., Elichegaray C. and Vie le Sage R. (1983) Application de l'analyse des correspondances à l'étude de la composition physicochimique de l'aérosol urbain. Atmos. Envir. 17, 73-78.

15) Tsalkani N., Perros P. and Toupance G. (1987) High PAN concentrations during nonsummer periods: A study of two episodes in Creteil (Paris). France. J. Atmos. Chem. 5, 291-299.

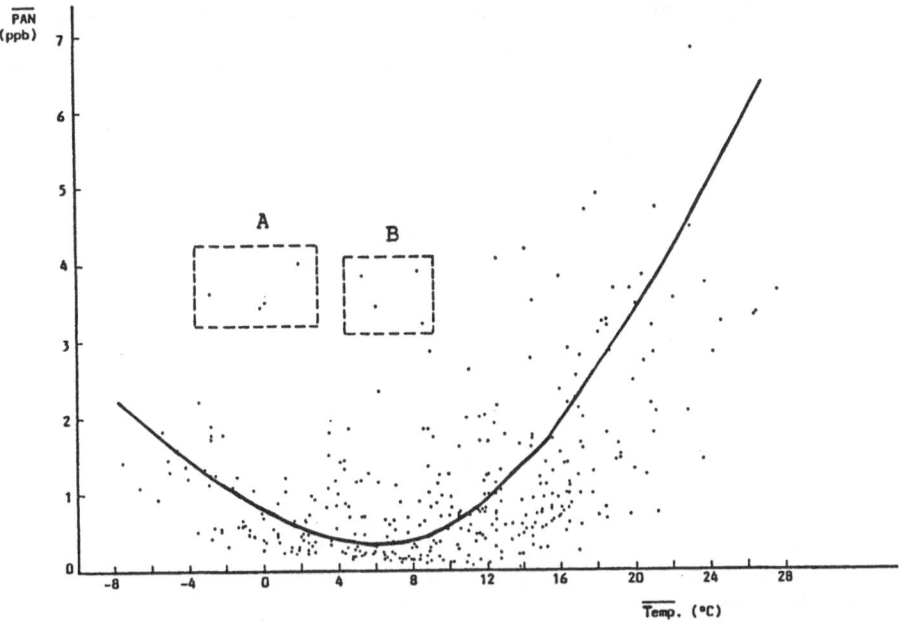

Figure 1 : Dependence of the daily average PAN concentrations on the daily average temperatures.

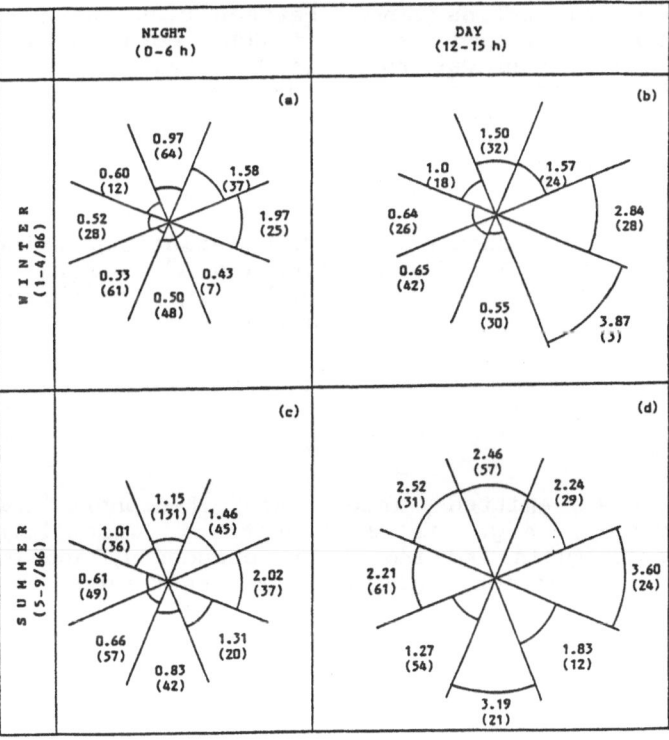

Figure 2 : Pollution roses for the winter and summer period during day- and night-time. PAN data in ppb. Figures in parentheses denote the number of hourly PAN measurements corresponding to each wind sector.

TEMPORAL VARIATIONS OF FORMALDEHYDE IN CONTINENTAL AIR MASSES

J. Slemr[1], W. Seiler[1], R. Teuber, and K.J. Rumpel[2],

[1] Fraunhofer-Institute for Atmospheric Environmental Research, Kreuzeckbahnstr. 19, 8100 Garmisch-Partenkirchen, F.R. Germany

[2] Umweltbundesamt, Meßstelle Deuselbach, 5509 Deuselbach, Mühlenstr., F.R. Germany

Summary

Concentrations of formaldehyde (HCHO) were determined in ambient air during 10 short measuring periods between January and September 1981 at the measuring site of the Umweltbundesamt (UBA) at Deuselbach, FRG (480 m elevation, 49.8° N, 7.2° E). HCHO levels were measured using the DNPH-method by applying impingers filled with aqueous DNPH solution. A total of 169 samples were taken and analysed in the laboratory. Concurrent mixing ratios of SO_2, CO_2, CO, and O_3 were measured and compared with the temporal variations of the HCHO mixing ratio. Meteorological data (solar radiation, wind direction and speed, temperature, precipitation etc.) were also obtained routinely at the station.

The HCHO mixing ratios ranged between 0.09 and 4.13 ppbv. The mean background value of HCHO obtained for clean marine air masses was found to be 0.28 ppbv. The highest HCHO mixing ratios were observed in polluted atmospheres, e.g. in January and February, indicating the influence of direct HCHO emissions by anthropogenic sources. Extremely low HCHO values below 0.10 ppbv were observed in clean marine air during winter conditions.

During late spring and summer, the HCHO concentrations showed distinct diurnal variations which correlated well with O_3, thus indicating photochemical HCHO production. Under winter conditions, anticorrelations between HCHO and O_3 were observed.

1. Introduction

Formaldehyde is emitted into the atmosphere by anthropogenic sources, e.g. as a result of incomplete combustion of fossil fuels and industrial processes, and by natural sources, e.g. direct emission from vegetation. The most important source of atmospheric HCHO is the photochemical degradation of hydrocarbons. Formaldehyde is mainly destroyed by reactions with HO_x radicals or by photolysis. Minor sinks are dry and wet deposition and heterogeneous reactions. The atmospheric residence time of HCHO is relatively short and depends strongly on the time-of-day, season, and latitude. At a latitude of 40°N at noon and under cloud-free conditions, the HCHO photolytic lifetime is estimated to be about 4 h on July 1 and 9 h on January 1 (Finlayson-Pitts and Pitts, 1986). Due to the short residence time, temporal and spatial fluctuations of HCHO mixing ratios should be expected.

The objective of this study was to measure HCHO mixing ratios at a rural station during different seasons and in different air masses. The results should contribute to our understanding of the distribution and behaviour of HCHO in tropospheric air within the PBL over continents.

2. Measuring Site and Experimental

Formaldehyde was measured at a monitoring station of the Umweltbundesamt near Deuselbach (49.8°N, 7.2°E), a small village situated in the low mountain range of the Hunsrueck. The station is located on a flat hill at an altitude of 480 m in a rural area and is surrounded by meadows, fields, and forests. The traffic density on the road nearby the measuring site was very low and the mainroad, a few kilometers south of Deuselbach, showed only a moderate traffic density so that a direct impact of exhaust emissions on the HCHO measurements can be excluded. Local emissions from Deuselbach (296 inhabitants) or from a few other villages and industrial plants within a radius of 30 km from the station may have influenced the measurements, as well as polluted air masses advected from industrial areas of Belgium and the Netherlands (100 - 300 km distance), Trier and Luxemburg (30 - 70 km distance), Neukirchen-Saarbrücken (40 - 60 km distance), Rhein-Main (90 - 100 km distance), Rhein-Neckar (100 - 120 km distance), und Cologne-Ruhrgebiet (130 - 200 km distance).

Air samples were taken 1.5 m above the surface at a site open to advection from all directions. HCHO was collected by drawing air through impingers filled with an acidic aqueous solution of 2,4-dinitrophenylhydrazine (DNPH) at a rate of 2 l/min for a sampling period of about 150 minutes. The formaldehyde hydrazone formed in the solution was subsequently extracted using CCl_4. The CCl_4 extract was washed with 2N HCl, evaporated to dryness, and then redissolved in CH_3OH for subsequent analysis by HPLC.

CO was monitored using the HgO-technique (Seiler at al., 1980). Data on O_3, SO_2 and CO_2 were provided by the Bundesumweltamt monitoring station and meteorological data by the adjacent German Weather Service station.

3. Results and Discussion

Ambient levels of HCHO were determined during ten short measuring periods, each 3 or 4 days long, from January through September 1981. During this time span, mixing ratios of HCHO ranged between 0.09 and 4.13 ppbv.

Generally low levels of HCHO were measured, with a minimum of 0.09 ppbv on January 15, during the first measuring period (January 13 - 16, 1981). At this time the station was under the influence of clean marine air masses, rapidly advected from the northwest over the Atlantic ocean and the North Sea to the measuring site. The simultaneously measured low mixing ratios of CO and SO_2 with values of 125 ppbv and 0.63 ppbv, respectively, indicate that these air masses were most likely not influenced by anthropogenic processes.

The highest HCHO value (4.13 ppbv) was observed during the measuring period between February 24 and 27. The maximum HCHO mixing ratio reached a value of 4.13 ppbv, which is more than ten times higher than the expected value of about 0.2 ppbv if the photochemical oxidation of CH_4 is the only HCHO source. During this measuring period, continental air masses were slowly advected from the south and southeast passing over highly industrialized areas. The high HCHO mixing ratios correlated well with the CO_2, SO_2 and CO mixing ratios. On two days, the formaldehyde mixing ratios exhibited diurnal variations which coincided well with the variation of solar radiation and the O_3 mixing ratios. The positive correlation between the HCHO mixing ratios with the O_3 mixing ratios and solar radiation at relatively high CO and SO_2 mixing ratios is indicative for photochemical production of HCHO in polluted air.

After March, the photochemical production of HCHO became more obvious. Observed diurnal variations correlated well with variations of O_3 mixing ratios and with solar radiation intensity. A good example is May 7 (Fig. 1). Maxima were generally observed during the early afternoon, whereas the minima occured during the early morning. The maximum HCHO values were associated with low levels of SO_2, CO and CO_2, indicating that under these circumstances the observed diurnal cycles of HCHO mixing ratios were not due to anthropogenic emissions. The positive correlation with O_3 indicates that these diurnal variations were caused by photochemical reactions, most likely by oxidation of non methane hydrocarbons.

HCHO mixing ratios, averaged over each measuring period, are plotted in Fig.2. The average values vary between 0.4 and 2.2 ppbv and are significantly higher than the average HCHO mixing ratio of 0.24 and 0.28 ppbv found by Teuber (1984) in clean marine air over the North Atlantic in March/April and October/November, respectively. The highest average HCHO values were observed during the second measuring period in January, during February and September. For these periods, the 48-h back trajectories indicated that the air masses had stayed for a long time over the continent until they reached the measuring station and were thus influenced by directly emitted anthropogenic HCHO.

Even during periods with lower formaldehyde levels (second measuring period in January and in March), the average HCHO values of at least 0.4 ppbv were still higher than the background values observed in the unpolluted air over the Northern Atlantic at similar latitudes. The positve correlation between the elevated HCHO mixing ratios and SO_2 and CO_2 mixing ratios found in Deuselbach indicate the influence of anthropogenically produced HCHO leading to elevated HCHO mixing ratios in polluted air masses. The influence of anthropogenic HCHO emissions is particularly significant in winter when HCHO residence times are considerably longer than in summer so that HCHO can be transported over greater distances and thus reach remote areas.

Calculated 48-h back trajectories demonstrate that highest HCHO mixing ratios were always associated with air masses which resided for long time periods over the continent and passed areas with high emission of pollutants. The correlation between HCHO and other pollutants indicate that during winter conditions the HCHO mixing ratios within the PBL over the European continent are dominated by anthropogenic emissions, whereas during summer the HCHO mixing ratio is strongly influenced by photochemical oxidation of NMHC, both of natural and anthropogenic origin.

References

Finlayson-Pitts, B.J. and Pitts, Jr., J.N. (1986) Atmospheric chemistry. Fundamentals and experimental techniques. New York. John Wiley & Sons, Inc.

Seiler, W., Giehl, H., and Roggendorf, P. (1980) Detection of carbon monoxide and hydrogen by conversion of mercury oxide to mercury vapor. Atmos. Technol. 12, 40-45.

Teuber, R. (1984) Räumliche und zeitliche Schwankungen von Formaldehyd in Reinluft. Diplomarbeit, Institut für Meteorologie, Universität Mainz

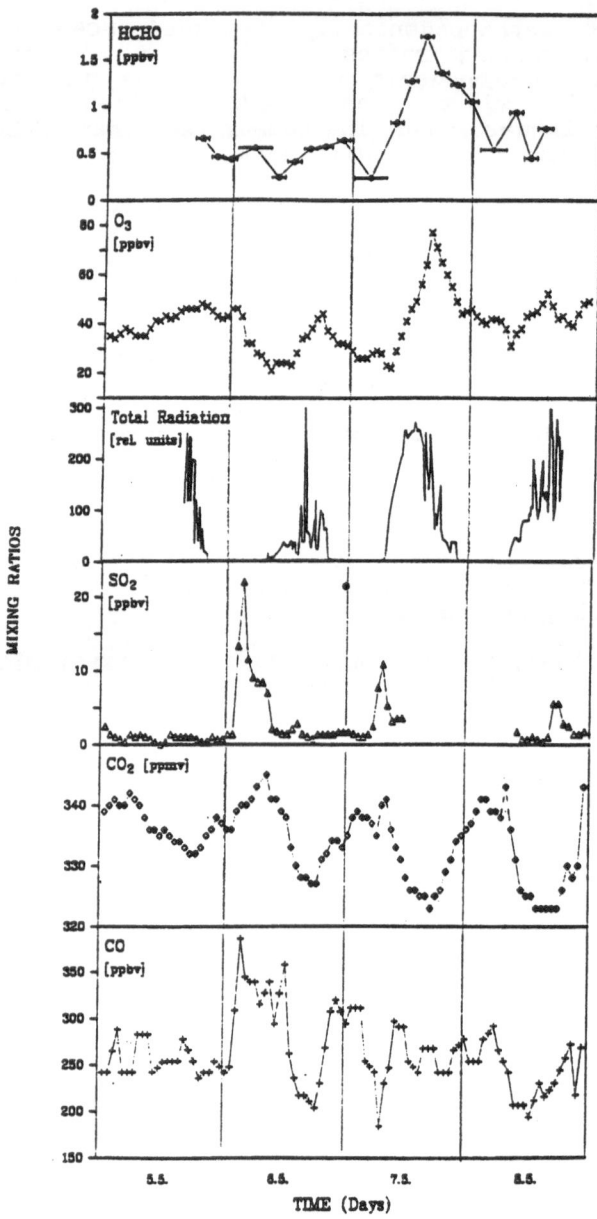

Fig. 1: Mixing ratios of HCHO, O_3, SO_2, CO_2, and CO and solar radiation intensity from May 5 through May 8, 1981

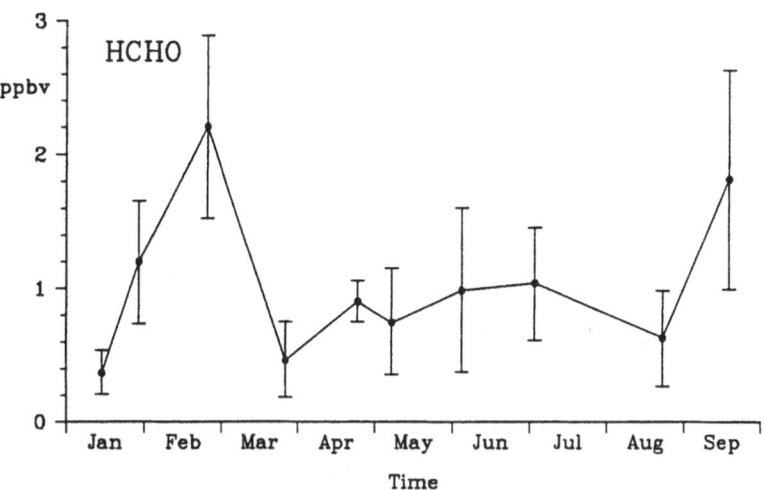

Fig. 2:   Averages of HCHO mixing ratios from January through
September 1981. The vertical bars show standard
deviations of mean values. Each measuring period
lasted 3 or 4 days.

Transport And Deposition Of S- And N- Compounds
In Central Europe Modelled As Annual Averages

W. Klug and B. Wortmann
Technische Hochschule Darmstadt

Summary

A long term model has been evaluated at the THD with the objective of
providing reasonable estimates of concentration and deposition
fields, as well as interregional fluxes of sulphur and nitrogen with
minimal expense of computer time and input data.

   Calculations have been carried out as annual averages for the
year 1982 in a grid covering the main industrial areas of Central and
Western Europe.

   In cases where sufficient data were available satisfactory
correspondence was found between calculated and measured values.

1. Introduction

   As a contribution to the PHOXA project (1) the long term transport
and deposition of acidifying substances has been simulated at the THD with
a simple Eulerian box model. The model used was originally designed to
calculate the transport of SO_x ($SO_2 + SO_4$) within the EMEP grid (2). In
order to process the PHOXA emission data this model was modified so that
the area of the boxes has the dimension of 1° longitude * 30' latitude
(ca. 60*60 km²). Furthermore, nitrogen was incorporated with the compounds
NO_2, HNO_3, NO_3 (NO_y) in a second, and NH_3, NH_4 (NH_x) in a third module.
All chemical reactions have been modelled using a pseudo first order
reaction rate without any coupling between SO_x, NO_y and NH_x (3).

   For the purpose of further model improvement a different chemical
scheme has been installed recently which considers the effect NH_3 has on
the formation of sulphate and nitrate.

2. The Model

   To calculate the concentration $C(i)$ of the species i, the following
equation has to be solved for every grid point:

$$C(i) = \frac{\dfrac{Q(i)}{H} + \dfrac{|U|}{DX} C_u(i) + \dfrac{|V|}{DY} C_v(i)}{\dfrac{|U|}{DX} + \dfrac{|V|}{DY} + R(i)}$$

with Q the source strength, C_u and C_v the concentrations of the neighbouring boxes which are advected with the wind velocity components U and V, DX and DY the meridional and lateral dimensions of the grid cell and R the total removal rate due to deposition and chemical conversion. Equations of the above form were solved for SO_2, H_2SO_4, NO_2, HNO_3, NO_3, NH_3, NH_4NO_3 and $(NH_4)HSO_4$ using the PHOXA data basis for SO_2, SO_4, NO, NO_2 and NH_3 from the year 1982.

The chemical mechanism used in the new THD model which is based on a scheme described by Derwent (4), is illustrated in Figure 1. Unlike the HARWELL model however, Ozone has not been considered and all emitted NO has been assumed to be converted immediately to NO_2.

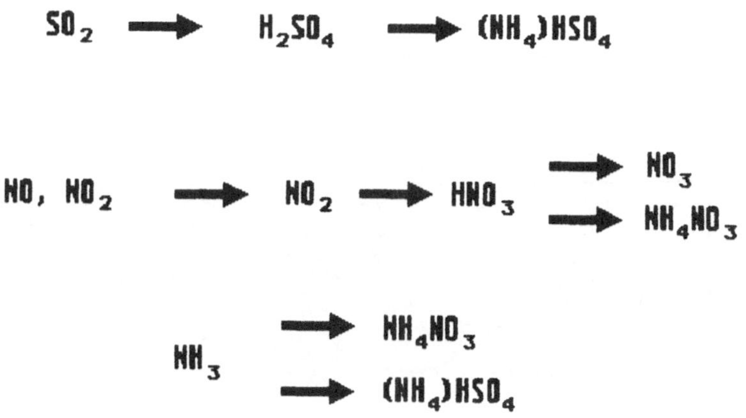

Figure 1. The chemical scheme employed in the THD long term model.

Other important assumptions are:
- constant mixing height,
- advection according to a windrose and a mean wind velocity,
- constant chemical transformation rates and dry deposition velocities with respect to time and space,
- wet deposition proportional to the mean annual amount of precipitation in each box ('constant drizzle') with different probabilities in each sector of the windrose.

A second version of the model described above, has been designed to provide boundary values for the PHOXA area processing SO_2, NO_2 and NH_3 emissions from neighbouring countries in the EMEP grid.

SO$_2$-CONCENTRATION $[\mu g/m^3]$

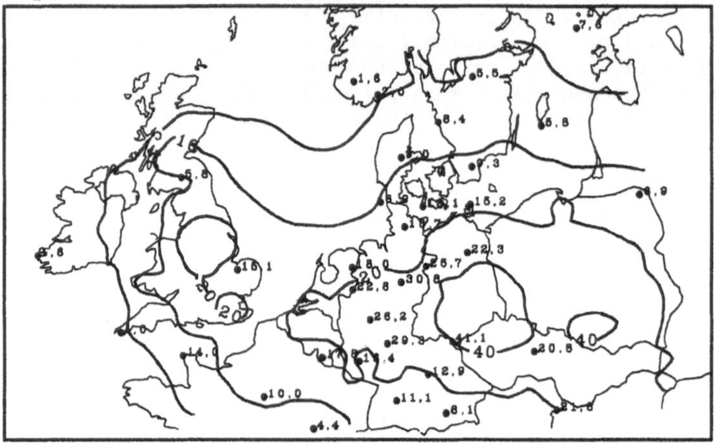

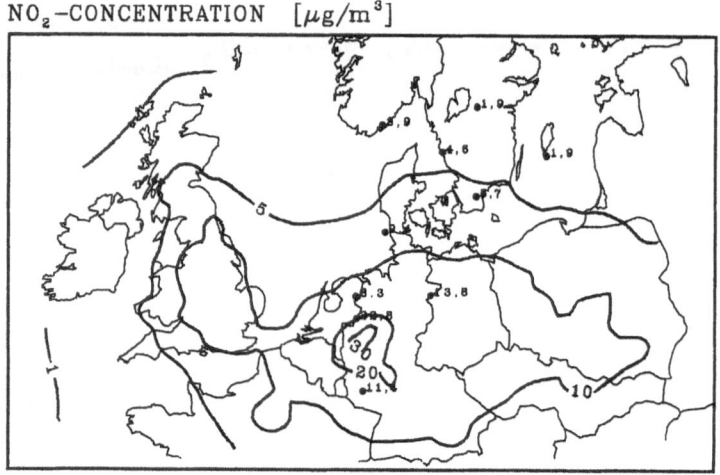

Figure 2. SO₂ concentrations in μg SO₂/m³. The numbers show measured values of the respective stations.

NO$_2$-CONCENTRATION $[\mu g/m^3]$

Figure 3. Same as Figure 2 but for NO₂ in μg NO₂/m³.

3. Results

Figures 2 and 3 show calculated mean concentrations for the year 1982 of SO₂ and NO₂ respectively, measured values for the same period have been added. The spatial distribution of both patterns is very similar, with high values along the 'industrial belt' from central England across the Rhine-Ruhr area and the southern parts of the GDR and Poland. However, the magnitude of the SO₂ concentrations exceeded those of NO₂ by roughly a factor of two. Also, the NO₂ maximum concentration can be found in the Rhine-Ruhr area, whereas the highest values for SO₂ have been calculated in the south-eastern part of the GDR.

A comparison with the observations indicates good correspondence with the calculated distribution and magnitude.

SO$_x$-WET DEPOSITION [g/m^2/a]

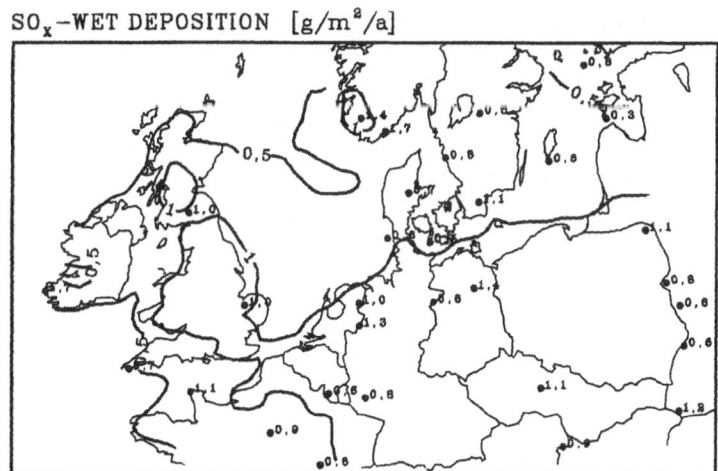

Figure 4. Wet deposition of SO$_x$ in g S/m^2.

NO$_y$-WET DEPOSITION [g/m^2/a]

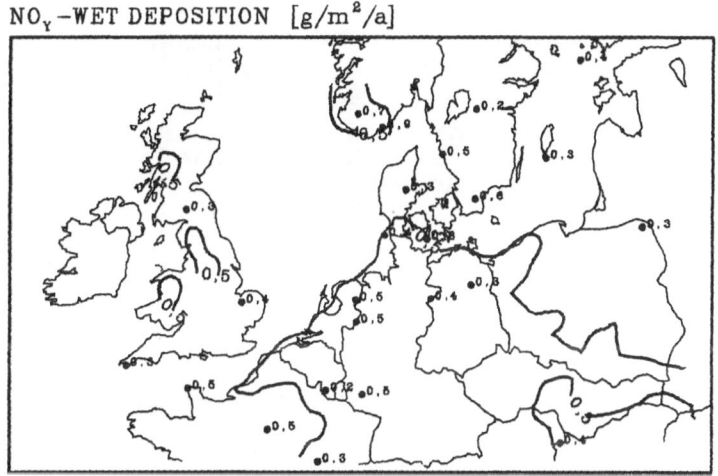

Figure 5. Wet deposition of NO$_y$ in g N/m^2.

Two examples of the simulation of wet deposition are illustrated in Figures 4 and 5, again measurements have been included. Compared to the calculated concentrations in Figures 2 and 3 wet deposition of SO$_x$ in Figure 4, as well as the corresponding results of NO$_y$ in Figure 5, are more uniformly distributed with values in both cases ranging roughly between .5 and 1 g/m^2 in units of sulphur or nitrogen respectively. The same holds true for the indicated observations.

As can be seen in both Figures, the model is able to reproduce secondary maxima of wet deposition caused by orographic rain for instance in Scotland and Norway. There, wet deposition turns out to be the most efficient removal process both for sulphur and nitrogen.

A statistical analysis of the measured and calculated data yielded correlation coefficients ranging from .63 to .91 for concentrations and .35 to .48 for wet depositions. With few exceptions, measurements and calculations differed by less than a factor of two.

Besides concentration and deposition fields, the model calculates budgets of SO_x, NO_x, NH_x and total nitrogen for every country in the model domain in the form of emitter-receiver matrices.

A more detailed picture of the budget of an individual country can be given if the fluxes across it's borders of a species emitted in another country are computed. In the case of the Federal Republic of Germany the budgets of SO_x and total nitrogen split up in foreign and indigenous sources are depicted in Figures 6 and 7. Similarly, any two areas (e.g. the North Sea, a certain industrial area etc.) with a resolution of ca. 30*30 km² (the original PHOXA grid size) may replace what is represented by 'Foreign' and 'Own' in the figures below.

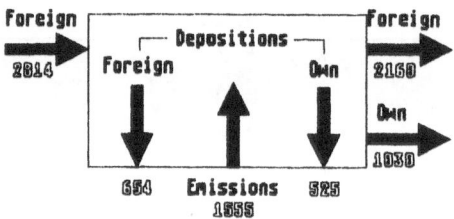

Figure 6. Sulphur budget of the FRG for the year 1982 in kt S/a.

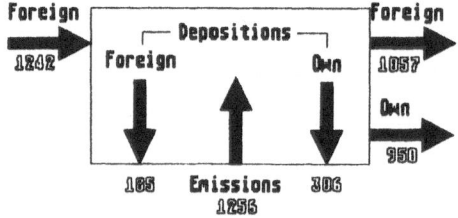

Figure 7. Same as Figure 6 but for nitrogen in kt N/a.

Acknowledgement

The project of which the results were presented here was sponsored by the Umweltbundesamt Berlin.

References

(1) Ludwig, C. and Meinl, H. (1987). Disperison Models as key elements in environmental decision making: the PHOXA programme as an example. 16th ITM on air pollution modelling and its application, Lindau.
(2) Klug, W. and Lüpkes, C. (1985). A comparison between long term interregional air pollution models. Final Report for Umweltbundesamt Berlin.
(3) Klug, W., Gömer, D. and Wortmann, B. (1989). The application of several interregional air pollution models for the simulation of acid deposition in the PHOXA programme. Final Report for Umweltbundesamt Berlin.
(4) Derwent, R.G. (1987). Modelling the long range transport of ammonia and ammonium compounds. Proceedings of the symposium "Ammonia and Acidification", pp 223-238, RIVM, Bilthoven.

DAILY PROFIL STUDY OF THE ATMOSPHERIC FORMALDEHYDE CONCENTRATIONS AT THE "POINTE DE PENMARC'H" UNDER PURE MARINE CONDITIONS

P. CARLIER, P. FRESNET, V. LESCOAT,
S. PASHALIDIS and G. MOUVIER
Laboratoire de physico-chimie de l'atmosphère
Université PARIS VII
2 place Jussieu - 75251 PARIS Cedex 05

SUMMARY

We have developed an atmospheric aldehyde measurement method, at the detection level of 0,15 ppbv, using a sampling by bubbling through an acidified 2,4-dinitrophenylhydrazine solution in acetonitrile at 2°C and a direct analysis of the produced 2,4-dinitrophenylhydrazones by HPLC with UV detection . By using this technique we measured the tropospheric formaldehyde concentrations at the "pointe de Penmarc'h" from July 4 to July 15 1988 at two levels (about 5 and 47 m high) with a sampling step of 2 hours.

As shown by the meteorological map analysis and the simultaneoously performed radon measurement ,the air masses reaching the breton coast have not travelled over a continent before at least a week . Consequently the performed measurements can be considered as representative of no perturbed air masses.

The average formaldehyde concentration was 0,4 ppbv which corresponds to the expected value from both the kinetic data for the oxidation of the oceanic methane background and for the formaldehyde photochimical depletion (photolysis, oxidation by OH radicals). By contrast with the observations made in continental areas, the daily variations were lightly pronounced. That can be explained by the rather long formaldehyde lifetime in such a medium and the very stable distribution of methane . Nevertheless it has been generaly observed slightly higher daily level at 5m than at 47m high and the opposite for night-time levels.

1. INTRODUCTION

The european project "OCEANO NOX" plans to study the oxidation mechanisms of trace compounds in the lower troposphere under the particular conditions of no perturbed marine atmosphere. For the reasons succintly recalled in the introduction of our paper related to the formaldehyde measurements during the Polarstern crossing (2), we have measured the atmospheric levels of this compound at the "Pointe

CONCENTRATIONS ATMOSPHERIQUES EN FORMALDEHYDE

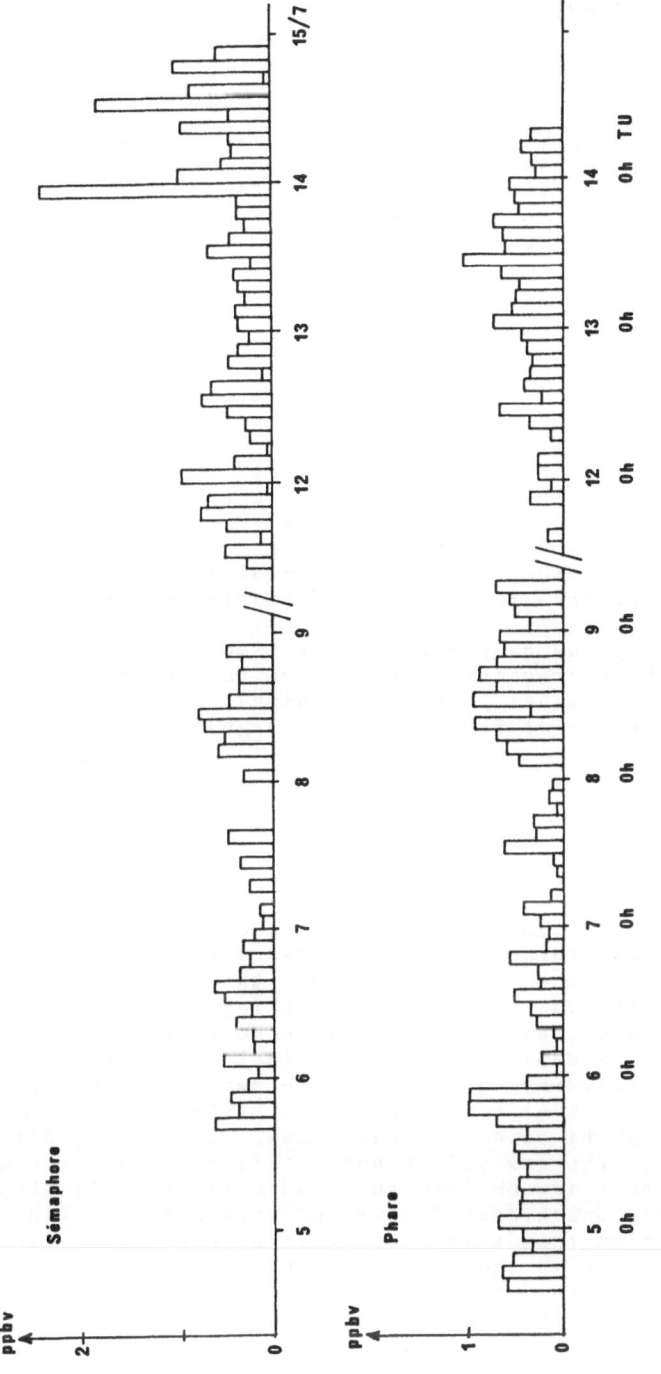

Figure 1 : Atmospheric level for formaldehyde at the "Pointe de Penmarc'h (July 1988)

de Penmarc'h". Some fuller information about the choise of this site and its avantages are available in the contribution related to the methyl iodide (3) and the DMS (4) also measured in the framework of the "OCEANO NOX" project.

This paper deals with the results obtained during the campaign of july 1988. Meteorological situations were propitious to the study of Atlantic air masses, which have not been under continental influences for at least a week as the ^{222}Rn measurements carried out by others participators in the campaign (5,6) have shown .

2. EXPERIMENTAL

The aldehydes are trapped by bubbling through an acidified solution of 2,4-dinitrophenylhydrazine in acetonitrile, at 2°C. A device constituted by 12 impingers held in a thermostatically cooled housing , by 12 electrogates drived by a camelock, by a membrane pump and by a gas volumeter get a 24 hours self sufficiency with a 2 hours sampling step. Afterwards the aldehydes are analysed as 2,4-dinitrophenyl-hydrazones by HPLC with UV detection (at 360 nm) within 48 hours. An automatic sampler enables to analyse up to 30 samples per day. For fuller information see references 7 and 8.

During the campaign of july 1988, as for the DMS (4), we have simultaneously sampled on the earth platform of the St Pierre signal station (5m high) and at the top of the old lighthouse of Penmarc'h (47m high).

3. RESULTS AND DISCUSSION

All the results are given on the figure 1. The obtained values are very low (an average of 0,4 ppbv for the two sampling points) and show fluctuations with an amplitude much weaker than in continental media ,specially on the top of the old lighthouse. As discussed talking of the results obtained on the Polarstern (4), this mean value (0,4 ppbv) agrees with the predicted one taking only into account the background methane oxidation induced by OH and the chemical formaldehyde depletion by OH radical and photolysis.

By analysing more subtly the data, it appears that this background level is clearly more disturbed at 5m high than at 47m high by secondary phenomena. It is very difficult to discuss on the raw values but, it is possible to deduce more easily some lesson from the mean daily profile study (fig. 2) because one breaks free from erratic fluctuations. Indeed at 47m high the daily profile is almost flat, with only a small maximum in the afternoon, whereas at 5m high a minimum (about 0,2 ppbv) at the night end and a maximum (about 0,6 ppbv) in the afternoon are observed.

Then, it emerges that the typical vertical gradient for the formaldehyde concentrations is positive during the night time (concentrations increase with the altitude) and nega-

PROFILS JOURNALIERS MOYENS DES CONCENTRATIONS DE HCHO A LA

POINTE DE PENMARC'H (07 - 1988).

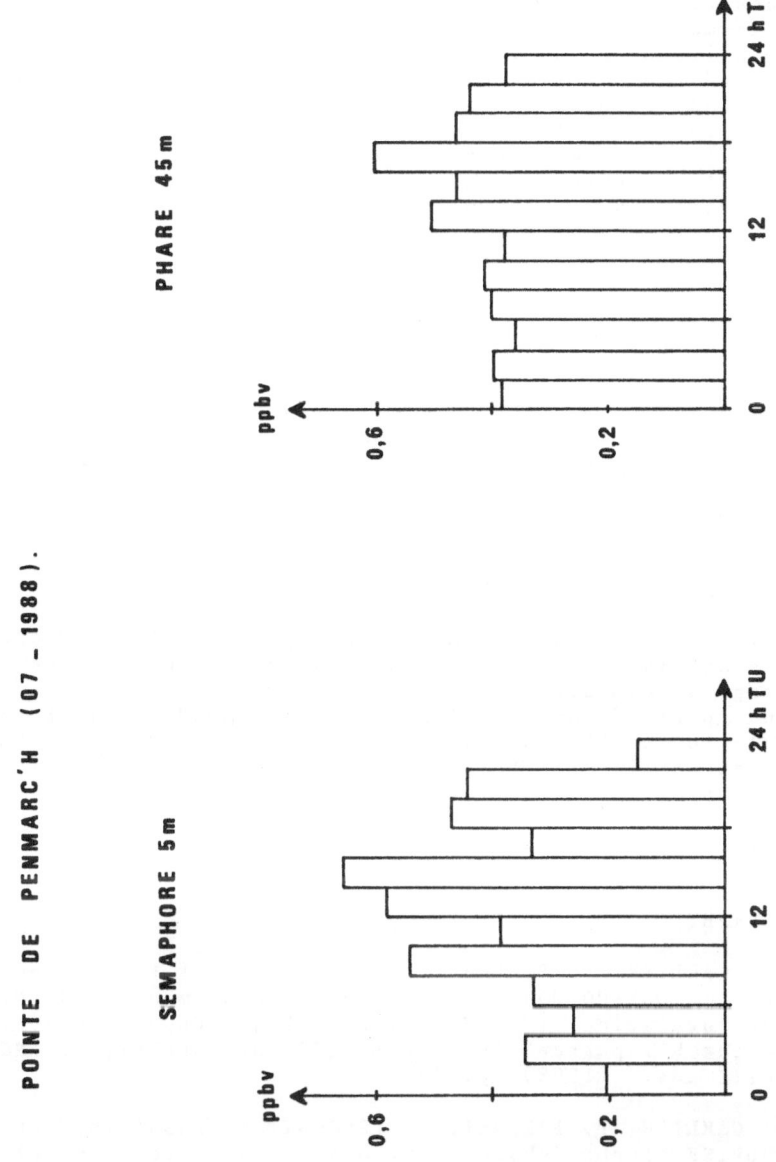

Figure 2 : Mean daily profiles of formaldehyde – a) at 5m high ; b) at 47m high

tive during the day time (concentrations decrease with the altitude). The positive night time gradient can be explained by the formaldehyde undersaturation of seawater (9 , 10) and consequently the ocean is a sink for this compound. On the other hand the negative day time gradient is very difficult to understand because it requires, in the first meters above the ocean surface, the being of an additional photochemical source sufficient to overcompensate for the oceanic sink. At the present statement of the researchs, we have not identified the compound(s) responsible of this formaldehyde formation.

4. CONCLUSION

It is now well established that the major phenomena controling the formaldehyde background level in no perturbed marine atmosphere are the chemical oxidations of methane as a source and of formaldehyde as a sink. Other revealed phenomena (dissolution in ocean, additional photochemical sources) are minor phenomena which affect only the first meters above the ocean surface.

ACKNOWLEDGMENTS

This work is a part of the european project "OCEANO NOX" which is financially supported by the Commission of the european Communities.
The participants in the "OCEANO NOX" project are grateful to the "Marine Nationale" and to the "Service des phares et balises" for the free access to the St Pierre signal station and to the old lighthouse of Penmarc'h . We would like to thank also all the staff of the establishments for their logistical support .

BIBLIOGRAPHY

1.   P. CARLIER ; Synthetic presentation of the european project OCEANO NOX. In : "Field measurements and their interpretation" ; COST 611 WP III ; CEE air pollution research report n° 14. S. BEILKE, J. MORELLI, G. ANGE-LETTI eds., (1988), p. 55.

2.   P. CARLIER, P. FRESNET, V. LESCOAT, S. PASHALIDIS et G. MOUVIER. Formaldehyde Background levels over the atlantic ocean during the Polarstern crossing from Bremerhaven to Rio Grande do sul. Ibiden.

3.   F. PETITET, M. TSETSI, P. CARLIER et G. MOUVIER . A Method for Methyl iodide measurement in the lower troposphere. Application to the marine atmosphere.Ibidem.

4.   S. PASHALIDIS, P. CARLIER et G. MOUVIER. Daily profile study of the atmospheric DMS concentrations near to an intense coastal source at the "Pointe de Penmarc'h". Ibidem.

5.   T. BRAUERS, H.P. DORN et U. PLATT . Private communication

6.   B. BONSANG et M. KANAKIDOU. Private communication .

7.   P. KALABOKAS. "Etude des facteurs sources et puits des composés carbonylés dans les masses d'air transitant au-dessus de la région parisienne". Thèse d doctorat - Paris VII - Septembre 1987 .

8.   V. LESCOAT . "Physicochimie des composés carbonylés dans les hydrométéores : métrologie et application. Etude des mécanismes d'oxydation. Thèse de doctorat - Paris VII - Février 1989.

9.   O.C. ZAFIRIOU, J. ALFORD, M. MERRERRA, E. PELTZER, A.M. THOMPSON et R.B. GAGOSIAN. Formaldehyde in remote marine air and rain : flux measurement and estimates ; J. geophys. Res. $\underline{7}$, (1980), 341.

10. P.S. LISS . "Gas transfer : Experiments and geochemical implications" in : Air-Sea Exchange of gases and particules - Nato ASI series C n° 108 - P.S. LISS et W.G.N. SLINN (Eds) ; R. Reidel Publ. Co. ; Dordrecht (1983) p 241.

MODELLING THE INFLUENCE OF METEOROLOGICAL AND OTHER FACTORS ON THE
CHEMISTRY OF OXIDIZED NITROGEN COMPOUNDS

Marke Hongisto
Finnish Meteorological Institute, Air Quality Department
Sahaajankatu 22 E, SF-00810 Helsinki, Finland

Summary

In long range transport models the choice of parametrization leads to
highly varying residence times of the compounds in the air. In this study
we have simulated the chemical transformation of some oxidized nitrogen
compounds above typical Finnish terrains in different meteorological
conditions. Our purpose was to collect information on how the simplifying
assumptions and variation of the model parameters influences the
transformation kinetics and deposition rates of the compounds. Simulations
are made for an instant release in clear sky conditions.

1. INTRODUCTION

Long or medium range transport models of chemical compounds are valuable in
tracing the origin of pollutants measured at environmental background
stations. However, for Finnish conditions the comparison of calculated and
measured concentrations has not led to a good compatibility. E.g.,
according to the latest EMEP-model calculations for sulphur and nitrogen
compounds, the modelled particulate sulphur concentrations were about half
of the measured values; SO_2 concentrations were even more underestimated
and the discrepancy was still greater for the nitrogen compounds (1). The
model parametrizations valid in Central European conditions are not
necessarily adequate for Finland. To estimate the influence of local
factors on the chemistry of NO_x derived compounds above typical Finnish
terrains, we have made case studies with the chemistry submodel of a
regional, Eulerian-type transport model, presently under construction in
the FMI. Simultaneously we have tested some simplifying assumptions that
could be appropriate in northern conditions.

2. MODEL STRUCTURE

In Lagrangian type models as in (2) the concentration c_i of compound i is
approximated as follows:

$$\frac{d}{dt}c_i + v\nabla c_i = -\sum_i k_i c_i c_k + \sum_j k_j c_j c_l + \frac{Q}{h} \frac{v_d}{h} c_i - \frac{WP}{h} c_i$$

where Q is the average emission rate of compound i per unit area, h is the
mixing height, k_i and k_j represent the transformation rates to other
substances via different chemical reactions with c_k or c_l, v_d is dry
deposition velocity and WP/h represents average wet deposition rate from
the volume element by scavenging coefficient W and precipitation amount P.
The emission is assumed to be instantaneously mixed to the total volume of
the grid element and advected out of the box by wind of velocity v.

By making a crude assumption that the average emission rate equals the advection rate, we can follow the transformation of average concentration of a box volume element as if it were a puff release. In the case studies described here, dry deposition represents the only means of exchange with the environment. The simulations are made in clear sky conditions. Another approach could be to follow the average concentration of a box with continuous emission and weather-type dependent dilution, but e.g. the available plume models are not adequate for long range transport, and the dilution by turbulent mixing could be much stronger than chemical transformation near the source; so the sensitivity analysis could fail.

In the troposphere the dominant interactions between oxidized nitrogen compounds are assumed to be the transformation of NO to NO_2 via hydrocarbon or O_3 reactions, NO_2 loss by photodissociation, or day-time gas phase reactions with OH or CH_3COO_2 radicals leading to HNO_3 or PAN formation, and night-time heterogeneous path to nitrate particles in water droplets or on aerosol surfaces. The structure of the model is shown in Fig.1., it is a modified version of the oxidized nitrogen part of the EMEP-NO_x chemistry model (2). The intercoupling reactions between the selected variables are linearized by giving the OH, RO_2 and CH_3COO_2 radical and ozone concentrations as diurnally and seasonally varying quantities.

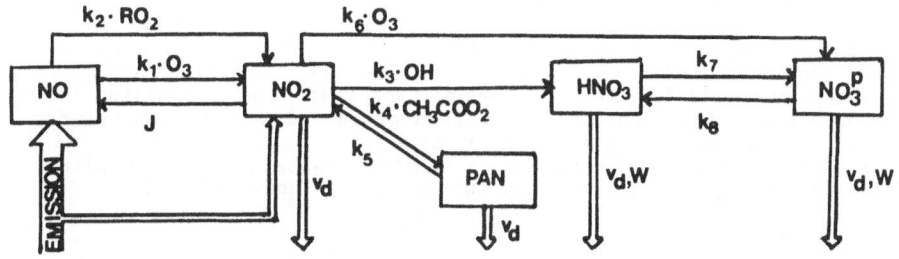

Fig.1. The structure of the chemical submodel. $k_1 = 2.3E-12exp(-1450/T)$, $k_2 = 7.6E-12$, $k_3 = 1.1E-11$, $K_4 = 3.2E-12$, $k_5 = 7.94E14exp(-12530/T)$, $k_6 = 1.2E-13exp(-2450/T)$, $k_7 = 1.5E-5$ (June),$0.5E-5$(December), $= 2 k_8$, $J = .01exp(-0.39sec:\Theta)(1-(CLF/2))$, where Θ = zenith angle of the sun and CLF = the cloudiness index. k_1-k_4 and k_6 in $cm^3molec^{-1}s^{-1}$,others in s^{-1}.

The differential equation system is solved by the quasi steady state approximation method (3). The reaction rates for the equations are given below Fig.1. Because HNO_3 is assumed to be scavenged inside a cloud to droplets in a few seconds (4), the gaseous nitric acid is assumed to exist only outside clouds, and the transformation rate of nitric acid to dissolved nitrate is assumed to be equal to the average formation rate of the clouds. In the higher parts of the clouds, and in the subsidence areas, the droplets can evaporate (evaporation happens in the clouds 10-25 times before they precipitate, see (5)). Then at the same time the acidity of the droplets increases and HNO_3 can be released unless it has formed a salt inside the droplet. Thus the reaction k_7 is bidirectional. In December, because of lower temperatures and ice clouds the formation rate of nitrate by scavenging is lower. Still, according to temperatures from the available sounding data, we have supercooled clouds below 2.5 km . The partition of nitric acid between the particle and gas phases is important because of their different deposition rates.

The ozone levels employed in the model were monthly average concentrations measured at Ähtäri background station, varying diurnally from 24 to 38 ppb in June and 18 to 23 ppb in December. Monthly average temperatures were -5...-7 °C in December, 10...16 °C in June. The mixing height varied from 200 to 1600 m in June, 250 to 500 m in December. The OH and CH_3COO_2 radical concentrations depend on the solar radiation; the $C(max)OH$ was $0.85*10^6$ cm^{-3} in summer and $0.3*10^6$ cm^{-3} in winter, the corresponding $C(max)CH_3COO_2$ values were $1.2*10^6$ and $0.8*10^6$ cm^{-3}. The organic radical concentration RO_2 was estimated to be $0.5*10^9$ cm^{-3} in winter and 10^9 cm^{-3} in summer. The initial concentration whose development is followed in the model, was 10^{10} molecules/cm^3, consisting of 95 % NO and 5 % NO_2 .The simulations were started at the shortest and longest days of the year at Helsinki latitude.

The dry deposition velocities were calculated by $v_d = (r_a + r_b + r_c)^{-1}$ where the aerodynamic resistance r_a was

$$r_a = (\ln(z/z_o) - \Psi(z/L))(0.4\ u_*)^{-1} \cdot$$

Here $\Psi(z/L)$ is the stability function for the vertical heat profile, z_o is the roughness parameter, and u_* the friction velocity. Values for the other resistances and surface roughness were calculated analogously, or taken from the rewiew article of Voldner (6). The stability functions needed in the calculations were taken from (7). The friction velocity, Monin-Obukhov length L, mixing height, as well as some other meteorological variables needed in the simulation runs, were calculated by the meteorological preprocessor developed in the Finnish Meteorological Institute by Pentti Vaajama. The deposition velocities calculated for each hour of a month for each compound were averaged to get the monthly diurnal values. From the average terrain types of southern Finland we have selected two representative areas called the coastal and inland sites, and the calculated v_d values for these two areas are shown in Fig.2.

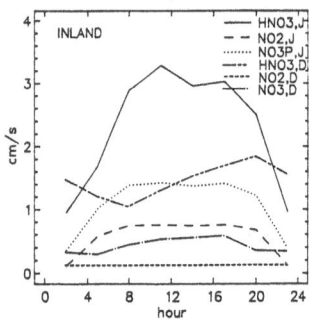 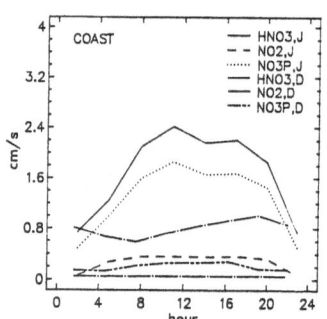

Fig 2. Dry deposition velocities, monthly averages, Southern Finland. I = inland site: 5 % lakes, 5 % fields, 70 % coniferous, 20 % deciduous forest, C = coastal site: 50 % sea, 10% fields, 20 % deciduous, 20% coniferous forest. d = December, j = June.

3. RESULTS

In table 1 we present the average residence times for nitrogen oxides in air as $T(30\%)NO_2$, the time needed for the concentration to drop to 30 % of the initial value. We also calculated the time required for the total concentration to drop 50 %, $T(50\%)TOT$, and the maximum values of the individual concentrations as a fraction of the initial value: NO_3 max, HNO_3 max and PAN max.

In figure 3 we present the total concentrations in June and December over sites consisting of just one or mixed terrain types. The development of the nitrate, NO_2 and HNO_3 concentrations are shown in figure 4.

The simulation is started at 7:40 local time, thus in winter the emitted NO-concentration transforms to NO_2 in a few time steps; the maximum NO-concentration reached after the photodissociation of NO_2 begins was about 5 % of the initial concentration. In light conditions of June 24th the NO_2 photodissociation rate is about six times greater at Helsinki latitude than December 24th, so we get a maximum of 27-28 % of NO at noon if we assume a photo-stationary state. The NO level is sensitive to the ozone and RO_2 concentrations and the temperature. In separate simulations (January 10th, Helsinki latitude) it has been shown that if the O_3 level is kept constant, the decrease of O_3 concentration from 60 to 15 ppb drops the ratio NO_2/NO from 30 to 7 in 0 °C temperature, and from 16 to 5 in -15 °C temperature. Also if we increase the RO_2 level to 10^{10} n/cm^3 from the values used in simulations, the maximum concentration of NO in summer is decreased by 80 % (with the same initial concentration of NO_x.) In stable, cold situations when turbulent mixing is low and ozone is consumed inside the plume, or near big natural or anthropogenic hydrocarbon emission sources, the oxidation of NO can be very slow.

TABLE 1

SUMMER

surface	coniferous forest	decid. forest	grass	water	coastal 50% see	inland site 90% forest
T(30%)NO_x	22:15	22:15	37:30	45:30	32:15	22:40
T(50%)TOT	20:30	20:15	36:15	41:40	24:30	20:40
NO3 max,%	14.8	13.7	20.4	17.5	16.3	14.8
HNO3 max	9.4	9.4	11.1	12.6	11.0	9.6
PAN max	1.6	1.6	1.8	1.8	1.8	1.7

WINTER

	coniferous forest	decid. forest	grass	ice water	coastal 50% see	inland site 90% forest
T(30%)NO_x	43:30	48:30	55:45	56:44	50:50	45:20
T(50%)TOT	41:15	>72:00	>72:00	>72:00	72:00	49:20
NO3 max,%	18.3	32.9	37.0	38.4	29.5	21.9
HNO3 max	2.0	2.6	4.6	8.2	3.4	2.3
PAN max	3.3	3.7	4.0	4.1	3.5	3.3

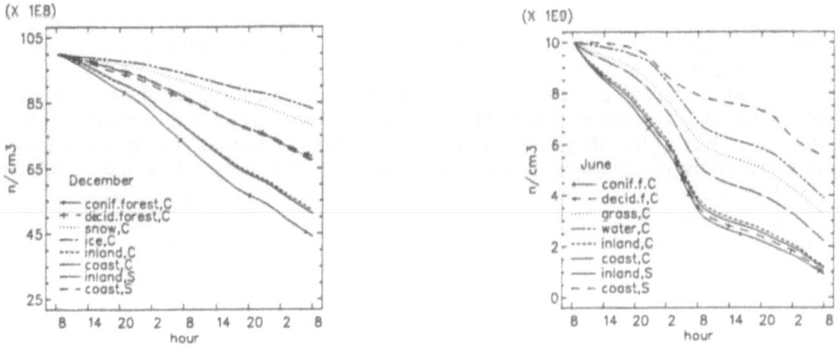

Fig 3. Development of total concentrations, June and December.

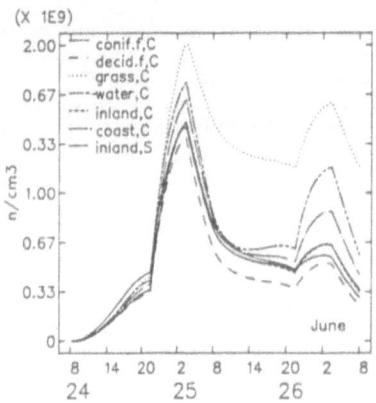 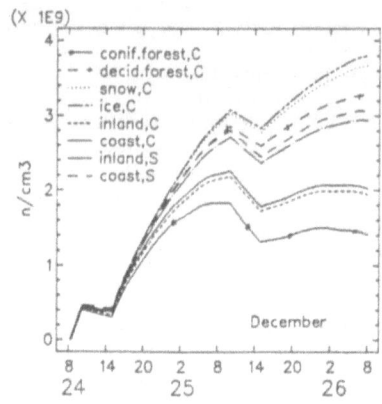

Fig.4a   Concentrations of nitrate, June and December

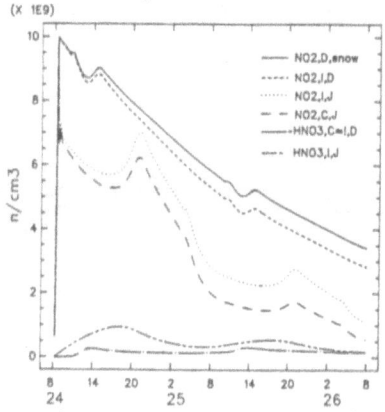

Fig.4b   Concentrations of NO$_2$ and HNO$_3$. C=coastal, I=inland, J=June,
           D = December.

The NO$_2$ concentration decreases quite steadily in winter.   In summer
it has a late evening maximum because of the NO decay.   At the simulation
dates the summer night lasted about 6 hours, and the winter night 19 hours,
which leads to much higher nitrate levels in winter, in spite of the two
times faster night-time formation rate in summer. At higher latitudes the
difference is more pronounced: if all other parameters except the sun
radiation level are kept constant, we get 89 % more nitrate in winter than
in summer at Helsinki latitude, but 216 % more at Rovaniemi latitude after
the first simulation night.
        Dry deposition influences the shape of the concentration curves,
especially in the morning, when the mixing height starts to rise (generally
after 8 AM), but the dry deposition velocity starts to grow after the
sunrise according to parametrization used.   Thus v_d/h is a very effective
sink at that time.   The HNO$_3$ concentration decreases strongly in summer
nights;   in winter nights it may even grow.   The explanation is partly the
variation of its precursor concentration NO$_2$, partly the lower OH levels at
night.   In the winter-time we get much less nitric acid than in summer.

4. Conclusions

We have shown, that even in a very simple air chemistry model the choice of
parametrization leads to quite different residence times of the emitted
pollutants in air.   In this work we have simulated the development of the

chemical composition of an air parcel passing over a single type of terrain. The examined simulation period may be too long, because the air parcel probably crosses several land types in three days. However, already from the first simulation hours it can be seen, that it is necessary to characterize the surface in detail in long range transport models; at least the ice/snow/water areas should be separated from the vegetated areas to get more reliable development of concentrations.

REFERENCES

(1) Tuovinen,J-P.,Kangas L.,Nordlund,G. Model calculations and nitrogen deposition in Finland. to be published in Acidification in Finland, ed. by P.Kauppi et all. Springer-Verlag, 1990.

(2) Eliassen,A.,Hov,O.,Iversen,T.,Saltbones,J.,Simpson,D.,1988. Estimates of airborne transboundary transport of sulphur and nitrogen over Europe. EMEP/MSC-W Report 1/88. The Norwegian Meteorological Institute, 79p.

(3) Hesstvedt,E. Hov,Ö.,Isaksen,S.A.,1978. Quasi-steady-state approximations in air pollution modelling: Comparison of two numerical shemes for oxidant prediction. Int. J. of chemical Kinetics, vol X, pp. 971-994.

(4) Chameides,W.L.,1984. The photochemistry of remote marine stratiform cloud. Journal of Geophysical Research, Vol.89,NO.D3,pp.4739-4755.

(5) Pruppacher,H.R.,1986. The role of cloud physics in atmospheric multiphase systems: Ten basic statements. In Chemistry of multiphase atmospheric systems, ed by W.Jaeschke. NATO ASI Series G,Vol 6, pp.133-190

(6) Voldner,E.C.,Barrie,L.A.,Sirois,A.,1986. A literature review of dry deposition of oxides of sulphur and nitrogen with emphasis on long-range transport modelling in North America. Atm.Env.Vol.20, No.11,pp.2101-2123.

(7) van Ulden,A.P.,Holtslag,A.A.M.,1985. Estimation of atmospheric boundary layer parameters for diffusion applications. Journal of Climate and Applied Meteorology, Vol.24,pp.1196-1207.

TEMPORAL VARIABILITY OF ATMOSPHERIC Pb, Cu AND Mn
CONCENTRATIONS AND FLUXES OVER THE NORTHWESTERN MEDITERRANEAN SEA.

E. REMOUDAKI, G. BERGAMETTI, R. LOSNO B. CHATENET and G. MOUVIER
Laboratoire de Physico-Chimie de l'atmosphère, Université Paris 7,
2 Place Jussieu, F-75251 Paris Cedex 05, FRANCE

Summary
Daily 24-h aerosol and total (wet+dry) atmospheric deposition samples (with a collection period of about 15 days) have been collected
at a coastal location in northwestern Corsica between February 1985
and October 1987. 3-D air mass trajectory analysis indicates that
atmospheric Pb, Cu, and Mn over the Western Mediterranean are long
range transported from continental source regions. The scavenging of
atmospheric Pb, Cu, and Mn by rains generates a seasonal cycle with
high atmospheric concentrations during the dry season and lower
concentrations during the wet season. Sporadic but intense Saharan
dust transport events are responsible for the highest atmospheric
concentrations of Mn. The ratios between the atmospheric deposition
of Pb, Cu and Mn and the corresponding mean daily precipitation rate
reach a maximum during the mediterranean summer. This is due to the
wet scavenging of a more loaded atmosphere during the dry season
than during the wet season. The precipitation frequency (Fp) is the
major factor influencing the seasonal variability of the atmospheric
Pb and Cu deposition. The relationship between the seasonal variability of Mn deposition and the Fp is less direct. This is due to
the occurrence of Saharan dust transport events which strongly
influence the variability of the Mn deposition.

1. INTRODUCTION
        The transport and input of elements originating from anthropogenic
and natural land-sources strongly influence the geochemistry of the
Mediterranean seawater (3),(4),(6). Previous studies (1), (5), (6), have
shown that the atmospheric trace metal concentrations are highly variable
on a daily time scale over the Northwestern Mediterranean. These studies
suggested that a continuous sampling strategy is needed to quantitatively
assess the inputs of trace elements into this basin. Such a sampling
strategy has been realized at a coastal site in Corsica between February
1985 and April 1988.
        We present here daily atmospheric concentrations and measured total
atmospheric deposition values for lead, copper and manganese. We finally
evaluate the major factors which control the transport and deposition of
Pb, Cu, and Mn to the above marine area.

2. SAMPLING AND ANALYSIS
        Sampling and analytical procedures are described elsewhere (8), (9).

3. RESULTS AND DISCUSSION
3.1. Temporal variability of the aerosol concentrations.
        Results on the daily concentration of Pb and Cu cover the period:
February 1985-April 1986. The daily measured concentrations of Mn cover

the period: February 1985-April 1988. The geometric mean atmospheric concentrations for the studied period are: 15.9 ng.m^{-3}, 2.1 ng.m^{-3} and 5 ng.m^{-3} for Pb, Cu and Mn respectively.

In previous papers (2), (3), (8), (9), we have shown that there is a high variability of the Pb, Cu, and Mn atmospheric concentrations occuring on a time scale of one day. Local precipitation events are responsible for the abrupt decreases of atmospheric lead, copper and manganese concentrations. Moreover, we observe strong variations of the atmospheric manganese concentrations which appear episodically during the sampling period: for example, atmospheric manganese concentration increases by a factor of 100 in less than 48 hours between December 28th and 30th 1985. These strong variations occur at the same time to Al and Si atmospheric concentrations and, are attributed to dust transport events from the arid or semiarid African regions (2), (3). These events are responsible for the highest measured concentrations of Mn, Al, and Si (3). Although manganese is an element of mixed origin (anthropogenic and crustal), when such events occur, the predominance of crustal Mn in the Med iterranean atmosphere becomes evident.

Our data also suggest a seasonal pattern of lead , copper, and manganese concentrations in the Western Mediterranean atmosphere. In table I are reported the geometric mean atmospheric concentrations for the three elements corresponding to each dry and wet season of the sampling period. Dry seasons are characterized by higher mean concentrations while wet seasons are characterized by lower mean concentrations. This pattern is inversely related to that of precipitation (table I). These results suggest that precipitation is the major factor controlling the seasonal pattern of the measured concentrations of Pb, Cu, and Mn.

A 3-D air mass trajectory analysis has been performed for the first year of measurements. The model of trajectory calculation reports the forecasted precipitation events greater than 0.1 mm.h^{-1}. The air mass trajectory study suggested that Pb, Cu and Mn are long range transported from continental source regions. This study also permitted to evaluate the influence of source regions and airflow patterns on the variations of the atmospheric concentrations of these elements (3), (8), (9). The main conclusion from the study of the source regions and airflow patterns is that precipitation occuring locally as well as during transport is the major factor controlling the variability of lead and copper atmospheric concentrations in the Western Mediterranean. The strong variations of the manganese concentrations are mainly due to the Saharan dust inputs and to the removal of the particles containing Mn by precipitation events.

3.2. Temporal variability of the Pb, Cu, and Mn atmospheric deposition.

The measured fluxes of Pb, Cu, and Mn range from 0.0009 to 0.0660 µg.cm^{-2}.d^{-1}, 0.0001 to 0.0070 µg.cm^{-2}.d^{-1} and 0.0002 to 0.0358 µg.cm^{-2}.d^{-1} respectively. In (8), (9) we have shown that the high deposition values of Pb, Cu and Mn result mainly from wet scavenging. Our results also suggest that there is a factor of 10 between the fluxes of Cu and Pb, the last one being always the higher. This proportion is in agreement with the amounts of Pb and Cu emitted by the anthropogenic European sources (7). Although the antropogenic emissions of Mn are of the same order of magnitude with those of Cu (7), manganese fluxes are much higher than those of Cu.

Moreover, the strong increases of the measured total (wet+dry) fluxes of Mn are in perfect agreement with the aluminium measured fluxes corresponding to the same samples (9). These increases are attributed to the occurrence of sporadic but intense events of saharan dust transport. The episodically increases of the Mn and Al fluxes are observed when such events are followed by precipitation. In these cases, manganese deposition values, due to the wet scavenging of high concentrations of mineral aerosol particles, are the highest observed.

3.2.a) Relationship between the Pb, Cu, and Mn deposition and the occurrence of the rainy days

In previous papers (8), (9) it has been shown that the ratios between the lead, copper and manganese fluxes and the corresponding mean daily precipitation rate (m.d.p.) are not constant. These ratios become higher during spring and summer. As discussed previously, lead, copper and manganese concentrations in the mediterranean atmosphere follow a seasonal pattern with high concentrations during the dry season and lower concentrations during the rainy season. We have shown that the high values of the ratios F(Pb)/m.d.p. and F(Cu)/m.d.p. during the dry season are mainly due to the wet scavenging of a more heavily loaded atmosphere. This is shown in figure 1. In this figure we have reported the Pb, Cu, Mn, and Al fluxes after application of a moving mean method to our data, with a time span of about 60 days. In the same figures we have reported the occurrence of the rainy days using the same method.

Indeed, Pb and Cu deposition are highest during the dry season and lowest during the wet season. The variations of Pb and Cu deposition are anticorrelated with those of the occurrence of the rainy days. We attribute this behaviour to the ability of Pb and Cu to be more efficiently transported from the source regions to the sampling area during the dry season. In contrast, the time needed for a sufficient reloading of the sampling site atmosphere is longer than the time interval between two rain events during the mediterranean " winter".

The above arguments are strongly supported by the comparison between the Pb (and Cu) and Na fluxes. We have shown that the variations of Na deposition are perfectly correlated with those of the occurrence of the rainy days. Na is primarily present as seasalt aerosol locally produced. Consequently, Na deposition is independent from transport processes. Its behaviour is explained by a very fast reloading of the local atmosphere, following a rain event, with locally produced seasalt aerosol.

Although manganese deposition also presents higher values during the dry season and lower values during the rainy season (fig. 1), it is mainly dominated by the occurrence of saharan dust transport events. This can be shown by the comparison between the Mn and Al fluxes (fig. 1). The temporal variability of these two curves is in good agreement.

These results suggest that two main factors control the seasonal variability of the manganese deposition:
 -The occurrence of precipitation locally as well as during transport.
 -The occurrence of saharan dust transport events.

3.2. b) Relationship between the frequency of the precipitation periods and the seasonal variability of Pb, Cu and Mn atmospheric deposition.

We divided the sampling period into dry and wet seasons. For both seasons, we defined the frequency of the precipitation periods Fp as the number of groups of precipitation events divided by the duration of the corresponding period in days. The groups of precipitation events are separated from each other by a period of 4 "dry" days at least. The choice of this time interval is in agreement with the estimated mean travel time of continental aerosol to Corsica (at least two days, (3)).

In table II we present the Pb, Cu, Mn fluxes as well as the frequency of the precipitation periods Fp corresponding to each season. Our results suggest that:

–High values of Fp mean that the atmosphere has not the time to be reloaded enough with long range transported Pb and Cu aerosol particles. So, Pb and Cu deposition presents low values during both rainy seasons (1985-1986, 1986-1987). The summer 1986 was frequently scavenged by rains and cannot classified as a "normal" dry season. The corresponding deposition values of Pb and Cu are lower than those observed during summer 1985.

–Low values of precipitation frequency Fp mean that dry deposition may become significant. The values of Pb and Cu fluxes corresponding to summer 1987 are ascribed to the lack of precipitation. Consequently, even if the atmosphere is heavily loaded with particles, the corresponding fluxes are low.

–An intermediate value of Fp allows sufficient reloading of the atmosphere with Pb and Cu aerosol particles as well as efficient scavenging by precipitation events. In this case (summer 1985), Pb and Cu fluxes are the highest observed and their seasonal variability becomes more pronounced.

The relationship between the manganese fluxes and the corresponding Fp appears less direct. The high total deposition value corresponding to the wet season 1985-1986 is ascribed to the occurrence of sporadic and intense saharan dust transport events. Indeed, 25 per cent of the total annual flux of Mn, corresponding to the first year of measurements, resulted from one single Saharan dust event which occurred between March 1st and 2nd, 1986. The relatively high value of Mn deposition during summer 1987 is also due to the occurrence of such events.

4. ACKNOWLEDGEMENTS

We thank the staff of the signal station of Capo Cavallo for their logistical support during the field experiments. We thank the French Marine Nationale for the free access to the signal station and the French Direction de la Météorologie Nationale for the use of the meteorological tower. We are grateful to A. Dutot, F. Dulac, D. Martin, B. Strauss and J. M. Gros. This work was supported by the Ministère de l'Environnement, the programme Flux Océanique of INSU CNRS (DYFAMED), by UNEP/WMO (MEDPOL) and, by a doctoral fellowship (Sectoral grant n° B/87000259, E. Remoudaki) of the Environmental Research Programme of the European Economic Communities.

5. REFERENCES

(1) ARNOLD M., A. Seghaier, D. Martin, P. Buat-Ménard, and R. Chesselet, Géochimie de l' aérosol marin au-dessus de la Méditerranée Occidentale, in Comptes Rendus des VI journées d' études sur les pollutions marines en Méditerranée. CIESM Monaco, 27-37, 1982.

(2) BERGAMETTI G., L. Gomes, E. Remoudaki, M. Desbois, D. Martin, and P. Buat-Ménard, Present transport and deposition patterns of African dusts to the Northwestern Mediterranean, NATO ASI series: Paleoclimatology and paleometeorology: Modern and past patterns of global atmospheric transport, 1989, KLUWER, DORDRECHT, in press.

(3) BERGAMETTI G., A. L. Dutot, P. Buat-Ménard, R. Losno, and E. Remoudaki, Seasonal variability of the elemental composition of atmospheric aerosol particles over the Northwestern Mediterranean, Tellus, 41B,

353-361, 1989.
(4) BUAT-MENARD P., J. Davies, E. Remoudaki, J. C. Micquel, G. Bergametti, C. E. Lambert, U. Ezat, C. Quetel, J. La Rosa, and S. W. Fowler, Non steady-state removal of atmospheric particles from Mediterranean surface waters, Nature, in press, 1989.
(5) CHESTER R., Sharples E. J., Sanders G. S., Saydam A. C., Saharan dust incursion over the Tyrrhenian Sea, Atmos. Environ., 18, 929-935, 1984.
(6) DULAC F., P. Buat-Ménard, M. Arnold, and U. Ezat, Atmospheric input of trace metals to the Western Mediterranean Sea: 1. Factors controlling the variability of atmospheric concentrations, J. Geophys. Res., 92, 8437-8453, 1987.
(7) PACYNA J., Estimation of the atmospheric emissions of trace elements from anthropogenic sources in Europe, Atmos. Environ., 18, 41-50, 1984.
(8) REMOUDAKI E., G. Bergametti, and P. Buat-Ménard, Temporal variability of atmospheric lead concentrations and fluxes over the Northwestern Mediterranean Sea, accepted for publication in J. Geophys. Res., 1989.
(9) REMOUDAKI E., G. Bergametti, and R. Losno, On the dynamic of the atmospheric input of copper and manganese into the Western Mediterranean Sea, submitted for publication in Atmos. Environ., June 1989.

Begin-End	m.d.p. (mm.d^{-1})	Pb (ng.m^{-3})	Cu (ng.m^{-3})	Mn (ng.m^{-3})
03/25/85-10/11/85 (dry season)	0.5	21.1	3.4	7.2
10/11/85-03/22/86 (wet season)	2.2	12.6	1.4	4.5
03/22/86-11/01/86 (dry season)	1.0	---	---	5.3
11/01/86-04/16/87 (wet season)	1.9	---	---	3.2
04/16/87-10/02/87 (dry season)	0.2	---	---	8.0

Table I: Geometric mean atmospheric concentrations and the mean daily precipitation rate m.d.p. for each dry and wet season.

Begin-End	Fp (d^{-1})	F(Cu) $(\mu\text{g.cm}^{-2}.\text{d}^{-1})$	F(Pb) $(\mu\text{g.cm}^{-2}.\text{d}^{-1})$	F(Mn) $(\mu\text{g.cm}^{-2}.\text{d}^{-1})$
03/25/85-10/11/85 (dry season)	0.030	0.0014	0.0118	0.0072
10/11/85-03/22/86 (wet season)	0.074	0.0004	0.0025	0.0062
03/22/86-11/01/86 (dry season)	0.058	0.0009	0.0042	0.0023
11/01/86-04/16/87 (wet season)	0.079	0.0005	0.0028	0.0023
04/16/87-10/02/87 (dry season)	0.018	0.0007	0.0024	0.0033

Table II: Seasonal Pb, Cu, and Mn deposition and the corresponding frequency of the precipitation periods Fp.

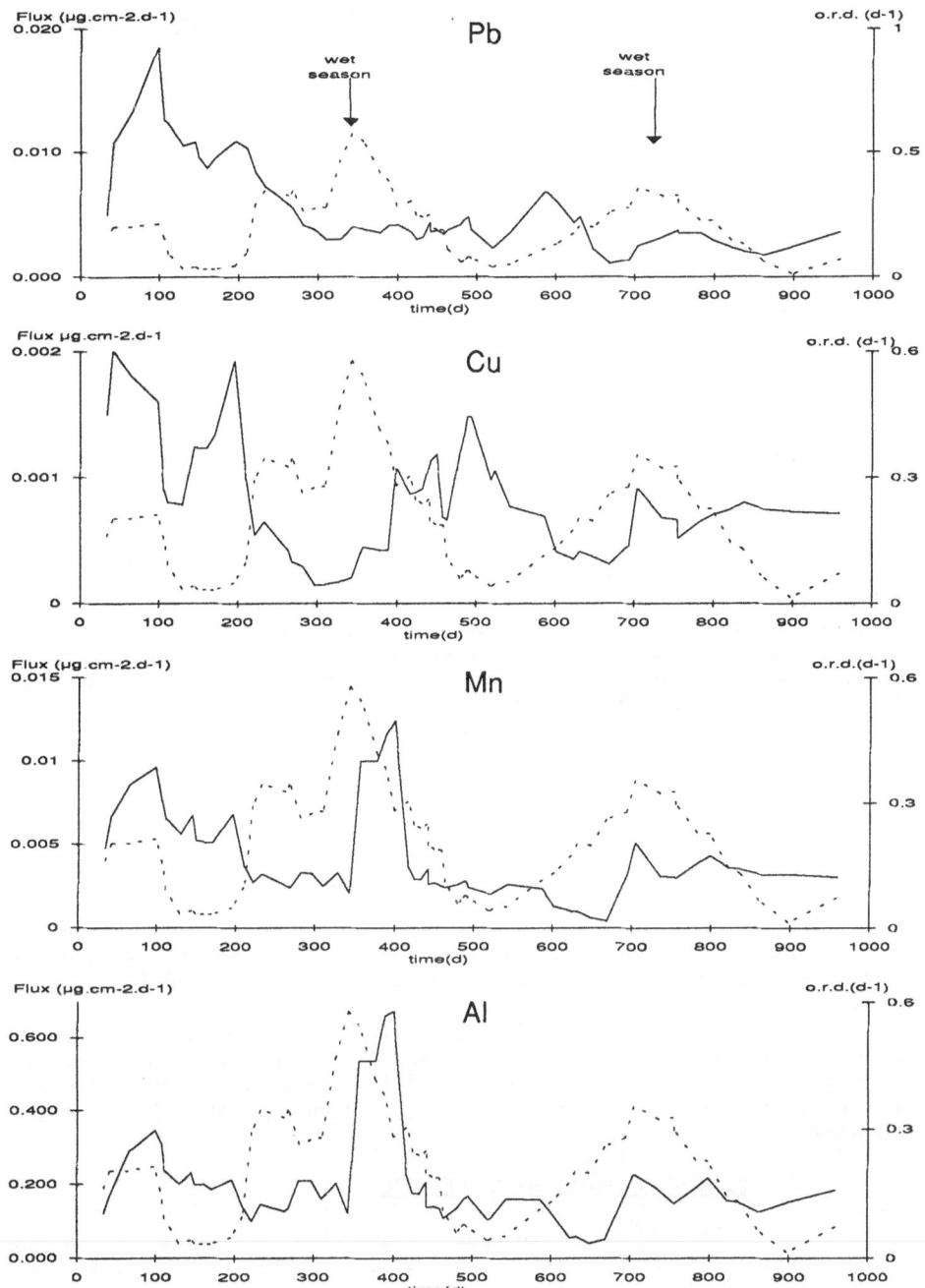

Figure 1: Solid lines: Pb, Cu, Mn, and Al deposition fluxes after application of a moving mean method with a time span of 60 days. Dotted lines: Occurrence of the rainy days (o.r.d.) after application of the same moving mean method.

IDENTIFICATION OF POLYCYCLIC AROMATIC NITRO DERIVATIVES

DURING A WINTER PHOTOCHEMICAL EPISODE IN PARIS

WORTHAM H.M. , MASCLET P. A. et MOUVIER G.

Laboratoire de Physico-Chimie de l'Atmosphère UA 717
Université PARIS VII, 2 Place Jussieu
PARIS , CEDEX 05 , FRANCE

RESUME

Une méthode analytique a été mise au point afin de doser les dérivés nitrés des Hydrocarbures Aromatiques Polycycliques (Nitro-HAP) présents dans l'atmosphère en phase particulaire et en phase gazeuse. Cette méthode a été expérimentée au cours d'un épisode photochimique pendant l'hiver 89 sur Paris. L'analyse comporte 4 étapes distinctes: le piégeage sur filtre pour la phase particulaire et sur XAD-2 pour la phase gazeuse; L'extraction au soxhlet; la préséparation de l'échantillon par HPLC préparative avec suivie du signal en UV; l'analyse des fractions par HPLC à détection par fluorescence après réduction catalytique. On a ainsi identifié et quantifié 9 Nitro-HAP dans chacune des deux phases (particulaire et gazeuse). Les concentrations de Nitro-1-Pyrène et de Nitro-2-Pyrène relevées sont les suivantes: 36 pg/m3 et 9 pg/m3 en phase gazeuse et 192 pg/m3 et 56 pg/m3 en phase particulaire. Ces valeurs ont été rapprochées des concentrations en HAP observées aux mêmes dates . L'évaluation des rapports N-HAP/HAP permet d'émettre des hypothèses sur la stabilité des nitro-HAP dans l'atmosphère.

1. INTRODUCTION

Les dérivés nitrés des Hydrocarbures Aromatiques Polycycliques (Nitro-HAP) sont les principaux produits de dégradation des HAP en phase gazeuse et particulaire. Ils sont formés de jour par réaction avec OH, O3 et les NOx et de nuit par réaction avec NO3 ou N2O5 (3).

Bien que présent à l'état de trace ces polluants secondaires ont un rôle déterminant dans la toxicité de l'atmosphère puisque le Nitro-1-Pyrène est, par exemple, tenu à lui seul pour responsable de plus de 30% de l'effet mutagène des émissions diésel (1)(2).

Pour évaluer la toxicité de l'air induite par cette famille de composés, il faut les doser dans l'atmosphère. Nous avons donc mis au point une méthode analytique, très sensible, permettant d'identifier et de doser 10 Nitro-HAP, testée au cours d'un épisode photochimique hivernal.

2. MISE AU POINT DE LA METHODE

Pour le piégeage comme pour l'extraction nous avons repris la méthode utilisée par PISTIKOPOULOS (4). La phase particulaire est recueillie sur des filtres en fibre de verre téflonnée et la phase gazeuse sur de la résine XAD-2. L'ensemble du protocole permet de limiter les artéfacts lors du prélèvement; ceux -ci, selon AREY J. et al. (5) ne doivent pas excéder les 3%. L'extraction est obtenue en 3 heures à l'aide d'un Soxhlet et par un mélange d'un tiers de Cyclohexane (C6H12) et de deux tiers de Dichlorométhane (CH2Cl2). L'extrait est repris dans 1 ml de Méthanol.

L'extrait est complexe puisque les pièges ne sont pas spécifiques. Il faut donc isoler les différentes familles de composés que l'on souhaite analyser. Cette pré-séparation a été obtenue par Chromatographie Liquide Haute Performance préparative (HPLC). Le suivi du signal est effectué en UV à 254 nm. Les conditions chromatographiques retenues sont les suivantes:

- La colonne semi-préparative utilisée est une MERCK Lichrosorb Si 60 (7 μm) de 25 cm de long et de 1 cm de diamètre intérieur. Le gradient d'élution est le suivant:

- 98% de C6H12 + 2% de CH2Cl2 pendant 5 min suivi d'un gradient linéaire qui abouti en 20 min à 95% de CH2Cl2 + 5% d'Ethanol a un débit de 2,5 ml/min.

Les deux fractions correspondant aux HAP et aux Nitro-HAP sont recueillies en sortie de colonne, les HAP de 6,3 à 8 min et les Nitro-HAP de 8,4 et 11,3 min (Fig I). Le rendement de cette opération de nettoyage de l'échantillon est excellent puisqu'il est de l'ordre de 90%.

FIG.I CHROMATOGRAMME DE LA PHASE GAZEUSE SUR LA COLONNE

SEMI - PREPARATIVE

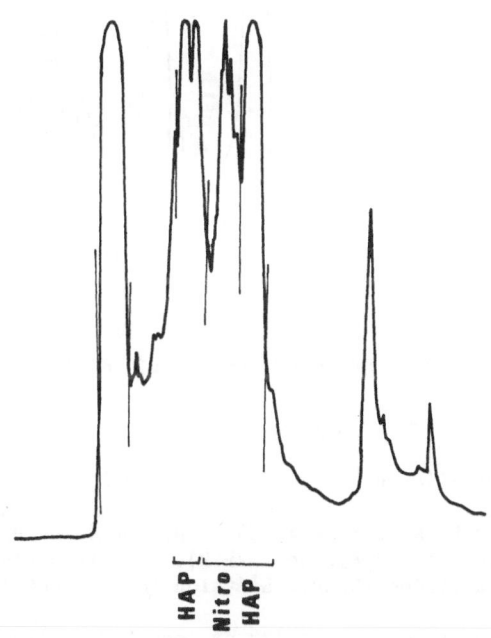

Les extraits "propres" d'HAP et de Nitro-HAP sont analysés par HPLC avec détèction fluorimétrique sur une colonne greffée en C18 (MERCK Lichrospher, 5 μm). La méthode d'analyse utilisée pour les HAP gazeux et particulaires a été décrite par ailleurs (4) mais la préséparation a été ajoutée.

Pour les Nitro-HAP le problème est plus complexe car ils ne sont pas fluorescents et la détection UV est insuffisante pour les concentrations observées. Il faut donc les réduire en Amino-HAP fluorescents et donc détectables à haute sensibilité. Cette réduction est obtenue à l'aide d'un catalyseur de Platine et de Rhodium adsorbé sur de l'Alumine (6)(7). Il est contenu dans un colonne de 12 cm de long afin que le temps de contact soit suffisant. Cette dernière est placée après la colonne analytique. Le chauffage à 75°C du catalyseur permet d'obtenir des rendements de réduction de 78 à 92% en fonction des Nitro-HAP considérés. Le gradient d'élution des Nitro-HAP est le suivant:

- 75% de MeOH + 25% d'H2O à 95% de MeOH + 5% d'H2O en 25 min avec un débit de 0.95 ml/min (Fig II).

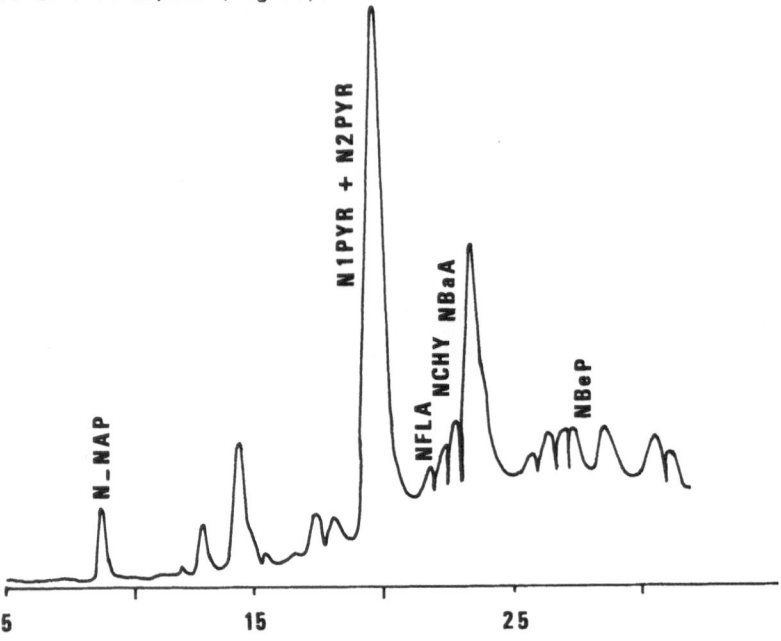

FIG.II CHROMATOGRAMME DE L'EXTRAIT DES NITRO-HAP PARTICULAIRE

$\lambda_{ex} = 360\,nm$ $\lambda_{em} = 430\,nm$

Pour identifier un maximum de Nitro-HAP on utilise 3 couples de longueurs d'ondes.
* λ exc = 366 nm et λ em = 530 nm permet de mesurer: le Nitro-1-Pyrène (N1PYR), le Nitro-2-Pyrène (N2PYR), le Nitro-Fluoranthène (NFLA) et le Nitro-Benzo(a)Pyrène (NBaP). L'étalon interne est le N-1-Naphtalène (N1NAP).
* λ exc = 360 nm et λ em = 430 nm permet d'analyser: le N1NAP, le Nitro-2-Naphtalène (N2NAP), les NPYR (N1PYR + N2PYR), le Nitro-Chrysène (NCHR), le Nitro-Benzo(a)Anthracène (NBaA) et le Nitro-Benzo(e)Pyrène (NBeP). On choisit ici l'Acridine comme étalon interne.
* λ exc = 313 nm et λ em = 400 nm permet d'analyser: le N1NAP, le N2NAP, le Nitro-Fluorène (NFLU). L'étalon interne est l'Acridine.

3. CONDITION DE L'ECHANTILLONNAGE

L'echantillonnage a été effectué du 30/01/89 au 03/02/89 à PARIS en haut d'une tour de 20 m de haut, loin de toute source d'émission. Ainsi on mesure bien les concentrations d'une source diffuse et homogène non influencée par les sources locales. Le prélèvement a duré 45 heures exclusivement pendant les heures de jour. Afin d'éviter le colmatage du filtre et la saturation de l'XAD-2, par dépassement du volume de retention, ces derniers sont changés quotidiennement. Les résultats des analyses des HAP et des Nitro-HAP sont respectivement répertoriés sur les histogrammes des figures III et IV. Ces résultats sont à rapprocher d'un certain nombre d'observations faites au cours de l'échantillonnage.

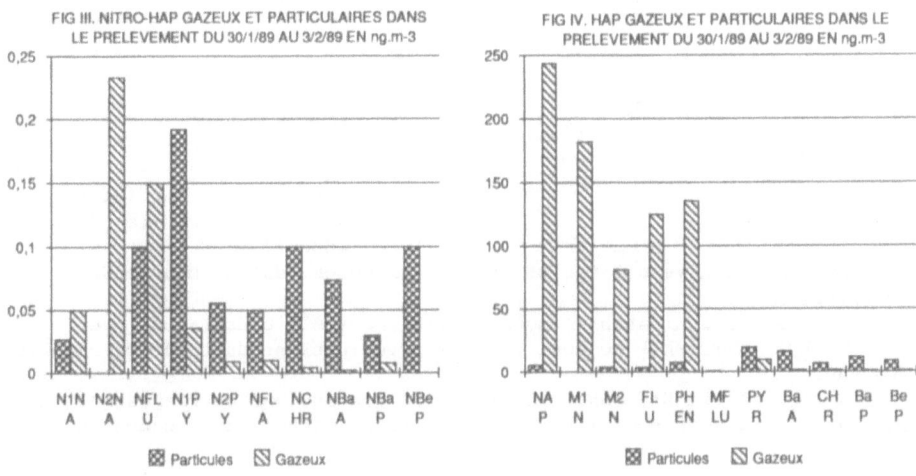

FIG III. NITRO-HAP GAZEUX ET PARTICULAIRES DANS LE PRELEVEMENT DU 30/1/89 AU 3/2/89 EN ng.m-3

FIG IV. HAP GAZEUX ET PARTICULAIRES DANS LE PRELEVEMENT DU 30/1/89 AU 3/2/89 EN ng.m-3

- Le dépôt particulaire recueilli sur les filtres était de 36 mg.
- La concentration moyenne des NOx relevée en milieu de journée était de 500 ppb.
- La température minimale enregistrée était de 1°C et l'amplitude thermique de 10°C. Le temps était sec et le ciel dégagé à partir de 13 ou 14 h après dissipation des brouillards. Ces divers paramètres montrent les fortes caractéristiques d'un épisode photochimique hivernal.

4. DISCUSSION

Tout d'abord, si l'on compare les concentrations des HAP, toutes phase confondues, avec les concentrations obtenues en Janvier et Février 86 (4) à Paris dans des conditions de température similaire, on s'aperçoit que les valeurs sont du même ordre de grandeur (Fig V). Mais les valeurs obtenues pendant l'hiver 89 sont des moyennes puisque le prélèvement a duré 45 heures, tandis que les analyses en 86 étaient effectuées sur des périodes de 3 heures en début et en fin de journée. D'aprés NIKOLAOU K. (5) ces deux périodes correspondent aux maxima journaliers. Cette constatation confirme la présence d'un épisode photochimique sur Paris pendant le prélèvement de 89. En effet les concentrations moyennes des HAP sont égales aux concentrations maximales des journées de l'hiver 86.

HAP GAZEUX + HAP PARTICULAIRES EN ng/m3 D'AIR			
DATE DU PRELEVEMENT	28/01/86 de 16h16 à 18h46	11/02/86 de 08h23 à 11h05	Moyenne diurne du 30/01/89 au 03/02/89
NAP	153.70	180.80	249.15
M1N	63.67	66.47	181.57
M2N	59.43	228.97	84.86
FLU	115.04	231.77	127.76
PHEN	79.97	158.67	142.54
PYR	27.48	35.89	16.47
BaA	10.74	31.79	17.51
CHR	4.64	13.63	7.67
BaP	3.80	12.83	12.61
BeP	11.60	28.24	9.57

FIG V. TABLEAU COMPARATIF DES CONCENTRATIONS OBTENUES PENDANT L'HIVER 86 (4) ET L'HIVER 89 (ce travail).

Etant donnée la météorologie exceptionnelle il faut s'attendre à des concentrations en Nitro-HAP particulièrement élevées. Mais nous ne possédons pour le moment aucune valeur de référence pour les deux phases. Il existe quelques valeurs dans la littérature pour la phase particulaire (10)(11)(13) mais elles n'ont pas été obtenues avec la même méthode et ne sont donc pas comparables.

Le rapport Nitro-HAP/HAP met en évidence la faible quantité des Nitro-HAP devant les HAP (Fig VI) (de 3.2% pour le Fluorène particulaire à 0.12% pour le Naphtalène et le Fluorène gazeux).

On peut à ce niveau formuler trois hypothèses qu'il conviendrait de vérifier.
- Les HAP sont stables pour la plus part et se dégradent peu.
- Les Nitro-HAP ne sont pas les principaux produits de dégradation des HAP.
- Les Nitro-HAP sont plus réactifs que les HAP et réagissent donc au fur et à mesure de leur formation pour donner des composés plus oxydés.

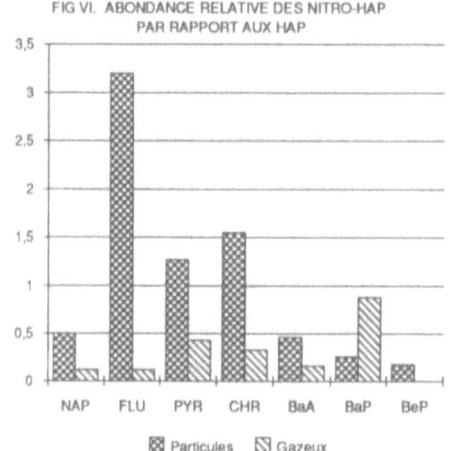

FIG VI. ABONDANCE RELATIVE DES NITRO-HAP PAR RAPPORT AUX HAP

NAP FLU PYR CHR BaA BaP BeP

⊠ Particules ◺ Gazeux

En exceptant le cas du BaP le rapport Nitro-HAP/HAP est plus de deux fois plus élevé en phase particulaire qu'en phase gazeuse. Ceci est d'autant plus remarquable que cette constatation se vérifie même lorsque la phase gazeuse représente la majorité des Nitro-HAP.

En comparant l'importance relative des deux phases (Fig IV), on remarque que la phase gazeuse est majoritaire pour le N-FLU ainsi que pour les Nitro-HAP de masses moléculaires inférieures et qu'elle est minoritaire pour les Nitro-HAP comportant au moins 4 cycles.

Toutefois les deux phases sont toujours présentes si l'on excepte le N-BeP et le N-2-NAP pour lesquels une seule phase a été mise en évidence dans notre travail. Le facteur déterminant l'importance relative des deux phases semble par conséquent être la masse moléculaire, comme pour les HAP.

Le cas du N-2-PYR est plus particulièrement significatif. En effet le N-2-PYR ne se forme pas à l'émission (14) et résulte donc de la transformation du Pyrène atmosphérique. Le seul mécanisme réactionnel diurne connu est l'addition radicalaire de OH (14). La présence du N-2-PYR dans un prélèvement diurne met donc en évidence cette réaction radicalaire et montre qu'au moins une fraction des HAP atmosphériques se transforme en Nitro-HAP.

5. CONCLUSION

La HPLC nous a donc permis de détecter et d'identifier 9 Nitro-HAP. Mais les méthodes de piégeage et de préparation de l'échantillon restent lourdes. Toutefois les résultats obtenus sont positifs à plusieurs points de vue. Tout d'abord la préséparation de l'échantillon a permis de totalement isoler les HAP des Nitro-HAP or ces deux familles de composés interfèrent lors des analyses. D'autre part cette première analyse de Nitro-HAP atmosphériques gazeux et particulaires effectuée en hiver peut servir de référence lors de prochains prélèvements. Et enfin si des variations journalières ne peuvent pas être détectées en raison de la durée de l'échantillonnage, les variations saisonnières pourront l'être.

REMERCIEMENTS
Nous tenons à remercier pour leurs efficaces collaborations Mlle E. BON NGUYEN et Mme S. MASCLET-BEYNE.

BIBLIOGRAPHIE
(1) TOKIWA H. ET OHNISKI Y. CRL crit. Reviews in toxicol. 17, 1986, 23
(2) ROSENKRANZ H.S. et MERMELSTEIN R. J. Environ. Sci. Health. C3 1985 23
(3) FINLAYSON-PITTS B. et PITTS J. "Atmospheric Chemistry" Eds: John Wiley et Sons Inc, New-York, 1986.
(4) PISTIKOPOULOS P. "Comportement physicochimique des HAP particulaires et gazeux dans l'atmosphère: mode de formation des aérosols, transport à méso-échelle, adaptation d'un modèle récepteur à des composés réactifs." Thèse de Docteur en science, université PARIS 7 1988.
(5) AREY J., ZIELINSKA B., ATKINSON R. et WINER A.M. Environ. Sci. Tecnol. 22, 1988, 457.
(6) Mac CREHAN W.A. et MAY W.E. Anal. Chem. 56, 1984, 625.
(7) TEJADA S., ZWEIDINGER R.B. et SIGSBY J.E. Anal. Chem. 58, 1986, 1827.
(8) Mac CREHAN W.A., MAY W.E., YANG S.D. et BENNER B. Anal. Chem. 60, 1988, 194.
(9) NIKOLAOU K. "Comportement physicochimique des HAP particulaires dans l'atmosphère et identification de leurs sources. " Thèse de 3eme cycle Université PARIS 7, 1983.
(10) KORFMACHER W.A., RUSHING L.G., AREY J., ZIELINSKA B. et PITTS J. J. HRC & CC. 1987, 641.
(11) ANG K.P., TAY B.T. et GUNASINGHAM H. Intern. J. Environmental Studies 163, 1987, 29.
(12) GREENBERG A. and DARAK F.B. Molecular structure and energetics 4, 1987, VCH Publishers, Inc.
(13) AREY J., ATKINSON R., ZLELINSKA B. et Mc ELROY P.A. Environ. Sci. Technol. 23, 1989, 321.
(14) PITTS J., AREY SWEETMAN J., ZIELINSKA B., WINER A. et ATKINSON R. Atmos. Environ. 19, 1985, 1601
(15) PITTS J. Atmos. Environ. 21, 1987, 2531.

Monitoring of Peroxides and other Trace Gas Constituents at a High Alpine Station

A. Neftel, B. E. Lehmann and M. S. Lehmann
Physics Institute, University of Bern
CH-3012 Bern, Switzerland

Summary

A number of atmospheric trace constituents is being monitored at the Swiss High Alpine Research Station at the Jungfraujoch (3450 m elevation) in the central Swiss alps.
Peroxides (H_2O_2 and organic peroxides) are measured with a two channel gas stripping system combined with Flow Injection Analysis (FIA). The system is based on a fluorometric method which uses peroxidase enzyme to catalyze the reaction in which hydroperoxides cause dimerization of 4-ethyl-phenol. The system has a time resolution of 2 minutes and a detection limit of 20ppt(v). A commercially available Differential Optical Absorption Spectroscopy (DOAS) instrument is used to measure SO_2, NO_2, O_3 and H_2O with high sensitivity and a time resolution of 15 minutes over an atmospheric path of 1000m. The DOAS system was installed in November 1988. During the winter 1988/1989 several prominent events have been observed, where SO_2 concentration raised from a background level around 1 mikrogram/m^3 up to 30 mikrograms/m^3 STP for periods of several hours to approximately two days.

1. Introduction

Recent reserach has shown, that H_2O_2 is an important trace species in polar ice samples and that it is preserved in ice up to thousand years old (Neftel 1984,1986, Stauffer 1988, Sigg 1988). In order to understand such polar records, we need to learn more about the atmospheric cycles of H_2O_2 . Last year, therefore, we initiated a program (within the COST 611 frame) to monitor peroxides and other trace constituents at the Jungfraujoch in the central Swiss alps. The high alpine research station at the Jungfraujoch (3450 m elevation) is a unique, well equipped facility for monitoring trace constituents in the free continental troposhere.
We are especially interested in determining the temporal and spatial scales over which peroxide concentrations are modulated.

2. Experimental setup

The peroxide analyzer is similar to the one developed at the National Center for Atmospheric Research (NCAR) in Boulder by Lazrus et. al. (1986). A scheme for the instrument is shown in Figure 1. Peroxides are measured through the enzymatic reaction of peroxides with 4-ethylphenol in the presence of the enzyme peroxidase. Intensively fluorescent dimers are formed in this reaction. For H_2O_2 , the reaction is very fast, and goes to completion after a few seconds. Organic peroxides react slower. We tested only Methylhydroperoxide (MeOOH), where a rate roughly 10 times slower than for H_2O_2 was found (Sigg, in preparation).
Peroxides are collected in a glass coil (10 spirals) that has an inner diameter of 2mm and a radius of approx. 1cm. The liquid flow rate was 0.4ml/min and the air flow was $2*10\text{-}3m3$ (STP)/min. After passing through the coil, the liquid is alternatively switched to the total peroxide channel or to the organic peroxide channel, where the enzyme catalase is added to destroy selectively H_2O_2 . Catalase decomposes H_2O_2 far more rapidly than organic peroxides. Again this was only tested for MeOOH, were H_2O_2 was removed 45 times more efficiently than was MeOOH at the flow conditions indicated above. With such a 2-channel analyzer only a rough discrimination between H_2O_2 and organic peroxides can be made,based on calibration using for example MeOOH. Another peroxide which will contribute to the organic peroxide signal is Hydroxymethyl-hydroperoxide. However, with increasing carbon chain length the system becomes less sensitive, because reaction rates are slower and the solubility is lower yielding lower uptake rates in the glass coil. For H_2O_2 a collection efficency of almost 100% was determined, whereas the collection efficiency for MeOOH was calculated to be only 60% (Lazrus, 1986) for the given flow conditions.

In addition, a Differential Optical Absorption Spectrometer (DOAS) from the Swedish OPSIS company was installed to monitor O_3, SO_2, NO_2 and H_2O. The system is shown schematically in Figure 2. A 968 m long path was installed between the research station and the PTT telecommunication station located on the Jungfraujoch mountain slope at 3667 m elevation. The emitter, placed at the PTT station, consists of a 15cm concave mirror with a 150 Watt Xenon high pressure lamp. A similar mirror in the receiver focusses the light into an optical fiber which is connected to the entrance slit of the spectrometer. The different compounds are measured sequentially at different wavelengths as indicated in Table 1. A 40nm interval is scanned at a 100 Hz frequency by means of a rotating disk with slits placed between the emergence slit of the monochromator and the photomultiplier. The system indicates the measured concentrations in g/m^3eff together with a deviation, which is a measure of the quality of the fit between the different reference spectra and the actual spectra that is being evaluated. Another parameter is indicated, which is proportional to the amount of light.

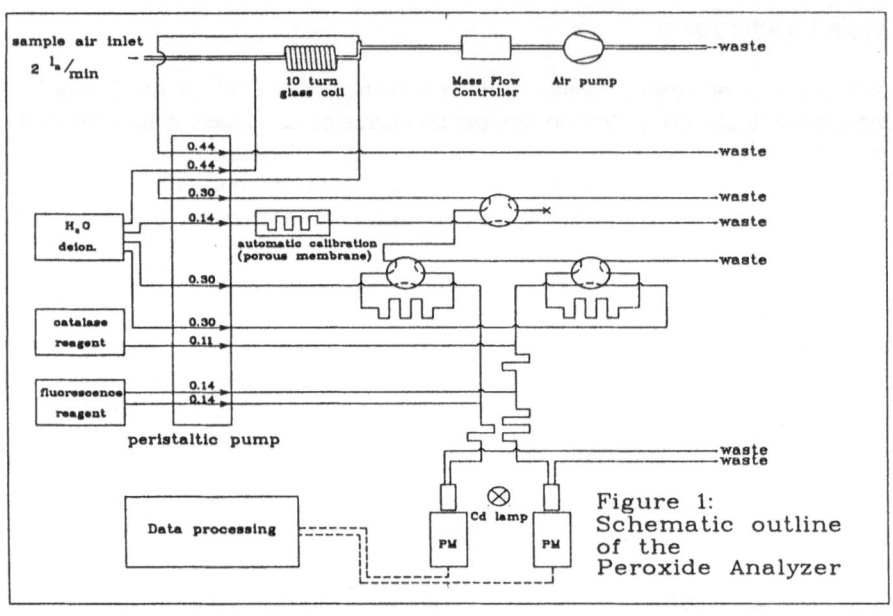

sample air inlet
2 $^{l_a}/min$

10 turn
glass coil

Mass Flow
Controller

Air pump

waste

waste

0.44
0.44
0.30
0.14

waste
waste

H₂O
deion.

automatic calibration
(porous membrane)

waste

0.30

waste

catalase
reagent

0.30
0.11

fluorescence
reagent

0.14
0.14

peristaltic pump

waste
waste

Data processing

Cd lamp

PM

PM

Figure 1:
Schematic outline
of the
Peroxide Analyzer

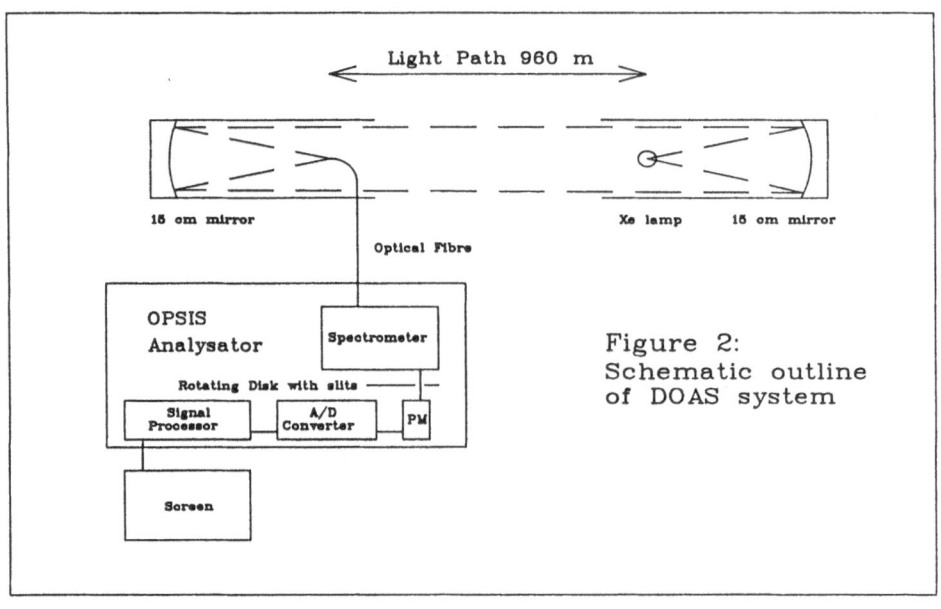

Light Path 960 m

15 cm mirror

Optical Fibre

Xe lamp

15 cm mirror

OPSIS
Analysator

Spectrometer

Rotating Disk with slits

Signal
Processor

A/D
Converter

PM

Screen

Figure 2:
Schematic outline
of DOAS system

Table 1

Component	Central Wavelength nm	Measuring Time min	Detection Limit ppb(V)
O_3	270	2	3
SO_2	300	2	0.1
HNO_2	350	2	0.1
NO_2	430	2	0.2
H_2O	720	2	$0.2 \ g/m^3$

a) Experience in operation with the DOAS system

The DOAS system was installed in November, 1988, and operated continuously until the beginning of August 1989 . It was then sent to Sweden for a general check up, because the grating setting in the spectrometer was no longer operating correctly.
DOAS measurements require good visibility. A special problem arises when the visibility changes during the measurement of one component. The system determines allways the light level and adjusts the gain of the photomultiplier at the beginning of a measurement . If the visibility increases during the measuring cycle the photomultiplier can be oversaturated, leading to unreastically high concentrations for SO_2 and NO_2. Thus prior to data interpretation the raw data must be evaluated carefully.

b) Experience in operation of the peroxide monitor

The peroxide system is not yet so far developed that it can run unmaintained for longer periods. We have so far measured different campaigns indicated in Table 2.

Table 2: Peroxide campaigns at the Jungfraujoch:

30th - 31st	January	1989
17th - 26th	Mai	1989
1st - 5th	June	1989
21st	August	1989

Problems arise mainly from contamination of the chemical reagents (which results in restricted flow through valves and T-pieces), and from air bubbles reaching the detector. Initially, we used MnO_2-coated AlO_2 grains in a small column to separate organic peroxides from H_2O_2. Such columns gave good results initially, but after only one day of continuous operation the quality of the separation decreased rapidly.

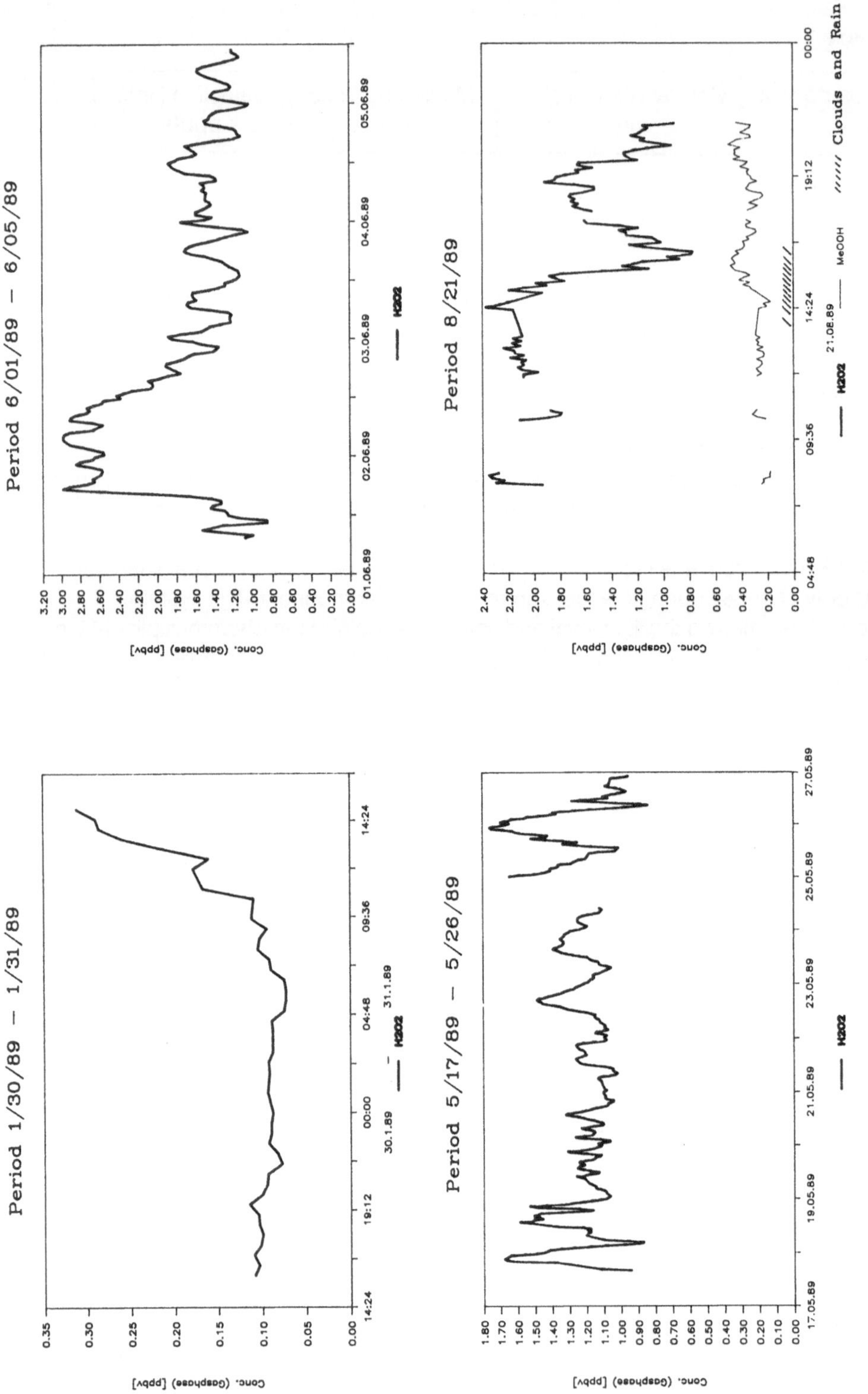

Figure 3: H_2O_2 record Jungfraujoch

3. Results and Discussion

a) H$_2$O$_2$ and organic peroxides

Figure 3 shows the recorded peroxide concentrations. Three characteristic patterns can be seen. (1) Small fluctuations, typically with an amplitude of the order of 10 - 20% of the mean value, on a time scale of a few hours; (2) diurnal variation with maximum values in the afternoon; and (3). large fluctuations with time scales of 1 day or more. Seasonal variations which have been predicted in various model calculations (e.g. Logan, 1981) and measured by Dollard (1988) for example, are also evident. Values below 1 ppb(V) occur in the winter and values up to 3 ppb(V) are recorded in summer months. A diurnal cycle is seen in almost all ground-based stations at low altitudes, with the lowest values occuring just before sunrise. In contrast to such stations, were the amplitude of the diurnal variations is almost 100% the amplitude at the Jungfraujoch is only about 15-30% of the mean value. Model calculations for the photochemical formation of peroxides in the boundary layer by Kleinmann (1986) fit the diurnal cycle for ground-based station well. In the absence of sunlight,peroxy radical formation is almost completely inhibited, so that the losses due to deposition are not compensated. With increasing altitude, deposition mechanisms become less effective, resulting in diurnal variations of lesser magnitude. The other important sink, photolytical destruction is also not active without sunlight. It is possible that the diurnal variation reflects the vertical air motion. Vertical H$_2$O$_2$ profiles show a "belly shape" type, with maximum concentrations above the planetary boundary layer (Heikes, 1987). The Jungfraujoch is situated above this maximum. During the day upwinds might transport H$_2$O$_2$ rich air masses to the Jungfraujoch and during the night the opposite situation occurs.

Figure 4:
500 hPa pressure maps 6/1/89 — 6/3/89

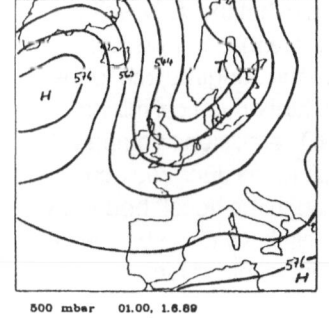
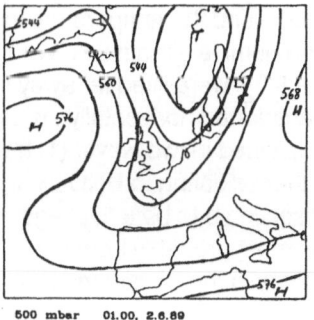
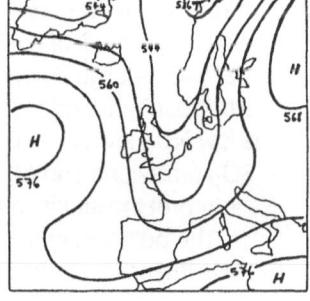

500 mbar   01.00, 1.6.89   500 mbar   01.00, 2.6.89   500 mbar   01.00 3.6.89

In addition, the peroxide concentration is affected by the trajectories of regional air masses over a time period corresponding to the residence time of peroxide in the atmosphere. An example of such a case occured on the 2nd and 3rd of June when a large H_2O_2 was recorded over more than 24 hours. The 500mb pressure map (Figure 4) for the 2nd of June shows a tendency for air masses to come from lower latitudes, i.e. from regions of higher H_2O_2 concentration. From these first results, we conclude that the concentration of H_2O_2 is mainly determined by two mechanisms. (1) the general meteorological situation which determines from which latitude the air masses are originating and (2) the diurnal behaviour of the vertical air mass transport, which is of course very specific to the Jungfraujoch.

The level of organic peroxides, expressed in MeOOH equivalents, could only be determined in the last campaign. During the measurements, a cloud passed the station. Consequently H_2O_2 was lowered, but the organic peroxide level increase slightly. This reflects the difference in solubility between H_2O_2 and organic peroxides, which is in the case of MeOOH two orders of magnitude. The organic fraction was determined to be 20% of the total peroxide level, which is in good agreement with the model calculations for nonurban conditions (e.g. Kleinmann, 1986).

b) DOAS measurements

Ozone and H_2O concentration at the Jungfraujoch are clearly above detection limits, whereas NO_2 and SO_2 are often at or below detection limits. Other components which could theoretically be measured with the OPSIS DOAS system, such as HNO_2 , CIO or Toluene are below the present detection limits.

For this first report we restrict ourselves to an interesting period at the end of January, when a large variability in the SO_2 signal was found. Figure 5 shows all of the measured concentrations together with the temperature record. These data can be divided into three different sections marked a,b and c. Section a is representative of typical winter background concentrations, section b is characterized by moderately elevated NO_2 and SO_2 levels; and section c marked by a peak in SO_2 and very low NO_2 concentrations. Just before the SO_2 event, a period with elevated O_3 levels occured. The H_2O concentration and the temperature indicate that the SO_2 spike is linked to the arrival of cold and humide air masses. We suppose that air masses, loaded with a large amount of SO_2, were lifted far away from the Jungfraujoch to the free troposphere and then transported horizontally , without further vertical mixing. The low NO_2 levels can be explained in two ways: (1) low NO_x emissions compared to SO_2 emissions and (2) further oxidation to HNO_3.in transit to the Jungfraujoch. We interpret the situation for section b to be a "leakage" through the "local" boundary layer. The pollution originates most probably from sources relativelly close by. Due to the less intensive photochemistry in the winter time NO_2 is not oxidized further.

Figure 5:
DOAS data for the period 1/18/89 - 1/27/89

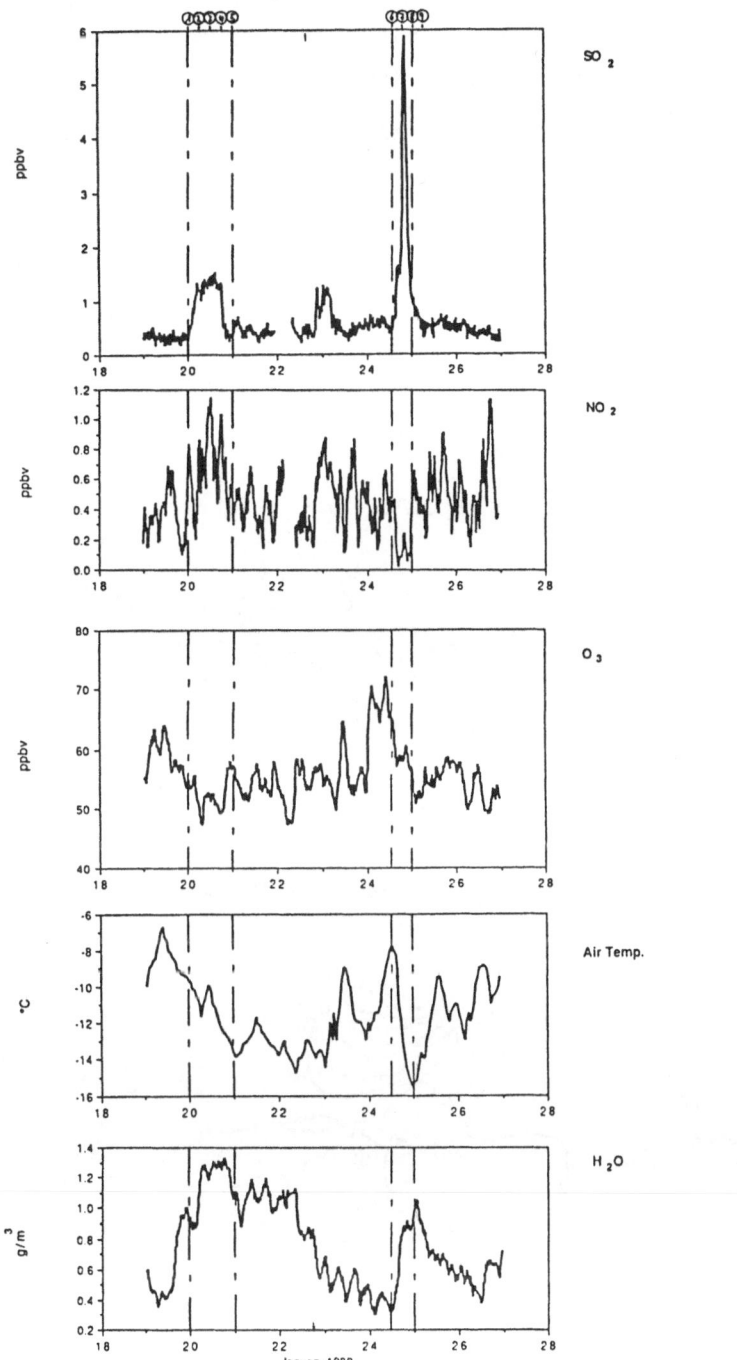

Trajectory analysis provides valuable information on air mass source regions. Figure 6 shows the 700 hPa 48 hour backtrajectories, which were calculated by Dr. H. Kolb (Institute for Meteorology and Geophysics, Vienna) within the Alptrack Program. It can be shown that the air arriving on the 20th of January passed over northern Italy, where pollution could have been incorporated into the air mass. The emission ratio of SO_2 to NO_x for this region is roughly 1:1. (EMEP Report 1/88, 1988). In contrast, the air causing the high SO_2 peak passed eastern europe, a region known to have high SO_2 emissions.

Similar events occured on the 23rd and 24th of November, 1988, and during the first week of February 1989, but these events remain to be analyzed in detail.

Figure 6:
48 hour back trajectories
(700 hPa level)
for the period
indicated in Figure 5

20.Jan.89	00.00	①
	06.00	②
	12.00	③
	18.00	④
21.Jan.89	00.00	⑤
24.Jan.89	12.00	⑥
	18.00	⑦
25.Jan.89	00.00	⑧
	06.00	⑨

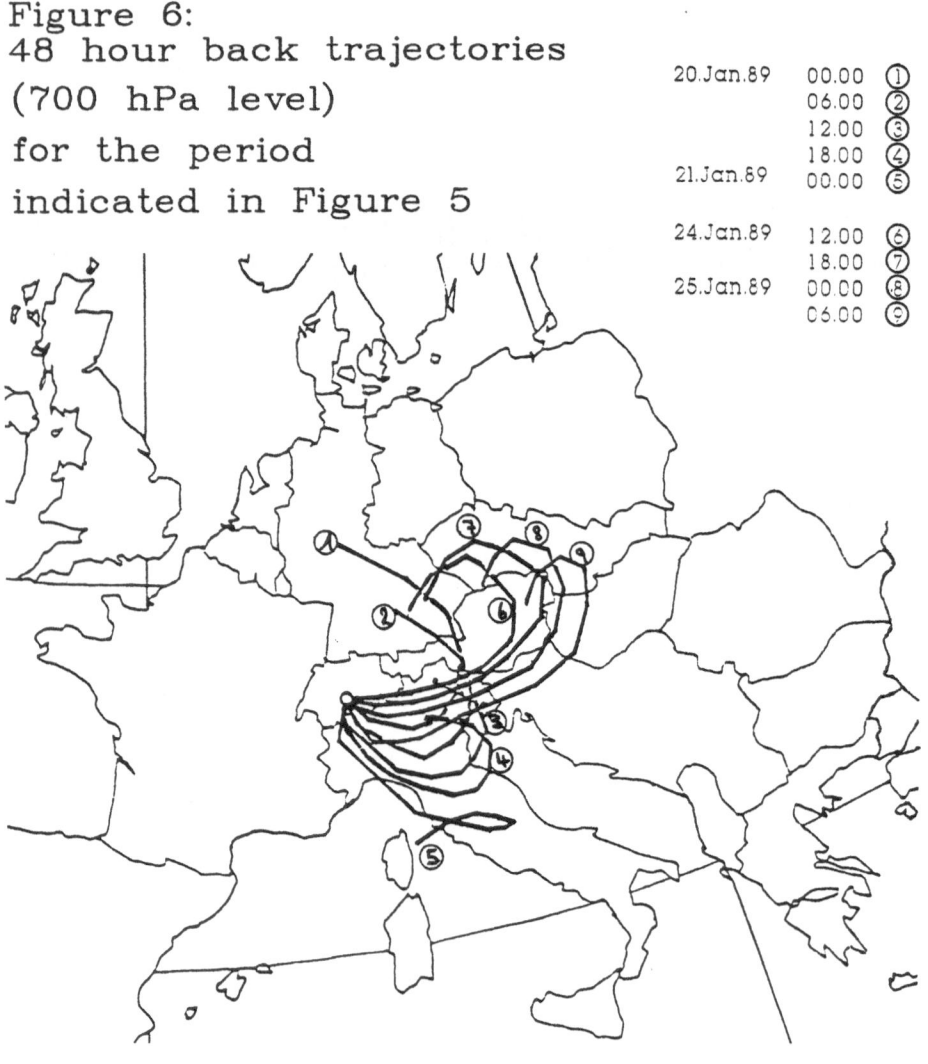

4. Conclusions and Outlook

Initial results indicate that the Jungfraujoch station is an interesting place to study both clean, free continental troposherical air, and pollution events that reach higher altitudes.

It is important to lower the detection limit for the DOAS SO_2 and NO_2 measuremets by at least a factor of four in order to trace their concentrations under clean air conditions. It may be possible to install a reflector at the PTT station, to double the path length. This would also require a stronger light source to compensate the loss in light transmitted over a greater distance.

For the peroxide monitor small improvements have to be made that it can be run unmaintained over longer periods. A better differentiation between organic peroxides and H_2O_2 must also be achieved.

The interpretation of initial results expressed herein is considered to be preliminary and must be confirmed by longer time series data combined with trajectory analysis, taking into account the vertical stability of air masses. Such efforts are underway within the Alptrack program, a subprogram of the Eurotrack program and within the new Swiss Pollumet program.

Acknowledgements:

This work was supported by the Swiss Federal Office for Education and Science within the COST 611 action. We thank Prof. Dr. H. Debrunner for the opportunity to work on the Jungfraujoch reserach station and Prof. Dr. Oeschger for his continuous support and interest. We are grateful to Dr. Anne Arquit for editorial assistance.

References:

Dollard G.J. and Davies T.J. (1988)
Measurements of gaseous hydrogen peroxide in southern england.
COST 611 Report No 14, Villefranche sur Mer,
Field measurements and their interpretation. 212-224

Heikes B.G., Kok G.L., Walega J.G. and Lazrus A.L. (1987) H_2O_2, O_3 and SO_2 measurements in the lower troposphere over the eastern United States during fall.
J. Geophys. Res. **92** 915-931

Kleinmann L.I. (1986)
Photochemical formation of peroxides in the boundary layer.
J. Geophys. Res. **91** 10889-10904

Logan J.A. (1981)
Tropospheric Chemistry: A global perspective.
J. Geophys. Res. **86** 7210-7254

Lazrus A.L., Kok G.L., Lind J.A. Gitlin S.N., Heikes B.G. and Shetter R.E. (1986)
Automated fluorometric method for hydrogen peroxide in air.
Analytical Chemistry **58** 594-597

Lehmann M.S., Lehmann B.E., Neftel A., Baltensperger U., Gäggeler H.W. and Jost D.T. (1989)
Long range transport of SO_2 and Aerosols to a high alpine station.
Poster presented at the symposium on meteorological aspects of mesoscale and long-range pollution transport IAMAP Reading UK

Neftel A., Jacob P. and Klockow D. (1984)
Measurements of hydrogen peroxide in polar ice samples.
Nature **311** 43-45

Neftel A., Jacob P. and Klockow D. (1986)
Long-term trend of Hydrogen Peroxide in polar ice cores.
Tellus **38B** 262-270

Neftel A., Blatter A. and Staffelbach T.
(1989, this conference)
Gas phase measurements of NH_3 and NH_4^+ with Differential Optical Absorption Spectrometer (DOAS) and Gas Stripping Scrubber in combination with Flow Injection Analysis

Sigg. A. and Neftel A. (1988)
Seasonal variations of hydrogen peroxide in polar ice cores.
Annual Glaciol. **10** 157-162

Sigg A. (1989 in preparation)
Wasserstoffperoxid in Eisbohrkernen aus Grönland und der Antarktis: Messreihen, ihre Interpretation und atmosphärisch-chemische Bedeutung.
Dissertation Universität Bern

Stauffer B. and Neftel A. (1988)
What have we learned from the ice cores about atmospheric changes in the concentrations of nitrous oxide, hydrogen peroxide, and other trace species.
The changing atmosphere eds. F. S. Rowland and I.S.A. Isaksen
John Wiley and Son 63-77 Dahlem Konferenzen 1988

Thompson A.M., Owens M.A. and Stewart R.W. (1989)
Sensitivity of tropospheric hydrogen peroxide to global chemical and climate change.
Geophysical Res. Letters **16** 53-56

FLUX MEASUREMENTS OVER PRAIRIE GRASSLANDS IN NORTHERN COLORADO

D. W. STOCKER[1], K. F. ZELLER[2], W.J. MASSMAN[2], D. HAZLETT[3], D. FOX[2], D. L. LUKENS[2] and D. H. STEDMAN[4]

1. EAWAG, Dubendorf, Switzerland, 2. US Forest Service, Fort Collins, Colorado, 3. Colo. State Univ. Fort Collins, Colorado, 4. Univ. Denver, Denver, Colorado.

SUMMARY

The fluxes of nitrogen oxides and ozone have been measured over the prairie grassland of Northern Colorado by eddy correlation. The deposition velocity of ozone varied diurnally. The ozone maximum deposition velocity occurred at around midday, $V_d = 0.4$ cm s^{-1}, and was zero overnight. The net flux of nitrogen oxides was upward, with a diurnal and a seasonal variation. In early spring there was an upward flux during the day, and zero or a slight downward flux overnight. The daytime NOx fluxes increased from the spring to early summer, as the soil temperature increased. During the summer, moderate rainfall after a few days drought initiated large bursts of upward NOx flux. The average NO$_2$ flux for the period March - July was 1.7 ng N m^{-2}s^{-1} (peak 8.8 ng N m^{-2}s^{-1}), and the average NOx flux for June and July was 5.1 ng N m^{-2}s^{-1} (peak 17.3 ng N m^{-2}s^{-1}).

1. INTRODUCTION

The emission of nitrogen oxides from microbiol processes in soils represents one of the major natural sources of nitrogen oxides to the troposphere, but its magnitude is still uncertain (1). Nitric oxide and nitrous oxide are produced by both nitrifying and denitrifying microbial organisms. NO is released under aerobic conditions and N$_2$O under anaerobic conditions. The production of NO varies with the soil temperature, soil moisture content, fertilization (amount and type), soil depth, vegetation cover and the atmospheric concentration (2).

Nitric oxide is rapidly oxidised in the troposphere, with a lifetime of about 1 minute, to nitrogen dioxide. This is the first step leading to the production of ozone and hydroxyl radicals (HO) which drive photochemical oxidation cycles leading to the formation of nitric acid (3).

In the present study the fluxes of NOx (NO + NO$_2$) were measured over the prairie of Northern Colorado by eddy correlation. The flux of ozone and supporting meteorological data were also measured. Eddy correlation uses a direct measurement of the covariance between the fluctuating vertical wind velocity and the fluctuating concentration of the pollutant to determine the flux. Although the flux is measured at a fixed point, it should be considered as the integration of the surface fluxes from some upwind area.

2. EXPERIMENTAL METHODS

The field site was located 50 km northeast of Fort Collins, CO, at the Central Plains Experimental Range, a research area adjacent to, and west of, the Pawnee National Grasslands. Vegetation is dominated by shortgrasses (64%), succulents (20%) and half shrubs (8%). A full description of the field site is given in reference (4). The chemical instruments were housed in an air-conditioned shelter. Air was brought to the instruments using a high-volume sample pump. The intake system consisted of a 26.5m long, 1.6 cm i.d. Teflon tube through which air was drawn at ca. 160 l/min. The residence time of a sample in the tube was ca 2s. The intake was mounted at a height of 6m on a meteorological tower.

Total NOx was measured using an NO + ozone chemiluminescent monitor, the higher oxides of nitrogen were reduced to NO using a hot molybdenum converter at 400°C (5). The principle species reduced were NO_2 and peroxyacetyl nitrate (PAN). The instrument response was ca. 0.3 nA/ppb with a detection limit of ca. 70 ppt. The response time was ca. 0.9s.

Ozone and nitrogen dioxide were also measured by chemiluminescence, ozone by reaction with eosin-y and NO_2 by reaction with a basic luminol solution (6,7). For the detection of ozone, the instrument response was ca. 0.12×10^{-6} A/ppb and the detection limit < 1 ppb. For the detection of NO_2 the response was ca. 0.3×10^{-6} A/ppb and the detection limit ca. 100 ppt. The response time of the instrument was ca. 0.7s for the analysis of both NO_2 and ozone.

The NOx and NO_2 analysers were calibrated twice per week. The output of the fast response ozone monitor was compared with that from a uv photometric ozone analyser. The baseline response of the chemical instruments was measured at the beginning of each sampling period by introducing "clean" air at the tower mounted intake to the chemical instruments.

Co-located on the tower with the intake for the chemical sensors was a 3-axis sonic anemometer. A platinum resistance thermometer was used to measure the temperature and a krypton hygrometer was used to measure the water vapour.. The tower mounted instruments were at a height of 6m. Nett radiation, the soil temperature and soil heat flux were also measured. Data was recorded using a personal computer, the software used for real-time data processing was adapted from McMillan's Basic flux programme for eddy correlation in non-simple terrain (8). The software rotates the coordinate reference frame to align with the mean wind velocity, hence the flux is measured perpendicular to the mean wind flow for the sampling period. Data is collected and stored as 30 minute averages.

The data was screened to eliminate those data points collected when the system was not operating under optimum conditions according to the following criteria:-

All data taken when the wind direction was between 195° and 225°. This precluded any interference by the tower and the instrument shelter on the measurements.

All momentum flux measurements larger than -5.5 cm^2s^{-2}. This removes all positive momentum fluxes and small negative stresses that are below the resolution of the instrumentation.

Fluxes measured when the roughness length (z_0) exceeded 25cm. Larger values of z_0 were associated with combinations of low wind speed and small negative stress measurements.

The screened data was then corrected for imperfect sensor response, sensor separation, sensor mismatching, aliasing, and spatial averaging (over a finite volume for the chemical sensors and line averaging for the temperature probe and sonic anemometer).

3. RESULTS

The diurnal average deposition velocity of ozone measured at the field site from May 26th to August 2nd, 1988 is 0.47 cm s^{-1}. The springtime deposition velocity is slightly less; 0.40 cm s^{-1} for the period March 15th to May 25th, 1988. The increased deposition velocity during the summer probably reflects the increased atmospheric turbulence and greater surface area as the plant cover increases. These values are comparable to those obtained at the same site during June and July 1987 using the same chemical instrumentation, but with a Gill UVW propellor anemometer. This demonstrates that, provided the appropriate correction factors are applied flux measurements can be carried out using either type of anemometer. These results are in good agreement with other recent measurements of the dry deposition of ozone to grassland (9,10).

The daytime concentration of NO_2 is ca. 1-2 ppb. The concentration gradually increases overnight, decreasing to daytime levels with the breakup of the nocturnal inversion layer. On some nights concentrations as high as 20 ppb are observed at around midnight. There are no local anthropogenic sources, so these air masses have probably been transported from the Front Range area of Colorado (Denver to Fort Collins) or Cheyenne (Wyo). In March the average daytime flux is (0.05 - 0.1) ng N m^{-2}s^{-1}. At night there is no flux, or a slight deposition flux. The daytime summer fluxes of NOx are much larger than those in the early spring, up to 0.4 ng N m^{-2}s^{-1}. Figure 1. shows the monthly average NO_2 flux alongside the soil temperature, total monthly rainfall and the total green leaf area index i.e. the area covered by living plants. The NO_2 flux increases as the soil temperature and amount of rainfall increase, and conditions are favourable for plant growth and soil microbial activity. The summertime fluxes are also influenced by rain storms, see Figure 2. After a period of a few days drought a moderate rainfall can stimulate a large burst of NOx flux (June 22nd). Continued wetting of the soil suppresses the flux, due to either; water logging, a decrease in soil temperature nutrient limitation or a combination of these. Likewise very heavy storms do not initiate NOx fluxes (July 7th), however, as the soil begins to dry out and warm up, increased NOx fluxes are observed (July 8th - July 10th). The fluxes of NOx and NO_2 are comparable for the period during which the NOx and NO_2 instruments were operating concurrently, see Figure 2. The results from this study fall within the range of previous measurements of NO and NOx fluxes over grasslands (11,12).

4. CONCLUSIONS

The flux of ozone has been successfully measured using different anemometers, yielding the same results. The prairie is a net source of NOx, which has probably been released at the surface as NO. The flux of NOx shows both a diurnal as well as a seasonal variation. Maximum upward fluxes of NOx occur during the summer when the soils are warm and have been moistened after a period of drought.

REFERENCES

(1) LOGAN, J.A., "Nitrogen oxides in the troposphere: Global and regional budgets", J. Geophys. Res., 88, 10785-10807, (1983).
(2) JOHANSSON, C., GRANAT, L., " Emission of nitric oxide from arable land", Tellus, 36B, 25-37, (1984).
(3) LOGAN, J.A., PRATHER, M.J., WOFSY, S.C. and MCELROY, M.B., "Tropospheric chemistry: A global perspective", J. Geophys. Res., 86, 7210-7254, (1981).
(4) ZELLER, K., MASSMAN, W., STOCKER, D., FOX, D.G., STEDMAN, D. and HAZLETT, D., " Initial results from the Pawnee eddy correlation system for dry deposition research", US Forest Service Research Paper, RM-282, (1989).
(5) DICKERSON, R.R., DELANY, A.C. and WARTBURG, A.F., "Further modification of a commercial NOx detector for high sensitivity", Rev. Sci. Instrum., 55, 1995-1998, (1984).
(6) RAY, J.D., STEDMAN, D.H. and WENDEL, G.J., "Fast chemiluminescent method for measurement of ambient ozone", Anal. Chem., 58, 598-600, (1986).
(7) WENDEL, G.J., STEDMAN, D.H., and CANTRELL, C.A. "Luminol-based nitrogen dioxide detector", Anal. Chem., 55, 937-940, (1983).
(8) MCMILLAN, R.T., "An eddy correlation technique with extended applicability to non-simple terrain", Bound. Layer Meteorol., 43, 231-245, (1988).
(9) DROPPO, J.G., "Concurrent measurements of ozone dry deposition using eddy correlation and profile flux methods", J. Geophys. Res., 90, 2111-2118, (1985).
(10) DELANY, A.C., FITZJARRALD, D.R., LENSCHOW, D.H., PEARSON, R., WENDEL, G.J., and WOODRUFF, B., "Direct measurements of nitrogen oxides and ozone fluxes over grassland", J. Atmos. Chem., 4, 429-444, (1986).
(11) PARRISH, D.D., WILLIAMS, E.J., FAHEY, D.W., LIU, S.C., and FEHSENFELD, F.C., "Measurement of nitrogen oxide fluxes from soils: Intercomparison of enclosure and gradient measurement techniques", J. Geophys. Res., 92, 2165-2171, (1987).
(12) WILLIAMS, E.J., PARRISH, D.D., and FEHSENFELD, F.C., "Determination of nitrogen oxide emissions from soils: Results from a grassland site in Colorado, United States", J. Geophys. Res., 92, 2173-2179, (1987).

Figure 1. A: The monthly average soil temperature (+). B: The monthly NO_2 flux measured during the spring and early summer ( ), total monthly rainfall (+) and total green leaf area index (◇).

Figure 2. A: The NOx ( ) and NO_2 (+) fluxes measured over the period June 16th to July 14th. The total daily rainfall (inches) ( ). B: Soil temperature measured over the same period.

Figure 1

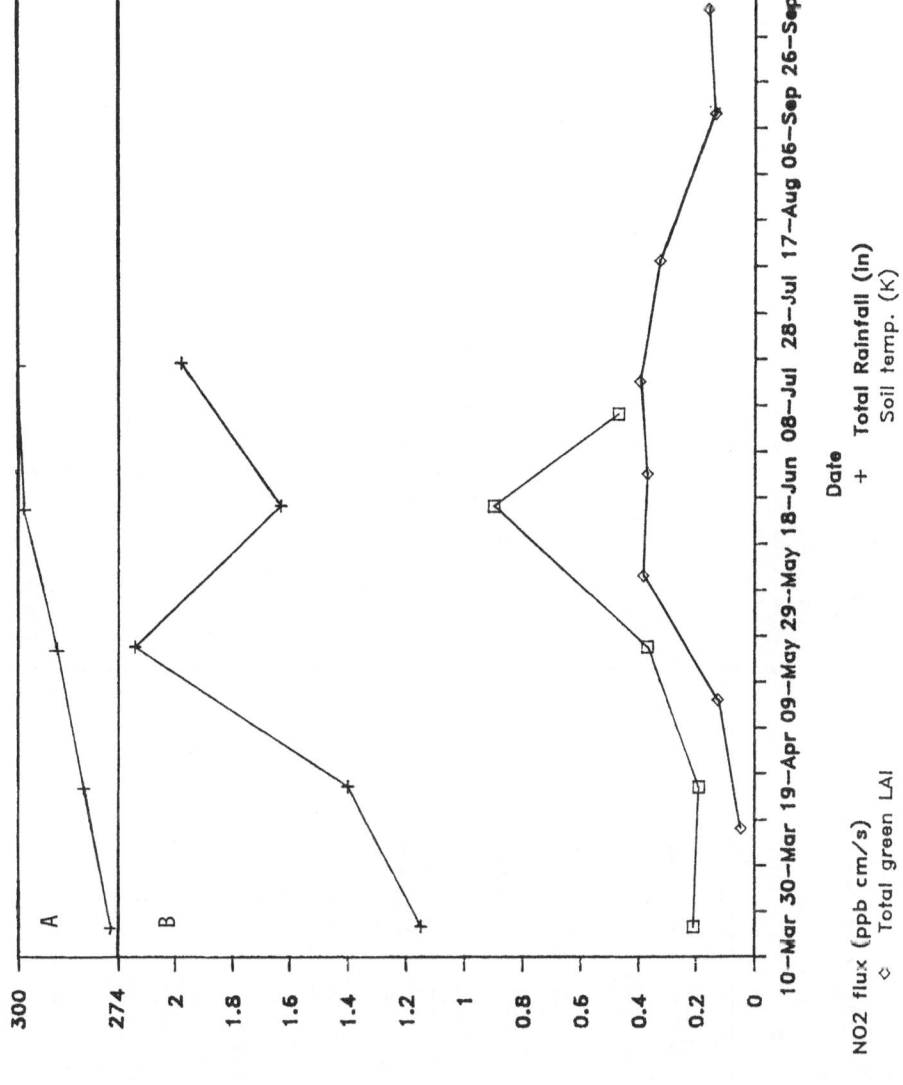

A

B

300

274

2

1.8

1.6

1.4

1.2

1

0.8

0.6

0.4

0.2

0

10–Mar 30–Mar 19–Apr 09–May 29–May 18–Jun 08–Jul 28–Jul 17–Aug 06–Sep 26–Sep

Date

□ NO2 flux (ppb cm/s)   + Total Rainfall (in)

◇ Total green LAI   Soil temp. (K)

Figure 2

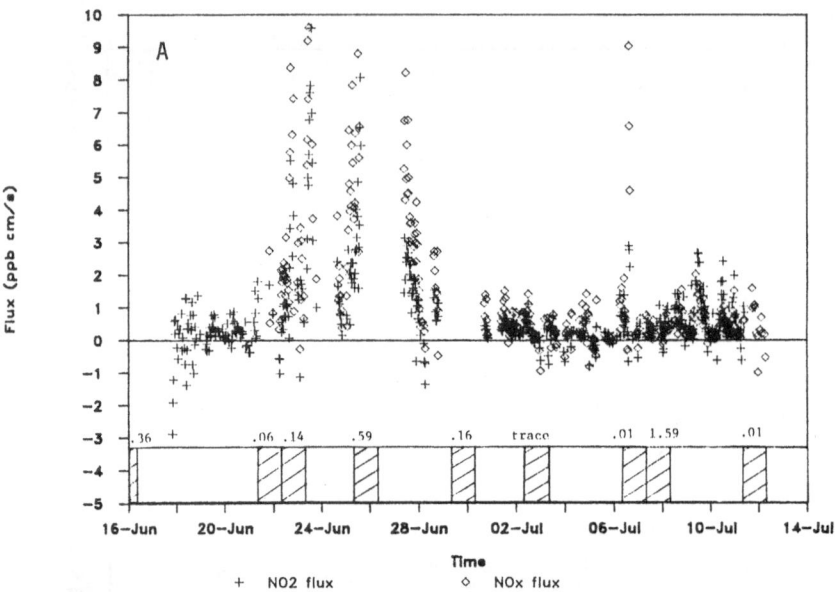

Figure 3

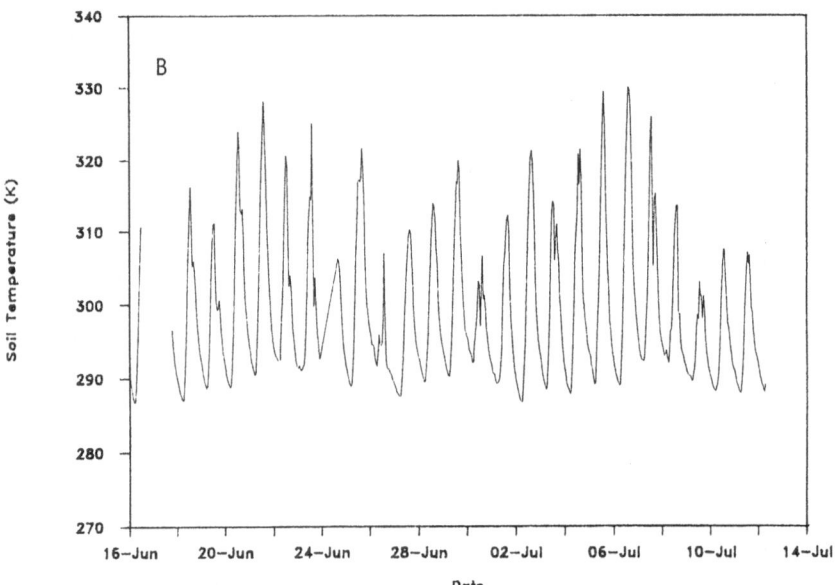

SESSION III/B

FIELD MEASUREMENT

AND

THEIR INTERPRETATION

1988 POLASTERN CRUISE ANT VII/1

MEASUREMENTS OF PHOTOLYSIS FREQUENCIES OF O_3 AND NO_2 AND AMBIENT O_3 CONCENTRATION ON THE POLARSTERN CRUISE ANT VII/1

J. Callies, U. Platt[*], and T. Brauers

Kernforschungsanlage Jülich GmbH, ICH 3, Atmosphärische Chemie
Postfach 1913, D-5170 Jülich, FRGermany

[*]Inst. für Umweltphysik, Universität Heidelberg
INF 366, D-6900 Heidelberg, FRGermany

Abstract

A comprehensive set of trace gas concentrations and meteorologic parameters were measured simultaneously during a cruise of the research vessel Polarstern from Bremerhaven (54°N, 8°E) to Rio Grande (32°S, 52°W) during Sept. 15 to Oct. 9 of 1988. This paper gives an overview of basic measurements made during that cruise: Meteorological parameters, photolysis frequencies of O_3 and NO_2, aerosol and ozone concentrations.

1. Introduction

Sources, sinks, and tranformation processes in marine air can contribute significantly to the global budgets of many trace species like reduced sulphur compounds, ozone, and hydrocarbons. Although more than two thirds of our planet are covered by oceans, little is known about concentrations and tranformation mechanisms of trace gases in the oceanic environment, for instance it is not clear, whether the remote marine troposphere is a net source or sink of ozone (1).

Given the fact that most of the trace gas cycles are chemically coupled it is of particular importance to measure a comprehensive set of species, simultaneously. Such a set will not only enhance our knowledge about the distribution of man-made and natural species, but could also be used to test our models of the chemical interactions in the undisturbed atmosphere.

A cruise of the research vessel Polarstern from Bremerhaven (54°N, 8°E) to Rio Grande (32°S, 52°W) during Sept. 15 to Oct. 9 of 1988 provided an opportunity to measure a wide variety of trace gases and physical parameters in the mid-atlantic (Fig.1). To allow a meridional cross section the cruise followed the 30°W meridian from (30°N, 30°W) to (30°S, 30°W).

A total of 15 research groups participated in the atmospheric chemistry program of this cruise. It is the purpose of this paper to give an overview of 'basic' measurements made during that cruise: Meteorological parameters and photolysis frequencies, which are a measure for the photochemical activity, as well as aerosol and ozone measurements. Those data form the basis and background information for the interpretation of the individual measurements reported seperately.

2. Experimental

Meteorological parameters (air and water temperatures, air presssure, relative humidity) were measured using standard meteorological instrumentation and recorded by the integrating data system (INDAS) of the Polarstern. The measured windspeed and winddirection were corrected for the movement of the vessel. Two photoelectric detectors observing the solar radiation fluxes in the wavelength ranges 290-315nm and 300-420nm, respectively, were used to determine the photolysis frequencies of ozone ($J(O_3)$) and nitrogen dioxide ($J(NO_2)$).

$$O_3 + h\nu \rightarrow O_2 + O(^1D) \quad \text{and} \quad NO_2 + h\nu \rightarrow NO + O(^3P)$$

Both detectors were equipped with scattering devices, giving less than 5% deviation from an uniform angular response for direct and scattered UV-radiation at zenith angles between 0 and 90° (2). In order to match the detector responses to the photoactinic spectra special filters were used. For absolute calibration $J(O_3)$ detectors had been exposed to solar radiation side by side with a chemical actinometer. While this procedure can yield an accuracy of better than ±16% (2) the data given in this paper my be somewhat less accurate due to aging of the reference instrument since its last actinometric calibration. The response of the $J(NO_2)$ detector was scaled to the empirical correlation of $J(NO_2)$ to the total radiation intensity at small to medium zenith angles reported by Bahe et al. (3). The uncertainty of the data is estimated at ±15%. The total radiation intensity was continuously recorded by a calibrated sensor (Kipp and Zonen CM6).

Aerosol concentrations were monitored by two instruments. First, a commercial condensation nuclei counter (TSI model 3020) was used to detect particles with radii greater than 0.01μm. Second, the total light scattering cross section of larger (r > 0.1μm) particles was recorded by an integrating nephelometer, which measured the amount of light scattered from an HeNe (λ = 633nm) laser beam. The instrument was calibrated by observing the Rayleigh scattering of aerosol free gases with different (known) Rayleigh scattering cross sections. Ozone concentrations were continuously recorded by a short-path UV absorption technique (DASIBI).

3. Results and Discussion

An outline of the cruise and the wind directions encountered is given in Fig. 1. During the first leg (Bremerhaven to 30°N, 30°W) variing directions were encountered. From 30°N, 30°W to the ITC (10°N, 30°W) the course crossed the NE passat region. In the ITC region (10°N, 30°W to 4°N, 30°W) again variing wind direction prevailed. The SE passat region extended from 4°N, 30°W to 25°S, 30°W. South of this point the wind was influenced by a strong low pressure system centered near 40°S, 47°W resulting in northerly winds.

The radiation measurements are shown in Fig. 2, maximum global radiation values reached 95 mW/cm². The UV-flux values, expressed as photolysis frequencies of NO_2 and O_3, respectively, reached maxima of $9.1*10^{-3}s^{-1}$ (with short term peaks up to $9.7*10^{-3}s^{-1}$) and $3.2*10^{-5}s^{-1}$ with half widths of 6 and 8 hours at (20°S, 30°W), respectively. Patchy clouds existed during the whole cruise, thus reducing the radiation below its possible maximum values. Particularly in the ITC region fog and clouds reduced the radiation levels.

The ozone mixing ratios (Fig. 3) at sea level in the northern hemisphere ranged from about 32 ppb (30°N) to 18 ppb (ITC). This is in general agreement with the expected meridional ozone distribution (1, 4). In contrast to that the observed ozone concentration in the southern hemisphere (0°N to 24°S) ranged from about 30 to 40ppb, roughly twice the amount expected (4).

CN concentrations in the northern hemisphere (with the exception of the ITC) were about 10-times higher than in the southern hemisphere, where the concentration was often lower than 50 CN/cm^3 (Fig. 3). The nephelometer data show relatively moderate levels of $(2 \text{ to } 5)*10^{-5}\text{m}^{-1}$. Assuming a monodisperse aerosol consisting of r = 0.2μm particles, a typical nephelometer reading of $3*10^{-5}\text{m}^{-1}$ would correspond to 240 particles/cm^3. However, the influence of sea spray (possibly generated by the ship) cannot be excluded, for instance only 10 particles of 1μm radius are required to cause the above reading.

Combining the measured photolysis frequency, ozone concentration, and humidity data it is possible to calculate OH production rates. Computed values range up to $8*10^6\text{cm}^{-3}\text{s}^{-1}$, due to the low NO_x mixing ratios encountered during most of the cruise (around 20ppt (5)) those production rates would correspond to noontime OH concentrations of only about 10^7cm^{-3}.

References

(1) J.A. Logan (1985) J. Geophys. Res. 90, 10463.
(2) W. Junkermann, U. Platt, A. Volz-Thomas (1989) J. Atm. Chem. 8, 203.
(3) F.C. Bahe, U. Schurath, and K.H. Becker (1980) Atm. Environm. 14, 711.
(4) S.C. Liu, D. Kley, and M. Mcfarland (1980) J. Geophys. Res. 85, 7546.
(5) D. Brüning and F. Rohrer, this volume.

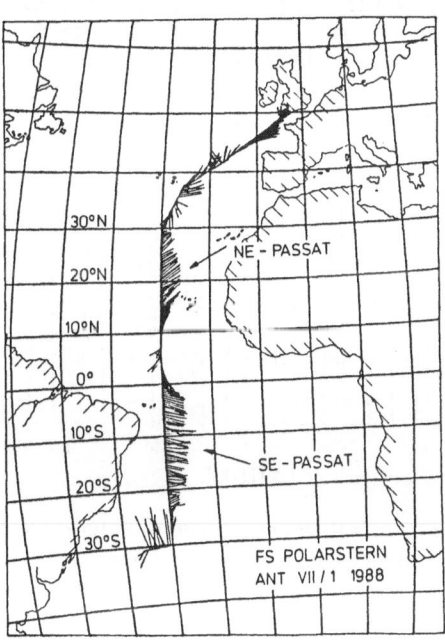

Fig. 1   Chart of Polarstern cruise ANT VII/1, lines at course indicate wind-direction and -speed.

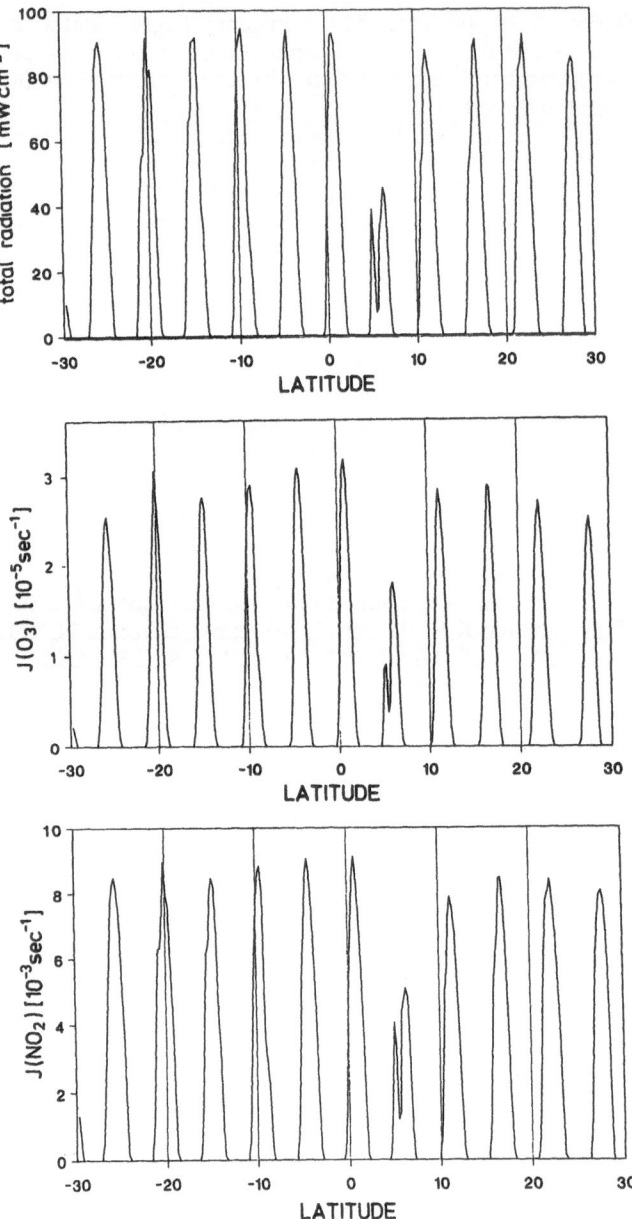

Fig. 2   Measured radiation fluxes as a function of latitiude from 30°S (-30) to 30°N (+30): Total radiation (upper panel), photolysis frequency of O_3 to yield $O(^1D)$ atoms (center panel), photolysis frequency of NO_2 (bottom panel).

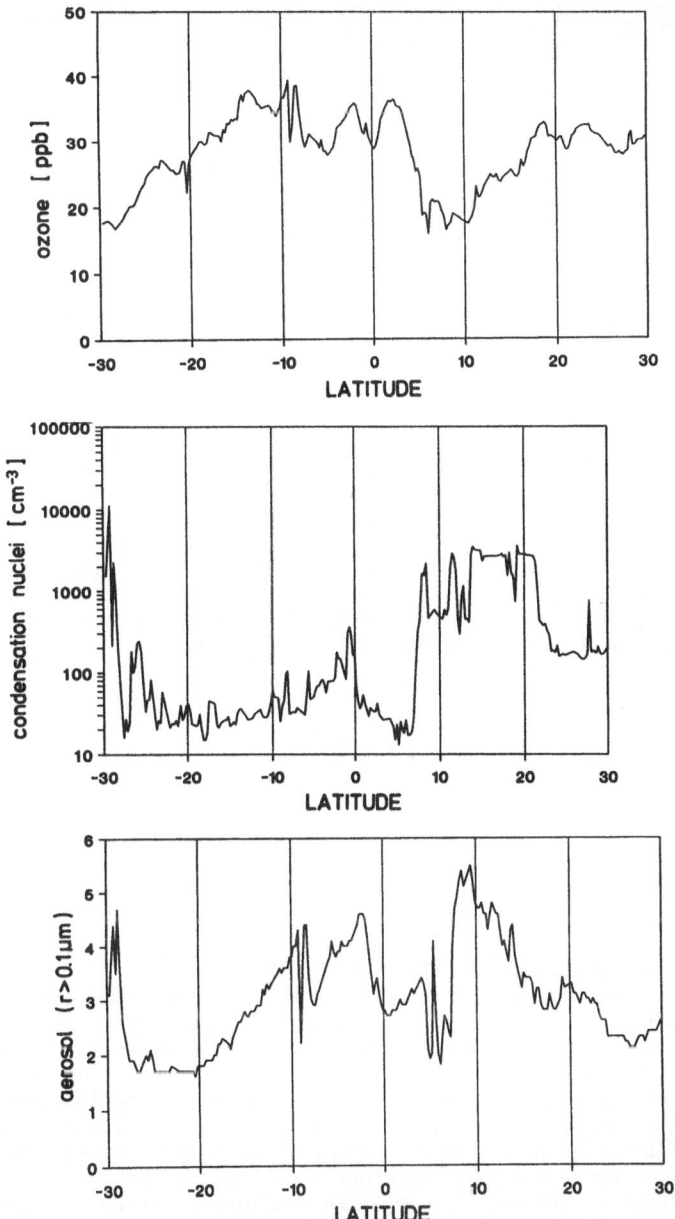

Fig. 3   Measured Ozone mixing ratio (upper panel), condensation nuclei concentration, and aerosol ligth scattering-signal (in units of $10^{-5}m^{-1}$, bottom panel). All data are plotted as function of latitude.

THE MERIDIONAL DISTRIBUTION OF OZONE AND WATER VAPOR OVER THE ATLANTIC OCEAN BETWEEN 30 °S AND 52 °N IN SEPTEMBER/OCTOBER 1988

H.G.J. SMIT, S. GILGE, D. KLEY,
Institut für Chemie 2 : Chemie der Belasteten Atmosphäre
Kernforschungsanlage Jülich, D–5170, FRG.

Abstract

During the scientific shipcruise (ANT VII/1) of the research vessel "Polarstern" we performed a sounding program for the measurement of tropospheric and lower stratospheric ozone and water vapor over the Atlantic Ocean between 30 °S and 52 °N. Tropospheric O_3 in the tropical region between 5 °N and 15 °N was characterized by relatively low values near the surface, extending up to the tropopause with the photolysis of O_3, followed by the reaction with water vapor, providing the major sink of ozone. In the subtropical regions, the O_3 showed a strong vertical gradient in the middle troposphere. In the SH relatively high levels were observed between 5 °S and 20 °S probably of photochemical origin due to biomass burning events in the tropics. From these data we derived a meridional distribution of ozone and watervapor. The discussion in this paper will be focused on the relation between ozone and watervapor in terms of long–range transport and of sources and sinks of ozone in the Northern and Southern Hemisphere.

1. Introduction

Although tropospheric ozone plays an important role in tropospheric chemistry as a precursor for reactive radicals, and as a significant absorber of infrared radiation, there is only little information about its distribution. In fact the ozone monitoring stations which obtain vertical ozone profiles operate mainly over the continents in the Northern Hemisphere (NH); the most well known station in Central Europe is at the Hohenpeissenberg in the FRG (Attmannspacher et al., 1984). Only a few stations operate in the Southern Hemisphere (SH) (Oltmans et al., 1989).

There is a particular dearth of information on the maritime distribution of tropospheric ozone over the Atlantic Ocean. It is particularly important to know the height resolved concentration of tropospheric ozone as the air masses approach the European continent with the prevailing winds coming from the West and, finally, from polluted regions of the American continent. This information is needed for two reasons: First, the ozone pollution over Europe can only be assessed by knowing the ozone concentration of the incoming air and, secondly, Eulerian models must have the proper boundary condition. We reported a spring–time meridional distribution of tropospheric ozone over the Central Atlantic from an ozone sounding campaign that took place between 37 °S and 46 °N in March/April 1987 during a shipcruise on the FS "Polarstern" (Smit et al., 1988). We observed that the concentration of tropospheric ozone in the NH was much higher than that in the SH and that the vertical gradient in the SH was much stronger than in the NH. In the tropics, the major sink of O_3 was near the surface due to photolysis of O_3 and the consecutive reaction of $O(^1D)$ with water vapor. Further, at 30 °N, strong influences from stratospheric/tropospheric exchange processes were observed.

In order to examine the seasonal differences of the meridional distribution of O_3

we performed a second ozone sounding campaign over the Atlantic Ocean aboard the FS "Polarstern", which took place in September/October 1988. In this note we will report on the meridional cross section of ozone over the Atlantic Ocean between 52 °N and 30 °S, mainly along the 30 °W–meridian, from the fall cruise.

2. Experimental details

The ozone sounding is a balloone borne experiment, consisting of a hydrogen filled latex balloon (Totex, Japan), a parachute and an ozone sonde plus radio sonde which measures the ozone concentration and simultaneously the meteorological parameters: pressure, temperature and relative humidity up to 30 km. Under the experimental conditions (1000 g payload, 1200 g free lift) the ascent velocity was around 5m/sec and the burst altitude near 30 km.

During ascent and descent, the measured parameters are transmitted by telemetry to the groundstation aboard the ship for further data processing. The ozone sonde and the radiosonde are coupled to a special interface for digital data transmission.

The ozone sensor, based on an electrochemical method (ECC–type, Scientific Pump Corporation, USA) after Komhyr (1969), generates an electrochemical current proportional to the flowrate of ozone. A small electrically driven gas sampling pump forces ambient air through the ozone sensor. By knowing the gas flow rate, its temperature and pressure , the measured electrical current due to ozone can be converted to the ozone concentration. The radiosonde (RS80–type, Vaisala, Finnland) has three sensors, one each for pressure, temperature and relative humidity which are all capacative devices and generate signal frequencies between 7 and 10 KHz.

The microcomputer controlled interface (TMAX, USA), digitizes all simultaneously measured signals such as the ozone current, the temperature of the gas sampling line and the analog frequencies of the RS80–radiosonde into a hexadecimal coded data frame of ASCII–characters. The digital data stream is modulated as a two tone signal (2.0 and 2.2 KHz) on the 403 MHz FM–transmitter of the radiosonde and telemetered to the groundstation.

The use of this interface electronics allows the simultaneous measurement of the concentrations of ozone and water vapor, so that its relationship enables us to investigate the O_3 varations in terms of transport of sources and sinks. Additionally, this interface has the advantage that the actual temperature of the air sampling is measured for a more accurate determination of the ozone concentration.

The groundstation consists of a conventional 403 MHz, FM–receiver, a modem for demodulating the two tone signal back into the hexadecimal coded data frame which is then fed into a personal computer for further data processing. About every 7 s a complete data frame is received such that for an ascent rate of about 5m/s the height resolution of the data transmission of a complete cycle of all parameters is equivalent to a 35 m height resolution.

Before each flight, the ozone sensors were carefully prepared and checked in the laboratory aboard the ship for accuracy and response time at ambient conditions (pressure ≈ 1000 hPa, temperature ≈ 22 °C and relative humidity ≈ 70%).

The checks for accuracy were performed with the use of an UV–photometer and an integrated ozone generator (1008 RS–type, Dasibi Corporation, USA) for ozone concentrations corresponding to atmospheric concentrations up to an altitude of 30 km. For all ozone sensors flown the accuracy was better than ± 5 %.

The response time was around 20 s at an in–flight temperature of the sensor of about 30 °C such that at an average ascent rate of 5m/s the effective height resolution for ozone measurement was about 100 m.

Prior to launch, an in situ comparison of sonde reading was made against a UV–photometer for monitoring the ozone concentration near sea surface (Callies,1989), our ozone sonde having the same location as the air intake for the photometer. The sensors agreed within to ± 3 pbbv.

The precision of the flight ozone sensors was investigated during the earlier campaign in 1987 (Smit et al, 1989) by twin soundings. It was better than ± 5 % for tropospheric ozone concentrations.

Altogether 40 soundings were made between 52 °N and 30 °S, with a latitudinal spacing of 1.5 to 2.5 degrees. Here, we will report the vertical profiles obtained during the ascent of the sondes.

On a few occasions, the balloon drifted through the ships engine exhaust plume after launch. This caused transient erratic readings that were discarded. The ozone sensor recovered within 1–2 minutes after passage through the exhaust fumes.

3. RESULTS

Presentation of all individual profiles of ozone and the simultaneously measured temperature and relative humidity would be beyond the scope of this article. They will be published as a technical report.

The latitudinal variation of the mixing ratio of ozone and water vapor within the atmospheric boundary layer (ABL) is shown in Figure 1 and was derived from the individual vertical profiles after determining the ABL–height from the slope of the potential temperature profile. The ABL–height was mostly around 1–2 km, whereas inside of the ITCZ there was no defined ABL due to the strong convective transport aloft.

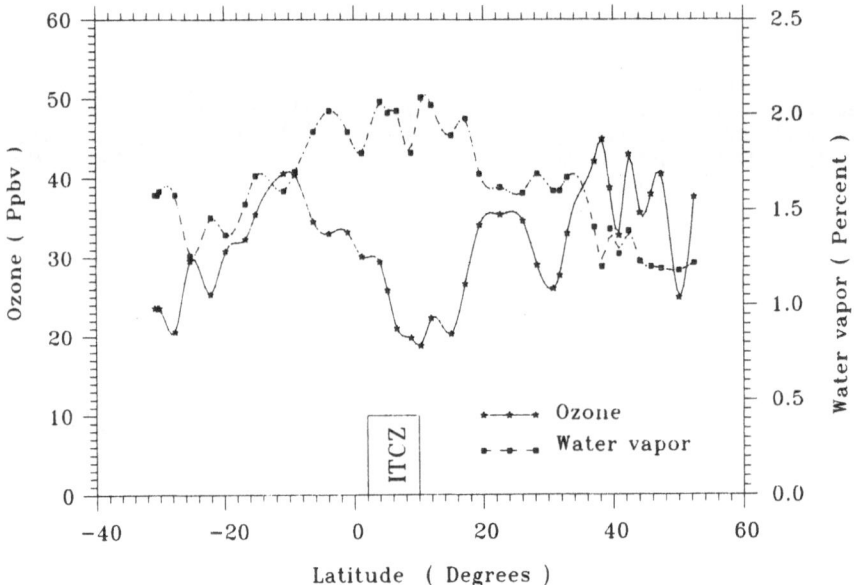

Figure 1: The mean mixingratios of ozone and water vapor within the atmospheric boundary layer (ABL).

In order to obtain a good representation of the large–scale meridional distribution of tropospheric ozone that was derived from the altogether 40 individual profiles of ozone. We used the following procedure:

First, the individual vertical profiles were made equidistant with a vertical spacing of 250 m by calculating the weighted mean of the measured ozone over the different height intervals. Secondly, equidistant profiles in the latitudinal direction were generated with a one degree spacing by linear interpolation. Thirdly, a low pass filter was used to smooth structures with length scales smaller than 500 m altitude and 2 degree latitude. The results are presented in Fig. 2.

The same procedure was applied to derive the meridional distribution of water vapor, shown in Figure 3.

Both meridional cross sections are each based on about 7000 independent measurements which are fairly evenly spaced in altitude and latitude. In both directions, altitude and latitude, the resolution of 100 m and 2 degree, respectively, is better than the width of the observed features. Therefore, the meridional distributions of ozone and water vapor, presented in Figure 2 and 3, respectivily, are well sampled and represent the large scale tropospheric distribution over the Central Atlantic Ocean in the period September/October 1988.

The tropopause heights, given in Figure 2 and 3 are determined from the measured temperature profiles following the guidelines of the WMO. The inter tropical convergence zone (ITCZ) was located between $2 ^0N$ and $10 ^0N$.

The general features of the ozone and water vapor distributions are discussed next:

Ozone in the atmospheric Boundary Layer:

ABL—ozone mixing ratios up to 1–2 km of altitude in the Northern Hemisphere (NH) and in the Southern Hemisphere (SH) varied between 20 and 40 ppbv with almost no vertical gradient. The lowest values of about 20 ppbv were measured between $7 ^0N$ and $15 ^0N$, mainly in the ITCZ. Northwards of this minimum we observed a sharp latitudinal gradient between $15 ^0N$ and $20 ^0N$, the ozone increasing from 20 ppbv to 35 pbbv. A similar increase was observed southwards of the ITCZ between $7 ^0N$ and the equator. In the NH, there was an increase between $20 ^0N$ and $35 ^0N$ to levels of 40 pbbv, while the highest values were measured northwards of $35 ^0N$. However, relatively high mixing ratios of O_3 were also found in the SH between the equator and $20 ^0S$ with a maximum of 40 ppbv at $15 ^0S$.

Ozone in the free Troposphere:

In the region of the ITCZ, the low mixing ratios of O_3 observed within the ABL, are also extending aloft up to the tropopause and even in the lower stratosphere. In the more subtropical regions, outside of the ITCZ, there was a strong vertical gradient in the lower troposphere above the ABL with ozone increasing aloft from around 30 ppbv at surface up to values of 60–70 ppbv at an altitude of 5 km. This effect was very strong in the SH between $15 ^0S$ and $20 ^0S$.

In the middle troposphere, there was only a weak vertical gradient, while in the upper troposphere the gradient increased, possibly due to the influence of stratospheric ozone. Further, the large scale distribution of ozone showed a strong negative correlation with the accompanying distribution of water vapor.

4. DISCUSSION

The large scale latitudinal variation of the height averaged O_3 in the ABL, shown in Figure 1, tracked well the latitudinal variations of O_3 in near surface air, which was also measured [Callies et al., (1989)].

The meridional cross section of O_3 in Figure 2 shows a very weak vertical gradient in the ABL, such that in general the O_3 is well mixed within this layer. Very little is known about the climatology of ozone in the ABL over the Atlantic. One of the few experimental investigations are reported by Winkler (1988), based on series of surface measurements over the Atlantic during the last decade. A comparison of the latitudinal variation given by Winkler for the September/October—period show the same general features for the NH as we have observed, but differs for the SH, especially between the ITCZ and $20 ^0S$. In the NH, Winklers data also show a strong latitudinal gradient between $20 ^0N$ and the ITCZ, where O_3 reaches its minimum. However, south of the ITCZ, Winkler's ozone climatology shows a very weak latitudinal gradient with O_3 reaching a local maximum of only 15 ppbv at $15 ^0S$. In contrast, we observe a maximum of 40 ppbv at $15 ^0S$.

The meridional cross section of Figure 2 demonstrates that the general features observed in the ABL extend well into the free troposphere.

The low mixing ratio of O_3 between $5 ^0N$ and $15 ^0N$ suggest the existence of a

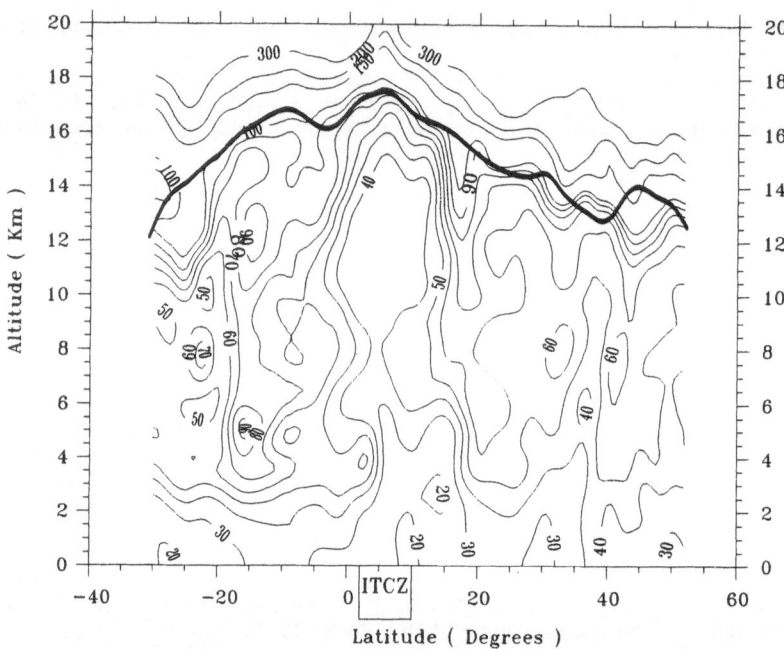

Figure 2:   Tropospheric meridional distribution of ozone, represented by isolines of mixingratios in pbbv. Tropopause heights are indicated by the fat line.

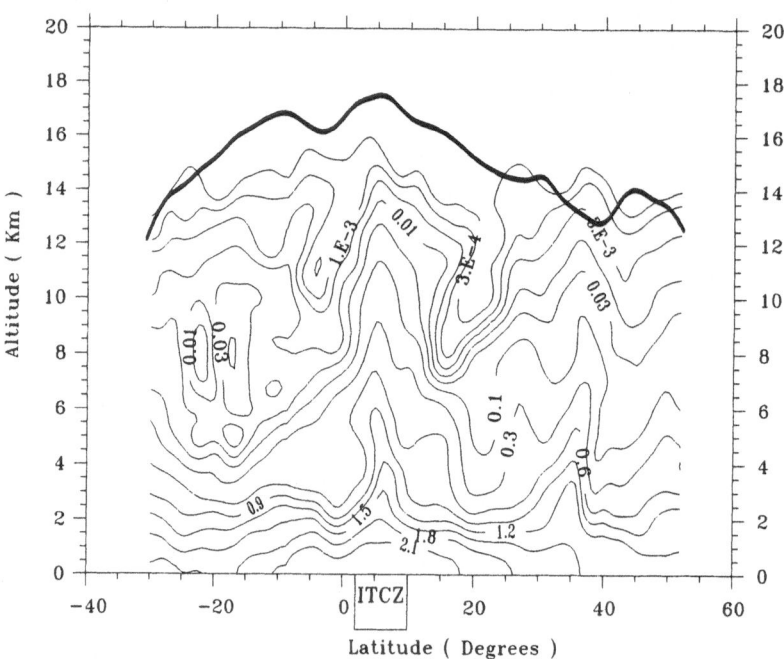

Figure 3:   Tropospheric meridional distribution of water vapor, represented by isolines of mixingratios in percents. Tropopause heights are indicated by the fat line.

sink in this region [Liu et al.,1983]. The accompanying high concentrations of water vapor, and maximum UV–radiation in the tropics provides a strong photochemical sink for ozone, by reactions (1) and (2)

$$O_3 + h\nu \longrightarrow O(^1D) + O_2 \qquad (1)$$
$$O(^1D) + H_2O \longrightarrow 2\,OH \qquad (2)$$

such that the relative O_3 loss rate increases nearly proportionally with increasing mixing ratio of water vapor. Figure 1 shows a high water vapor content within the ABL in the central tropics with the consequence that this photochemical sink will be most effective in this region [Smit et al., 1988]. The shapes of the H_2O–isolines in figure 3 imply that the influence of this photochemical loss mechanismen is extending aloft into the middle troposphere through dynamical effects. Due to the uplifting branch of the Hadley cell, air is advected upwards, carrying ozone deficient air into the upper troposphere and probably even in to the lower stratosphere.

From the shape of the O_3 isolines it appears that the ITCZ region between 2 ^{0}N and 10 ^{0}N near the surface can be contoured aloft by the 40–50 pbbv isolines, and that this vertical transport dominates the vertical distribution of O_3. This means that there is almost no latitudinal confluence of ozone rich air from higher latitudes into this region.

In the upper–and middle troposphere, just outwards of the ITCZ, and centered around 15 ^{0}N and 10 ^{0}S, ozone and water vapor are negatively correlated and the shape of the isolines of O_3 and H_2O (Figures 2 and 3) implies a downward transport of ozone rich air into the middle troposphere. This effect provides relatively high O_3 levels of about 70 ppbv to the middle troposphere at around 15 ^{0}N and 10 ^{0}S, thus inducing a strong latitudinal gradient north–and southwards of the ITCZ in the middle troposphere. The strong vertical gradient observed in the lower troposphere above the ABL at these latitudes is probably due to a confluence of ozone rich air from the middle troposphere above and ozone poor air in the lower troposphere transported by the dominating passat winds towards the ITCZ. In addition, the latitudinal gradient will become steeper by the photochemical sink of O_3 in the tropical ABL.

In the next section, we will focus the attention to the region of the middle troposphere centered around 5 km of altitude and 15 ^{0}S to 25 ^{0}S latitude. It is here were we observe a substantial air mass, characterized by large ozone mixing ratios of up to 90 ppbv. From the shape of the tropopause height and the isolines of ozone at this altitude (see Figure 2) it can be concluded that there was no recent intrusion of stratospheric air although from the simultaneously measured water vapor (see Figure 3) the indications are that this air mass originated in the upper troposphere. One major question in the debate on the origin of tropospheric ozone is related to photochemical production versus stratospheric intrusion. As measurements of NO_x, HNO_3, CO, NMHC in surface near air underlying these air masses indicate, these are aged air masses that some time ago possessed the potential for photochemical ozone production, as seen in the enhanced HNO_3/NO_x ratio, large concentration of CO and low NMHC concentration [Brüning et al., (1989), Papenbrock, (1989), Mathieu, (1989), Koppmann et al., (1989)]. In–situ production of ozone was also discounted.

The origin of the enhanced ozone levels in surface near air at these latitudes is clearly related to the air mass with high ozone concentration, sitting above, as discussed above. Transport to the surface should have occurred through a slow diffusion process as indicated by the spacing of the ozone isolines above.

The ozone being photochemical in origin and having been transported down from the upper troposphere implies that the air must have experienced emissions of ozone precursors at least once in its photochemical history.

Biomass burning, followed by a rising plume to the upper troposphere somewhere on the continents, followed by long–range transport in the upper troposphere and final descent to the ocean where it was measured seems to be an attractive hypothesis for the high ozone levels in the middle troposphere and ABL at around -10 ^{0}S. However, adiabatic and diabatic air trajectories would need to be constructed and analyzed before definite conclusions can be drawn.

5. CONCLUSIONS
   The meridional cross section of tropospheric ozone derived from 40 vertical profiles during the ANT VII/1–cruise aboard the FS "Polarstern", provides an excellent view of the latitudinal variations of ozone. Combined with the simultaneously measured meridional cross section of water vapor we are able to explain some features of the ozone varations in terms of transport, and sources and sinks of this gas.
   In the tropical regions, low values of ozone were observed in the ABL which, together with the high levels of water vapor, implies a maximum of photochemical losses. Due to strong vertical advective transport in the ITCZ, the influence of the sink located in the ABL on the distribution of ozone extends up to the tropopause.
   In the SH and NH, the subtropical regions were characterized by a downward transport of ozone rich air from the upper–into the middle–troposphere, inducing a negative latitudinal gradient towards the ITCZ. The lower troposphere in these regions exhibits a strong vertical gradient above the ABL, probably due to a vertical confluence of ozone rich air and the dominating horizontal transport by the passat winds, combined with photochemical losses of O_3 in the ABL.
   Relatively high ozone levels were found between 5 ^{0}S and 20 ^{0}S in the middle and lower troposphere. There where no indications for stratospheric intrusion, while surface measurements of photo–oxidants involved in the production of O_3 [Brüning et al., (1989), Papenbrock, (1989), Mathieu, (1989), Koppmann et al., (1989)] suggest a photochemical origin. which is probably due to biomass burning events over the continent combined with a convectiv transport aloft to the upper troposphere followed by long range transport and final descent to the ocean.

6. ACKNOWLEDGEMENT

   The authors wish to thank the Alfred Wegener Institute, Bremerhaven, FRG, for the oppurtunity to participate in the ANT VII/1 scientific expedition aboard the FS "Polarstern". Further, we want to thank the crew aboard the ship for their expert assistance during the cruise.

REFERENCES

(1)   ATTMANNSPACHER, W., HARTMANNSGRUBER, R. and LANG, P. (1984). Long period tendencies of atmospheric ozone based on ozone measurements started in 1967 at the Hohenpeissenberg Meteorological Observatory. Meteorol.Rdsch.,37, 193–199.
(2)   BRÜNING, D., ROHRER, F. (1989). Surface NO– and NO_2–mixing ratios between 30 ^{0}N and 30 ^{0}S in the atlantic region. Submitted to Proc. fifth European Symposium on physico–chemical behaviour of atmospheric pollutants, Varese (It.), September 1989.
(3)   CALLIES, J., PLATT, U. and BRAUERS, T. (1989). Measurements of photolysis frequencies of O_3 and NO_2 and ambient O_3 concentration on the Polarstern ANT VII/1. Submitted to Proc. fifth European Symposium on physico–chemical behaviour of atmospheric pollutants, Varese (It.), September 1989.
(4)   KOMHYR, W.D. (1969). Electrochemical concentration cells for gas analysis. Ann.Geoph.,25,203–210.
(5)   LIU, S.C., MCFARLAND, M., KLEY, D., ZAFIRIOU, O., and HUEBERT, B. (1983). Tropospheric NO_x and O_3 budgets in the Equatorial Pacific. J.Geophys.Res.,88,1360–1368.
(6)   OLTMANS, S.J. and KOMHYR, W.D. (1989). Ozone in the remote troposphere from surface and ozone sonde observations. Submitted to Int. Conf. on the Generation of oxidants on regional and global scales, Norwich, July 1989.
(7)   MATHIEU, B. (1989). Measurements of formaldehyde, acetaldehyde and carbonmonoxide over the mid–atlantic from 40 ^{0}N to 30 ^{0}S. Submitted to Proc. fifth European Symposium on physico–chemical behaviour of atmospheric

pollutants, Varese (It.), September 1989.

(8)   PAPENBROCK, Th. and STUHL, F. (1989). Detection of nitric acid in air by a laser—photolysis fragment fluorescence (LPFF) method. Submitted to Proc. fifth European Symposium on physico—chemical behaviour of atmospheric pollutants, Varese (It.), September 1989.

(9)   SMIT, H.G.J., KLEY, D., MCKEEN, S., VOLZ, A., and GILGE, S. (1988). The latitudinal and vertical distribution of tropospheric ozone over the Atlantic Ocean in the Southern and Northern Hemisphere. Submitted to Proc. Quadrennial Ozone Symposium,Göttingen, Aug.1988.

(10) WINKLER, P. (1988). Surface ozone over the Atlantic Ocean, J.Atm.Chem.,7,73—91.

HYDROGEN PEROXIDE CONCENTRATION VARIATIONS IN MARINE TROPOSPHERIC ATMOSPHERE

P. JACOB and D. KLOCKOW

Institut für Spektrochemie und angewandte Spektroskopie (ISAS)
Dortmund, F.R.G.

Summary

Hydrogen peroxide, one of the key compounds in multiphase atmospheric chemistry, was measured on an Atlantic cruise of the German research vessel "Polarstern", in rain, seawater and ambient air by a chemiluminescence technique, gas phase H_2O_2 after cryogenic sampling.
    The presented results show an increase of gas phase mixing ratios of about 45 pptv per degree latitude between 50°N and 0°, and a maximum of 3.5 ppbv around the equator. Generally higher mixing ratios were observed in the southern hemisphere, with a clear diurnal variation.
    Correlations with meteorological parameters and concentrations of other atmospheric trace gases are given, as well as comparisons with model calculations.

1. INTRODUCTION
    The importance of hydrogen peroxide in atmospheric chemistry arises from his oxidizing potential in the liquid phase and from the fact being involved in gas- and liquid phase radical chemistry (1, 2).
    H_2O_2 production is mainly controlled by mixing ratios of O_3, CO, NO_x, H_2O, and UV-radiation intensity (3). The most important sinks are heterogeneous loss (wet and dry deposition), liquid phase conversion, and photolysis.
    Models of global tropospheric chemistry predict H_2O_2-mixing ratios between a few pptv to about 5 ppbv as a function of latitude, altitude and season (4, 5).
    These calculations could be confirmed by measurements within the past 3 years (6, 7, 8). The published data show a positive gradient of the hydrogen peroxide mixing ratios with altitude as well as from higher latitudes to the equator, and clear seasonal variations with maxima in summer. These results were obtained in continental air masses; on board of the "Polarstern" we had the opportunity to measure hydrogen peroxide in marine troposphere.

2. EXPERIMENTAL
    Gaseous hydrogen peroxide was sampled by a cryogenic technique (9). Ambient air was passed with a flux of 0.7 m^3/h through a cooled (−45 °C) glass tube, in which H_2O_2 was trapped together with water vapor in the ice phase. The determination followed immediately after melting of the ice by peroxyoxalate chemiluminescence (10). Sampling time was one hour, except during the early morning from 2 to 8, where a 6 hour sample was taken.

Rainwater was collected with a polypropylene funnel in polypropylene vessels and analyzed immediately also by peroxyoxalate chemiluminescence.

3. RESULTS AND DISCUSSION

Figure 1 shows a profile of the H_2O_2 mixing ratio during the cruise
of the "Polarstern", starting at Bremerhaven (52°N 8°O) and ending the
south Atlantic (30°S 36°W), exhibiting some remarkable in features.

1. As predicted the H_2O_2 mixing ratio increased from higher latitudes towards the equator and decreased again with higher latitudes in the
southern hemisphere. This is more obvious in figure 2, where the 10 degree average of the H_2O_2 mixing ratio is plotted. From this figure one
can calculate a H_2O_2-gradient of about 45 pptv per degree latitude, a
value, which compares well with that from van Valin et al. (8), valid
for an altitude at 2000 m.

2. The general level seems to be higher in the southern hemisphere
at this time of the year (see also figure 2).

3. The maximum around the equator reaches 3.5 ppbv H_2O_2. This compares well with measurements in Brazil at 13°S (11) and in the marine
troposphere of the eastern part of the Atlantic close to the equator
(12), with H_2O_2-mixing ratios between 3 and 4 ppbv.

4. Very low concentrations were measured on sept. 17, 1988, passing
the English Channel. Here polluted air masses with high NO mixing ratios
limited the H_2O_2-production.

5. A diurnal variation with a maximum close to midnight could be
observed, more distinctly in the southern than in the northern hemisphere. Compared to measurements in continental air, the amplitude of the
diurnal variation in marine air is lower and the maximum is shifted from
the later afternoon to midnight. This phenomenon most probably is a consequence of differences in vertical mixing over continental and ocean
areas, respectively, during daytime and nighttime. As mentioned above,
the H_2O_2-mixing ratio increases with altitude, and ground level concentrations are controlled by vertical fluxes, which in turn are related to
temperature differences between air and ground or air and water surface,
respectively. In contrast to a stable atmospheric layer above continental areas during nighttime, the vertical mixing above the ocean surface
reaches a maximum around midnight, as illustrated in figure 3. This compares well with the averaged diurnal variation of the H_2O_2-mixing ratio
shown in figure 4.

6. During rain events – with maximum H_2O_2-concentrations of 0.12 mM
– gasphase H_2O_2 mixing ratio decreased due to wash out processes, provided that the hydrogen peroxide content in rain was lower than the equilibrium concentration, given by Henry's Law constant.

Especially the rain collected on 9/28 within the intertropical convergence zone was far from equilibrium (less than 5 % of the equilibrium
concentration), with the consequence, that gaseous H_2O_2 was reduced from
1.4 to 0.80 pptv. The reason for the low hydrogen peroxide concentration
in the precipitation could be traced back by V. Lescoat (13) to an input
of S(IV)-rich continental air in high altitudes, thus consuming H_2O_2
within the clouds. Besides this, huge amounts of Sahara dust, observed
in rainwater samples and in a one stage impactor, may have led to a catalytic destruction of H_2O_2.

7. Correlations with other parameters. Global radiation, O_3 frequency, and the H_2O-mixing ratio are well correlated with H_2O_2. This is
also the case for O_3 and CO mixing ratios in the southern hemisphere,
whereas in the nothern hemisphere the relation seems to be more complex.

Except the first 4 days, the NO_x mixing ratio was generally to low (< 50 pptv) to be important for the H_2O_2 –chemistry.

ACKNOWLEDGEMENTS
This work was supported by Alfred Wegener Institut.

REFERENCES

(1) PENKETT, S.A., JONES, B.M.R., BRICE, K.A. and EGGLETON, A.E.J. (1979). The importance of atmospheric O_3 and H_2O_2 in oxidizing SO_2 in cloud- and rainwater. Atmos. Environ. **13**, 123 - 137.

(2) CHAMEIDES, W.L. and DAVIS, D.D. (1982). The free radical chemistry of cloud droplets and its impact upon the composition of rain. J. Geophys. Res. **87**, 4863 - 4867.

(3) McELROY, W.J. (1986). Sources of hydrogen peroxide in cloud-water. Atmos. Environ. **20**, 427 - 438.

(4) LOGAN, J.A., PRATHER, M.J., WOFSY, S.C. and McELROY, M.B. (1981). Tropospheric chemistry: A global perspective. J. Geophys. Res. **86**(C8), 7210 - 7254.

(5) KLEINMANN, L. (1986). Photochemical formation of peroxides in the boundary layer. J. Geophys. Res. **91**, 10889 - 10904.

(6) HEIKES; B.G., KOK, G.L., WALEGA, J.G. and LAZRUS, A.L. (1987). H_2O_2, O_3, and SO_2 measurements in the lower troposphere over the eastern U.S.A. during fall. J. Geophys. Res. **92**, 915 - 931.

(7) van VALIN, C.C., RAY, J.D., BOATMAN, J.F. and GUNTER, R.L. (1987). Hydrogen peroxide in air during winter over the south-central United States. Geophys. Res. Lett. **14**, 1146 - 1149.

(8) JACOB, P., NEFTEL, A. and KLOCKOW, D. (1986). Die Peroxyoxalat-Chemilumineszenz und ihre Anwendung zur Bestimmung von Wasserstoffperoxid in Niederschlägen und Außenluft. in: VDI-Berichte Nr. 608 "Meßtechnik in der Luftreinhaltung", VDI-Verlag, Düsseldorf, S. 377 -399.

(9) JACOB, P., TAVARES, T.M. and KLOCKOW, D. (1986). Methodology for the determination of gaseous hydrogen peroxide in ambient air. Fres. Z. Anal. Chem. **325**, 359 - 365.

(10) KLOCKOW, D. and JACOB, P. (1986). The peroxyoxalate chemiluminescence and its application to the determination of hydrogen peroxide in precipitation. in: Chemistry of multiphase atmospheric systems, Ed. Jaeschke, W., Springer Verlag, Berlin, Heidelberg, 117 - 130.

(11) JACOB, P., TAVARES, T.M., ROCHA, V.C. and KLOCKOW, D. (1989). Atmospheric H_2O_2 field measurements in a tropical environment. Bahia, Brazil, submitted to Atmos. Environ.

(12) HARRIS, G. Pers. communication.

(13) LESCOAT, V. (1989). Physico-Chimie des composes carbonyles dans les hydrometeores: metrologie et application. Etudes de mecanismes d'oxidation. Thesis, Paris

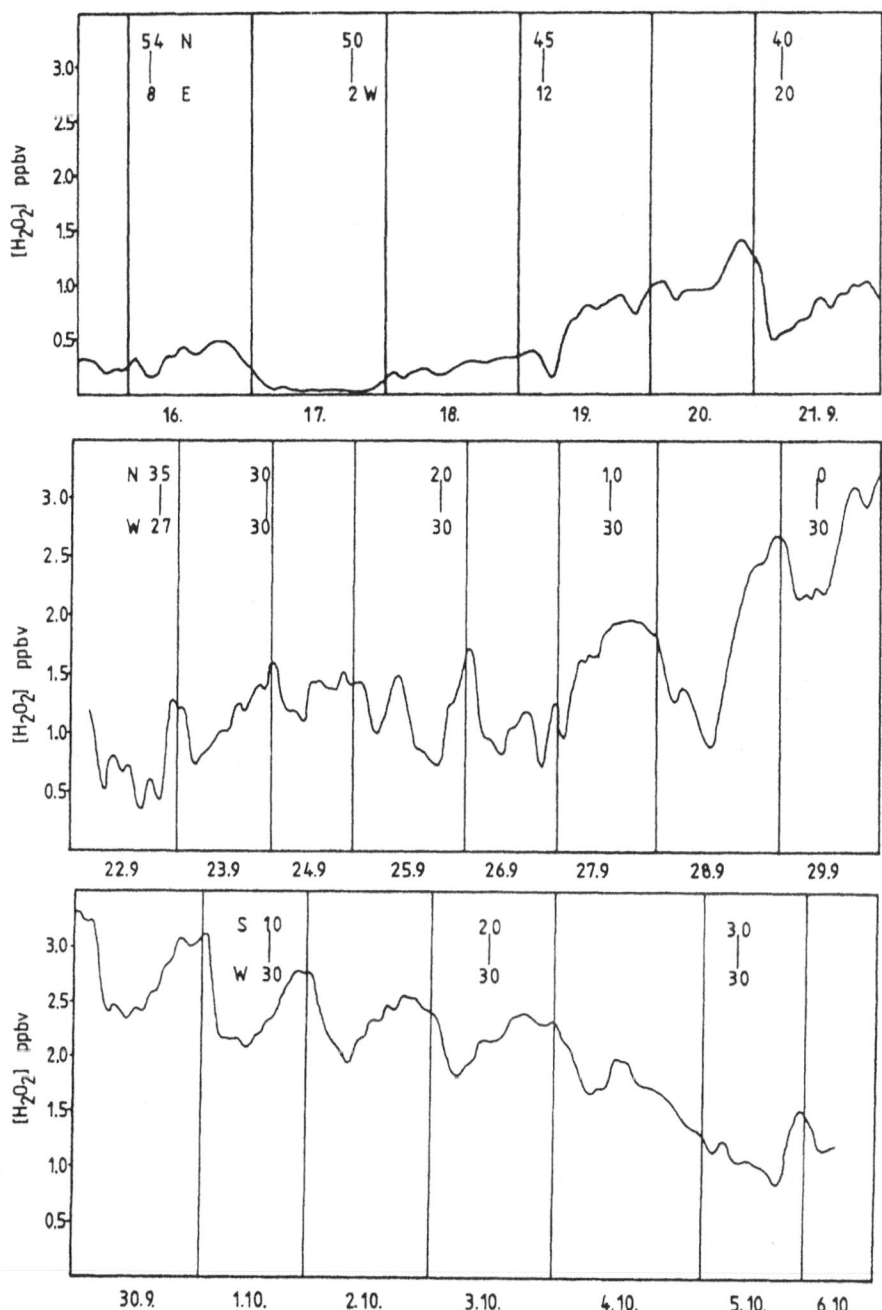

Figure 1: Hydrogen peroxide mixing ratio in the marine atmosphere measured during the cruise of the "Polarstern"
R: Rain events

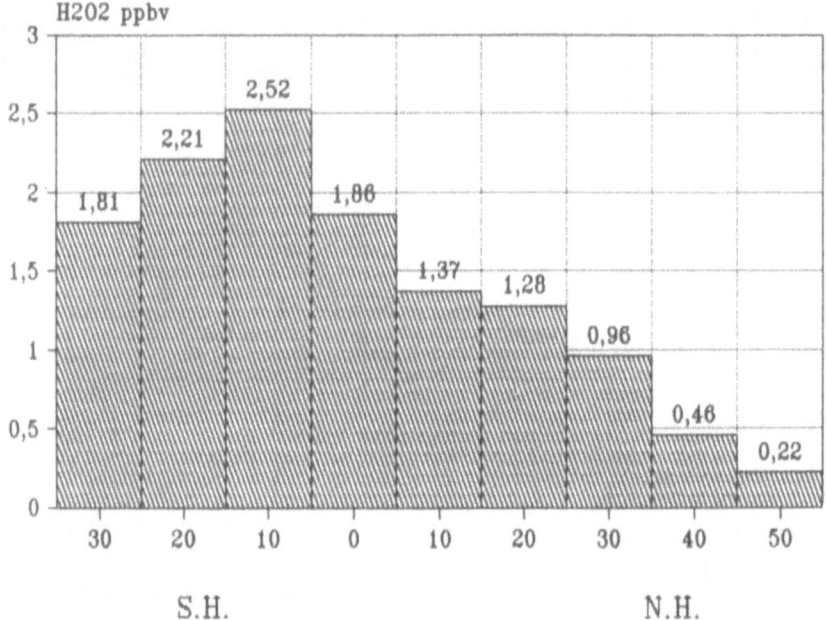

Figure 2: Latitude profile of hydrogen peroxide mixing ratio from 54°N to 32°S along the 30 degree meridian from 9/23 to 10/5, 1988

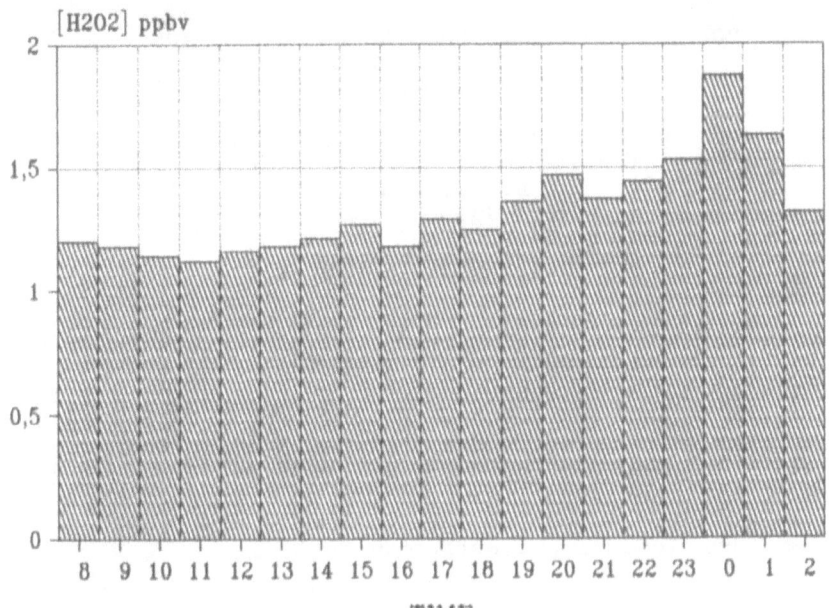

Figure 3: Averaged diurnal variation of the hydrogen peroxide mixing ratio during the cruise of the "Polarstern"

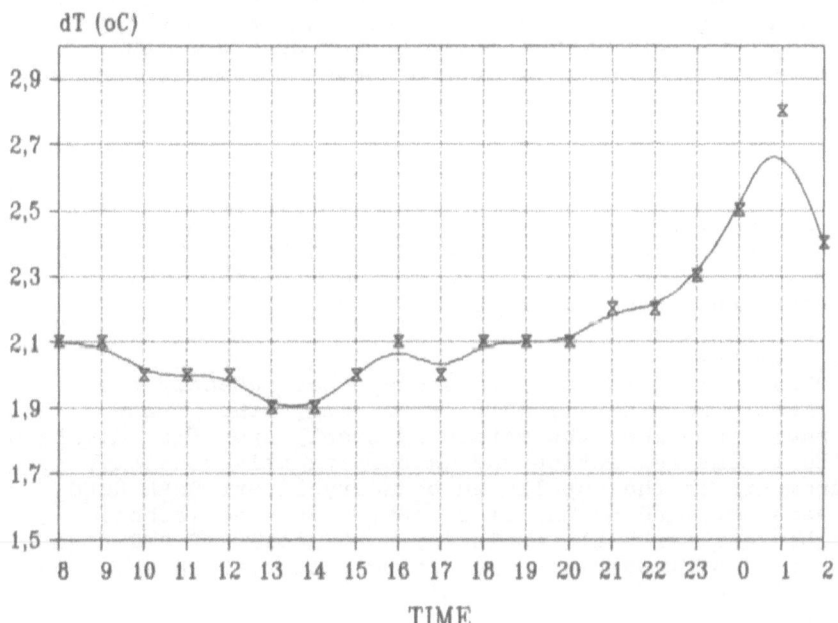

Figure 4: Averaged diurnal variation of the temperature difference seawater surface / air (18 m above sea surface) during the cruise of the "Polarstern"

POLARSTERN 1988: MEASUREMENTS OF TRACE GASES USING
TUNABLE DIODE LASERS
AND INTERCOMPARISONS WITH OTHER METHODS.

G.W. Harris, D. Klemp, T. Zenker and J.P. Burrows,
Max Planck Institute for Chemistry, Mainz, FRG.,
B. Mattieu,
KFA Julich, FRG
P. Jacob,
University of Dortmund, FRG

Summary

Measurements of NO_2, HCHO, H_2O_2 and HCl were made by the
highly specific method of mid infra-red absorption
spectroscopy using tunable diode lasers (TDLAS) during
the 1988 Polarstern expedition. S.H. NO_2 levels suggest
net photochemical destruction of O_3. N.H. HCHO averaged
0.47+/-0.2 ppbv and indicates OH noontime maxima of 2-
$4x10^6$ cm^{-3}. An upper limit for HCl of 0.25 ppbv was
measured in oceanic air. The TDLAS data are compared to
those from less direct methods.

1. Introduction

We report measurements of NO_2, HCHO, H_2O_2 and HCl
carried out aboard F.S. Polarstern during September-October
1988 using tunable diode laser absorption spectroscopy.
Details of the equipment and procedures are reported
elsewhere (1,2); the apparatus is capable of monitoring four
species simultaneously, however TDLAS measurements of all of
the above molecules were not carried out throughout the
entire cruise. NO_2 was measured continuously from 30^ON to
30OS, HCHO and H_2O_2 data were obtained overlapping with
other measurements by potentially less specific techniques,
which aid in the interpretation of these other data, and a
study of HCl by TDLAS was performed at 38^ON, 28^OW.

2. Measurement intercomparisons

2.1 Formaldehyde

TDLAS applied to HCHO measurement has recently been
described in some detail (3). As for many other trace
atmospheric species, the method is highly specific, displays
excellent time resolution and is sufficiently sensitive for
measurements in the unpolluted boundary layer. DNPH HCHO
data were obtained by one of us (BM) using the methods
described earlier (4), employing both cooled (2-4°C), and
uncooled (~25°C) samplers containing aqueous DNPH solutions
at pH 2.4. Most samples were collected with the cooled
samplers to allow the simultaneous determination of
acetaldehyde, which is collected with only low efficiency in
the uncooled samplers (4). However, comparison of the HCHO
data from the cooled samplers with that from both the

uncooled DNPH samplers and the TDLAS, indicated significant positive interferences or artifacts for cooled samplers. Similar effects have previously been reported (4) for summertime conditions, but are apparently not observed during winter or other periods of low photochemical activity. The data on HCHO and acetaldehyde obtained with the cooled DNPH samplers during this cruise are considered unreliable, and are not reported here.

Standards intercomparison for the TDLAS and DNPH systems agreed to within 1%, and a total of eight fully overlapping measurements of atmospheric HCHO by TDLAS and the uncooled DNPH method are presented in Table I. The TDLAS data are the means of the five minute values measured during the DNPH sampling periods (usually approximately one hour).

Table I. Directly overlapping HCHO measurements made by tunable diode laser spectroscopy and the DNPH – HPLC method

TDLAS ppbv	+/-	DNPH ppbv	+/-	Ratio DNPH/TDLAS
0.48	0.09	0.44	0.17	0.92
0.39	0.09	0.41	0.07	1.05
0.62	0.17	0.57	0.05	0.92
0.57	0.25	0.48	0.04	0.84
0.29	0.24	0.40	0.05	1.38
0.48	0.19	0.28	0.05	0.58
0.29	0.06	0.33	0.03	1.14
0.50	0.14	0.21	0.10	0.42

The mean ratio of the DNPH data to the spectroscopic measurement is 0.91+0.28, with only one measurement pair differing by more than the combined estimated uncertainty. If this latter measurement pair is rejected, a mean ratio (DNPH/TDLAS) of 0.98+0.23 (N = 7) is obtained, suggesting that the DNPH method with an uncooled sampler does not suffer from interferences or artifacts, and that both methods are sufficently sensitive to make formaldehyde measurements in the remote marine boundary layer, as previously reported, (1-5).

2.2 Hydrogen Peroxide.

The chemiluminesence technique used by one of us (PJ) to determine H_2O_2 has been described previously (6) and the measurements obtained during Polarstern 88 are reported seperately (7). Overlapping TDLAS measurements were made from late evening on the 21st September until shortly after dawn on September 23rd 1988 in the Azores region of the north Atlantic. Light rain occurred during the 6 h. overnight sample collection for the chemiluminesence method (7). After the rain event, both methods indicated a

parallel increase in the H_2O_2 mixing ratio during the
morning of the 22nd Sept. to values of 1.2-1.4 ppbv near
local noon. Both methods then showed rather steady values of
H_2O_2 until midnight, which decreased to 0.6 - 0.8 ppbv
throughout the rest of the night. The agreement between the
two techniques was satisfactory during this period. However,
during the light rain event of the night of the 21st-22nd,
and immediatetly prior to the event, the TDLAS showed values
of H_2O_2 which were about half those from the
chemiluminesence method (TDLAS 0.6-0.2 ppbv, CL 1.2-0.5
ppbv), although the trends were parallel. The reason for the
discrepency is not clear, but possibly either the
chemiluminesence samples became contaminated by aqueous
phase H_2O_2, or the throughput of the TDLAS inlet system was
lower for H_2O_2 under conditions very near 100% relative
humidity, than for the calibration H_2O_2 carried by dry
nitrogen. We are investigating the latter possibilty in
laboratory tests.

3. Discussion of the TDLAS measurement data
3.1 HCHO
    The mean mixing ratio of formaldehyde as measured by
TDLAS was 0.47+0.2 ppbv between 40°N and 17°N, exhibiting no
systematic diurnal variation, but with a gradient towards
lower values in the south of the measurement region. Weak
diurnal HCHO variation in the remote boundary layer has
been observed previously (4,5,8). In the atmosphere over the
ocean the predominant source of HCHO is methane oxidation,
especially in areas of low biological activity as in the
present case (9). Since the gas phase source (OH attack on
methane), and the gas phase sinks, (OH attack on HCHO and
photolysis of HCHO) vary similarily throughout the day, the
mean diurnal variation is determined largely by deposition
of HCHO. Previous TDLAS studies of HCHO in the marine
boundary layer (5) have been used to estimate the deposition
velocity for HCHO to the ocean surface as ˜0.3 cm.sec^{-1}, and
with this information it is possible to construct a simple
model of the dependance of stationary state HCHO on the
prevailing hydroxyl radical concentrations.
    With the aid of hydrocarbon measurements carried out on
board (10), appropriate HCHO photolysis rates, and this
photostationary state model, we estimate the mean daily
maximum OH radical concentrations to have been 2-4x10^6 cm^{-3}
during the measurement period.

3.2 Hydrogen peroxide
    The TDLAS H_2O_2 data was obtained in the region north of
the Azores and showed a daytime maximum value of ˜1.4 ppbv.
Our earlier, much more extensive measurements measurements
of H_2O_2 aboard F.S. Meteor, (5), suggest that such values
may be quite typical for the northern tropical regions of
the Atlantic ocean in autumn. The present restricted TDLAS
data set does not allow us to comment on the diurnal
behaviour or latitudinal distribution of H_2O_2.

3.3 Nitrogen dioxide

NO_2 was measured throughout the passage from 30°N, 30°W
to 30°S, 30°W. Much of the northern hemisphere data was
rejected because of contamination from an air exhaust on
Polarstern when the relative wind direction was 285+15°. All
data for which the relative wind velocity was less than 2
$m.sec^{-1}$ were also rejected. We believe that the remaining
NO_2 measurements were not influenced by local emissions.

Observed northern hemisphere NO_2 mixing ratios in air
free from local contamination were between 30 and 60 pptv,
with a trend apparent towards lower values nearer the ITCZ.
South of the ITCZ the NO_2 fell to values usually below 20
pptv, which was the TDLAS system detection limit for 5
minute signal averaging. Co-addition of the individual 5
minute spectra, and numerical reprocessing to reduce the
effects noise and drift, improved detection limits at the
expense of time resolution and revealed mean NO_2 mixing
ratios for each 2° latitude interval shown in Table 2.

In the table, the third column shows 95% confidence
intervals on the central values calculated by consideration
of the residuals after a least squares fit of the
calibration spectra to the averaged, background corrected,
ambient air spectra. It can be seen that in about one half
of the cases, and especially in the southern hemisphere, the
confidence intervals include zero NO_2 mixing ratios.
However, the method used to derive ambient mixing ratios
from the spectra is equally likely to produce positive and
negative results if NO_2 is truly absent, while only one of
the 22 calculated NO_2 mixing ratios in table 2 is negative.
Thus, although the NO_2 was very near to the TDLAS system
detection limits, the data indicate that the southern
hemisphere NO_2 levels were 10 - 20 pptv from the equator to
about 8°S and may be taken to suggest even lower levels
between 10° and 28°S.

These data may be compared with the measurements using
chemiluminesence (CL) and a photolytic converter by Bruening
and Rohrer (11). For the southern hemisphere they report
mean NO_2 as 15+10 pptv; the TDLAS result is 9+13 pptv.
While these average figures are not in disagreement, there
are indications from individual data points that the CL
results were sometimes higher than those from the TDLAS. It
is clear, however, from the combined TDLAS and CL (11) data
sets, that the NO_x levels in the southern hemisphere were
predominantly below those required to promote photochemical
production of O_3 in the boundary layer.

3.4 Hydrogen chloride

There is general agreement that gas phase HCl may be
released from sea-salt aerosol following acidification by
strong acids such as H_2SO_4 and HNO_3, (12) and references
therein. HCl may also be liberated by methanesulphonic acid
(13). The potential importance of these sources of chlorine

Table 2. Mean mixing ratios of NO_2 in 2° latitude intervals as measured by TDLAS.

Centre of latitude interval (N+,S-)	Mean NO_2 (pptv)	95% confidence interval (pptv)	No. of spectra in avg.
29	56.6	+ 12.0	12
27	47.7	+ 18.5	16
17	40.2	+ 14.5	5
13	31.2	+ 11.0	20
11	31.4	+ 22.0	23
5	10.8	+ 12.6	71
3	16.7	+ 12.7	84
1	11.8	+ 14.8	105
-1	17.6	+ 14.8	41
-3	15.1	+ 15.5	90
-5	19.7	+ 9.5	54
-7	10.0	+ 11.0	51
-9	5.8	+ 7.4	116
-11	7.6	+ 8.0	90
-13	5.2	+ 7.3	66
-15	13.7	+ 10.6	94
-17	13.8	+ 8.3	74
-19	-1.5	+ 8.9	60
-21	2.3	+ 5.7	45
-23	4.6	+ 10.1	84
-25	11.1	+ 9.6	63
-27	2.8	+ 12.7	86

in the troposphere has been pointed out by Singh and Kasting (14), who noted that Cl atoms, released after reaction of OH with HCl, are very much more reactive towards many non-methane hydrocarbons than are OH radicals themselves. Using a detailed one dimensional model, Singh and Kasting (14), showed that if the mixing ratio of HCl in the marine boundary layer is 0.5 - 2 ppbv, then the fractional amount of NMHC oxidized by Cl atoms relative to the total amount oxidized by Cl and OH in the troposphere, lies between 20 and 40%.

It is however, uncertain what a reasonable average value for the boundary layer HCl mixing ratio may be. Several measurements of "Gas-phase Inorganic Chlorine", interpreted as HCl, are summarized in references 12 and 13, and suggest the range used in the model (14), while two spectroscopic determinations are in disagreement with each other, Farmer et al., (15) reporting 1 to 2 ppbv in both marine and continental air, in contrast to values of 1 to 100 pptv from Marché et al, (16). Moreover, measurements by an HCl-specific derivatization GC technique (17), yielded values of 50 to 100 pptv for Atlantic airmasses uninfluenced by continents.

To obtain further information on HCl in oceanic air, we made exploratory measurements using the unambiguous TDLAS method. An appropriate laser diode was made available to us by the Fraunhofer-Institut für Physikalische Meßtechnik, Freiburg, FRG., for which we are grateful.

The system was calibrated by addition of HCl from a permeation device at the ambient air inlet, the output of the device being determined by timed acid-base titration of a NaOH solution in an impinger. The calibration spike was 2.5 ppbv, and the reproducibilty of this signal implied a detection limit of ~50 pptv.

The measurements aboard Polarstern were carried out on Sept. 24th at position $28^{\circ}N$, $30^{\circ}W$, under cloudless conditions, a wind speed of 5 m.sec^{-1}, relative humidity 65% and air temperature $23^{\circ}C$, the latter parameters being measured by the ship's sensors ~6 m above our sample inlet and ~18 m above the ocean. Gas phase HNO_3 and aerosol NO_3^- measured by other investigators on Polarstern (this symposium) suggest values on the order of 100 pptv and ~0.5 x10^{-6} g.m^{-3} respectively.

Only an upper limit for the HCl mixing ratio of 0.25 ppbv could be determined from the TDLAS data. The upper limit arises because of severe memory effects associated with the desorption of the HCl calibration gas from the teflon walls of the inlet system and/or from the surfaces inside the White cell. The observed HCl signal decreased to half its initial value of 2.5 ppbv within 5 minutes of the removal of the calibration spike, but thereafter dropped much more slowly, reaching an apparently stable level of ~0.25 ppbv only after three hours had elapsed. This constant signal was observed for a further two hours, and may represent the ambient HCl mixing ratio at the site. However tests suggested that much of the remaining signal was still due to surface off-gassing, rather than to ambient HCl.

Our data thus tend to support the lower group of measurements of gas phase HCl in the marine boundary layer, and show that the HCl at our measurement site was well below the range considered in (14). Based on the results of Singh and Kasting (14), we estimate that if the upper limit of 0.25 ppbv HCl measured here is typical, then less than 5% of NMHC near the surface is oxidized through chlorine chemistry, and less than ~10% in the troposphere as a whole. Clearly, however, further direct measurements of HCl in the marine atmosphere are needed to resolve this important question.

REFERENCES

(1) HARRIS, G.W., BURROWS, J.P., KLEMP D. and ZENKER T., (1989). A high sensitivity, multi-laser Tunable Diode Laser instrument for Trace Gas Measurements in the Remote Troposphere. Paper in preparation.
(2) ZENKER T., (1989) PhD Thesis, University of Mainz, in preparation.

(3) HARRIS, G.W., MACKAY, G.I., IGUCHI, T., MAYNE, L.K.
     and SCHIFF, H.I. (1989) Measurements of Formaldehyde in
     the Troposphere by Tunable Diode Laser Absorption
     Spectroscopy. J. Atmos. Chem., 8, 119-137.
(4) SCHUBERT, B., SCHMIDT, U., and EHHALT, D.H., (1988)
     Untersuchungen zum Nachweis und zur Chemie von
     Formaldehyde und Acetaldehyde in der unteren
     Troposphaere. KFA Bericht Jul-2257, December 1988
(5) HARRIS, G.W.,BURROWS, J.P., KLEMP, D. and ZENKER, T.
     (1989). Tunable Diode Laser Measurements in the
     Tropical Atlantic Boundary Layer. International
     Conference on the Generation of Oxidants on Regional
     and Global Scales, Norwich UK, July 3-7 1989.
(6) JACOB, P., TAVARES, T.M., and KLOCKOW, D., (1986)
     Methodology for the determination of gaseous hydrogen
     peroxide in ambient air. Fres. Z. Anal. Chem., 325,
     359-364, 1986.
(7) JACOB, P. and KLOCKOW, D., (1989) Hydrogen peroxide
     concentration variations in the Marine Troposphere,
     (This symposium).
(8) LOWE D.C. and SCHMIDT, U., (1983) Formaldehyde (HCHO)
     measurements in the nonurban atmosphere, J. Gephys.
     Res., 88, 10884 - 10858.
(9) PLASS, Ch., JOHNEN,F.J. KOPPMANN, R. and RUDLOPH, J.
     The latitudinal distribution of NMHC in the Atlantic
     and their fluxes into the atmosphere. This Symposium.
(10) KOPPMANN, R., JOHNEN, F.J., PLASS, Ch.and RUDOLPH,J.
     (1989) The latitudinal distribution of light non-
     methane hydrocarbons over the mid Atalantic between
     $40^{\circ}N$ and $30^{\circ}S$. This symposium.
(11) BRUENING, D and ROHRER, F. (1989) Surface NO and NO_2
     mixing ratios measured between $30^{\circ}N$ and $30^{\circ}S$ in the
     Atlantic region. (This symposium).
(12) CICERONE, R.J., (1981)Halogens in the Atmosphere.
     Rev. Geophys. and Space Phys., 19(1), 123-129
(13) BRIMBLECOMBE, P., and CLEGG, L., (1988)
     The solubility and behaviour of Acid Gases in the
     Marine Aerosol., J. Atmos. Chem., 7, 1-18.
(14) SINGH, H. B. and KASTING, J.F., (1988)
     Chlorine-hydrocarbon Photochemistry in the Marine
     Troposphere and Lower Stratosphere, Journal of
     Atmospheric Chemistry, 7, 261-285, 1988.
(15) FARMER, C.B., RAPER, O.F., NORTON, R.H., (1976)
     Spectroscopic detection and Vertical Distribution of
     HCl in the Troposphere and Stratosphere, Geophys. Res.
     Letters. 3(1), 13-16.
(16) MARCHE, P., BARBE, A., SECROUN, C., CORR, J
     and JOUVE, P., (1980). Ground based Spectroscopic
     measurement of HCl, Geophys. Res. Let., 7(11), 869-872
(17) VIERKORN-RUDOLPH, B., RUDOLPH, J., MEIXNER, F.J.,
     BACHMANN, K. and SCHWARZ, B. (1984) Vertical and
     Horizontal profiles of Hydrogen Chloride in the
     Mediterranean Region. In "Physico-Chemical Behaviour of
     Atmospheric Pollutants". Varese 433-440

DETECTION OF NITRIC ACID IN AIR BY A LASER-PHOTOLYSIS FRAGMENT-FLUORESCENCE (LPFF) METHOD

Th. PAPENBROCK and F. STUHL
Physikalische Chemie I, Ruhr-Universität,
D-4360 Bochum, Federal Republic of Germany

Summary
Gaseous nitric acid has been detected in ambient air by a novel laser-photolysis fragment-fluorescence (LPFF) method. Besides measurements at various continental locations we have recently determined HNO_3 on the Atlantic Ocean. The measurements are continuous with a typical resolution of up to 15 min. A detection limit of about 30 pptv was reached for a 1 h time constant. Mixing ratios in the ppbv-range were observed on the continent while it was usually lower than 100 pptv on the ocean.

1. Introduction

Nitric acid (HNO_3) is one of the dominant end products of the important atmospheric NO_x-chemistry. It is produced by reactions such as

$$OH + NO_2 + M \rightarrow HNO_3 + M$$
$$NO_3 + H_2CO \ (CH_3CHO) \rightarrow HNO_3 + HCO \ (CH_3CO)$$
$$N_2O_5 + H_2O \rightarrow 2 \ HNO_3.$$

It contributes significantly to the acidity of rain and is one of the components of the equilibrium

$$HNO_3 + NH_3 \longleftrightarrow NH_4NO_3.$$

Most of the previous detection methods such as filters and denuders require relatively long collection times. It has been therefore difficult to detect diurnal trends by those methods. We have recently developed a laser-photolysis fragment-fluorescence (LPFF) method to detect HNO_3 at low mixing ratios with a reasonable time resolution. For a thorough understanding of this method we have previously studied the ArF-laser photolysis of HNO_3 to yield excited OH(A) in a two-photon, two-step mechanism[1-3]. We have furthermore studied the ArF laser photolysis of HONO[4] which appears to be an important intermediate in this photolysis mechanism[3,5] (it is also important in the atmosphere) and properties of the excited OH[5] being used for detection in this system. In this paper we wish to present some results of our first field measurements.

2. Experimental

HNO_3 is detected by $OH(A^2\Sigma^+ \rightarrow X^2\pi)$-fluorescence which is generated in the ArF-laser photolysis. A status report on the feasibility for the detection of HNO_3 in air by LPFF including a number of experimental details have been reported previously[6]. The apparatus used is shown schematically in Fig. 1. The excimer laser (Lambda Physik, EMG 101 MSC) emits light pulses of about 15 ns width at 193 nm. The pulse energy is kept constant at about 130 ± 10 mJ by a modified power lock unit. After shaping the cross section of the laser beam, the laser light traverses a

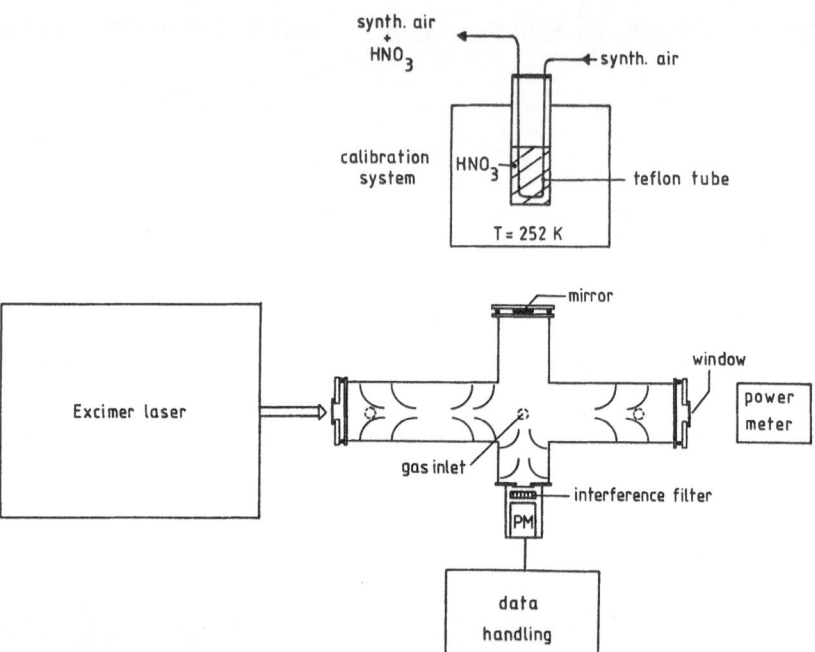

Fig. 1 Schematic diagram of the laser-photolysis fragment-fluorescence apparatus.

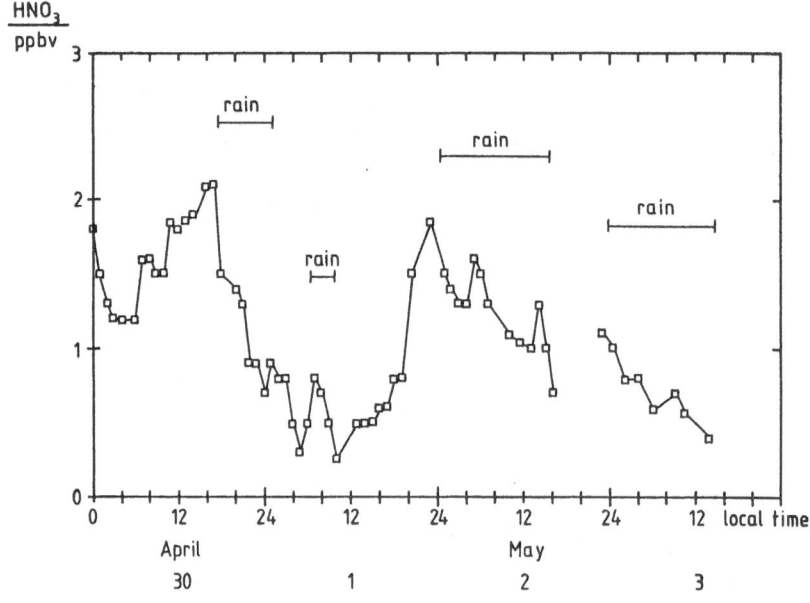

Fig. 2 Diurnal HNO₃ mixing ratios from April 30 to May 3, 1987 at Ruhr-Universität Bochum during the begin of a rainy period.

glass tube of 13 cm diameter and 92 cm length. The tube contains a number of apertures to reduce stray light of the laser and from the windows of the photolysis cell. A photomultiplier (EMI 9813 QGB) observes the OH fluorescence at right angles to the laser beam through additional apertures and an interference filter (λ_o = 308.8 nm, $\Delta\lambda$ = 4.4 nm, 52 % transmission). In spite of the above precautions some scattered light reached the photomultiplier cathode. The signal from the photomultiplier was therefore monitored by a gated integrator and processed by a personal computer. An aperture of 510 ns delayed 50 ns after the laser allowed efficient discrimination from stray light. At the same time, the signal was registered by a chart recorder as a back up system. OH(A,v=0) has a radiative lifetime of about 690 ns, which can be very efficiently shortened by quenching by atmospheric constituents. The pressure of the photolyzed air was therefore reduced to 8 mbar to lengthen the effective decay of the OH(A) fluorescence to be captured by the gate.

The flow of the incoming air sample was led directly to the observed photolysis volume which has the size of about 1 x 0.5 x 1 cm^3. This minimises contact of the sample with the walls of the vessel before HNO_3 is photolyzed. During measurements the vessel was pumped by a rotary pump. To remove traces of HNO_3 from the vessel, the system could be evacuated to less than 5 x 10^{-4} mbar by a turbomolecular pump. This procedure became repeatedly necessary to obtain a useful "zero" signal. The calibration of the system was regularly performed during the measurements using permeation of HNO_3 through teflon. A teflon tube was immersed into liquid HNO_3 (99.5 % purity) at 253 K. This temperature was kept constant at all times by a Lauda cryostat. Before use, this permeation device was calibrated by titration using a NaOH solution. The calibration method was confirmed using ion chromatography in the laboratories of UBA, Schauinsland, and ECN, Petten. Mixing air saturated with HNO_3 together with another well defined flow of air gave the same results. It was observed that such permeation devices age. Therefore the titration was performed regularly about twice a week.

The sampling line usually was 3 to 8 m long and consisted of a teflon tube of 4 mm inner diameter and 1 mm wall thickness. A needle valve made of teflon, which was adjusted by a stepping motor to keep the pressure constant in the photolysis cell, was attached at the outer end of this tube. The flow through the cell was estimated to be about 600 sccm. The needle valve was usually located not too close to surfaces. On "RV Polarstern" the valve was positioned in the fast flow intake (at about atmospheric pressure) of a constantly running commercial vacuum cleaner. Particles were separated from the flow by a virtual impactor. The whole measuring device was located on port side behind the bridge to receive air from the dominant wind direction before it gets in contact with the ship. Those measurements which are thought to be contaminated by the ship were deleted.

3. Results and Discussion

Measurements of HNO_3 using the LPFF-method were performed since 1986. During this time the calibration method, detection limit and data handling were gradually improved. Measuring sites were
   (a) Ruhr-Universität Bochum,
   (b) UBA-Meßstation, Schauinsland, Black Forest,
   (c) Kernforschungsanlage Jülich,
   (d) ECN, Petten, Netherlands, and
   (e) Research vessel "Polarstern", 30° West, from 30°N to 30°S, cruise across the Atlantic Ocean from Bremerhaven to Rio do Sul, Brazil.
   In addition, an international intercomparison test was performed in

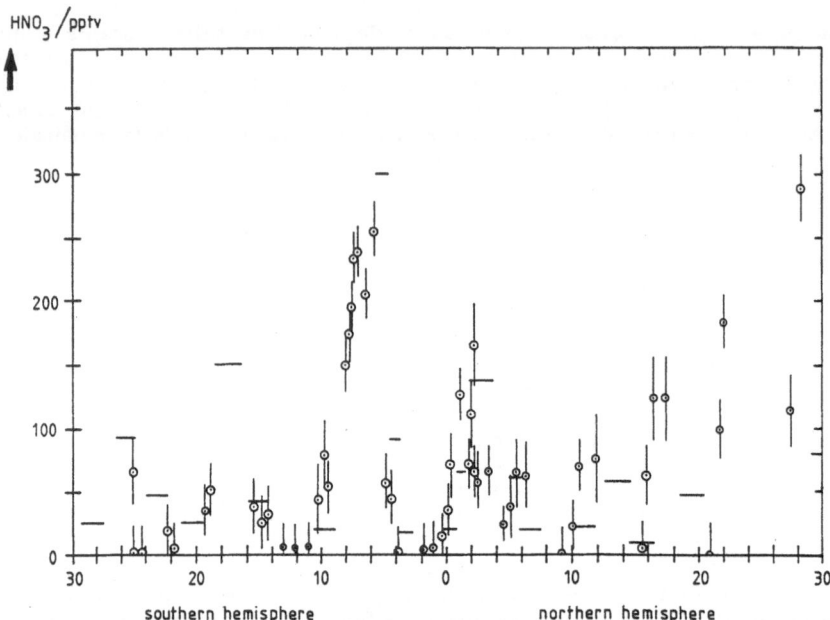

Fig. 3 HNO$_3$ mixing ratios on the Atlantic Ocean during a cruise of RV
Polarstern from 30°N (Sept. 24, 1988) to 30°S (Oct. 5, 1988) along
30°W. The horizontal bars represent measurements by nylon filters[7].

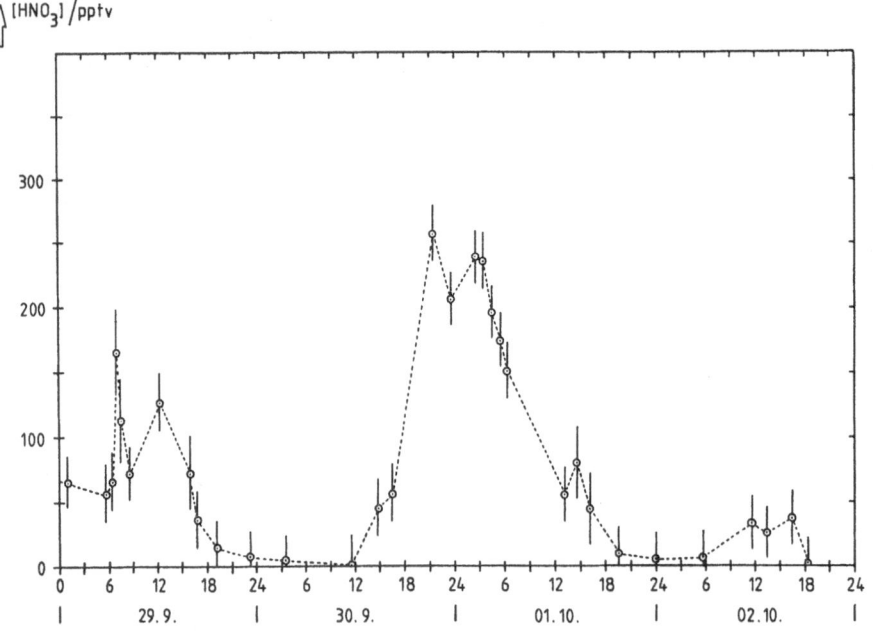

Fig. 4 Measurements of Fig. 3 from Sept. 29 to Oct. 2 on an enlarged time
scale.

the laboratories of the Landesanstalt für Immissionsschutz (LIS), Essen.

A few examples of these measurements will be presented here. Fig. 2 displays a period of 4 days sampling at our home laboratory (Ruhr-Universität). It shows the HNO_3 mixing ratios after the ending of a period of relatively warm and sunny spring weather. The change in weather pattern started with a heavy thunderstorm at 4 p.m. on April, 30. Thereafter there were showers off and on. Obviously, rain gradually decreases the HNO_3 concentration; the washout lasts for several hours. On the occasions of later rainfalls the washout repeats. In between, during the periods without precipitation, the HNO_3 mixing ratio grows. The highest mixing ratio observed on April 30 occurs late afternoon. This observation was repeatedly made during days with sunshine. Note, that the mixing ratio can be as high as several ppbv and down to the tenth ppbv range. During a number of other days we have observed that the mixing ratio was larger during noon than at midnight and larger in summer than in winter.

A number of experiments in the Black Forest and at the North Sea Coast (Netherlands) were performed parallel to measurements by denuders. They show that, in general, LPFF deviates no more from denuder measurements than denuders deviate from each other.

In contrast to these continental measurements, the values obtained on the ocean are almost two orders of magnitude lower. Fig. 3 displays all data gathered from 30°N to 30°S. Included in this figure are measurements obtained during the same cruise by Müller and Rudolf[7] using nylon filters. With a few exceptions the agreement between the two very different methods is good. From the data of the Northern Hemisphere we find a gradient of HNO_3 decrease from North to South of about 4.4 pptv/degree. In the Southern Hemisphere an average concentration of less than 30 pptv was deduced from the data available, if the elevated mixing ratios at roughly 6°S are excluded.

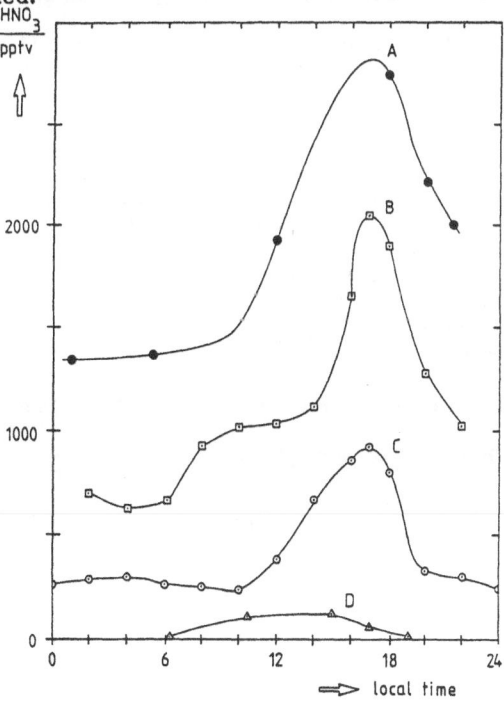

Fig. 5 Selected diurnal mixing ratios for sunny days and different sites. A, Ruhr-Universität Bochum, Aug. 13, 1986; B, Schauinsland, Black Forest, July 14, 1987; C, Petten, Netherlands, Aug. 12, 1987; D, RV Polarstern, 30°W, 19.8° to 14.4°N, Sept. 26, 1988.

Two episodes of relatively high concentrations take place at 6°S and about 2°N. These two episodes are expanded in Fig. 4 and put on the local time scale. The first two increased data points coincide with crossing the plumes of two passing ships. Elevated NO concentrations were registered at the same time by Rohrer[8]. We therefore presume that the ship engines (possibly Diesel) have emitted the measured pollutants.

The data of Sept. 29 and Oct. 2 show that, during the day, the mixing ratio is higher than during night as often observed during this trip. A notable exception with remarkably high concentration is the night from Sept. 30 to Oct. 1, the second episode. Shortly before or during this episode elevated concentrations of C_2H_2[9], K^+[10], and CO[11] and slightly increased concentrations of SO_4^{2-} and NO_3^- were observed[10]. We believe, that perhaps biomass burning has generated these trace gases. Long range transport of anthropogenic odd nitrogen, HNO_3 in particular, has been recently shown to influence measurements at Mauna Loa, Hawaii.[12,13]

Finally, Fig. 5 displays a feature frequently observed during sunny days: low concentrations at night and a maximum late afternoon. The data in this figure provide a preliminary comparison of this feature for different sites. Note the very different mixing ratios encountered in polluted and (almost) not polluted locations.

4. Acknowledgement

We gratefully acknowledge financial support by BMFT, Grant Nr. 0743131/2. We thank Dipl. Ing. Graul (UBA Schauinsland), Dr. Slanina and M.P. Keuken (ECN), Dr. Rohrer and Prof. Ehhalt (KFA) and the crew of Polarstern for the hospitality we enjoyed in the respective institutions.

5. References

1. Th. Papenbrock, H.K. Haak and F. Stuhl, Ber. Bunsenges. Phys. Chem. **88**, 675 (1984).
2. R.D. Kenner, F. Rohrer and F. Stuhl, Chem. Phys. Letts. **116**, 374 (1985).
3. R.D. Kenner, F. Rohrer, Th. Papenbrock and F. Stuhl, J. Chem. Phys. **90**, 1294 (1986).
4. R.D. Kenner, F. Rohrer and F. Stuhl, J. Phys. Chem. **90**, 2635 (1986).
5. R.D. Kenner, F.P. Capetanakis and F. Stuhl, submitted to J. Phys. Chem.
6. Th. Papenbrock and F. Stuhl, Physico-Chemical Behaviour of Atmospheric Pollutants, Proceedings of the Fourth European Symposium, Stresa (Italien), Sept. 1986, G. Angeletti and G. Restelli, Eds., D. Reidel, Dordrecht, 1987.
7. K.P. Müller and J. Rudolph, private communication.
8. F. Rohrer, private communication.
9. R. Koppmann, private communication.
10. S. Bürgermeister, private communication.
11. R. Bauer, private communication.
12. J. F. Galasyn, K. L. Tschudy, and B. J. Huebert, J. Geophys. Research **92**, 3105 (1987).
13. H. Levy II and W. J. Moxim, Nature **338**, 326 (1989).

THE LATITUDINAL DISTRIBUTION OF LIGHT NON-METHANE HYDROCARBONS OVER THE MID-ATLANTIC BETWEEN 40°N AND 30°S

R. Koppmann, F.J. Johnen, Ch. Plass and J. Rudolph

Institut für Atmophärische Chemie
Kernforschungsanlage Jülich GmbH
D-5170 Jülich, Postfach 1913, F.R.G.

SUMMARY

During the cruise ANT VII/1 (September/October 1988) of the German research vessel "Polarstern" the latitudinal distributions of several non-methane hydrocarbons were measured by in-situ gas chromatography. The measurements covered the latitude range between 40°N and 30°S.

In the southern hemisphere the mixing ratios of propane and n-butane were in the range of 50-100 ppt, the mixing ratios of acetylene and ethane, which are longerlived, were around 100 ppt and 500 ppt, respectively. The mixing ratios of the shortlived alkenes, ethene and propene, were <30 ppt.

All alkanes exhibit a considerable increase from the southern to the northern hemisphere. Compared with the cruise ANT V/5 (March/April 1987) the mixing ratios in the northern hemisphere in September/October 1988 proved to be a factor of 3 lower than in March/April 1987, probably due to seasonal effects. The average south/north increase of ethene and propene was a factor of 2 for both measurement series.

The relative pattern of the hydrocarbons near the intertropical convergence zone was very similar for both sets of measurements. The results indicate that the mixing ratios of the longerlived species are primarily due to long range transport from continental areas, whereas the shortlived species may be dominated by oceanic emissions.

1. INTRODUCTION

In the past years a number of investigations of light nonmethane hydrocarbons (NMHC) have been published (Rudolph and Johnen, 1989, and references therein). The results of these measurements indicate that most light NMHC have substantial oceanic sources (cf. Bonsang and Lambert, 1985; Bonsang et al., 1988; Rudolph and Ehhalt, 1981; Singh and Salas, 1982). This assumption is supported by the presence of dissolved light NMHC in the surface water of the ocean (cf. Lamontagne et al., 1974; Bonsang et al., 1988).

In order to improve our knowledge on the dependence of atmospheric mixing ratios on the primary production we carried out parallel measurements of light NMHC in the air and in the surface water of the Atlantic during the cruise ANT VII/1 of R.V. Polarstern in September/October 1988.
In this paper the results of the atmospheric measurements of ethene, ethane, n-butane and acetylene are presented.

2.   EXPERIMENTAL

The light NMHC were measured by in-situ gas chromatography. The instrument was installed in a container on the port side of the navigation deck. The air intake line extended about 4 m over the structure of the ship in order to avoid contamination by ship emission. The stainless steel inlet line was permanently flushed with outside air at a flow rate of $\approx$ 30 dm^3min^{-1} to avoid wall losses and contaminations.
The samples were preconcentrated from 2 - 4 dm^3 (STP) of air at liquid nitrogen temperature. The preconcentration technique is described in detail by Rudolph et al. (1989). The sample was injected into a gas chromatograph, where the light fraction (C_2-C_4) was separated on a packed column (6 m, 2 mm ID, Porapack Q, 100/120 mesh) and the heavy fraction (C_5-C_{10}) on a fused silica capillary column (DB5, 60 m, 0.32 mm ID). The hydrocarbons were measured by a flame ionization detector. The instrument is very similar to the one described by Rudolph et al. (1986). The mixing ratios were calculated by comparing the sample with a reference air of known composition. The mixing ratios of the different hydrocarbons in the reference air were in the range of a fraction of a ppb to a few ppb.

3.   RESULTS AND DISCUSSION

More than 90 measurements of light NMHC were made during this cruise of R.V. Polarstern. The measurements covered a latitude range from 40°N to 30°S. The cruise track is described in detail by Callies et al. (1989). Most of these measurements were made exactly at 30°W longitude. The latitudinal distributions of ethene, ethane, n-butane and acetylene are plotted in Figs. 1 - 4.
Between 35°N and 20°N ethane, ethene and n-butane showed relatively low mixing ratios. These are obviously due to the meteorological situation. The air masses were advected by a stable high pressure region over the Atlantic and had no contact with continents for several days (K. Arpe, 1988).
South of 20°N the situation changed. The weather maps indicated a recent contact of the air masses with the African continent resulting in elevated mixing ratios of all hydrocarbons except acetylene.
Possible alkane sources over Africa are anthropogenic activities, such as natural gas losses and evaporation losses from oil fields. Increasing alkene mixing ratios are due to oceanic emissions from the upwelling areas along the north west coast of Africa or emission from vegetation. Preliminary analysis of the wind trajectories show that the transport time of the airmasses from the coast line to the ship was in the order of 2 - 3 days which corresponds to the mean atmospheric lifetime of ethene in this latitude.

Table 1. Mixing ratios (ppt) of some light NMHC in the marine atmosphere

	northern hemisphere		southern hemisphere	
	ANT V/5 March/April (1987)	ANT VII/1 Sept./Oct. (1988)	ANT V/5 March/April (1987)	ANT VII/1 Sept./Oct. (1988)
Ethane	1810 ± 460	593 ± 194	278 ± 104	497 ± 118
n-Butane	94 ± 80	73 ± 72	20 ± 9	14 ± 9[1]
Ethene	58 ± 26	42 ± 32	25 ± 18	22 ± 9
Acetylene	-	112 ± 47	-	79 ± 29

[1] several measurements were below the lower limit of detection

In the intertropical convergence zone (ITCZ), the mixing ratios dropped by a factor of 2, but showed only a slight gradient towards southern latitudes for the alkanes south of the ITCZ, and almost no gradient for ethene. This indicates that in the southern hemisphere the measurements were less influenced by recent continental impact.

In Table 1 our average hemispheric mixing ratios of ethene, ethane and n-butane (September/October 1988) are compared with values given by Rudolph and Johnen, 1989 for the same cruise track, but for a different season (March/April 1987).

The data for ethane indicate that there is a seasonal cycle for the alkanes in both hemispheres with higher values in spring. In March/April the average mixing ratio in the northern hemisphere is a factor of 3 higher than in September/October, while it is a factor of 2 lower in the southern hemisphere.

Acetylene, with a mean atmospheric lifetime of one month, shows a different behaviour than the alkanes and alkenes.
In the northern hemisphere we found a north to south gradient of about 4 ppt per degree latitude. This decrease with latitude is comparable to the gradient reported by Rudolph and Ehhalt (1981) for Jan./Feb. However, the absolute mixing ratios differ by a factor of 2, which probably indicates a seasonal variation in the northern hemisphere. We also have to consider dilution effects since most of the measurements were done more than 1500 km from the nearest coast line.

South of the ITCZ the acetylene mixing ratio increased from 50 to 150 ppt and decreased towards southern latitudes with a similar gradient as in the northern hemisphere. These relatively high mixing ratios may be the result of biomass burning in southern hemispheric winter. Measurements of acetylene in Antarctica showed that the highest mixing ratios occur between August and October (Rudolph et al., 1989).

ACKNOWLEDGEMENTS

We thank the Alfred-Wegener-Institut für Polar- und Meeresforschung for the opportunity to participate in the ANT VII/1 cruise of research vessel Polarstern.
This work was supported financially by the Bundesminister für Forschung und Technologie of the Federal Republic of Germany under grant No. 0 744 102 5.

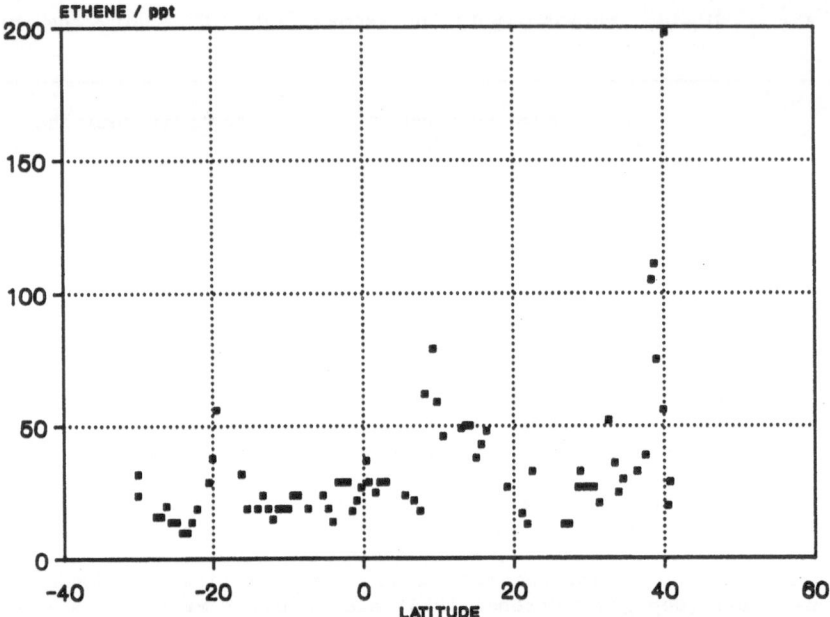

Fig. 1.　　Latitudinal variation of the C_2H_4 mixing ratio.

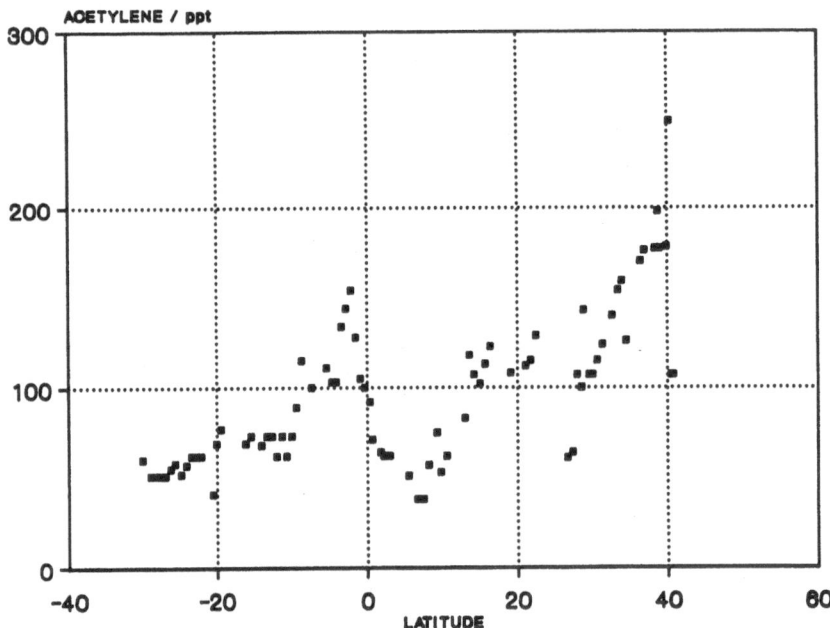

Fig. 2.　　Latitudinal variation of the C_2H_2 mixing ratio.

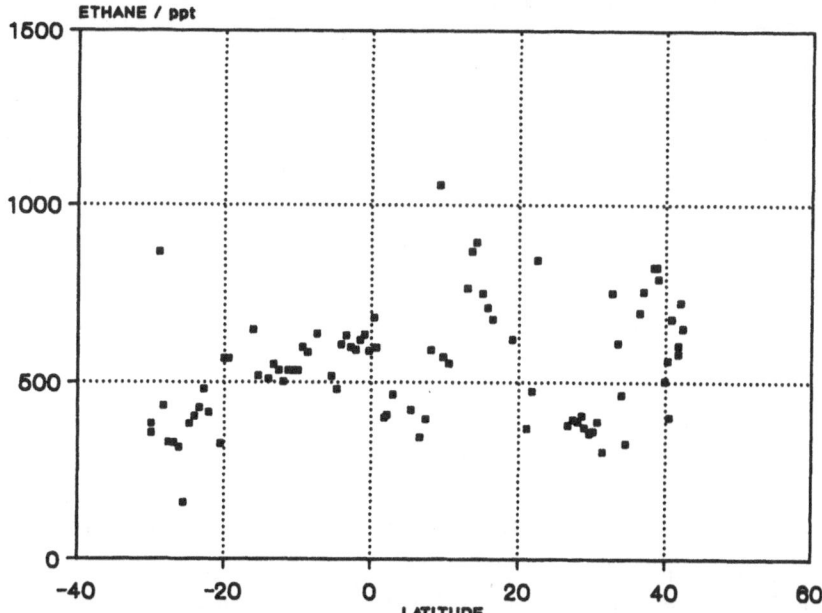

Fig. 3.   Latitudinal variation of the C_2H_6 mixing ratio.

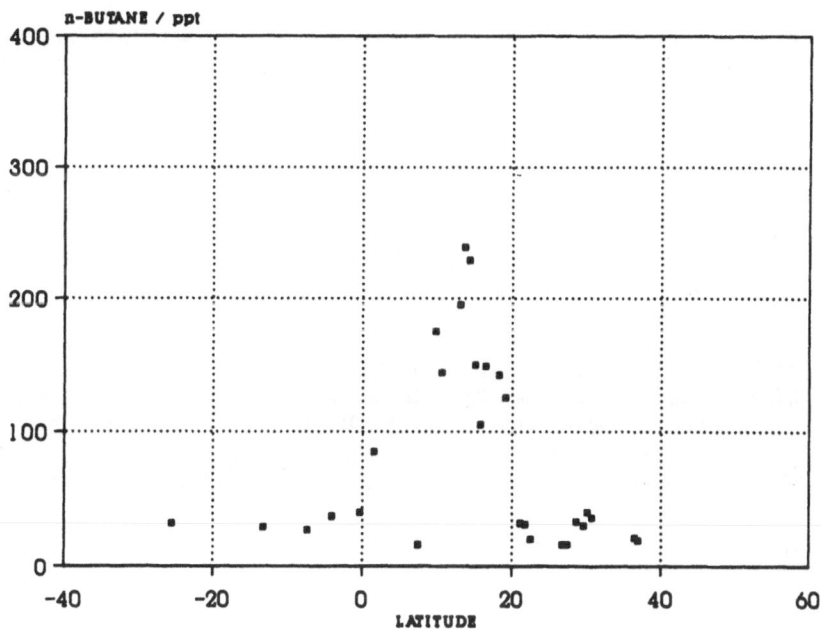

Fig. 4.   Latitudinal variation of the n-C_4H_{10} mixing ratio.

REFERENCES

(1)   Arpe, K., 1988, private communication.

(2)   Bonsang, B., M. Kanakidou, G. Lambert, and P. Monfray (1988). The marine source of C_2-C_5 aliphatic hydrocarbons. J. Atmos. Chem. 6, 3-20.

(3)   Bonsang, B., and G. Lambert (1985). Nonmethane hydrocarbons in an oceanic atmosphere. J. Atmos. Chem. 2, 257-271.

(4)   Callies, J., U. Platt, and T. Brauers (1989). Measurements of photolysis frequencies of O_3 and NO_2 and ambient O_3 concentration on the Polarstern cruise ANT VII/1. This volume.

(5)   Ehhalt, D.H., and J. Rudolph (1984). On the importance of light hydrocarbons in multiphase atmospheric systems. Ber. Kernforschungsanlage Jülich, JÜL-1942, pp. 1-43.

(6)   Lamontagne, R.A., J.W. Swinnerton, and V.J. Linnenbom (1974). C_1-C_4 hydrocarbons in the North and South Pacific. Tellus 26, 71-77.

(7)   Rudolph, J., and D.H. Ehhalt (1981). Measurements of C_2-C_5 hydrocarbons over the North Atlantic. J. Geophys. Res. 86, 11.959-11.964.

(8)   Rudolph, J., F.J. Johnen, and A. Khedim (1986). Problems connected with the analysis of halocarbons and hydrocarbons in the non-urban atmosphere. Intern. J. Environ. Anal. Chem. 27, 97-122.

(9)   Rudolph, J., F.J. Johnen, A. Khedim, and G. Pilwat (1989). The use of automated on line gaschromatography for the monitoring of organic trace gases in the atmosphere at low levels. Int. J. Environ. Anal. Chem. (in press).

(10) Rudolph, J., and F.J. Johnen (1989). Measurements of light atmospheric hydrocarbons over the Atlantic in regions of low biological activity. J. Geophys. Res. (in press).

(11) Rudolph, J., A. Khedim, and D. Wagenbach (1989). The seasonal variation of light NMHC in the Antarctic Troposphere. J. Geophys. Res. (in press).

(12) Sing, H.B., and L.J. Salas (1982). Measurement of selected light hydrocarbons over the Pacific Ocean: Latitudinal and seasonal variations. Geophys. Res. Lett. 9, 842-845.

THE LATITUDINAL DISTRIBUTION OF NMHC IN THE ATLANTIC AND THEIR FLUXES INTO THE ATMOSPHERE

Ch. Plass, F.J. Johnen, R. Koppmann, and J. Rudolph

Institut für Atmophärische Chemie
Kernforschungsanlage Jülich GmbH
D-5170 Jülich, Postfach 1913, F.R.G.

SUMMARY

During Polarstern cruise ANT VII/1 in September 1988 over the Atlantic the concentrations of light nonmethane hydrocarbons (NMHC) were determined in-situ, both in the mixed layer of the ocean and in the atmosphere.
The oceanic concentrations of the C_2-C_4 species ranged from several nl gaseous hydrocarbon per liter seawater to below the detection limit of 0.05 nl/l. With increasing carbon number the abundances of the compounds decreased. The alkenes generally predominated the alkanes. Concentrations decreased from 30°N to 30°S by factor of 10 to 100. The pattern in the northern hemisphere showed much more variability whereas structures in the southern hemisphere were smoother.
Hydrocarbons in all seawater-samples were supersaturated with respect to the atmosphere by 2 or 3 orders of magnitude. Calculated emission rates are in the range of 10^8 molec $cm^{-2}s^{-1}$ and show a similar pattern to the seawater concentrations.
The atmospheric removal of ethene seems to be balanced by the oceanic emissions. This supports the assumption that the ocean is the dominant ethene source in remote marine atmosphere.

1. INTRODUCTION

Light NMHC concentrations in oceanic surface water have been reported by Swinnerton et al. (1974) and Bonsang et al. (1988). They all found a high supersaturation in ocean water with respect to atmosphere. Estimates of the oceanic emissions of hydrocarbons to the atmosphere have been published (Rudolph and Ehhalt, 1981; Bonsang et al., 1988), but these values were based on small or incomplete data sets.
During the cruise ANT VII/1 of the R.V. Polarstern over the mid Atlantic in September/October 1988 we simultaneously measured the concentrations of light NMHC in surface seawater and in air.
In this paper the seawater concentrations for C_2-C_4 hydrocarbons are presented. The fluxes of these hydrocarbons to the atmosphere are calculated and for ethene a simplified atmospheric balance is discussed.

2. EXPERIMENTAL

The NMHC concentrations in seawater were measured by gas chromatography aboard the ship. The samples were taken from 11 m depth by means of an inlet extending about 0.5 m below the hull of the ship. The water was pumped through a stainless steel manifold to the laboratory at a flowrate of about 100 l/min. Samples were split off from the stream and passed through a glass microfibre filter (Whatman GF/C). Volumes of 870 ml were transferred to a stripping chamber similar to that described by Swinnerton et al. (1967). The volatile hydrocarbons were stripped from solution by purging with ultrapure Helium at a flowrate of 100 ccm/min for 30 minutes. The stripping efficiency was more than 85 % for the C_2-C_4 hydrocarbons.

The preconcentration and analytical technique is described by Koppmann et al. (this volume) and Rudolph et al. (1989). The hydrocarbon concentrations in seawater were calculated by comparing the sample from 870 ml of seawater with reference air of known composition. The lower limit of detection for the C_2-C_4 hydrocarbons was about 0.05 nanoliter of dissolved hydrocarbon per liter of seawater. The accuracy of the method was 20 % for C_2, 25 % for C_3, and 40 % for C_4 hydrocarbons, the differences are due to different stripping efficiencies and calibration accuracies.

3. RESULTS AND DISCUSSION

During cruise ANT VII/1 (Callies et al., this volume) 65 measurements of hydrocarbons in seawater were performed between 35°N and 30°S. The latitudinal distributions of ethene and propene are shown in Figure 1.

Ethene showed maximum concentrations of 10 nl/l at 30°N. The concentrations decreased by a factor of 10 towards 30°S with lower variability in the southern hemisphere compared to the northern hemisphere.

The propene concentration followed the same pattern with a factor of 2 lower concentration values. The different C_2-C_4 alkenes were linearly correlated with correlation coefficients of >0.95.

The ethane seawater concentration (Figure 2) peaked at 30°N with about 12 nl/l and decreased to 30°S roughly by a factor of 100.

The correlation between the concentrations of different alkanes is excellent with coefficients better than 0.98.

The correlation is not as good if we plot ethene versus ethane concentrations (Figure 3). The correlation coefficient is 0.9 and the linear regression curve intercepts the ethene-axis at (1.4 ± 0.14) nl/l.

The alkene-axis intercept is a typical feature for alkene versus alkane concentration plots, whereas it is not observed in the alkene-alkene and alkane-alkane comparison.

This indicates a different production mechanisms for alkenes and alkanes.

In Table 1 oceanic concentrations and emission rates averaged for the regions from 35°N to the equator and from the equator to 30°S are given. The emissions were calculated according to Liss and Merlivat (1988) from the seawater concentrations, the one hour averages of the wind velocity and the seawater temperature. Since the ocean is supersaturated in C_2-C_4 hydrocarbons by 2-3 orders of magnitude relative to atmospheric mixing ratios (Koppmann et al., this volume), the latter have negligible impact on the oceanic emissions.

The NMHC concentrations showed a pattern with alkenes dominating and abundances decreasing with increasing carbon numbers in both hemispheres. The alkene to alkane ratios increased from north to south. The average concentrations found here are comparable to the average baseline concentrations reported by Swinnerton and Lamontagne (1974) but lower than the values found by Bonsang et al. (1988).

Table 1:

The C_2-C_4 hydrocarbon seawater concentrations and the oceanic emission rates. The values are averages of our data from the northern and southern hemisphere.

	Seawater concentrations		Emission rates	
	Northern Hem.	Southern Hem.	Northern Hem.	Southern Hem.
	nl/l	nl/l	10^7 molec $(cm^{-2}s^{-1})$	
Ethane	3,4	0,36	31	5,5
Propane	0,98	0,16	8,9	2,3
i-Butane	0,08	0,02[1]	0,7	0,3[1]
n-Butane	0,21	0,03[1]	1,8	0,5[1]
Ethene	4,3	1,4	40	18
Propene	2,2	0,64	21	8,5
1-Butene	1,2	0,38	12	5,0
Acetylene	0,09	0,06[1]	0,9	0,8[1]

[1] Several measurements were below the lower limit of detection.

The mean emission flux of ethene calculated here for the open ocean area of 4.8×10^{-11} g $cm^{-2}h^{-1}$ is comparable to estimates by Rudolph and Ehhalt (1981) of 6×10^{-11} g $cm^{-2}h^{-1}$ for the Atlantic region of 70 - 77°N and 10°E - 5°W but considerably lower than the value of 6×10^{-10} g cm^{-2} h^{-1}, which corresponds to Bonsang's global ocean ethene-emission of about 15 Mt C/year (Bonsang et al., 1988).

For the shortlived alkenes the local emission must be largely balanced by local destruction. An atmospheric balance for the hydrocarbons allows an estimate of the oceanic contribution to the local atmospheric hydrocarbon concentration. This is shown for the example of ethene in figure 4. The oceanic emissions were calculated as described before. To estimate atmospheric destruction we assumed a well mixed boundary layer of 2 km height. The atmospheric ethene (Koppmann et al., this volume) is mostly removed by reaction with OH. The removal by reaction with ozone is neglected here as it contributes with only 20 % to the total ethene destruction. We used an OH concentration calculated by Callies (private communication) and a rate constant of the OH-ethene reaction of $8,54 \times 10^{-12}$ $cm^3 molec^{-1}s^{-1}$ at 298 k (Atkinson, 1986). The OH concentration was based on the measurements of O_3 and NO_2 photolysis frequencies and ozone mixing ratios (Callies et al., this volume), the NO_x measurements from Brüning et al. (this volume) and our CO measurements from grab samples.

In general the oceanic emissions balanced the atmospheric loss of ethene. In the regions around 10°N, the equator and 20°S the atmospheric removal considerably exceeded the emissions possibly due to advection. This assumption is confirmed for the region 15°N to ITCZ, where the airmasses were in contact with the African continent 2 - 3 days before they passed the Polarstern (Koppmann et al., this volume). From this estimated balance we conclude, that for marine airmasses the dominant source of ethene is the oceanic emission.

ACKNOWLEDGEMENTS

We thank the Alfred-Wegener-Institut für Polar- und Meeresforschung for the opportunity to participate in the ANT VII/1 cruise of research vessel Polarstern.
This work was supported financially by the Bundesminister für Forschung und Technologie of the Federal Republic of Germany under grant No. 0 744 102 5.

REFERENCES

(1) Atkinson, R. (1986). Kinetics and mechanisms of the gas-phase reactions of the hydroxyl radical with organic compounds under atmospheric conditions. Chem. Rev. 86, 69-201.

(2) Bonsang, B., M. Kanakidou, G. Lambert, and P. Monfray (1988). The marine source of C_2-C_5 aliphatic hydrocarbons. J. Atmos. Chem. 6, 3-20.

(3) Brüning, D., and F. Rohrer (1989). Surface NO and NO_2 mixing ratios measured between 30°N and 30°S in the atlantic region. This Volume.

(4) Callies, J., U. Platt, and T. Brauers (1989). Measurements of photolysis frequencies of O_3 and NO_2 and ambient O_3 concentration on the Polarstern cruise ANT VII/1. This Volume.

(5) Koppmann, R., F.J. Johnen, Ch. Plass, and J. Rudolph (1989). The latitudinal distribution of light non-methane hydrocarbons over the mid-Atlantic between 40°N and 30°S. This volume.

(6) Rudolph, J., and D.H. Ehhalt (1981). Measurements of C_2-C_5 hydrocarbons over the North Atlantic. J. Geophys. Res. 86, 11.959-11.964.

(7) Rudolph, J., F.J. Johnen, A. Khedim, and G. Pilwat (1989). The use of automated on line gaschromatography for the monitoring of organic trace gases in the atmosphere at low levels. Int. J. Environ. Anal. Chem. (in press).

(8) Swinnerton, J.W., and R.A. Lamontagne (1974). Oceanic distribution of low-molecular weight hydrocarbons. Environm. Sci. Technol. 8, 657-663.

(9) Swinnerton, J.W., and V.J. Linnenbom (1967). Determination of the C_1 to C_4 hydrocarbons in sea water by gas chromatography. J. Gas Chromatography 5, 570-573.

(10) Liss, P.S., and L. Merlivat (1986). Air-sea gas exchange rates: introduction and synthesis. In: The Role of Air-Sea Exchange in Geochemical Cycling, P. Buat-Menard (eds.), 113-127, Reidel Publ. Comp., Dordrecht.

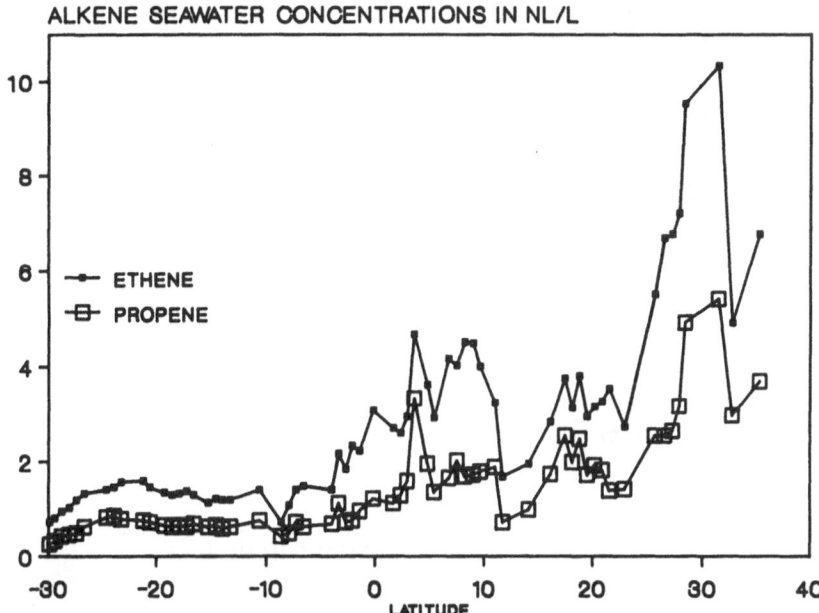

Fig. 1:   The latitudinal distribution of ethene and propene in seawater.

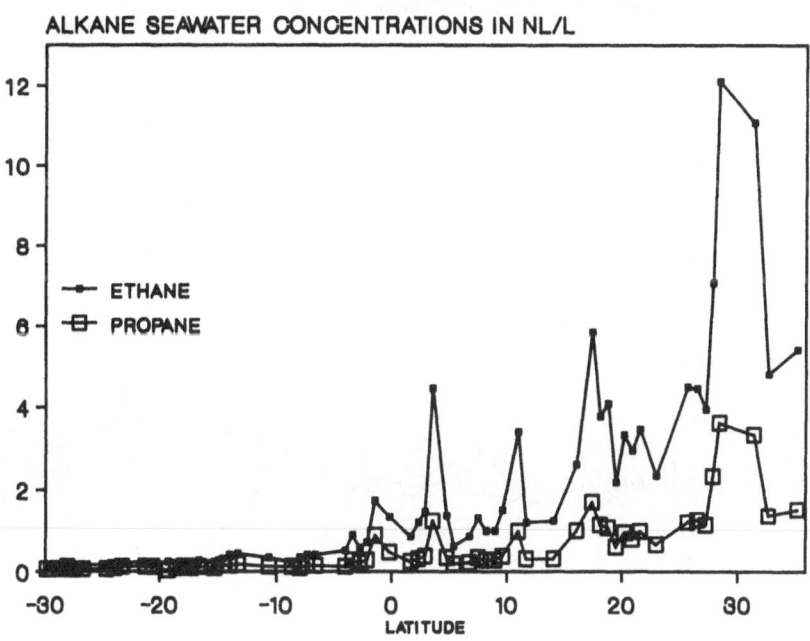

Fig. 2:   The latitudinal distribution of ethane and propane in seawater.

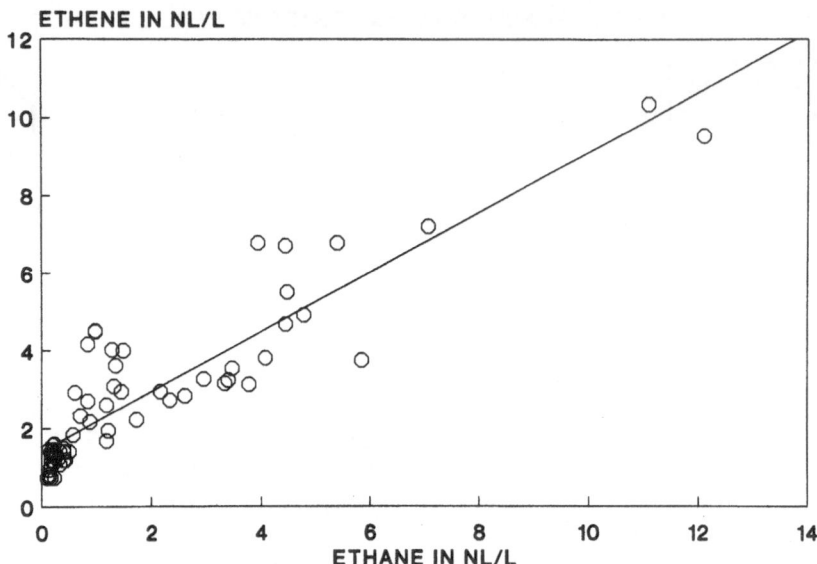

Fig. 3: Plot of ethene versus ethane concentration in seawater. The solid line represents the linear regression:
[ethene] = (1,4 ± 0,14)nl/l + (0,77 ± 0,05) x [ethane]

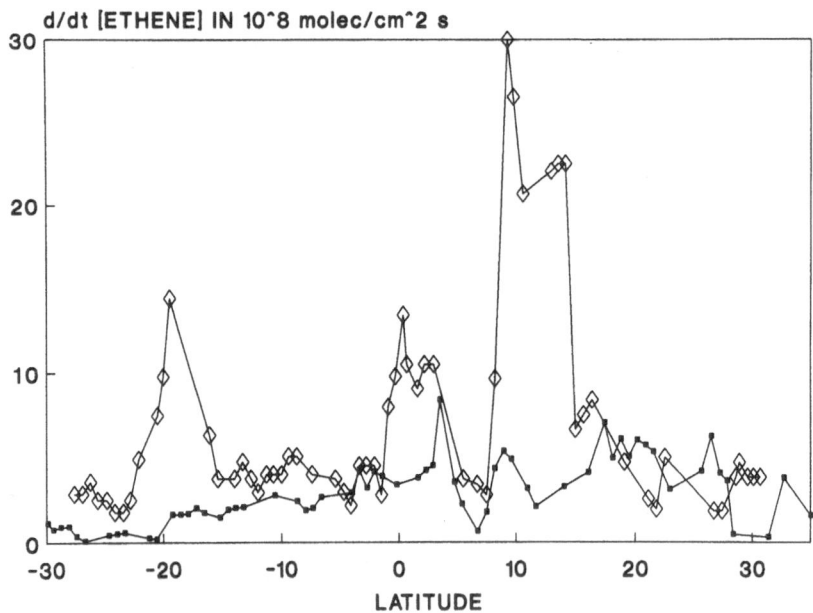

Fig. 4: Comparison of oceanic emissions of ethene (dots) with atmospheric removal in a 2 km high atmospheric column by reaction with OH radicals (diamonds).

FORMALDEHYDE BACKGROUND LEVELS OVER THE ATLANTIC OCEAN DURING THE POLARSTERN CROSSING FROM BREMERHAVEN TO RIO GRANDE DO SUL

P. CARLIER, P. FRESNET, V. LESCOAT, S. PASHALIDIS
and G. MOUVIER
Laboratoire de physico-chimie de l'atmosphère
Université PARIS VII
2 place Jussieu - 75251 PARIS Cedex 05

SUMMARY

During the POLASTERN crossing from Bremerhaven to Rio Grande do sul (15 september to 9 october 1988) we have measured the formaldehyde atmospheric concentrations at about 30 m above the ocean. The used experimental technique is the very same that we have used during the campaign at the "pointe de Penmarc'h" in July 1988. Outside of the area under continental influence the average levels are 0,4 ppbv in the north hemisphere , 0,3 ppbv in the south hemisphere and about 2 ppbv in the intertropical convergence zone. While the main values observed outside the ITCZ are in agreement with those expected from the methane oxidation and the formaldehyde photochimical depletion , these obtained in the ITCZ can only be explained by taking into account the relative high concentrations of alcenes found in this area and their fast oxidation.

1 - INTRODUCTION

Aldehydes and more particularly the formaldehyde, play a key role for the comprehension of the physical chemistry of the low troposphere and this for well-known reasons today (1). If studies about their behaviour in the continental environment are passably numerous (for exemple 2-4), those connected with no perturbed marine atmospheres are yet very few (5-10) and do not allow to have a representative idea of the space-time formaldehyde distribution. That justifies the experiments about this compound that we have carried out both within the framework of the European project OCEANO NOX (11) and during the Polarstern crossing ANT VII/1.

2 - EXPERIMENTAL

The methods of sampling and analysis are the very same with those used for the campaign at the "pointe de Penmarc'h" (11) and which are discribed in details in another paper (12-13). The samplings have been carried out with a step of two hours on the flat-roof of the watchtower at

about 30 meters above the floating line. The analysis have
been carried out aboard the day after the samplings at the
latest.

3 - RESULTS AND DISCUSSION

All the results from the measurements two hours by two
hours of the formaldehyde are represented on the figure 1.
In order to interpret the observed levels, it is fitting to
distinguish five periods which overlap those resulting from
the wind direction analysis (14) :
    1°) as far as the Azores
    2°) from the Azores to the 10th parallel
    3°) from the 10th parallel to the Equator (ITCZ)
    4°) from the Equator to the 30th southern parallel
    5°) after the 30th southern parallel.

* Periods 1 and 5 *
    During these periods, the study of winds has showed
that the analysed air was under continental influences coming from either Occidental Europe or Brazil and dating from
less than 3 or 4 days. This result is in another way confirmed by study of nitrogen oxides (15) and nitric acid (16).
From both the formaldehyde lifetime (a few days in a such
environment) and the possibility of regeneration of this
compound from other organic compounds, it is perfectly logi-
cal to find here some formaldehyde concentrations (up to 2,8
ppb) clearly upper than either these mentioned in the literature for no perturbed marine atmospheres (5-10) or measured by ourselves at the "Pointe de Penmarc'h" within the
framework of the european project "OCEANO NOX" (11).

* Periods 2 and 4 *
    The study of both the meteorological situation (14) and
the other compounds in the air samples (14, 13) has showed
that in these regions the air has not been under the influence of continental sources for probably at least a week.
Therefore we can consider that the air is representative of
a no perturbed oceanic environment.
    Consequently the concentrations have been clearly
lower (an average of 0.3 ppbv in the southern hemisphere and
0.4 ppbv in the northern hemisphere) and have showed very
low daily variations. It is remarkable that these values
correspond with those that we can calculate with the hypothesis of stationary concentrations and only by considе-
ring both the oxidation of the background methane by $OH^{\cdot}$
radical as source and the chemical depletion of the formaldehyde (by $OH^{\cdot}$, NO_3 or $h\nu$) as sink.
    By considering a sufficient long time (that means a
long time in comparison with the lifetime of the formaldehyde) we can neglect the variation of the contents of the
atmospheric reservoir in comparison with the input and the
output fluxes. We deduce of this that the average concentration is the one that equilibrates the sources and the

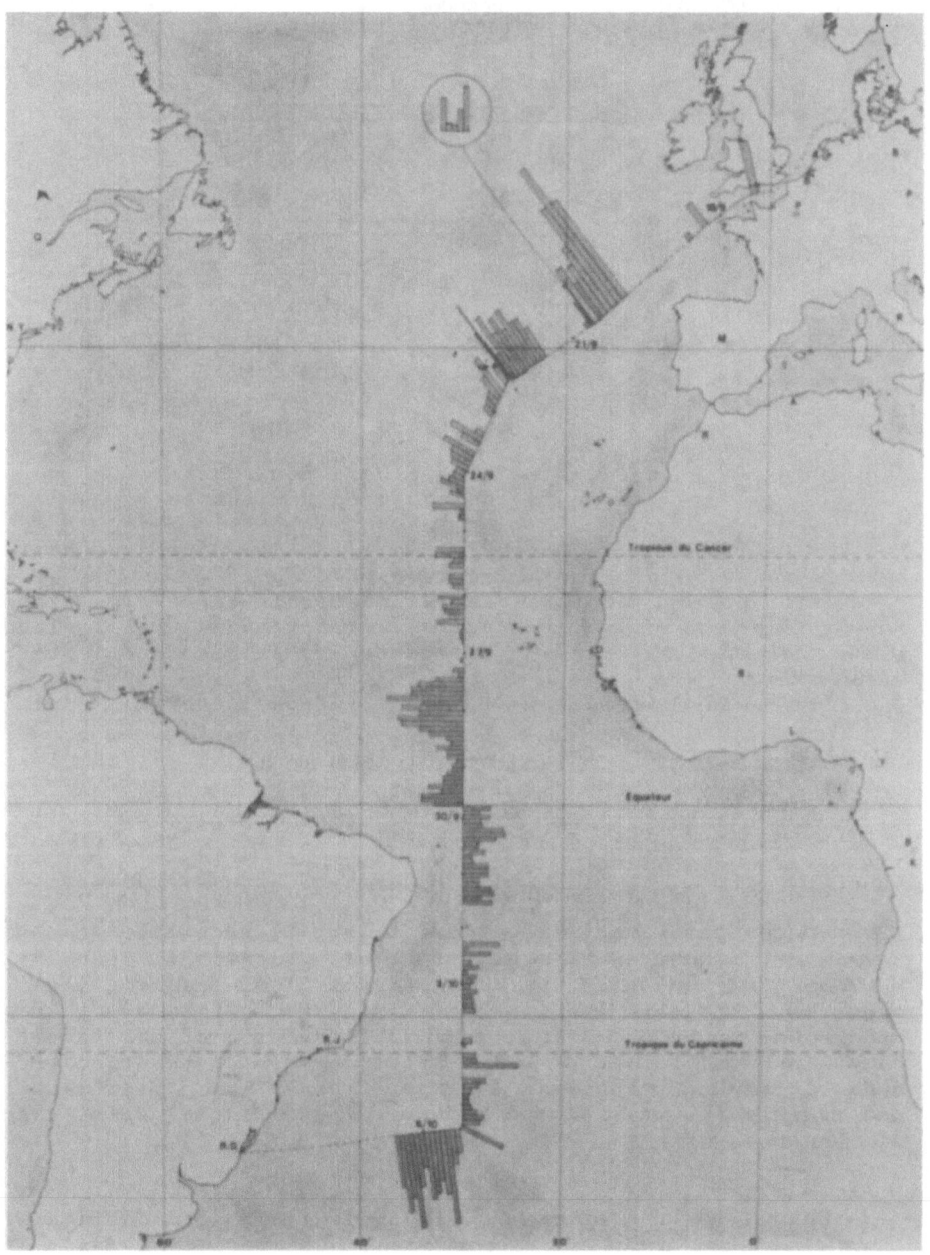

Figure 1 : The atmospheric formaldehyde concentrations during the PULARSTERN crossing ANT VII/1 (Sept.-Oct. 1988) . ▄▄▄ = 1 ppbv

sinks of formaldehyde. Then :

$$k_{CH_4+OH}\left[\overline{CH_4}\right]\left[\overline{OH}\right] = k_{HCHO+OH}\left[\overline{HCHO}\right]\left[\overline{OH}\right] + k_{HCHO+NO_3}\left[\overline{NO_3}\right]\left[\overline{HCHO}\right] + \overline{\emptyset}\left[\overline{HCHO}\right]$$

or

that we can write.

$$\left[\overline{HCHO}\right] = \cfrac{k_{CH_4+OH}\left[\overline{CH_4}\right]\left[\overline{OH}\right]}{k_{HCHO+OH}\left[\overline{OH}\right] + k_{HCHO + NO_3}\left[\overline{NO_3}\right] + \overline{\emptyset}_{HCHO}}$$

$$\left[\overline{HCHO}\right] = \frac{k_{CH_4+OH}\left[\overline{CH_4}\right]}{k_{HCHO+OH}} \times \cfrac{1}{1 + \cfrac{k_{HCHO+}\left[\overline{NO_3}\right]}{k_{HCHO+OH}\left[\overline{OH}\right]} + \cfrac{\overline{\emptyset}_{HCHO}}{k_{HCHO+CH}\left[\overline{OH}\right]}}$$

The first factor is well known. With the following values :

$$k_{CH_4+OH} = 8 \times 10^{-15} \text{ cm}^3\text{s}^{-1}$$

$$k_{HCHO+OH} = 1,1 \times 10^{-11} \text{ cm}^3\text{s}^{-1}$$

and $[CH_4]$ = 1.6 ppmv in the northern hemisphere
= 1.2 ppmv in the southern hemisphere

This factor is worth 1.2 ppbv in the northern hemisphere and 0.9 ppbv in the southern hemisphere.
The second factor is an adimensional corrective term
which depends only on the relative importances, on a long
time, of OH, NO_3 and hγ considered as sinks for the formaldehyde.
By taking some extremal values (strong with NO_3, small

with OH) it appears that $k_{HCHO+NO_3}\left[\overline{NO_3}\right]/k_{HCHO+OH}\left[\overline{OH}\right]$ is certainly

lower than 0.1 and can be neglected.

About the term $\overline{\emptyset}_{HCHO} / k_{HCHO+OH}\left[\overline{OH}\right]$ the varied data of

literature unfortunately allow only to estimate to be
between 1 and 4 and consequently the corrective term is
between 0.2 and 0.5. The terminal result in perfect agreement with the obtained measurement : the oxidation of the
background methane and the chemical depletion of the formaldehyde are sufficient to explain the levels of the formaldehyde in such pure oceanic environments. These conclusions
are roughly the very same of those carried out by WARNECK et
al. on this subject (17).

* Period 3 *
Here, we have observed ,at first, upper concentrations
than in the areas corresponding to the periods 2 and 4.
Afterwards we have noticed a sudden decrease (the morning of
sept. 28). This fall is tied to the single significant rain
which has affected our measurements. Consequently, it is
tied to the scavenging of the formaldehyde (slightly soluble

compound).

The increase of the formaldehyde levels outside of the precipitation in this area can be correlated with the values of hydrocarbons measured by PLASS et al. (18). Indeed in this area, they recorded NMHC's concentrations clearly upper than in areas 2 and 4 and more particularly for alcenes which are more reactive than alcanes. Then we have not more the right to consider that the background methane is the single source of the formaldehyde in a no perturbed oceanic environment. If the origin of these NMHC is not clearly identified nevertheless it is sure that they are responsible for the increase of the formaldehyde in this area.

4 - CONCLUSION

The comparison of our results with both meteorological data and results about others compounds (NO_x, HNO_3,...) allow eventually to end at a very coherent global explanation about the observed formaldehyde values.

It appears that the formaldehyde background values are very strongly modified by either even far perturbations (areas 1 and 5) or additionnal sources of hydrocarbons (area 3). Consequently, it is also a good criterion to control the purely oceanic origin of an air mass .

ACKNOWLEDGEMENT

The autors thank the professors D. H. EHHALT and U. PLATT as well the "Alfred Wegener Institut" (Prof. Dr. G. KRAUSE) for their invitation to participe in this campaign of atmospheric measurements.

BIBLIOGRAPHY

(1) P. CARLIER, H. HANNACHI et G. MOUVIER ; The chemistry of carbonyl compounds in the atmosphere - A review. Atmos. Environ., 20, (1986), 2079.

(2) H. PLATT, D. PERNER et H.W. PATZ ; Simultaneous measurement of atmospheric SO_2, O_3 and NO_2 by DOAS. J. Geophys. Res., 84C, (1975), 6329.

(3) D. GROSJEAN, R. SWANSON et C. ELLIS ; Carbonyls in Los Angeles air : contribution of direct emission and photochemistry. Sci. total Environ., 29, (1983), 65.

(4) P. KALABOKAS, P. CARLIER, P. FRESNET, G. MOUVIER et G. TOUPANCE ; Field studies of aldehyde chemistry in the Paris Area . Atmos. Environ., 22, (1988) 147.

(5) U. PLATT et D. PERNER ; Direct measurement of atmospheric CH_2O, HNO_2, O_3, NO_2 and SO_2 by DOAS. J. Geophys. Res., 85C, (1980), 7453.

(6)   K.   FUSHIMI et Y.   HIYAKE ; Contents of formaldehyde in
       the   air above the surface of the   ocean.   J.   Geophys.
       Res., 85C, (1980), 7533.

(7)   D.C.   ZAFIRIOU,   J. ALFORD, H.   HERRERA, G.T.   PECTZER et
       R.B.   GAGOSIAN ; Formaldehyde in remote marine are and
       rain :   flux measurements and estimates.   Geophys. Res.
       Lett., 7, (1980), 341.

(8)   D.C.   LOWE, U.   SCHMIDT et D.H.   EHHALT ; A new technique
       for   measuring tropospheric formaldehyde Geophys.   Res.
       Lett., 7, (1980), 825.

(9)   V.   NEITZERT et W.   SEILER ; Measurements of formaldehyde
       in clean air. Geophys. Res. Lett. 8 (1981), 79.

(10)   D.C. LOWE et U.   SCHMIDT ; Formaldehyde in the non-urban
       atmosphere. J. Geophys. Res., 88C, (1983), 10844.

(11)   P. CARLIER, P. FRESNET, V. LESCOAT, S. PASHALIDIS et G.
       MOUVIER ;   Daily profile study of the atmospheric for-
       maldehyde   concentrations at the "Pointe de   Penmarc'h"
       under pure marine conditions. Ibidem.

(12)   P.   KALABOKAS ; Etude des facteurs sources et puits des
       composés   carbonylés dans les masses   d'air   transitant
       au-dessus de la région parisienne.
       Thèse de Doctorat, Paris VII, Septembre 1987.

(13)   V.   LESCOAT ;   Physico-chimie des composés   carbonylés
       dans   les hydrométéores :   métrologie et application   ;
       études des mécanismes d'oxydation.
       Thèse de Doctorat, Paris VII, Février 1989.

(14)   J.   CALLIES,   U.   PLATT et T.   BRAUERS ; Measurement of
       photolysis   frequencies   of O_3 and NO_2 and   ambient O_3
       concentration   on   the   Polarstern   cruise   ANT   VII/1.
       Ibidem.

(15)   D.   BRUENING et F.   ROHER ;   Surface NO and NO_2 mining
       ratios measured between 30° N and 30° S in the atlantic
       region. Ibidem.

(16)   T. PAPENBROCK et F. STUHL ; Detection of nitric acid in
       air by a laser photolysis fragment fluorescence method.
       Ibidem.

(17)   P.   WARNECK,   W. KLIPPEL et G.K. MOORTGAT ; Formaldehyd
       in troposphärischen Reinluft. Ber.   Bunsenges.   Phys.
       Chem. 82, (1978), 1136.

(18)   C.   PLASS, F.J. JOHNEN, R. KOPPMANN et J. RUDOLPH ; The
       latitudinal   distribution   of NMHC in   the   atlantic
       and their fluxes into the atmosphere. Ibidem.

METHANESULFONATE AND NON-SEA-SALT SULFATE IN THE MARINE AEROSOL AND PRECIPITATION

Bürgermeister, S., Georgii, H.-W. and Staubes, R.
Inst. f. Meteorologie und Geophysik, J.W.Goethe - Universität,
Feldbergstr. 47, 6000 Frankfurt 1

Summary

The distribution of non-sea-salt (nss) sulfate and methanesulfonate (MSA) in the marine aerosol and precipitation was investigated during two passages with the research vessel "Polarstern" across the Atlantic between 50 °N and 30 °S in March/April 1987 and Sept./Oct. 1988. The aerosol concentrations over the tropical and subtropical Atlantic variied in the range 2 - 15 ng $S(MSA)/m^3$ and 150 - 900 ng $S(SO_4^=)/m^3$. Higher concentrations up to 25 ng $S(MSA)/m^3$ and 3500 ng $S(SO_4^=)/m^3$ were found approaching the European continent. The ratio of nss sulfate and MSA was determined to be 21.5 on an average in remote marine airmasses. Concentrations of 0.5 - 25 μg $S(MSA)/l$ and 50 - 1500 μg $S(SO_4^=)/l$ were measured in rainwater with the lowest values found near the equator (10 °S - 10 °N). On the basis of the observed concentrations of MSA and nss sulfate in the aerosol and rainwater the dry and wet deposition of these compounds to the Atlantic was calculated.

1. INTRODUCTION

The emission of dimethylsulfide (DMS, CH_3SCH_3) from the oceans to the atmosphere is one of the most important natural sources of atmospheric sulfur with an estimated amount of 40 Tg $S(DMS)/yr$ (1), about half of the global anthropogenic sulfur emissions. On the basis of several laboratory studies it is suggested that the major products of the atmospheric DMS oxidation by OH- and NO_3-radicals are sulfur dioxide (SO_2) and methanesulfonic acid (CH_3SO_3H) with informations about the yield of these compounds still being discrepant (2, 3, 4). Final atmospheric oxidation products are sulfate ($SO_4^=$) and methanesulfonate (MSA, $CH_3SO_3^-$) particles which are very effective as cloud condensation nuclei and hence influence cloud albedo and climate (5). Moreover both compounds are contributing to the acidity of rain.

In this study the concentrations of nss sulfate and methanesulfonate were determined in the aerosol and precipitation over the Atlantic Ocean comprising a latitude range from 50 °N to 30 °S. These measurements should give information about the distribution of MSA especially over the tropical and subtropical Atlantic where to our knowledge no measurements of this compound were conducted so far. Moreover the importance of methanesulfonate in comparison to nss sulfate as part of the marine atmospheric sulfur cycle should be investigated.

2. EXPERIMENTAL

2.1. Methods

Aerosol was sampled in a height of 20 m above sea level by pumping 30 m^3 of ambient air over a period of 12 hours through Microsorban 98 - filters (Ø 47 mm). The pumps were controled by a wind vane to minimize the influence of sources located on board the ship. Sampling of rainwater was carried out near by in a polyethylene funnel

(ϕ 25 cm) which could be directed according to the wind. The aerosol filters were flushed with 4 * 5 ml of deionized water to extract the soluble compounds. These solutions as well as the rainwater samples were analyzed by ion chromatography using a Dionex HPIC AS-3 column for separation of anions. For the analysis of the anions a 0.8 mM $NaHCO_3$ (MSA) and a 2.4 mM Na_2CO_3/3 mM $NaHCO_3$ (sulfate) eluant was used. Sodium and potassium were also analysed by ion chromatography using a Dionex column CS 1 and a 5 mM HCl eluant. The amount of nss sulfate and nss potassium in the aerosol was calculated by assuming the ratio of sea salt sulfate (potassium) and sodium in the aerosol to be the same as in seawater ($S(SO_4^-)/Na^+$ = 0.0825, K^+/Na^+ = 0.037).

2.2. Sampling sites

The measurements over the Atlantic were carried out during two cruises with the research vessel "Polarstern". The first passage (ANT V/5, 19.3. - 18.4.87) started in Puerto Madryn (Argentina, 42 °S, 65 °W), extended along 30 °W from 30 °S to 30 °N and was continued from the Azores to the Channel into the North Sea to finish in Bremerhaven (F.R.G., 54 °N, 9 °E). In Fig. 1 the three-dimensional-48h-backward trajectories (arriving at 12h local time in 950/700 hPa) are plotted for ANT V/5. They were calculated by the European Medium Range Weather Forecast Center in Reading (GB). These trajectories verify that the investigated air masses were predominantly of marine origin and not in contact with continents for at least two days. The route of ANT VII/1 was very similar, it started in Bremerhaven (14.9.88) and finished in Rio Grande (Brazil, 32 °S, 52 °W).

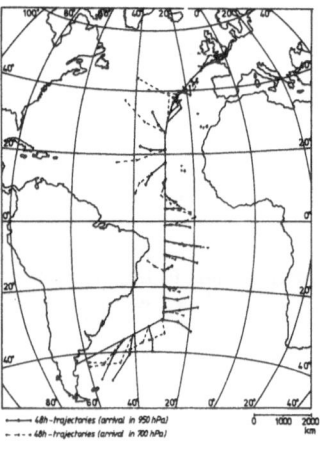

Fig.1: Cruise track of ANT V/5

3. RESULTS AND DISCUSSION

3.1. Distribution of methanesulfonate and nss sulfate over the Atlantic

3.1.1. Aerosol

The concentration of methanesulfonate in the aerosol during the first passage (ANT V/5) is plotted in Fig. 2. The lowest values (< 2 ng $S(MSA)/m^3$) were measured between 5 °S and 25 °S in the region of the oligotrophic Brazil Current. The highest concentrations (about 10 - 25 ng $S(MSA)/m^3$) were observed north of 40 °N approaching the European continent and traversing the Channel. They were caused by a remarcable increase of the chlorophyll and DMSP concentration in the surface water and a consequently higher emission rate of DMS (6). The secondary maxima of the MSA concentration at 35 - 30 °S and 5 - 25 °N were also correlated with an increase of the DMSP and DMS concentration in the ocean water.

During ANT VII/1 the concentrations of MSA were higher on the average and showed a smaller variability with values of 5 - 10 ng $S(MSA)/m^3$ being observed in the region 25 °S - 45 °N (Fig. 3). A concentration maximum with values of 15 - 20 ng $S(MSA)/m^3$ was again measured in the Channel. The most significant deviation compared to the results of ANT V/5 was found in the region 10 °S - 25 °S where the MSA concentrations were higher by a factor of about 4 during ANT VII/1. During both passages eastern wind directions (trade winds) dominated in this area. To some extend the different MSA concentrations can be explained by a relative high wind speed of 8.5

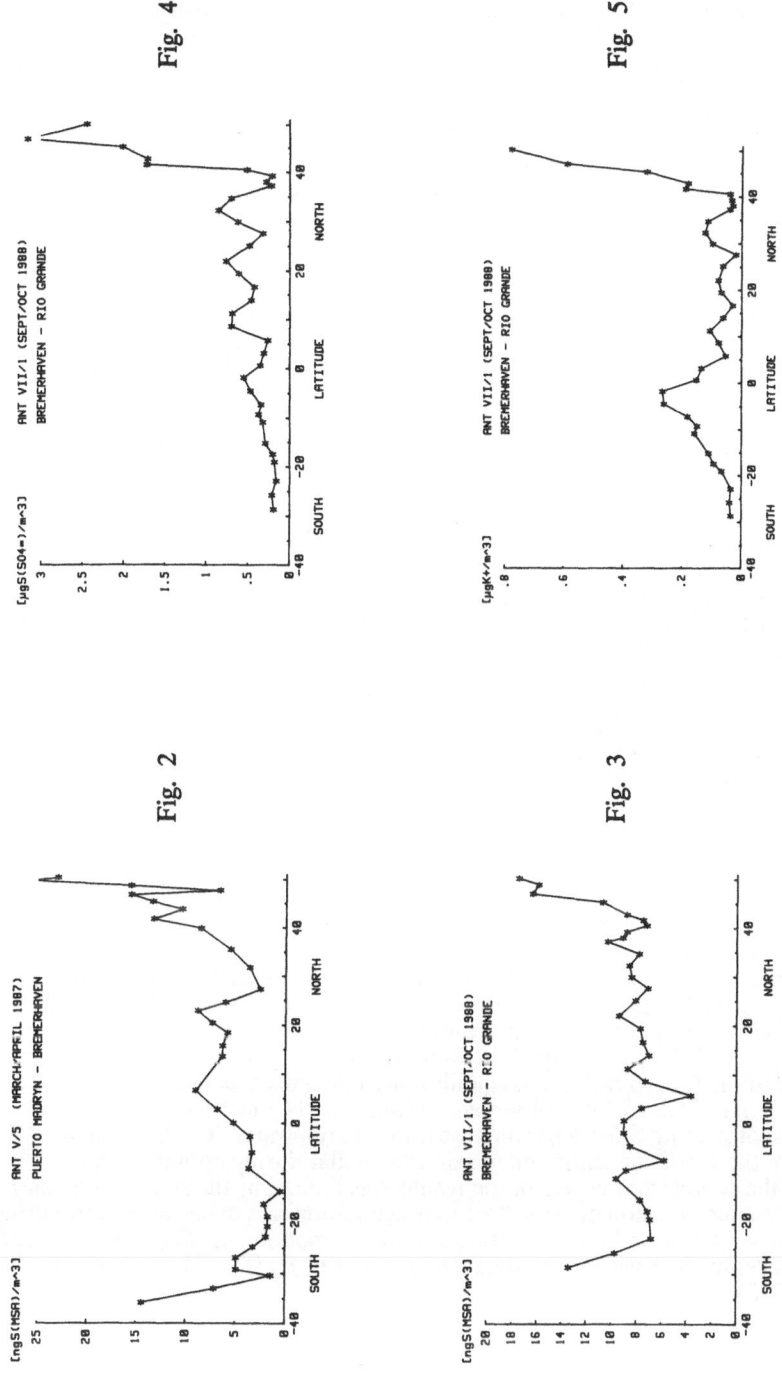

Fig. 2 - 5: Latitudinal dependence of the methanesulfonate (2,3), nss sulfate (4) and nss potassium (5) aerosol concentration over the Atlantic Ocean (35 °S - 50 °N)

m/s in October 88 (March 87: 5 m/s) causing a stronger air-sea exchange and therefore favouring the emission of DMS. Another possible reason are seasonal changes in the characteristics of the surface ocean (3.4 °C lower water temperature in October 1988) influencing the DMS production in this region.

Fig. 4 shows the latitudinal course of the nss sulfate concentration which was only determined during ANT VII/1. North of 40 °N the measurements were obviously influenced by anthropogenic sources located in Europe resulting in concentrations of 1.5 - 3.5 μg $S(SO_4^=)/m^3$. Between 40 °N and the Innertropical Convergence Zone (ITCZ) nss sulfate concentrations of 200 - 900 ng $S(SO_4^=)/m^3$ were found. The maxima observed at 32 °N, 22 °N and 8 - 12 °N were caused by advection of continental in-fluenced air masses from North Africa by the trade winds. In the southern hemisphere a constant decrease of nss sulfate was observed from 500 ng $S(SO_4^=)/m^3$ at 0 - 5 °S to 150 - 200 ng $S(SO_4^=)/m^3$ in the region 17 - 29 °S. A maximum of the nss potassium concentration determined in the area 0° - 17 °S (Fig. 5) verifies that the investigated air masses in this region were influenced by continental sources as advection of dust particles or more likely biomass burning. Therefore it is assumed that during ANT VII/1 remote marine conditions in the South Atlantic were limited to the section 17° - 30 °S where the mean ratio of nss sulfate and methanesulfonate was determined to be 21.5.

3.1.2. Rainwater

In Fig. 6 the MSA concen-tration in rainwater measured during ANT V/5 and ANT VII/1 is plotted as a function of latitude. If several samples of rainwater were taken during one precipitation event the values were summarized with respect to the volume. The highest concentrations of 16 - 25 μg S(MSA)/l were measured between 27 °S and 30 °S, in a local shower as well as in precipitation caused by a cold front advecting air masses from South America. In the ITCZ (3 °N - 6 °N) and in tropical showers at 2 °S - 8 °S the concentration decreased to 0.5 - 2.0 μg S(MSA)/l,

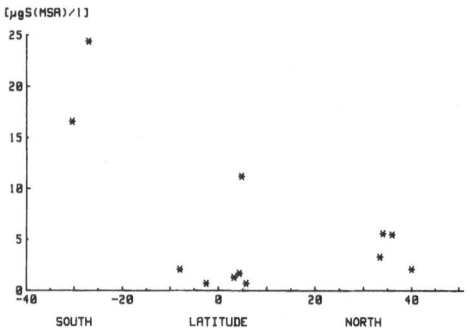

Fig. 6: MSA concentration in rainwater (ANT V/5, ANT VII/1)

with the exception of a slight shower (amount of rain: < 0.5 mm) just south of the ITCZ (11.2 μg S(MSA)/l). The low values in the ITCZ were influenced by the high amount of precipitation in this area with consequently stronger dilution of dissolved compounds. In the ITCZ the acidity of the rainwater was very low, its pH (5.67) corresponded to the CO_2/bicarbonate equilibrium in deionized water, while in the other regions a pH of 4.5 - 5.2 was observed. Between 33 °N and 40 °N MSA concentrations of 2 - 6 μg S(MSA)/l were measured in rainwater. The latitudinal dependence of the MSA concentration in rain was similar during both passages. Nevertheless the representativeness of the results was limited by the sparse frequency of rainfall. The concentration of nss sulfate in precipitation was determined only during ANT VII/1 (n = 6). The ratio of nss sulfate and MSA amounted to 92 (61.8 - 112) with nss sulfate concentrations variing between 0.06 mg $S(SO_4^=)/l$ (ITCZ) and 1.63 mg $S(SO_4^=)/l$ (30 °S).

3.2. Deposition of methanesulfonate and nss sulfate to the Atlantic

The mean dry deposition velocity of methanesulfonate and nss sulfate particles to the ocean is supposed to be 0.4 cm/s according to Varhelyi and Gravenhorst (7) and Galloway and Whelpdale (8). It is assumed that the deposition velocity of both compounds is equal because of their similar size distribution (9) and physical properties. The uncertainty of this value is quite high ($\pm$ 0.3 cm/s (7)) due to its dependence on wind speed and particle size distribution as well as different model estimations parameterizing the process of deposition (10), (11). Our measurements during ANT V/5 and ANT VII/1 resulted in a mean concentration of 6.5 ng S(MSA) /m^3 and 410 ng S($SO_4^=$)/m^3 over the Atlantic (35 °S - 40 °N). With these assumptions a mean dry deposition of 0.82 mg S(MSA)/(m^2*yr) and 51.7 mg S($SO_4^=$)/(m^2*yr) was calculated.

The wet deposition of nss sulfate and methanesulfonate was calculated by taking into account the annual mean precipitation of the regions where the measurements were conducted (12) which variied between 530 l/(m^2*yr) (27 °S) and 2020 l/(m^2*yr) (4.5 °N). The result of the precipitation at 30 °S (5.10.1988) was not taken into account because in this case it was obvious that continental influenced airmasses were advected from South America. Summarizing the results of ANT V/5 and ANT VII/1 a mean annual wet deposition of 3.90 mg S(MSA)/(m^2*yr) and 193 mg S($SO_4^=$)/(m^2*yr) was determined. Comparing the estimations for dry and wet deposition of MSA and nss sulfate to the Atlantic ocean it is proved that wet deposition represents the dominant sink for both compounds with a percentage of about 80 % of the total deposition.

The flux of DMS from the Atlantic ocean to the atmosphere was ascertained during ANT V/5 to be 67.8 - 78.8 mg S(DMS)/(m^2*yr) (6). A similar value of 68.8 mg S(DMS)/(m^2*yr) was estimated by Andreae (1) for oligotrophic areas as the tropical North Atlantic and the Brazil Current. Comparing this emission rate with the calculated total deposition of MSA (4.72 mg S(MSA)/(m^2*yr)) it is suggested that about 6 - 7 % of atmospheric DMS is oxidized to MSA. The total deposition of nss sulfate estimated for the course of ANT VII/1 is remarkably higher than the suggested DMS emission. This confirms the assumption that advection of sulfate from continental sources influenced the measurements during this passage to a significant amount.

4. CONCLUSIONS

The concentration of the MSA-aerosol variied between 2 and 20 ng S(MSA)/m^3 in the boundary layer of the Atlantic (30 °S - 50 °N). Aerosol concentrations of nss sulfate between 150 and 900 ng S($SO_4^=$)/m^3 were observed over the Atlantic (30 °S - 40 °N) with the lowest values (150 - 200 ng S($SO_4^=$)/m^3) being typical for the remote South Atlantic where the concentration ratio of nss sulfate and methanesulfonate was found to be 14 - 25. On the basis of this ratio it is concluded that 4 -7 % of atmospheric DMS is converted to MSA over the South Atlantic presuming that the DMS oxidation was the only source of MSA and nss sulfate particles and other products could be neglected. The concentration of MSA and nss sulfate in rainwater was highly variable (0.5 - 25 μg S(MSA)/l and 60 - 1700 μgS($SO_4^=$)/l) with a minimum observed near the equator. The dry and wet deposition of MSA and nss sulfate to the Atlantic was calculated taking into account the concentrations of these compounds observed in the aerosol and rain. The wet deposition proved to be the dominant sink representing about 80 % of the total deposition. The dry deposition of MSA was determined to be 0.82 mg S(MSA) /(m^2*yr) and its wet deposition 3.90 mg S(MSA)/(m^2*yr). Comparing these results with the flux of DMS from the Atlantic to the atmosphere determined in the same region it is estimated that 6 - 7 % of the sulfur emitted as DMS is deposited as methanesulfonate to the Atlantic.

ACKNOWLEDGEMENTS

We thank the captain and the crew of "Polarstern" for their co-operation. Our work was financed by "Bundesministerium für Forschung und Technik" and "Deutsche Forschungsgemeinschaft".

REFERENCES

(1) Andreae, M.O. (1985). Sulfur emissions to the atmosphere in remote areas, in "The biogeochemical cycling of sulfur and nitrogen in the remote atmosphere", ed. by J.N. Galloway et al., D. Reidel Publ. Comp., Dordrecht, 5 - 25

(2) Grosjean, D. and Lewis, R. (1982). Atmospheric photooxidation of methylsulfide Geophys. Res. Letters, 9, 1203 - 1206

(3) Hatakeyama, S., Izumi, K. and Akimoto, H. (1985). Yield of SO_2 and formation of aerosol in the photooxidation of DMS under atmospheric conditions, Atm. Env., 19, 135 - 141

(4) Barnes, I., Bastian, V. and Becker, K.H. (1986). Products and kinetics of the OH initiated oxidation of SO_2, CH_3SH, DMS, DMDS and DMSO, Proc. of the 4th Europ. Symp. "Physico-chemical behaviour of atmospheric pollutants" in Stresa, ed. by Angeletti, G. and Restelli, G., D. Reidel Publ. Company, Dordrecht, 327 - 337

(5) Charlson, R.J., Lovelock, J.E., Andreae, M.O. and Warren, S.G. (1987). Oceanic phytoplankton, atmospheric sulfur, cloud albedo and climate, Nature, 326, 655 - 661

(6) Bürgermeister, S., Zimmermann, R.L., Georgii, H.-W.,Bingemer, H.G., Kirst, G.O., Janssen, M. and Ernst, W. (1989). On the biogenic origin of dimethylsulfide: relation between chlorophyll, ATP, organismic DMSP, phytoplankton species and DMS distribution in Atlantic surface water and atmosphere, submitted to J. Geophys. Res.

(7) Varhelyi, G. and Gravenhorst, G. (1983). Production rate of airborne sea-salt sulfur deduced from chemical analysis of marine aerosols and precipitation, J. Geophys. Res., 88, 6737 - 6751

(8) Galloway, J.N. and Whelpdale, D.M. (1987). Watox-86 overview and western North Atlantic Ocean S and N atmospheric budgets, Global Biogeochem. Cycles, 1, 261 - 281

(9) Saltzman, E.S., Savoie, D.L., Prospero, J.M. and Zika, R.G. (1986). Elevated atmospheric sulfur levels off the Peruvian coast, J. Geoph. Res., 91, 7913 - 7918

(10) Slinn, S.A. and Slinn, W.G.N. (1980). Predictions for particle deposition on natural waters, Atm. Env., 14, 1013 - 1016

(11) Williams, R.M. (1982). A model for the dry deposition of particles to natural water surfaces, Atm. Env., 16, 1933 -1938

(12) Hantel, M. and Fischer, G. (1989). Landoldt-Börnstein, new series, ed. by Hellwege, K.-H. and Madelung, O., V/4 c 2, 322

DMS DISTRIBUTION STUDY OVER THE ATLANTIC OCEAN DURING THE POLARSTERN CROSSING FROM BREMERHAVEN TO RIO GRANDE DO SUL

S. PASHALIDIS, P. CARLIER, G. MOUVIER

Laboratoire de physico-chimie de l'atmosphère
Université Paris VII , 2 place Jussieu , 75251 Paris cedex 05

SUMMARY

During the crossing of the Atlantic ocean on the POLARSTERN from Bremerhaven to Rio Grande do Sul, for 24 days, from 15 September to 9 October 1988, we have measured the DMS concentrations in the troposphere at a height of 30 m over the sea-level.
The organosulphur compounds were trapped by adsorption in a TENAX cartridge and they were analysed, after thermodesorption and preconcentration in a microtrap, by gas chromatography with flame photometric detection. Generally the observed DMS levels were extremely low (in the range of 20 to 300 ng(DMS)m^{-3}) which are in agreement with the fact that the crossed areas were biologicaly very poor. The highest values (west of the Cape Verda Archipelago) correspond to a slightly less poor area.

1. INTRODUCTION

Several studies on the role of dimethylsulfide (DMS) in the physical chemistry of the lower marine troposphere are justified by the fact that this molecule is the principal source of reduced sulphur in such environment and is consequently one of the foundamental precursors of the background acidity of the atmosphere not disturbed by man. The different aspects of the problem : biogenic processes of emission, spatio-temporal distribution , mecanisms of photooxidation, processes of the depositions etc... have been recently the object of a very complete book edited by the ACS.(1) Furthermore CHARLSON and al. (2) gave light to the fact that one of the products of the oxidation of DMS, methanesulfonic acid is a very efficient precursor for the condensation nuclei and they brought up the question if the aerosols generated by this pathsway play a significant role in the global radiative budget.Thus it appears that the DMS can not only have an impact on the chemical quality of the air but also an impact on the climate.
The previous studies showed an extreme variability

of the DMS productivity by the sea and the oceans (3-5) and of its atmospheric level. The crossing of the Polarstern supplied an excellent opportunity to improve our knowledge about the production capacity of varied atlantic ocean areas (see article of U. PLATT).

2. EXPERIMENTAL

The organosulphur compounds (particularly DMS) were sampled on small columns (80mm x 4mm) on a refrigerated adsorbant (0,1g of TENAX T.A. at -24°C). The precolumn packed with iron (II)sulfate ($FeSO_4$) was behind the column in order to destroy the atmospheric oxidants (NO_2, O_3...) and to preserve the organosulphur compounds already retained from an indesirable oxidation. This phenomenon was already mentioned by other authors (4). The columns were put in an automatic system of desorption (DANI STP 33.50) and the organosulphur compounds analyzed by CPG/FPD (VARIAN 3700) (see fig. 1).

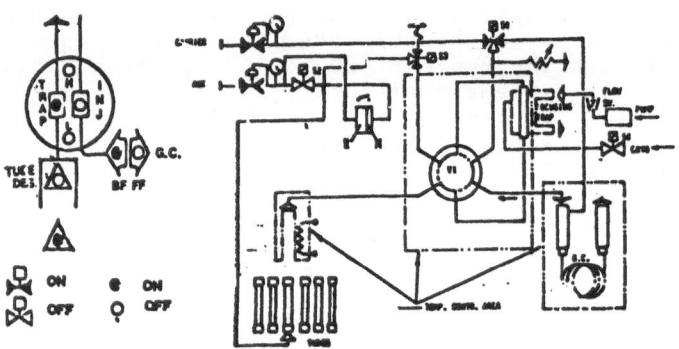

fig. 1 Sequential tube desorber

The detection limit was 20 ng $(DMS)/m^3$. The sampling was performed on the platform of the Polarstern, situated 30 m above the sea-level.

The DMS was also measured during this campaign, by STAUBES (6). Their trapping method consists of a cryogenic trap by liquid argon and they mesured the organosulphur compounds on the Polarstern platform about 18 m above the sea level. In spite of these differences we have observed an acceptable concordance between the two sets of measurements.

3. RESULTS AND DISCUSSION

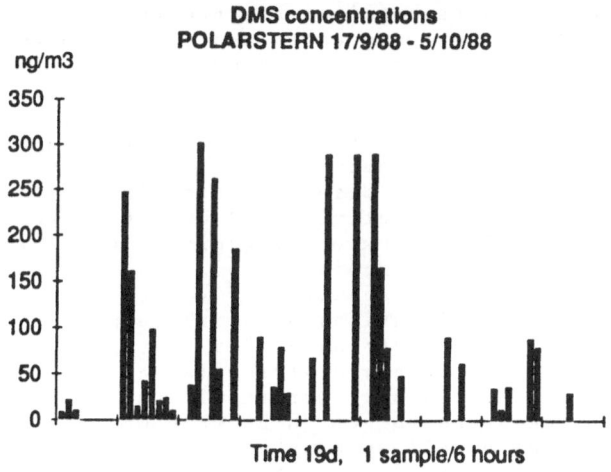

Fig. 2 DMS concentrations from 17/9 to 5/10/88

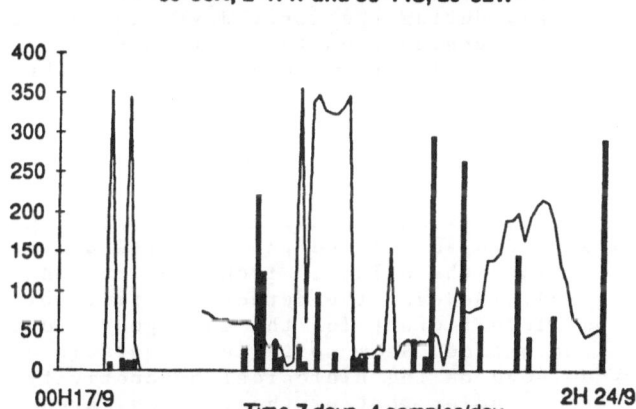

Fig. 3 DMS concentrations from 17/9 to 23/9/88

All the results obtained are presented in the fig.
2. The values that we present are often between the
lowest values cited in the literature. Maximum 300 ng
(DMS)/m3 and also sometimes less than our detection limit
(20 ng/m3).
However we expected to find stronger values. Those
values can be explained by the fact that we were
crossing a really liquid desert. According to the map of
primary biological productivity (8) we always sampled in
oceanic areas with a flux of biogenic carbon lower than
100 g(C)/m2/yr. (oligotrophic zones).
The highest observed values correspond
systematically to situations when the wind was blowing
from East to the Polarstern and was carrying air masses
from zones near to western Africa (area of Islands of
Cape Verda Archipelago). Our results are in agreement
with the measurements of MIHALOPOULOS et al. (7)
referring to the upwelling zone near to Somali. These
authors mentioned that the DMS productivity does not
follow exactly the global biological productivity.
Furthermore they showed that the central zone of maximum
productivity which is dominated by diatom algae was
rather poor in DMS relatively to the peripheric zones of
less global biological productivity but dominated by
dinoflagels which are plainly more rich in DMS. This
fact can explain that our DMS values are distributed in
a range greater than one order of magnitude.
Except DMS carbon sulfide (CS2) was detected in
our analyses only during the first days. The air analyzed
was strongly influenced from the western Europe and the
anthropogenic origin fom this compounds is probable.This
is confirmed from the study of the other compounds
(NO_x(9), aldehydes(10) for instance).

4. CONCLUSION

The technical analysis of DMS in low concentrations
that we have developed allowed permited to take a profil
of DMS crossing the Atlantic from Bremerhaven to Rio
Grande do Sul. However the detection limit of 20 ng
(DMS)/m3 was insufficient for the more poor zones. The
observed fluctuations seem to be in acceptable agreement
with our knowledge on the biological productivity of the
crossing areas and/or that of the proximal areas able to
influence the DMS concentrations in sampled air.

5. ACKNOWLEDGEMENTS

The authors would like to thank the professors D.H.
EHHALT and U. PLATT as well the "Alfred Wegener Institut"
(Pr Dr G. KRAUSE) for their invitation to participate in
this campaign of atmospheric measurements.

REFERENCES

1.E. S. SALTZMAN (ed)
"Biogenic sulfur in the environment". ACS Symposium
series Washington D.C. (1989)

2.R. J. CHARLSON ,J.E. LOVELOCK ,M.O. ANDREAE,S.G. WAREN
Oceanic phytoplankton, atmospheric sulphur, cloud albedo
and climate .
Nature, 326,655,1987

3.B.C.NGUYEN,S. BELVISO ,B. BONSANG ,G. LAMBERT
Dimethyl sulfide lifetime in the marine atmosphere
D Reidel publishing company, 1984

4.M.O.ANDREAE et al
J. Geoph. Res., 20, 12891, 1985.

5.B.C.NGUYEN, B. BONSANG ,A. GAUDRY
J.Geoph. Res., 88, 10903, 1983

6.R. STAUBES, H.W. GEORGII, S. BURGERMEISTER
Biogenic sulphur compounds in seawater and the marine
atmosphere. ibidem)

7.N. MIHALOPOULOS, B.C. NGUYEN and S. BELVISO S
Dimethyl sulfide in the Somali upweling area
International congress of geochemistry and
cosmochemistry, Paris, June 1988.

8.H. LIETH
"Versuch einer kartographischen Darstellung der
produktivitat der Pflanzendecke auf der Erde"
in :Geographisches Tachenbuch. Wiesbaden 1964-65

9.T. BRAUER, H.P. DORN, U.PLATT
Spectroscopic measurement of NO2,O3,SO2,IO and NO3 in
maritime air. ibidem

10.P. CARLIER, P.FRESNET, V.LESCOAT, S.PASHALIDIS and G.
MOUVIER
Daily profile study of the atmospheric formaldehyde
concentrations at the "Pointe de Penmarc'h" under pure
marine conditions. ibidem

BIOGENIC SULFUR COMPOUNDS IN SEAWATER AND THE MARINE ATMOSPHERE

R. Staubes, H.-W. Georgii, S. Bürgermeister
Inst. f. Meteorology and Geophysics
University of Frankfurt

SUMMARY

Dimethyl Sulfide (DMS), Carbonyl Sulfide (COS) and Carbon Disulfide (CS_2) were determined in surface water and the overlying atmosphere during a cruise between Bremerhaven, FRG (52°N) and Rio Grande do Sul, Brasil (32°S) on board the polar research vessel 'Polarstern'. The cruise track included both oligotrophic and coastal waters, the air samples included both remote marine air and air masses influenced by terrestrial or coastal inputs. The ranges of measured atmospheric concentrations for all samples were 5 - 100 pptv DMS, 420 - 670 pptv COS and <3 - 160 pptv CS_2. Both atmospheric DMS and CS_2 in the marine boundary layer were characterized by a significant spatial variability, in particular an increase of DMS and a rapid decrease of CS_2 with distance from land has been found. Due to its long tropospheric lifetime COS was distributed fairly homogeneous along the cruise track showing a mean of 550 pptv. With respect to atmospheric DMS, COS and CS_2 seawater was found to be supersaturated with dissolved Dimethyl Sulfide, Carbonyl Sulfide and Carbon Disulfide. The results show DMS to be the dominant sulfur gas in all the waters examined, with less amounts of COS and CS_2. Our measurements reveal daytime maxima in dissolved COS and CS_2 concentrations, indicating photochemical influenced sources of these sulfur compounds in seawater.

1. INTRODUCTION

Oceanic emissions of biogenic sulfur gases including Dimethyl Sulfide, Carbonyl Sulfide and Carbon Disulfide constitute a major flux of sulfur to the atmosphere, a source which is believed to be responsible for the background levels of SO_2, non sea salt sulfate and methansulfonic acid. These compounds are important factors in cloud chemistry and global climate as contributors to cloud condensation nuclei (1). The dominant sulfur compound released from the oceans and thus the most important precursor of non sea salt sulfate is considered to be DMS which is produced from metabolic processes in certain algae (2). Surface layers in different regions of the World's oceans are always observed to be supersaturated with DMS relative to the atmosphere, implying a net flux to the atmosphere (3,4,5). Andreae and Raemdonck (6) estimated a flux of 40±20 TgS(DMS)/yr into the atmosphere which accounts for 80 - 95% of the gaseous sulfur flux from the oceans and about one half of the estimated natural sulfur flux per year. Released to the atmosphere DMS is rapidly oxidized by several mechanisms, the most important one is the reaction with OH radicals (7). Over the open ocean very few measurements exist for sulfur gases other than DMS. Most of the existing measurements centered on coastal environments (8,9). CS_2 measurements in open ocean seawater (10,11) indicate that the ocean is supersaturated with respect to the atmosphere. In the atmosphere CS_2 is oxidized through reactions with OH radicals producing SO_2 and COS in equimolar quantities (12). COS is the most abundant sulfur gas in the atmosphere on a global scale, one important natural source for COS in the atmosphere is the emission from seawater (13). Due to its long tropospheric lifetime of 2 - 7 years (14), COS can diffuse into the stratosphere, where it is photolyzed and oxidized to sulfate. Therefore COS is thought to be a major contributor to the stratospheric sulfate aerosol layer which affects the Earth's radiation balance and climate (15).

In this paper we present simultaneous determinations of DMS, COS and CS_2 in surface seawater and the marine atmosphere performed during a cruise track of RV 'Polarstern' in the Atlantic Ocean from 52°N to 30°S in 1988.

2. EXPERIMENTAL

Air concentrations of DMS, COS and CS_2 were measured four times a day at 01, 07, 13 and 19h local time on the upper deck of the ship, 21m above the sea level, forward of potential contamination from stack emissions. The air sampled was enriched cryogenically at the temperature of liquid argon (-185°C) in U-shaped glass sample loops (30cm x 6mm I.D.) sealed with high-vacuum glass valves with PTFE Teflon sealings. Typical sample volumes were 10 L of air at a rate of 500 ml/min measured by mass flow controllers. Due to co-trapping of atmospheric oxidants or oxidant precursors which can cause significant DMS losses in the sample loops it was necessary to use prefilters. We employed glass tubes filled with 5% NA_2CO_3 on Anakrom C22 upstream of the cryotraps. The efficiency of these scrubber tubes have been tested during the field measurements by DMS standard additions to natural air.

Simultaneously to the determination of atmospheric DMS, COS and CS_2 the concentrations of these compounds in surface seawater have been examined. Seawater was obtained from the ship's continuous seawater pumping system. Samples were injected through a glass fiber filter to remove algal cells into a gas stripping column. A glass frit on the bottom of the column allows the sulfur-free purge gas nitrogen to bubble through the sampled water. Thus the volatile compounds are stripped from the seawater into the nitrogen stream, which passes through a dryer tube filled with $MgClO_4$ and subsequently through a cryotrap immersed in liquid argon to enrich the sulfur compounds. No detectable sulfur blanks had been shown when reanalyzing purged seawater. To avoid storage artifacts surface water samples were analyzed immediately after collection.

Analysis of both air and seawater samples were performed using a gaschromatographic system equipped with a flame photometric detector. The sensitivity of this system is 3 - 4 pptv in atmospheric samples and 0.5 - 1 ngS/l in seawater samples. Further details on the analytical system are given by Staubes et al. (16).

3. RESULTS

DMS, COS and CS_2 in surface seawater
The latitudinal distribution of DMS, COS and CS_2 in seawater is given in Fig. 1. DMS (Fig.1a) was found always to be present in surface water far in excess of concentrations expected at atmospheric equilibrium. The mean DMS concentration for all samples was 30 ngS(DMS)/l (standard deviation: 28,6 ngS(DMS)/l), which was evidently lower than the average of 100 ngS(DMS)/l for the world's oceans estimated by Andreae (17), confirming the oligotrophic character of the Atlantic. Generally the surface DMS values in the North Atlantic were higher than those measured in the South Atlantic. The distinct fluctuations of the observed DMS concentrations might be related to phytoplankton speciation and their stage of growth and senescence. Estimates of the latitudinal distribution of certain algae species and bacteria provided a rough classification of the observed water masses into 4 characteristic sections (H. Kuosa, pers. comm.), which separates regions with different phytoplankton population and

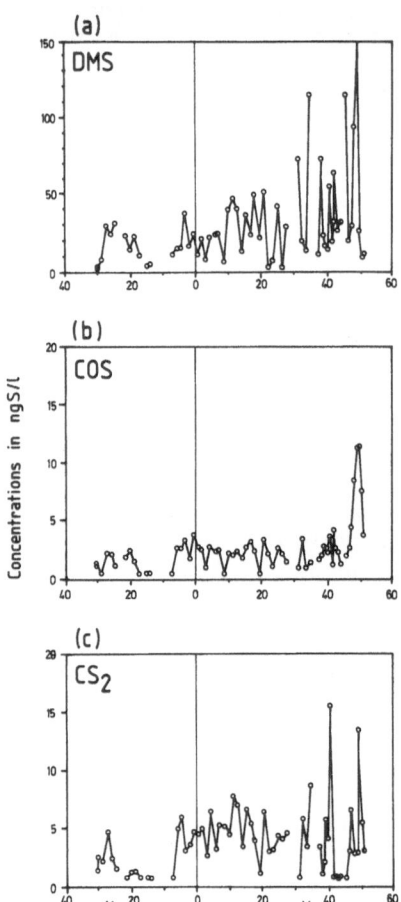

(a)

DMS

(b)

COS

Concentrations in ngS/l

(c)

CS$_2$

south   latitude   north

Fig.1: Latitudinal distribution of DMS, COS and CS$_2$ in surface seawater along the cruise of RV Polarstern

primary productivity rates. As shown in Table 1 highest DMS concentrations with maximum values of 150 ngS(DMS)/l (mean: 50 ngS(DMS)/l) occured north of 44°N in a continental shelf area containing the highest biomass amounts of the area traversed. Between 42°N and 2°S the averaged DMS concentration decreased to 30 ngS(DMS)/l with a range of 3 to 73 ngS(DMS)/l coinciding with a reduced biomass content and an alteration of phytoplankton composition in seawater. A further reduction of DMS was observed between 2°S and 30°S with an average of 16 ngS(DMS)/l. A comparison of the DMS values in seawater with the amount of different algae species showed the best agreement of DMS concentrations with the number of Dinoflaggelates. This type of algae was the most abundant of the phytoplankters observed during the cruise. A regular diurnal variation of DMS in seawater was not observed indicating that the DMS production is not influenced by photochemical pro- cesses. Also, no correlation between dissolved DMS and water temperature was found. The concentration of COS and CS$_2$ in surface water (Fig. 1b,1c) showed evidently lower values than the DMS concentrations. Both COS and CS$_2$ contribute about one tenth to the total sulfur amount observed in surface sea- water. COS concentrations fell in the range from less than 0,5 ngS(COS)/l (detection limit) to 11 ngS(COS)/l with a mean of 2,6 ngS(COS)/l. CS$_2$ concentra- tions were slightly enhanced compared to the observed COS values with an average of 3,7 ngS(CS$_2$)/l ranging from less than 0,8 ngS(CS$_2$)/l (detection limit) to 15 ngS(CS$_2$)/l. Similar to the latitudinal DMS distribution pattern, COS and CS$_2$ showed maximum values north of 40°N and minima in the South Atlantic (Tab.1). The peak concentrations of COS and CS$_2$ observed around 40°N exceeded the average values of both compounds by a factor of 4 - 5. COS and CS$_2$ values measured south of 40°N revealed only a rather weak amplitude with a range of 2 ngS(COS)/l and 3 ngS(CS$_2$)/l, respectively. The daily variations of both dissolved COS and CS$_2$ showed daytime maxima between 13h and 19h local time indicating a photochemical influenced source of both compounds in seawater. The daytime/nighttime ratios varied between 1,1 and 3. Strongest differences between daytime and nighttime concentrations occured in the southern hemisphere where enhanced insolation values had been observed. Previously, Ferek and Andreae (18) found a strong diurnal variation of COS in surface seawater and suggested a COS production from dissolved organic sulfur compounds by photochemical reactions.

Region	Characteristics	Mean concentrations in ngS/L		
		DMS	COS	CS_2
from 50.1°N, 1.6°W to 44.1°N, 13,2°W (9 samples)	high biomass content, ca.150 µgC/l, mainly diatoms; moderate amounts of dinoflagellates	53.8	6.05	4.49
from 44.1°N, 13,2°W to 17.3°N, 30,0°W (22 samples)	low biomass content, ca. 20 µgC/l, mainly dinoflagellates and chrysochromulina	33.9	2.2	3.72
from 17.3°N, 30,0°W, to 3.4°S, 30,0°W (12 samples)	low biomass content, ca. 20 µgC/l, mainly dinoflagellates and chryso- chromulina, moderate amounts of diatoms and tryochodesmium	25.2	2.3	4.96
from 3.4°S,30,0°W to 30.0°S,30,0°W (16 samples)	low biomass content, mainly dinoflagellates and chrysochromulina	16.4	1.6	2.25

Tab.1: Mean concentrations of DMS, COS and CS_2 summarized for regions of different biological activity along the cruise track of RV Polarstern in 1988

Atmospheric DMS, COS and CS_2
The results of DMS, COS and CS_2 measurements in the atmosphere are shown in Fig.2. The distribution of atmospheric DMS showed strong fluctuations over the whole sampling period with concentrations ranging from 6 to 130 ngS(DMS)/m^3 (Fig. 2a). The total mean DMS concentration was 45,5 ngS(DMS)/m^3 (34 pptv), slightly higher than the average concentration of 30 ngS(DMS)/m^3 reported by Bürgermeister et al. (19) for a South-North Atlantic transect in 1987, but much lower than the global average over the world's oceans of 150 ngS(DMS)/m^3 observed by Andreae et al. (20). The difference may be due to the fact that measurements by Andreae et al. were made over biologically much more productive waters resulting in a higher flux of DMS into the atmosphere.
Our data show no significant correlation between atmospheric and seawater DMS values. This is caused by the fact that the DMS concentration in marine air is not only determined by the rate of sea-to-air transfer, but also controlled evidently by the rate of removal processes. Only in a few limited regions of the cruise there was a tendency to simultaneously occuring enhanced atmospheric and dissolved DMS concentrations (10°N - 20°N, Fig.2a) and coinciding low DMS values in atmosphere and seawater (5°N - 5°S). The sharp increase in atmospheric values around 30°S to a maximum of 130 ngS(DMS)/m^3 which was not reflected in surface water values may be caused either by transport mechanisms out of regions of high DMS productivity or by an efficient gas stripping of the ocean's surface by strong winds (11 m/s, highest values during the whole

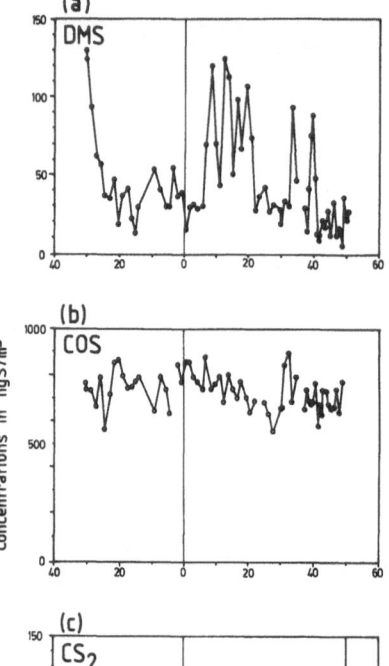

(a)
DMS

(b)
COS

Concentrations in ngS/m³

(c)
CS₂

south   north
latitude

Fig.2: Latitudinal distribution of atmospheric DMS, COS and CS$_2$ along the cruise track of RV Polarstern

cruise). Minimum DMS concentrations were found in continentally influenced air masses north of 40°N which might be the result of increased values of oxidants reacting faster with DMS than OH. A decrease of atmospheric DMS concentrations in polluted air had been previously shown by Andreae et al. (17) and Saltzman and Cooper (21). No systematic diurnal trend of DMS concentrations could be observed in this region. A regular diurnal variation of atmospheric DMS was found between 40°N and 30°S over the Atlantic with a minimum between 13h and 19h and a nighttime maximum between 1h and 7h local time (Fig.3). Minimum and maximum values differed by a factor of 1,7. The observed weak diurnal variation in the range between 20 and 30 ngS(DMS)/m³ could be explained by the oxidation of DMS by NO$_3$ radicals at night. The function of NO$_3$ radicals as an effective sink for atmospheric DMS during the night has been previously reported by Andreae et al. (17). In the southern hemisphere the nighttime concentration of DMS as well as the amplitude of the diurnal DMS pattern were lower compared to the northern hemisphere indicating an enhanced rate of DMS oxidation by NO$_3$ radicals over the South Atlantic.

The COS concentration measured in the boundary layer over the Atlantic (Fig.2b) showed an average of 733 ngS(COS)/m³ (550 pptv) with only minor variations (standard deviation: 77 ngS(COS)/m³). About 10% lower mean atmospheric COS values of 670 and 685 ngS(COS)/m³, respectively, were found by Johnson and Harrison (14) and Torres et al. (22) over the Pacific Ocean. Our measurements showed slightly increased COS values on the average in the southern hemisphere compared to the northern hemisphere (752 ngS(COS)/m³ and 725 ngS(COS)/m³), indicating that the observed air masses over the South Atlantic were highly influenced by continental inputs. This assumption is supported by measurements of CO and ethane on board the ship, showing also enhanced concentrations in the southern hemisphere (R.Bauer, R.Koppmann, pers. comm.). The latitudinal distribution of atmospheric CS$_2$ is shown in Fig. 2c. The CS$_2$ concentrations showed a clear difference between coastal and open ocean areas. North of 40°N concentrations fell in the range of 40 to 420 ngS(CS$_2$)/m³ (15 - 160 pptv), increasing with higher latitudes and further approach to the European continent. South of 40°N most measurements showed CS$_2$ values less than 8 ngS(CS$_2$)/m³ (detection limit) with only few exceptions predominantly occuring in the northern hemisphere.

Sea-to-air flux of DMS, COS and CS$_2$
We found surface seawater to be supersaturated with DMS, COS and CS$_2$ relative to

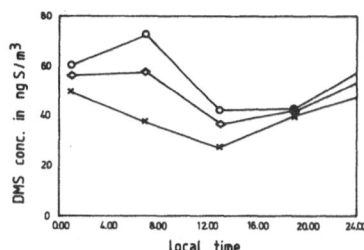

Fig.3: Mean diurnal variations of the atmospheric DMS concentration in the northern hemisphere ($40°N - 0°$, circles), in the southern hemisphere ($0° - 30°S$, crosses) and between $40°N$ and $30°S$ (rhombs)

the atmosphere throughout the study region. To calculate fluxes from the ocean to the atmosphere from our concentration data, we use the thin-layer model of Liss and Slater (23). They described the flux as

$$F = -v(c_g/H - c_l)$$

where v is the transfer velocity, c_g and c_l the concentrations in the atmosphere and in seawater, respectively, and H the dimensionless Henry's law constant. The windspeed-dependent transfer velocity for DMS, COS and CS_2 could be calculated from the actual wind data given from the ship's registration system and by using the diffusion coefficient given for radon at certain windspeeds by Smethie et al. (24). The Henry's law constants of DMS, COS and CS_2 used for the estimation of flux rates are $H_{DMS} = 0,074$ (25), $H_{COS} = 1,0 - 1,5$ (26) and $H_{CS2} = 1,2$ (11). The mean fluxes of DMS from the ocean to the atmosphere estimated in the North and South Atlantic are 93 ngS(DMS)/m²min and 74 ngS(DMS)/m²min, respectively. Andreae et al. averaged their data obtained from a variety of oceanic regions to a mean global DMS flux of 100 ngS(DMS)/m²min. The COS flux we observed in the Atlantic Ocean averaged to about 5 ngS(COS)/m²min. CS_2 fluxes were calculated to be in the range of 5,8 - 11,4 ngS(CS_2)/m²min. These values exceed CS_2 flux estimates made by Kim and Andreae (11) in the western regions of the North Atlantic by a factor of 5. From our data it is evident that the oceans represent a substantial sulfur source in particular for atmospheric DMS, but also for atmospheric COS and CS_2.

ACKNOWLEDGEMENT

The research in this paper has been supported by the Deutsche Forschungs-gemeinschaft. We thank the crew of the research vessel Polarstern for their collaboration. We gratefully acknowledge the help of H.Kuosa for providing his data on algae distribution in seawater.

REFERENCES

(1)   Charlson, R.J.; Lovelock, J.E.; Andreae, M.O.; Warren, S.G.   (1987): Oceanic phytoplankton, atmospheric sulphur, cloud albedo and climate; Nature, Vol.326, 655 - 661

(2)   Dacey, J.W.H.; Wakeham, S.G. (1986): Oceanic dimethylsulfide: Production during zooplankton grazing on phytoplankton; Science, 233, 1314 - 1316

(3)   Barnard, W.R.; Andreae, M.O.; Watkins, W.E.; Bingemer, H.; Georgii, H.W. (1982): The flux of dimethylsulfide from the oceans to the atmosphere; J.Geophys.Res., Vol.87, 8787 - 8793

(4)   Andreae, M.O.; Ferek, R.J.; Bermond, F.; Byrd, K.P.; Engstrom, R.T.; Hardin, S.; Houmere, P.D.; LeMarrec, F.; Raemdonck, H. (1985): Dimethyl Sulfide in the Marine Atmosphere; J.Geophys.Res., Vol.90, 12891 - 12900

(5)   Bates, T.S.; Cline, J.D.; Gammon, R.H., Kelly-Hansen, S.R. (1987): Regional and Seasonal Variations in the Flux of Oceanic Dimethylsulfide to the Atmosphere; J.Geophys.Res., Vol.92, 2930 - 2938

(6)   Andreae, M.O. and Raemdonck, H. (1983): Dimethylsulfide in the surface ocean and the marine atmosphere, A global view; Science, 233, 1314 - 1316

(7)   Graedel, T.E. (1982): Reduced sulfur emission from the open oceans; Geophys.

Res. Lett., Vol.6, 1203 - 1206

(8)   Jorgensen, B.B.; Okholm-Hansen, B. (1985): Emissions of biogenic sulfur gases from a Danish Estuary; Atm. Environ., 19, 1737 - 1749

(9)   Turner, S.M.; Liss, P.S. (1985): Measurements of various sulphur gases in a coastal environment; J.Atmos.Chem. 2, 223 - 232

(10) Lovelock, J.E. (1974): CS_2 and the natural sulphur cycle; Nature 248, 625 - 626

(11) Kim, K.H.; Andreae, M.O. (1987): Carbon Disulfide in Seawater and the Marine Atmosphere over the North Atlantic; J.Geophys.Res., Vol.92, 14733 - 14738

(12) Jones, B.M.R.; Cox, R.A.; Penkett, S.A. (1983): Atmospheric chemistry of carbon disulphide; J.Atmos.Chem. 1, 65 - 86

(13) Toon,O.B.; Kasting, J.F.; Turco, R.P.; Liu, M.S. (1987): The Sulfur Cycle in the Marine Atmosphere; J.Geophys.Res., Vol.92 943 - 963

(14) Johnson, J.E.; Harrison, H. (1986): Carbonyl sulfide concentrations in the surface waters and above the Pacific Ocean; J.Geophys.Res, Vol.91, 7883 - 7888

(15) Crutzen; P.J. (1976): The possible importance of CSO for the sulfate layer of the stratosphere; Geophys.Res.Lett., Vol.3, 73 - 78

(16) Staubes, R.; Georgii, H.W.; Ockelmann, G. (1989): Emissions of COS, DMS and CS_2 from various soils in Germany; Tellus, Vol.41B, 305 - 313

(17) Andreae, M.O. (1985): The emission of sulfur to the remote atmosphere, in: The Biogeochemical Cycling of Sulfur and Nitrogen in the Remote Atmosphere; edited by J.N. Galloway, R.J. Charlson, M.O. Andreae, H. Rohde, D. Reidel Publ. Comp., Dordrecht

(18) Ferek, R.J. and Andreae, M.O. (1984): Photochemical Production of Carbonyl Sulfide in Marine Surface Waters, Nature 307, 148 - 150

(19) Bürgermeister, S.; Zimmermann, R.L.; Georgii, H.W.; Kirst, G.O., Janssen, M.; Ernst, W. (1989): On the biogenic origin of Dimethylsulfide: Relation between chlorophyll, ATP; organismic DMSP; Phytoplankton species and DMS distribution in Atlantic surface water and atmosphere, submitted to J.Geophys.Res.

(20) Andreae, M.O.; Barnard, W.R.; Ammons, J.M. (1983): The Biological Production of Dimethylsulfide in the Ocean and its Role in the Global Atmospheric Sulfur Budget; Ecol.Bull. 35, Stockholm, 167 - 177

(21) Saltzman, E.S., Cooper, D.J. (1988): Shipboard measurements of Atmospheric Dimethylsulfide and Hydrogen Sulfide in the Caribean and Gulf of Mexico; J.Atmos.Chem. 7, 191 - 209

(22) Torres, A.L.; Maroulis, P.J.; Goldberg, A.B.; Bandy, A.R. (1980): Atmospheric OCS measurements on Project Gametag; J.Geophys.Res., Vol.85, 7357 - 7360

(23) Liss, P.S.; Slater, P.G. (1974): Flux of gases across the air-sea interface; Nature 247, 181 - 184

(24) Smethie, W.M.; Takahashi, T.; Chipman, D.W. (1985): Gas exchange and CO_2 flux in the tropical Atlantic Ocean determined from ^{222}Rn and pCO_2 Measurements; J.Geophys.Res., Vol.90, No.C4, 7005 - 7022

(25) Bingemer, H. (1984): Dimethylsulfid in Ozean und mariner Atmosphäre - Experimentelle Untersuchung einer natürlichen Schwefelquelle für die Atmosphäre, Ph.D.thesis, J.W.Goethe Univ., Frankfurt

NET TOTAL RADIATION AT THE SEA SURFACE

OF THE ATLANTIC OCEAN BETWEEN 30°N AND 30°S

H.D. Behr

Deutscher Wetterdienst, Meteorologisches Observatorium Hamburg
Frahmredder 95, D-2000 Hamburg 65

ABSTRACT

The three-weeks cruise ANT VII/1 of RV Polarstern in September/
October 1988 through the Atlantic Ocean along 30°W between 30°N
and 30°S offered a good opportunity to obtain a meridional distribution of solar and longwave radiation components out of
this area. The knowledge of the spatial and temporal distribution of these quantities at the sea surface is important for
numerous meteorological, oceanographical, and physico-chemical
investigations. This work is the continuation of the measurements made during the cruise ANT V/5 of RV Polarstern in March/
April 1987. As the cruise extended across the both subtropics
and tropics of the Atlantic Ocean characteristic daily courses
and meridional distributions of the radiation components and
atmospheric turbidity could be worked out. Special attention
will be given to the ultraviolet component of global radiation.

1. INTRODUCTION

The following quantities were recorded by commercially produced
instruments:

Global solar radiation (G), reflected global radiation (R),
direct solar radiation (I), sunshine duration (S), longwave atmospheric radiation (A), and longwave sea-surface radiation (E).

The UV-B global radiation of sun and sky between 280 nm and
315 nm had been recorded by a non-commercially produced instrument,
described by Dehne (1). An improved version had been used on board.
The instrument takes account for one of the important effects of
UV-B radiation; its calibrated output is proportional to the efficiency of global radiation for minimal erythema (sunburning effect)
on human skin.

During the cruise over the Atlantic Ocean along 30°W between
30°N and 30°S a complete data set was obtained every minute. As this
cruise was a continuation of ANT V/5 details about the sites of the
instruments and their maintenance and data processing as well may be
taken from Behr (2).

2. TOTAL RADIATION BALANCE

Figure 1 shows the meridional distribution of the daily sums of
the net total radiation Q on a horizontal surface and its components.
Q is given by:

$$Q = (G-R) - (E-A) \tag{1}$$

A remarkable increase in A and decrease in G reveals the position of
the Intertropical Convergence Zone (ITCZ) at 6°N. G and A are inverse

related at other latitudes as well according to cloudiness of those areas. Information about this quantity may be taken from the meridional distribution of direct solar radiation I, which is strictly correlated with the relative sunshine duration S/S_0, where S_0 is the astronomical sunshine duration. During the cruise I and therefore S/S_0 where larger in the southern hemisphere (SH) than in the northern hemisphere (NH). This causes higher values of G in SH and lower ones in NH, A runs vice versa. The sea-surface radiation E shows a steady course through the latitudes, the maximum lays at 6°N, this corresponds to a sea-surface temperature of 29°C. The reflected global radiation R does not show any variation with latitude, except in the vicinity of the ITCZ where it decreases according to G. As G is larger in SH than in NH as already mentioned the net total radiation Q is larger in SH as well. Q shows a similar trend with latitude as discussed by Kondratyev (3): a maximum at both sides of the ITCZ and declines to higher latitudes.

3. UV-B GLOBAL RADIATION

UV-B global radiation is the radiation of sun and sky on a horizontal surface in the wavelength range from 280 nm to 315 nm. Though it takes only a small fraction (< 0.4 %) of the energy of the total global radiation spectrum, its contribution to may physico-chemical processes is considerable. Furthermore UV-B global radiation shows great variations with sun elevation angle and ozone content of the atmosphere as well.

Figure 2 shows the meridional distribution of the daily sums of global solar radiation G and UV-B global radiation respectively. The daily sums of UV-B reach 1.45 Wh/m^2 during the cruise, this value had been confirmed by theoretical calculations made by Dehne (4). Furthermore the quotient UV-B/G is plotted, it gives an information about the fraction of UV-B on global solar radiation. During the cruise UV-B was larger in SH than in NH according to lower values of ozone content in SH than in NH, which is not typical for this area. As cloudiness reduces G to a higher degree than UV-B, the quotient UV-B/G increases in the vicinity of the ITCZ, but remains small at higher latitudes.

The influence of the sun elevation angle γ on UV-B is shown in Figure 3. The daily sums had been computed in five different ways: UV(0) = daily sums of the time between sunrise and sunset, UV(γ) = daily sums for the time intervals with γ exceeding 50°, 60°, 70°, or 80°, respectively. A first look at the curves reveals that the daily sums decrease according to neglecting low sun elevation angles. Though the cruise took place during the time of the equinox this decrease was not symmetric in both hemispheres. The different meridional distributions of the ozone content in both hemispheres may cause this behaviour. Other influences on UV-B as different cloudiness and atmospheric turbidity along 30°W could not analysed with the help of the data of this cruise till now.

In the following the contribution of the irradiance of UV-B to the daily sums will be discussed under another point of view. In Figure 4 the maximum persistence in minutes for given classes of UV-B irradiances is plotted versus the corresponding class values; the width of each class is 1 µW/cm^2. The irradiances between 5 µW/cm^2 and 14 µW/cm^2 occur in both hemispheres seldomly, but beyond 14 µW/cm^2 the longer persistences of higher UV-B classes in SH exceed

– 694 –

those in NH on account of the variability of the cloudiness, depending mainly on the synoptic situation. These longer persistences ac-
cumulates to higher daily sums in SH than in NH.

Figure 5 shows the daily courses of global solar radiation G
and UV-B global radiation respectively at a nearly cloudless day
(04.10.1988; 25°S 30°W). In addition the quotient UV-B/G is plotted.
G and UV-B are given in relative units related to their maxima at
noon. The time used is true local time (TLT). This figure reveals
three properties of UV-B: (1) UV-B vanishes at low sun elevation
angles earlier than G does, (2) with increasing sun elevation angle
UV-B increases more rapidly than G, (3) clouds influence G more than
UV-B. Figure 6 shows the dependency of UV-B/G on sun elevation angle.
As data of sunshine duration S are available, seven different cases
had been taken into consideration: (1) all cases, (2) S=100 % should
last for 10 min continuously, (3) same, but for 15 min, (4) same,
but for 20 min, (5) same, but for 25 min, (6) same, but for 30 min,
(7) overcast sky should last for 10 min at least. All curves coincide for lower sun elevation angles, but spread open for higher
angles with a triparation as follows: overcast sky – all cases –
sunny cases. This may be interpreted that at cloudy conditions UV-B
will be intensified by scattering at clouds.

4. ATMOSPHERIC TURBIDITY

The reduction of the transmittance of the clean and dry atmosphere by extinction by atmospheric trace substances is called turbidity of the atmosphere. Aerosol particles are the solid and fluid
trace substances, while the gaseous components are the atmospheric
trace gases especially water vapor. The extraterrestrial solar radiation I_0 is reduced by the turbidity of the atmosphere to direct
solar radiation $I(\gamma)$, reaching the surface; $I(\gamma)$ depends on the sun
elevation angle γ. Kasten (5) proposed a formula to compute the Linke
turbidity factor T_L by:

$$I(\gamma) = I_0 \cdot \exp\left(-T_L/(0.9 + 9.4 \cdot \sin\gamma)\right) \qquad (2)$$

$I(\gamma)$ had been recorded during the cruise, while I_0 and γ can be computed according to the ship position and time (TLT). In order to
eliminate all measurements of $I(\gamma)$ reduced by clouds the following
restriction had been used: S=100 % should last at least for 10 min.
T_L was computed according to eq. (2) under the restriction already
mentioned.

Figure 7 shows the meridional distribution of the Linke turbidity factor obtained from the data of both cruises, solid lines:
ANT VII/1 (Sep.–Oct. 1988), dashed lines: ANT V/5. (March–April 1987).
Shading depicts the area of the ITCZ. Both lines reveal the same
peculiarities: (1) a maximum in the vicinity of the ITCZ at 6°N which
was caused by the high water vapor content essentially. The high
values of T_L between 10°N and 15°N were caused in both cases by sand-
storms originating from the southwest of the Sahara. On account of
minor sources of dust and sand in SH T_L did not reach similar high
values at the corresponding latitudes, (2) a minimum near 10°S
($T_L \approx 3.0$) gives a hint at clean air masses reaching this position
from areas south of Africa. No dust or sand influenced the turbidity
of the air in this area. During ANT VII/1 south of 25°S T_L increased
again according to the synoptic situation: a low transported dusty

air masses from southern Brasil eastwards, details may be taken from Behr (6).

In order to illustrate the daily course of the Linke turbidity factor the data of 04.10.1988 (25°S, 30°W) are plotted in Figure 8. The relative sunshine duration S/S_0 was 90.5 %. T_L oscillates between 3.0 near sunrise/sunset and 5.5 near noon caused by swelling of aerosol during the day. This phenomenon was firstly mentioned by Linke (7).

5. SUMMARY

The data obtained during an intense three-weeks-cruise accross different climatic zones of the Atlantic Ocean reveal regional differences in the radiation energy budget of that area.

There is an obvious increase of all radiation quantities from higher latitudes up to the vicinity of the ITCZ. The high cloud cover connected with ITCZ itself reduces all short wave radiation and causes an increase of longwave atmospheric radiation.

The quotient UV-B/G increases with sun elevation angle. The increase is less steep for cloudless conditions.

The Linke turbidity factor varies between 7.0 near the ITCZ and $\approx$ 3.0 at 25°S.

REFERENCES

(1) Dehne, K. (1977). Design and performance of a new instrument for measuring UV-B global radiation. In: papers presented at the WMO-Technical Conference on Instruments and Methods of Observation (TECIMO), Hamburg, 27-30 July 1977, WMO-No. 480, 173-178, Genève.

(2) Behr, H.D. (1989). Radiation balance at the sea surface of the Atlantic Ocean between 40°S and 40°N. Submitted to Journal of Geophysical Research.

(3) Kondratyev, K.Ya. (1969). Radiation in the atmosphere. Academic press, New York, London, pp 912.

(4) Dehne, K. (1983). Neue Berechnungen zur Klimatologie der erythemwirksamen UV-Globalstrahlung. Annalen der Meteorologie (N.F.), 20, 115-116.

(5) Kasten, F. (1980). A simple parameterization of the pyrheliometric formula for determining the Linke turbidity factor. Meteorol. Rdsch., 33, 124-127.

(6) Behr, H.D., H. Köhler, R. Schmidt, D. Winterkemper (1989). Weather conditions during ANT VII/1. Berichte zur Polarforschung, 62, 9-15.

(7) Linke, F. (1924). Ergebnisse von Messungen der Sonnenstrahlung Über dem Atlantischen Ozean und Argentinien. Meteor. Z, 41, 42-46.

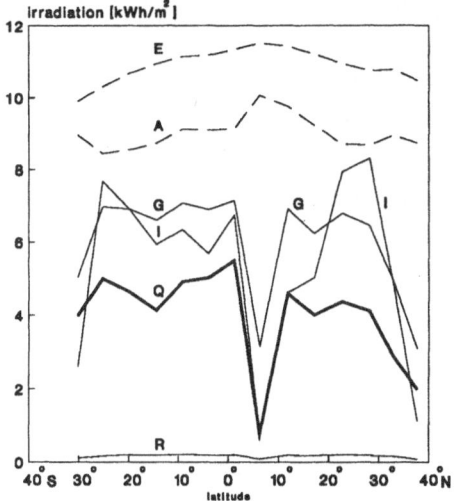

Fig. 1 Meridional distribution of the daily sums
G, R, I, A, E, and Q

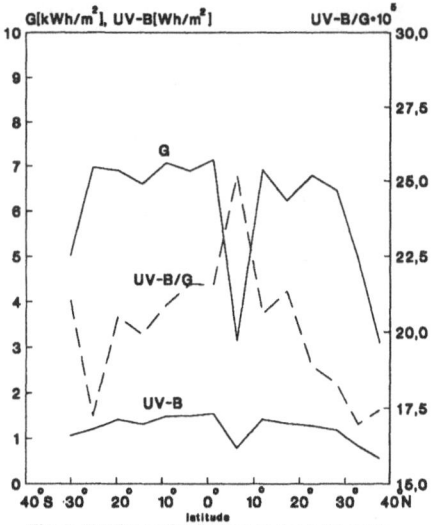

Fig. 2 Meridional distribution of the daily sums
G, UV-B, and UV-B/G

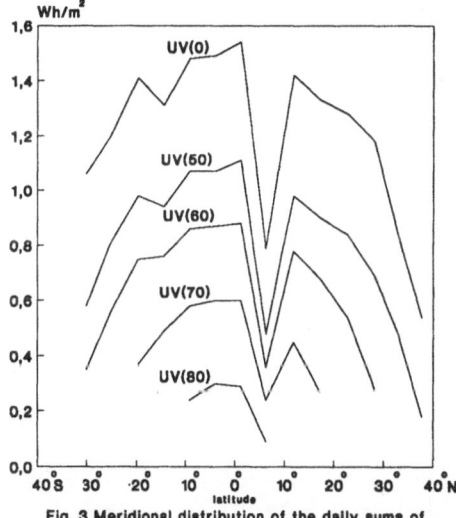

Fig. 3 Meridional distribution of the daily sums of
UV-B summed up above different sun elevations angles

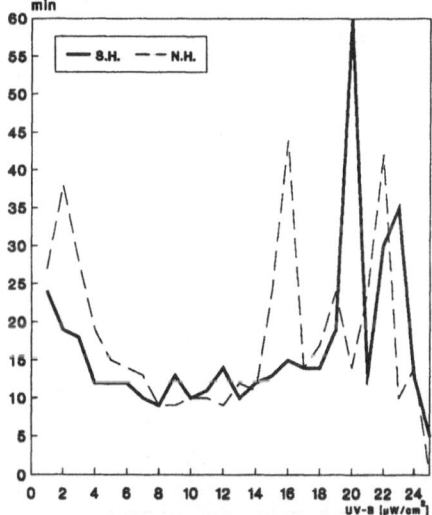

Fig. 4 Maximum duration of UV-B

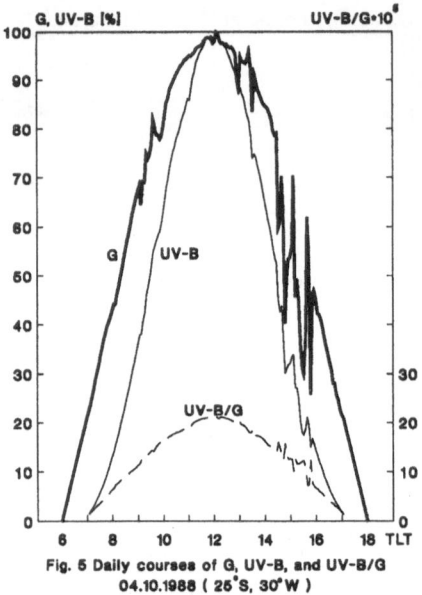

Fig. 5 Daily courses of G, UV-B, and UV-B/G
04.10.1988 (25°S, 30°W)

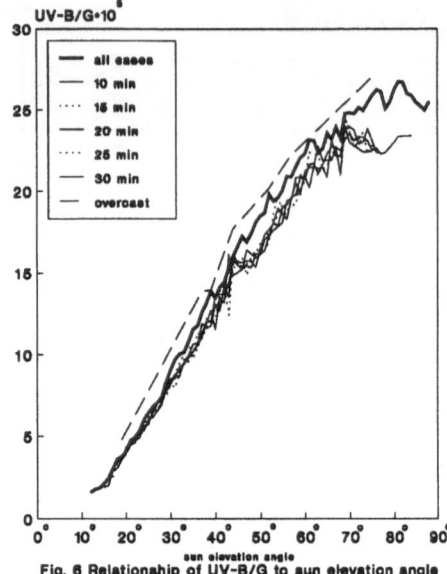

Fig. 6 Relationship of UV-B/G to sun elevation angle
for different sky conditions

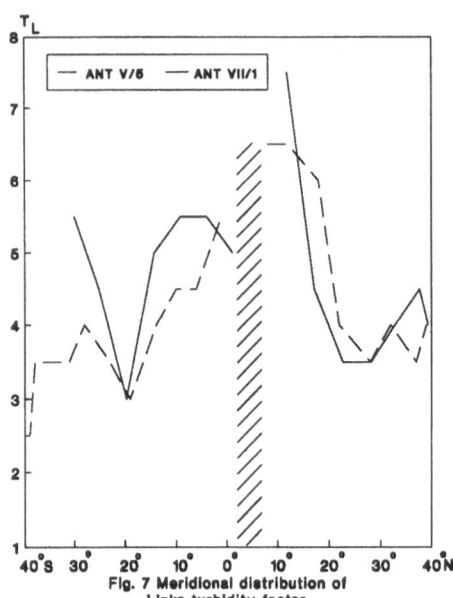

Fig. 7 Meridional distribution of
Linke turbidity factor

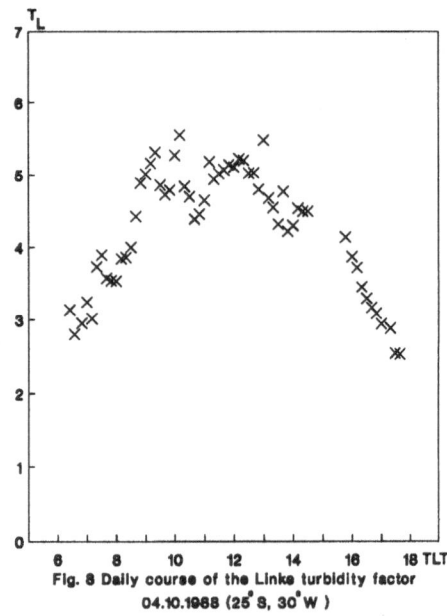

Fig. 8 Daily course of the Linke turbidity factor
04.10.1988 (25°S, 30°W)

SURFACE NO- AND NO₂-MIXING RATIOS

MEASURED BETWEEN 30°N AND 30°S IN THE ATLANTIC REGION

D. BRÜNING and F. ROHRER

Institut für Atmosphärische Chemie
Kernforschungsanlage Jülich GmbH
Postfach 1913, D-5170 Jülich, F.R.G.

SUMMARY

Surface NO- and NO_2-mixing ratios have been measured aboard the research vessel 'POLARSTERN' during the mission ANT VII/1 from September 24. to October 5.,1988. Measurements were taken along 30°W over the Atlantic region covering latitudes between 30°N and 30°S. The average mixing ratios were about 10 pptv NO and 30 pptv NO_2. We measured elevated mixing ratios at 13°N (probably due to air masses originating from the west-african continent) and in the region of the ITCZ between 8°N and 5°N. Because of probable contamination by the ship, the measured mixing ratios could be only used as upper limits.

1. INTRODUCTION

Nitrogen oxide plays a controlling role in the chain of catalytic reactions which produce tropospheric ozone. This role is most easily illustrated by the oxidation of carbon monoxide. In the presence of sufficient NO it takes the path,

CO	+ OH	-> H	+ CO_2		R1
H	+ O_2 + M	-> HO_2	+ M		R2
HO_2	+ NO	-> OH	+ NO_2		R3
NO_2	+ hv	-> NO	+ O		R4
O	+ O_2 + M	-> O_3	+ M		R5

CO	+ 2 O_2	-> CO_2	+ O_3

which results in a net production of O_3. In the absence of NO, the oxidation of CO takes a different path:

CO	+ OH	-> H	+ CO_2		R1
H	+ O_2 + M	-> HO_2	+ M		R2
HO_2	+ O_3	-> OH	+ 2 O_2		R6

CO	+ O_3	-> CO_2	+ O_2

In this case, ozone is destroyed. At the "critical" volume mixing ratio of NO of 10 pptv, production (by R1 to R5) and destruction (by R1, R2, R6) are equal. Above this value, the oxidation of CO leads to a photochemical production of ozone.

To estimate the global tropospheric production of O_3, the global background concentration of NO is required. We have therefore recently started to measure the NO volume mixing ratio on a global scale (1). Here, we report the results of the NO-, NO_2- and NO_y'-measurements aboard the research vessel 'POLARSTERN' during the mission ANT VII/1 from September 24. to October 5.,1988. The measurements were taken at 30°W in the atlantic region covering latitudes between 30°N and 30°S.

A number of other components (CO, CH_4, CH_2O, O_3,non methane hydrocarbons, PAN, $J(NO_2)$, $J(O^1D)$, H_2O_2, HNO_3, NO_3 and sulphur-compounds) were measured by other groups during the campagne.

2. EXPERIMENTAL

NO , NO_2 and NO_y' were measured by a chemiluminescence detector described earlier (2). We define here NO_y' as the sum of NO, NO_2, NO_3, $2 \times N_2O_5$, HNO_2, HNO_4, PAN and nitrate-aerosol . NO_y' does not include HNO_3 in contrast to the usual definition of NO_y which includes HNO_3 (3).

The instrument used is shown schematically in Fig. 1. It consisted of two detectors operated in parallel. NO was detected by the chemiluminescence reaction between NO in the sample air and ozone added from an ozone generator.

| NO | $+ O_3$ | $-> NO_2^*$ | $+ O_2$ | | R7 |
| NO_2* | | $-> NO_2$ | $+ hv$ | | R8 |

$$NO \quad + O_3 \qquad -> NO_2^* \quad + O_2 \qquad\qquad R7$$
$$NO_2^* \qquad\qquad -> NO_2 \quad + hv \qquad\qquad R8$$

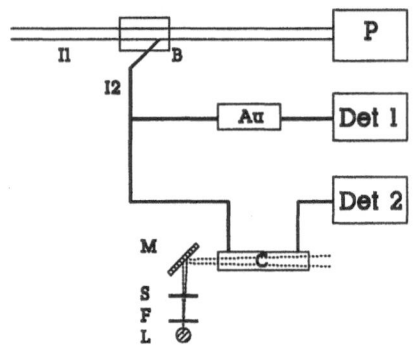

Fig.1 Schematic drawing of the chemiluminescence detector system

I1   teflon inlet tube,d$=14$mm,l$=$ 6m,100 l min^{-1}
I2   teflon inlet tube,d$=$ 4mm,l$=$15m,6 l min^{-1}
B   salt separator
P   pump , 100 l min^{-1}
Au NO_y'-converter
C   photolytic converter
M   UV-mirror
S   shutter
F   filter
L   300 W Xe-arc-lamp
Det1 NO_y'-detector
Det2 NO/NO_2-detector

The fluorescence was detected by a photomultiplier tube (EMI 9658 R, cooled to -40°C) through a red filter (RG 610). To increase the sensitivity, the reaction chamber consisted of gold-plated aluminium. The dark current of the PMT and the interfering chemiluminescence reactions were measured in a zero-mode by passing the sample-air/ozone mixture through a teflon-coated relaxation volume. The NO-signal was calculated from the difference between the measure-mode (without passage through the relaxation volume) and the average of the two neighboring zero modes. The signal from the PMT's was fed to photon counters, which were controlled by a computer. For ambient measurements, the counters were read out each 30 seconds with a mode switching time of 180 seconds. The flow and the size of the relaxation volume were chosen, so that the residence time in the relaxation volume was 2 times the lifetime of NO against ozone. This experimental setup discriminated against longlived interfering chemiluminescences and instrumental artefacts.

The reaction mixture was pumped through the reaction vessel at 15 mbar pressure and a volume flow rate of 100 l min^{-1}. The flow was controlled by measuring the pressure drop across a capillary in the inlet tube. The signal, which was obtained with the help of differential pressure gauges, was used to drive a

Teflon needle valve which regulated the flow to a constant rate. The pressure drop was calibrated against the flow controlers of the gas calibration system each day. With this equipment, the flow rate was constant to better than 2 % .

The inlet system consisted of three parts (see fig.1) : a 6 m teflon tube with an inner diameter of 14 mm. The air sample was pumped through this tube with a flow of 100 l min^{-1}. The second part was a teflon salt separator for separation of sea salt and reduction of pressure. The third part was a 15 m Teflon tube with an inner diameter of 4 mm.The sample air was pumped through this tube with a flow of 6 l min^{-1} at a pressure of 800 mbar. The residence time of the sample air in the inlet system has been 2 s. The NO mixing ratio has been corrected for the reaction of ozone with NO in the inlet system. The correction has been 2% for 30 ppb ozone. The inlet point was mounted 2 m apart from the side of the ship at the port side of the upper deck, 20 m in front of the funnel and 15 m above sea level.

NO and NO$_2$ were measured by detector 2. The air sample was passing a photolytic converter, which was build out of a glass tube (volume 580 ml) with quartz-windows and a 300 W Xe-arc-lamp with integrated back-reflector. The light of the Xe-lamp was passing a filter, cutting wavelength below 350 nm, a shutter and a UV-mirror which was only reflecting between 325 - 475 nm. If the shutter was open, light of the wavelength region 350 - 475 nm could enter the converter and photolytically dissociate NO$_2$ (reaction R4). With the shutter closed, detector 2 was measuring NO.

With the shutter open, the signal was proportional to E*NO$_2$.(E is the efficiency of the photolytic converter.) This efficiency was determined each day by measuring a known amount of NO$_2$ in synthetic air. Normally, the efficiency was 0.5. The efficiency of the converter for measuring sample air was depending on the ozone, NO and NO$_2$ concentrations of the sample. This was corrected by model calculations of the chemical system in the converter. The ozone mixing ratio during the campagne was variing between 20 - 30 ppb. For 20 - 30 ppb ozone, 12 ppt NO, 25 ppt NO$_2$ and a residence time of 20 s, the efficiency was 0.49 - 0.48. The model calculation was also correcting for the reaction of NO with ozone inside the converter.

NO$_y$' was measured after conversion to NO in a gold converter (3). The converter consisted of a heated (400°C) gold tube (l = 0.3 m, d = 6 mm). A low amount of CO (0.1 %) was added to the sample air for a complete reduction of the various NO$_y$'-components. CO was giving a fake-NO-signal of 50 ppt when directed through the converter. The size of this fake signal was determined each day at the time of calibration and subtracted from the NO$_y$' signal. NO$_y$' was not including HNO$_3$, because HNO$_3$ was sticking onto the walls of the 15 m Teflon inlet tube. This fact was tested experimentally. After 12 hours, no signal of HNO$_3$ could be detected for an addition of 2 ppb HNO$_3$ at the beginning of the inlet tube. Because of the fake signal of CO and the missing HNO$_3$ response, the NO$_y$'-results will be not discussed.

The sensitivity of the NO/NO$_y$ detectors was determined each day with a calibrated NO/N$_2$-mixture (AIRCO) and synthetic air. The calibration mixture was prepared by using two thermostatted mass flow controlers (Advanced Semiconductor Materials), which had been calibrated by soap bubble flow meters. The NO/NO$_2$-detector showed a sensitivity of 2 counts s^{-1} ppt^{-1} (NO) and 1 count s^{-1} ppt^{-1} (NO$_2$), the NO$_y$-detector 2 counts s^{-1} ppt^{-1} (NO$_y$). The calibration-factors showed day to day variations of less than 5% . The relative calibration of both detectors was changing less than 2% per day. The mixing ratio of the NO calibration mixture was 7500 ppt. With that mixing ratio, both detectors showed standard deviations of 50 ppt for a 10 s integration interval.

The detection limit was calculated from the frequency distribution of the measure-mode signals of synthetic air. These distributions were gaussian and showed standard deviations of 6 ppt NO and 15 ppt NO$_y$' for a 10 second integration time. The signal-strength was 600 counts s^{-1} for NO and 1600 counts s^{-1} for NO$_y$' originating from the dark current of the photomultipliers and additional chemiluminescences caused by the ozone-generator. With 2 counts s^{-1} ppt^{-1} , the standard deviations are roughly a factor 2 higher than those calculated by photon-counting statistics. For 180 s integration time, one can calculate the detection limit (2 times the standard deviation of the background signal) of 3 ppt NO, 6 ppt NO$_2$ and 7 ppt NO$_y$'. For easier handling, the data of the 30 second intervalls were averaged for 10 minutes. For graphical presentation, those data were then averaged over half a degree of latitude (generally 200 minutes). Estimated from the detection limit, those half-degree-means should have a standard deviation of 1.5 ppt NO, 3 ppt NO$_2$ and 3 ppt NO$_y$'.

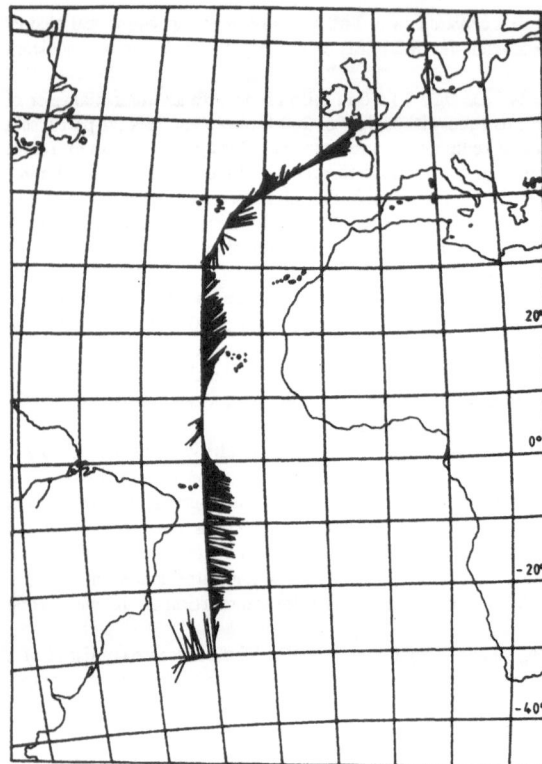

Auxiliary meteorological information namely wind direction, wind speed, position, temperature and pressure was supplied by the navigation system of the ship. Fig.2 showes the wind direction and speed along the track of the ship. The ITCZ can be clearly recognized between 5°N and 8°N.

Fig.2 Wind direction and track of the ship. The lines are pointing from the direction of the wind to the position of the ship. The size of the lines is proportional to the wind speed. Generally, the wind speed has been around 10 m s^{-1}.

3.RESULTS AND DISCUSSION

Fig. 3 to 5 are showing the results for the NO-, NO$_2$- and NO$_y$'-mixing ratios. The results with signals which could be clearly attached to ships passing nearby were taken out. The whole data set can be divided into three parts. The first part covers the region between 30°N and 15°N. Here, NO showed mixing ratios around 10 ± 4 ppt and NO$_2$ around 25 ± 8 ppt. The wind direction was north-east, from port and aft of the ship (north-east trade wind). The deviations are roughly a factor 2 higher than those calculated out of the detection limit and are reflecting the variability of the signal.

The region between 13°N and 0°S was different. The mixing ratios and deviations are a factor 2 higher than before. Those high signals were accompanied by the arrival of wind carried sand and grasshoppers at the ship at 15°N. The air masses passing the ship must have had their origin at the african continent and could not have been transported for a long period of time. The same conclusion could be drawn out of wind charts of that region. At this time, the ship was also crossing the ITCZ between 8°N and 5°N (see Fig.2). The wind speed has been occasionally very low, and the wind direction was changing more than 360°.

At 0°, the conditions were changing again. The mixing ratios at 0° to 30°S were around 5 ± 10 ppt NO and 15 ± 10 ppt NO$_2$. The wind direction was south-east (south-east trade wind). So those air masses must have been transported a long way over the South-Atlantic Ocean. At this region, the wind was coming from the front of the ship. The variability of the data was very high, although the signal was very near to the detection limit. The deviations were a factor 4-6 higher than calculated for the detection limit.

The stars in Fig.3 point out the highest light intensity each day. There are only a few locations, where one can detect a marked diurnal variation of NO and NO$_2$(at 12°N, 6°N, 14°S and at 27°S). Especially at 12°N and 6°N where the NO$_2$-mixing ratios were highest, the maximum of NO at noon was correlated with a minimum of NO$_2$. But the NO-mixing ratio was not going down to zero at night, although the error bars normally included zero at night.

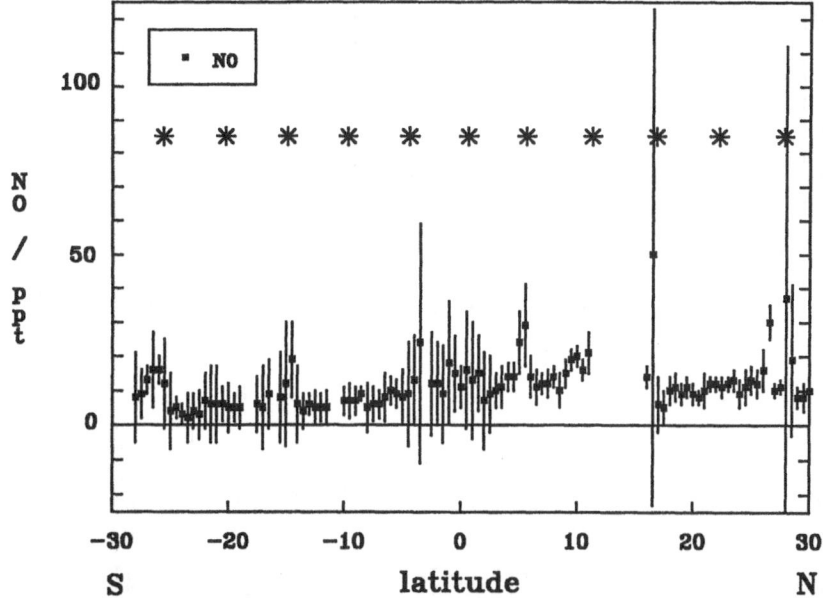

Fig. 3   NO-mixing ratios measured at 30°W, 30°N - 30°S , 24.9.1988 - 5.10.1988, error bars represent one
time the standard deviation, stars indicate the maximum of light intensity each day

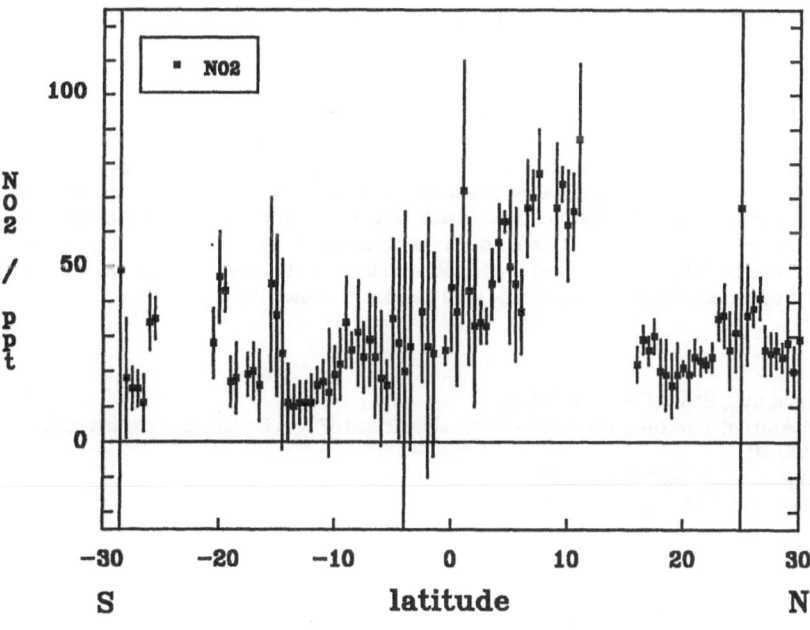

Fig. 4   NO$_2$-mixing ratios measured at 30°W, 30°N - 30°S , 24.9.1988 - 5.10.1988, error bars represent one
time the standard deviation

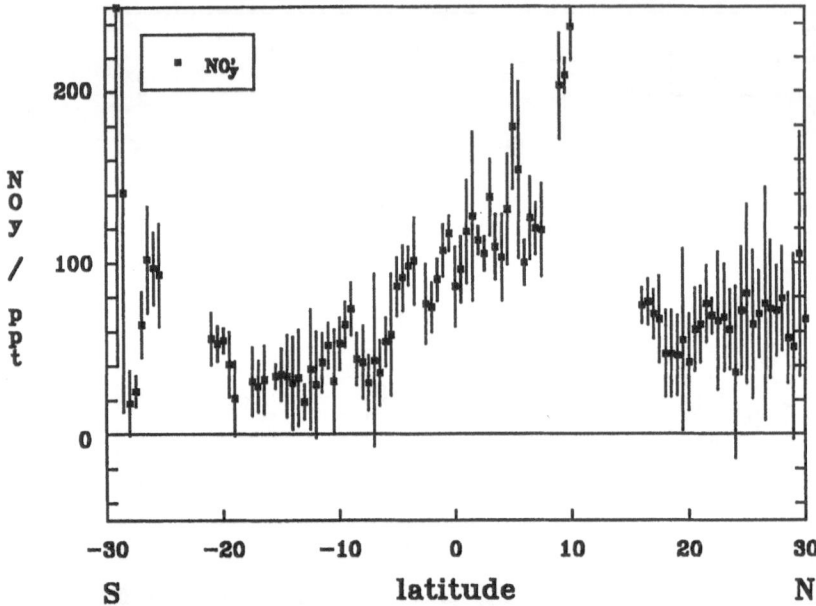

Fig. 5 NO_y'-mixing ratios measured at 30°W, 30°N - 30°S , 24.9.1988 - 5.10.1988
NO_y' does not include HNO_3 (see text), error bars represent one time the standard deviation

There can be several possible explanations for this observation:

- There could be signal from interfering chemiluminescences. This possibility could not be ruled out completely, but the experiment was designed especially to discriminate against interferences(2).
- The ocean could be a weak source of NO, but the source strenght required to maintain the observed NO_x concentrations at 15m above sea level exceeds by far that accepted for the ocean.
- The inlet line could be contaminated, evaporating NO_x. This possibility could be ruled out, because the inlet system was completely changed during the campaign showing no effect on the signal.
- The most probable explanation was contamination from the ship itself. The mixing ratios were high at the beginning of the campaign when the wind came from port and aft of the ship (with higher probability of contamination by the ships smoke-stack) and they were low at the end of the cruise when the winds came from the front of the ship with lower probability for a pickup of contamination. Nevertheless even then 2ppt of NO were observed at night indicating the continued presence of contamination probably due to a number of smaller ventholes of the ship. This considerations were supported by the observation of very strange wind trajectories aboard the ship.

REFERENCES

(1) Drummond,J.W., Ehhalt,D.H., and Volz,A.
    Measurements of nitric oxide between 0 - 12 km altitude and 67°N - 60°S latitude obtained during STRATOZ III.
    J. Geophys. Res. 93, 15831-15849, 1988.
(2) Drummond,J.W., Ehhalt,D.H., and Volz,A.
    An optimized chemiluminescence detector for tropospheric NO measurements.
    J. Atmos. Chem. 2, 287-306, 1985.
(3) Fahey,D.W., Eubank,C.S., Hübler,G., and Fehsenfeld,F.C.
    Evaluation of a catalytic reduction technique for the measurement of total reactive odd-nitrogen NO_y in the atmosphere.
    J. Atmos. Chem. 3, 435-468, 1985.

MEASUREMENTS OF PEROXYACETYL NITRATE IN THE MARINE ATMOSPHERE

K.P. Müller, J. Rudolph, and K. Wohlfart

Institut für Atmosphärische Chemie
Kernforschungsanlage Jülich GmbH
D-5170 Jülich, Postfach 1913, F.R.G.

Summary

Peroxyacetylnitrate was measured during the Polar-stern cruise ANT VII/1 from Bremerhaven to Rio Grande/Brazil (52°N - 31°S). A preconcentration system in combination with ECD gas chromatography allowed a lower detection limit of 0.4 ppt. Mixing ratios of PAN ranged over 3 orders of magnitude. Several systematic fluctuations were observed between the highest ratio of 2000 ppt in the English Channel and less than 0.4 ppt south of the Azore Islands (38° N). The observed wind directions indicate that long-range transport from continental Europe was the major factor. In tropical latitudes, the mixing ratio of PAN dropped below the detection limit of 0.4 ppt. At 30°S off the eastern coast of South America, PAN mixing ratios of 10 to 100 ppt were detected in continentally influenced air masses.

I. Introduction

Peroxyacetyl nitrate (PAN) is a photochemical oxidant and a component of urban photochemical smog with concentrations ranging up to 2 - 10 ppb in urban regions. In nonurban continental Europe mixing ratios are typically lower than 2 ppb, in general only a fraction of a ppb.

Recent investigations have shown that PAN also is present in the upper troposphere (Rudolph et al., 1987) and the marine troposphere (Singh and Salas, 1986, Goudena et al., 1980, Lovelock, 1974). Due to its strongly temperature dependent thermal decomposition, PAN has a variable life time ranging from about 30 minutes at ground level to several months at high altitudes. Consequently, PAN may be transported over large distances through the cold upper troposphere into the "clean" atmosphere (Crutzen, 1979). In this paper we investigate the distribution of PAN over the remote Atlantic.

II. Experimental

        The PAN measurements were made during the cruise ANT VII/1 of the German research vessel RV Polarstern in September/October 1988. Fig. 1 shows the cruise track from Bremerhaven to Rio Grande. The wind vectors in fig. 1 indicate the wind directions and the wind velocities.

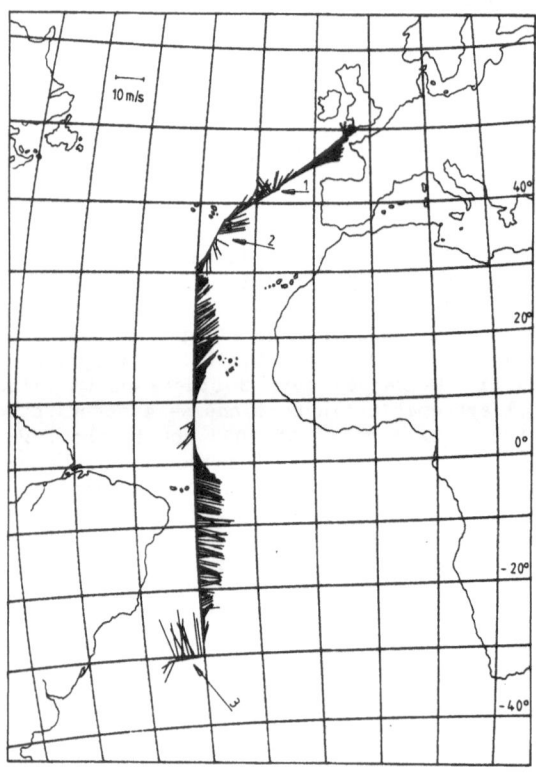

Fig. 1:
Cruise track of the RV POLARSTERN during leg ANT VII/1 from Bremerhaven to Rio Grande do Sul (Brazil) from Sept. 15 to Oct. 9, 1988. Significant changes in mixing ratios of PAN are indicated by the arrows 1-3.

        Using gas chromatography combined with an ECD-detector, a lower detection limit of 5 - 10 ppt can be obtained. This detection limit has been improved by the implementation of a cryogenic preconcentration step. At temperatures lower than 195 K, PAN is quantitatively adsorbed onto a glass tube and desorbed at temperatures between 278 - 288 K. The details of the cryogenic preconcentration method in connection with ECD gas chromatography are described by Müller and Rudolph (1989). With cryogenic preconcentration the lower limit of detection was 0.4 ppt. The reproducibility of the method was about 10 %.

        The PAN gaschromatograph was installed in a container on the port side of the navigation deck. The PTFE inlet line extended about 3 meters over the structure of the ship. The inlet line was continuously flushed by a pump with a flow rate of about 50 ml/sec.

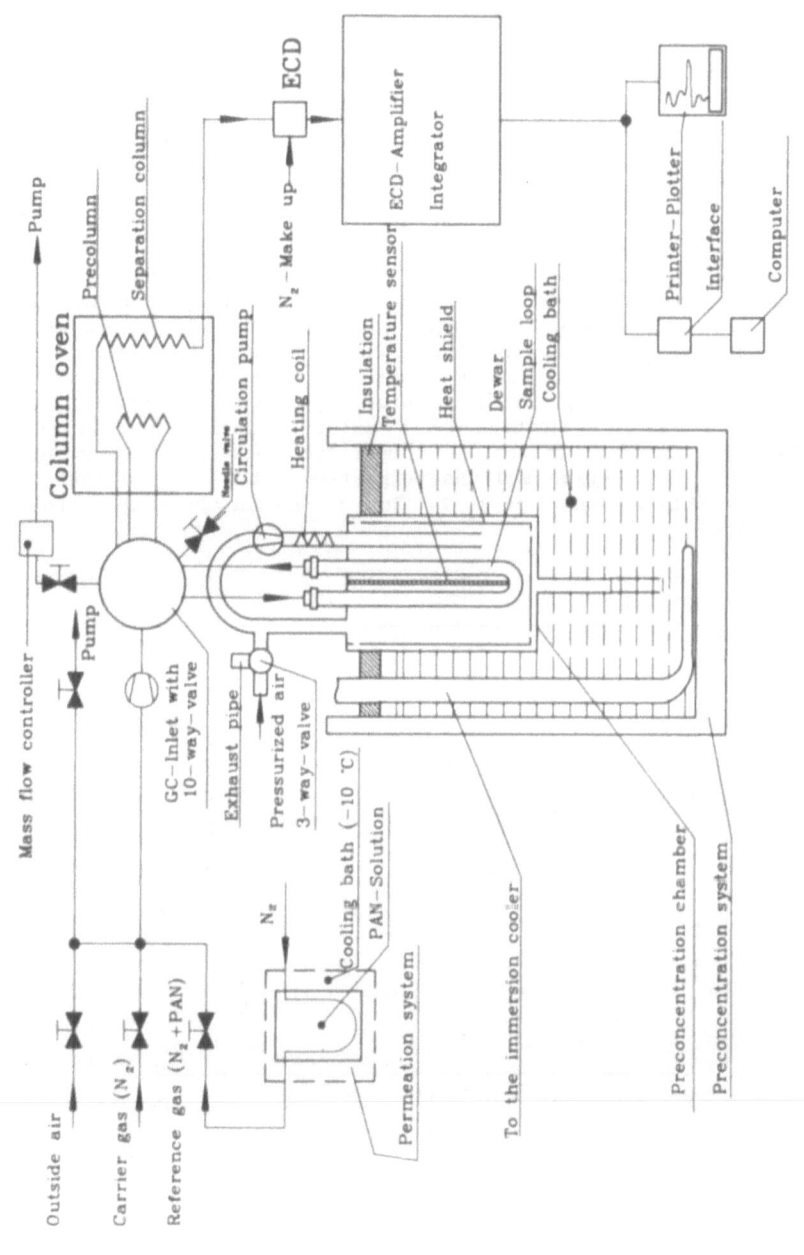

Fig. 2: Schematic drawing of the experimental setup including calibration and cryogenic preconcentration system.

A schematic drawing of the instrument is shown in Figure 2. In the GC-inlet system the air is passed through a glass sample loop. A massflow meter controlled the flow rate of the sampled air. During preconcentration the sample loop is immersed in a cooling liquid at 193 K. At this temperature PAN is quantitatively adsorbed onto the glass surface of the sample loop. For desorption of PAN the temperature of the preconcentration loop is raised to about 280 K. Then the sample is injected by means of a 10-way valve.

After injection the air sample is separated on the GC-precolumn (300 * 2 mm, glass, packed with 5 % PEG 400 on Chromosorb WHP 80/100 mesh). In this precolumn, PAN is separated from low-volatile substances and transfered to the separation column (same as precolumn, but 1000 mm long). The low-volatile constituents on the precolumn are backflushed.

For calibration a constant gas flow of 30 cm^3/min of N_2 is passed through a permeation device. The permeation device consists of a PTFE tube immersed in a dilute solution of PAN in n-heptane. In order to obtain a low and stable permeation rate and to reduce the rate of PAN decomposition in the solution, the permeation device is permanently kept in a cryostat at 263 K. The PAN mixing ratio in the calibration gas is analyzed by passing it through a dilute aqueous sodiumhydroxide solution. In alkaline solution, PAN decays quantitatively to NO_2^-, which is subsequently measured by spectrophotometry after derivatisation. Additionally the liquid injection of a dilute solution of PAN in n-heptane is used for calibration. A comparison between calibration by injection of PAN in solution and the permeation method showed that both methods agree within an error of 10 %. This is within the total error of the measurements.

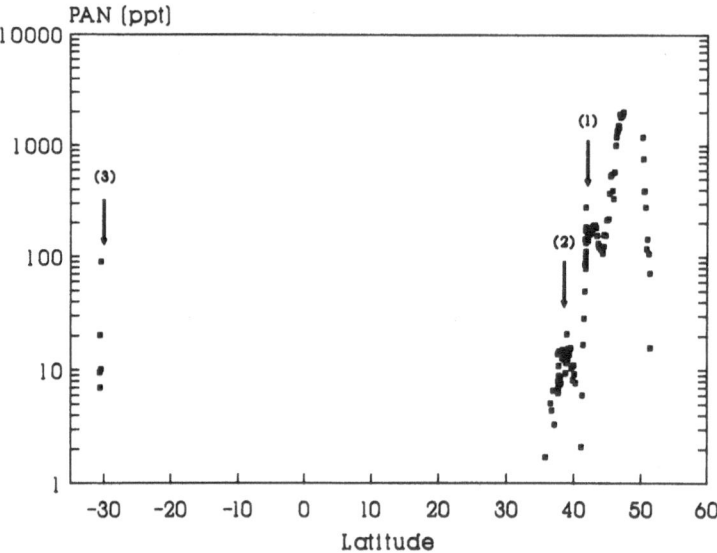

Fig. 3: PAN mixing ratios in the marine troposhere, latitudinal profile (52°N - 31°S) measured during ANT VII/1. Arrows 1-3 correspond to changes in wind direction, (see also Fig. 1).

III. Results and Discussion

Fig. 3 shows the latitudinal profile of PAN. The mixing ratios of PAN varied from about 2000 ppt in the English Channel to less than 0.4 ppt (detection limit) south of the Azore Islands. The variations of the mixing ratio during the cruise are indicated by the arrows 1-3 in figure 1. In tropical latitudes at 30°W between 30°N - 30°S the mixing ratio was below the detection limit. At 31°S/34°W PAN was detected again in continentally influenced air with concentrations between 10 - 100 ppt. It is possible to correlate the strong variations in mixing ratio with the changes in the main wind directions coresponding to different air masses (see Fig.1 and Figure 3).

In the latitude range 30°N-30°S the air temperatures at sealevel were in general around 25°C, sometimes even higher. At these temperatures the thermal lifetime of PAN is only about half an hour. Since the concentrations of the precursors of PAN (NMHC and NO_x) were quite low in this latitude range (Koppmann et al., 1989; Brüning and Rohrer, 1989) we expect a negligible in situ production of PAN. In addition horizontal transport is not fast enough to maintain substantial PAN mixing ratios against a decay time of half an hour over the remote ocean. Consequently the very low PAN mixing ratios we found over the open Atlantic in tropical and subtropical latitudes are not surprising.

We can compare our results with those from other authors. Goudena et al. (1980) measured the latitudinal distribution of PAN during seven ship cruises between Amsterdam or Rotterdam (52°N, Netherlands) and Paramaribo (6°N, Surinam) or Rio de Janeiro (23°S, Brasil). Their experimental technique allowed only measurements of PAN mixing ratios above 50 ppt. Thus most of their measurements only gave an upper limit. Between 45°N and 53°N they found PAN mixing ratios ranging from less than 50 ppt to several hundred ppt, in the English Channel sometimes even more than one ppb. Towards lower latitudes the PAN mixing ratios decreased rapidly, between 33°N and 45°N only occasionally the PAN mixing ratios were above the detection limit and never exceeded 250 ppt. South of 33°N Goudena et al. never found detectable levels of PAN.

Lovelock and Penkett (1974) reported considerably higher PAN concentrations for the open Atlantic between 53°N and 18°N. During a ship cruise from Hamburg to Santo Domingo in October 1973 they measured PAN mixing ratios with daily averages ranging from 0.1 to 2.5 ppb. But it should be noted, that Singh and Viezee (1988) report the formation of PAN artefacts by sampling procedures similar to those used by Lovelock and Penkett.

There are also measurements of the latitudinal distribution of PAN over the Pacific. Singh et al. (1986) measured PAN concentrations around 50 ppt between 50°N and 30°N, decreasing with decreasing latitudes to about 5 ppt at the equator. Their southern hemispheric average was around 5 ppt with no systematic latitude dependence. Most of these measurements were made not far from the American west-coast and thus continental influence on the sampled air masses cannot be excluded. However, no data about wind direction and wind velocity or origin of the air masses are reported.

Our PAN measurements at tropical and subtropical latitudes over the open Atlantic are much lower than the previous measurements and indicate that at sealevel and lower latitudes no significant background levels of PAN exist. This is compatible with the short atmospheric PAN lifetime and the low concentrations of PAN precursors.

Acknowledgements

We thank the Alfred-Wegener-Institut für Polar- und Meeresforschung for the opportunity to participate in the ANT VII/1 cruise of the research vessel POLARSTERN. This work was supported financially by the Bundesminister für Forschung und Technologie of the Federal Republic of Germany under grant No.07441025.

References

Baulch, D. L., C. A. Cox, R. F. Hampson Jr., J. A. Kerr, J. Troe, and R.T. Watson; Evaluated kinetic and photochemical data for atmospheric chemistry: Supplement II; J. Phys. chem. Ref. Data, 4, 1259-1273, 1984.

Brüning D., and F. Rohrer; Surface NO- and NO_2-mixing ratios measured between 30°N and 30°S in the atlantic region, this volume.

Crutzen, P. J.; The role of NO and NO_2 in the chemistry of the troposphere and stratosphere; Ann. Rev. Earth. Planet. Sci., 7, 443-472, 1979.

Garland, J. A., and S. A. Penkett; Absorption of peroxyacetyl nitrate and ozone by natural surfaces; Atmospheric Environment, 10, 1127-1131, 1976.

Goudena, E. J. G., R. Guicherit, and K.D. den Hout,; Peroxyacetylnitrate, ozone and some halocarbon measurements over the atlantic; Instituut voor milieuhygiene en gezondheidstechnick TNO, Publicatie no. 725, Postbus 214, 2600 AE Delft, Netherlands, 1980.

Koppmann, R., F. J. Johnen, Ch. Plass and J. Rudolph; The latitudinal distribution of light non-methane hydrocarbons over the mid-Atlantic between 40°N and 30°S; this volume.

Lovelock, J. E., and S.A. Penkett; PAN over the Atlantic and the smell of clean linen; Nature, 249, 434, 1974.

Müller, K. P., and J. Rudolph; An automated technique for the measurement of peroxyacetyl nitrate in ambient air at ppb and ppt levels; Int. Journ. of Environment. Anal. Chem.; in press, 1989.

Rudolph, J., B. Vierkorn-Rudolph, and F. X. Meixner; large-scale distribution of peroxyacetylnitrate, results from the STRATOS III flights; Journal of Geophysical Research, Vol. 92, No. D6, 6653-6661, 1987.

Schurath, U., U. Kortmann , and S. Glavas; properties, formation and detection of peroxyacetyl nitrate; in: Proceedings of the 3rd European Symposium on Physico-Chemical Behavior of Atmospheric Pollutants, Varese, Italy, pp. 27-37, 1984.

Singh, H. B., and L. J. Salas; The global distribution of peroxyacetyl nitrate; Nature, 321, 588-591, 1986.

Singh, H.B., and W. Viezee; Enhancement of PAN abundance in the Pacific marine air upon contact with selected surfaces; Atmospheric Environment, Vol. 22, 419-422, 1988.

CONCENTRATION AND SOURCES OF TRACE ELEMENTS OVER THE ATLANTIC OCEAN

R.LOSNO, P.CARLIER, L.GOMES and G.BERGAMETTI
Laboratoire de Physico-Chimie de l'Atmosphère
Université Paris 7, 2 Place Jussieu, 75251 PARIS CEDEX 05, France.

Summary

Measurements of solid inorganic aerosol were performed over the Atlantic Ocean during the cruise of the Polarstern (ANT VII/1) in October and November 1988, between Bremerhaven (F.R.G.) and Rio Grande do Sul (Brazil).
A large decrease in concentrations of pollutants (S and Zn) is observed when the ship has lived the coast of North Europe, leading to low background levels in the tropical and equatorial areas. On the other hand, crustal material as aluminium, silicium or iron exhibits strong variations in concentration during the cruise especially when an air flow is originated from North African desert regions. Elements as phosphorus and manganese show changes in concentrations which can be interpreted as the addition of both crustal and anthropogenic contributions. A general decrease between the beginning and the end of the cruise appears, similar to those observed in the case of sulfur or zinc. However, a significant crustal input, strongly correlated with the hight levels of aluminium and silicium is observed in the North Atlantic Tropical area.

1. INTRODUCTION

The aerosol studies may be very useful to understand the transport processes through the atmosphere, and then the geochemistry of the continent-ocean interactions. Some elements are emitted by the human activities and pollution, others by the eolian abrasion of the earth crust. Remote areas are very useful in these cases, because the sources are far from the sampling location and then the air masses are enough homogeneous in space and time in front of the duration of the sampling and the availability of meteorological data. Moreover, it is now well known that, in open ocean, the atmosphere is the major source for several element such as Al, Si and Mn which can act as limitants in the marine life cycle.

We are going to present here some preliminary results about the variations of Al, Si, P, S, Mn and Zn concentrations, followed over the North Sea and the Atlantic Ocean during a four weeks field experiment on the German polar ship "Polarsten" ANT VII/1.

2. SAMPLING AND ANALYSIS

Aerosol was sampled by pumping air through a Nuclepore polycarbonate filter of 0.4 μm pore size. The filters were placed at the top of a 3-meter long pole fixed in front of the iceberg guard plateform of the ship, which is 25 m over the sea level and about 10 m over the higher deck. To prevent contamination by the smoke emitted by the ship at about 40 m back, the pumps where stopped when local wind conditions were not propicious (w.s.<1.5m/s and wind direction no more than 80° of the front). The lock of pumps operated few seconds after bad wind speed and immediatly when the wind direction becomes bad. By opposite, a ten-to twenty seconds delay was performed before the restart of pumps after better wind conditions.

The figure 1 shows the route of the ship and the sampling period from Bremerhaven (FRG) to Rio Grande Do Sul (Brazil). The filters were changes each day and half or at each weather change. Sampled volume varies between 5 and 35 cubic meters of filtered air.

Analyses were performed with a X-Ray Fluorescence spectrometer, in thin layer condition (1) directly on the filter. Analytical conditions are printed on table 1.

Element	Crystal	Primary beam	Type of counter	Detection limit	Uncertainty
Al	PET 002	chrome	CFG	75 ng	7 %
Si	PET 002	chrome	CFG	112 ng	6 %
P	GE 111	chrome	CFG	20 ng	4 %
S	GE 111	chrome	CFG	100 ng	6 %
Mn	LiF 100	cuivre	CFG	14 ng	4 %
Zn	LiF 220	or	CS	30 ng	4 %

Table 1 : Analytical conditions by XRF.
CFG : Gaseous flux counter. CS : Scintillator.

3. RESULTS AND DISCUSSIONS:

The data are plotted on figure 2, with trace metal concentrations as a function of latitude.

Particulate anthropogenic elements exhibit high atmospheric concentrations in the North Sea and in the Channel. Concentrations are high as 50 ng.m^{-3} for Zn and 3500 ng.m^{-3} for S. These values are in agreement with those observed by Dedeurwarder et al. (2) and Cambray et al. (3). These high concentrations primarily result from land based sources in North European industrial countries. A strong decrease is then observed to the Middle North Atlantic. Concentrations are in the range of 1 to 4 ng.m^{-3} for Zn and 200 to 800 ng.m^{-3} for S with no significant changes between 40°N and 30°S. These concentrations are similar to those observed by Bonsang (4) for SO_4^{2-} in remote marine areas and by Buat-Ménard (5) for Zn over the Atlantic Ocean. One order of magnitude separates data obtained in polluted marine areas and those from remote marine area. This confirms the stress produced by human activities on some regional marine regions.

Southern hemisphere concentrations of Si and Al show very lower values (13 to 48 ng.m^{-3} for Al) than observed in the North Sea, probably resulting on the closest proximity of continents. But the main feature for these elements is the major peak in concentration observed between 20°N-10°N. They are higher than 2200 and 5500 ng.m^{-3} for Al and Si respectively. As shown by air mass trajectories (figure 3), an African desertic origin can be attributed to this dust, and underlines the major role played by this source on oceanic biogeochemical cycle and sedimentation.

P and Mn result in the atmosphere from both natural (crustal) and anthropogenic emissions. Their concentration profile clearly suggest a major anthropogenic contribution in the North Sea with a similar behaviour as observed for Zn or S and a major crustal origin between 20°N and 10°N while these elements exhibit the same peak as Al and Si.

4. CONCLUSIONS:

These preliminary results on particulate matter collected over North Sea and Atlantic Ocean confirm the spatial variability of all the elements whatever their source. For each element, at last one order of magnitude exists in concentration levels, underlining the necessity of long duration regional studies.

5. ACKOWLEDGEMENTS:

We thank the "Alfred Wengener Institut" (FRG, Bremerhaven), to have invited us on bord of the ship "Polarstern".

6. REFERENCES:

(1) LOSNO R., BERGAMETTI G. and MOUVIER G. (1987), Determination of optimal conditions for atmospheric aerosol analysis by X-Ray fluorescence. Environmental Technology Letters, Vol 8, 77-86.
(2) DEDEURWAERDER H.L., BAYENS W.F. and DEHAIRS F.A. (1985), Estimates of dry and wet deposition of several trace metals in the Southern Bight of the North Sea. Proc. of the 5th Int. Conf. on Heavy Metals in the Environment, Athens, Sept. 1985.
(3) CAMBRAY R.S., JEFFERIES D.F. and TOPPING G. (1979), The atmospheric input of trace elements to the North Sea. Marine Sci. Communications, Vol 5, 175-194.
(4) BONSANG B. (1980), Cycle atmosphérique du soufre d'origine marine. Thèse d'état, Université de Picardie, 231 p.
(5) BUAT-MENARD P. (1979), Influence de la retombée atmosphérique sur la chimie des métaux en trace dans la matière en suspension de l'Atlantique Nord. Thèse d'état, Université Paris 7, 233 p.

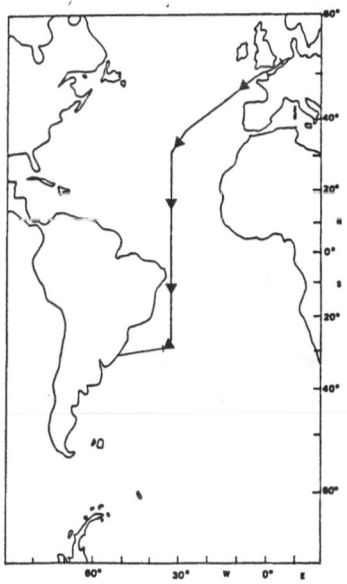

FIGURE 1:route of the ship

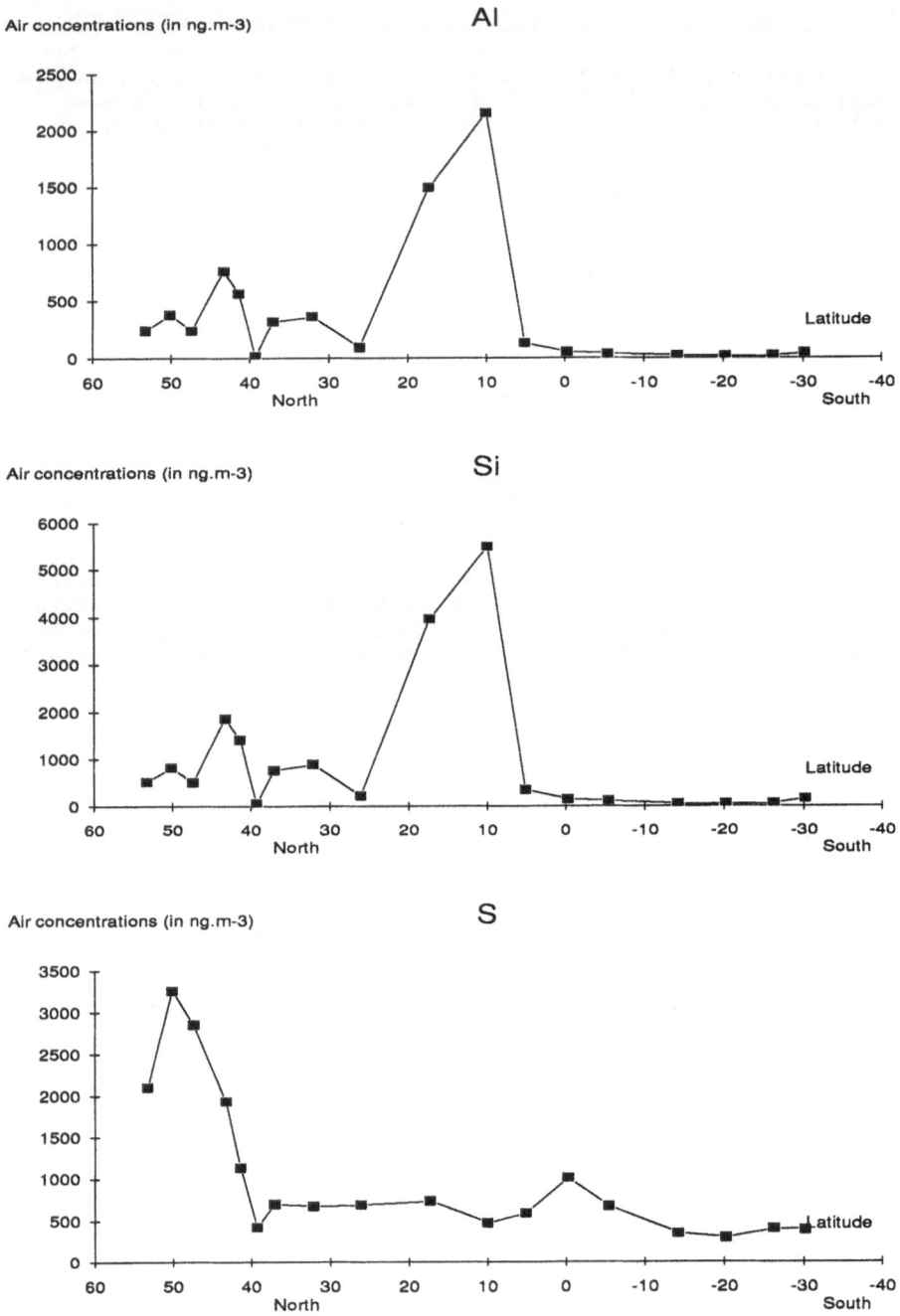

FIGURE 2: Concentration profiles of trace elements.

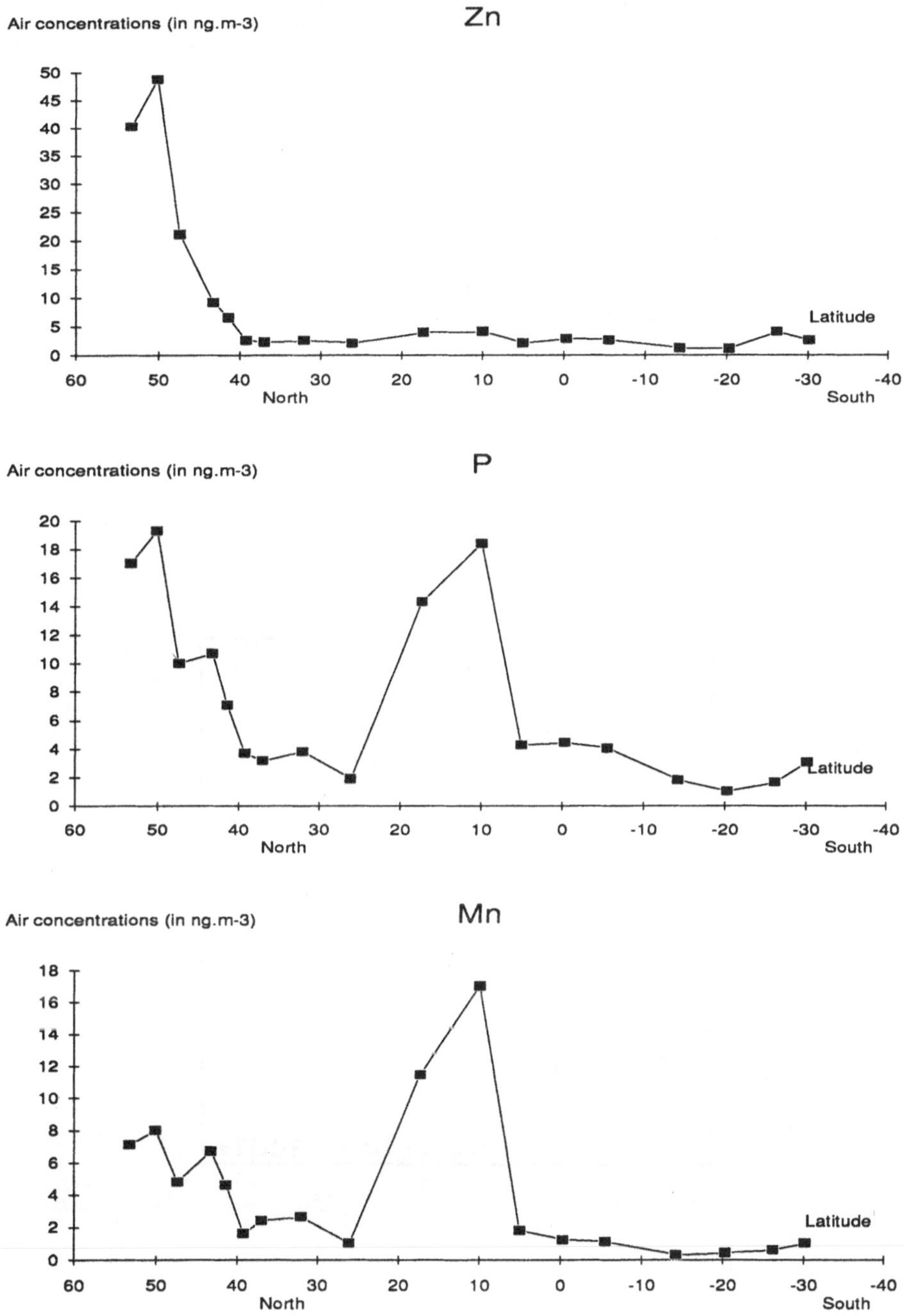

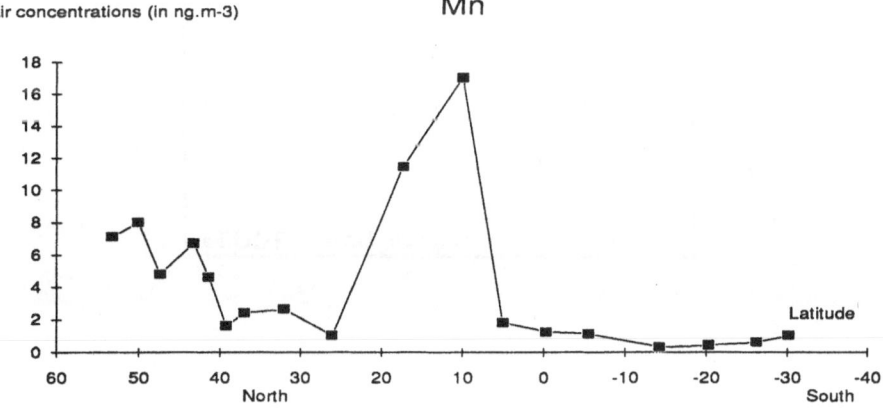

FIGURE 2 (Continued): Concentration profiles of trace elements.

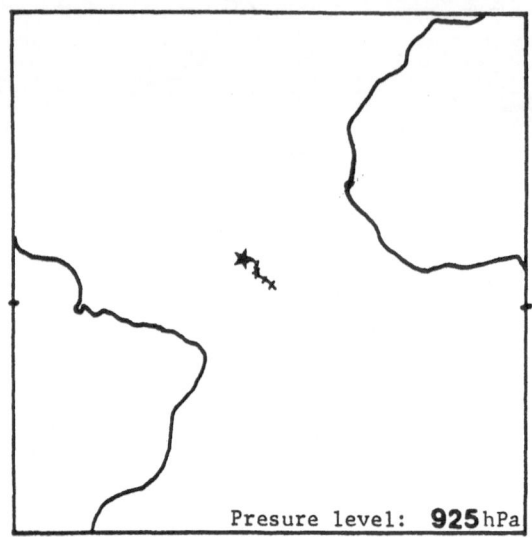

Presure level: **925** hPa

FIGURE 3: Air masses trajectories.

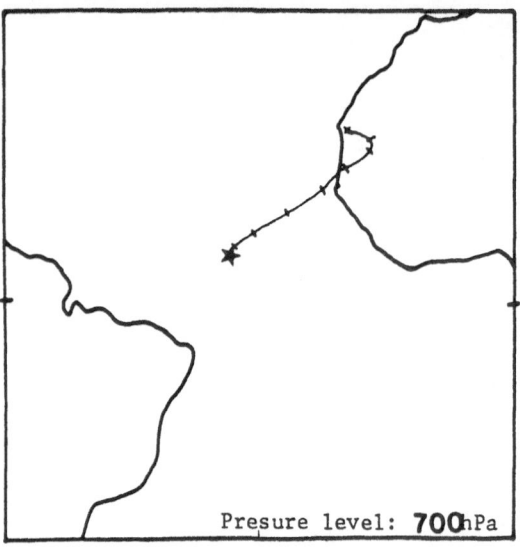

Presure level: **700** hPa

O. HOV, Project Leader of COST Project 611

A condensed view of atmospheric chemistry is given in the Figure.
For the first time we have seen significant contributions in all the main areas of atmospheric chemistry, which shows that COST611 is an important meeting place for atmospheric scientists in Europe. The strengthening of COST611 comes as a result of intensified effort in the field of atmospheric chemistry research in many European countries and within the CEC, an effort which is expected to gain further momentum as the STEP programme comes into action. Appropriate links have been established with Eurotrac, which means that the forces now can be channelled to the scientific rather than organizational work.

The emphasis has shifted over the years from the study of local pollution problems in Europe to the regional and global aspects of air chemistry, and how man is perturbing the chemical cycles between the earth and the atmosphere. This emphasis on regional and global problems is important, because those will remain as local problems are being cleared away. In Europe there are at least two major pollution regimes; the one prevailing in the Mediterranean countries, and the one that we are accustomed to in Central and Northern Europe. We still know quite little about the pollution regimes in the Mediterranean and how it influences the quality of life there and the composition of the northern hemisphere, although a few contributions on this topic have been presented by our Spanish and Portuguese colleagues.

For the first time a major part of the scientific programme of a COST611 meeting has been devoted to the reporting of joint projects which cross over institutional and national barriers. I think that it can already be stated that the emphasis on creating joint projects has been a success, because it educates people, instrumentation and analytical methods become harmonized, and the scientific quality of the work is enhanced. This is in my opinion well reflected in the contributions at this symposium.

In the wake of the joint international projects within COST611, one activity has been taken up which is of particular significance:
Instrument intercomparisons under controlled conditions to improve the comparability of measurements, and to establish primary standards against which other methods can be compared. A range of different instrumentation will still be needed since the purposes of measurements can be many, but a primary standard can provide absolute reference points for other methods. At this meeting, the instrument intercomparison of nitric acid and nitrate in Rome one year ago has been reported, and this has helped putting in perspective the difficult task of separating HNO_3 and nitrate in the air, and we have also been informed about intercalibrations within EMEP and about a comparison of wet only precipitation collectors.

Other joint projects that have reported their results include :
- the role of the marine atmosphere in the determination of the composition of air coming into Europe from the Atlantic. This project organizes both field and laboratory studies in a fruitful way.
- Composition of the remote marine atmosphere as measured in the Polarstern cruise one year ago
- The Halipp-projects with their emphasis on heterogeneous chemistry
- Lactoz, devoted to chemical processes related to tropospheric ozone with special emphasis on dark reactions

- Fog composition experiment
- Instrumentation for dry deposition measurements

Stratospheric work has been reported with a paper on possible chemical processes underlying the release of atomic chlorine in the Antarctic stratosphere, and on reactions involving the ClO radical. Also measurements have been reported on the tropospheric and stratospheric concentration of decomposition products of halogenated organic species, work which will be of increasing importance in the future as the emissions of CFCs continue and replacements also are introduced.

In summary, a lot of effort has been devoted to analytical methods to measure atmospheric trace species and reaction processes.
- Particularly strong has been the emphasis on the marine atmosphere, both with respect to halogenated species, S, NO_X, O_3, NHMC, H_2O_2, oxygenated organic species, and including reaction kinetics of their decomposition and transformation
- There has also been an emphasis on NO_X & NH_3 chemistry and measurements of species derived from NO_X and NH_3 over land, and how these species contribute to acid deposition
- Tropospheric O_3, including chemical processes where there has been an emphasis on nighttime chemistry which is an important part of the ozone problem
- Peroxides in air and rain water

Which areas have been given less attention? Going back to the Figure, it seems that a lot of effort has been done on instrumentation and process studies, while
- models to link emissions, atmospheric transformation and transport to concentrations or deposition, have been presented only to a small extent.
- Dry removal processes have not been discussed at the symposium
- The natural S-cycle has been paid considerable attention, much less effort has been devoted to the natural cycle of NMHC and NO_X, of these natural NMHCs are of particular significance in Europe and especially in the southern part.
- The cycles of pollutants in the Mediterranean countries and their influence on the composition of the northern hemisphere atmosphere, need more attention
- Particulate matter is an important part of the atmospheric pollution burden, both locally and regionally, and the physics and chemistry of aerosols is important to understand in connection e.g. with the nitrate budget, since some size fractions are more enriched in nitrate than others.
- As emission controls are introduced for many species, it is important to keep track of the changes in atmospheric composition which follow. One point here is e.g. the increases N_2O emissions due to the introduction of catalytic convertors
- "organic compounds" need further definition and specification

As the STEP programme gains momentum, I am sure that the future meetings of COST611 will be the focal points of the reporting of significant progress in atmospheric chemistry research in Europe, and this will also include stratospheric chemistry.

Feedbacks in atmospheric chemistry

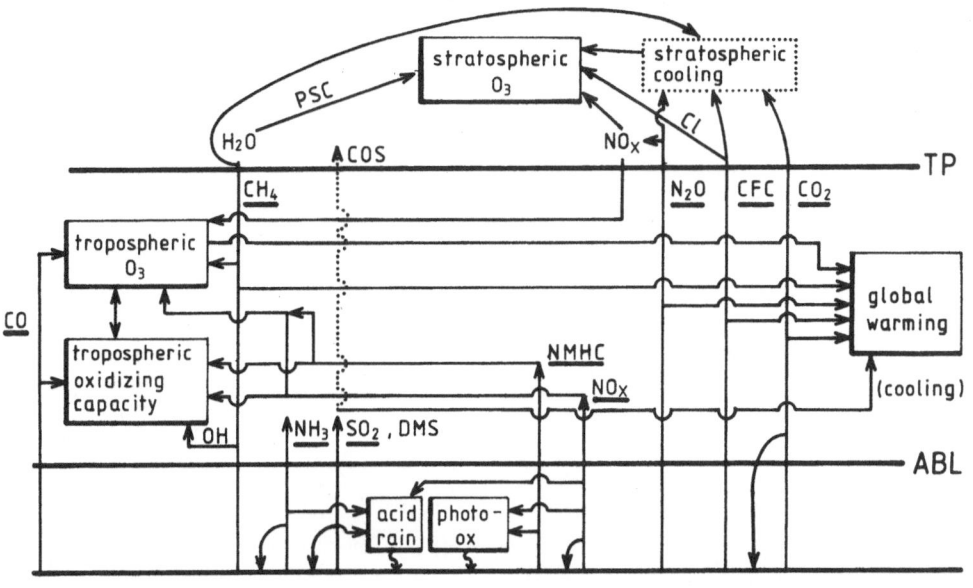

LIST OF PARTICIPANTS

ALLEGRINI I.
CNR-Ist. Inquinamento Atmosferico
CP-10
I-00016 MONTEROTONDO Stazione Roma

ANDRES M.D.
CIEMAT-Ipryma
22, Av. Complutense
E-28040 MADRID

ANGELETTI G.
Commission of the E.C.
DG XII/E1
200, Rue de la Loi
B-1049 BRUXELLES

ARLANDER D.
Institut für Chemie-3
KFA Jülich
D-5170 JÜLICH

ARTINANO B.
CIEMAT-Ipryma
22, Av. Complutense
E-28040 MADRID

ATKINS D.H.F.
JRC-Environment Institute
Ispra Site
I-21020 ISPRA (VA)

BACHMANN K.
Technische Hochschule Darmstadt
Fachbereich 8
Hochschulstrasse, 4
D-6100 DARMSTADT

BANGE P.
NV-KEMA
Env. Res. Dept.
P.O. Box 9035
NL-6800 ET ARNHEM

BARNES I.R.
Bergische Universitat
Physik. Chemie FB 9
Gauss Strasse, 20
D-5600 WUPPERTAL 1

BECKER K.H.
Bergische Universitat
Phys. Chemie FB 9
Gauss Strasse, 20
D-5600 WUPPERTAL 1

BEHNKE W.
Fraunhofer Inst. f. Toxikologie
Nikolai Fuchs Strasse
D-3000 HANNOVER 61

BEHR H.D.
Deutscher Wetterdienst
P.O. Box 650 150 e Jussieu
D-2000 HAMBURG 65

BEILKE S.
Umweltbundesamt
Pilot Station Frankfurt
Frankfurterstrasse, 135
D-6050 OFFENBACH

BENTER T.
Universitat Kiel
Phys. Chemie
Ludewig Meyn Strasse, 8
D-2300 KIEL 1

BIGGS P.
Oxford Univ.
Physical Chemistry Lab.
South Parks Road
OXFORD OX1 3QZ - UK

BOHM R.
FG Luftreinhaltung
Strasse des 17 Juni, 135
D-BERLIN

BONNET P.
Commission of the E.C.
DG XI
200, Rue de la Loi
B-1049 BRUXELLES

BORREGO C.
Univ. de Aveiro
Dep. de Ambiente
P-3800 AVEIRO

BORRELL P.
EUROTRAC-ISS
Fraunhofer IAU
Hindenburg Strasse, 43
D-8100 GARMISCH-PARTENKIRCHEN

BOULAUD D.
C.E.A.
DPT/ISPIN
P.O. Box 6
F-92265 FONTENAY AUX ROSES

BRAUERS T.
KFA-Jülich
ICH-3
P.O. Box 1913
D-5170 JÜLICH

BROCKMAN K.
Bergische Universitat
Phys. Chemie FB 9
Gauss Strasse, 20
D-5600 WUPPERTAL 1

BROWN A.C.
Oxford University
Physical Chemistry Lab.
South Parks Road
OXFORD OX1 3QZ - UK

BRUCKMANN P.
Umweltbehorde
Dept. Luftuntersuchungen
Neuer Kamp 25
D-2000 HAMBURG 36

BRUNING D.
KFA-Jülich
ICH-3
P.O. Box 1913
D-5170 JÜLICH

BURGERMEISTER S.
Goethe Universitat
Inst. f. Meteorologie u. Geophysik
Feldbergstrasse, 47
D-6000 FRANKFURT 1

BUTTINI P.
ENI Ricerche
Via Ramarini, 32
I-00015 MONTEROTONDO Stazione Roma

BUXTON G.
Leeds University
Cookridge Radiation Centre
LEEDS LS16 6QB - UK

CANOSA-MAS C.E.
Oxford University
Physical Chemistry Lab.
South Parks Road
OXFORD OX1 3QZ - UK

CARLIER P.
Univ. Paris VII
LPCA
2, Place Jussieu
F-75251 PARIS CEDEX 05

CASADO J.
Ecole Centrale de Lyon
BP 163
F-69131 ECULLY CEDEX

CECINATO A.
CNR Ist. Inquinamento Atmosferico
CP-10
I-00016 MONTEROTONDO Stazione Roma

CICCIOLI P.
CNR Ist. Inquinamento Atmosferico
CP-10
I-00016 MONTEROTONDO Stazione Roma

CORTIELLO M.
CNR Ist. Inquinamento Atmosferico
CP-10
I-00016 MONTEROTONDO Stazione Roma

COUTINHO M.
Univ. de Aveiro
Dep. de Ambiente
P-3800 AVEIRO

COX R.
UKAEA
British Antartic Survey
High Cross
Madingley Road
CAMBRIDGE

CROWLEY J.
Max Planck Inst. f. Chemie
Saarstrasse, 23
D-6500 MAINZ

DEISTER U.
Max Planck Inst. f. Chemie
Saarstrasse, 23
D-6500 MAINZ

DEVOLDER P.
Univ. de Lille
Lab. Chimie de la Combustion
F-59655 VILLENEUVE D'ASCQ

DORN H.P.
KFA-Jülich
ICH-3
P.O. Box 1913
D-5170 JÜLICH

DUBOIS J.
Secretariat d'Etat, Environment
14, Bd. du General Leclerc
F-92524 NEULLY

ELLERMANN T.
Risø National Laboratory
Chemistry Dept.
DK-4000 ROSKILDE

ELSHOUT A.
N.V.-KEMA
Env. Res. Dept.
P.O. Box 9035
NL-6800 ET ARNHEM

FACCHINI M.C.
CNR Ist. FISBAT
Via dei Castagnoli, 1
I-40126 BOLOGNA

FEBO A.
CNR Ist. Inquinamento Atmosferico
CP-10
I-00016 MONTEROTONDO Stazione Roma

FENGER J:
Min. of the Environment
Risø National Lab.
DK-4000 ROSKILDE

FERM M.
Swedish Environmental Res. Inst.
P.O. Box 47086
S-40258 GOTHENBURG

FILBY G.
K.F.K. - Karlsruhe
P.O. Box 3640
D-7500 KARLSRUHE

FINK E.H.
Universitaet
Gausse Strasse, 20
D-5600 WUPPERTAL

FOSTER P.
Univ. J. Fourier
GRECA
1, Rue F. Raoult
F-38000 GRENOBLE

FUHRER K.
Univ. of Bern
Physics Inst. C14 Lab.
Sidlerstrasse, 5
CH-3012 BERN

FUZZI S.
CNR Ist. FISBAT
Via de Castagnoli, 1
I-40126 BOLOGNA

GEISS F.
JRC Environment Institute
Ispra Site
I-21020 ISPRA (VA)

GEORGII H.W.
Goethe Universitat
Inst. f. Meteorologie u. Geophysik
Feldbergstrasse, 47
D-6000 FRANKFURT

GERLACH R.
Institut für Chemie-3
KFA Jülich
D-5170 JÜLICH

GLAVAS S.
University of Patras
Department of Chemistry
GR-26110 PATRAS

GRUNDAHL L.
NERI
Frederiksborgvej, 399
DK-4000 ROSKILDE

GUILLARD CH.
Ecole Centrale de Lyon
BP 163
F-69131 ECULLY CEDEX

HAAKS D.
Aero-Laser Gmbh
Am ETS Stadion 1
D-8100 GARMISCH-PARTENKIRCHEN

HANSSEN J.E.
Norwegian Inst. for Air Research
P.O. Box 64
N-2001 LILLESTROM

HARRIS G.
Max Planck Inst. f. Chemie
Saarstrasse, 23
D-6500 MAINZ

HARTMANN D.
Univ. Gottingen
Inst. f. Phys. Chemie
Tammanstrasse, 6
D-3400 GOTTINGEN

HARTMANN W.R.
Max Planck Inst. f. Chemie
Saarstrasse, 23
D-6500 MAINZ

HAUPTMANN J.
Tech. Hochsch. Darmstadt
Fachbereich 8
Hochschulstr., 4
D-6100 DARMSTADT

HAYMAN G.
Harwell Laboratory
Chemical Physics Group
B551 Harwell
DIDCOT OXON OX11 ORA - UK

HEIMANN G.
Max Planck Inst. f. Chemie
Saarstrasse, 23
D-6500 MAINZ

HELAS G.
Max Planck Inst. f. Chemie
Abt. Biogeochemie
P.O. Box 3060
D-6500 MAINZ

HERRMANN H.
IPC GOTTINGEN
D-3400 GOTTINGEN

HJORTH J.
JRC Environment Institute
Ispra Site
I-21020 ISPRA (VA)

HOEFKEN K.
GSF München
Ingolstadtes Landstrasse 1
D-8042 NEUHERBERG

HOFZUMAHAUS A.
KFA-Jülich
ICH-3
P.O. Box 1913
D-5170 JÜLICH

HONGISTO M.
Finnish Meteorological Inst.
Sahaajankatu 22E
SF-00810 HELSINKI

HOV O.
Norwegian Inst. for Air Research
P.O. Box 64
N-2001 LILLESTROM

JACOB P.
Univ. for Spectrochemistry and
Applied Spectroscopy
Busen Kirchoff Strasse, 11
D-4600 DORTMUND

JACOB V.
Univ. J. Fourier
GRECA
1, Rue F. Raoult
F-38000 GRENOBLE

JAESCHKE W.
Univ Frankfurt
Zentrum f. Umweltforschung
R. Mayer Strasse 7-9
D-6000 FRANKFURT 1

JENKIN M.
Harwell Laboratory
Chemical Physics Group
B551 Harwell
DIDCOT OXON OX11 ORA - UK

JOOS E.
Electricité de France
6, Quai Watier
F-78400 CHATOU

JUNKERMANN W.
Fraunhofer IAU
Kreuzeckbahnstrasse, 18
D-8100 GARMISCH-PARTENKIRCHEN

KANAKIDOU M.
M.PMI. for Chemistry Atmospheric Div.
D-6500 MAINZ

KARLSSON V.
Finnish Meteorological Inst.
Sahaajakatu 22E
SF-00810 HELSINKI

KINS L.
Univ. München
Meteorologisches Inst.
Barbarastrasse, 16
D-8000 MÜNCHEN 4C

KLOSE A.
Commission of the E.C.
DG XII/G1
200, Rue de la Loi
B-1049 BRUXELLES

KNISPEL R.
Fraunhofer Inst. f. Toxikologie
Nikolai Fuchs Strasse, 1
D-3000 HANNOVER 61

KOCH R.
Fraunhofer Inst. f. Toxikologie
Nikolai Fuchs Strasse, 1
D-3000 HANNOVER 61

KOPPMANN R.
KFA-Jülich
ICH-3
P.O. Box 1913
D-5170 JÜLICH

KOSKINEN T.
Finnish Meteorological Inst.
Sahaajankatu 22E
SF-00810 HELSINKI

KOTZIAS D.
JRC Environment Institute
Ispra Site
I-21020 ISPRA (VA)

KUTSENOGIY P.
Max Planck Inst. für Chemie
Saarstrasse, 23
D-6500 MAINZ

LAGRANGE J.
EHICS
URA 405 au CNRS
Rue Blaise Pascal, 1
F-67000 STRASBOURG

LAGRANGE P.
EHICS
URA 405 au CNRS
Rue Blaise Pascal, 1
F-67000 STRASBOURG

LAMMEL G.
KFK-Karlsruhe
Lab. f. Aerosolphysik und Filtertech
P.O. Box 3640
D-7500 KARLSRUHE

LANGHANS I.
Katholieke Univ. Leuven
Celestinijnenlaal 200F
B-3030 LEUVEN

LANGROVA S.
Univ. of Goteborg
Dept. of Inorganic Chemistry
S-41296 GOTEBORG

LAURILA S.
Finnish Meteorological Inst.
Sahaajankatu 22E
SF-00810 HELSINKI

LAVERDET G.
CNRS
CRCCHT
1C, Av. Recherche Scientifique
F-45071 ORLEANS

LE BRAS G.
CNRS
CRCCHT
1C, Av. Recherche Scientifique
F-45071 ORLEANS

LEHMANN M.
Univ. of Bern
Physics Inst. C14 Lab.
Sidlerstrasse, 5
CH-3012 BERN

LESCLAUX R.
Univ. de Bordeaux I
Lab. Photophysique
F-33405 TALENCE CEDEX

LEVSEN K.
Fraunhofer Inst. f. Toxikologie
Nikolai Fuchs Strasse, 1
D-3000 HANNOVER 61

LIBERTI A.
Univ. di Roma
Dip. di Chimica
P.le Aldo Moro, 5
I-00185 ROMA

LJUNGSTROM E.
Univ. of Goteborg
Dept. of Inorganic Chemistry
S-41296 GOTEBORG

MAC LEOD H.
CNRS
CRCCHT
1C, Av. Recherche Scientifique
F-45071 ORLEANS

MADELAINE G.
CEA - DPT ISPIN
P.O. Box 6
Av. de la Division Leclerc
F-92256 FONTENAY AUX ROSES

MALICET J.
Univ. de Reims
Lab. Chimie-Physique
P.O. Box 347
Moulin de la Housse
F-51062 REIMS

MARTIN M.
CIEMAT - Ipryma
Av. Complutense, 22
E-28040 MADRID

MASNIERE P.
Electricité de France
6, Quai Watier
F-78401 CHATOU

McELROY J.
Central Electricity Generat. Board
CERL
Kelvin Av.
LEATHERHEAD Surrey KT22 75E - UK

MIGON C.
Lab. de Physique et Chimie Marines
LPCM - LA DARSE
P.O. Box 8
F-06230 VILLEFRANCHE SUR MER

MIHALOPOULOS N.
Bergische Universitaet
Physik. Chemie FB 9
Gauss Strasse, 20
D-5600 WUPPERTAL 1

MILLAN M.
CIEMAT-Ipryma
Av. Complutense, 22
E-28040 MADRID

MIRABEL P.
Univ. L. Pasteur
1, Rue Blaise Pascal
F-67000 STRASBOURG

MONKS P.
Oxford Univ.
Physical Chemistry Lab.
South Parks Road
OXFORD OX1 3QZ - UK

MOORTGAT G.
Max Planck Inst. f. Chemie
Air Chemistry Div.
Saarstrasse, 23
D-6500 MAINZ

MORELLI J.
Ecole Normale Superieure
Inst. de Biogeochimie Marine
1, Rue M. Arnoux
F-92120 MONTROUGE

MULLER J.
Umweltbundesamt
Pilot Station Frankfurt
135, Frankfurterstrasse
D-6050 OFFENBACH

MULLER M.
Ministère de l'Environnement
SRETIE
14, Bd. General Leclerc
F-92524 NEUILLY

NEFTEL A.
Univ. of Bern
Physics Inst. C14 Lab.
Sidlerstrasse, 5
CH-3012 BERN

NEUROTH R.
KFA
Institut für Atmosphärische Chemie 3
D - JÜLICH

NIELSEN C.J.
Univ. of Oslo
Chemistry Dept.
N-0315 BLINDERN OSLO 3

NIELSEN O.J.
Risø National Laboratory
Chemistry Dept.
DK-4000 ROSKILDE

NODOP K.
Goethe Univ.
Inst. f. Meteorologie
Feldbergstrasse, 47
D-6000 FRANKFURT 1

NOLTING F.
Fraunhofer Inst. f. Toxikologie
und Aerosolforschung
Nikolai Fuchs Strasse 1
D-3000 HANNOVER 61

OCKELMANN G.
Goethe Univ.
Inst. f. Meteorologie und Geophysik
D-6000 FRANKFURT 1

OLSEN I.
Univ. of Oslo
Chemistry Dept.
N-0315 BLINDERN OSLO 3

OTJES R.
E.C.N.
P.O. Box 1
NL-1755 ZG PETTEN

OTT H.
Commission of the E.C.
DG XII/E1
200, Rue de la Loi
B-1049 BRUXELLES

PAFFRATH G.
DVLR
German Aerospace Research Establ.
Inst. Atmospheric Physics
D-8031 OBERPFAFFENHOFEN

PALMGREN JENSEN F.
NERI
Frederiksborgvey, 399
DK-4000 ROSKILDE

PAPENBROCK T.
Ruhr Univ. Bochum
Physikalische Chemie I
P.O. Box 102148
D-4630 BOCHUM

PARR D.
Univ. of Oxford
Physical Chemistry Laboratory
South Park Road
OXFORD OX1 3QZ - UK

PASHALIDIS S.
Univ. de Paris VII
LPCA
2, Place Jussieu
F-75252 PARIS CEDEX 05

PEETERS J.
Katholieke Univ. Leuven
Celestijnenlaan 200 F
B-3030 HEVERLEE

PERNER D.
Max Planck Inst. f. Chemie
Saarstrasse, 23
D-6500 MAINZ

PERRINO C.
CNR Ist. Inquinamento Atmosferico
CP-10
I-00016 MONTEROTONDO Stazione Roma

PERROS P.
Univ. Paris Val de Marne
LPCE
Av. du General De Gaulle
F-94000 CRETEIL

PICHAT J.C.
CNRS Ecole Centrale de Lyon
Photocatalyse et Environnement
P.O. Box 163
F-69131 ECULLY CEDEX

PLASS C.
KFA-Jülich
ICH-3
P.O. Box 1913
D-5170 JULICH

PLATT U.
Univ. Heidelberg
Inst. f. Umweltphysik
D-6900 HEIDELBERG

PLAZA J.
CIEMAT-Ipryma
Av. Complutense, 22
E-28040 MADRID

POLZER J.
Tech. Hochschule Darmstadt
Fachbereich 8
D-6100 DARMSTADT

POSSANZINI M.
CNR
Via Salaria km 29.300
I-00185 ROMA

POULET G.
CNRS
CRCCHT
1C, Av. Recherche Scientifique
F-45071 ORLEANS CEDEX 2

PRUSS A.
Ecole Polytechnique Fédérale
de Lausanne
Lasen
CH-1015 LAUSANNE

RABER T.
Max Planck Inst. f. Chemie
Chemistry Dept.
Saarstrasse, 23
D-6500 MAINZ

REINHOLDT K.
Max Planck Inst. f. Chemie
P.O. Box 3060
Saarstrasse, 23
D-6500 MAINZ

REMOUDAKI E.
Univ. de Paris VII
LPCA
2, Place Jussieu
F-75005 PARIS CEDEX 05

RESTELLI G.
JRC Environment Institute
Ispra Site
I-21020 ISPRA (VA)

RINDONE B.
Univ. di Milano
Dip. Chimica Organica e Industriale
Via Venezia, 21
I-20133 MILANO

ROHRER F.
KFA-Jülich
ICH-3
P.O. Box 1913
D-5170 JÜLICH

ROMER F.
N.V.-KEMA
Emv. Res. Dept.
P.O. Box 9035
NL-6800 ET ARNHEM

ROSSET P.
Univ. de Clermont F.
LAMP
12, Av. de Landais
F-63000 CLERMONT-FERRAND

SALMON G.A.
Univ. of Leeds
Cookridge Hospital
LEEDS LS16 6QB - UK

SAWERYSYN J.P.
Univ. de Lille 1
Lab. Cinetique et Chimie de la Combustion
F-59655 VILLENEUVE D'ASCQ CEDEX

SCHURATH U.
Univ. Bonn
Physikalische Chemie
Wegelerstrasse, 12
D-5300 BONN

SCOTT J.A.
Dublin Univ. College
Physics Dept.
Belfield
IRL-DUBLIN 4

SIDEBOTTOM H.
Univ. College Dublin
Chemistry Dept.
Belfield
IRL-DUBLIN 4

SIESE M.
Fraunhofer Inst. f. Toxikologie
und Aerosolforschung
Nikolai Fuchs Strasse, 1
D-3000 HANNOVER 61

SKOV H.
JRC Environment Institute
Ispra Site
I-21020 ISPRA (VA)

SLEMR J.
Fraunhofer IAU
Kreuzeckbahnstrasse, 19
D-8100 GARMISCH-PARTENKIRCHEN

SMIT H.J.G.
KFA-Jülich
ICH-2
P.O. Box 1913
D-5170 JÜLICH

STAUBES R.
Goethe Univ.
Inst. f. Meteorologie und Geophys.
Feldbergstrasse, 47
D-6000 FRANKFURT 1

STOCKER D.
EAWAG
CH-8600 DUBENDORF (ZURICH)

STUHL F.
Ruhr Univ. Bochum
Physikalische Chemie 1
P.O. Box 102148
D-4630 BOCHUM

TSETSI M.
Univ. de Paris VII
LPCA
2, Place Jussieu
F-75251 PARIS CEDEX 05

VAREY R.H.
NPTEC
Kelvin Avenue
LEATHERHEAD SURREY, U.K.

VERSINO B.
JRC Environment Institute
Ispra Site
I-21020 ISPRA (VA)

VINCKIER C.
Katholieke Univ. Leuven
Celestijnenlaan 200 F
B-3030 HEVERLEE

VOGT R.
Univ. Kiel
Inst. f. Phys. Chemie
Olshausen Strasse, 40
D-2300 KIEL

VOLKWEIN S.
Univ. Frankfurt
Zentrum für Umweltforsch.
Robert-Mayer-Strasse, 7-9
D-6000 FRANKFURT am Main 1

WALDEN J.
Finnish Meteorological Inst.
Sahaajankatu 22E
SF-00810 HELSINKI

WALSH J.J.
Univ. College Dublin
Dept. Chemical Engineering
Upper Marrion Str.
IRL-DUBLIN 2

WANGBERG I.
Univ. of Goteborg
Dept. of Inorganic Chemistry
S-41296 GOTEBORG

WARNECK P.
Max Planck Inst. f. Chemie
Saarstrasse, 23
D-6500 MAINZ

WAYNE R.P.
Univ. of Oxford
Physical Chemistry Laboratory
South Parks Road
OXFORD OX1 3QZ - UK

WEBER E.
Bundesminister f. Umwelt
Naturschutz und Reaktorsicherheit
P.O. Box 120629
D-5300 BONN 1

WEBER M.
KFA
Institut für Atmosphärische Chemie 3
D - JÜLICH

WILLE U.
Univ. Kiel
Inst. Phys. Chemie
Olshausenstrasse, 40
D 2300 KIEL

WINKLER P.
Deutscher Wetterdienst
Meteorologisches Observatorium
Frahmredder, 95
D-2000 HAMBURG 65

WOOD N.
Univ. of Leeds
Cookridge Radiation Res. Centre
Cookridge Hospital
LEEDS LS16 6QB - UK

WORTHAM H.
Univ. de Paris VII
LPCA
2, place Jussieu
F-75251 PARIS CEDEX

WORTMANN B.
Technische Hochschule Darmstadt
Inst. f. Meteorologie
Hochschulestrasse, 1
D-6100 DARMSTADT

WYERS G.P.
E.C.N.
P.O. Box 1
NL-1755 ZG PETTEN

ZABEL F.
Bergische Univ.
Pysikalische Chemie FB9
Gaussstrasse, 20
D-5600 WUPPERTAL 1

ZETZSCH R.
Fraunhofer Inst. f. Toxikologie
und Aerosolforschung
Nikolai Fuchs Strasse, 1
D-3000 HANNOVER 61

INDEX OF AUTHORS

COX, R.A. , 155, 196, 385
CROWLEY, J.N. , 343, 373

DEISTER, U. , 263
DEREXEL, Eph. , 433
DEVOLDER, P. , 317, 322
DIOURI, M. , 12
DI PALO, V. , 75
DONLON, M. , 311, 361
DORN, H.-P. , 103, 237
DUANE, M. , 396
DUTOT, A.L. , 566
DÜWEL, U. , 114

ELEGANT, L. , 499
ELLERMANN, T. , 225
ELSEN, L. , 172
ENDERLE, K.H. , 57
ESLICK, I.C. , 75
EXNER, M. , 302

FACCHINI, M.C. , 57, 507
FEBO, A. , 69, 140
FOSTER, P. , 291
FOX, D. , 619
FRESNET, P. , 584, 671
FUZZI, S. , 57, 507

GEORGII, H.W. , 44, 512, 677, 688
GILGE, S. , 632
GLAVAS, S. , 184
GOMES, L. , 713
GOURMI, A. , 317

HANSSEN, J.E. , 38
HARRIS, G.W. , 646
HARTMANN, D. , 354
HARTMANN, W.R. , 560
HAUPTMANN, J. , 98
HAYMAN, G.D. , 196
HAZLETT, D. , 619
HELAS, G. , 441, 555, 560
HERRMANN, H. , 302
HERMANN, J.-M. , 283
HJORTH, J. , 177, 231, 402
HOFZUMAHAUS, A. , 103
HONGISTO, M. , 590
HORIE, O. , 343
HOV, O. , 51, 719

ISRAEL, G.W. , 145

JACOB, P. , 640, 646
JACOB, V. , 291
JAESCKE, W. , 57

JANSSEN, L.H.J.M. , 446, 485
JENKIN, M.E. , 385
JENSEN, N. , 177
JOHNEN, F.J. , 659, 665
JOOS, E. , 492
JOSEPH, M. , 410
JOURDAIN, J.L. , 160
JUNKERMANN, W. , 524

KARBACH, M.P. , 75
KARTHÄUSER, J. , 354
KEMA, N.V. , 446, 485
KEUKEN, M.P. , 6, 92
KINS, L. , 537
KIRCHNER, W. , 270
KLEMP, D. , 646
KLEY, D. , 632
KLOCKOW, D. , 640
KLUG, W. , 578
KOCH, R. , 322
KOPPMANN, R. , 659, 665
KOTZIAS, D. , 396
KRANZ, E. , 128
KROLL, G. , 518

LAFFOND, M. , 291
LAGRANGE, J. , 109, 257
LAGRANGE, P. , 109, 257
LAMMEL, G. , 128, 471
LANCAR, I.T. , 190
LANGROVA, S. , 336
LAVERDET, G. , 160, 190
LAURILA, T. , 465
LE BRAS, G. , 160, 190, 373
LEHMANN, B.E. , 608
LEHMANN, M.S. , 608
LESCLAUX, R. , 349
LESCOAT, V. , 584, 671
LIND, J. , 507
LITTLE, M.R. , 330
LJUNCSTRÖM, E. , 336
LOHSE, C. , 177
LOIRAT, H. , 349
LOSNO, R. , 596, 713
LUKENS, D.L. , 619

MAC LEOD, H. , 160
MADELAINE, G. , 12
MARTIN, M. , 543
MASCLET, P.A. , 602
MASNIERE, P. ,433
MASSMAN, W.J. , 619
MATTIEU, B. , 646
McELROY, W.J. , 251
MELLOUKI, A. , 160

METZIG, G. , 134
MIGON, C. , 499
MILLAN, M. , 543
MINESHOS, G. , 184
MONKS, P.S. , 204, 410
MOORTGAT, G.K. , 343, 366, 373
MORELLI, J. , 499
MOUVIER, G. , 32, 459, 584, 596, 602, 671, 683
MÜLLER, J. , 210
MÜLLER, K.P. , 707
MUNK, J. , 225
MURA, S. , 75

NEFTEL, A. , 83, 608
NELSEN, W. , 172
NICOLAS, E. , 499
NICOLLIN, B. , 396
NIELSEN, C.J. , 231
NIELSEN, O.J. , 220, 225, 311
NIEUWSTADT, F.T.M. , 485
NODOP, K. , 44

OCKELMANN, G.E.F. , 512
O'FARRELL, D.J. , 361
OLSEN, I.M.W. , 231
OTJES, R.P. , 6, 92
OTTOBRINI, G. , 231
ORSI, G. , 57, 507

PAFFRATH, D. , 531
PAGSBERG, P. , 225
PAPENBROCK, Th. , 653
PARCHATKA, U. , 75
PARR, A.D. , 330
PASHALIDIS, S. , 459, 584, 671, 683
PAUWELLS, J.-F. , 317
PEETERS, J. , 379
PERNER, D. , 75, 471
PERRAUD, R. , 291
PERRINO, C. , 69, 140
PERROS, P. , 426, 566
PETITET, F. , 32
PICHAT, P. , 283
PLASS, Ch. , 659, 665
PLATT, U. , 103, 237, 627
PLAZA, J. , 543
POLZER, J. , 215
POSSANZINI, M. , 75
POULET, G. , 160, 190, 373
PROYOU, A. , 426

RABER, W. , 366
RATTIGAN, O. , 220
RAYEZ, M.-T. , 349
REICH, Th. , 114

REINHOLDT, K. , 366
REISCHL, G. , 57
REMOUDAKI, E. , 596
RESTELLI, G. , V, 177, 402
RICHARD, E. , 454
RINDONE, B. , 402
ROHRER, F. , 549, 701
RÖMER, F.G. , 446
RÖSLER, F.M. , 531
ROSSET, R. , 454
RUA, J. , 479
RUDOLPH, J. , 659, 665, 707
RUMPEL, K.J. , 572

SALMON, G.A. , 245
SAWERYSYN, J.-P. , 317
SCHLITT, H. , 396
SCHURATH, U. , 184, 270
SCHUSTER, G. , 441, 555
SEILER, W. , 572
SIDEBOTTOM, H.W. , 220, 311, 361
SIESE, M. , 322
SKOV, H. , 177
SLANINA, J. , 6, 92
SLEMR, F. , 524
SLEMR, J. , 572
SMIT, H.G.J. , 632
SMITH, S.J. , 330
STAFFELBACH, T. , 83
STARCKE, J. , 172
STAUBES, R. , 677, 688
STEDMAN, D.M. , 619
STOCKER, D.W. , 619
STUHL, F. , 653

TEUBER, R. , 572
TE WINKEL, B.H. , 446
TOUPANCE, G. , 426, 566
TREACY, J.J. , 220, 311, 361
TSALKANI, N. , 566
TSETSI, M. , 32

VEYRET, B. , 349

WALDEN, J.A. , 121
WÄNBERG, I. , 336
WARNECK, P. , 263, 471
WAYERS-IJPELMANN, A. , 6
WAYGOOD, S.J. , 251, 330
WAYNE, R.P. , 204, 330, 410
WELTER, F. , 270
WILSON, S. , 441, 555
WINKLER, P. , 63, 518
WITTE, F. , 322
WOHLFART, K. , 207